EARTH: PORTRAIT OF A PLANET

Earth
Portrait of a Planet

STEPHEN MARSHAK

University of Illinois

With contributions from

Donald Prothero

Occidental College

W. W. Norton & Company

NEW YORK   LONDON

The text of this book is composed in Stone Serif, with the display
    set in Frutiger.
Composition by TSI Graphics
Manufacturing by Courier

Editor: Jack Repcheck
Developmental editor: Susan Gaustad
Director of manufacturing: Roy Tedoff
Photography editors: Ruth Mandel, Neil Ryder Hoos, Sarah Chamberlin
Project editors: JoAnn Simony, Kim Yi
Layout artists: JoAnn Simony, Roberta Flechner, Cathy Lombardi
Editorial assistant: Garrett Michaels
Book designer: Joan Greenfield

Library of Congress Cataloging-in-Publication Data

Marshak, Stephen, 1955–
    Earth : portrait of a planet / by Stephen Marshak ; with
contributions by Donald Prothero.
        p. cm.
    Includes bibliographical references and index.
    ISBN 0-393-97423-5 (pbk.)
    1. Geology. I. Prothero, Donald R. II. Title.

QE26.2.M365 2001
550—dc21   00-048031

W. W. Norton & Company, Inc., 500 Fifth Avenue, New York, N.Y.
10110
www.wwnorton.com

W. W. Norton & Company Ltd., Castle House, 75/76 Wells Street,
London W1T 3QT

2 3 4 5 6 7 8 9 0

DEDICATION

To Kathy, David, and Emma, who helped in (and put up with!) this

endeavor in so many ways over the years it resided in our home

PART V
Earth Resources

PART VI
Processes and Problems at the Earth's Surface

The modern science of geology (or geoscience), the study of the Earth, began in the late eighteenth century. So in comparison with other sciences, geology is a young subject. But over the past two centuries, thousands of geologists have provided answers to a wide range of questions: Why do earthquakes and volcanoes happen? What causes mountains to rise? How is the ocean floor different from the land surface? Where do rocks form, and what are they made of? How do Earth's varied landscapes develop and change through time? How has the climate changed through time? When did our planet form, and by what process? Has the map of Earth's surface remained the same through time? What do fossils tell us about the evolution of life? Where do we dig to find valuable ore, and where do we drill to find oil? How does the land react with air and water? Indeed, a look at almost any natural feature leads to a new question, and new questions fuel the need for new research. Thus, geoscience remains an active and exciting field today.

Before the mid-twentieth century, geoscientists studied each of the questions listed above on its own, without considering its relation to other issues. But since 1960, there have been two "paradigm shifting" advances that have unified thinking about the Earth and its features. The first, called the theory of plate tectonics, shows that the Earth's outer shell, rather than being static, consists of discrete plates that constantly move very slowly, so that the map of our planet constantly changes. We now understand that plate interactions cause earthquakes and volcanoes, build mountains, provide gases for the atmosphere, and affect the distribution of life on Earth. The second advance, called the concept of Earth systems science, emphasizes that the planet's water, land, atmosphere, and living inhabitants are dynamically interconnected. Earth materials constantly cycle among various living and nonliving reservoirs on, above, and within the planet, and the history of life is intimately linked to the history of the physical Earth.

Earth: Portrait of a Planet is an introduction to geology that weaves the theory of plate tectonics and the concept of Earth systems science into its narrative from the beginning, and thus strives to create a modern, coherent image—a portrait—of our planet. This book will help students understand the origin of the Earth and its internal structure, the processes that cause plates to move and the consequences of this movement, the origin of Earth's diverse landscapes (from deserts to glaciers to coasts), the materials that make up the Earth (including the resources they contain), the envelope of air and water that surrounds and interacts with the solid Earth, the long history of the Earth, and the ways in which the planet changes through time. In addition, students will acquire the background they need to address such practical issues as investing in property, as they learn about the hazards posed by earthquakes, unstable slopes, and floods. And they will gain the knowledge to understand political issues such as global warming and nuclear-waste disposal.

NARRATIVE THEMES

Although students will learn plenty of geological terminology in this book, to develop a more complete understanding of the Earth, they must go beyond vocabulary and be aware of fundamental concepts, or narrative themes, that explain how the Earth works. These themes provide a Peg-Board on which to hang observations and ideas, and allow students to make connections between them. Several narrative themes (discussed more fully in the Prelude) are emphasized throughout the text.

1. The Earth is a complex system in which the solid Earth, the oceans, the atmosphere, and life are interconnected to yield a planet unique in the solar system.

2. Most geological processes can be understood in the context of plate tectonics theory.

3. The Earth is a planet, formed like other planets from dust and gas, but a constantly changing one.

4. The Earth is very old—about 4.6 billion years old. During this time, the map of the planet and its surface features have changed, and life has evolved.

5. Internal processes (driven by Earth's internal heat) and external processes (driven by heat from the Sun) interact at the Earth's surface to create our landscapes.

6. Natural hazards—earthquakes, volcanoes, landslides, floods—and processes such as the depletion of oil and gas reserves are of vital interest to us all.

7. Physical features of the Earth are linked to life processes.

8. Science comes from observation, and people make scientific discoveries.

9. The study of geology can increase science literacy in such areas as chemistry and physics.

ORGANIZATION

The topics covered in this book have been arranged so that students can build their knowledge of geology on a foundation of basic concepts. Thus, the book starts with cosmology and the formation of the Earth, and then introduces the architecture of our planet, from surface to center. With this background, we can delve into plate tectonics theory. Plate tectonics appears early, a departure from standard practice in introductory geology texts, so that students can relate all subsequent chapters to this concept. Knowing about plate tectonics, for example, helps students understand the next suite of chapters on minerals, rocks, and the rock cycle. A knowledge of plate tectonics and rocks together then provides a basis for learning about volcanoes, earthquakes, and mountains. And with this background, we can see how the map of the Earth and the life on it have changed through the vast expanse of geologic time, and how energy and mineral resources have developed.

The final part of the book addresses processes and problems occurring at or near the Earth's surface, from the unstable slopes of hills, down the course of rivers, to the shores of the sea and beyond. This section also includes a summary of atmospheric science and concludes with a topic of growing concern in society—global change, particularly climate change.

SPECIAL FEATURES

Broad Application

Earth: Portrait of a Planet provides complete coverage of topics used in a traditional physical geology or introduction to geology course. But, recognizing that many instructors cover aspects of historical geology and Earth systems science as well, the book also provides chapters that address Earth history, the atmosphere, the oceans, and global change.

Flexible Organization

Though the sequence of chapters was chosen for a reason, this book is designed to be flexible enough for instructors to choose their own strategies for teaching geology. Thus, each chapter is largely self-contained, reiterating relevant material or at least referring to other chapters where certain topics can be reviewed. This apparent redundancy is intentional, for geology is a nonlinear subject: the individual topics are so interrelated that there is not always a single best way to order them.

Societal Issues

Geology's practical applications are addressed in chapters on energy resources, mineral resources, global change, and

mass wasting. Further, chapters on earthquakes, volcanoes, and landscapes highlight geological hazards. And students are encouraged to apply their geological understanding to environmental issues, where relevant.

Boxed Inserts

Throughout the text, boxes expand on specific topics by giving further scientific background, additional detail, or related information that's just plain interesting.

Detailed Illustrations

It's hard to understand features of the Earth system without being able to see them. To help students visualize topics, this book is lavishly illustrated, with figures that attempt to give a realistic context for a geologic feature without overwhelming students with extraneous detail. The talented artists who worked on the book have "pushed the envelope" of modern computer graphics, and the result is the most realistic pedagogical art ever provided by a geoscience text.

Photographs from around the world have been assembled for this book. Where appropriate, they are accompanied by annotated sketches labeled "What a geologist sees," to help students discover what the photos show.

Featured Paintings

In addition to individual figures, British painter Gary Hincks has provided spectacular two-page spreads for most chapters. These paintings illustrate key concepts introduced in the chapters and visually emphasize the relationships between components of the Earth system.

SUPPLEMENTS

For Instructors

1. *Norton Media Library with PowerPoint Slides*
Included on this CD-ROM (dual platform) are approximately 100 photographs, 300 state-of-the-art illustrations from the text, and 49 unique and dynamic Flash animations. Developed by Stephen Marshak in collaboration with Precision Graphics, and by Declan DePaor, these animations illustrate key geologic principles that are difficult to convey through static images. Some examples:

- Transform faulting
- Plate boundaries
- Hot-spot volcanoes
- Subduction
- Rifting
- Mineral growth
- The formation of oceanic crust
- The formation of cross beds

- Transgression and regression
- Types of faults
- Seismic-wave motion
- How a seismograph works
- Types of unconformity
- Folding
- Geologic history

- Oil formation and trapping
- The evolution of a meandering stream
- Glacial advance and retreat
- Milankovitch cycles

Designed for lecture display or student use, these animations can be enlarged to full-screen view, and feature VCR-like controls that allow you to pause, fast-forward, or rewind for more effective use in the classroom.

2. Overhead Transparency Set

The text illustrations are featured in a complete set of transparency acetates.

3. Test-Item File

Prepared by Stephen Marshak, Terry Engelder of Pennsylvania State University, and John Werner of the University of Illinois, this test bank contains over 1,200 multiple-choice and true-false test questions. It is available in printed form or in Norton TestMaker (MicroTest III), a flexible electronic testing system for IBM-compatible or Macintosh computers. The computerized test-item file includes approximately 700 additional multiple-choice and true-false questions from the Study Guide.

4. Instructor's Resource Manual

This manual, prepared by John Werner, contains useful material to assist instructors as they prepare their lectures.

For Students

1. Portrait of a Planet Website

This resource features nearly fifty interactive animations of dynamic processes, with an emphasis on plate tectonics, geologic hazards, and Earth systems science concepts. Overviews, key terms and definitions, crossword puzzles, and multiple-choice quizzes test students' understanding of chapter content. Biweekly Earth Science News updates from Newswise.com and specially commissioned articles by science writer Elisabeth Mason help them to apply their knowledge and further highlight the relevancy and inherent interest of geologic concepts.

2. Portrait of a Planet CD-ROM

Selected contents from the student website are also available on a free CD-ROM packaged with every copy of the text.

3. Study Guide

Written by Rita Leafgren of the University of Northern Colorado, this thorough review provides summaries and study advice for each chapter, recall and matching exercises, short-answer questions, figure-labeling exercises, and practice tests.

ACKNOWLEDGMENTS

I am very grateful for the assistance of many people in bringing this book from the concept stage to the shelf. First and foremost, I wish to thank my family. My wife Kathy helped throughout in the overwhelming task of keeping track of text and figures. In addition, she edited text, copied drafts, and provided invaluable advice. My daughter Emma spent many nights helping to organize figures, and my son David helped me keep the project in perspective. During the early stages, I benefited greatly from discussions with Philip Sandberg, who contributed ideas that helped establish the organization and tone of the book. Donald Prothero assisted at a later stage by contributing text and editorial comments, and providing end-of-chapter questions and suggested readings. I also wish to thank Fernando Alkmim, who helped me figure out the best way to explain some of the complex topics in the book.

The publisher, W. W. Norton, has been incredibly supportive and generous in their investment in this project. I am particularly grateful to the editorial staff. Steve Mosberg helped lay out the basic framework of the book and provided valuable suggestions for content, and Rick Mixter refined the framework and set it on track to its final form. Jack Repcheck bulldozed aside obstacles to bring the project to completion, has been a constant source of encouragement, and has provided numerous innovative ideas that have strengthened the book. Susan Gaustad has been an outstanding copy editor, who has come up with wonderful suggestions for wordings. Susan's sharp eye caught many of those little errors that are so easy to miss. April Lange expertly coordinated the development of ancillaries, particularly the CD-ROM. Ruth Mandel and, later, Neil Hoos and Sarah Chamberlin have been outstanding photo researchers, incorporating my own photographs and locating spectacular photos from contributors. JoAnn Simony ably managed the Herculean task of overseeing production for the book.

The illustrations have involved many artists. Precision Graphics, of Champaign, Illinois, has done a phenomenal job of producing overwhelming volumes of high-quality figures on short notice. Joanne Bales of PG has been particularly helpful in working out the details of figure design with me and in coordinating the artists who worked on the book. Stan Maddock helped create the style of the figures, and produced a great many of them. George Kelvin drew a number of the figures in the earlier chapters. And Gary Hincks has produced the incredible two-page spreads, in part using his own designs and geological insight.

I also much appreciate insightful reviews or discussions of the manuscript by the following geologists. The careful comments by Barbara Tewksbury were particularly valuable.

Jack C. Allen, Bucknell University
David W. Andersen, San Jose State University
Philip Astwood, University of South Carolina
Keith Bell, Carleton University
Mary Lou Bevier, University of British Columbia
George S. Clark, University of Manitoba
Patrick M. Colgan, Northeastern University
John W. Creasy, Bates College
Robert T. Dodd, State University of New York at Stony Brook
James E. Evans, Bowling Green State University
William D. Gosnold, University of North Dakota
Bryce M. Hand, Syracuse University

Donna M. Jurdy, Northwestern University
Robert Lawrence, Oregon State University
John A. Madsen, University of Delaware
Charlie Onasch, Bowling Green State University
Lisa M. Pratt, Indiana University
Bob Reynolds, Central Oregon Community College
Kevin G. Stewart, University of North Carolina at Chapel Hill
Barbara Tewksbury, Hamilton College
Thomas M. Tharp, Purdue University
Kathryn Thorbjarnarson, San Diego State University
Robert T. Todd, SUNY Stony Brook
Jon Tso, Radford University
William E. Sanford, Colorado State University
Alan Whittington, University of Illinois
Lorraine W. Wolf, Auburn University
Christopher J. Woltemade, Shippensburg University

ABOUT THE AUTHOR

Stephen Marshak is currently professor and head of the Department of Geology at the University of Illinois, Urbana-Champaign. He holds an A.B. from Cornell University, an M.S. from the University of Arizona, and a Ph.D. from Columbia University (Lamont-Doherty Earth Observatory). Steve's research interests lie in the fields of structural geology and tectonics. He has served as chair of the Division of Structural Geology and Tectonics of the Geological Society of America, and as a member of the National Science Foundation Panel for Tectonics. Over the years, he has had the opportunity to explore geology in the field on several continents. Recently, his work has focused on understanding the development of mountain belts, particularly in the ancient crust of Brazil (the accompanying photograph was taken in Serra do Espinhaço). Steve has been on the faculty of the University of Illinois since 1983, teaching courses in introductory geology, structural geology, tectonics, field geology, and petroleum geology, and has won the university's highest teaching award. He shares authorship on two previous books: *Basic Methods of Structural Geology* and *Earth Structure: An Introduction to Structural Geology and Tectonics*.

THANKS!

I greatly appreciate your selection of this book as your entrée into the science of geology. This is a first edition, and as such can certainly benefit from input by users. I welcome your comments, especially if you find text or figures that are in error or not clear. Please contact me at: smarshak@uiuc.edu.

Stephen Marshak

EARTH: PORTRAIT OF A PLANET

And Just What Is Geology?

Civilization exists by geological consent, subject to change without notice.
—WILL DURANT (1885–1981)

P.1 IN SEARCH OF IDEAS

In the glare of the midnight sun, our C-130 Hercules transport plane rose from a smooth ice runway on the frozen sea surface at McMurdo Station, Antarctica, and we were off to spend a month studying unusual rocks exposed on a cliff about 250 kilometers (km) away. As we climbed past the smoking summit of Mt. Erebus, Earth's southernmost volcano, we had one nagging thought: no aircraft had ever landed at our destination, so the ground conditions there were unknown; if deep snow covered the landing site, the massive plane might get stuck and not be able to return to McMurdo. Because of this concern, the flight crew had added a crate of rocket canisters to the pile of snowmobiles, sleds, tents, and food in the plane's cargo hold. "If the turboprops can't lift us, we can clip a few canisters to the tail, light them, and rocket out of the snow," they claimed.

For the next hour, we flew along the Transantarctic Mountains, a ridge of rock that divides the continent into two parts, East Antarctica and West Antarctica (▶Fig. P.1). A vast ice sheet, in places over 3 km thick, covers East Antarctica—the surface of this ice sheet forms a high plain known as the Polar Plateau. Rivers of ice from the Polar Plateau slowly flow down valleys cut through the Transantarctic Mountains. (Ice sheets and ice rivers are called **glaciers.**) From the plane's windows, we admired long stripes of rock debris that had been shed from mountains onto the glaciers in these valleys. Suddenly, we heard the engines slow.

As the plane descended, it lowered its landing gear, which was equipped with skis. The loadmaster shouted a reminder of the emergency alarm code: "If you hear three short blasts of the siren, hold on for dear life!" Roaring toward the ground, the plane touched the surface of our first choice for a landing spot, the ice at the base of the rock cliff we wanted to study. *Wham, wham, wham, wham!!!!*

We can see the Earth system at a glance near the Maroon Bells, a row of mountains in Colorado. Here, sunlight, air, water, rock, and life all interact.

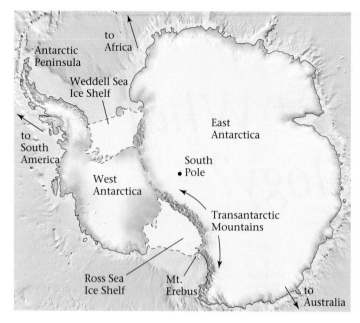

FIGURE P.1 Map of Antarctica.

FIGURE P.3 Geologists sledding to a field area in Antarctica. The sleds contain a month's worth of food, sample bags, rock hammers, and notebooks as well as tents and clothes (and a case of frozen beer).

Sastrugi (frozen snow drifts) rippled the ice surface, and as the plane's skis slammed into them at about 180 km an hour, it seemed as though a fairy-tale giant was shaking the plane. Seconds later, the landing aborted, we were airborne again, looking for a softer runway above the cliff. Finally, we landed in a field of deep snow, unloaded, and bade farewell to the plane (▶Fig. P.2). The Hercules trundled for kilometers through the snow before gaining enough speed to take off, but fortunately did not need to use the rocket canisters. When the plane passed beyond the horizon, the silence of Antarctica hit us—no trees rustled, no dogs barked, and no traffic rumbled in this stark land of black

FIGURE P.4 A geologist studying exposed rocks on a mountain slope on the desert island of Zabargad, in the Red Sea, off the coast of Egypt.

FIGURE P.2 Geologists unloading a cargo of tents, sleds, and snowmobiles from the tail of a C-130 Hercules transport plane that has just landed in a snowfield. Note the large skis over the wheels.

rock and white ice. It would take us a day and a half to haul our sleds of food and equipment down to our study site (▶Fig. P.3). All this to look at a few dumb rocks?

Geologists, scientists who study the Earth, explore remote regions like Antarctica almost routinely. Such efforts often strike people in other professions as a strange way to make a living. Here's how Scottish poet Walter Scott (1771–1832) described geologists at work: "Some rin uphill and down dale, knappin' the chucky stones to pieces like sa' many roadmakers run daft. They say it is to see how the warld was made!" Indeed—to see how the world was made, to see how it continues to evolve, to find its valuable resources, to prevent contamination of its waters and soils, and to predict its sudden movements. That is why geologists spend months at sea drilling holes in the ocean floor, why they scale mountains (▶Fig. P.4), camp in humid jungles, and trudge through desert winds. That is why geologists use electron microscopes to examine the atomic structure of minerals, use mass spectrometers to define the composition of rock and water, and use supercomputers to model the paths of earthquake waves. For over two centuries, geologists have pored over the Earth—in search of ideas to explain how our planet operates.

curred today off Japan," or "landslides will threaten the city," or "contaminants from the proposed toxic waste dump are destroying the town's water supply," or "there's only a limited supply of oil left," the scientists referred to are geologists.

The fascination of geology attracts many people to careers in this science. Thousands of geologists work for oil, mining, water, engineering, and environmental companies, while a smaller number work in universities, government geological surveys, and independent laboratories. Nevertheless, since the majority of students reading this book will not become professional geologists, it's fair to ask the question "Why should people, in general, study geology?"

First, geology may be one of the most practical subjects you can learn, for geologic phenomena and issues affect our daily lives, sometimes in unexpected ways:

- Landslides, volcanoes, floods, and earthquakes destroy communities.
- The discovery of a new oil field lowers energy costs, which in turn saves consumers money.

P.2 THE NATURE OF GEOLOGY

Geology, the study of the Earth, focuses on describing our planet's composition, behavior, and history. As is true with most sciences, geology consists of numerous subdisciplines, listed in Table P.1. In fact, because geology covers such a diversity of subjects, researchers commonly substitute the term **geoscience.**

Not only do geologists address academic questions, such as the formation and composition of the Earth, the causes of earthquakes (▶Fig. P.5) and ice ages, and the evolution of life, they also address practical problems, such as how to prevent groundwater contamination, how to find oil and minerals, and how to stabilize slopes. And in recent years, geologists have participated in the study of global climate change, for the long-term record of climate change lies in layers of sediment and rock. When news reports begin, "Scientists say . . ." and then continue with "an earthquake oc-

TABLE P.1 Principal Subdisciplines of Geology (Geoscience)

Name	Subject of Study
Engineering geology	The stability of geologic materials at the Earth's surface, for such purposes as controlling landslides and building tunnels.
Environmental geology	Interactions between the environment and geologic materials, and the contamination of geologic materials.
Geochemistry	Chemical compositions of materials in the Earth and chemical reactions in the natural environment.
Geochronology	The age (in years) of geologic materials, the Earth, and extraterrestrial objects.
Geomorphology	Landscape formation and evolution.
Geophysics	Physical characteristics of the whole Earth (such as Earth's magnetic field and gravity field) and of forces in the Earth.
Hydrogeology	Groundwater, its movement, and its reaction with rock and soil.
Mineralogy	The chemistry and physical properties of minerals.
Paleontology	Fossils and the evolution of life as preserved in the rock record.
Petrology	Rocks and their formation.
Sedimentology	Sediments and their deposition.
Seismology	Earthquakes and the Earth's interior as revealed by earthquake waves.
Stratigraphy	The succession of sedimentary rock layers.
Structural geology	Rock deformation in response to the application of force.
Tectonics	Regional geologic features (such as mountain belts) and plate movements and their consequences.

FIGURE P.5 Human-made cities cannot withstand the vibrations of a large earthquake. These apartment buildings collapsed during an earthquake in Turkey.

- Concerns about the availability of energy or metallic resources sometimes trigger conflicts.
- The loss of a town's freshwater supply discourages a factory from locating in the town.
- Your state or province proposes to locate a nuclear-waste-disposal facility near you.
- A storm destroys houses built along a nearby beach, so your insurance rates go up.

Clearly, all citizens of the twenty-first century, not just professional geologists, will need to make decisions concerning Earth-related issues. And they will be able to make more reasoned decisions if they have a basic understanding of geologic phenomena. History is full of appalling stories of people who ignored geological insight and paid a horrible price for their ignorance. Your knowledge of geology may help you to avoid building your home on a hazardous floodplain or fault zone, on an unstable slope, or along a rapidly eroding coast. With a basic understanding of groundwater, you may be able to save money when drilling an irrigation well, and with knowledge of the geologic controls on resource distribution, you may be able to invest more wisely in the resource industry.

Second, the study of geology gives you a perspective on the planet that no other field can. As you will see, the Earth is a complicated system; its living organisms, climate, and solid rock are all interrelated. Geologic study reveals Earth's antiquity (it's about 4.6 billion years old) and demonstrates how the planet has changed profoundly during its existence. What was the center of the Universe to our ancestors becomes, with the development of geological perspective,

our "island in space" today, and what was an unchanging orb originating at the same time as humanity becomes a dynamic planet that existed long before people did.

Third, the study of geology puts human achievements and natural disasters in context. On the one hand, our cities seem to be no match for the power of an earthquake, and a rise in sea level may swamp all major population centers. But on the other hand, we are now changing the face of the land worldwide at rates that far exceed those resulting from natural geologic processes. By studying geology, we develop a frame of reference for judging the impact of these changes.

Finally, when you finish reading this book, your view of the world will be forever colored by geological curiosity. When you next go on a road trip, the rock exposures next to the highway will no longer be gray, faceless cliffs, but will present complex puzzles of texture and color telling a story of Earth's history. When you walk in the mountains, you will think of the many forces that shape and reshape the Earth's surface. And when you hear about a natural disaster, you will have insight into the processes that brought it about.

P.3 THEMES OF THIS BOOK

A number of narrative themes appear (and reappear) throughout the text. These themes, listed below, can be viewed as the book's "take home" message.

1. ***The Earth is a unique, evolving system.*** Geologists increasingly recognize that the Earth is a complicated system; its interior, solid surface, oceans, atmosphere, and life forms interact in many ways to yield the landscapes and environment in which we live. Within this **Earth system,** chemical elements pass in cycles between different types of rock, between rock and sea, between sea and air, and between all of these entities and life. Aside from the residue of occasional collisions with fragments of rock or ice (asteroids or comets), all the material involved in these cycles originates in the Earth itself—our planet is truly an island in space.

2. ***Plate tectonics is a unifying idea that explains Earth processes.*** Like other planets, Earth is not a homogeneous ball, but rather consists of concentric layers: from center to surface, Earth has a core, mantle, and crust. We live on the surface of the crust, where it meets the atmosphere and the oceans. In the 1960s, geologists recognized that the crust together with part of the underlying mantle form a 100–150-km-thick semirigid shell. Large cracks separate this shell into discrete pieces, called **plates,** which move very slowly relative to one another (▶Fig. P.6). The theory that describes this movement and its consequences is now known as the

- - - - Trench or collision zone ———— Ridge ———— Transform ⬅ 5 cm/yr

FIGURE P.6 This map shows the Earth's principal plates. The arrow on each plate indicates the direction the plate moves, and the length of the arrow indicates the plate's velocity.

theory of plate tectonics, and it serves as the foundation for understanding most geologic phenomena. Although plates move very slowly, generally less than 10 centimeters (cm) a year, their movements yield earthquakes, volcanoes, and mountain ranges, and cause the distribution of continents to change over time.

3. *The Earth is a planet.* The subtitle of this book, *Portrait of a Planet,* highlights the view that despite the uniqueness of Earth's system and inhabitants, Earth fundamentally can be viewed as a planet, formed like the other planets of the solar system from dust and gas that encircled the newborn Sun. Though Earth resembles the other inner planets (Mercury, Venus, and Mars), it differs from them in having plate tectonics, an oxygen-rich atmosphere and liquid-water ocean, and abundant life. Further, because of the dynamic interactions among various aspects of the Earth system, our planet is constantly changing; the other inner planets are static.

4. *The Earth is very old.* Geological data indicate that the Earth formed 4.6 billion years ago—plenty of time for geologic processes to generate and destroy features of the Earth's surface, for life forms to evolve, and for the map of the planet to change. Plate-movement rates of only a few centimeters

per year, if those movements continue for hundreds of millions of years, can move a continent thousands of kilometers. In geology, we have time enough to build mountains and time enough to grind them down many times over. To define intervals of this time, geologists have invented a time scale, known as the **geologic time scale** (▶Fig. P.7). Geologists call the last 545 million years the **Phanerozoic eon,** and all time before that the **Precambrian.** They further divide the Precambrian into three main intervals named, from oldest to youngest, the **Hadean,** the **Archean,** and the **Proterozoic** eons, and the Phanerozoic eon into three main intervals named, from oldest to youngest, the **Paleozoic,** the **Meso-zoic,** and the **Cenozoic** eras (Chapter 13 provides further details about geologic time).

5. *Internal and external processes interact at the Earth's surface.* **Internal processes** are those phenomena that ultimately are driven by heat from inside the Earth. Plate movement is an example, and since plate movements cause mountain building, earthquakes, and volcanoes, we call all of these phenomena internal processes as well. **External processes** are those phenomena that ultimately are driven by heat supplied by radiation coming to the Earth from the Sun. This heat drives the movement of air and water, which grinds and sculpts the Earth's surface and transports the de-

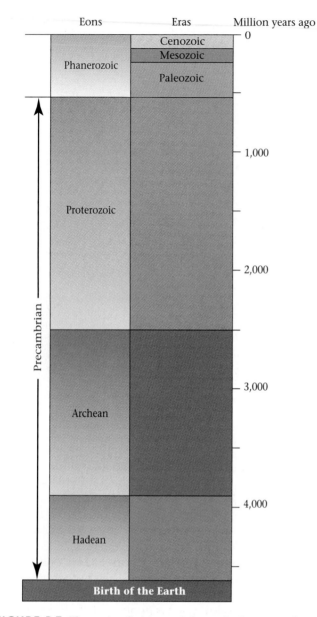

FIGURE P.7 The major divisions of the geologic time scale.

processes such as groundwater flow and contamination or depletion of oil and gas reserves are of vital interest to every inhabitant of this planet. They are often a matter of life and death. Linkages between geology and the environment are therefore stressed throughout the book.

7. **Physical aspects of the Earth system are linked to life processes.** All life on this planet depends on such physical features as the minerals in soil, the temperature, humidity, and composition of the atmosphere, and the flow of surface and subsurface water. And life in turn affects and alters these same physical features. For example, the atmosphere's oxygen comes primarily from plant photosynthesis, a life activity, and it is this oxygen that permits complex animals to survive. The oxygen also affects chemical reactions between air, water, and rock. Without the physical Earth, life could not exist, but without life, this planet's surface might have become a frozen wasteland like that of Mars, or enshrouded in acidic clouds like that of Venus.

8. **Science comes from observation, and people make scientific discoveries.** Science is not a subjective guess or an arbitrary dogma, but rather a consistent set of objective statements resulting from the application of the **scientific method** (see Box P.1). Every scientific idea must be constantly subjected to testing and possible refutation, and can be accepted only when supported by documented observations. Further, scientific ideas do not appear out of nowhere, but are the result of human efforts. Wherever possible, this book shows where geologic ideas came from, and tries to answer the question "How do we know that?"

9. **The study of geology can increase general science literacy.** Studying geology provides an ideal opportunity to learn basic concepts of chemistry and physics, because these concepts can be applied directly to understanding tangible phenomena. Thus, where appropriate, basic concepts of physical science are introduced in "Science Toolboxes." Also, Appendix A provides a systematic introduction to matter and energy, for those readers who have not learned this information previously.

As you read this book, please keep these themes in mind. Don't view geology as a list of terms to memorize, but rather as an interconnected set of concepts to digest. Most of all, enjoy yourself as you learn about what may be the most fascinating planet in the Universe.

bris to new locations, where it accumulates. The interaction between internal and external processes form the landscapes of our planet.

6. **Geologic phenomena affect our environment.** Volcanoes, earthquakes, landslides, floods, and even more subtle

The Scientific Method

Sometime during the past 200 million years, a large block of rock or metal (a meteor), which had previously been orbiting the Sun, crossed the path of Earth's orbit. In seconds, it pierced the atmosphere and slammed into our planet (thereby becoming a meteorite) at a site in what is now the central United States, a landscape of flat cornfields. The impact released more energy than a nuclear bomb—a cloud of shattered rock and dust blasted skyward, and once-horizontal layers of rock from deep below the ground sprang upward and stood on end in the gaping hole left by the impact. When the dust had settled, a huge crater surrounded by debris marked the surface of the Earth at the impact site. Later in Earth history, running water and blowing wind wore down this jagged scar, and some 15,000 years ago, sediments (sand, gravel, and mud) carried by a vast glacier buried what remained, hiding it entirely from view (▶Fig. P.8a, b). Wow! So much history beneath a cornfield. Have you ever wondered how geologists come up with such a story? It takes scientific investigation.

The movies often portray science as a dangerous tool, capable of creating Frankenstein's monster, and scientists as warped or nerdy characters with thick glasses and poor taste in clothes. In reality, **science** is simply the use of observation, experiment, and calculation to explain how nature operates, and **scientists** are people who study and try to understand natural phenomena. Scientists carry out their work using the **scientific method,** a sequence of steps for systematically analyzing scientific problems in a way that leads to verifiable results. Let's see how geologists employed the steps of the scientific method to come up with the meteorite-impact story.

1. *Recognizing the problem:* Any scientific project, like any detective story, begins by identifying a mystery. The cornfield mystery came to light when water drillers discovered limestone, a rock typically made of shell fragments, just below the 15,000-year-old glacial sediment. In surrounding regions, the rock at this level consists of sandstone, made of cemented-together sand grains, which differs greatly in composition from limestone. Since limestone can be used to build roads, make cement, and produce the agricultural lime used in treating soil, workers stripped off the glacial sediment and began excavating the limestone. They were amazed to

FIGURE P.8 (a) The site of an ancient meteorite impact in the American Midwest, before impact. Note the horizontal layers of rock below the ground surface. The thin lines represent boundaries between the successive layers. During impact, a large crater, surrounded by debris, forms. (b) The site of the impact today. The crater and the surface debris were eroded away. Relatively recently, the area was buried by glacial till (sediments). Underground, the impact disrupted layers of rock by tilting them and by generating faults (fractures on which sliding occurs).

Crater Debris

Fault

Glacial till

Layer of rock

(a)

(b)

find that rock layers exposed in the quarry tilted steeply and had been shattered by large cracks. In the surrounding regions, all rock layers are horizontal, like the layers in a birthday cake, the limestone layer lies underneath the sandstone, and the rocks contain relatively few cracks. Curious geologists came to investigate, and soon realized that the geologic features of the land just beneath the cornfield presented a problem to be explained: What phenomena had brought limestone up close to the Earth's surface, tilted the layering in the rocks, and shattered the rocks?

2. *Collecting data:* The scientific method proceeds with the collection of observations or clues that point to an answer. Geologists studied the quarry and determined the age of its rocks, measured the orientation of rock layers, and "documented" (made a written or photographic record of) the fractures that broke up the rocks.

3. *Proposing hypotheses:* A scientific **hypothesis** is merely a possible explanation, involving only naturally occurring processes, that can explain the set of observations. Scientists propose hypotheses during or after their initial data collection. The geologists working in the quarry came up with two alternative hypotheses. First, the features in this region could result from a volcanic explosion; and second, they could result from a meteorite impact.

4. *Testing hypotheses:* Since a hypothesis is no more than an idea that can be either right or wrong, scientists must put hypotheses through a series of tests to see if they work. The geologists at the quarry compared their field observations with published observations made at other sites of volcanic explosions and meteorite impacts, and studied the results of experiments designed to simulate such events. They learned that if the geologic features visible in the quarry were the result of volcanism, the quarry should contain certain rock types that appear only in volcanoes. No such rocks were found. If, however, the features were the consequence of an impact, the rocks should contain **shatter cones,** small, cone-shaped fractures formed only by meteorite impact (▶Fig. P.9). Shatter cones can easily be overlooked, so the geologists returned to the quarry specifically to search for them, and found them in abundance. The impact hypothesis passed the test!

Theories are scientific ideas supported by an abundance of evidence; they have passed many tests and failed none. Scientists have more confidence in a theory than they do in a hypothesis. Continued study in the quarry eventually yielded so much evidence for impact that the impact hypothesis came to be viewed as a theory. Keep in mind, however, that theories are not *necessarily* correct—scientists continue to test theories over a long time. Successful theories withstand the test of time and are supported by so many observations that they come to be widely accepted. (As you will discover in Chapters 3 and 4, geologists consider the idea that continents drift around the surface of the Earth to be a theory, because so much evidence supports it.) However, some theories may eventually be disproven, to be replaced by better ones.

Some scientific ideas must be considered absolutely correct, for if they were violated, the natural Universe as we know it would not exist. Such ideas are called **scientific laws,** and examples include the law of gravity, the laws of motion, and the laws of thermodynamics.

FIGURE P.9 Shatter cones in limestone. These cone-shaped fractures, formed only by severe impact, open up in the direction away from the impact. At this locality, the cones open up downward, indicating that the impact came from above.

Our Island in Space

When you look out toward the horizon from a mountain top, the Earth seems endless; and before the modern era, many people thought it was. But to astronauts flying to the Moon, the Earth is merely a small, shining globe—they can see half the planet in a single glance. We all ride on an island in space as it circles the Sun and, together with the Sun, rush through space at an immense speed.

Earth may not be endless, but it is a very special planet: its temperature and composition, unlike those of the other planets in the solar system, make it habitable. In this part, we first learn scientific ideas about how the Earth, and the Universe that contains it, came to be. Then we take a quick tour of the planet, looking particularly at its composition and various layers. With this background, we're ready to encounter the twentieth-century revolution in geology that yielded the set of ideas we call the theory of plate tectonics. We'll see that this theory, which proposes that the outer layer of the Earth is divided into "plates" that move with respect to one another, provides a rational explanation for a great variety of geologic features—from the formation of continents to the distribution of fossils. In fact, geologists now realize that plate movements are even responsible for generating the gases from which the atmosphere and oceans formed, and without which life could not exist.

CHAPTER 1

Cosmology and the Birth of Earth

1.1 INTRODUCTION

Sometime in the distant past, perhaps more than 100,000 generations ago, humans developed the capacity for complex, conscious thought. This amazing ability, which distinguishes our species from all others, brought with it the gift of curiosity, an innate desire to understand and explain the workings of ourselves and all that surrounds us—our **Universe.** Questions that we ask about the Universe differ little from questions a child asks of a playmate: Where do you come from? How old are you? Such musings first spawned legend and lore, with heroes, gods, and goddesses performing supernatural activities. Increasingly, **science,** the systematic analysis of natural phenomena, has provided insight into these questions. However, the development of **cosmology,** the study of the overall structure of the Universe, has proven to be a rough one, booby-trapped with tempting but flawed approaches and littered with discarded prejudices.

In this chapter, we begin with a brief historical sketch of cosmological thought. For brevity, the discussion is restricted to the Western tradition, though an equally rich history of ideas developed in other cultures. We then look at currently accepted ideas of modern cosmology and the key discoveries that led to scientific ideas about how our planet fits into the fabric of a changing Universe. The chapter concludes with a description of Earth's formation 4.6 billion years ago.

1.2 AN EVOLVING IMAGE OF THE EARTH AND SPACE

Three thousand years ago (1000 B.C.E.; "before the Common Era"), the Earth's human population totaled only several million, the pyramids of Egypt had already been weathering in the desert for 1,600 years, and Homer, the great Greek poet, was compiling the *Iliad* and the *Odyssey*. In Homer's day, astronomers of the Mediterranean region knew the difference between stars and planets. They had

When the Hubble Space Telescope looks into what, to the naked eye, appears to be the black void of the night sky, it reveals a spectacle of disks and spirals of hazy light. Each of these is a distant galaxy, a cluster of as many as 300 billion stars. This is the fabric of space.

observed that the positions of stars remained fixed relative to one another, but that the whole star field slowly revolved around a fixed point, while the planets moved relative to the stars and to each other, etching seemingly complex paths across the night sky (▶Fig. 1.1). In fact, the word "planet" comes from the Greek word *planés,* which means "wanderer." Despite their knowledge of the heavens, though, astronomers of Homer's day did not realize that Earth itself is a planet. Greek philosophers believed the Earth to be a flat disk with land toward the center and water around the margins; this disk lay at the center of a **celestial sphere,** beneath a dome to which the stars were attached and above an underworld governed by the fearsome god Hades. Placing the Mediterranean region at the center of the Universe must have made the ancient Greeks feel quite important, indeed! Philosophers of the day toyed with numerous explanations for the Sun: to some, it was a burning bowl of oil, and to others, a ball of red-hot iron. Most philosophers favored the notion that movements of celestial bodies represented the activities of gods and goddesses.

Then, beginning around 600 B.C.E. (the golden age of ancient science), Greek philosophers began to argue about the structure of the Universe. Perhaps the most contentious point in this debate concerned the position of the Earth in the Universe. Some pictured the Earth as a sphere that, along with all other celestial objects including the Sun, followed circular orbits around a "central fire." Others popularized the **geocentric Universe concept** (▶Fig. 1.2a), in which the Earth sat motionless in the center of the heavens while other bodies made perfectly circular orbits around it. Notably, most philosophers assumed that orbits had a circular shape, because they considered circles to be the most perfect of geometric forms and believed that the Universe was a perfect creation.

Around 250 B.C.E., Greek philosophers proposed the **heliocentric Universe concept,** in which all heavenly objects including the Earth orbited the Sun (▶Fig. 1.2b). But this idea found little favor, and three centuries later Ptolemy (100–170 C.E.), an influential Egyptian mathematician, once again championed the geocentric concept. Ptolemy bolstered his view with complex calculations that seemed to predict the wanderings of the planets. The church hierarchy adopted this hypothesis as dogma, because it certified that the Earth occupied the most important place in the Universe and implied that humans were thus the Universe's most important creatures. With the fall of Rome in 476 C.E., Europe and the Middle East entered the Middle Ages. For the next millennium, most scientific study in the Western world ceased, and those who disagreed with the Ptolemaic view of the Universe, with a flat Earth in its center, risked accusations of heresy.

Then came the Renaissance. The very word means rebirth and revitalization, and in fifteenth-century Europe bold thinkers spawned a new age of exploration and discovery. One

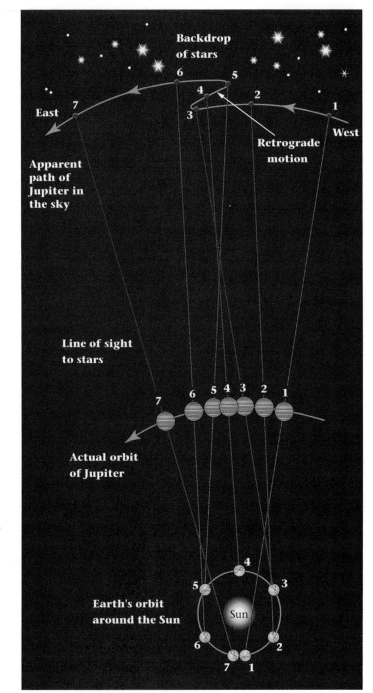

FIGURE 1.1 Each line represents a line of sight from the Earth through Jupiter to the backdrop of stars. The numbers indicate successive positions of Earth and Jupiter. Notice that the retrograde motion (position 4, for example) occurs because the Earth's orbit around the Sun is smaller than Jupiter's. Thus, for part of the year, Earth is actually moving in a direction opposite to that of Jupiter.

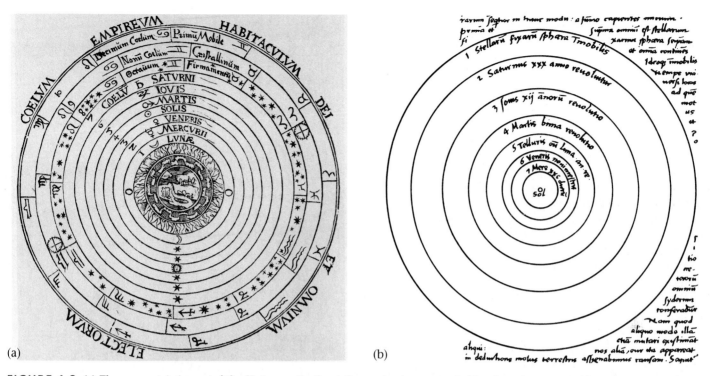

FIGURE 1.2 (a) The geocentric image of the Universe. Earth, at the center, is surrounded by air and fire and the Moon, Mercury, Venus, the Sun, Mars, Jupiter, and Saturn. Everything lies within the globe of the stars. (b) The heliocentric view of the Universe, as illustrated in this woodcut from Copernicus's *De revolutionibus*.

of the most influential was an Italian geographer and astronomer, Paulo Toscanelli (1397–1482), who reintroduced the idea of a spherical Earth around 1430, with profound consequences for exploration. In the 1480s, Toscanelli received a letter from an unknown sea captain named Christopher Columbus, inquiring about the Earth's shape. Toscanelli responded with a brief description of his ideas, and Columbus, armed with Toscanelli's report, proposed to the King and Queen of Spain that on a spherical Earth he would be able to reach the riches of India by sailing west, across the Atlantic. The monarchs agreed to fund his voyage to test the hypothesis. Of course, Columbus hit a fairly large roadblock, now known as America, and never made it to India. But two decades after this trip, Ferdinand Magellan did succeed in circumnavigating the planet, thus making it possible to create a globe, a map drawn on a sphere, the first realistic image of our planet.

As Renaissance thinking spread across Europe, scientists began to attack the geocentric Universe concept. Nicolaus Copernicus (1473–1543) reintroduced and justified the heliocentric concept in a book called *De revolutionibus* (Concerning the Revolutions) but, perhaps fearing the wrath of church officials, published the book just days before he died (Fig. 1.2b). *De revolutionibus* did indeed spark a bitter battle that pitted astronomers like Johannes Kepler (1571–1630)

and Galileo (1564–1642) against the establishment. Kepler showed that the planets follow elliptical, not circular, orbits, and thus demonstrated that Ptolemy's calculations, the bulwark of the geocentric hypothesis, were wrong. Galileo, using the newly invented telescope, observed that Venus has phases like our own Moon (a characteristic that would only be possible if Venus orbited the Sun) and that moons orbit Jupiter. His discoveries demonstrated that all heavenly bodies do not revolve around the Earth, and thus contradicted the geocentric hypothesis. To reward Galileo for this immensely important conceptual leap, officials charged him with heresy, and he spent the last ten years of his life under house arrest.

In the year of Galileo's death, Isaac Newton (1642–1727), perhaps the greatest scientist of all time, was born in England. Newton derived mathematical laws governing gravity and basic mechanics (the movement of objects in space), and thus provided the tools for explaining how physical processes operate in nature. Thus, as the seventeenth century came to a close, people for the first time possessed a clear image of the movements of planets. Unfortunately for human self-esteem, Earth was demoted from its place of prominence at the center of the Universe to become merely one of many planets circling the Sun.

1.3 A SENSE OF SCALE

We use enormous numbers to describe the size of the Earth, the distance from the Earth to the Sun, the distance between stars, and the distance between galaxies. Who came up with these numbers? In this section, we find out.

Circumference of the Earth

Eratosthenes (c. 276–194 B.C.E.) was in charge of the library in Alexandria, one of the great centers of learning in the Mediterranean region. One day while filing papyrus scrolls, he came across a report noting that in the southern Egyptian city of Syene, the Sun lit the base of a deep vertical

BOX 1.1

THE REST OF THE STORY

Earth's Rotation

If you gaze at the night sky for a long time, you'll see that the stars move in a circular path around the North Star. This movement suggests either that the Earth spins on its axis (an imaginary line connecting the North and South Poles) with respect to the stars, or that the stars orbit the Earth (▶Fig. 1.3). Curiously, it was not until the middle of the nineteenth century that Jean Bernard Léon Foucault (1819–1868), a French physicist, proved that the Earth spins on its axis. He made this discovery by setting a heavy pendulum, attached to a long string, in motion. As the pendulum continued to swing for many days, Foucault noted that the plane in which it oscillated (a plane perpendicular to the Earth's surface) appeared to rotate around a vertical axis (a

line perpendicular to the Earth's surface). If Newton's first law of motion—objects in motion remain in motion, objects at rest remain at rest—was correct, then the only explanation for this phenomenon was that the Earth rotated under the pendulum while the pendulum continued to swing in the same plane (▶Fig. 1.4a, b). Foucault displayed his discovery beneath the great dome of the Pantheon in Paris, to much acclaim.

We now know that, in fact, the Earth's spin axis is not fixed in orientation; rather, it wobbles. This wobble, known as **precession,** corresponds to the wobble of a toy top as it spins. We'll see later in this book that the precession of the Earth's axis, which takes 26,500 years, may affect the planet's climate.

FIGURE 1.3 Time exposure of the night sky over an observatory. Note that the stars appear to be fixed relative to one another, but that they rotate around a central point, now the North Star. This motion is actually due to the rotation of the Earth on its axis.

FIGURE 1.4 Foucault's experiment. (a) An oscillating pendulum at a given time. (b) The same pendulum at a later time. The pendulum stays in the same plane, but the Earth, and hence the frame, rotates.

(a)   (b)

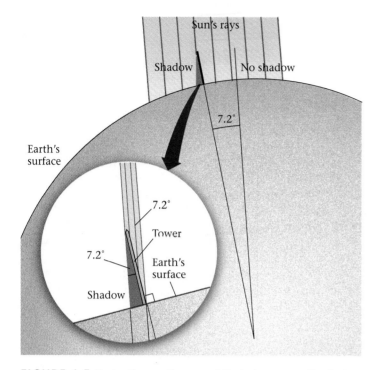

FIGURE 1.5 Eratosthenes discovered that at noon on the first day of summer, the Sun's rays were perpendicular to the Earth's surface (that is, parallel to the radius of the Earth) at Syene, but made an angle of 7.2° with respect to a vertical tower at Alexandria. Thus, the distance between Alexandria and Syene represented 7.2°/360° of the Earth's circumference. Knowing the distance between the two cities, therefore, allowed him to calculate the circumference of the Earth.

well precisely at noon on the first day of summer. Eratosthenes deduced that the Sun's rays at noon on this day must be exactly perpendicular to the Earth's surface at Syene, and that if the Earth was spherical, then the Sun's rays could *not* simultaneously be perpendicular to the Earth's surface at Alexandria, 800 km to the north (▶Fig. 1.5). So on the first day of summer, Eratosthenes measured the shadow cast by a tower in Alexandria at noon. The angle between the tower and the Sun's rays, as indicated by the shadow's length, proved to be 7.2°. He then commanded a servant to pace out a straight line from Alexandria to Syene. The sore-footed servant found the distance to be 5,000 stadia (1 stadium = 0.1572 km). Knowing that a circle contains 360°, Eratosthenes then calculated the Earth's circumference as follows:

$$360°/x = 7.2°/5,000 \text{ stadia, so}$$
$$x = 360° \ (5,000 \text{ stadia})/7.2° = 250,000 \text{ stadia, or}$$
$$39,300 \text{ km (24,421 miles).}$$

Thus, twenty-two centuries ago, Eratosthenes determined the circumference of the Earth to within 2% of today's accepted value (40,008 km, or 24,865 miles) without the aid of any sophisticated surveying equipment—a truly amazing feat. The slight error in his calculation came in measuring the distance to Syene, for the servant occasionally stumbled and miscounted his pace.

The Distance from Earth to Celestial Objects

Around 200 B.C.E., Greek mathematicians, using ingenious geometric calculations, determined that the distance to the Moon was about thirty times the Earth's diameter, or 382,260 km; the true distance, on average, is actually 381,555 km (about 237,100 miles). But it wasn't until the seventeenth century that astronomers figured out that the mean distance between the Earth and the Sun is about four hundred times the distance between the Earth and the Moon, or 149,600,000 km (about 93,000,000 miles). As for the stars, the ancient Greeks realized that they must be much farther away than the Sun in order for them to appear as a fixed backdrop behind the Moon and planets, but they had no way of calculating the actual distance. Our current notion of the vastness of the Universe, in fact, only arose in the past two centuries. In 1838, astronomers first applied the **parallax method** to determine the distance to a nearby star (see Box 1.2) and found that the nearest star to Earth, Alpha (α) Centauri, lies 40.85 trillion km away.

Since it's hard to fathom the distances to planets and stars without visualizing a more reasonably sized example, imagine that the Sun is the size of an orange. At this scale, the Earth would be a grain of sand at a distance of 10 meters (m) (30 feet) from the orange. Pluto, which on average is the planet farthest from the Sun, would lie about 650 m (about 2,100 feet) from the orange, and Alpha Centauri 2,000 km (about 1,243 miles) from the orange.

When astronomers realized that light travels at a constant (that is, unchanging) speed of about 300,000 km (about 186,000 miles) per second, they realized that they had a way to describe the huge distances between objects in space conveniently. Since the distance light travels in a specified time is constant, they defined a large distance by stating how long it takes for light to traverse that distance. For example, it takes light about 1.3 seconds to travel from the Earth to the Moon, so we can say that the Moon is about 1.3 light seconds, rather than 381,555 km, away. Similarly, we can say that the Sun is 8.3 light minutes away, instead of 150 million km away. A **light year,** the distance that light travels in one Earth year, equals about 9.5 trillion km (about 6 trillion miles). When you look up at Alpha Centauri, 4.3 light years distant, you see light that started on its journey to Earth about 4.3 years ago.

The parallax method can't be used for measuring distances beyond about 30 light years. In fact, astronomers didn't develop techniques for measuring the distance to very distant stars and galaxies until the twentieth century. With these techniques, they determined that the farthest celestial objects that can be seen with the naked eye are

BOX 1.2

SCIENCE TOOLBOX

The Parallax Method

To understand the parallax method, hold up your thumb at arm's length, and close your left eye. Then align your thumb with some object on a distant wall. Without moving your arm, close your right eye and open your left eye. Notice that your thumb now aligns with a different part of the wall. The apparent movement of your thumb across a distant background is called **parallax** (▶Fig. 1.6a), and the angle $\alpha/2$ is the parallax angle. Parallax occurs because your eyes are a few centimeters apart from each other, so each eye sees your finger from a different perspective. If you know the distance between your eyes (about 6.5 cm for most people), you can calculate the distance to your thumb. In this example, the distance between your eyes is the **baseline**. Notice that the parallax decreases as you move your thumb farther away from your face.

To determine the distance to a nearby star, we need a very long baseline. Fortunately, because the Earth orbits the Sun, we have a convenient one (▶Fig. 1.6b)—the diameter of Earth's orbit, about 300,000,000 km. Astronomers observe the position of a nearby star relative to a backdrop of very distant stars at two times that are six months apart (that is, after Earth has covered half its orbit). They can then measure the angle $\alpha/2$. Knowing this angle, astronomers can use trigonometry to complete their calculation of the distance.

FIGURE 1.6 (a) If you number a scale along a distant wall, hold up a thumb or finger, and close each eye in turn, you will notice that the number behind your finger depends on which eye you look through (an effect known as parallax). The distance between your eyes is the baseline. Your brain is able to synthesize the differing images to create a 3-D image in your mind. (b) Astronomers use parallax to determine the distance to a nearby star, Alpha (α) Centauri. The two positions of Earth six months apart provide the baseline.

over 2.2 million light years away. Powerful telescopes allow us to see much farther. The edge of the visible Universe lies 15 billion light years away, which means that light traveling to Earth from the edge of the Universe began its journey over 10 billion years before the Earth even existed. When such numbers became available by the middle of the twentieth century, people came to the realization that the dimensions of the Universe must be truly staggering.

1.4 THE MODERN IMAGE OF THE UNIVERSE

We've seen that the burst of discovery during the Renaissance forced people to change their view of Earth's central place in the Universe. Eventually, they realized that the Earth is but one of nine planets in the **solar system** (the Sun, and the planets and other objects that travel around it). Stars are not randomly scattered through the Universe; gravity pulls them together to form immense systems, or groups, called **galaxies.** The Sun is but one of over 300 billion stars clustered together to form the Milky Way galaxy, and the Milky Way is but one of the 100 billion galaxies that comprise the visible Universe. Galaxies are so far away that to the naked eye they look like stars in the night sky. The nearest galaxy to ours, Andromeda, lies over 2.2 million light years away. Overall, galaxies are not evenly distributed through the Universe; there are some regions where galaxies cluster together and others that have no galaxies.

FIGURE 1.7 An image of what the Milky Way galaxy might look like if viewed from outside. Note that the galaxy consists of spiral arms around a central cluster. Our Sun lies at the edge of one of these arms.

If we could view the Milky Way from a great distance, it would look like a flattened spiral, 100,000 light years across, with great curving arms gradually swirling around a glowing, disk-like center (►Fig. 1.7). Presently, our solar system lies near the outer edge of one of these arms and rotates around the center of the galaxy about once every 250 million years. Relative to the galactic center, we hurtle through space at about 200 km per second.

1.5 FORMING THE UNIVERSE

Do galaxies move with respect to other galaxies? Does the Universe become larger or smaller with time? Has the Universe always existed? An answer to these fundamental questions came from an understanding of a phenomenon called the Doppler effect. Though the term may be unfamiliar, the phenomenon it describes is an everyday experience. To understand the phenomenon, you need to be familiar with the nature of waves: please study Box 1.3.

The Doppler Effect

When you hear the scream of a train whistle, the sound you hear has moved through the air from the whistle to your ear in the form of sound waves. As each wave passes, air alternately compresses, then expands. The **pitch** of the sound, meaning its note in the musical scale, depends on the **frequency** of the sound waves, or the number of waves that pass a point in a given time interval. Now imagine that as you are standing on the station platform, the train moves toward you. The sound of the whistle gets louder as the train approaches, but its pitch remains the same. Then, the instant the train passes, the pitch abruptly changes; it sounds like a lower note in the musical scale. An Austrian physicist, C. J. Doppler (1803–1853), first interpreted this phenomenon, and thus it is now known as the **Doppler effect.** When the train moves toward you, the sound has a higher frequency (the waves are closer together), because the sound source, the whistle, has moved slightly closer to you between the instant that it emits one wave and the instant that it emits the next (►Fig. 1.9a, b). When the train moves away from you, the sound has a lower frequency (the waves are farther apart), because the whistle has moved slightly farther from you between the instant it emits one wave and the instant it emits the next.

Light energy also moves in the form of waves, though light waves are different from sound waves in that they do not need a medium (like air or water) in order to exist. In shape, light waves resemble water waves. Visible light comes in many colors—the colors of the rainbow. The color of light you see depends on the frequency of the light waves, just as the pitch of a sound you hear depends on the

BOX 1.3

SCIENCE
TOOLBOX

Transmitting Energy by Waves

Imagine that you throw a pebble into a pond. As it enters the water, the pebble briefly depresses the water surface, but an instant later, once the pebble is below the surface, water rushes back to fill in the gap. The impact has transferred energy to the surface of the water. This energy then moves from the site of impact in all directions on the surface of the water in the form of waves. As you watch the waves, they seem to move horizontally across the water. But as they pass under a floating stick, the stick bobs up and down. In other words, **waves** are disturbances that travel from one point to another by causing periodic back-and-forth movements, but without physically transferring material from the wave source to another location. Note that as the wave passes the stick, the energy that was originally transferred from the pebble to the surface of the water now causes the stick to move.

If the back-and-forth motion as a wave passes lies at right angles to the direction the wave travels, we refer to the wave as a **shear wave** (▶Fig. 1.8a). The highest point on a wave is its **crest,** and the lowest point is its **trough.** When describing waves, we specify the **wavelength,** or the distance between successive crests or successive troughs; the **frequency,** or the number of crests (or troughs) that pass a fixed reference point in a given time; and the **amplitude,** or the distance between the crest and the trough.

Now imagine that you are standing next to a parked train, and the engineer blows the whistle. The sound consists of energy passing through the air from the whistle to your ear in the form of waves traveling at about 1,240 km an hour (about 760 mph). As sound energy passes through a body of air, the air first squashes together, then relaxes (▶Fig. 1.8b). Sound waves differ from shear waves in that the direction in which the air molecules move back and forth is parallel (rather than at right angles) to the direction that the wave moves. We call waves with this geometry **compressional waves.** Sound waves transfer energy from the whistle to your ear and cause your eardrum to vibrate, just as the water waves made the stick bob up and down. The vibrations of your eardrum, in turn, stimulate nerves that send information about the sound to your brain.

The sound you hear from the whistle has a specific pitch. To most people, **pitch** refers to the note in the musical scale that the sound makes. Pitch depends on the frequency of the sound waves. For example, the A above middle C on a piano has a frequency of 440 Hz. ("Hz" is an abbreviation for "hertz," the unit of frequency; 1 Hz = 1 cycle, or complete vibration, per second.) You perceive sounds of different pitches as different musical notes.

FIGURE 1.8 (a) Cross section of a series of water waves (a type of shear wave). (b) An image of sound waves (a type of compressional wave). Sound waves consist of alternating regions where air is compressed (closely spaced lines) and regions where air has dilated (widely spaced lines).

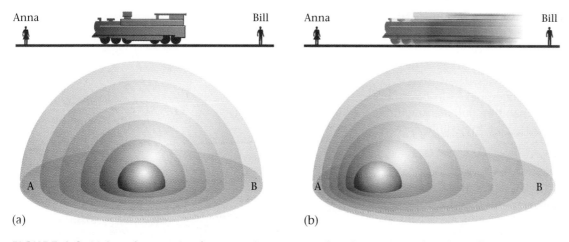

FIGURE 1.9 (a) Sound emanating from a stationary source has the same wavelength in all directions (the circles represent the waves), and observers at points A and B (Anna and Bill) hear the same pitch. (b) If the source is moving toward Anna, the wavelength seems shorter to Anna than to Bill. Therefore, Anna hears a higher-pitched (higher-frequency) sound than Bill does.

frequency of sound waves. Red light has a longer wavelength (lower frequency) than blue light (▶Fig. 1.10a, b). The Doppler effect also applies to light, but is only noticeable if the light source moves at over half the speed of light. If a light source moves away from you, the light will become redder (as the light shifts to lower frequency), and if the source moves toward you, the light will become bluer (as the light shifts to higher frequency). We call these changes the **red shift** and the **blue shift,** respectively.

FIGURE 1.10 Shapewise, light waves look like ocean waves in that they are shear waves, but physically, they are quite different. (a) Blue light has a relatively short wavelength (higher frequency). (b) Red light has a relatively long wavelength (lower frequency).

Blue light (high frequency)

(a)

Red light (low frequency)

(b)

Red Shifts and the Expanding Universe Theory

No one realized that the Doppler effect applied to light from stars until one frosty night in 1929, when, at the Mt. Wilson Observatory high in the mountains of California, Edwin Hubble and his colleague Milton Humason began studying light from a distant galaxy. The light was so faint that in order to see it, they had to use a clockwork mechanism to move their telescope so that it could stay focused on the galaxy while the Earth rotated. Hubble and Humason studied the light's wavelengths, using a special prism that divides light into a spectrum of different colors, and found that the wavelengths were shifted toward the red direction (toward longer wavelengths, or lower frequencies) (▶Fig. 1.11a–c). In other words, light from the distant galaxy exhibited a red shift.

Hubble and Humason immediately recognized that the red shift resulted from the Doppler effect. Their observation could mean only one thing: the distant galaxy must be moving away from Earth at an immense velocity. They began to look at light from other galaxies to find out which were moving toward Earth and which were moving away. To their surprise, they found that the light from *all* distant galaxies, regardless of their direction from Earth, exhibited a red shift (the galaxies were moving away).

How can all galaxies be moving away from us, regardless of which direction we look? Hubble puzzled over this question and finally came up with a solution. The whole Universe must be expanding! To picture the expanding Universe, imagine a ball of bread dough with raisins scattered throughout. As the dough bakes and expands into a loaf, each raisin moves away from its neighbors, in every direction (►Fig. 1.12a, b). Hubble's proposal came to be known as the **expanding Universe theory**.

Hubble's expanding Universe theory marked a revolution in thinking. No longer could we view the Universe as being fixed in dimension, with galaxies locked in position. Now we see the Universe as an expanding bubble, in which galaxies race away from each other at incredible speeds.

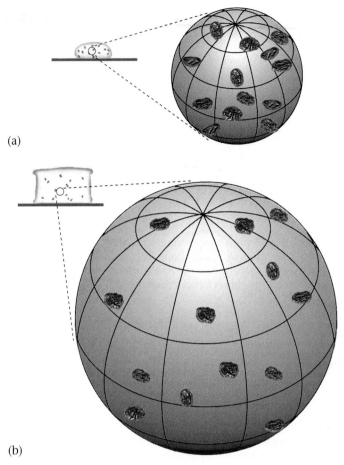

(a)

(b)

FIGURE 1.11 (a) Light, radio waves, and X-rays are all examples of electromagnetic energy. This energy is transmitted by electromagnetic waves. A spectrum represents the range of wavelengths for electromagnetic energy; very long wavelength energy (radio waves) lies at one end of the spectrum, while very short wavelength energy (X-rays and gamma rays) lies at the other. Visible light, which occupies a small portion of the spectrum, spans colors of the rainbow, from red (longer wavelengths) to violet (shorter wavelengths). (b) Starlight collected by a telescope, when passed through a prism, divides into the colors of the rainbow, which can be printed on a photographic plate to create a spectrograph. However, certain specific wavelengths, which were absorbed by atoms in the star, do not appear (they are black stripes, or absorption lines, on the spectrograph). (c) Simplified spectrographs from two different galaxies. The absorption lines for calcium shift toward red (to the left) for the distant galaxy.

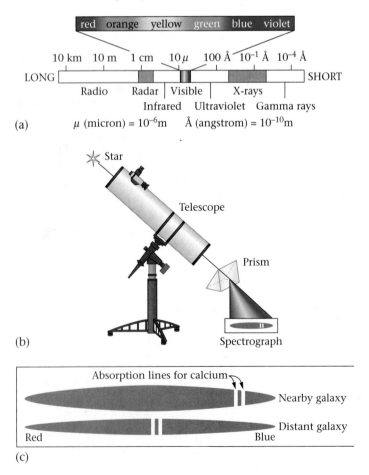

(a)

(b)

(c)

FIGURE 1.12 (a) At the dough stage, raisins in raisin bread are relatively close to one another. (b) During baking, the bread expands, and the raisins (like galaxies in the expanding Universe) have all moved away from each other. Notice that *all* raisins move away from their neighbors, regardless of direction.

This image immediately triggers the key question of cosmology: Has the Universe always been expanding, or did the expansion begin at some specific time in the past?

The Big Bang

Most astronomers have concluded that expansion did indeed begin at a specific time, with a cataclysmic explosion called the **big bang.** According to one version of the big bang theory, all matter and all energy—everything that now comprises the Universe—was initially packed into one point, a point so small that it occupied almost no volume at all. For reasons that no one understands, the point exploded between 10 and 20 billion years ago (an estimate based on the age of the most distant objects in the visible Universe). To learn what may have happened next, you first need to review some terms from basic physics and chemistry; please study Appendix A.

1.6 GROWING SOLAR SYSTEMS OUT OF CHAOS

Aftermath of the Big Bang

According to calculations, at the instant of the big bang, everything that was to become the Universe was so hot that matter could not exist in any form. Thermal energy (energy caused by the vibration and movement of particles) kept even the smallest atomic pieces, quarks, from sticking together. A second later, however, the Universe had expanded and cooled to about 5 billion degrees Celsius, five times the temperature of the center of the Sun. Under these conditions, protons and neutrons, the basic building blocks of atoms, could form. About 300,000 years later, the Universe had cooled enough for whole atoms (hydrogen and helium) to form. On its 1 millionth birthday, the Universe still looked totally different than it does today. Instead of stars, galaxies, and planets, the Universe consisted of a hot cloud of hydrogen (98%) and helium (2%) atoms. Astronomers refer to such a cloud of gas or dust in space as a **nebula.**

Astronomers suggest that the initial push of the big bang sent atoms gyrating outward in a wild, random dance. Eventually, the gas distribution became heterogeneous, meaning that the **density** (the mass, or amount of matter, in a given volume) of the nebula varied from place to place. The initial nebula began to clump into separate nebulae, separated from one another by emptier space (▶Fig. 1.13). The distribution of matter in the Universe perhaps began to resemble the distribution of mist in the sky on a day when there are scattered clouds.

Star Formation

In the context of the big bang theory, gas in the nebulae eventually accumulated into revolving balls and began to pull in even more matter from their surroundings—a grand case of "the rich getting richer." As gravity caused a gas ball to collapse inward, the ball began to spin progressively faster, just like a spinning figure skater who speeds up when she pulls her arms close to her side. **Centrifugal force,** the familiar outward-directed force that sends you flying off a spinning merry-go-round if you don't hold on, caused the spinning cloud to flatten into a disk with a bulbous core. The central core of the disk, where most of the mass concentrated, continued to collapse inward because of gravitational attraction. The resulting body, known as a **protostar,** became dense inside. As density progressively increased, temperature also increased, for the compression of a gas makes its temperature rise. Eventually, the core of a protostar became incredibly dense, and its temperature reached millions of degrees (see art on pp. 26–27).

Under the high pressures and temperatures reached within a protostar, atoms of hydrogen zip around at high velocity and bash into neighboring atoms. The violence of the collisions causes the nuclei of the hydrogen atoms to fuse (stick) together, thereby creating new, larger atoms of helium. Physicists refer to this process as **nuclear fusion** (or **nucleosynthesis**). The fusion of smaller atoms to produce larger atoms releases vast amounts of energy—the energy of a star! When the first nuclear furnace fired in a protostar, perhaps a couple of million years after the big bang, it became a true star, and the first starlight pierced the primordial blackness of the Universe. Soon, many protostars fired up, and a first generation of stars came to life. As we see later in this chapter, the disk-like nebulae surrounding these newborn stars eventually developed into planets.

FIGURE 1.13 Gases clump to form distinct nebulae, which look like clouds in the sky. In this Hubble Space Telescope picture, stars that have already formed light up the nebulae.

1.7 WE ARE ALL MADE OF STARDUST

Element Factories

The Universe today contains ninety-two naturally occurring elements (Table 1.1; see Appendix A for a discussion of elements). Where do these elements come from? They didn't all form during the big bang, because the newborn Universe consisted only of hydrogen and helium. They formed later, during the life cycle of a star.

The nebula condenses into a swirling disc, with a central ball surrounded by rings.

Forming the solar system, according to the nebula hypothesis: A second- or third-generation nebula forms from hydrogen and helium left over from the big bang, as well as from heavier elements that were produced by fusion reactions in stars or during explosions of stars.

Gravity reshapes the proto-Earth into a sphere. The interior of the Earth separates into a core and mantle.

Forming the planets from planetesimals: Planetesimals grow by continuous collisions. Gradually, an irregularly shaped proto-Earth develops. The interior heats up and becomes soft.

Soon after Earth forms, a small planet collides with it, blasting debris that forms a ring around the Earth.

The Moon forms from the ring of debris.

The ball at the center grows dense and hot enough for fusion reactions to begin. It becomes the Sun. Dust (solid particles) condenses in the rings.

Dust particles collide and stick together, forming planetesimals.

The Birth of the Earth-Moon System

Eventually, the atmosphere develops from volcanic gases. When the Earth becomes cool enough, moisture condenses and rains to create the oceans.

TABLE 1.1 Some Common Naturally Occurring Elements[a]

Name	Symbol	Atomic Number	Atomic Weight
Hydrogen	H	1	1.0
Helium	He	2	4.0
Carbon	C	6	12.0
Nitrogen	N	7	14.0
Oxygen	O	8	16.0
Fluorine	F	9	19.0
Neon	Ne	10	20.2
Sodium	Na	11	23.0
Magnesium	Mg	12	24.3
Aluminum	Al	13	27.0
Silicon	Si	14	28.1
Sulfur	S	16	32.1
Chlorine	Cl	17	35.5
Potassium	K	19	39.1
Calcium	Ca	20	40.1
Chromium	Cr	24	52.0
Iron	Fe	26	55.9
Cobalt	Co	27	58.9
Nickel	Ni	28	58.7
Copper	Cu	29	63.6
Zinc	Zn	30	65.4
Silver	Ag	47	107.9
Tin	Sn	50	118.7
Gold	Au	79	197.0
Mercury	Hg	80	200.6
Lead	Pb	82	207.2
Radon	Rn	86	222.0
Radium	Ra	88	226.0
Uranium	U	92	238.0

[a]See Appendix A for a complete periodic table.

With the discovery of nuclear fusion, physicists proposed just before World War II that stars are element factories, constantly fashioning heavier (large atomic weight) atoms out of lighter (small atomic weight) atoms through the process of nuclear fusion. When a star the size of our Sun first forms, hydrogen makes up essentially all the fuel. Hydrogen atoms fuse to form helium, and, once the hydrogen has been used up, helium atoms fuse to form still larger atoms. When the helium of the star has been consumed, then other fusion reactions take place, leading to the formation of still other, heavier, elements such as carbon. Nitrogen and oxygen atoms form when hydrogen atoms

fuse to carbon. Fusion reactions to form even larger atoms require temperatures higher than can occur in a star like our Sun, but which can occur in the core of very massive stars (ten to thirty times the mass of the Sun). In these massive stars, large atoms like iron (fifty-six times the size of hydrogen) can develop.

Because stars have a limited mass, they contain a limited amount of fuel and thus can only burn for a limited time.* Large stars, because of their greater mass, become denser and therefore hotter than small stars, so fusion reactions happen at a faster rate in bigger stars than in smaller ones. In other words, large stars burn out faster than do small stars. A star the size of the Sun will burn for about 10 billion years, whereas a star that is twenty-five times bigger may burn only for a few million years. When a star runs out of fuel, it dies, ejecting gases into space in the process. These gases contribute heavier elements into surrounding nebulae. A very massive star becomes a **supernova** when it dies, meaning that it explodes cataclysmically, ejecting large quantities of matter into space to form new nebulae (▶Fig. 1.14). (The Latin word *nova,* or "new," emphasizes that when such explosions occur, a previously invisible star temporarily becomes so bright that it's visible in the night sky.) The formation of elements larger than iron can only take place under conditions achieved during a supernova explosion.

To sum up the workings of the element factories: Hydrogen, the lightest element in the Universe, formed during the big bang. Helium, the next lightest element, formed partly during the big bang and partly by subsequent fusion reactions in stars. Larger atoms, up to iron, developed as a result of fusion reactions in stars, with larger stars producing heavier elements. Supernova explosions, in particular, produce very heavy elements.

In effect, when the first generation of stars died, they left a legacy of new elements that mixed with residual gas from the big bang. A second generation of stars and associated planets formed out of the new, compositionally more diverse nebulae. Second-generation stars lived and died, and contributed elements to third-generation stars. Succeeding generations of stars and planets contain a greater proportion of heavier elements. Because different stars live for varied periods of time, at any given moment the Universe contains many different generations of stars, including small stars that have been living for a long time and large stars that have only recently arrived on the scene. The mix of elements we find on Earth includes relics of primor-

*This kind of "burning," the energy-producing nuclear reaction in stars, is a very different process from burning wood in your fireplace. During the burning of wood, a chemical reaction, the chemical bonds holding atoms in the wood's molecules together break, and atoms simply rearrange into new chemical compounds.

FIGURE 1.14 Very heavy elements form during supernova explosions. Here, we see the rapidly expanding shell of gas ejected into space by an explosion.

dial gas from the big bang as well as the disgorged guts of dead stars. Think of it—elements that make up your body once formed inside a star!

Forming Our Solar System

According to calculations, the star we now know as the Sun developed less than 5 billion years ago, perhaps 10 billion years after the big bang. It's likely a third- or fourth-generation star, created from a nebula that contained all ninety-two elements (including elements formed in previous stars and during supernova explosions). The planets of the solar system developed from the residual ring of gas and dust, known as a **protoplanetary nebula,** that surrounded the Sun. Astronomers refer to the concept that planets grow out of planetary nebula as the **nebula theory of planet formation** (see art on pp. 26–27).

Initially, the protoplanetary nebula around the Sun consisted of a gas and dust cloud. Some atoms of heavier elements in the cloud combined into molecules, and these collected to form more solid dust specks. The dust, in turn, collected into tiny, solid pieces of rock and metal called **planetesimals.** Gases like hydrogen and helium, however, remained free. As it orbited the proto-Sun, the planetary nebula began to segregate into discrete rings. Planetesimals within a ring attracted one another gravitationally and collided. With each collision, some matter was ejected but some stayed attached, so gradually, over millions of years, the planetesimals clumped together into larger bodies. When smaller planetesimals collided with larger ones, the smaller body stuck to the larger. Finally, planet-sized bodies developed, and these **protoplanets** acted like vacuum cleaners, sucking in residual planetesimals that crossed their orbits. This residual matter

fell as **meteorites** (solid chunks) onto the protoplanets. Gradually, the protoplanets grew to become planets.

When the Sun began to produce fusion energy, it also produced a **solar wind,** a stream of particles (such as protons and electrons) flowing outward into space—these particles had enough energy to escape from the Sun's gravity (▶Fig. 1.15). As the planets formed, solar wind blew light, gaseous elements such as hydrogen and helium out of the inner portion of the solar system, leaving the inner planets (Mercury, Venus, Earth, and Mars) relatively free of these elements. Thus, the inner planets are composed primarily of rock and metal. In the outer parts of the solar system, however, the wind was weaker, and the gravitational attraction of the rocky protoplanets was strong enough to hold on to the lighter elements. These grew into the outer planets (Jupiter, Saturn, Uranus, and Neptune), large balls of gas surrounding a core of metal and rock (▶Fig. 1.16). Jupiter, for example, has a rocky core perhaps ten to twenty times the size of Earth, but this core comprises only 0.5% of the volume of the planet. Astronomers commonly refer to the inner planets as the **terrestrial** (Earth-like) planets, and the outer as the **Jovian** (Jupiter-like), or **gas-giant,** planets.

One of the prime reasons astronomers like the nebula theory of planet formation is that it explains why the **ecliptic** (the elliptical, or oval, plane traced out by a planet's orbit) of each planet except Pluto is nearly the same, and why all planets orbit the Sun in the same direction (▶Fig. 1.17). These observations make sense if all the planets formed out of a flattened disk of gas moving in the same direction around a central mass. Pluto, whose ecliptic is inclined relative to other planets' ecliptics, may be an errant moon, knocked out of orbit from around another planet.

FIGURE 1.15 The solar wind is made up of streams of particles flowing into space from the Sun.

FIGURE 1.16 The relative sizes of the planets of our solar system.

Forming the Earth's Moon

Without a doubt, the Moon, dominating the night sky, has affected humanity in innumerable ways. Its gravitational pull attracts the water in the sea and creates the tides, and its phases have governed the calendars of most societies since civilization began, and even before. But though we could see the Moon clearly with telescopes, we did not know how it came into being until after the *Apollo* astronauts returned samples to Earth during the late 1960s and early 1970s.

Before the *Apollo* mission, astronomers speculated that the Moon had once been another planet that became trapped by the Earth's gravity field, or that the Moon had been pulled out of the Earth by the gravitational attraction of an unknown body passing nearby. But after analyzing the *Apollo* samples, geologists concluded that the Moon formed when a Mars-sized protoplanet collided with Earth perhaps tens of millions of years after Earth first developed, blasting a substantial part of the Earth's rocky mantle into orbit around the Earth, and transferring so much energy that much of the Earth melted. The ring of debris

left from this collision quickly coalesced to form the Moon (see pp. 26–27). Once formed, the Moon and Earth continued to be pummeled by the barrage of meteors that all planets endured while their orbits still contained countless tiny planetesimals. The early meteor impacts on the Earth have long since been erased by continuing geologic activity, but on the static (unchanging) Moon the craters remain as stark testimony to the early stage of the solar system's evolution.

Why Is the Earth Round?

Planetesimals are jagged or irregular in shape. Planets, on the other hand, are nearly spheres. Why are planets spherical? Suppose you take a block of cheese outside on a hot summer afternoon and place it on a table. As the Sun warms the cheese, it gets softer and softer, and eventually gravity causes the cheese to spread out in a pancake-like blob on the table. What this model shows is that gravitational force alone can cause material to change shape if the material is soft enough. Now let's apply this model to planetary growth.

FIGURE 1.17 The planets' ecliptics. Note that the orbits of all planets except Pluto lie in the same plane. Pluto may be an errant moon; some astronomers do not call it a planet.

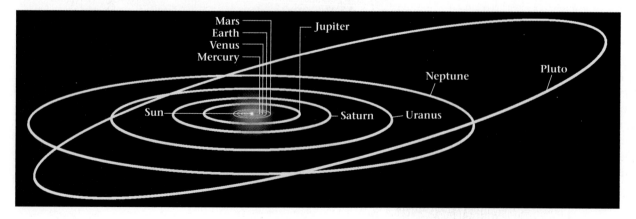

Cool rock is quite strong, as you know if you've ever been hit in the head by a stone. The rock composing a small planetesimal is cool and strong enough so that the force of gravity is not sufficient to cause the rock to flow. Thus, planetesimals maintain the irregular shapes they originate with or that develop when collisions lop off large chunks from them. But once a planetesimal grows beyond a certain critical size (somewhat smaller than our Moon), the insides of the planet become warm, and the rock becomes soft enough to flow in response to gravity; also, the gravitational force has become strong. As a consequence, protrusions are pulled inward toward the center, and the planetesimal reforms into a shape that permits the force of gravity to be the same at all points on its surface. This special shape is a sphere. (In reality, the centrifugal force created by the rotation of the planets causes them to be slightly flattened at the poles; see pp. 26–27.)

CHAPTER SUMMARY

• Most Greek philosophers favored a geocentric Universe concept, which placed the Earth at the center of the Universe, with the planets and Sun orbiting around the Earth within a celestial sphere speckled with stars. The heliocentric (Sun-centered) Universe concept was not widely accepted.

• Ptolemy's mathematical formulations led to the acceptance of the geocentric Universe concept throughout the Middle Ages.

• The Renaissance brought a revolution in scientific thought. The idea of a spherical Earth returned, and Copernicus reintroduced the heliocentric concept, which was then proven by Galileo. Newton introduced physical laws that allowed people to understand motion in the Universe.

• Eratosthenes was able to measure the size of the Earth in ancient times, but it was not until fairly recently that astronomers accurately determined the distance to the Sun, planets, and stars. Distances in the Universe are so large that they must be measured in light years.

• The Earth is one of nine planets orbiting the Sun, and this solar system lies on the outer edge of a slowly revolving galaxy, the Milky Way, which is composed of about 300 billion stars. The Universe contains at least hundreds of billions of galaxies.

• The red shift of distant galaxies, a manifestation of the Doppler effect, indicates that all distant galaxies are moving away from the Earth. This observation leads to the expanding Universe theory. Most astronomers agree that this expansion began after the big bang, a cataclysmic explosion about 10–20 billion years ago.

• The first atoms (hydrogen and helium) of the Universe developed about 1 million years after the big bang. These atoms formed vast gas clouds, called nebulae.

• According to the nebula theory of planet formation, gravity caused clumps of gas in the nebulae to coalesce into dense, revolving balls. As these balls of gas collapsed inward, they evolved into flattened disks with a bulbous center. The protostars at the center of these disks eventually became dense and hot enough that fusion reactions began. When this happened, they became true stars, emitting heat and light.

• Heavier elements form during fusion reactions in stars; the heaviest are probably made during supernova explosions. The Earth and the life forms on it contain elements that could only have been produced during the life cycle of stars. Thus, we are all made of stardust. Several generations of stars have formed and have died since the big bang.

• Planets developed from the rings of gas and dust, the planetary nebulae, that surrounded protostars. The gas condensed into planetesimals that then clumped together to form protoplanets, and finally true planets. In our solar system, solar wind blew lighter elements outward, where they accumulated and formed the Jovian, or gas-giant, planets. The rocky and metallic balls in the inner part of the solar system did not acquire huge gas coatings; they became the terrestrial planets.

• The Moon formed from debris blasted free from Earth when our planet collided with a Mars-sized planet early during the history of the solar system.

• A planet the size of the Moon or larger will assume a near-spherical shape because the warm rock inside is so soft that gravity smoothes out irregularities.

KEY TERMS

amplitude (p. 22)	nuclear fusion (p. 25)
big bang theory (p. 24)	parallax (p. 19)
celestial sphere (p. 16)	pitch (p. 21)
centrifugal force (p. 25)	planet (p. 29)
compressional waves (p. 22)	protoplanetary nebula
cosmology (p. 15)	(p. 29)
Doppler effect (p. 21)	planetesimals (p. 29)
ecliptic (p. 29)	protoplanet (p. 29)
expanding Universe theory (p. 24)	protostar (p. 25)
frequency (p. 21)	red shift (p. 23)
galaxy (p. 21)	shear waves (p. 22)
geocentric Universe concept (p. 16)	solar wind (p. 29)
heliocentric Universe concept (p. 16)	supernova (p. 28)
Jovian planets (p. 29)	terrestrial planets (p. 29)
light year (p. 19)	Universe (p. 15)
meteorite (p. 29)	wavelength (p. 22)
nebula (p. 25)	

REVIEW QUESTIONS

1. Why do the planets appear to move with respect to the stars?

2. Contrast the geocentric and heliocentric Universe concepts.

3. How did Galileo's observations support the heliocentric Universe concept?

4. Describe how Foucault's pendulum demonstrates that the Earth is rotating on its axis.

5. How did Eratosthenes calculate the circumference of the Earth?

6. Describe how the parallax method can be used to estimate the distance to far objects.

7. You hear the main character in a cheap science-fiction movie say he will "return ten light years from now." What's wrong with his usage of the term "light year"? What are light years actually a measure of?

8. Describe how the Doppler effect works.

9. If you hear a train whistle's pitch growing higher and higher, is it moving toward you or away from you?

10. What does the red shift of the galaxies tell us about their motion with respect to the Earth?

11. Briefly describe the steps in the formation of the Universe and the solar system.

12. How is a supernova different from a normal star?

13. Why do the inner planets consist mostly of rock and metal, but the outer planets mostly of gas?

14. Why are all the planets in the solar system (except Pluto) orbiting the Sun in the same direction and in the same plane?

15. Describe how the Moon was formed.

16. Why is the Earth round?

SUGGESTED READING

Adams, F., and G. Laughlin. 1999. *The Five Ages of the Universe.* New York: Touchstone.

Ahrens, T. J. 1994. The origin of the Earth. *Physics Today* 47: 38–45.

Allegre, C. 1992. *From Stone to Star.* Cambridge, Mass.: Harvard University Press.

Beatty, J. K., and A. Chaikin, eds. 1990. *The New Solar System.* Cambridge, Mass.: Sky.

Chown, M. 1993. *Afterglow of Creation.* New York: Arrow.

Clark, S. 1995. *Towards the Edge of the Universe.* New York: Springer Verlag.

Hartmann, J., and R. Miller. 1991. *The History of the Earth: An Illustrated Chronicle of an Evolving Planet.* New York: Workman.

Hawking, S. 1988. *A Brief History of Time.* New York: Bantam.

Kuhn, K. F. 1994. *In Quest of the Universe.* St. Paul, Minn.: West.

Silk, J. 1988. *The Big Bang: The Creation and Evolution of the Universe.* New York: Freeman.

———. 1994. *A Brief History of the Universe.* New York: Freeman.

Van den Bergh, S., and J. E. Hesser. 1993. How the Milky Way formed. *Scientific American* 268 (1): 72–81.

Weinberg, S. 1996. *The First Three Minutes.* New York: Flamingo.

Wetherill, G. W. 1990. Formation of the Earth. *Annual Reviews of Earth and Planetary Sciences* 18: 205–56.

CHAPTER 2

Journey to the Center of the Earth

2.1 INTRODUCTION

For most of human history, people perceived the other planets of our solar system to be nothing more than bright points of light that moved in relation to each other and in relation to the stars. In the seventeenth century, when Galileo first aimed his telescope skyward, they became hazy spheres, and through the nineteenth and early twentieth centuries, as more powerful telescopes were developed, our image of the planets continued to change. But now that we have actually sent space probes out to investigate them, we have exquisitely detailed pictures displaying the texture and color of the planetary surfaces, as well as basic data about their composition. From this information, geoscientists have come up with hypotheses to explain the history of the planets.

What if we turned the tables and became explorers from another planet visiting Earth for the first time? What would we see? Even without touching the planet, we could detect its magnetic field and atmosphere, and could characterize its surface. We could certainly distinguish regions of land, sea, and ice. We could also get an idea of the nature of Earth's interior, though we could not see the details. In the first part of this chapter, we rocket to Earth and study its external characteristics. In the second part, we build an image of the planet's interior, based on a variety of data. (Of course, no one can see the interior directly, because high pressures and temperatures would crush and melt any visitor.) This high-speed tour of Earth will provide a frame of reference for the remainder of the book.

2.2 WELCOME TO THE NEIGHBORHOOD

Our journey begins in interplanetary space, far outside the orbit of the Moon. Compared with the air we breathe at sea level, interplanetary space is a **vacuum,** meaning that it contains very little matter in a given volume. Here, there are only 5,000 (that is, 5×10^3) atoms per liter, while in air at sea level, there are 27,000,000,000,000,000,000,000 (2.7×10^{22}) atoms per liter. (Note that it's more

Miners digging for gold 3.5 km below the Earth's surface in South Africa. Even at this depth, the miner has traversed only 0.0005 of the Earth's radius.

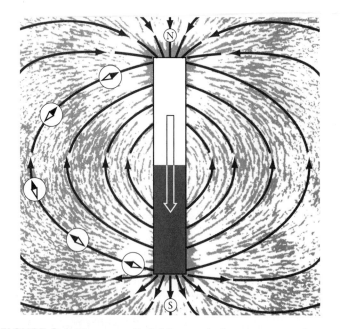

FIGURE 2.1 The magnetic field around a bar magnet can be displayed by sprinkling iron filings on a sheet of paper lying over the magnet. The filings define curving trajectories—these are the field lines. Note that the magnet has a north pole and a south pole.

convenient to write very large numbers using scientific notation.) Some of the atoms in interplanetary space are left over from the nebulae out of which the solar system formed, some escaped from the atmospheres of planets, and some, ejected from the Sun, comprise the solar wind (see Chapter 1). Interestingly, we can actually see the solar wind

in action, in the familiar tail of a comet. **Comets** are small balls of ice and dust, probably remaining from the formation of the solar system, that follow highly elliptical orbits around the Sun. When they approach the Sun, they warm up and start to evaporate. Solar wind blows the resulting stream of gas outward, creating the glowing streak that we recognize as the tail.

As our rocket approaches the Earth, its instruments detect the planet's magnetic field, like a signpost shouting, "Approaching Earth!" A **magnetic field** is the region affected by the force emanating from a magnet. This force, which grows progressively stronger as you approach the magnet, attracts or repels electrically charged or magnetic particles. (Of note, humans have been aware of magnetism since at least 500 B.C.E., and have used compasses since around 500 C.E. William Gilbert, physician to England's Queen Elizabeth I around 1600, seems to have been the first scientist to suggest that compasses work because the Earth itself behaves like a magnet.) Earth's magnetic field, like the familiar magnetic field around a bar magnet, is largely a **dipole:** it has a north pole and a south pole. We can portray the magnetic field by drawing **magnetic field lines,** the trajectories along which magnetic particles would align, or charged particles would flow, if placed in the field (▶Fig. 2.1). Since the solar wind contains electrically charged particles, the wind interacts with Earth's magnetic field, distorting it into a huge teardrop pointing away from the Sun. Fortunately, the magnetic field deflects the wind, so that most of the particles do not reach the Earth's surface. In this way, the magnetic field acts like a shield against the solar wind; the region inside this magnetic shield is called the **magnetosphere** (▶Fig. 2.2).

FIGURE 2.2 The magnetic field of the Earth interacts with the solar wind—the wind distorts the field so that it tapers away from the Sun, and the field isolates the Earth from most of the wind. Note the Van Allen belts near the Earth.

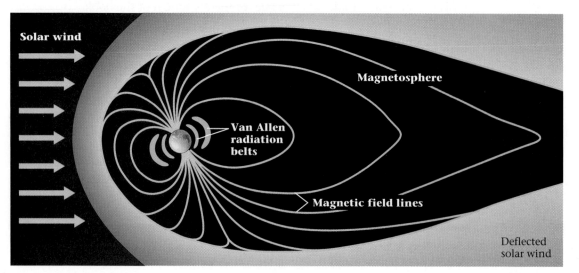

Though it protects the Earth from the solar wind, the magnetic field does not stop our rocket ship, and we continue to speed toward the planet. At distances of about 3,000 km and 10,500 km out from the Earth, we encounter two distinct belts called the **Van Allen radiation belts,** named for the physicist who first recognized them in 1959. These consist of solar wind particles as well as **cosmic rays** (nuclei of atoms, especially of hydrogen, that bombard the Earth from deep space; they may have been emitted from supernova explosions) that were moving so fast they were able to penetrate the weaker outer part of the magnetic field and were then trapped by the stronger magnetic field closer to the Earth. By trapping cosmic rays, the Van Allen belts protect life on Earth from dangerous radiation.

2.3 THE ATMOSPHERE

Approaching still closer to the Earth, we enter its **atmosphere,** an envelope of gas consisting overall of 78% nitrogen (N_2) and 21% oxygen (O_2), with minor amounts (1% total) of argon, carbon dioxide (CO_2), neon, methane, ozone, carbon monoxide, and sulfur dioxide, that surrounds

the Earth (▶Fig. 2.3a, b). Of note, other terrestrial planets have atmospheres, but none of these is like Earth's. Venus's atmosphere, dense enough to hide the planet's surface, consists almost entirely of carbon dioxide. Mercury has only a trace of an atmosphere: because of the planet's high temperature, its atmosphere boiled away and escaped to space long ago. Mars has a thin atmosphere that, like Venus's, consists almost entirely of carbon dioxide. It's almost arbitrary where we place the outer limit of Earth's atmosphere, for the density of gas comprising the atmosphere gradually decreases with altitude until it's the same as that of interplanetary

FIGURE 2.3 (a) Sunset view from the space shuttle. The gases and dust of the atmosphere reflect light and absorb certain wavelengths of light, creating a glowing palette of color. The vacuum of space is always black. (b) Nitrogen and oxygen comprise most of the gas in the atmosphere.

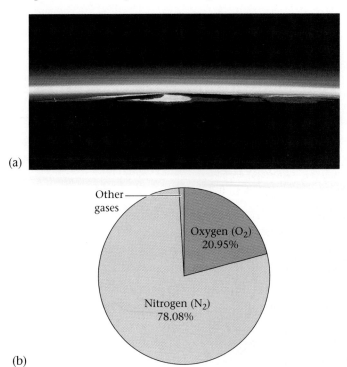

(a)

(b)

FIGURE 2.4 (a) A graph displaying the variation of air pressure with elevation shows that by an elevation of 30 km, atmospheric pressure is less than 1% of the atmospheric pressure at sea level. (b) Atmospheric density increases toward the base of the atmosphere because the weight of the upper atmosphere squeezes together gas molecules in the lower atmosphere. (c) By analogy, if you place a spring on a table in a gravity field, the weight of the upper part of the spring pushes down on the lower part and causes it to squeeze together.

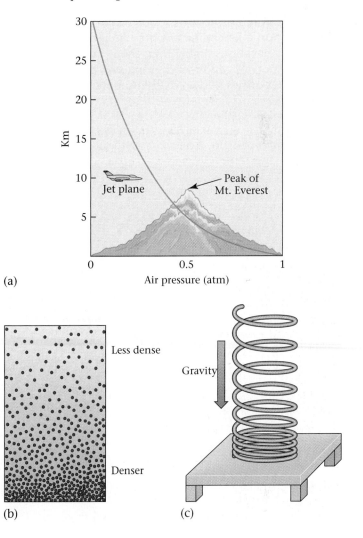

(a)

(b)   (c)

space (this occurs about 10,000 km from Earth). But 99% of the gas in the atmosphere lies below 50 km, and most of the remaining 1% lies between 50 and 500 km.

The density of the atmosphere at a given elevation reflects the pressure caused by the weight of the overlying column of air: this overlying air decreases with increasing altitude, so pressure also decreases with increasing altitude. By definition, **pressure** is the "push" acting on a material (pressure = force per unit area); in this case, it squeezes molecules in the air closer together. At sea level, air pressure on average is 1 atmosphere (atm; 1 atm = 1.03 kilograms per square centimeter = 14.7 pounds per square inch; and 1 atm = 1.01 bars, another unit of pressure); at an elevation of 5.6 km, air pressure is 0.5 atm (50% of the air lies below 5.6 km); while at the peak of Mt. Everest, 8.85 km high, air pressure is only 0.3 atm (▶Fig. 2.4a–c). Where the space shuttle orbits the Earth, an altitude of about 400 km (about 250 miles), air pressure is only 0.0000001 atm. Keep in mind that humans cannot live for long at elevations greater than about 4.5–5.5 km. If your hike to the top of Mt. Everest and down again lasts more than a few hours, you'll need to breathe bottled oxygen to survive. Commercial jet airplanes fly at an altitude of about 11 km, where there is enough oxygen to run a jet engine but not enough for a person to breathe. At elevations higher than about 25 km, jet engines no longer function, and rocket propulsion becomes a necessity (a rocket carries its own oxygen supply).

The nature of the atmosphere changes with distance from the Earth. Because of these changes, atmospheric scientists divide the atmosphere into layers. Notably, most winds and clouds develop only in the lowest layer, the **troposphere,** where air undergoes convection (hence the prefix *tropos,* Greek for "turning"). **Convection** refers to the circulation that occurs when different parts of a fluid (a liquid or gas) have different temperatures—warmer portions of a fluid are less dense and thus rise, while cooler portions are denser and thus sink (see Appendix A). Winds develop when sinking cool air moves in to replace rising warm air. Clouds, mists of tiny droplets, form when the rising air contains moisture, for as this air rises, it cools, and the moisture condenses. (The other atmospheric layers, from above the troposphere to the top, are the stratosphere, mesosphere, and thermosphere; these will be discussed in Chapter 20.)

2.4 LAND AND OCEANS

Now we've gone into orbit around the Earth, and we set about mapping the planet—what obvious features should we put on the map? To start with, Earth's surface looks totally different from that of any other terrestrial planet or moon in the solar system. The surfaces of other planets are rocky regions, pockmarked by craters. In contrast, Earth has

FIGURE 2.5 This map of the Earth shows variations in elevation of both the land surface and the sea floor. Darker blues are deeper water in the ocean. Greens are lower elevation on land.

FIGURE 2.6 A hypsometric diagram shows the proportions of the Earth's solid surface at different elevations. Two principal zones—the continents and adjacent continental shelf areas (the submerged margins of continents), and the ocean floor—account for most of Earth's area. Mountains and deep trenches cover relatively little area.

three distinctly different types of surface: dry land (continents and islands), the oceans, and ice-covered areas. Land (including ice-covered land) covers about 30% of the Earth's surface, while the oceans cover the remaining 70%. Lakes and rivers cover a small portion of the land. Geologists sometimes refer to the surface water of Earth (lakes, rivers, and oceans), along with **groundwater** (water that fills openings underground) and water vapor in the atmosphere, as the **hydrosphere.** Craters or their relicts do exist, but they are relatively rare.

While orbiting the Earth, our instruments record variations in elevation—the **topography**—of the solid Earth, and show diverse landscapes such as mountains, valleys, and plains on the continents (▶Fig. 2.5). Our instruments are so sensitive that they also detect variations in the elevation of the sea floor, and we recognize submarine plains, oceanic ridges, and deep trenches, or troughs. The results of such measurements can be portrayed on a special kind of graph called a **hypsometric curve,** which plots surface elevation on the vertical axis and the percentage of the Earth's surface on the horizontal axis (▶Fig. 2.6). Notice that a relatively small proportion of the Earth's surface occurs at very high elevations (mountain ranges) or at great depths (oceanic trenches). In fact, most of the land surface lies just within a kilometer of the sea level, while most of the sea floor is between 2.5 and 4.5 km deep. Clearly, a slight change in the sea level would dramatically change the amount of dry land—a rise, for example, would flood much of the land.

Even without taking samples of the Earth, sensors in our orbiting spaceship find that the temperature of the atmosphere varies with latitude (distance from the equator)

FIGURE 2.7 On the Earth, temperature varies with latitude. These temperature variations, along with moisture variations, create different climatic belts.

and that Earth has complex weather patterns, with huge atmospheric storms tracking across its surface (▶Fig. 2.7). We might also be able to determine that the oceans contain saltwater, which moves in vast currents; that rivers drain the land, indicating that water cycles between the land, the atmosphere, and the sea; and that the Earth supports vegetation and animal life, which vary in character with location. Finally, we discover that the Earth has various climatic belts, which differ from one another in terms of air temperature, moisture content, and vegetation type. From our vantage point in space, we clearly recognize the complexity of the Earth system.

2.5 WHAT IS THE EARTH MADE OF?

Elemental Composition

At this point, we leave our fantasy space voyage and turn our attention to the materials that make up the Earth, which we need to know about before proceeding to the Earth's interior. In Chapter 1, we learned that the atoms that comprise the Earth consist of a mixture of elements left over from the big bang as well as elements produced by fusion reactions in stars and during supernova explosions. During the birth of the solar system, solar wind blew lighter elements away from the region in which Earth was forming, like wind separating chaff from wheat. Earth formed from the heavier elements left behind: iron (35%), oxygen (30%), silicon (15%), and magnesium (10%) make up most of Earth's matter (▶Fig. 2.8). The remaining 10% consists of the other eighty-eight naturally occurring elements.

FIGURE 2.8 The proportions of major elements on the Earth. Note that iron and oxygen are the most abundant elements.

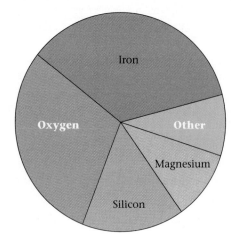

Categories of Earth Materials

A great variety of materials make up the Earth. Here are some of the main categories.

- *Organic chemicals:* Carbon-containing compounds that occur in living organisms, or that resemble such compounds, are called **organic chemicals.** Examples include oil, protein, plastic, fat, and rubber. Certain simple carbon-containing materials, however—such as pure carbon (C), carbon dioxide (CO_2), carbon monoxide (CO), lime (CaO), and calcium carbonate ($CaCO_3$)—are *not* considered organic.

- *Minerals:* A solid inorganic (not organic) substance in which atoms are arranged in an orderly pattern is called a **mineral.** Minerals grow either by the cooling and freezing of a liquid or by precipitation out of a water solution. **Precipitation** occurs when atoms that had been dissolved in water come together and form a solid. (For example, solid salt forms by precipitation out of seawater when the water evaporates.) A single coherent sample of a mineral that grew to its present shape and has smooth, flat faces is called a **crystal,** while an irregularly shaped sample, or fragment, derived from a once-larger crystal or group of crystals is a **grain.** Familiar minerals include quartz, diamond, feldspar, and calcite.

- *Glasses:* A solid in which atoms are not arranged in an orderly pattern is called a **glass.** Glass forms when a liquid freezes so fast that atoms do not have time to organize into an orderly pattern. Common window glass, for example, is made when molten (liquid) quartz freezes quickly.

- *Rocks:* Aggregates of mineral crystals or grains, and masses of natural glass, are called **rocks.** Geologists recognize three main groups of rocks. (1) **Igneous rocks** develop when hot molten rock cools and freezes solid. (2) **Sedimentary rocks** form from grains that break off preexisting rock and become cemented together, or from minerals that precipitate out of a water solution; an accumulation of loose mineral grains (grains that have not stuck together) is called **sediment.** (3) **Metamorphic rocks** are created when preexisting rocks undergo changes including the growth of new minerals in response to heat and pressure. (See Chapters 6–8.)

- *Metals:* Solids composed almost entirely of metal atoms (such as iron, aluminum, copper, and tin) are called **metals.** An **alloy** is a mixture containing more than one type of metal atom (for example, bronze is a mixture of copper and tin).

- *Melts:* **Melts** form when solid materials become hot and transform into a liquid. Molten rock is a type of melt—

geologists distinguish between **magma,** which is molten rock beneath the Earth's surface, and **lava,** molten rock that has flowed out onto the Earth's surface.

• *Volatiles:* Materials that easily transform into a gas at the relatively low temperatures found at the Earth's surface are called **volatiles.** Examples include water, carbon dioxide, and the other gases that comprise the atmosphere.

Notably, the most common minerals in the Earth contain the compound **silica** (SiO_2) mixed in varying proportions with other elements (typically iron, magnesium, aluminum, calcium, potassium, and sodium). These are called **silicate minerals,** and, not surprisingly, rocks composed of silicate minerals are called **silicate rocks.** Geologists distinguish four classes of igneous silicate rocks based on the proportion of silica-rich minerals within them, relative to the proportion of iron- and magnesium-rich minerals within them. In order, from greatest to least proportion of silica-rich minerals, these classes are **silicic** (abundant silica, relatively little iron and magnesium), **intermediate** (between silicic and mafic), **mafic** (relatively little silica, abundant iron and magnesium), and **ultramafic** (large amounts of iron and magnesium, very little silica). As the proportion of silica-rich minerals in a rock increases, the **density** (mass per unit volume) decreases. Thus, silicic rocks are less dense than mafic rocks.

Within each class, there are many different rock types, each with a name, that differ from one another in terms of composition (chemical makeup) and crystal size. These will be discussed in detail in Part II, but for now we need to know four rock names: **granite** (a silicic rock with large grains), **basalt** (a mafic rock with small grains), **gabbro** (a mafic rock with large grains), and **peridotite** (an ultramafic rock with large grains).

2.6 DISCOVERING THAT THE EARTH HAS LAYERS

To find out what's inside the Earth, our rocket ship won't be much use. In fact, the world's deepest mine shaft penetrates gold-bearing rock that lies only about 3.5 km (2 miles) beneath South Africa. Though miners seeking this gold must begin their workday by plummeting straight down a vertical shaft for almost ten minutes aboard the world's fastest elevator, the shaft is little more than a scratch on Earth's surface when compared with the planet's radius (the distance from the center to the surface, 6,370 km). Even the deepest well ever drilled, a 12-km-deep puncture in northern Russia, penetrates only the upper 0.03% of the Earth. We literally live on the thin skin of our planet, its interior forever inaccessible to our wanderings.

FIGURE 2.9 A literary image of the Earth's insides: *The Fallen Angels Entering Pandemonium, from* [Milton's] *"Paradise Lost," Book 1,* by English painter John Martin (1789–1854).

People have wondered about the Earth's interior since ancient times, knowing that it was the source of incandescent lavas spewed from volcanoes, of precious gems and metals, of sparkling spring water, and of mysterious powers that were strong enough to shake the ground and topple buildings. Without the ability to observe the Earth's interior firsthand, pre-twentieth-century authors dreamed up fanciful images of what Earth's interior looked like. For example, the English poet John Milton (1608–1674) portrayed the underworld as a "dungeon horrible, on all sides round, as one great furnace flamed"—a perception widely held in the West (▶Fig. 2.9). Perhaps the molten rock and sulfur-rich smoke that spewed from volcanoes in the Mediterranean inspired such images. In the eighteenth and nineteenth centuries, European writers thought that the Earth's interior resembled a sponge, containing open caverns variously filled with molten rock, water, or air. In this way, the interior could provide both the water that bubbled up at springs *and* the lava that erupted at volcanoes. In fact, in French author Jules Verne's popular 1864 novel *Journey to the Center of the Earth,* three explorers find a route through interconnected caverns to the Earth's center. Our present image of the Earth's interior, one made up of distinct layers, is the end product of many discoveries made during the past two hundred years (see art on pp. 42–43). The discoveries were made with the help of certain clues.

Clues from Measuring Earth's Density

The first key to understanding the Earth's interior came from studies that provided an estimate of the planet's den-

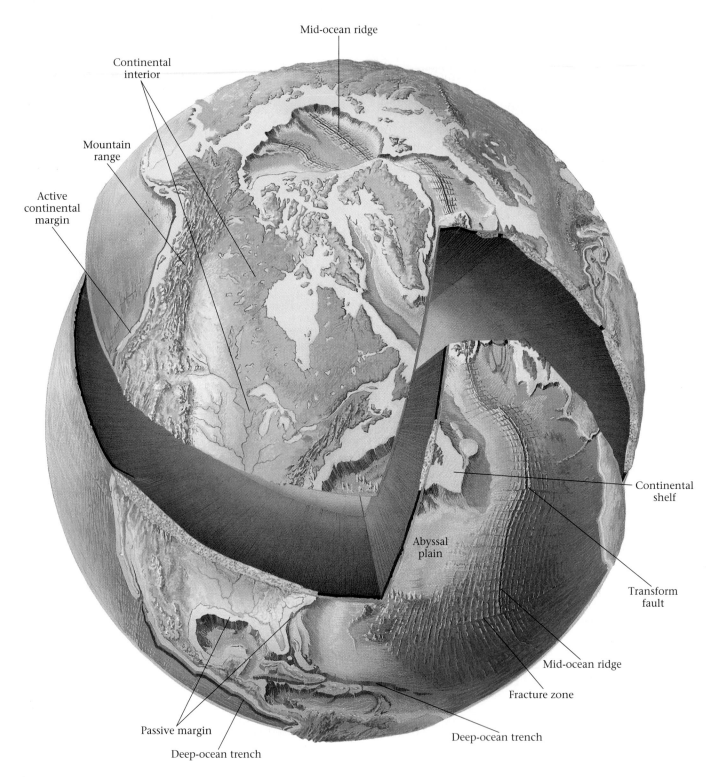

Mid-ocean ridge

Continental interior

Mountain range

Active continental margin

Continental shelf

Abyssal plain

Transform fault

Mid-ocean ridge

Fracture zone

Passive margin

Deep-ocean trench

Deep-ocean trench

The Interior of the Earth

If we could break open the Earth, we would see that its interior consists of a series of concentric layers, called (in order from the surface to the center) the crust, the mantle, and the core. The crust is a relatively thin skin (7–10 km beneath oceans, 25–70 km beneath the land surface). Oceanic crust consists of basalt (mafic rock), while the average continental crust is intermediate to silicic. The mantle, which overall has the composition of ultramafic rock, can be divided into three layers: upper mantle, transition zone, and lower mantle. The core can be divided into an outer core of liquid iron alloy and an inner core of solid iron alloy. Temperature increases

progressively with depth, so at the Earth's center the temperature may approach that of the Sun's surface.

Within the mantle and outer core, there is swirling, convective flow. Flow within the outer core creates the Earth's magnetic field.

When discussing plate tectonics, it is convenient to call the outer part of the Earth, a relatively rigid shell composed of the crust and uppermost mantle, the lithosphere and the underlying warmer, more plastic portion of the mantle the asthenosphere. These are not shown in this painting.

Geographic
north pole

Lines of magnetic force

Magnetic
north pole

North America

Mantle
plume

Mantle

Inner core
(solid metal alloy)

Outer core
(liquid metal alloy)

Magnetic
south pole

Geographic
south pole

Upper mantle

Transition zone

Lower mantle

Crust

Liquid outer core

Solid inner core

Ultramafic rock

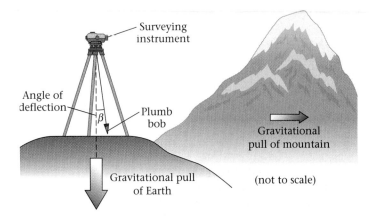

FIGURE 2.10 A surveyor noticed that the plumb line he was using to level his surveying instrument did not hang exactly vertically near a mountain; it was deflected by an angle ß, owing to the gravitational attraction of the mountain. The angle of deflection represents the ratio between the mass of the mountain and the mass of the whole Earth.

sity (mass per unit volume). To determine Earth's density, you must first determine Earth's mass (the amount of matter making up the Earth). The first realistic estimate of Earth's mass came in 1776 from the work of a British astronomer, Nevil Maskelyne. Maskelyne suggested weighing the Earth by examining the deflection of a plumb bob attached to a surveying instrument. (A plumb bob is a weight at the end of a line used to determine the orientation of a vertical line.) He proposed that the angle of deflection (ß) of a plumb bob caused by the gravitational attraction of a mountain would indicate the magnitude of a gravitational attraction exerted by the mountain's mass relative to the gravitational attraction of the Earth's mass. Maskelyne tested his hypothesis at Schiehallion Mountain in Scotland (▶Fig. 2.10). His results led to an estimate of 4.5 grams per cubic centimeter (g/cm^3) for Earth's density (this is the number after its conversion into the modern metric system, for the metric system hadn't been invented yet). In 1778, English physicist Henry Cavendish, using a different method for the experiment, arrived at a density estimate of 5.45 g/cm^3 (more than five times the density of water), fairly close to modern estimates. As typical rocks (like granite and basalt) at the surface of the Earth have a density of only 2.2–2.5 g/cm^3), the average density of Earth exceeds that of its surface rocks. Certainly the open voids that Jules Verne described could not exist!

Clues from Measuring Earth's Shape

Once they had determined that the density of the Earth's interior was greater than that of its surface rocks, nineteenth-century geologists asked, "Does this density increase *gradually* with depth, or does the Earth consist of a less dense shell surrounding a denser core?" If the Earth's density increased only gradually with depth, most of its mass would lie far from the center, and centrifugal force from the spinning of the planet would cause the Earth to flatten into a disk. Since this doesn't happen, geologists concluded that much of Earth's mass must be concentrated close to the center. The dense center came to be known as the core. Eventually, geologists found that the density of this core approaches 13 g/cm^3.

To further determine the nature of Earth's interior, geologists measured tides, the rise and fall of the Earth's surface in response to the gravitational attraction of the Moon and Sun. If the Earth were composed of a liquid surrounded by only a thin solid crust, then the surface of the land would rise and fall daily like the surface of the sea. We don't observe such behavior, so the Earth's interior must be largely solid. Thus, the image conveyed by the popular press that the outer skin of the Earth floats on a sea of molten rock is simply not true.

By the end of the nineteenth century, geologists had recognized that the Earth resembled a hard-boiled egg, in that it had three principal layers: a not-so-dense **crust** (like an eggshell; composed of rocks like granite, basalt, and gabbro), a denser solid **mantle** in between (the white; composed of a then-unknown material), and a very dense **core** (the yolk; also composed of a then-unknown material). Clearly, many questions remained. How thick are the layers? Are the boundaries between layers sharp or gradational? And what exactly are the layers composed of?

Clues from the Study of Earthquakes: Refining the Image

One day in 1889, a physicist in Germany noticed that the pendulum in his lab began to move without having been touched. He reasoned that the pendulum was actually standing still, because of its inertia (the tendency of an object at rest to remain at rest, and of an object in motion to remain in motion), and that the Earth was moving under it. A few days later, he read in a newspaper that a large earthquake had taken place in Japan minutes before the movement of his pendulum began. The physicist knew that an **earthquake** is a vibration caused by the sudden breaking of rocks in the Earth, and quickly deduced that the energy responsible for jiggling his laboratory in Germany had traveled through the Earth from Japan. This energy moves in the form of waves, called either **seismic waves** or **earthquake waves,** that resemble the shock waves you feel with your hands when you snap a stick (▶Fig. 2.11). (Of note, the breaking of rock during an earthquake either produces a new fracture in the Earth or causes sliding on a preexisting fracture. A fracture on which sliding occurs is called a **fault.**)

Geologists soon realized that the study of seismic waves traveling through the Earth might provide a tool for explor-

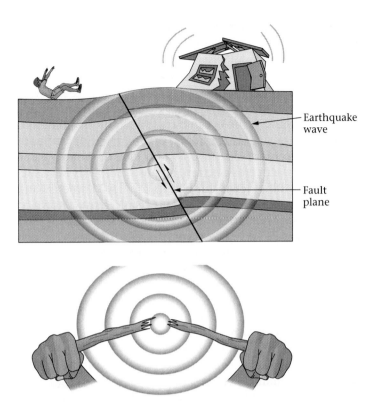

FIGURE 2.11 When the rock inside the Earth suddenly breaks and slips, forming a fracture called a fault, it generates shock waves that pass through the Earth and shake the surface (creating an earthquake), much as the sound waves from a stick snapping travel to you and make your eardrum vibrate.

ing the Earth's insides (much like doctors use ultrasound to study a patient's insides). They eventually discovered that **seismic velocity,** the speed at which seismic waves travel, changes abruptly at specific depths beneath the surface. A boundary at which the velocity changes, called a **seismic-velocity discontinuity,** represents a pronounced change in the physical nature (such as density and compressibility) of the material comprising the Earth. Major discontinuities, for example, define the boundaries between the crust and the mantle, and between the mantle and the core (see Interlude C). As more data became available, geoscientists identified subtler seismic-velocity discontinuities, which subdivide the mantle into three layers and the core into two.

Studies of earthquakes allowed geologists to put the dimensions of Earth's layers in perspective. The crust, which varies from 10 km thick beneath the oceans to 35 km thick beneath the interiors of continents and 70 km thick beneath mountain ranges, accounts for less than 1% of the Earth's radius. The mantle, which spans the region from the base of the crust to a depth of 2,885 km, accounts for most of the volume of the Earth. And the core itself has a radius of 3,486 km, about twice that of the Moon.

Pressure and Temperature Inside the Earth

In order to keep underground tunnels from collapsing under the pressure created by the weight of overlying rock, mining engineers must design sturdy support structures. Not surprisingly, deeper tunnels require stronger supports: the downward push from the weight of overlying rock increases with depth, simply because the mass of the overlying rock layer increases with depth. At the Earth's center, pressure probably reaches about 3,600,000 atm.

Temperature also increases with depth in the Earth. Even on a cool winter's day, miners who chisel away at gold veins exposed in tunnels 3.5 km below the surface swelter in temperatures of about 53°C (127°F). We refer to the rate of change in temperature with depth as the **geothermal gradient.** In the upper part of the crust, the geothermal gradient averages between 15° and 50°C per km. At greater depths, the rate decreases (to 10°C per km or less), so that at a depth of about 35 km below the surface of a continent, the temperature hovers "only" between 350° and 550°C. No one has ever directly measured the temperature at the Earth's center, but recent calculations suggest it may reach 4,300°C, slightly less than the temperature at the surface of the Sun.

2.7 WHAT ARE THE LAYERS MADE OF?

We saw earlier that the material composing the Earth's insides must be much denser than familiar surface rocks like granite and basalt. To discover what this material was, geologists

- conducted laboratory experiments to determine what kinds of materials inside the Earth could serve as a source for magma;

- studied unusual chunks of rock that may have been carried up from the mantle in magma;

- conducted laboratory experiments to measure seismic velocities in samples of known rock types, so that they could compare these velocities with observed velocities in the Earth; and

- estimated which elements would be present in the Earth if the Earth had formed out of planetesimals similar in composition to **meteorites** (chunks of rock and/or metal alloy that fell from space and landed on Earth).

As a result of this work, we now have a pretty clear sense of what the layers inside the Earth are made of, though this picture is constantly being adjusted when new findings become available. Let's now look at the properties of individual layers, starting with the Earth's surface.

The Crust

We know more about the crust, the outermost layer of the Earth, than we do about the underlying mantle and core, because we can directly observe the rocks that it consists of. The crust's thickness comprises only 0.2 to 0.6% of the Earth's radius—if the Earth were the size of a balloon, the crust would be thinner than the balloon's skin—yet it is our home and the source of all our resources.

Geologists distinguish between two fundamentally different types of crust—**oceanic crust,** which underlies the sea floor, and **continental crust,** which underlies continents (▶Fig. 2.12a). The crust is not simply cooled mantle, like the skin on chocolate pudding, but rather consists of a variety of rocks that differ in composition (chemical makeup) from mantle rock. Of note, the atoms making up crustal rock were originally part of the mantle; they were extracted to form the crust when these parts of the mantle melted and generated magma that rose to or near the Earth's surface.

FIGURE 2.12 (a) This simplified cross section illustrates the differences between continental crust and oceanic crust. Note that the thickness of continental crust can vary greatly. (b) Oceanic crust is denser than continental crust.

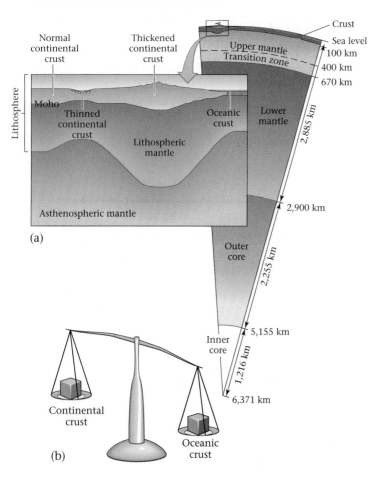

The base of the crust (the crust-mantle boundary) is defined by a seismic-velocity discontinuity that was first recognized by a Croatian seismologist, Andrija Mohorovičić, in 1909, and thus is referred to as the **Moho.**

Oceanic crust is only 7 to 10 km thick. At highway speeds (100 km per hour), you could drive a distance equal to the thickness of the oceanic crust in about five minutes. (It would take sixty-three hours, driving nonstop, to reach the Earth's center.) We have a good idea of what oceanic crust looks like in cross section, because geologists have succeeded in drilling down through its top few kilometers and have found places where slices of oceanic crust, known as **ophiolites,** have been incorporated in mountains and therefore have been exposed on dry land. Studies of such examples show that oceanic crust consists of fairly uniform layers. At the top, we find a blanket of sediment, generally less than 1 km thick, composed of clay grains and plankton (tiny floating animal and plant life) shells that have settled like snow. Beneath this blanket, the oceanic crust consists of a layer of basalt and, below that, a layer of gabbro (both mafic igneous rocks).

Most continental crust is about 35 to 40 km thick on average—about four times the thickness of oceanic crust—but its thickness varies much more than does oceanic crust. Plate motions can cause continents to stretch like taffy in narrow bands called rifts (here, the crust may be only 25 km thick) or to squash and thicken in mountain belts (here, the crust may become 70 km thick). In contrast to oceanic crust, continental crust contains a great variety of rock types, ranging from mafic to silicic in composition. Overall, continental crust is less mafic than oceanic crust—it has a silicic (granite-like) to intermediate composition—so a block of average continental crust weighs less than a same-size block of oceanic crust (▶Fig. 2.12b).

Finally, it is important to note that most rock in the crust (both continental and oceanic) contains **pores** (tiny open spaces), and in much of the upper several kilometers of the crust the pores are filled with water. (Earth's liquid-water supply, therefore, does not all lie on the surface.) This subsurface water, or **groundwater,** provides the water that farmers pump out of wells for irrigation and that cities pump out for their water supplies.

The Mantle

The mantle of the Earth forms a 2,885-km-thick layer surrounding the core. In terms of volume, it is the largest part of the Earth. In contrast to the crust, the mantle consists of ultramafic rock, peridotite, which is very rich in iron and magnesium and very poor in silica. This means that peridotite, though rare at the Earth's surface, is actually the most abundant rock in our planet. Overall, density in the mantle increases from about 3.5 g/cm^3 at the top to about 5.5 g/cm^3 at its base, but as noted before, this increase does not occur gradually. Rather, the mantle consists of three layers, defined by jumps in seismic velocity: the **upper mantle,** down to a depth of 400 km, the **transition zone,** from there down to a depth of 670 km, and the **lower mantle,** from there down to the core-mantle boundary.

Almost all of the mantle is solid rock. But even though it's solid, mantle rock is so hot that it's soft enough to flow extremely slowly—at the rate of less than a few centimeters a year, roughly the rate at which your fingernails grow. "Soft" here does not mean liquid, it simply means that over long periods of time mantle rock can change shape without breaking. As an analogy, think of soft wax, which can change shape without breaking. Between 100 and 200 km deep in the mantle beneath ocean basins, a region called the **low-velocity zone,** up to 2% of the mantle has melted and exists as droplets of molten rock between solid grains.

Though overall the temperature of the mantle increases with depth, it varies significantly with location even at the same depth. The warmer regions are less dense, while the cooler regions are denser. The blotchy pattern of warmer and cooler mantle demonstrates that the mantle is convecting, like water in a simmering pot (see art on pp. 42–43).

Warm mantle gradually flows upward, while cooler, denser mantle sinks.

The Core

Early calculations suggested that the core had the same density as gold, so for a number of years people held the false impression that vast riches lay at the heart of our planet. Alas, geologists eventually concluded that the core consists of a far less glamorous material, iron alloy (iron mixed with lesser amounts of oxygen, nickel, silicon, or sulfur). They arrived at this conclusion, in part, by comparing the properties of the core with the properties of metallic (iron) meteorites (see Box 2.1).

Studies of how seismic waves bend (refract) as they pass through the Earth, along with the discovery that certain types of earthquake waves cannot pass through the outer part of the core (see Interlude C), led geoscientists to divide the core into two parts, the **outer core** (between 2,900 and 5,155 km deep) and the **inner core** (from a depth of 5,155 km down to the Earth's center at 6,371 km). The outer core

---

BOX 2.1

THE REST OF THE STORY

Meteorites

All kinds of matter orbit the Sun, from the planets and their moons to the **asteroids** (fragments formed when two planet-sized bodies collided soon after the birth of the solar system) that now occupy a belt between the orbits of Mars and Jupiter, to smaller chunks and dust specks. Some of the smaller chunks, and even a few asteroids, follow elliptical orbits that cross the path of Earth. When the smaller pieces get close enough, the Earth's gravity pulls them in and they fall to Earth. An object falling to Earth is a **meteoroid.**

When a meteoroid passes through the atmosphere, friction between the air and the meteoroid in the mesosphere heats it up so that its surface melts and begins to evaporate. When this happens, the meteoroid emits a tail of glowing gas, which we see as a bright streak across the sky—we call this a **meteor,** or shooting star. Most meteoroids are sand-size or smaller and burn up before they fall to an elevation of about 25 km. Only larger meteoroids survive their fiery plunge and strike the Earth. Once a meteoroid hits the Earth, we call it a **meteorite.** On average, meteorites add about 100 tons of matter to the Earth every year. Curiously, scientists did not believe that objects we now consider meteorites came from space until the end of the eighteenth century, after a spectacular meteor shower during which people saw a large number

of meteorites land near Siena, Italy. The explanation that meteorites came from space finally gained wide acceptance in 1803, when a shower dropped 3,000 meteorites in Normandy, France.

Scientists distinguish three classes of meteorites: stony, iron, and stony-iron. About 93% of all meteorites that have been recovered are stony, meaning that they are made of rock. Most of these may also be referred to as chondrites, because they contain tiny silica-rich spheres called chondrules. About 6% of meteorites are iron, so named because they consist entirely of iron alloy. Stony-iron meteorites contain both rock and iron alloy.

Some meteorites appear to be chunks of the Moon or Mars that were blasted into space during large meteoroid impacts long ago. Others are relicts of primordial planetesimals. But most meteorites appear to come from the asteroid belt. The existence of distinct stony and iron meteorites is evidence that the asteroids formed from the collision of two planetary objects, each big enough to have separated internally into a metal core (the source of iron meteorites) and a rocky mantle (the source of stony meteorites). Clearly, the study of meteorites provides a fascinating window through which to observe the early history of the solar system.

is a *liquid* iron alloy with a density of 10 to 12 g/cm³. It can exist as a liquid because the temperature in the outer core is so high that even the great pressures squeezing the region cannot lock atoms into a solid framework. Because it is a liquid, the iron alloy of the outer core can flow; as we will see in Chapter 3, this flow generates Earth's magnetic field.

The inner core, with a radius of about 1,220 km and a density of 13 g/cm³, is a *solid* iron alloy, which may reach a temperature of over 4,300°C. Even though it is hotter than the outer core, the inner core is a solid because it is deeper and subjected to even greater pressure. The pressure keeps atoms from wandering freely, so they pack together tightly in very dense materials. The inner core probably grows through time, at the expense of the outer core, as the Earth slowly cools. Interestingly, recent data suggest that the inner core rotates faster than the rest of the Earth because of the electromagnetic force applied to it by the outer core.

2.8 THE LITHOSPHERE AND ASTHENOSPHERE

So far, we have divided the Earth into layers based on their seismic velocity, which depends on their composition and mineral structure. In the past few decades, geoscientists have also found it convenient to divide the outer several hundred kilometers of our planet into two layers according to how easily the rock comprising each layer can flow. That is, geoscientists now distinguish between the **lithosphere,** the relatively rigid, nonflowable layer that lies between the surface and a depth of about 100–150 km, and the **asthenosphere,** the relatively soft, flowable layer that lies below a depth of 100–150 km. Another term used to describe these layers is **viscosity,** meaning the resistance of a material to flow. Materials with low viscosity (like water) flow easily, while materials with high viscosity (like molasses) flow slowly. The lithosphere thus has a higher viscosity than the asthenosphere.

We can picture the lithosphere as Earth's rigid outer shell. Notice that it consists of the entire crust *plus* the uppermost part of the mantle. Thus, the boundary between lithosphere and asthenosphere lies within the upper mantle, significantly below the Moho. The **oceanic lithosphere,** topped by oceanic crust, reaches a thickness of about 100 km, while the **continental lithosphere,** topped by continental crust, reaches a thickness of about 150 km. (These numbers are approximations.) The asthenosphere is mostly solid, but small amounts of melt occur in the upper 100 km or so of the asthenosphere below ocean basins. We can't really define a specific base to the asthenosphere, for the entire mantle below a depth of about 100–150 km is soft enough to flow. Some geologists place the base of the asthenosphere at around the upper mantle/transition zone boundary, just for convenience.

The boundary between the lithosphere and asthenosphere occurs at a temperature of about 1,280°C, for at this temperature mantle rock becomes soft enough to flow. That's because the viscosity of a material depends on its temperature. To see how, take a cube of candle wax and place it in the freezer. The wax becomes very rigid and can maintain its shape for long periods of time; in fact, if you were to drop the cold wax, it would shatter. But if you take another block of wax and place it in a warm (not hot) oven, it will become soft (its viscosity decreases), so that you could easily mold it into another shape. In fact, the force of gravity alone may cause the warm wax to flow and assume the shape of a pancake. Rock behaves somewhat similarly to the wax blocks. When rock is cool, it is quite rigid—certainly, the rocks exposed in the cool temperatures of Earth's surface are very hard and will shatter when smashed with a hammer. But at high temperatures, rocks become soft and can flow, though much more slowly than wax. This ability to flow slowly occurs at a temperature much lower than is necessary to cause rocks to melt. Rock of the lithosphere is cool enough to behave rigidly, whereas rock of the asthenosphere is warm enough to flow easily. In Chapter 4, we will learn how Earth's rigid lithosphere has broken into plates that move with respect to each other.

CHAPTER SUMMARY

- The Earth has a magnetic field, which shields it from solar wind. Closer to Earth, the field creates the Van Allen belts, which trap cosmic rays.

- A layer of gas surrounds the Earth. This atmosphere (78% nitrogen, 21% oxygen, 1% other) can be subdivided into distinct layers. All weather occurs in the troposphere, the layer we live in. Air pressure decreases with elevation, so 50% of the gas in the atmosphere is below 5.5 km.

- The Earth consists of organic chemicals, minerals, glasses, rocks, metals, melts, and volatiles. Most rocks on Earth contain silica (SiO_2). We distinguish between silicic, intermediate, mafic, and ultramafic rocks based on the proportion of silica to iron and magnesium within them. Silicic rocks have the greatest proportion of silica-rich minerals and are the least dense, while ultramafic rocks have the greatest proportion of iron and magnesium-rich minerals and are the most dense.

- The Earth's interior can be divided into three compositionally distinct layers, named in sequence from the surface down: the crust, the mantle, and the core. The first recognition of this division came from studying the density and shape of the Earth.

- Pressure and temperature both increase with depth in the Earth. At the center, pressure is 3.6 million times greater than at the surface, and temperature reaches 4,300°C, nearly as hot as the surface of the Sun. The rate at which temperature increases is the geothermal gradient.

• Studies of seismic waves revealed the existence of sub-layers in the core (outer core and inner core) and mantle (upper mantle, transition zone, and lower mantle).

• The crust is a thin skin that varies in thickness from 7–10 km (beneath the oceans) to 25–70 km (beneath the continents). Oceanic crust is mafic in composition, while average continental crust is silicic to intermediate. The mantle is composed of ultramafic rock. The core consists of iron alloy.

• The crust plus the upper part of the mantle comprise the lithosphere, a relatively rigid shell that lies over the asthenosphere, the soft layer of the mantle, below.

KEY TERMS

alloy (p. 40)
asthenosphere (p. 48)
atmosphere (p. 37)
comet (p. 36)
continental crust (p. 46)
convection (p. 38)
core (p. 44)
crust (p. 44)
dipole (p. 36)
earthquake (p. 44)
fault (p. 44)
geothermal gradient (p. 45)
glass (p. 40)
groundwater (p. 39)
hypsometric curve (p. 39)
inner core (p. 47)
lithosphere (p. 48)
lower mantle (p. 47)
low-velocity zone (p. 47)
mafic (p. 41)
magnetic field (p. 36)
magnetic field lines (p. 36)
magnetosphere (p. 36)
mantle (p. 44)

melt (p. 40)
metal (p. 40)
meteorite (p. 45)
meteoroid (p. 47)
mineral (p. 40)
Moho (p. 46)
oceanic crust (p. 46)
ophiolite (p. 46)
organic chemical (p. 40)
outer core (p. 47)
pores (p. 46)
pressure (p. 38)
rock (p. 40)
seismic-velocity discontinuity (p. 45)
seismic waves (p. 44)
silicic (p. 41)
topography (p. 39)
transition zone (p. 47)
troposphere (p. 38)
upper mantle (p. 47)
vacuum (p. 35)
Van Allen radiation belts (p. 37)
volatiles (p. 41)

REVIEW QUESTIONS

1. Describe the shape of the magnetic field of the Earth. What does it do to the solar wind?

2. What are the Van Allen radiation belts? How do they protect the Earth?

3. With every breath you take, what are you breathing?

4. In what layer of the atmosphere does most weather occur?

5. Why don't jet engines work above 25 km?

6. What is the proportion of land and sea on the Earth's surface?

7. What is a hypsometric curve?

8. What is the elevation of most of Earth's surface that is above sea level? What is the elevation of most of the sea floor?

9. What are the main categories of materials that make up the Earth?

10. How deep is the deepest well to ever penetrate the Earth's crust? How does this compare with the distance to the center of the Earth?

11. What does the average density of the Earth, and of the rocks of the crust, tell us about the density of the Earth's interior?

12. How do we know that the Earth's mass is mostly concentrated at the center? How do we know that Earth must be largely solid?

13. What is the geothermal gradient? Why does it get hotter toward the center of the Earth?

14. Contrast continental and oceanic crust in terms of thickness and composition.

15. What is the Moho, and how was it first recognized?

16. What is the mantle composed of? How do its density and temperature change with depth?

17. What is the core composed of? How do we know this?

18. Contrast the lithosphere and asthenosphere. How are the criteria for defining lithosphere and asthenosphere different from the criteria used to recognize the crust, mantle, and core?

SUGGESTED READING

Bolt, B. A. 1982. *Inside the Earth*. San Francisco: Freeman.

Brown, G. C., and A. E. Mussett. 1993. *The Inaccessible Earth*. London: Chapman and Hall.

Dawson, J. 1993. CAT scanning the Earth. *Earth* 2(3): 36–41.

Ernst, W. G. 1990. *The Dynamic Planet*. New York: Columbia University Press.

Jeanloz, R., and T. Lay. 1993. The core-mantle boundary. *Scientific American* 268(5): 48–55.

Song, X., and P. G. Richards. 1996. Seismological evidence for differential rotation of the Earth's inner core. *Nature* 382: 221–24.

Wysession, M. 1995. The inner workings of the Earth. *American Scientist* 83(2): 134–46.

Drifting Continents and Spreading Seas

3.1 INTRODUCTION

In September 1930, fifteen explorers led by a German meteorologist, Alfred Wegener, set out across the endless snowfields of Greenland to resupply two weather observers stranded at a remote camp. The observers were planning to spend the long polar night recording wind speeds and temperatures in Greenland's polar plateau. At the time, Wegener was somewhat infamous (▶Fig. 3.1): fifteen years earlier, he had published a small book, *The Origin of the Continents and Oceans,* in which he had dared to challenge geologists' long-held assumption that the continents had remained fixed in position through **geologic time** (the time since the formation of the Earth). Instead, Wegener proposed that the present distribution of continents and ocean basins had evolved only recently. According to Wegener, the continents had once fit together like pieces of a giant jigsaw puzzle, to make one vast supercontinent. He suggested that this supercontinent, which he named **Pangaea** (pronounced Pan-jee-ah; Greek for "all land"), later fragmented into separate continents that then drifted apart, moving slowly to their present positions (▶Fig. 3.2). Drifting continents? Absurd! Or so proclaimed the leading geologists of the day.

At a widely publicized 1926 geology conference in New York City, Wegener presented arguments, in imperfect English, for his **continental-drift hypothesis** to a phalanx of celebrated American professors. The Americans scoffed: "What force could possibly be great enough to move the immense mass of a continent?" Wegener couldn't provide a good answer, and despite all the observations he had made that favored continental drift, he failed to convince his audience.

Now, four years later, Wegener faced an even greater challenge. As he headed into the interior of Greenland, the weather worsened, and most of his party turned back. But Wegener felt he could not abandon the isolated observers, and with two companions he trudged forward. On October 30, 1930, Wegener reached the observers and dropped off enough supplies to last the winter. Wegener and one companion set out on the return trip the next day, but they never made it home.

A fossil of Glossopteris *from an exposure in Antarctica. The presence of this fossil on many continents was one of the observations that led to the proposal of continental drift.*

FIGURE 3.1 Alfred Wegener, the German meteorologist who proposed a comprehensive model of continental drift and presented geologic evidence in support of the idea.

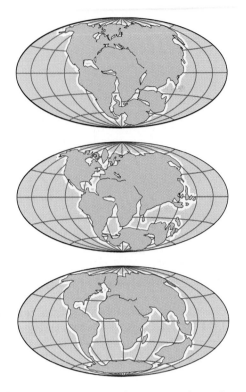

FIGURE 3.2 Wegener's image of Pangaea and its subsequent breakup and dispersal.

It took four decades, but today geologists have accepted Wegener's hypothesis and take it for granted that the map of the Earth constantly changes; continents have indeed waltzed around its surface, variously combining and breaking apart through geologic time. The revolution in thought began in 1960, when Harry Hess, a Princeton University professor, proposed that continents drift apart because new ocean floor forms between them by a process that his contemporary Robert Dietz labeled **sea-floor spreading,** and that continents move toward each other when the old ocean floor between them sinks back down into the Earth's interior, a process called **subduction.** By 1968, geologists had developed a fairly complete model describing continental drift, sea-floor spreading, and subduction. In this model, Earth's lithosphere, its outer, relatively rigid shell, actually consists of about twenty distinct pieces, or **plates,** that slowly move relative to one another. Because we can confirm this model by many observations, it has gained the status of a theory (see Prelude), which we now call the **theory of plate tectonics,** from the Greek word *tekton,* which means "builder"; plate movements "build" regional geologic features. Geologists view plate tectonics as the grand unifying theory of geology, because it can successfully explain a great many geologic phenomena, as we will see.

In this chapter, we learn about the observations that led Wegener to propose his continental-drift hypothesis. Then

we look at paleomagnetism, the record of Earth's past magnetic field preserved in rock, which provides a key proof of continental drift. Finally, we learn how observations about the sea floor made by geologists during the mid–twentieth century led Harry Hess to propose the concept of sea-floor spreading. In Chapter 4, we will build on these concepts and become acquainted with modern plate tectonics theory. The knowledge of continental drift and sea-floor spreading that we gain in this chapter and of plate tectonics in Chapter 4 will prove vital in understanding the geologic phenomena we encounter in the rest of the book.

3.2 WEGENER'S EVIDENCE FOR CONTINENTAL DRIFT

Before Wegener, geologists viewed the continents and oceans as immobile—fixed in position throughout geologic time. According to Wegener, however, the positions of continents change through time. He suggested that a vast supercontinent he called Pangaea existed until the middle of the **Mesozoic** era (the interval of geologic time, commonly known as the "Age of Dinosaurs," between 245 and 65 million years ago), when it broke apart to form the continents we see today, and that these continents then drifted away from each other. (Geologists now realize that supercontinents formed and dispersed at least a few times during Earth's history—the name Pangaea applies only to the most recent supercontinent.) Let's look at some of Wegener's arguments and see why he came to this conclusion.

The Fit of the Continents

Almost as soon as maps of the Atlantic coastlines became available in the 1500s, scholars noticed the fit of the continents. If you cut out the continents from a map, they do roughly fit together. For example, the northwest coast of Africa tucks in against the eastern coast of North America, and the bulge of eastern South America nestles cozily into the indentation of southwestern Africa. Australia, Antarctica, and India all connect to the southeast of Africa, while Greenland, Europe, and Asia pack against the northeastern margin of North America. In fact, all the continents can be joined, with remarkably few overlaps or gaps, to create Pangaea. Modern plate tectonics theory can now even explain the misfits. Wegener concluded that the fit was too good to be coincidence.

Locations of Past Glaciations

Wegener was an Arctic meteorologist by training, so not surprisingly, he had a strong interest in **glaciers,** rivers or sheets of ice that slowly flow across the land surface. He realized that glaciers form mostly at high latitudes, and thus that by studying the past locations of glaciers, he might be able to determine the past locations of continents. We'll look at glaciers in detail in Chapter 22, but we need to know something about them here, to understand Wegener's arguments.

When a glacier moves, it scrapes **sediment** (pebbles, boulders, sand, silt, and mud) off the ground and carries it along. The sediment freezes into the base of the glacier, so the glacier then becomes like sandpaper and grinds exposed rock beneath it. In fact, rocks protruding from the base of the ice carve scratches (**striations**) into the underlying rock, and these scratches indicate the direction in which the ice has flowed. When the glacier eventually melts, the sediment collects on the ground and creates a distinctive layer of **till,** a mixture of mud, sand, pebbles, and larger rocks. Later on, the till may be buried and preserved. As we'll see in Chapter 12, the age of a till layer, or any layer of sediment, can be figured by studying its **fossils** (preserved relicts of organisms). Today, glaciers are found only in polar regions and in high mountains. But by studying the distribution and age of ancient till, geoscientists have determined that at several times during Earth's history, glaciers covered large areas of continents. We refer to these times as **ice ages.** One of the major ice ages occurred about 260–280 million years ago, near the end of the **Paleozoic** era (the interval of geologic time between 545 and 245 million years ago).

Why was the study of ancient glacial deposits important to Wegener? When Wegener plotted the locations of till deposited by late Paleozoic glaciers, he found that in this time interval glaciers occurred in southern South America, southern Africa, southern India, Antarctica, and southern Australia, places that are all now widely separated from one another and, with the exception of Antarctica, do not currently lie in cold polar regions (▶Fig. 3.3a). But to Wegener's amazement, all the late Paleozoic glaciated areas lie adjacent to each other on a map of Pangaea (▶Fig. 3.3b). Furthermore, when he plotted the glacial striations, they all pointed roughly outward from a location in southeastern Africa, just as would be expected if an ice sheet comparable to the present-day Antarctic polar ice cap developed in southeastern Africa and spread outward from its origin. In other words, Wegener determined that the distribution of

FIGURE 3.3 (a) The distribution of late Paleozoic glacial deposits on a map of the present-day Earth. The arrows indicate the orientation of striations. (b) The distribution of these glacial deposits on a map of the southern portion of Pangaea. Note that the glaciated areas fit together to define a polar ice cap.

(a)

(b)

glaciations at the end of the Paleozoic era could easily be explained if the continents were united in Pangaea, with the southern part of Pangaea located over the South Pole, but could not be explained if the continents had always been in their present positions. Thus, observations on the distribution of late Paleozoic glaciation provided strong support for the existence of Pangaea.

The Distribution of Equatorial Climatic Belts

If the southern part of Pangaea straddled the South Pole at the end of the Paleozoic era, then during this same time interval southern North America, southern Europe, and northwestern Africa would have straddled the equator and would have had tropical or subtropical climates. So Wegener searched for evidence that this was so by studying sedimentary rocks that were formed at this time, for the material comprising these rocks can reveal clues to the climate. Specifically, in the swamps and jungles of tropical regions,

FIGURE 3.4 Map of Pangaea, showing the distribution of coal deposits and reefs (indicating tropical environments), and sand-dune deposits and salt deposits (indicating subtropical environments). Note how deposits now on different continents align in distinct belts.

thick deposits of plant material accumulate, and when deeply buried, this material transforms into coal. Further, in the clear shallow seas of tropical regions, large reefs built by coral-like organisms develop offshore. Finally, subtropical regions, on either side of the tropical belt, contain deserts, an environment for sand-dune formation and the accumulation of salt from evaporating seawater or salt lakes. Wegener thought that the distribution of late Paleozoic coal, sand-dune deposits, and salt deposits could define climate belts on Pangaea.

Sure enough, in the belt of Pangaea that Wegener expected to be equatorial, late Paleozoic sedimentary rock layers included abundant coal and the relics of reefs; and in the portions of Pangaea that Wegener predicted would be subtropical, late Paleozoic sedimentary rock layers included relics of desert dunes and of salt (▶Fig. 3.4). On a present-day map of our planet, these deposits are scattered around the globe in discontinuous fragments; further, many of the deposits now occur at high latitudes, where they cannot have formed. However, in Wegener's Pangaea, the deposits align in continuous bands that occupy appropriate latitudes. Thus, the distribution of late Paleozoic coal, reef, salt, and sand-dune deposits supported Wegener's proposal.

The Distribution of Fossils

Today, different continents provide homes for different species. Kangaroos, for example, live only in Australia, while zebras roam the plains of Africa. Similarly, many kinds of plants grow only on one continent and not on others. Why? Because land-dwelling species of animals and plants cannot swim across vast oceans and thus evolve independently on different continents. During a period of Earth history when all continents were in contact, however, land animals and plants conceivably could have dispersed, so the same species might have appeared on many continents.

With this concept in mind, Wegener plotted locations of fossils of land-dwelling species that lived during the late Paleozoic and early Mesozoic eras (between about 300 and 210 million years ago) and found that they had indeed existed on several continents (▶Fig. 3.5). For example, an early Mesozoic land-dwelling reptile called *Cynognathus* lived in both southern South America and southern Africa. *Glossopteris,* a species of seed fern, flourished in regions that now comprise South America, Africa, India, Antarctica, and Australia. *Mesosaurus,* a freshwater reptile, inhabited portions of what is now South America and Africa. *Lystrosaurus,* another land-dwelling reptile, wandered through present-day Africa, India, and Antarctica. None of these species could have traversed a large ocean. Thus, Wegener argued, the distribution of these species required the continents to have been adjacent to one another in the late Paleozoic and early Mesozoic eras.

Considering that paleontologists found fossils of species such as *Glossopteris* in Africa, South America, and India, Wegener suspected that they might also be found in Antarctica. The tragic efforts of Captain Robert Scott and his party of British explorers, who reached the South Pole in 1912, confirmed this proposal. On their return trip, the party died of starvation and cold, only 11 km from a food cache. When their bodies were found, their sled loads included *Glossopteris* fossils that they had hauled for hundreds of kilometers, in the process burning valuable calories that could possibly have kept them alive long enough to reach the cache. In 1969, paleontologists found fossils of *Lystrosaurus* in Antarctica, providing further confirmation that Wegener was right and that the continents were once connected.

Matching Geologic Units

In the same way that an art historian can identify a Picasso painting and an architect a Victorian design, a geoscientist can identify a distinctive group of rocks. Wegener found that the same distinctive **Precambrian** (the interval of geologic time between Earth's formation and 545 million years ago) rock assemblages occurred on the east coast of South America and the west coast of Africa, regions now separated by an ocean (▶Fig. 3.6a). If the continents were joined to create Pangaea in the past, then these matching rock groups would have been adjacent to one another, and thus could have composed continuous blocks. Wegener also noted that belts of rocks in the Appalachian Mountains of the United States and Canada closely resembled belts of rocks in mountains of southern Greenland, Great Britain, Scandinavia, and northwest Africa (▶Fig. 3.6b), regions that lay adjacent to each other in Pangaea. Wegener thus demonstrated that not only did the coastlines of continents match—their component rocks did too.

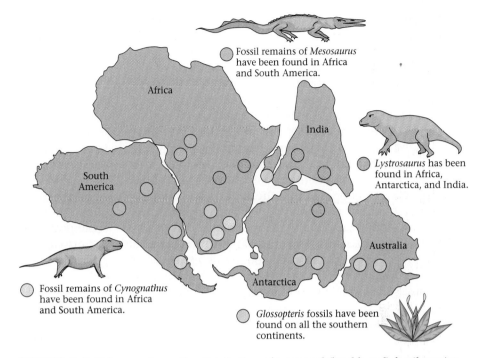

FIGURE 3.5 This map shows the distribution of terrestrial (land-based) fossil species. Note that creatures like *Lystrosaurus* could not have swum across the Atlantic to reach Africa.

FIGURE 3.6 (a) Distinctive areas of rock units on South America link with those on Africa, as if they were once connected and later broke apart. (b) If the continents are returned to their positions in Pangaea by closing the Atlantic, mountain belts of the Appalachians lie adjacent to similar-age mountain belts in Greenland, Great Britain, Scandinavia, and Africa.

(a)   (b)

Criticism of Wegener's Ideas

Wegener's model of a supercontinent that later broke apart explained the distribution of glaciers, coal, sand dunes, distinctive rock units, and fossils we find today. Clearly, he had compiled a strong case for continental drift. But Wegener, as noted earlier, could not adequately explain how or why continents drifted. In his writings, Wegener suggested that the centrifugal force created by the rotation of the Earth could cause a supercontinent centered at a pole to break up into pieces that would move toward equatorial latitudes. He proposed that the continental crust (he didn't refer to the lithosphere, which includes the crust and the uppermost part of the mantle) moved by "plowing" through oceanic crust as a ship plows through water. But other geologists of the time found his explanation wholly unsatisfactory. Experiments showed that the relatively weak rock making up continents cannot plow through the relatively strong rock making up the ocean floor, and that centrifugal force generated by Earth's spin is a million times too small to move a continent.

Wegener left on his final expedition to the ice cap of Greenland having failed to convince his peers, and he froze to death in the icy wasteland never knowing that his ideas would smolder for decades before being reborn as the basis of the broader theory of plate tectonics. During these decades, a handful of iconoclasts continued to champion Wegener's notions. Among these was Arthur Holmes, a highly respected British geologist who argued that huge convection cells existed inside the Earth, slowly transporting hot rock from the deep interior up to the surface. Holmes suggested that continents might be split and the pieces dragged apart in response to convective flow in the mantle. But in general, the validity of continental drift could not be tested until many more discoveries had been made about the nature of the Earth, using twentieth-century instruments and techniques. Key among these was the discovery of a phenomenon called paleomagnetism.

3.3 PALEOMAGNETISM AND APPARENT POLAR-WANDER PATHS

Mid-twentieth-century studies of magnetism in rocks provided data that effectively proved the ideas of continental drift and, later, sea-floor spreading. These studies showed that rocks preserve a record of Earth's past magnetic field, from which we can deduce lithosphere plate movements. The record of ancient magnetism preserved in rock is called **paleomagnetism.** In order to understand paleomagnetism and how it enabled geologists to test the continental-drift hypothesis, and later the sea-floor-spreading hypothesis, we must first look at the nature of Earth's magnetic field in a bit more depth than we did in Chapter 2. If you need to review background information on magnetism in general, please see Box 3.1.

Earth's Magnetic Field, Revisited

In Chapter 2, we learned that Earth has a magnetic field, which deflects the solar wind and traps cosmic rays. Why does this field exist? Geologists do not yet have a complete answer, but they have hypothesized that the field results from the circulation of liquid iron alloy, an electrical conductor, in the Earth's outer core—in other words, the outer core behaves like an electromagnet (▶Fig. 3.8a; see Box 3.2). For convenience, however, we can picture the planet as a giant bar magnet, with a north magnetic pole and a south magnetic pole (▶Fig. 3.8b). The "north-seeking end" of a compass points toward the north magnetic pole, while the "south-seeking end" points toward the south magnetic pole. We define the **dipole** of the Earth as an imaginary arrow that points from the north magnetic pole to the south magnetic pole, and passes through the planet's center.

Presently, Earth's dipole tilts at about 11° to the planet's **rotational axis** (the imaginary line through the center of the Earth around which Earth spins). Therefore, the **geographical poles** of the planet, the places where the rotational axis intersects the Earth's surface, do not coincide exactly with the magnetic poles. For example, the north magnetic pole currently lies in arctic Canada. As a consequence, the north-seeking end of a compass needle in New York points about 14° west of north. The angle between the direction that a compass needle points at a given location and the direction to "true" (geographic) north is called the **magnetic declination** (▶Fig. 3.9). Measurements over the past couple of centuries show that magnetic poles migrate very slowly through time, never straying more than about 15° of latitude from the geographic pole. In fact, the magnetic declination of a compass changes by only 0.2° to 0.5° per year. Notably, when averaged over about 10,000 years, the magnetic poles are thought to coincide with the geographic poles.

▶Figure 3.10 illustrates the magnetic field lines in space around the Earth, as seen in cross section (without the warping caused by solar wind). At the equator, the lines parallel Earth's surface; at mid-latitudes, the lines tilt at an angle to the surface; and at the magnetic poles, the lines are perpendicular to the surface. Thus, if we traveled to the equator and set up a magnetic needle such that it could pivot up and down freely, the needle would be horizontal. If we took the needle to mid-latitudes, it would tilt at an angle to Earth's surface; and at a magnetic pole, the needle would point straight down. The needle's angle of tilt (which, as Fig. 3.10 shows, depends on latitude) is called the **magnetic inclination.** Note that a familiar compass needle

BOX 3.1

*SCIENCE
TOOLBOX*

The Fundamentals of Magnetism

If you hold a magnet over a pile of steel paper clips, it will lift the paper clips against the force of gravity. The magnet exerts an attractive force that pulls on the clips. A magnet can also create a repulsive force that pushes an object away. For example, when oriented appropriately, one ring magnet can levitate another. The push or pull exerted by a magnet is a **magnetic force;** this force creates an invisible **magnetic field** around the magnet. Magnetic forces can be created by a **permanent magnet,** a special material that behaves magnetically for a long time all by itself. Magnetic forces can also be produced by an electric current passing through a wire. An electrical device that produces a magnetic field is an **electromagnet.** The stronger the magnet, the greater its **magnetization.** The strength of the pull that an object feels when placed in a field depends on the magnet's magnetization and on its distance from the object.

When other magnets, special materials (such as iron), or electric charges enter a magnetic field, they feel a magnetic force. **Compass needles,** for example, are simply magnetic needles that can pivot freely and that align with Earth's magnetic field. Recall from Chapter 2 that you can symbolically represent a magnetic field by a pattern of curving lines, known as magnetic field lines. You can see the form of these lines by sprinkling iron filings on a sheet of paper placed over a bar magnet; each filing acts like a tiny magnetic compass needle and aligns itself with the magnetic field lines.

All magnets have two **magnetic poles,** a north pole at one end and a south pole at the other. Opposite poles attract, but like poles repel. The imaginary line through the magnet that connects one pole to another is the magnet's **dipole.** Physicists specify the dipole by an arrow that points from the north to the south pole. The **polarity** of a magnet refers to the direction the arrow points; the dipoles of magnets with opposite polarity are represented by arrows with arrowheads at opposite ends. Because of the dipolar nature of magnetic fields, we can draw arrowheads on magnetic field lines oriented to form a continuous loop through the magnet.

Individual atoms behave like tiny electromagnets, because the orbiting of the electrons around the nucleus creates an electric current that, in turn, generates a magnetic field (▶Fig. 3.7a). Each atom, in effect, looks like a little dipole. But even though all materials consist of atoms, not all materials behave like strong, permanent magnets. In fact, most materials (wood, plastic, glass, gold, tin, etc.) are essentially nonmagnetic. That's because the atomic dipoles in the materials are randomly oriented, so overall the dipoles of the atoms cancel each other out (▶Fig. 3.7b). In a permanent magnet, however, all atomic dipoles lock into alignment with one another. When this happens, the magnetization of each atom adds to that of its neighbor, so the material as a whole becomes magnetic (▶Fig. 3.7c).

(a)

(b) Magnetization = 0 (because ↑ + ↓ = 0)

(c) Strong magnetization

FIGURE 3.7 (a) A magnetic dipole created by the rotation of an electron around an atom's nucleus. The dipole can be represented by an arrow that points from north to south. (b) In a nonmagnetic material, atoms tilt all different ways, so the dipoles cancel each other out, yielding a net magnetization of 0. (c) In a permanent magnet, the dipoles lock into alignment, so that they add to each other and produce a strong cumulative magnetization.

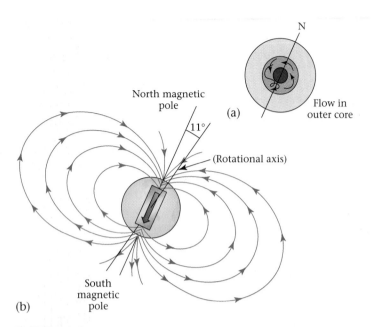

FIGURE 3.8 (a) The convective flow of liquid iron alloy in the Earth's outer core creates an electric current that in turn generates a magnetic field. Earth's magnetism creates magnetic lines of force in space. (b) We can picture Earth's magnetism by imagining that it contains a giant bar magnet. The dipole of this magnet points presently from the north magnetic pole to the south magnetic pole, and it pierces the Earth at the magnetic poles. Today, the magnetic poles do not coincide exactly with the Earth's geographic poles.

does not indicate inclination, because it cannot tilt—a compass needle aligns parallel to the "projection" of the magnetic field lines on the Earth's surface. (You can think of the projection as the shadow of a magnetic field line.)

How Do Rocks Develop Paleomagnetism?

More than 1,500 years ago, Chinese sailors discovered that an elongated piece of **lodestone** suspended from a thread "magically" pivots until it points north, and thus this rock could help guide their voyages. We now know that lodestone exhibits this behavior because it consists of **magnetite,** an iron-rich mineral that acts like a permanent magnet (see Box 3.1). Small crystals of magnetite or other magnetic minerals occur in many other rocks as well, including basalt. Each crystal produces a tiny magnetic force. The sum of the magnetic forces produced by all the crystals makes the rock as a whole magnetic. The magnetism of a rock, though, is generally too weak to affect a compass; it wasn't until the twentieth century that geologists developed instruments sensitive enough to detect the magnetism.

To see how magnetic rocks preserve a record of Earth's past magnetic field, let's examine one kind of rock, basalt. Basalt is an iron-containing igneous rock that forms when lava, flowing out of a volcano, cools and solidifies. When lava first comes out of a volcano, it is very hot (up to about 1,200°C), and thermal energy makes its atoms wobble and

Generating Earth's Magnetic Field

Explanations for how the Earth generates a magnetic field rely on the **theory of electromagnetism:** an electric current generates magnetism, and the movement of an electrical conductor (such as metal wire) in a magnetic field produces electricity. Because iron alloy is a good electrical conductor (electrons flow through it easily), the flow of iron alloy makes the Earth's outer core into an electromagnet.

To better understand the generation of Earth's magnetic field, consider an electric power plant. In a power plant, water or wind power spins a wire coil (an electrical conductor) around an iron bar (a permanent magnet). This apparatus is a **dynamo.** The motion of the wire in the bar's magnetic field generates an electric current in the wire, which in turn generates more magnetism. Similarly, in the Earth, convection (see

Appendix A) and the Earth's rotation cause liquid iron alloy (an electrical conductor) to flow in Earth's magnetic field. This flow generates an electric current, which in turn generates more magnetism.

The Earth differs from an electric power plant in that there is no permanent magnet in the center of the Earth; at the very high temperatures of the core, permanent magnets can't exist, because thermal energy makes atoms vibrate and tumble so fast that their tiny magnetic dipoles can't lock into alignment (see Box 3.1). Instead, the Earth is a "self-exciting dynamo," meaning that the magnetism produced by electric currents in the outer core is the magnetism that led to the generation of an electric current in the flowing iron alloy in the first place—this system perpetuates itself.

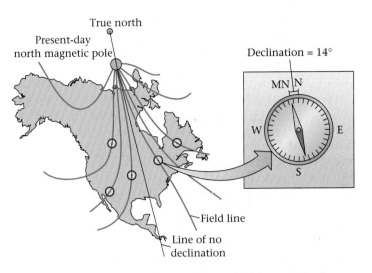

FIGURE 3.9 The projection of magnetic field lines in North America. Recall that lines of longitude run north-south, so in most places a compass needle will not parallel longitude. For example, a compass needle at New York would make an angle of 14° to the west of true north. Note that along the circumference that passes through both magnetic north and geographic north, the magnetic declination = 0°.

FIGURE 3.10 An illustration of magnetic inclination. A magnetic needle that is free to rotate around a horizontal axis aligns with magnetic field lines (here depicted in cross section). Because magnetic field lines curve in space, this needle is horizontal at the equator, tilts at an angle at mid-latitudes, and is vertical at the magnetic pole. Therefore, the angle of tilt depends on the latitude.

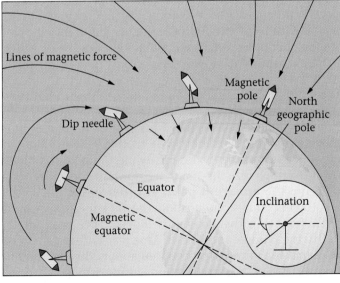

tumble chaotically. Each atom acts like a mini-dipole; the mini-dipoles of the wildly dancing atoms point in all different directions. When this happens, the magnetic force exerted by one atom cancels out the force of another, so the lava as a whole is not magnetic (▶Fig. 3.11a). However, as the temperature of the lava decreases, basalt rock starts to solidify. As the magnetite crystals in the basalt form and cool (that is, as thermal energy decreases), their iron atoms stop wobbling. The dipoles of all the atoms gradually become parallel with each other and with the magnetic field lines at the location where the basalt cools. Finally, at temperatures below 350°–550°C, the dipoles lock into position

FIGURE 3.11 The formation of paleomagnetism. (a) At high temperatures (greater than 350°–550°C), thermal vibration makes atoms move randomly; the dipoles thus cancel each other out, and the sample has no overall magnetic dipole. (b) As the sample cools below 350°–550°C, the atoms slow down, and their dipoles lock into alignment with the Earth's field.

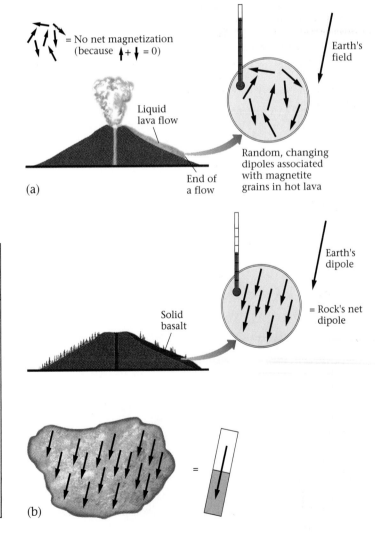

(pointing in the direction of the magnetic pole), and the basalt becomes a permanent magnet (▶Fig. 3.11b). Since this alignment is permanent, the basalt provides a record of the orientation of the Earth's magnetic field lines, relative to the rock, at the time the rock cooled. This record is paleomagnetism.

Interpreting *Apparent* Polar-Wander Paths: Evidence for Continental Drift

When geologists measured paleomagnetism in samples of basalt that had formed millions of years ago, they were surprised to find that the dipoles representing this paleomagnetism did not point to the present-day magnetic poles of the Earth (▶Fig. 3.12). At first, they interpreted this observation to mean that the magnetic poles of Earth were in different locations in the past, and they introduced the term **paleopole** to refer to the supposed position of the Earth's magnetic pole in the past. In some basalt samples, paleopoles were more than 90° away from present-day magnetic poles (see Box 3.3 to learn about locating paleopoles). Geologists then deduced that the positions of Earth's magnetic poles must change progressively through time, and they called this phenomenon **polar wander.** As we'll see, this initial interpretation of polar wander was quite wrong.

FIGURE 3.12 A rock sample can maintain paleomagnetization for millions of years. The dipole representing the paleomagnetism in this rock sample, from the village of Nono on the equator, does not parallel the Earth's present field. Note that I is not 0°, as it would be today for rock forming near the equator. (See Fig. 3.13a, b.)

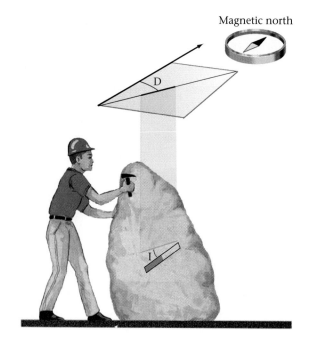

Geologists next decided to track what they thought was the "pole position" of Earth's magnetic field through time. So they measured the magnetic field preserved in rocks of many different ages from about the same location. ▶Figure 3.14 shows how this would be done for a locality called X on an imaginary continent. In Figure 3.14b, the paleopole for rock sample 1, which is 600 million years old, lies at position 1 on the map. In other words, position 1 indicates the location of Earth's magnetic pole, *relative to locality X,* 600 million years ago. In sample 2, which is 500 million years old, the paleopole lies at position 2 on the map, and so on. When all the points are plotted, the resulting curving line shows the progressive change in the position of the Earth's magnetic pole, relative to locality X, assuming that the position of X on Earth has been fixed through time. This curve was called a **polar-wander path.** Note that the polar-wander path ends near the present North Pole, because recent rocks became magnetized when Earth's magnetic field was supposedly close to its position today.

In the late 1950s, geologists determined what they thought was the polar-wander path for Europe. When this path was plotted, they did not accept the notion of continental drift, and assumed that the position of the continents was fixed. Thus, they interpreted the path to represent how the position of Earth's north magnetic pole migrated through time. Were they in for a surprise! When geologists then figured the polar-wander path for North America, they found that North America's path differed from Europe's (▶Fig. 3.15a). In fact, when paths were plotted for all continents, they all turned out to be different from one another (▶Fig. 3.15b). The hypothesis that the continents are fixed and the magnetic poles move simply cannot explain this observation. If the magnetic poles really moved while the continents stayed fixed, then all continents would have the same polar-wander paths.

Geologists suddenly realized that they were looking at polar-wander paths in the wrong way. It's not the pole that moves with respect to fixed continents, but rather the continents that move with respect to a fixed pole (▶Fig. 3.16a, b). As a continent moves, its position in relation to the Earth's magnetic field changes, so rocks formed at different times record different *apparent* positions of the pole. Further, if the pole is fixed, then in order for each continent to have its own unique polar-wander path, the continents must move, or "drift," in relation to each other. Poles do *not* wander, but the continents do drift while the pole stays fairly fixed. To emphasize this point, we now call curves like those of Figure 3.15 **apparent polar-wander paths.**

Having finally understood the meaning of apparent polar-wander paths, geologists looked once again at the paths for Europe and North America, and realized that they had the same shape between about 280 and 180 million years ago, the time period during which the continents supposedly linked together to form part of Pangaea (▶Fig. 3.17). During times when continents move with respect to one another, they develop different apparent polar-wander paths, but when continents are stuck together, they develop the same apparent polar-wander path.

The discovery that apparent polar-wander paths for different continents differ from one another can be explained by continental drift. But how could continental drift take place? Until that question could be answered, most geologists were not fully comfortable with the concept, and most still would not accept Wegener's idea. But the stage was set for the proposal of the sea-floor-spreading hypothesis, an idea that ultimately explains how continental drift occurs. The sea-floor-spreading concept culminated decades of sea-floor exploration, which we now review.

BOX 3.3

THE REST OF THE STORY

Finding Paleopoles

How do you find a paleopole position from the orientation of a paleomagnetic dipole in a rock sample? Keep in mind that the paleomagnetic dipole points to the location of the magnetic pole at the time the rock sample cooled. The horizontal projection of the dipole arrow on the Earth's surface (picture the projection as the shadow cast on the Earth's surface by the arrow if the Sun were directly overhead) is like a compass needle that points in the direction of the paleopole; that is, the projection defines an imaginary circle around the Earth that passes through the paleopole and the sample (▶Fig. 3.13a, b). Note that when drawing the circle, we assume that the declination at the time the sample was magnetized equals 0 (because we assume that averaged over time, the magnetic pole coincides with the geographic pole). To find the position of the paleopole on this circle, we must look at the inclination of the paleomagnetic dipole in the rock. Recall that inclination depends on latitude (Fig. 3.10). Thus, the inclination of the paleomagnetic dipole defines the **paleolatitude** of the sample with respect to the paleopole, and paleolatitude simply represents the distance (measured in degrees) from the pole along the circle to where the sample formed.

(a)

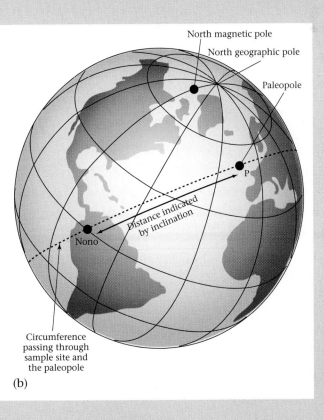

(b)

FIGURE 3.13 (a) Relative to present-day magnetic north, the sample of Figure 3.12 has a declination angle D in map view, and an inclination angle I in a vertical plane. (b) The paleomagnetic dipole preserved in the rock indicates that, relative to the village of Nono, the north magnetic pole sat at point P when the rock formed. The dipole defines the orientation of a circumference (dashed line) that passes between the sample locality and the paleopole. The distance between the two along this line is defined by the inclination.

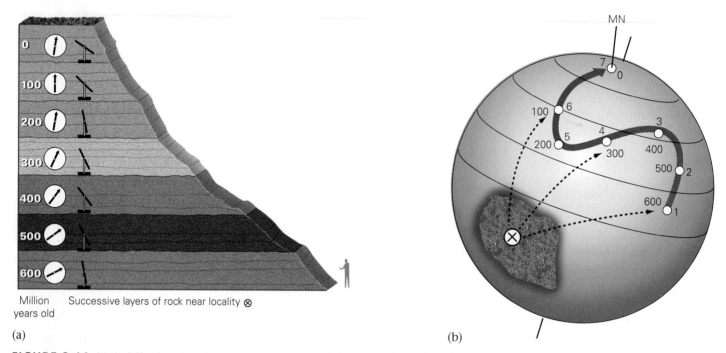

(a)   (b)

FIGURE 3.14 (a) A cliff at location X exposes a succession of dated lava flows. A geologist measures the orientation of the dipole in the rock. (The arrowheads aren't shown, as they don't matter here.) The paleopoles are the places where the dipole intersects the surface of the Earth. (b) The succession of paleopole positions through time for location X defines the polar-wander path for the location. The path ends at the position of the present-day magnetic pole, near the North Pole.

FIGURE 3.15 (a) Apparent polar-wander paths for North America and Europe for the past 300 million years, plotted on a present-day map of the Earth. (b) A map of apparent polar-wander paths for four continents, plotted on the present-day Earth.

(a)   (b)

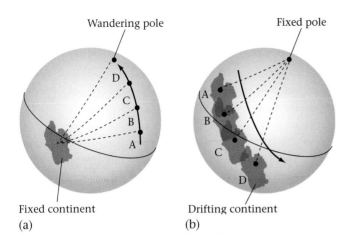

FIGURE 3.16 The two alternative explanations for an apparent polar-wander path. (a) In a "true polar-wander" model, the continent is fixed, so to explain polar-wander paths, the magnetic pole must move substantially. (In reality, the magnetic pole does move a little, but it never strays very far from the geographic pole.) (b) In a continental-drift model, the magnetic pole is fixed near the geographic pole, and the continent drifts relative to the pole.

FIGURE 3.17 The apparent polar-wander paths for North America and Europe would have coincided with each other from about 280 to 180 million years ago, because Europe and North America moved together as a unit when both were part of Pangaea. When Pangaea broke up, the two began to develop separate apparent polar-wander paths.

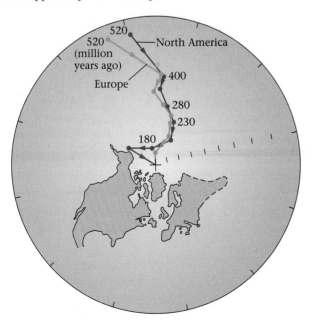

3.4 SETTING THE STAGE FOR THE DISCOVERY OF SEA-FLOOR SPREADING

New Images of Sea-Floor Bathymetry

Before World War II, we knew less about the shape of the ocean floor than we did about the shape of the Moon's surface. After all, we could at least see the surface of the Moon, and could use a telescope to map its craters and mares (flat areas). But our knowledge of sea-floor **bathymetry** (the shape of the sea-floor surface) came only from scattered "soundings" of the sea floor. To sound the ocean depths, a surveyor let out a length of cable with a heavy weight attached. When the weight hit the sea floor, the length of the cable indicated the depth of the floor. Needless to say, it took many hours to make a single measurement, and not many could be made. Nevertheless, soundings carried out between 1872 and 1876 by a British research vessel named the *Challenger* did hint at the existence of submarine mountain ranges and deep troughs.

Military needs gave a boost to sea-floor exploration, for as submarine fleets grew, navies required detailed maps showing variations in the depth of the sea floor. During the middle decades of the twentieth century, the invention of **echo sounding** (sonar) permitted such maps to be made. Echo sounding works on the same principle that a bat uses to navigate and find insects. A sound pulse emitted from a ship travels down through the water, bounces off the sea floor, and returns up as an echo through the water to a receiver on the ship. Since sound waves travel at a known velocity, the time between the sound emission and the detection of the echo defines the distance between the ship and the sea floor (velocity = distance/time, so distance = velocity × time). As the ship moves, echo sounding permits observers to obtain a continuous record of the depth of the sea floor; the resulting cross section showing depth plotted against location is called a **bathymetric profile** (▶Fig. 3.18a, b). By cruising back and forth across the ocean many times, investigators obtained a series of bathymetric profiles and from these constructed maps of the sea floor. Bathymetric maps revealed several important features of the ocean floor.

- *Mid-ocean ridges:* The floor beneath all major oceans includes two provinces: **abyssal plains,** the broad, relatively flat regions of the ocean that lie at a depth of about 4.5 km below sea level; and **mid-ocean ridges,** elongate submarine mountain ranges that lie only about 2–2.5 km below sea level. The Mid-Atlantic Ridge, for example, runs down the middle of the Atlantic Ocean, the East Pacific Rise cuts across the South Pacific Ocean, and the Southeast Indian Ocean Ridge bisects the Great

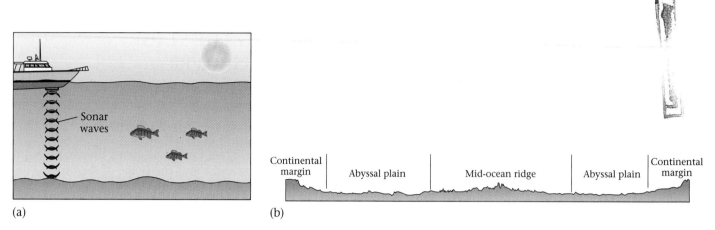

(a)

(b)

FIGURE 3.18 (a) To make a bathymetric profile, researchers use sonar. (b) An east-west bathymetric profile of the Atlantic Ocean.

Southern Ocean between Australia and Antarctica (▶Figs. 3.19, 3.20a). Geologists call the crest (top) of the mid-ocean ridge the **ridge axis.** All mid-ocean ridges are roughly symmetrical—bathymetry on one side of the axis is nearly a mirror image of bathymetry on the other side. Some, like the Mid-Atlantic Ridge, include steep escarpments (cliffs) as well as a distinct **axial trough,** a narrow depression that runs along the ridge axis; others have an overall smooth surface.

• *Deep-ocean trenches:* Along much of the perimeter of the Pacific Ocean, and in a few other localities as well, the ocean floor reaches astounding depths of 8–12 km—deep enough to swallow Mt. Everest. These deep areas define elongate troughs that are now referred to as **trenches.** Trenches border **volcanic arcs,** curving chains of active volcanoes.

• *Seamount chains:* Numerous islands poke up from the ocean floor: for example, the Hawaiian Islands lie in the middle of the Pacific. In addition to islands that rise above sea level, echo sounding has detected many **seamounts** (isolated submarine mountains). Oceanic islands and seamounts, which began as volcanoes,

FIGURE 3.19 The mid-ocean ridges, fracture zones, and principal deep-ocean trenches of today's oceans.

(a)

(b)

FIGURE 3.20 (a) Profile of a mid-ocean ridge, deep-ocean trench, and seamount chain. (b) Block diagram illustrating a fracture zone.

typically occur in chains, but in contrast to the volcanic arcs that border deep-ocean trenches, only one island at the end of a seamount chain is actively erupting today.

- *Fracture zones:* Surveys reveal that the ocean floor is diced up by narrow bands of vertical fractures. In each ocean, these **fracture zones** lie parallel to one another and roughly at right angles to mid-ocean ridges, effectively segmenting the ridges into small pieces (▶Fig. 3.20b).

New Observations on the Nature of Oceanic Crust

By the mid–twentieth century, geologists had discovered many important characteristics of the sea-floor crust. These discoveries led them to realize that oceanic crust is quite different from continental crust, and further that bathymetric features of the ocean floor provide clues to the origin of the crust. Specifically:

- Much of the ocean floor is covered by a layer of sediment composed of clay and the tiny shells of dead plankton. This layer varies in thickness—no sediment covers the mid-ocean ridge axis, while thicker sediment is found toward the margin of the ocean. But even at its thickest, the sediment layer is too thin to have been accumulating for the entirety of Earth history.

- By dredging up samples, geologists learned that oceanic crust contains no granite and no metamorphic rock, common rock types on continents. Rather, oceanic crust contains only basalt and gabbro. Thus, it is fundamentally different in composition from continental crust.

- **Heat flow,** the rate at which heat rises from the Earth's interior up through the floor of the ocean, is not the same everywhere in the oceans. Rather, more heat seems to rise beneath mid-ocean ridges than elsewhere (▶Fig. 3.21a, b). This observation led geologists to speculate that magma might be rising into the crust just below the mid-

ocean ridge axis, because this hot molten rock could transfer heat into the crust.

- When maps showing the distribution of earthquakes in oceanic regions became available in the years after World War II, geologists realized that earthquakes in these regions do not occur randomly, but rather define distinct

FIGURE 3.21 (a) In a mid-ocean ridge, heat from the mantle flows up through the crust; heat flow decreases away from the ridge axis. (b) The graph shows heat flow as a function of distance from the ridge axis. Heat flow is measured in heat-flow units, or HFU (1 HFU = 10^{-6} calories/cm/s).

(a)

(b)

FIGURE 3.22 A 1953 map showing the distribution of earthquakes in the ocean basins.

belts (▶Fig. 3.22). Some belts follow trenches, some mid-ocean ridge axes, and others lie along portions of fracture zones. Since earthquakes define locations where rocks break and move, geologists realized that these bathymetric features are places where movements of the crust take place.

Now let's see how Harry Hess used these observations to come up with the hypothesis of sea-floor spreading.

3.5 HARRY HESS AND HIS "ESSAY IN GEOPOETRY"

In the late 1950s, Harry Hess, after studying the observations described above, realized that the thinness of the sediment layer on the ocean floor meant that the ocean floor might be much younger than the continents, and that the progressive increase in thickness of the sediment away from mid-ocean ridges could mean that the ridges themselves were younger than the deeper parts of the ocean floor. If this was so, then somehow new ocean floor must be forming at the ridges, and the ocean basins (the water-filled depressions that make an ocean) must be getting wider with time. But how? The association of earthquakes with mid-ocean ridges suggested to him that the sea floor was cracking and splitting apart at the ridge. The discovery of high heat flow along mid-ocean ridge axes provided the final piece of the puzzle, indicating as it did the presence of molten rock beneath the ridges. In 1960, Hess suggested that molten rock rose upward beneath mid-ocean ridges and that this material solidified to create oceanic crust. The new sea floor then moved away from the ridge, a process we now call **sea-floor spreading.** Hess realized that old ocean floor must be consumed somewhere, or the Earth would have to grow. He suggested that deep-ocean trenches might be places where the sea floor sank back into the mantle, and that earthquakes at trenches were evidence of this movement, but he didn't understand how the movement took place (▶Fig. 3.23).

Hess and his contemporaries realized that the sea-floor-spreading hypothesis instantly provided the long-sought explanation of how continental drift occurs. Rather than plowing through oceanic crust as Wegener suggested, an impossibility given that rocks comprising continental crust are weaker than rocks in oceanic crust, continents passively move apart as the sea floor between them spreads at mid-ocean ridges, and passively move together as the sea floor between them sinks back into the mantle at trenches. Thus, sea-floor spreading proved to be an important step on the route to plate tectonics—the idea seemed so good that Hess referred to his description of it as "an essay in geopoetry." But other key discoveries would have to take place before the whole theory of plate tectonics came together.

3.6 MARINE MAGNETIC ANOMALIES: EVIDENCE FOR SEA-FLOOR SPREADING

In 1963, two discoveries provided strong evidence for sea-floor spreading. The first discovery was that the measured strength of Earth's magnetic field is not the same everywhere

FIGURE 3.23 Harry Hess's basic concept of sea-floor spreading. New sea floor forms at the mid-ocean ridge axis. As a result, the ocean grows wider. Old sea floor sinks into the mantle at a trench. Earthquakes occur at ridges and trenches.

in the ocean basins; the variations are now called marine magnetic anomalies. The second was that Earth's dipole reverses direction every now and then; such sudden reversals of the Earth's polarity are now called magnetic reversals. To understand why geologists find the concept of sea-floor spreading so appealing, we first need to learn about anomalies and reversals.

Marine Magnetic Anomalies

Geologists can measure the strength of Earth's magnetic field with an instrument called a **magnetometer.** At any given location on the surface of the Earth, the magnetic field that you measure includes two parts: one that is created by the main dipole of the Earth (which is caused in turn by the flow of liquid iron in the outer core) and another that is created by the magnetism of near-surface rock. The part resulting from the flow of the outer core is called the **dipole field** of the Earth. A **magnetic anomaly** is the difference between the expected strength of the dipole field at a certain location and the actual measured strength of the magnetic field at that location. Places where the field strength is stronger than expected are **positive anomalies,** and places where the field strength is less than expected are **negative anomalies.**

On continents, the pattern of magnetic anomalies is very irregular, with an irregularly shaped area of positive anomaly in one place and an irregularly shaped area of negative anomaly in another, depending on the composition of rocks at or near the Earth's surface. A body of iron-rich rock, for example, causes a positive anomaly. But magnetic anomalies on the sea floor yield a surprisingly different pattern.

In the mid–twentieth century, geologists first began to tow magnetometers back and forth across the ocean to map variations in magnetic field strength over the ocean floor.

FIGURE 3.24 (a) A ship sailing through the ocean dragging a magnetometer detects first a positive anomaly then a negative one, then a positive one, then a negative one. (b) Magnetic anomalies on the sea floor off the northwest coast of the United States. The dark bands are positive anomalies, the uncolored bands negative anomalies. Note the distinctive stripes of alternating anomalies. A positive anomaly overlies the crest of the Juan de Fuca Ridge (a small mid-ocean ridge).

(a)

(b)

Magnetic field lines

The Earth behaves like a giant magnet, and thus is surrounded by a magnetic field. The magnetism is due to the flow of liquid iron alloy in the outer core.

The age of oceanic crust varies with location. The youngest crust lies along a mid-ocean ridge, and the oldest along the coasts of continents. Here, the different color stripes correspond to different ages of oceanic crust. Red is youngest, purple is oldest.

The rock of oceanic crust preserves a record of the Earth's magnetic polarity at the time the crust formed. Eventually, a symmetric pattern of polarity stripes develops.

Marine magnetic anomalies are stripes representing alternating bands of oceanic crust that differ in the measured strength of the magnetic field above them. Stronger fields are measured over crust with normal polarity, while weaker fields are measured over crust with reversed polarity.

Normal polarity

Reversed polarity

Earth's magnetic field can be represented by a dipole that points from the north magnetic pole to the south. Every now and then, the magnetic polarity reverses.

Magnetic reversals are recorded in a succession of lava flows. Here, lavas with normal polarity are red, while lavas with reversed polarity are yellow.

Lava flows at a volcano.

By dating successive lava flows, geologists can determine the timing and duration of magnetic reversals.

The red stripes indicate rock with normal polarity, and the yellow stripes rock with reversed polarity.

Normal polarity   Reversed polarity   Mid-ocean ridge (normal polarity)

Paleomagnetism

They expected to find few, if any, anomalies, because they had observed that ocean crust consisted of the same rock types (basalt and gabbro) everywhere. But their measurements revealed distinct anomalies. As a boat cruised along a single traverse (course), the magnetometer's gauge would detect strong signals (a positive anomaly) and then weak signals (a negative anomaly). A graph of signal strength versus distance along the traverse, therefore, has a sawtooth shape (▶Fig. 3.24a). When the data from many traverses was compiled on a map, these **marine magnetic anomalies** defined distinctive, alternating bands. And if we color positive anomalies black and negative anomalies white, the pattern made by the anomalies resembles the stripes on a candy cane (▶Fig. 3.24b). The mystery of the marine magnetic anomaly pattern, however, remained unsolved until geologists recognized the existence of magnetic reversals.

Magnetic Reversals

Soon after geologists began to study the phenomenon of paleomagnetism, they decided to see if the magnetism of rocks changed as time passed. To do this, they measured the paleomagnetism of many successive basalt layers, cooled from lava, that had been erupted by a volcano over a long period of time. To their surprise, they found that the **polarity** (which end of a magnet points north and which end points south; see Box 3.1) of the paleomagnetic field of some layers was the same as that of Earth's present magnetic field, while in other layers it was the opposite. Recall that Earth's magnetic field can be represented by an arrow (dipole) that

points from north to south; in some of the lava flows, the paleomagnetic dipole pointed south (these layers showed **normal polarity**), but in others the dipole pointed north (these layers had **reversed polarity**) (▶Fig. 3.25).

At first, observations of reversed polarity were largely ignored, thought to be the result of lightning strikes or of chemical reactions between rock and water. But when repeated measurements from around the world revealed a systematic pattern of alternating normal and reversed polarity in rock layers, geologists realized that reversals were a global, not local, phenomenon. At various times during Earth history, the polarity of Earth's magnetic field has suddenly reversed; in other words, sometimes the Earth has normal polarity, as it does today, and sometimes it has reversed polarity (▶Fig. 3.26a, b). Times when the Earth's field flips from normal to reversed polarity, or vice versa, are

FIGURE 3.26 The magnetic field of the Earth has reversed polarity at various times during Earth history. (a) If the dipole points from north to south, Earth has normal polarity. (b) If the dipole points from south to north, Earth has reversed polarity.

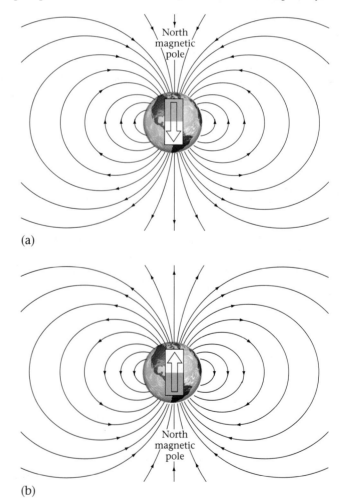

(a)

(b)

FIGURE 3.25 In a succession of lava flows on land, different flows exhibit different polarity (indicated here by whether the arrow points up or down). When these reversals are plotted on a time column, we have a magnetic-reversal chronology.

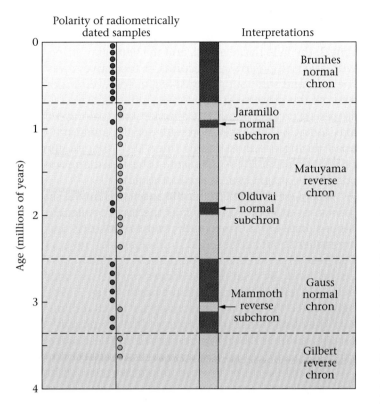

Polarity of radiometrically dated samples

Interpretations

Brunhes normal chron

Jaramillo normal subchron

Matuyama reverse chron

Olduvai normal subchron

Gauss normal chron

Mammoth reverse subchron

Gilbert reverse chron

Age (millions of years)

FIGURE 3.27 Radiometric dating of lava flows allows us to determine the age of magnetic reversals during the past 4.5 million years. Major intervals of a given polarity are referred to as polarity chrons, and are named after scientists who contributed to the understanding of Earth's magnetic field. Shorter-duration reversals are called subchrons.

In the 1950s, about the same time geologists discovered polarity reversals, they developed a technique that permitted them to define the age of a rock in years, by measuring the rate of decay of radioactive elements in the rock. The technique is called **radiometric dating** (and will be discussed in detail in Chapter 12). Geologists applied the technique to date the layers of rock in which they obtained their paleomagnetic measurements, and thus determined *when* the magnetic field of the Earth reversed. With this information, they constructed the history of magnetic reversals, now called the **magnetic-reversal chronology.** A diagram representing the Earth's magnetic-reversal chronology (▶Fig. 3.27) shows that reversals do not occur regularly, so the lengths of different **polarity chrons,** the time intervals between reversals, are different. For example, we have had a normal-polarity chron for about the last 700,000 years. Before that, there was a reversed-polarity chron that lasted about 1.75 million years. Geologists named the youngest four polarity chrons (Brunhes, Matuyama, Gauss, and Gilbert) after scientists who had made important contributions to the study of rock magnetism. As more measurements became available, investigators realized that there were some short-duration reversals (less than 200,000 years long) within the chrons, and called these shorter reversals **polarity subchrons.** Radiometric dating methods are not accurate enough to date reversals that are older than about 4.5 million years, because for rocks older than that, the uncertainty in a date exceeds the duration of a polarity subchron. (Dates have uncertainties. For example, a radiometric date of 5 million years may have an uncertainty of 300,000 years. We express this date as 5.0 ± 0.3 million years; that means the date is in the range of 4.7 to 5.3 million years.)

The Interpretation of Marine Anomalies

Armed with our knowledge of magnetic reversals, we can now understand the explanation for marine magnetic anomalies. A graduate student in England, Fred Vine, working with his adviser, Drummond Mathews, and a Canadian geologist, Lawrence Morley (working independently), discovered a solution to this riddle. Simply put, the three suggested that a positive anomaly occurs over areas of sea floor where basalt has normal polarity. In these areas, the magnetic force produced by the basalt *adds* to the force produced by Earth's dipole and creates a stronger magnetic signal than expected, as measured by the magnetometer. A negative anomaly occurs over regions of sea floor where basalt has reversed polarity. Here, the magnetic force of the basalt *subtracts* from the force produced by the dipole and results in a weaker magnetic signal (▶Fig. 3.28a).

Vine, Matthews, and Morley pointed out that marine magnetic anomalies could have formed as a consequence of sea-floor spreading—sea floor yielding positive anomalies

called **magnetic reversals.** When the Earth has reversed polarity, the south magnetic pole lies near the north geographic pole, and the north magnetic pole lies near the south geographic pole. If you were to use a compass during periods when the Earth's magnetic field was reversed, the north-seeking end of the needle would point to the south geographic pole.

Note that magnetic reversals are not related to the gradual change in the inclination and declination of the paleomagnetic dipoles we find in successive layers of rock that results from continental drift (as represented by apparent polar-wander paths; see Fig. 3.14a). Also, magnetic reversals are not related to the small change in declination caused by the slight movement of the dipole with respect to Earth's rotational axis. Indeed, reversals do *not* occur by the slow migration of the dipole around the Earth and past the equator. Rather, geologists have found evidence indicating that reversals take place quickly, perhaps in tens to thousands of years. Still, though magnetic reversals have now been well documented, the mechanism by which they occur remains uncertain and continues to be a subject of research.

(a)

(b)

FIGURE 3.28 (a) The explanation of marine anomalies. The sea floor beneath positive anomalies has the same polarity as Earth's field and therefore adds to it. The sea floor beneath negative anomalies has reversed polarity and thus subtracts from Earth's field. (b) The symmetry of the magnetic anomalies measured across the Mid-Atlantic Ridge south of Iceland. Note that individual anomalies are somewhat irregular, because the process of forming the sea floor in detail happens in discontinuous pulses along the length of the ridge.

developed at times when the Earth had normal polarity, while sea floor yielding negative anomalies formed when the Earth had reversed polarity. If this was so, then the pattern of anomalies should be symmetric with respect to the axis of a mid-ocean ridge. With this idea in mind, geologists set sail to measure magnetic anomalies near mid-ocean ridges (▶Fig. 3.28b). By 1966, the story was complete. In the examples they studied, the magnetic anomaly pattern on one side of a ridge was nearly a mirror image of the anomaly pattern on the other.

Let's look more closely at how marine magnetic anomalies are formed. Please refer to ▶Figure 3.29a. At time 1 (sometime in the past), a time of normal polarity, the dark stripe of sea floor forms. The tiny dipoles of magnetite grains in basalt making up this stripe align with the Earth's

field. As it forms, the rock in this stripe migrates away from the ridge axis, half to the right and half to the left. Later, at time 2, the field has reversed, and the light-gray stripe forms with reversed polarity. As it forms, it too moves away from the axis, and still younger crust begins to develop along the axis. As the process continues over millions of years, many stripes are created. A positive anomaly exists along the ridge axis today, because this represents sea floor that has developed during the most recent interval of time, a chron of normal polarity. The magnetism of the rock along the ridge adds to the magnetism of the Earth's field.

Geologists realized that if the anomalies on the sea floor formed by sea-floor spreading as the Earth's magnetic field flipped between normal and reversed polarity, then the anomalies should correspond to the magnetic reversals that had been discovered and radiometrically dated in basalt layers on land. By relating the stripes on the sea floor with magnetic reversals found in dated basalt (▶Fig. 3.29b), geologists dated the sea floor back to an age of 4.5 million years. And they found that the relative widths of anomaly stripes on the sea floor exactly corresponded to the relative durations of polarity chrons in the magnetic-reversal chronology.

The relationship between anomaly-stripe width and polarity-chron duration provides the key for determining the rate (velocity) of sea-floor spreading, for it indicates that the rate of spreading has been constant for the last 4.5 million years. Remember that velocity = distance/time. In the North Atlantic Ocean, 4.5-million-year-old sea floor lies 45 km away from the ridge axis. Therefore, the velocity at which the sea floor moves away from the ridge axis = 45 km / 4.5 million years. This means that the crust moves away from the Mid-Atlantic Ridge axis at a rate of 1 cm per year, or that a point on one side of the ridge moves away from a point on the other side by 2 cm per year. We call this number the **spreading rate.** In the Pacific Ocean, sea-floor spreading occurs at the East Pacific Rise (geographers named this a "rise" because it is not as rough and jagged as the Mid-Atlantic Ridge). The anomaly stripes bordering the East Pacific Rise are much wider, and 4.5-million-year-old sea floor lies about 225 km from the rise axis. This requires the sea floor to move away from the rise at a rate of about 5 cm per year, so the spreading rate for the East Pacific Rise is about 10 cm per year.

Geologists also realized that if they could assume that the rate of sea-floor spreading has remained constant for a long time, then they could date the ages of magnetic-field reversals further back in history, simply by measuring the distance of successive magnetic anomalies from the ridge axis (time = distance/velocity). Such analysis thus defined the magnetic-reversal chronology back to about 80 million years based on the magnetic anomaly stripes of the Atlantic Ocean (▶Fig. 3.29c). Eventually, by examining stripes in all the world's oceans, geologists defined the reversal chronol-

FIGURE 3.29 (a) The progressive development of stripes of alternating polarity in the ocean floor. Each time represents a successive stage of new sea floor forming at a mid-ocean ridge, while Earth's field undergoes magnetic reversals. (b) The observed stripes correlate with the polarity chrons and subchrons measured in lava flows on land. (c) The reversal chronology for the last 76 million years, based on marine magnetic anomalies.

ogy back to around 180 million years ago. The spectacular correspondence between the record of marine magnetic anomalies and the magnetic-reversal chronology can only be explained by the sea-floor-spreading hypothesis. Thus, the discovery and explanation of marine magnetic anomalies served as a proof of sea-floor spreading, and allowed geologists to measure rates of spreading. At a rate of 5 cm per year, sea-floor spreading produces a 5,000-km-wide ocean in 100 million years (only about 2% of Earth history).

3.7 DEEP-SEA DRILLING: FURTHER EVIDENCE

Soon after geologists around the world began to accept the idea of sea-floor spreading, an opportunity arose to really put the concept to the test. In the late 1960s, a drilling ship called the *Glomar Challenger* set out to sail around the ocean drilling holes into the sea floor. This amazing ship could lower enough drill pipe to drill in 5-km-deep water and could continue to drill until the hole reached a depth of about 1.7 km (1.1 miles) below the sea floor. Drillers brought up cores of rock or sediment that geoscientists then studied on board.

On one of its early cruises, the *Glomar Challenger* drilled a series of holes through sea-floor sediment to the basalt layer, spaced at progressively greater distances from the axis of the Mid-Atlantic Ridge. If the model of sea-floor spreading was correct, then the sediment layer should have gotten progressively thicker away from the axis, and the age of the oldest sediment just above the basalt should have become progressively older away from the axis. When the drilling and the analyses were complete, the predictions were confirmed. Eventually, drilling would also show that the oldest ocean floor is only around 200 million years old.

CHAPTER SUMMARY

• Alfred Wegener proposed that continents had once been stuck together to form a single huge supercontinent (Pangaea) and had subsequently drifted apart. He called this idea the continental-drift hypothesis.

• Wegener drew from several different sources of data to support his hypothesis: (1) coastlines on opposite sides of the ocean match up; (2) the distribution of late Paleozoic glaciation can be explained if the glaciers comprised a polar ice cap over the south end of Pangaea; (3) the distribution of late Paleozoic/early Mesozoic equatorial climatic belts is compatible with the concept of Pangaea; (4) the distribution of fossil species suggests the existence of a supercontinent; (5) distinctive rock units that are now on opposite sides of the ocean formed continuous belts on Pangaea.

• Despite all the observations that supported continental drift, most geologists did not initially accept the idea, because no one could explain how continents could move.

• Rocks retain a record of the Earth's magnetic field at the time they formed. This record is called paleomagnetism. By measuring paleomagnetism in successively older rocks, geologists found that the apparent position of the Earth's magnetic pole relative to the rocks changes through time. Successive positions of the pole define an apparent polar-wander path.

• Polar-wander paths are different for different continents. This observation can be readily explained by continental drift: continents move with respect to one another, while the Earth's magnetic poles remain fixed.

• The invention of echo sounding permitted explorers to make detailed maps of the sea floor. These maps revealed the existence of mid-ocean ridges, deep-ocean trenches, seamount chains, and fracture zones. Heat flow is generally greater near the axis of a mid-ocean ridge.

• Around 1960, Harry Hess proposed the hypothesis of sea-floor spreading. According to this hypothesis, new sea floor forms at mid-ocean ridges, above a band of upwelling mantle, then spreads symmetrically away from the ridge axis. As a consequence, ocean basins get progressively wider with time, and the continents on either side of the ocean basins drift apart. Eventually, the ocean floor sinks back into the mantle at deep-ocean trenches.

• Magnetometer surveys of the sea floor revealed marine magnetic anomalies. Positive anomalies, where the magnetic field strength is greater than expected, and negative anomalies, where the magnetic field strength is less than expected, are arranged in alternating stripes.

• During the 1950s, geologists documented that the Earth's magnetic field reverses polarity every now and then. The record of reversals, dated by radiometric techniques, is called the magnetic-reversal chronology.

• The proof of sea-floor spreading came from the interpretation of marine magnetic anomalies. According to Vine, Mathews, and Morley, the sea floor that forms when the Earth has normal polarity creates positive anomalies, and the sea floor that forms when the Earth has reversed polarity creates negative anomalies. Anomalies are symmetric with respect to a mid-ocean ridge axis, and their widths are proportional to the duration of polarity chrons, facts that can only be explained by sea-floor spreading. Study of anomalies allows us to calculate the rate of spreading.

• Drilling of the sea floor confirmed its age and served as another proof.

KEY TERMS

apparent polar-wander path (p. 60)	magnetic inclination (p. 56)
bathymetric profile (p. 63)	magnetic pole (p. 57)
bathymetry (p. 63)	magnetic-reversal chronology (p. 71)
continental drift (p. 51)	magnetometer (p. 67)
deep-ocean trench (p. 64)	mid-ocean ridge (p. 63)
dipole (p. 56)	normal and reversed polarity (p. 70)
electromagnetism (p. 58)	
fracture zone (p. 65)	paleomagnetism (p. 56)
geographical pole (p. 56)	paleopole (p. 60)
heat flow (p. 65)	Pangaea (p. 51)
ice age (p. 53)	plate tectonics (p. 52)
magnetic anomaly (p. 67)	polarity (p. 57)
magnetic declination (p. 56)	sea-floor spreading (p. 52)
magnetic dipole (p. 56)	seamount (p. 64)
magnetic field (p. 57)	subduction (p. 52)
magnetic force (p. 57)	

REVIEW QUESTIONS

1. What was Wegener's continental-drift hypothesis?

2. How does the fit of the coastlines around the Atlantic support continental drift?

3. How does the evidence of past glaciations support continental drift?

4. How does the evidence of equatorial climatic belts support continental drift?

5. How does the distribution of ancient fossils support continental drift?

6. Why were geologists initially skeptical of Wegener's theory of continental drift?

7. Describe how the angle of inclination of the Earth's magnetic field varies with latitude. How could this information be used to determine the ancient latitude of a continent?

8. How does a basalt rock become magnetized?

9. Why did the discovery of apparent polar-wander paths show that the continents, rather than the poles, had moved?

10. Describe the characteristics of mid-ocean ridges, deep-ocean trenches, and seamount chains.

11. How did the observations of heat flow and seismicity support the hypothesis of sea-floor spreading?

12. What is a marine magnetic anomaly? How is it detected?

13. Describe the pattern of marine magnetic anomalies across a mid-ocean ridge. How is this explained?

14. How were the reversals of the Earth's magnetic field discovered? How did they corroborate the sea-floor-spreading hypothesis?

SUGGESTED READING

Allegre, C. 1988. *The Behavior of the Earth: Continental and Seafloor Mobility.* Cambridge: Harvard University Press.

Bird, J. M., ed. 1980. *Plate Tectonics.* Washington, D.C.: American Geophysical Union.

Butler, R. F. 1992. *Paleomagnetism: Magnetic Domains to Geologic Terranes.* Boston: Blackwell.

Erikson, J. 1992. *Plate Tectonics: Unraveling the Mysteries of the Earth.* New York: Facts on File.

Glen, W. 1982. *The Road to Jaramillo: Critical Years of the Revolution in Earth Sciences.* Palo Alto, Calif.: Stanford.

LeGrand, H. E. 1988. *Drifting Continents and Sifting Theories.* Cambridge, Eng.: Cambridge University Press.

McPhee, J. A. 1998. *Annals of the Former World.* New York: Farrar, Straus and Giroux.

Moores, E. M., ed. 1990. *Plate Tectonics: Readings from "Scientific American."* New York: Freeman.

Sullivan, W. 1991. *Continents in Motion: The New Earth Debate.* 2nd ed. New York: American Institute of Physics.

CHAPTER 4

The Way the Earth Works: Plate Tectonics

4.1 INTRODUCTION

Thomas Kuhn, an influential historian of science working in the 1960s, argued that scientific thought evolves in fits and starts. According to Kuhn, scientists base their interpretation of the natural world on an established line of reasoning, a **scientific paradigm,** for many years. But once in a while, a revolutionary thinker, or group of thinkers, proposes a radically new point of view that invalidates the old paradigm. Almost immediately, the scientific community scraps old hypotheses and formulates others consistent with the new paradigm. Kuhn called such abrupt changes in thought a **scientific revolution.** Some new paradigms work so well that they become scientific law and will never be replaced.

In physics, Isaac Newton's mathematical description of moving objects established the paradigm that natural phenomena obey physical laws; older paradigms suggesting that natural phenomena followed Greek philosophy had to be scrapped. In biology, Charles Darwin's proposal that species evolve by natural selection required biologists to rethink hypotheses based on the older paradigm that species never change. And in geology, a scientific revolution in the 1960s yielded the new paradigm that the outer layer of the Earth, the lithosphere, consists of separate pieces, or plates, that move with respect to each other. This idea, which we now call the **theory of plate tectonics,** or simply **plate tectonics,** required geologists to cast aside hypotheses rooted in the paradigm of fixed continents, and thus led to a complete restructuring of how geologists thought about Earth history. Compare this book with a geology textbook from the 1950s, and you will instantly see the difference.

Alfred Wegener planted the seed of plate tectonics theory with his proposal of continental drift in 1915, but between 1915 and 1960 this seed lay dormant while

Astronauts in the space shuttle Endeavor *could see the consequences of plate tectonics right outside their window. This photo from space shows the Sinai Peninsula, separated from Egypt to the west and the Arabian Peninsula to the east by rifts, narrow belts where the crust has stretched and broken apart. Note the green stripe in the middle—this is the Nile Valley, which flows into the southern Mediterranean Sea. The triangle of green is the Nile Delta.*

geoscientists focused on collecting new data about the Earth. Discoveries about sea-floor bathymetry, heat flow, marine sediments, and apparent polar wander led to the germination of the seed in 1960, with Harry Hess's proposal of sea-floor spreading. The roots took hold three years later when marine magnetic anomalies supplied strong evidence for sea-floor spreading. During the next five years, the study of geoscience turned into a feeding frenzy, as many investigators dropped what they'd been doing and turned their attention to testing the hypothesis of sea-floor spreading and describing its broader implications; by 1968, thanks primarily to the work of perhaps two dozen different investigators, the sea-floor-spreading hypothesis had bloomed into plate tectonics theory. Geologists clarified the concept of a plate, described the types of plate boundaries, calculated plate motions, related plate tectonics to earthquakes and volcanoes, showed how plate motions generate mountain belts and seamount chains, and defined the history of past plate motions. In the 1970s, after excited investigators had presented their new ideas to standing-room-only audiences at many conferences, the geoscience community, with few exceptions, embraced plate tectonics theory and has built on it ever since.

This chapter outlines the key elements of plate tectonics theory. We first learn about lithosphere plates, the types of plate boundaries, and the nature of geologic activity that occurs at each. We then look at hot spots and other special locations on plates. Finally, we see how continents break apart and how they collide, and learn what makes plates move. Because plate tectonics theory is geology's grand unifying theory, it is now an essential foundation for the discussion of all geology.

4.2 WHAT DO WE MEAN BY PLATE TECTONICS?

The Concept of a Lithosphere Plate

As we learned in Chapter 2, geoscientists divide the interior of the Earth into layers. If we want to distinguish layers according to the speed at which seismic waves pass through them, we speak of the crust, upper mantle, lower mantle, outer core, and inner core, and define the boundaries between these layers by abrupt changes in the speed of earthquake waves. But if, instead, we want to distinguish layers based on whether or not they flow under certain conditions, we use the names "lithosphere" and "asthenosphere." The **lithosphere** consists of the crust plus the top (cooler) part of the upper mantle. It behaves relatively rigidly, meaning that overall it does not flow easily, but bends and flexes,

or breaks, when acted on by a force. The lithosphere floats on a relatively soft layer called the **asthenosphere,** composed of mantle that can flow (though very slowly) when acted on by force, because it is warmer than lithospheric rock (hotter than about 1,280°C). The asthenosphere can convect, because of its ability to flow, but the lithosphere cannot.

Continental lithosphere and oceanic lithosphere differ markedly in their thickness, because the rate at which temperature increases with depth in continental lithosphere differs from the rate in oceanic lithosphere. On average, continental lithosphere has a thickness of 150 km, while old oceanic lithosphere has a thickness of about 100 km; for reasons discussed later in this chapter, very young oceanic lithosphere thins to about 10 km near the crest of mid-ocean ridges. Recall also that continental crust has an average thickness of 35–40 km and contains relatively low-density rock (granite and many other rock types), while oceanic crust reaches a thickness of only 10 km and consists of relatively high-density rock (basalt and gabbro).

The surface of continental lithosphere lies at a higher elevation than the surface of oceanic lithosphere. To picture why, imagine that we have two blocks of oak (a high-density hardwood), one 15 cm thick and one 10 cm thick. On top of the thicker block, we glue a 4-cm-thick layer of cork (a low-density bark), and on top of the thinner block, we glue a 1-cm-thick layer of pine (a medium-density softwood). Now place the two blocks in water (▶Fig. 4.1). The total mass of the cork-covered block exceeds the total mass of the pine-

FIGURE 4.1 We can picture continental lithosphere as a thick oak block (lithospheric mantle) overlain by a layer of cork (continental crust), and oceanic lithosphere as a thinner block of oak (lithospheric mantle) overlain by a layer of pine (oceanic crust). The pine layer is thinner than the cork layer. If both blocks float in a tub of water, the surface of the thick cork/oak block lies at a higher elevation than that of the pine/oak block. Thus, the ocean floor lies 4 km below the surface of the continents, on average, because lithosphere, like the wood blocks, floats on the asthenosphere.

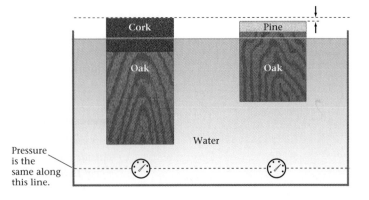

covered block, so the base of the cork-covered block sinks deeper into the water. But because the cork-covered block is thicker and has a lower overall density, it floats higher. In our analogy, the cork-covered block represents continental lithosphere, with its thick crust of low-density rock, while the pine-covered block represents oceanic lithosphere, with its thin crust of medium-density rock. The oak represents the mantle part of the lithosphere, thicker for the continent than for the ocean. Our analogy emphasizes that ocean basins, the low areas that fill with water to form oceans, exist because continental lithosphere floats higher than oceanic lithosphere (see Box 4.1).

The lithosphere forms the Earth's relatively rigid shell. But unlike the shell of a freshly laid hen's egg, the lithosphere shell contains a number of major "breaks," which

separate the lithosphere into distinct pieces.* We call the pieces **lithosphere plates,** or simply **plates,** and we call the breaks **plate boundaries.** Geoscientists distinguish

---

*Note that the above definition tacitly equates the base of a plate with the base of the lithosphere. This definition works well for oceanic plates, because the top 100 km of asthenosphere directly beneath oceanic lithosphere is particularly weak, owing to the presence of 2–4% melt along the surfaces of grains that serves to lubricate movement. This weak region is known as the seismic low-velocity zone, because seismic waves travel more slowly through a fluid. But some geologists now think that the upper 100–150 km of the asthenosphere moves with the continental lithosphere. Thus, the base of continental plates—the base of the layer that moves during plate motion—may actually lie within the asthenosphere. To simplify the discussion, however, we ignore this complication.

BOX 4.1

SCIENCE TOOLBOX

Archimedes' Principle of Buoyancy

Archimedes (c. 287–212 B.C.E.), a Greek mathematician and inventor, left an amazing legacy of discoveries. He described the geometry of spheres, cylinders, and spirals, introduced the concept of a center of gravity, and was the first to understand buoyancy. **Buoyancy** refers to the upward force acting on an object immersed or floating in a fluid. According to legend, Archimedes recognized this concept suddenly, while taking a bath, and was so inspired that he jumped out of the bath and ran home naked, shouting, "Eureka!"

Archimedes realized that when you place a solid object in water, the object displaces a volume of water equal in mass to the object (▶Fig. 4.2a, b). An object

denser than water, like a stone, sinks through the water, because even when completely submerged, the stone's mass exceeds the mass of the water displaced. When submerged, however, the stone weighs less than it does in air. An object less dense than water, such as an iceberg, sinks only until the mass of the water displaced equals the total mass of the iceberg. This condition happens while part of the iceberg still protrudes up into the air. Put another way, an object placed in a fluid feels a buoyancy force that tends to push it up. If the object's weight is less than the buoyancy force, the object floats, but if its weight is greater than the buoyancy force, the object sinks.

FIGURE 4.2 Archimedes' principle of buoyancy. (a) Ice is less dense than water, so it is buoyant. When an iceberg floats, 80% of the ice lies underwater. (b) The ice sinks until the total mass of the water displaced equals the total mass of the iceberg. Since water is denser, the volume of the water displaced is less than the volume of the iceberg, so the iceberg protrudes above the water.

Hot-spot
volcano

Transform
plate
boundary

Volcanic arc

Trench

Continental rift

Convergent plate
boundary

Subducting oceanic
lithosphere

Collisional mountain belt

Continental crust

Continental lithosphere

Lithospheric mantle

Asthenosphere

The Theory of Plate Tectonics

The outer portion of the Earth is a relatively rigid layer called the lithosphere. It consists of the crust (oceanic or continental) and the uppermost mantle. The mantle below the lithosphere is relatively plastic (it can flow) and is called the asthenosphere. The difference in behavior (rigid vs. plastic) between lithospheric mantle and asthenospheric mantle is a consequence of temperature—the former is cooler than the latter. Continental lithosphere is typically about 150 km thick, while oceanic lithosphere is about 100 km thick. (Note: they are not drawn to scale in this image.)

According to the theory of plate tectonics, the lithosphere is broken into about twenty plates that move relative to one another. Most of the motion is accommodated by sliding along plate boundaries (the edges of plates); plate interiors stay relatively undeformed. There are three kinds of plate boundaries.

1. Divergent boundaries: Here, two plates move apart by a process called sea-floor spreading. Divergent boundaries are marked by a mid-ocean ridge. Asthenospheric mantle rises

Triple junction

Sea-floor spreading

Mid-ocean ridge

Divergent plate boundary

Transform plate boundary

Occanic lithosphere

Inactive (extinct) hot-spot volcano

Active hot-spot volcano

Oceanic crust

Litho-spheric mantle

Asthenosphere

Mantle plume

beneath a mid-ocean ridge and partially melts, forming magma. The magma rises to create new ocean crust. The lithospheric mantle thickens progressively away from the ridge axis as the plate cools.

2. Convergent boundaries: Here, two plates move together, and one plate subducts beneath another (it sinks down into the mantle). Only oceanic lithosphere can subduct. At the Earth's surface, the boundary between the two plates is marked by a deep-ocean trench. During subduction, melting above the downgoing plate produces magma that rises to form a volcanic arc.

3. Transform boundaries: Here, one plate slides sideways past another, without the creation of a new plate or the subduction of an old one. The boundary is marked by a fault, a fracture on which sliding occurs. Transform boundaries link segments of mid-ocean ridges. They may also cut through continental lithosphere.

A point at which three plate boundaries meet is called a triple junction. This figure shows a triple junction where three mid-ocean ridges meet. Where two continents converge, they collide and form a collisional mountain range. This happens because continental crust is too buoyant to be subducted. At continental rifts, a continent stretches and may break in two. Rifts are marked by the existence of many faults. If a continent rifts apart, a new mid-ocean ridge develops. Hot-spot volcanoes form above plumes of hot mantle rock that rise from near the core-mantle boundary. As a plate drifts over a hot spot, it leaves a chain of extinct volcanoes.

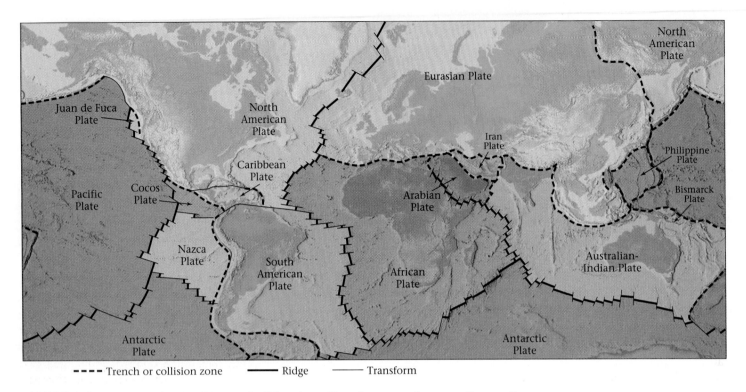

---- Trench or collision zone —— Ridge —— Transform

FIGURE 4.3 The major plates making up the lithosphere. Note that some plates are all ocean floor, while some contain both continents and oceans. Thus, some plate boundaries lie along continental margins (coasts), while others do not. For example, the eastern border of South America is not a plate boundary, but the western edge is.

twelve major plates and numerous microplates (eight to twenty, depending on the criteria used to define the boundaries). Some plates have familiar names (for example, the North American Plate, the African Plate), while some do not (the Cocos Plate, the Juan de Fuca Plate).

As illustrated in ►Figure 4.3, some plate boundaries follow **continental margins,** or coasts, while others do not. For this reason, we distinguish between **active continental margins,** which are plate boundaries, and **passive margins,** which are not plate boundaries. Along passive margins, continental crust is thinner than normal, and the upper (higher) part has broken into wedge-shaped slices (►Fig. 4.4). Thick (10–15 km) accumulations of sediment cover this thinned crust. The surface of this sediment layer is a broad, shallow (less than 500 m deep) region of the continent called the **continental shelf,** home to the major fisheries of the world. Note that some plates consist entirely of oceanic lithosphere or entirely of continental lithosphere, while some plates consist of both. For example, the Nazca Plate is made up entirely of ocean floor, while the North American Plate consists of North America plus the western half of the North Atlantic Ocean.

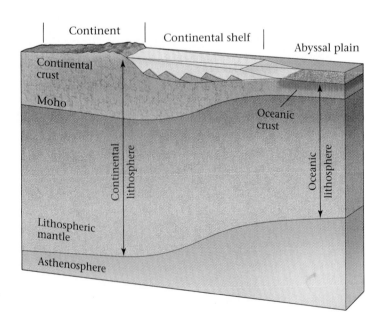

FIGURE 4.4 In this cross section of a passive margin, note that the continental crust thins along the boundary (see section 4.7, "Continental Rifting"). The sediment pile that accumulates over this thinned crust underlies the continental shelf.

The Basic Premise of Plate Tectonics

We can now restate plate tectonics theory as follows. The Earth's lithosphere is divided into plates that move relative to one another and relative to the underlying asthenosphere. Plate movement occurs at rates of 1 to 15 cm per year. As a plate moves, its internal area remains largely rigid and intact, but rock along the plate's boundaries undergoes **deformation** (cracking, sliding, bending, stretching, and squashing) as it grinds or scrapes against its neighbors or pulls away from its neighbors. As plates move, so do the continents that form part of the plates, resulting in continental drift. Because of plate tectonics, the map of Earth's surface constantly changes.

Identifying Plate Boundaries

How do we recognize the location of a plate boundary? The answer becomes clear from looking at a map showing the locations of earthquakes (▶Fig. 4.5). Recall from Chapter 2 that earthquakes are vibrations caused by shock waves that are generated where rock breaks and suddenly shears (slides) along a **fault** (a fracture on which sliding occurs). The **focus** of the earthquake is the spot where the fault begins to slip, and the **epicenter** marks the point on the surface of the Earth directly above the focus. Earthquake epicenters do not speckle the Earth's surface randomly, like buckshot on a target. Rather, the majority are located in relatively narrow, distinct belts. These **earthquake belts** define the position of plate boundaries, because the faulting (fracturing and slipping) that occurs along plate boundaries as plates move generates earthquakes. (We will learn more about this process in Chapter 10.) **Plate interiors,** regions away from the plate boundaries, remain relatively earthquake free because they are stronger and do not accommodate much movement.

While earthquakes serve as the most definitive indicator of plate boundaries, other prominent geologic features may also be found along plate boundaries. By the end of this chapter, we will see that each type of plate boundary is associated with a diagnostic group of geologic features (such as volcanoes, trenches, or mountains), generated by movement at the plate boundary. Note that earthquakes happen frequently along active continental margins, such as the Pacific coast of the Americas, but not along passive continental margins, like the east coast of the Americas.

Geologists define three types of plate boundary, based simply on the relative motions of the plates on either side of the boundary (▶Fig. 4.6a–c). A boundary at which two plates move apart from each other is called a **divergent plate boundary.** A boundary at which two plates move toward each other so that one plate sinks beneath the other is called a **convergent plate boundary.** And a boundary at which one plate slips laterally past (along the side of) an-

FIGURE 4.5 The locations of most earthquakes fall in distinct bands. These earthquake belts define the positions of the plate boundaries.

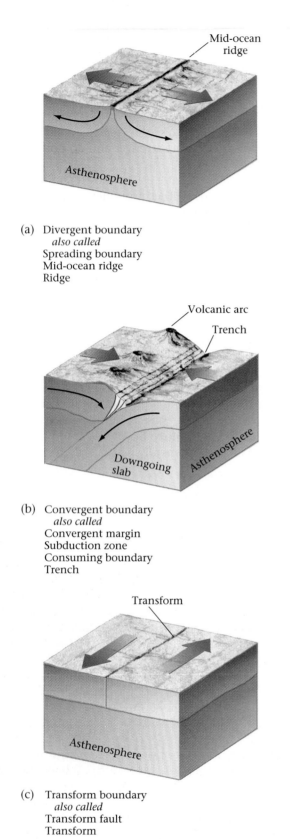

(a) Divergent boundary
   also called
   Spreading boundary
   Mid-ocean ridge
   Ridge

(b) Convergent boundary
   also called
   Convergent margin
   Subduction zone
   Consuming boundary
   Trench

(c) Transform boundary
   also called
   Transform fault
   Transform

FIGURE 4.6 (a) At a divergent boundary (its other names are listed below), two oceanic plates move away from each other. The lithosphere thickens with increasing distance from the ridge. (b) At a convergent boundary, one oceanic plate bends and sinks into the mantle beneath another plate. (c) At a transform boundary, two plates slide past each other along a vertical fault surface.

other plate is called a **transform plate boundary.** Each type looks and behaves differently from the others, as we will now see.

4.3 DIVERGENT PLATE BOUNDARIES AND SEA-FLOOR SPREADING

At divergent boundaries, or **spreading boundaries,** two oceanic plates move apart by the process of sea-floor spreading. Note that an open space does not develop between diverging plates. Rather, as the plates move apart, new oceanic lithosphere forms along the divergent boundary (▶Fig. 4.7). This process takes place at the linear submarine mountain ranges called mid-ocean ridges (such as the Mid-Atlantic Ridge, the East Pacific Rise, and the Southeast Indian Ocean Ridge), which rise 2 km above the adjacent abyssal plains of the ocean. Thus, geologists also commonly call a divergent boundary a **mid-ocean ridge,** or simply a **ridge.**

Characteristics of a Mid-Ocean Ridge

To get a better idea of a divergent boundary, let's look at one mid-ocean ridge, the Mid-Atlantic Ridge, in more detail. The Mid-Atlantic Ridge extends from the waters between northern Greenland and northern Scandinavia (latitude 80°N) southward across the equator to a latitude of 55°S, opposite the southern tip of South America. At its northern end, the ridge connects to the divergent boundary that underlies the Arctic Ocean, and at its southern end, it connects to the divergent boundary that runs between Africa and Antarctica. For most of its length, it's about 1,500 km wide.

Geologists have mapped segments of the Mid-Atlantic Ridge in detail, using sonar from ships and observations from research submarines. They have found that the formation of new sea floor takes place only across a remarkably narrow band—less than a few kilometers wide—along the **axis** (centerline) of the ridge. The axis lies at water depths of about 2–2.5 km. Along ridges (like the Mid-Atlantic) where sea-floor spreading occurs slowly, the axis lies in a narrow trough about 500 m deep and less than 10 km wide, bordered on either side by steep cliffs. Roughly speaking, the Mid-Atlantic Ridge is symmetrical—its eastern half looks like a mirror image of its western half. Along its length, the ridge consists of short segments (tens to hun-

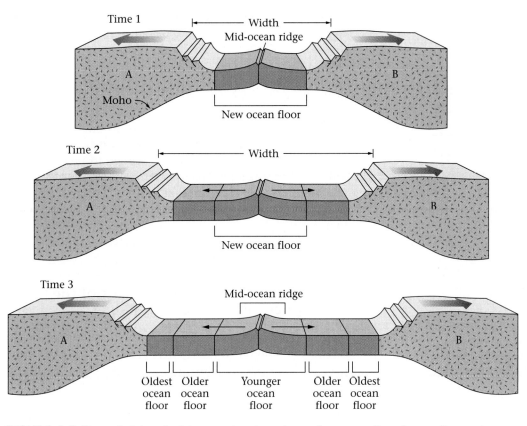

FIGURE 4.7 These sketches depict successive stages in sea-floor spreading along a divergent boundary (mid-ocean ridge); only the crust is shown. The top figure represents an early stage of the process, after the mid-ocean ridge formed but before the ocean grew very wide. With time, as seen in the next two figures, the ocean gets wider and continent A drifts way from continent B. Note that the youngest ocean crust lies closest to the ridge.

dreds of km long). If you walked north or south along the ridge axis in one of these segments, you would have to jog east or west every time you reached the end of the segment in order to reach the next one. The segments are linked to each other by fracture zones, narrow bands of broken-up crust, which lie at almost 90° to the ridge axis.

Not all mid-ocean ridges look just like the Mid-Atlantic. For example, ridges at which spreading occurs rapidly, such as the East Pacific Rise, do not have the axial trough we see along the Mid-Atlantic Ridge. Also, faster-spreading ridges are much wider.

The Formation of Oceanic Crust at a Mid-Ocean Ridge

As noted above, sea-floor spreading does not create an open space between diverging plates. Rather, as each increment of spreading occurs, new sea floor develops in the space. How does this happen?

As sea-floor spreading takes place, hot asthenosphere (the soft flowable part of the mantle) rises beneath the ridge (▶Fig. 4.8). As this asthenosphere rises, it begins to melt,

producing molten rock, or **magma.** (Molten rock underground is magma, while molten rock on Earth's surface is lava.) Magma has a lower density than solid rock, so it behaves buoyantly and rises. It eventually fills a space, or **magma chamber,** in the crust below the ridge axis. Some of the magma solidifies along the side of the chamber, to make the coarse-grained, mafic igneous rock called gabbro. Some of the magma rises still higher to fill vertical cracks, where it solidifies and forms wall-like sheets, or **dikes,** of basalt. Finally, some magma rises all the way to the surface of the sea floor at the ridge axis and spills out of small submarine volcanoes. The resulting lava cools to form a layer of basalt blobs, called **pillow basalt,** on the sea floor. We can't easily see the submarine volcanoes because they occur at depths of more than 2 km beneath sea level, but they have been observed by the research submarine *Alvin* (see Box 4.2). *Alvin* has also detected chimneys spewing hot, mineralized water that rose through cracks in the sea floor, after being heated by magma below the surface. These chimneys are called **black smokers** (▶Fig. 4.9).

As soon as it forms, new oceanic crust moves away from the ridge axis, and as this happens, more magma rises from

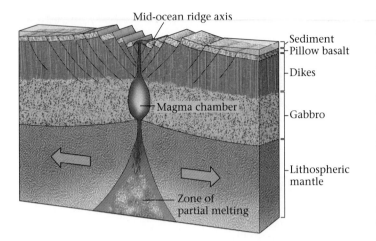

FIGURE 4.8 How new lithosphere forms at a mid-ocean ridge. Rising hot asthenosphere partly melts underneath the ridge axis. The molten rock, magma, rises to fill a magma chamber in the crust. Some of the magma solidifies along the side of the chamber, to make coarse-grained mafic rock called gabbro. Some magma rises still farther to fill cracks, solidifying into basalt that forms wall-like sheets of rock called dikes. Finally, some magma rises all the way to the surface of the sea floor at the ridge axis. This magma, now called lava, spills out and forms a layer of basalt. As sea-floor spreading continues, the oceanic crust breaks along faults. Also, as a plate moves away from a ridge axis and cools, the lithospheric mantle thickens.

FIGURE 4.9 A column of superhot water gushing from a vent (known as a "black smoker") along the mid-ocean ridge. The water has been heated by magma (molten rock) just below the surface. The cloud of "smoke" actually consists of tiny mineral grains; these minerals are first dissolved in the hot water, but when the hot water mixes with cold water of the sea, they precipitate. Many exotic species of life, such as giant worms, found nowhere else on Earth, live around these vents.

below, so still more crust forms. In other words, like a vast, continuously moving conveyor belt, magma from the mantle rises to the Earth's surface at the ridge, solidifies to form oceanic crust, and then moves laterally away from the ridge. Because all sea floor forms at mid-ocean ridges, the youngest sea floor occurs on either side of the ridge, and sea floor becomes progressively older away from the ridge. In the Atlantic Ocean, the oldest sea floor lies adjacent to the passive continental margins on either side of the ocean (▶Fig. 4.10).

The tension (stretching force) applied to newly formed solid crust as spreading takes place breaks this new crust, resulting in the formation of faults. Slip on the faults causes divergent-boundary earthquakes and creates numerous cliffs that parallel the ridge axis. Fortunately, mid-ocean ridge earthquakes occur far from populated areas and rarely cause damage.

The Formation of the Mantle Part of the Lithosphere at a Mid-Ocean Ridge

So far, we've seen how oceanic crust forms at mid-ocean ridges. What about the formation of the mantle part of the oceanic lithosphere? Remember that this part consists of the cooler uppermost area of the mantle, in which temperatures are less than about 1,280°C. At the ridge axis, such temperatures occur almost at the base of the crust, because of the presence of rising hot asthenosphere and hot magma, so the lithospheric mantle beneath the ridge axis effectively doesn't exist. But as the newly formed oceanic crust moves away from the ridge axis, the crust and the uppermost mantle directly beneath it gradually cool as they lose heat to the ocean above. As soon as this mantle cools below 1,280°C, it becomes, by definition, part of the lithosphere.

As oceanic lithosphere continues to move away from the ridge axis, it continues to cool, so the lithospheric mantle, and therefore the oceanic lithosphere as a whole, grows progressively thicker (▶Fig. 4.11a). This thickening doesn't change the thickness of the oceanic crust, for the crust formed entirely at the ridge axis. The rate at which cooling and thickening occur decreases with distance from the ridge axis. In fact, by the time the lithosphere is about 80 million years old, and resides at 800 to 1,500 km from the ridge axis (depending on the spreading rate), it has just about reached its maximum thickness (▶Fig. 4.11b).

The Reason Mid-Ocean Ridges Are High

Why does the surface of the sea floor rise to form a mid-ocean ridge along divergent plate boundaries? The answer once again comes from considering the buoyancy of oceanic lithosphere (see Box 4.1). As sea floor ages, the asthenosphere below cools enough to become part of the lithosphere, and the lithospheric mantle thickens. Cooler

Ma	0-5	5-21	21-38	38-52	52-65	65-145	145-160
Age	Pleistocene to Pilocene	Miocene	Oligocene	Eocene	Paleocene	Cretaceous	Jurassic

FIGURE 4.10 This map of the world shows the age of the sea floor. Note how the sea floor grows older with increasing distance from the ridge axis. (Ma = million years ago.)

rock is denser than warmer rock, so the process of cooling and thickening the lithosphere, like adding ballast to a ship, causes the lithosphere to sink deeper into the asthenosphere (▶Fig. 4.11c). Hot young lithosphere, on the other hand, is less dense and floats higher; this constitutes the mid-ocean ridge. Because lithosphere cools and thickens as it grows older, the depth of the sea floor reflects its age (Fig. 4.11b).

4.4 CONVERGENT PLATE BOUNDARIES AND SUBDUCTION

At convergent plate boundaries, or **convergent margins,** two plates, at least one of which is oceanic, move toward each other. But rather than butting each other like angry rams, one oceanic plate bends and begins to sink down into the asthenosphere beneath the other plate. Geologists refer to the sinking process as **subduction,** so convergent boundaries are also known as **subduction zones.** Because subduction at a convergent boundary consumes old ocean lithosphere and thus closes (or "consumes") oceanic basins, geologists also refer to convergent boundaries as **consuming boundaries,** and because they are delineated by deep-ocean trenches, they are sometimes simply called **trenches.** The amount of oceanic plate consumption worldwide, averaged over time, equals the amount of sea-floor spreading worldwide, so the surface area of the Earth remains constant.

Subduction occurs for a simple reason: oceanic lithosphere, once it has aged at least 10 million years, is denser than asthenosphere (because lithosphere cools and contracts as it ages), and thus can sink through the asthenosphere. When it lies flat on the surface of the asthenosphere,

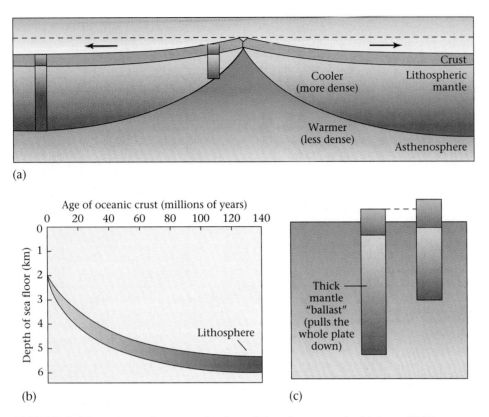

(a)

(b)   (c)

FIGURE 4.11 (a) As sea floor ages, the dense lithospheric mantle thickens. (b) The thickness of the lithosphere and the depth of the sea floor both increase as a plate moves away from the ridge and grows older. (c) Like the ballast of a ship, thicker lithosphere sinks deeper into the mantle.

oceanic lithosphere can't sink because the resistance of the asthenosphere to flow is too great; however, once the end of the convergent plate is pushed into the mantle, it begins to sink like an anchor falling to the bottom of a lake (▶Fig. 4.12a, b). As the lithosphere sinks, asthenosphere flows out of the way, just as water flows out of the way of an anchor. Because the asthenosphere resists flow, oceanic lithosphere can sink only very slowly, at a rate of less than 10–15 cm per year.

Note that the **downgoing plate** (or **slab**), the plate that has been subducted, *must* be composed of oceanic lithosphere. The **overriding plate** (or **slab**), which does not sink, can consist of either oceanic or continental lithosphere. Continental lithosphere cannot be subducted because it is too buoyant; the low-density rocks of continental crust act like a life preserver keeping the continent afloat. If continental crust moves into a convergent margin, subduction stops. Most oceanic lithosphere eventually sinks—in fact, as noted above, all ocean floor on the planet is less than about 200 million years old. But because continental lithosphere cannot subduct, some continental crust has persisted at the surface of the Earth for over 3.8 billion years.

Earthquakes and the Fate of Subducted Plates

At convergent plate boundaries, the downgoing plate grinds along the base of the overriding plate, a process that generates large earthquakes. These earthquakes occur fairly close to the Earth's surface, so some of them trigger massive destruction in coastal cities. But earthquakes also happen in downgoing plates at greater depths, deep below the overriding plate. In fact, geologists have detected earthquakes within downgoing plates to a depth of 670 km; the belt of earthquakes in a downgoing plate is the **Wadati-Benioff zone,** named for its two discoverers (▶Fig. 4.13).

What causes earthquakes in a Wadati-Benioff zone is not totally understood—they may result from the breaking of rock during movement on faults within the plate, or they may be due to the sudden collapse of mineral crystals in response to great pressures. (A crystal collapses when atoms in the crystal rearrange so as to occupy a smaller volume. The process, somewhat like clapping your hands together, generates shock waves.) At depths greater than 670 km, conditions leading to earthquakes evidently do not occur. Recent evidence, however, indicates that downgoing plates do continue to sink below a depth of 670 km—they just do so without generating earthquakes. In fact, studies suggest that the lower mantle may be a graveyard for old subducted plates. Most likely, these old plates slowly absorb heat from their surroundings until they become warm and soft enough to start flowing along with the rest of the mantle. When this happens, the material of the old plate may mix in with the rest of the mantle.

Geologic Features of a Convergent Boundary

To become familiar with the various geologic features that occur along a convergent plate boundary, let's look at an example, the boundary between the west coast of the South American Plate and the eastern edge of the Nazca Plate (a portion of the Pacific Ocean floor). A deep-ocean trench, the Peru-Chile Trench, delineates this boundary. Such trenches form where the plate bends as it starts to sink into the as-

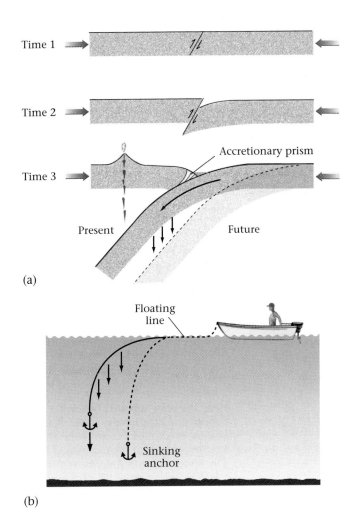

FIGURE 4.12 (a) The concept of subduction. A plate bends, and one piece pushes over the other. Oceanic lithosphere is denser than the underlying asthenosphere, but when it lies flat on the surface of the asthenosphere, it can't sink because the resistance of the asthenosphere to flow is too great. However, once the end of the plate is pushed into the mantle, the lithosphere begins to sink. (b) The process of sinking is like an anchor pulling a floating anchor line down. As a consequence, the bend in the plate (or in the anchor line) progressively moves with time.

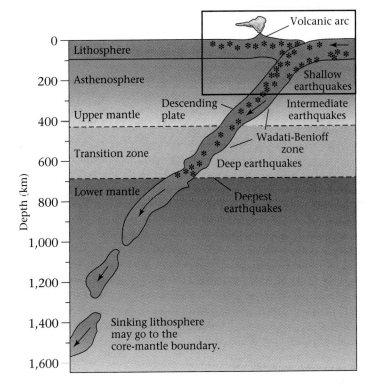

FIGURE 4.13 The Wadati-Benioff zone is a band of earthquakes that occur in subducted oceanic lithosphere. The discovery of these earthquakes led to the proposal of subduction.

thenosphere. Most trenches are named for a nearby geographic feature. For example, the Peru-Chile Trench is named for the countries that border it, and the Aleutian Trench is named for the adjacent Aleutian Islands.

In the Peru-Chile Trench, as the downgoing plate slides under the overriding plate, sediment (clay and plankton) that had settled on the surface of the downgoing plate, as well as sand that fell into the trench from the shores of South America, gets scraped up and incorporated in a wedge-shaped mass known as an **accretionary prism** (▶Fig. 4.14a). An accretionary prism forms in basically the same way as a pile of snow in front of a plow, and like the

snow, the sediment tends to be squashed and contorted during the formation of the prism (▶Fig. 4.14b).

A chain of volcanoes known as a **volcanic arc** develops behind the accretionary prism. As we will see in Chapter 6, the magma that feeds these volcanoes forms at or just above the surface of the downgoing plate when the plate reaches a depth of about 150 km below the Earth's surface. If the volcanic arc forms where an oceanic plate subducts beneath continental lithosphere, the resulting chain of volcanoes grows on the continent and forms a **continental volcanic arc.** If, however, the volcanic arc forms where one oceanic plate subducts beneath another oceanic plate, the resulting volcanoes form a chain of islands known as a **volcanic island arc** (▶Fig. 4.15a). In some cases, sea-floor spreading occurs behind an island arc, creating a small ocean basin called a **marginal sea** (▶Fig. 4.15b). The Japan Sea, for example, formed in this way.

4.5 TRANSFORM PLATE BOUNDARIES

We saw earlier that the spreading axis of a mid-ocean ridge consists of short segments linked by fracture zones. The geometric relationship of fracture zones to ridge segments,

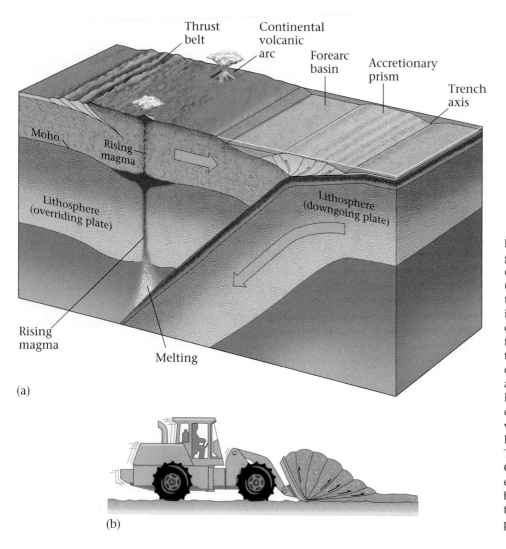

(a)

(b)

FIGURE 4.14 (a) This model shows the geometry of subduction along an active continental margin. The trench axis (lowest part of the trench) roughly defines the plate boundary. Numerous faults form in the accretionary prism, which is composed of material scraped off the sea floor. Behind the prism lies a basin of trapped sediment. A volcanic arc is created from magma that forms at or just above the surface of the downgoing plate. Here, the plate subducts beneath continental lithosphere, so the chain of volcanoes is called a continental arc. Faulting occurs on the backside of the arc. The Andes in South America and the Cascades in the United States are examples of such continental arcs. (b) A bulldozer pushing snow or soil is similar to the development of an accretionary prism.

FIGURE 4.15 (a) The Aleutian Islands in Alaska exemplify an island arc. (b) Subduction along an island arc. Here, the volcanoes build on the sea floor. Behind some island arcs, a marginal sea forms. This sea resembles a small ocean basin, with a spreading ridge that is created when the plate behind the arc moves away from the arc.

(a)

(b)

and evidence indicating that fracture zones are made up of broken-up crust, originally led geoscientists to conclude that fracture zones were faults. They then incorrectly assumed that sliding on faults in fracture zones broke an originally continuous ridge into segments and displaced the segments sideways (▶Fig. 4.16a). This interpretation implies that one segment moves with respect to its neighbor, as shown by the arrows in ▶Figure 4.16b. But soon after Harry Hess proposed his model of the sea-floor spreading in 1960, a Canadian named J. Tuzo Wilson realized that if sea-floor spreading really occurred, then the notion that fracture zones offset an originally continuous ridge could not be correct.

In Wilson's alternative interpretation, the fracture zone formed *at the same time* as the ridge axis itself, and thus the ridge consisted of separate segments to start with. These segments were *linked* (not offset) by fracture zones. With this idea in mind, he drew a sketch map showing two ridge-axis segments linked by a fracture zone, and drew arrows to indicate the direction that ocean crust was moving, relative to the ridge axis, as a result of sea-floor spreading (▶Fig. 4.16c). Look at Wilson's arrows. Clearly, the movement direction on the fracture zone must be opposite to the movement direction that geologists originally thought occurred on the structure. Further, in Wilson's model, slip occurs *only* along the segment of the fracture zone between the two ridge segments. Plate A moves with respect to plate B as sea-floor spreading occurs on the mid-ocean ridge. This movement results in slip along the segment of the fracture zone between points X and Y. But to the west of point X, the fracture zone continues merely as a boundary between two different parts of plate A. The portion of plate A at point 1, just to the north of the boundary (▶Fig. 4.16d), must be younger than the portion at point 2 just to the south, because point 1 lies closer to the ridge axis; but since points 1 and 2 move at the same speed, this segment of the fracture zone does not slip.

Wilson introduced the term **transform fault** for the actively slipping segment of a fracture zone between two ridge segments, and he pointed out that transform faults made a third type of plate boundary. Geologists now also call them transform boundaries, or simply **transforms.** At a transform boundary, one plate slides sideways past another, but no new plate forms and no old plate is consumed. Transform boundaries are therefore defined by a vertical fault on which slip parallels the Earth's surface.

As soon as Wilson published his idea, Lynn Sykes, of Columbia University, realized that the study of earthquakes could serve as a test. Sykes examined earthquakes that occurred along the Mid-Atlantic Ridge, and was able to tell precisely where they took place. He also determined which way rock slipped during the earthquakes. Just as Wilson had predicted, earthquakes only occurred along ridges and

along the transform faults, the portions of fracture zones between ridge segments (▶Fig. 4.16e). The sliding direction on the transform faults was also what Wilson had predicted. Only sea-floor spreading could explain the observed movement. So Wilson's proposal and Sykes's confirmation served as additional evidence for sea-floor spreading.

Most transform boundaries are fairly short, and serve only to link segments of mid-ocean ridges. But not all transforms link ridge segments. Some, such as the Alpine Fault of New Zealand, link trenches, while others link a trench to a ridge segment. Further, not all transform faults occur in oceanic lithosphere; a few cut across continental lithosphere. The San Andreas Fault, for example, which cuts across California, defines part of the plate boundary between the North American Plate and the Pacific Plate—the portion of California that lies to the west of the fault (including Los Angeles) is part of the Pacific Plate, while the portion that lies to the east of the fault is part of the North American Plate (▶Fig. 4.17a, b). On average, the Pacific Plate moves about 6 cm north, relative to North America, every year. If this motion continues, Los Angeles will become a suburb of Anchorage, Alaska, in about 100 million years.

The grinding of one plate past another along a transform fault generates frequent earthquakes, including some huge ones. Fortunately, most of these earthquakes occur out in the ocean basin, far from people. But large earthquakes along the transform faults that cut across continental crust can be very destructive. A huge earthquake on the San Andreas Fault, for example, destroyed San Francisco in 1906. Smaller earthquakes in more recent years, such as the 1989 earthquake that once again hit the San Francisco Bay area, have killed dozens of people and have caused billions of dollars of damage.

4.6 SPECIAL LOCATIONS IN THE PLATE MOSAIC

Triple Junctions

So far, we've focused attention on boundaries—divergent (mid-ocean ridge), convergent (trench), and transform—between *two* plates. But there are several places where *three* plate boundaries intersect at a point. Geologists refer to these points as **triple junctions.** We name triple junctions after the types of boundaries that intersect. For example, the triple junction formed where the Southwest Indian Ocean Ridge intersects two arms of the Mid–Indian Ocean Ridge (this is the triple junction of the African, Antarctic, and Australian Plates) is a ridge-ridge-ridge triple junction

FIGURE 4.16 (a) In this incorrect interpretation of an oceanic fracture zone, the fault forms and cuts across an originally continuous ridge. (b) After slip on the fault, indicated by the arrows, the ridge consists of two segments. (c) In Wilson's correct interpretation, the ridge initiates at the same time as the transform, and thus was never continuous. Note that the way in which the fault slips (along the fracture zone between points X and Y) makes sense if sea-floor spreading takes place, but contrasts with the slip in (b). (d) Even though the ocean grows, the transform can stay the same length. Point 1 on plate A is younger than point 2 because it lies closer to the ridge axis. (e) The fracture zone beyond the ends of the transform does not slip, and thus is not a plate boundary. It does, however, mark the boundary between portions of the plate that are different in age.

(a)

(b)

FIGURE 4.17 (a) The San Andreas Fault is a transform plate boundary between the Pacific Plate to the west and the North American Plate to the east. At its southeast end, the San Andreas connects to spreading ridge segments in the Gulf of California. (b) The San Andreas Fault where it cuts across a dry landscape. The fault trace is the narrow valley.

BOX 4.2

THE HUMAN ANGLE

So You Want to See a Plate Boundary?

Plate boundaries are so important that geologists scrambled to study them when they were first recognized. But where do you go if you want to see one for real? First, it's important to realize that plate boundaries are not a single line on the surface of the Earth, but are actually zones of faulting perhaps 40–200 km wide. Still, although you can't straddle a plate boundary the way you can a painted stripe, there are in fact places where you can see a plate-boundary zone.

Let's start with a transform boundary. Most transform boundaries are on the sea floor, but a few cross continental crust. One of the most accessible of these is the San Andreas Fault in California. North of San Francisco, a visitor to Earthquake Park near Point Reyes will see the trace of a segment of the fault that

ruptured during the great 1906 earthquake. A subtle depression on the ground marks the fault trace. This depression formed because rupturing weakened a band of rock and soil, which then collapsed. Small ponds and marshes also occur along the fault, marking places where a slight amount of stretching caused the surface of the ground to sink. In the San Francisco area, there are numerous faults related to the San Andreas. Some of these have moved slowly in recent decades, and have offset curbstones bordering streets. Farther south, in central California, the fault passes across an orchard, offsetting rows of trees and, beneath a winery, offsetting wine casks in the basement. Where the fault crosses a highway near the town of Palmdale, the layering of rock in road cuts (exposures) adjacent to the highway has been contorted wildly. Farther south still, movement on the fault has squeezed up ridges of broken rock.

Convergent boundaries prove more elusive. You can get close to one by standing in the adjacent volcanic arc. If you visit Mt. Rainier or any of the other volcanoes in Washington or Oregon (▶Fig. 4.18a), for example, you are not far from the convergent boundary between the Juan de Fuca Plate and the North American Plate. Likewise, when you hike in the Andes of South America, you are close to the convergent boundary between the Nazca Plate and the South American Plate. But to actually see the boundary, you would have to descend through 7 to 12 km of water to reach the deep darkness at the floor of the Peru-Chile Trench. And after all that effort, you would find that the boundary lies buried by sediment.

There are over 40,000 km of divergent plate boundaries on our planet, marked by mid-ocean ridges. But despite their height, almost all of the mid-ocean ridge system lies beneath nearly 2.5 km of water. Geologists in the research mini-submarine *Alvin*

have explored the mid-ocean ridge of the North Atlantic Ocean. Because ocean crust has only just formed at the ridge axis, features of the plate boundary have not been buried by sediment, and stand out in the glare of *Alvin*'s headlights. Peeking through the portholes of the submarine, geologists have seen molten rock oozing up through cracks in the crust, chimneys spewing out clouds of boiling water saturated with dissolved minerals, and steep cliffs created by movement on faults along the plate boundary.

Most of us will never have the chance to squeeze into *Alvin*'s cramped quarters for a claustrophobic drop to the sea floor (even though it sinks like a stone, *Alvin* takes almost two hours to reach bottom). But amazingly, there is one place in the world where we can stand on a divergent boundary: Iceland (▶Fig. 4.18b). This island straddles the Mid-Atlantic Ridge. A walking or driving tour of Iceland will take you across fault-bounded troughs and to fissures that frequently spew fountains of lava. When you stand in eastern Iceland, you are on the Eurasian Plate, but when you stand in western Iceland, you are on the North American Plate.

(a)   (b)

FIGURE 4.18 (a) A volcanic arc in the western United States; (b) the Mid-Atlantic Ridge exposed in Iceland.

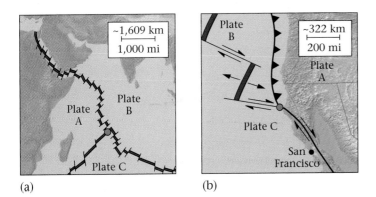

(a)   (b)

FIGURE 4.19 (a) A ridge-ridge-ridge triple junction (at the dot). (b) A trench-transform-transform triple junction.

(►Fig. 4.19a). The triple junction north of San Francisco is a trench-transform-transform triple junction (►Fig. 4.19b).

Hot Spots

Most **subaerial** (above sea level or the land surface) volcanoes are situated in the volcanic arcs that border trenches. Small volcanoes also lie along mid-ocean ridges, but ocean water hides most of them. The volcanoes of volcanic arcs and mid-ocean ridges are **plate-boundary volcanoes,** formed as a consequence of movement along the boundary. Not all volcanoes on Earth are plate-boundary volcanoes, however. For example, Hawaii, a huge active volcano, lies in

the middle of the Pacific Plate, and Yellowstone National Park, site of a recent volcano, lies in the northwest corner of Wyoming, in the interior of the United States. Worldwide, geoscientists have identified about one hundred volcanoes that exist as isolated points and are not a consequence of movement at a plate boundary; these are called **hot-spot volcanoes,** or simply **hot spots** (►Fig. 4.20). Most hot spots are located in the interiors of plates, away from the boundaries, but a few happen to lie on mid-ocean ridges.

What causes hot-spot volcanoes? J. Tuzo Wilson, the discoverer of transform faults, was one of the first to come up with an explanation. After examining hot spots in the interiors of ocean plates, Wilson noted that an erupting hot-spot volcano formed an island at the end of a chain of now-dead (no-longer erupting) volcanic islands and seamounts. (A volcano along a convergent plate boundary, in contrast, is one of many in a chain that are all active at about the same time.) Wilson noted further that all the hot-spot volcanoes in the Pacific lie at the southeast end of a northwest-southeast-lying chain of dead volcanoes (►Fig. 4.21). With this image in mind, Wilson suggested in 1963 that a hot-spot volcano develops over a heat source in the mantle that is fixed relative to the moving plate; an active volcano represents the location of the heat source. The chain of seamounts and inactive (dead) volcanic islands linked to the hot-spot volcano represent locations on the plate that were once over the source but have since moved off. Subsequently, the heat source came to be associated with a **mantle plume,** a column of very hot rock rising up through the mantle.

FIGURE 4.20 The dots represent the locations of hot-spot volcanoes. The tails represent hot-spot tracks.

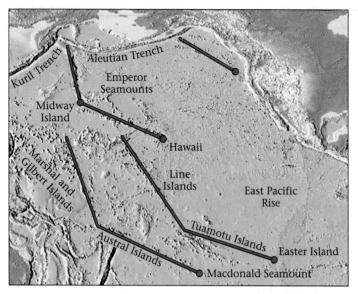

FIGURE 4.21 Hot-spot tracks in the Pacific Ocean. The small dots represent islands or seamounts. The straight lines indicate the geometry of the tracks. Note that the chains have a 40° bend in them, resulting from a change in the direction of motion of the Pacific Plate about 40 million years ago.

Let's look more closely at how Wilson's model works. Mantle plumes are thought to originate just above the core-mantle boundary, where heat from the Earth's core warms the base of the mantle. Because it expands as it heats, the hot rock above the core-mantle boundary becomes less dense than overlying mantle and begins to stream upward in a column-shaped mass—at the rate of only a few centimeters per year. When the hot rock reaches the base of the lithosphere, it partly melts, for reasons discussed in Chapter 6. The magma formed by melting then rises through the lithosphere and erupts at a volcano on the Earth's surface (▶Fig. 4.22a). All the while, the plate on which the volcano grows continues to shift, so eventually the volcano moves off the plume and dies, or "goes extinct." Meanwhile, a new volcano grows over the plume (whose presence creates a hot spot). If the plume lasts for a long time in the same place, it will spawn a long chain of dead (extinct) volcanoes; again, at any one time, only the volcano directly over the plume can be active. As the plate beneath the extinct volcano grows older, it sinks, and the volcano sinks with it, eventually becoming submerged. Once submerged, the volcano can be called a seamount. In some cases, a coral reef grows on top of the extinct volcano as it submerges, so that the seamount has a flat top; a flat-topped seamount is called a **guyot.** (Charles Darwin, during the voyage of the *Beagle* in the 1830s, was the first to recognize the slow sinking of dead volcanoes.)

The Hawaiian chain provides a clear example of the volcanism associated with a hot spot. Volcanic eruptions occur

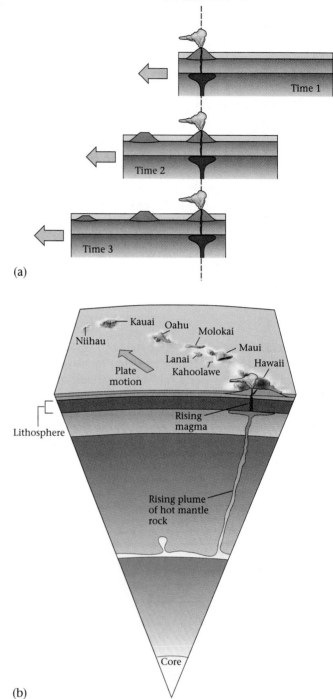

FIGURE 4.22 (a) A mantle plume causes a hot spot to form at the base of a plate, leading to the growth of a volcano on the surface of the plate. As the plate moves, the volcano is carried off the hot spot; it then dies (goes extinct), and a new volcano forms above the hot spot. As the process continues, a chain of extinct volcanoes develops, with the oldest one farthest from the hot spot. The extinct volcanoes gradually sink below sea level and become seamounts. Only the volcano above the hot spot erupts. (b) The plume that forms a hot spot arises from the base of the mantle. Here, we see the plume that underlies the Hawaiian Islands.

today only on the big island of Hawaii (▶Fig. 4.22b). Other islands to the northwest are remnants of dead volcanoes. The oldest of these is Kauai, the island farthest from the big island. To the northwest of Kauai, still older volcanic remnants are found, and these continue in a straight line to Midway Island, site of a major battle during World War II. To the northwest of Midway Island, the chain of volcanic remnants no longer pokes above sea level, and so we refer to them as seamounts. As shown on Figure 4.21, the Emperor seamount chain extends northwest of Midway all the way to the junction between the Aleutian and Kuril Trenches just east of the Kamchatka Peninsula.

Wilson discovered how hot spots develop by looking at oceanic hot-spot volcanoes. But some hot spots lie within continents. For example, several have been active in the interior of Africa, and one now underlies Yellowstone National Park. The famous **geysers,** natural steam and hot-water fountains, of Yellowstone exist because hot magma, formed above the Yellowstone hot spot, lies not far below the surface of the park. Yellowstone lies at the northeast end of the Snake River Plain, a valley covered with the products of volcanic eruptions in the past. The Snake River Plain volcanics represent the track left as North America drifted westward over the fixed Yellowstone hot spot.

As mentioned earlier, a few hot spots lie on mid-ocean ridges. Where this happens, a volcanic island protrudes above sea level, because the hot spot produces far more magma than does a normal mid-ocean ridge. Iceland, for example, formed where a hot spot underlies the Mid-Atlantic Ridge. The extra volcanism of the hot spot built up the island of Iceland so that it rises almost 3 km above other places on the Mid-Atlantic Ridge. Hot-spot volcanoes astride mid-ocean ridges eventually split in half, so that two symmetrical seamount chains form to either side of the ridge axis.

4.7 THE BIRTH AND DEATH OF PLATE BOUNDARIES

The configuration of plates and plate boundaries visible on our planet today has not existed for all of geologic history, and will not exist indefinitely into the future. Because of plate motion, oceanic plates

form and are later consumed, while continents merge and later split apart. How does a new divergent boundary come into existence, and how does a convergent boundary cease to exist? Most new divergent boundaries form when a continent splits and separates into two continents. We call this process **continental rifting.** A convergent boundary ceases to exist when a piece of buoyant lithosphere, such as a continent or an island arc, moves into the subduction zone. We call this process **collision.**

Continental Rifting

A **continental rift** is a linear belt in which continental lithosphere pulls apart (▶Fig. 4.23). The lithosphere stretches horizontally, so it thins vertically, much like a piece of taffy you pull between your fingers. Near the sur-

FIGURE 4.23 When a continent stretches during continental rifting, the upper part of the crust breaks up into a series of faults. The lower part of the crust stretches more like soft plastic. The region that has stretched is the rift. With continued stretching, the crust becomes much thinner, and the asthenosphere that rises beneath the rift partly melts. As a consequence, volcanoes form in the rift. Eventually, the continent breaks in two, and a new mid-ocean ridge forms. With time, an ocean develops. The relics of the stretched and broken crust of the rift underlie the thick sediment wedge of the passive margins.

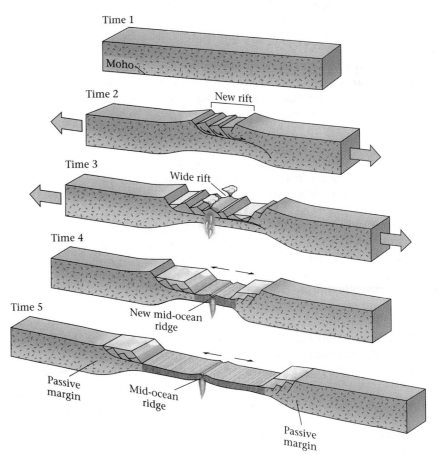

face of the continent, where the crust is cold and brittle, the stretching causes rock to break and faults to develop. The faulting then makes blocks of rock slide down, and as a result leads to the formation of a low area that gradually becomes buried by sediment. Lower in the crust, where the rock is warmer and softer, stretching may take place in a plastic manner, without breaking the rock. The whole region that stretches is the rift.

As lithosphere thins, hot asthenosphere rises beneath the rift and partly melts. This molten rock erupts at volcanoes along the rift. If rifting continues for a long enough time, the continent breaks in two, a new mid-ocean ridge (a divergent boundary) forms, and sea-floor spreading begins. The relict of the rift evolves into a passive margin (Fig. 4.4). In some cases, however, rifting stops before the continent splits in two. Then, the rift remains as a permanent scar in the crust, recognized by a belt of faults, volcanic rocks, and a thick layer of sediment.

Perhaps the most spectacular example of a rift today occurs in eastern Africa; geoscientists aptly refer to it as the East African Rift (▶Fig. 4.24). To astronauts in orbit, the rift looks like a giant gash in the crust. On the ground, it consists of a deep trough bordered on both sides by high cliffs that were made by faulting. Along the length of the rift, several major volcanoes smoke and fume; these include snow-crested Mt. Kilimanjaro, towering over 6 km above the savanna. Another major rift, known as the Basin and Range Province, breaks up the landscape of the western United States between Salt Lake City, Utah, and Reno, Nevada. Here, movement on numerous faults tilted blocks of crust to form narrow mountain ranges, while sediment that eroded from the blocks filled the adjacent basins (the low areas between the ranges).

Collision

India was once a small, separate continent that lay far to the south of Asia. But subduction consumed the ocean between India and Asia, and India moved northward, finally slamming into the southern margin of Asia 40–50 million years ago. Continental lithosphere, unlike oceanic lithosphere, is too buoyant to subduct. So when India collided with Asia, the attached oceanic plate broke off and sank down into the deep mantle. But India pushed hard into Asia, squashing the rocks and sediment that once lay between the two continents into the 8-km-high welt that we know as the Himalayan Mountains. During this process, not only did the surface of the Earth here rise in this mountain belt, but the crust became thicker. Generally, the crust beneath a mountain range is 60–70 km thick, about twice the thickness of normal continental crust. (We will learn more about mountain building in Chapter 11.)

Geoscientists refer to the process of two buoyant pieces of lithosphere converging and smashing together as **colli-**

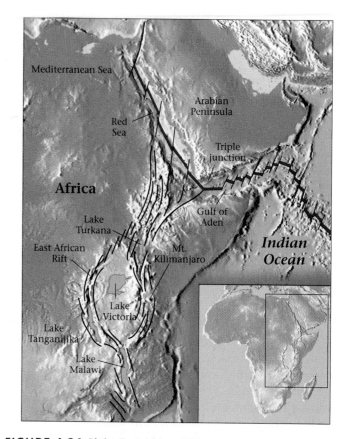

FIGURE 4.24 If the East African Rift were to continue growing, part of Africa would break off, forming a continental fragment. Note that the East African Rift intersects the Red Sea and the Gulf of Aden at a triple junction. The Red Sea and Gulf of Aden started as rifts but are now narrow oceans, bisected by new mid-ocean ridges. Many of the most important fossils of early man have come from rocks within the East African Rift; the region's lakes, formed because the axis of the rift drops down, provide an environment in which these hominids could survive.

sion (▶Fig. 4.25a, b). Some collisions involve two continents, some involve continents and an island arc. When a collision is complete, the convergent plate boundary that once existed between the two colliding pieces ceases to exist. Collisions yield some of the most spectacular mountains on the planet, such as the Himalayas and the Alps. They also yielded major mountain ranges in the past, which subsequently eroded away so that today we see only their relicts. For example, the Appalachian Mountains in the eastern United States were created as a consequence of three collisions. After the last one, a collision between Africa and North America around 280 million years ago, North America became part of the Pangaea supercontinent. The rifting apart of Pangaea later formed the Atlantic Ocean.

The relicts of ancient collisional mountain ranges like the Appalachians demonstrate that Pangaea was not the only supercontinent to exist during Earth's history. Before Pangaea, there were smaller continents, separated from one

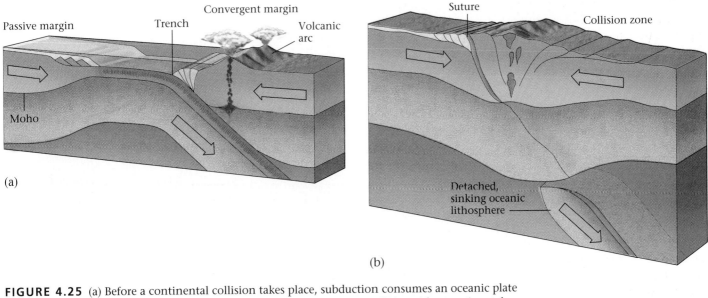

FIGURE 4.25 (a) Before a continental collision takes place, subduction consumes an oceanic plate until it collides with another plate. Here, a passive continental margin collides with a continental volcanic arc. (b) After the collision, the oceanic plate detaches and sinks into the mantle. Rock caught in the collision zone gets broken, bent, and squashed, and forms a mountain range. Slivers of oceanic crust may be trapped along the boundary, or suture, between what once had been two continents. As the crust squashes horizontally, it thickens vertically.

another by ocean basins that no longer exist. In fact, growing evidence suggests that continents have combined and dispersed perhaps several times in the course of the planet's existence.

4.8 WHAT DRIVES PLATE MOTION?

When geoscientists first proposed plate tectonics, they thought the process occurred simply because convective flow in the asthenosphere actively dragged plates along, as if they were rafts on a flowing river. Thus, early images depicting plate motion showed simple convection cells—elliptical flow paths of convecting asthenosphere—beneath mid-ocean ridges (▶Fig. 4.26). At first glance, this hypothesis sounded pretty good, but on closer examination, it failed. Among other reasons, it is impossible to draw a global arrangement of convection cells that can explain the complex geometry of plate boundaries on Earth. Gradually, geoscientists came to the conclusion that convective flow within the asthenosphere does not drive all plate motion. Hot asthenosphere does rise in some places and sink in others because of temperature contrasts, but the specific direc-

FIGURE 4.26 In an early hypothesis for the forces that drive plates, convection cells in the asthenosphere, indicated by the arrows that show the presumed flow direction in the asthenosphere, dragged the plates along. While convection certainly occurs, such simple geometries are now thought to be unlikely.

(a)

(b)

FIGURE 4.27 (a) A simplified profile (not to scale) of a mid-ocean ridge. Note that along the flanks of the ridge, the sea floor slopes. The elevation of the ridge causes an outward ridge-push force that drives the lithosphere plate away from the ridge. A similar situation exists in a glass containing honey and water. If the boundary between the honey and water tilts, the honey exerts an outward force at its base. (b) In this cross section illustrating slab-pull force, the oceanic plate is denser than the asthenosphere, so it sinks into the asthenosphere like a stone into water, only much more slowly. (c) In addition to ridge push and slab pull, the plates feel shear forces along their base as they move into the asthenosphere. Their movement may also be resisted by shear along transform faults or at the base of a continent at a subduction zone.

F_{RP} = Ridge-push force
F_{SP} = Slab-pull force
F_{B} = Shear force at the base of the plate

F_{CS} = Shear force between two plates
F_{TR} = Shear force along a transform fault

(c)

tions of this flow do not necessarily coincide with the directions of plate motion. Today, geoscientists favor the hypothesis that plates move primarily in response to two forces, ridge-push force and slab-pull force.

Ridge-push force develops because mid-ocean ridges lie at a higher elevation than the adjacent abyssal plains of the ocean (▶Fig. 4.27a). To understand ridge-push force, imagine you have a glass containing a layer of water over a layer of honey. By tilting the glass momentarily and then returning it to its upright position, you can create a temporary slope in the boundary between these substances. While the boundary has this slope, gravity causes the elevated honey to push against the side of glass adjacent to the

honey at the lower elevation. If the glass were to suddenly disappear, this force would push the honey out over the table. The geometry of a mid-ocean ridge resembles this situation. The surface of the sea floor is higher along a mid-ocean ridge axis than in adjacent abyssal plains, because the lithosphere underlying the ridge is thin and warm and thus less dense and more buoyant than the lithosphere beneath the abyssal plains. (Note that the elevation of the ridge is *not* due to an upward push of rising magma from below.) Thus, the surface of the sea floor overall slopes away from the ridge axis. Gravity causes the elevated lithosphere at the ridge axis to push on the lithosphere that lies farther from the axis (much like the tilted honey layer pushes on the side of the glass), making it move away. As lithosphere moves away from the ridge axis, new hot asthenosphere rises to fill the gap; it then moves away, cools, and itself becomes lithosphere. During this process, some of the rising asthenosphere melts, generating magma that then rises, solidifies, and forms the rock of the oceanic crust (see Chapter 6). Note that the upward movement of asthenosphere beneath a mid-ocean ridge is a *consequence* of sea-floor spreading, not the cause.

Slab-pull force, the force that downgoing plates (also called downgoing slabs) apply to oceanic lithosphere at a convergent margin, arises simply because lithosphere that was created more than 10 million years ago is denser than asthenosphere, so it can sink into the asthenosphere (▶Fig. 4.27b). Thus, once an oceanic plate starts to sink, it gradually pulls the rest of the plate along behind it, like an anchor pulling down the anchor line. This "pull" is the slab-pull force.

Now let's summarize our discussion of forces that drive plate motions. Plates move away from ridges—in other words, sea-floor spreading occurs—in response to the ridge push force, and as this happens, new asthenosphere rises to fill the space that opens between the plates. Magma derived from this rising asthenosphere floats still farther up and solidifies to form new oceanic crust. Old, cool oceanic lithosphere, being denser than asthenosphere, sinks down into the asthenosphere, creating slab-pull force that tows the rest of the plate along with it. But ridge push and slab pull are not the only forces acting on the plate—shear (sliding) of the convecting asthenosphere exerts a force on the base of the plate, just as flowing water exerts a force on the base of a boat tied to a dock. If this shear happens to be in the same direction the plate is already moving, it can speed up the motion, but if the shear is in the opposite direction, it might slow the plate down. Also, where one plate grinds against another, as occurs along a transform fault or at the base of an overriding plate at a convergent margin, friction (the resistance to sliding on a surface) may slow the plate down (▶Fig. 4.27c). Since plate motion does occur, ridge-push and slab-pull forces must be greater than all the resistance forces combined.

4.9 THE VELOCITY OF PLATE MOTIONS

How fast do plates move? It depends on your frame of reference. To illustrate this concept, imagine two cars speeding in the same direction down the highway. From the viewpoint of a tree along the side of the road, car A zips by at 100 km an hour, while car B moves at 80 km an hour. But relative to car B, car A moves at only 20 km an hour. Likewise, geologists use two different frames of reference for describing plate velocity (velocity = distance/time). If we describe the movement of plate A with respect to plate B, then we are talking about **relative plate velocity.** But if we describe the movement of both plates relative to a fixed point in the mantle, then we are speaking of **absolute plate velocity.**

We've already seen one method of determining relative plate motions. Geoscientists measure the distance of a known magnetic anomaly from the axis of a mid-ocean ridge, and calculate the velocity of a plate relative to the ridge axis by applying the equation: plate velocity = dis-

---

Does Plate Tectonics Occur on Other Planets?

Our solar system contains four terrestrial planets and numerous terrestrial-like moons. Do any of these planets or moons display the consequences of plate tectonics? Planetary geologists, who have studied images of these planets taken through telescopes or from exploratory satellites, say no. The Earth appears to be the only body in the solar system to experience plate tectonics. Why? Because Earth's interior has remained warm enough for flow to take place in its mantle. For this reason, asthenosphere can rise at mid-ocean ridges, and can move out of the way of subducting plates. The mantles of Mercury, Mars, the Moon, and Venus apparently cannot flow like that of Earth, so these planets have no active volcanoes, no continental drift, and no sea-floor spreading. Earlier in solar-system history, however, when they were warmer, Mars and Venus did have volcanoes, perhaps comparable to hot-spot volcanoes on Earth. Indeed, the pattern of faults on Venus suggests that plate-tectonics-like movement struggled to begin but never quite succeeded.

The Jovian planets consist mostly of gas, so they cannot possibly have a rigid lithosphere. Curiously, though, some of the moons of Jupiter and Saturn appear to exhibit tectonic features. In fact, the *Galileo* satellite in 1996 photographed a volcano in the act of erupting on Io, a moon of Jupiter. But these volcanoes reflect some other type of melting process, not plate tectonics.

Convergent boundary ▲▲▲   Ridge ━━   Transform ——   Absolute plate motions ←   Relative plate motions ←→
(5.5 cm per year)

FIGURE 4.28 Relative plate velocities: the light-blue arrows show the rate and direction at which the plate on one side of the boundary is moving with respect to the plate on the other side. Outward-pointing arrows indicate spreading (divergent boundaries), inward-pointing arrows indicate subduction (convergent boundaries), and parallel arrows show transform motion. The length of an arrow represents the velocity. Absolute plate velocities: the dark-blue arrows show the velocity of the plates with respect to a fixed point in the mantle.

tance from the anomaly to the ridge axis / age of anomaly. The velocity of the plate on one side of the ridge relative to the plate on the other is twice this value. We've also seen a way to determine absolute plate motions. If we assume that the position of a mantle plume does not change much for a long time, then the track of hot-spot volcanoes on the plate moving over the plume provides a record of the plate's absolute velocity and indicates the direction of movement (▶Fig. 4.28). The Hawaiian-Emperor seamount chain, for example, defines the absolute velocity of the Pacific Plate. Note that the Hawaiian chain runs northwest, while the Emperor chain curves north-northwest (the bend occurs at Midway Island) (Fig. 4.21). Radiometric dates of volcanic rocks from Midway indicate that they formed about 40 million years ago. Thus, the direction in which the Pacific Plate moved changed significantly at this time.

Based on the calculations described above, geologists have determined that relative plate motions on Earth today occur at rates of about 1–15 cm per year. But these rates, while small, can yield large displacements given the immensity of geologic time; in a million years, a plate can move 100 km. Can we detect such slow rates? Until the last decade, the answer was no. Now the answer is yes. Satellites now orbiting the Earth are providing us with the **global positioning system** (GPS). Sailors and pilots use a GPS receiver to navigate; automobile drivers can now use a GPS receiver to reach their destinations; and geologists use an array of GPS receivers to monitor plate motions. If done carefully enough, we can detect displacements of millimeters per year. In other words, we can now see the plates move!

4.10 THE DYNAMIC PLANET

Now, having completed our two-chapter introduction to plate tectonics, we can see more easily why plate tectonics holds the key to understanding most of geology. To start with, plate tectonics explains the origin and distribution of earthquakes, major sea-floor features (mid-ocean ridges, deep-ocean trenches, seamount chains, and fracture zones), and volcanoes (▶Fig. 4.29). It also tells us why mountain

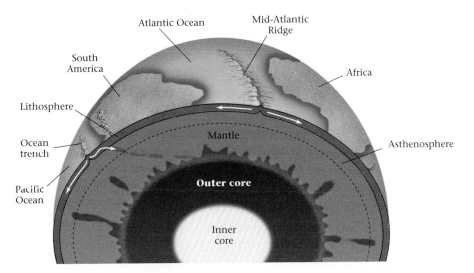

FIGURE 4.29 Plate tectonics involves the transfer of material from the mantle to the surface and back down again.

belts form. Finally, plate tectonics explains the drift of continents and why the distribution of land changes with time, a change that significantly affects the evolution of life on Earth. In coming chapters, we will explore these consequences and others in more detail.

CHAPTER SUMMARY

• The lithosphere, the rigid outer layer of the Earth, is broken into discrete plates that move relative to one another. Plates consist of the crust and the uppermost (cooler) mantle. Lithosphere plates effectively float on the underlying soft asthenosphere. Continental drift and sea-floor spreading are manifestations of plate movement.

• Some continental margins are plate boundaries, but many are not. A single plate can consist of continental lithosphere, oceanic lithosphere, or both.

• Most plate interactions occur along plate boundaries; the interior of plates remain relatively rigid and intact. Earthquakes delineate the position of plate boundaries.

• There are three types of plate boundaries—divergent, convergent, and transform—distinguished from each other by the movement the plate on one side of the boundary makes relative to the plate on the other side.

• Divergent boundaries are marked by mid-ocean ridges. At divergent boundaries, sea-floor spreading takes place, a process that forms new oceanic lithosphere.

• Convergent boundaries, also called convergent margins or subduction zones, are marked by deep-ocean trenches and volcanic arcs. At convergent boundaries, oceanic lithosphere of the downgoing plate is subducted beneath an overriding plate. The overriding plate can consist of either continental or oceanic lithosphere. An accretionary prism forms out of sediment scraped off the downgoing plate as it subducts.

• Subducted lithosphere sinks back into the mantle. Its existence can be tracked down to a depth of about 670 km by a belt of earthquakes known as the Wadati-Benioff zone.

• Transform boundaries, also called transform faults, are marked by large faults at which one plate slides past another. No new plate forms and no old plate is consumed at a transform boundary.

• Triple junctions are points where three plate boundaries intersect.

• Hot spots are places where a plume of hot mantle rock rises from just above the core-mantle boundary and causes anomalous volcanism at an isolated volcano. As a plate moves over the mantle plume, the volcano moves off the hot spot and dies, and a new volcano forms over the hot spot. As a result, hot spots spawn seamount/island chains.

• A large continent can split into two smaller ones by the process of rifting. During rifting, continental lithosphere stretches and thins. If it finally breaks apart, a new mid-ocean ridge forms and sea-floor spreading begins. Not all rifts go all the way to form a new mid-ocean ridge.

• Convergent plate boundaries cease to exist when a buoyant piece of lithosphere (a continent or an island arc) moves into the subduction zone. When that happens, collision occurs. The collision between two continents yields large mountain ranges.

• Ridge-push force and slab-pull force drive plate motions. Plates move at rates of about 1–15 cm per year. Modern satellite measurements can detect these motions.

• As the remainder of this book shows, plate tectonics theory provides a basis for understanding many geologic phenomena.

KEY TERMS

absolute plate velocity (p. 101)
accretionary prism (p. 89)
active (convergent, subducting) margin (p. 82, 87)
Archimedes' principle of buoyancy (p. 79)
asthenosphere (p. 78)
black smoker (p. 85)
buoyancy (p. 79)

collision (p. 98)
continental margin (p. 82)
continental rift (p. 97)
continental shelf (p. 82)
convergent plate boundary (p. 83)
divergent plate boundary (p. 83)
downgoing plate (slab) (p. 88)

earthquake belt (p. 83)
hot spot (p. 95)
lithosphere (p. 78)
lithosphere plates (p. 79)
magma (p. 85)
magma chamber (p. 85)
mantle plume (p. 95)
overriding plate (slab)
  (p. 88)
passive margin (p. 82)
plate boundary (p. 79)
plate tectonics (p. 77)
plate-boundary volcanoes
  (p. 95)

relative plate velocity (p. 101)
ridge-push force (p. 100)
scientific paradigm (p. 77)
scientific revolution (p. 77)
slab-pull force (p. 101)
subduction zone (p. 87)
transform plate boundary
  (p. 84)
transform fault (p. 91)
triple junction (p. 91)
volcanic arc (island and
  continental) (p. 89)
Wadati-Benioff zone
  (p. 88)

REVIEW QUESTIONS

1. What is a scientific revolution? How is plate tectonics an example of a scientific revolution?

2. What are the characteristics of a lithosphere plate?

3. How does oceanic crust differ from continental crust in thickness, composition, and density?

4. Describe how Archimedes' principle of buoyancy can be applied to continental and oceanic lithosphere.

5. Contrast active and passive margins.

6. What are the basic premises of plate tectonics?

7. How do we identify a plate boundary?

8. Describe the three types of plate boundaries.

9. How does crust form along a mid-ocean ridge?

10. What happens to the mantle beneath the mid-ocean ridge?

11. Why are mid-ocean ridges high?

12. Why is subduction necessary on a nonexpanding Earth with spreading ridges?

13. What is a Wadati-Benioff zone, and how does it help to define the location of subducting plates?

14. Describe the major features of a convergent boundary.

15. Why are transform plate boundaries required on an Earth with spreading and subducting plate boundaries?

16. What are two examples of famous transform faults?

17. What is a triple junction?

18. Explain the processes that form a hot spot.

19. How can hot spots be used to track the past motions of the overlying plate?

20. How is a seamount chain produced?

21. Describe the characteristics of a continental rift, and give examples of where this process is occurring today.

22. Describe the process of continental collision, and give examples of where this process has occurred in the past.

23. Discuss the two major forces that move lithosphere plates.

SUGGESTED READING

Allegre, C. 1988. *The Behavior of the Earth: Continental and Seafloor Mobility.* Cambridge: Harvard University Press.

Condie, K. C. 1997. *Plate Tectonics and Crustal Evolution.* 4th ed. Boston: Butterworth–Heinemann.

Cox, A., and R. B. Hart. 1986. *Plate Tectonics: How It Works.* Palo Alto, Calif.: Blackwell.

Kearey, P., and F. J. Vine. 1996. *Global Tectonics.* 2nd ed. Cambridge, Mass.: Blackwell.

Moores, E. M., ed. 1990. *Shaping the Earth: Tectonics of Continents and Oceans.* Readings from *Scientific American.* New York: Freeman.

Moores, E. M., and R. J. Twiss. 1995. *Tectonics.* New York: Freeman.

Earth Materials

W hat is the Earth made of? There are four basic components: the solid Earth (the crust, mantle, and core), the biosphere (living organisms), the atmosphere (the envelope of gas surrounding the planet), and the hydrosphere (the liquid and solid water at or near the ground surface). In this part of the book, we focus on the materials that make up the crust and mantle of the solid Earth. We will find that these consist primarily of rock. Most rock, in turn, contains minerals, so minerals are, in effect, the building blocks of our planet.

We therefore begin in Chapter 5 by learning about minerals and how they develop. Then we see, in Interlude A, how geologists distinguish three categories of rock—igneous, sedimentary, and metamorphic—according to how the rocks form. In each of the next three chapters (6, 7, and 8), we look at one of these rock categories. Finally, Interlude B shows us how, on our dynamic planet, Earth materials pass through a rock cycle, as atoms comprising one rock type end up being incorporated into a succession of other rock types.

CHAPTER 5

Patterns in Nature: Minerals

5.1 INTRODUCTION

Zabargad Island rises barren and brown above the Red Sea, about 70 km off the coast of southern Egypt. Nothing grows on Zabargad, except for scruffy grass and a few shrubs, so no one lives there now. But in the days of the pharaohs, dozens of slaves toiled on this 5-square-km patch of desert, gradually chipping their way into the side of its highest hill, seeking glassy green, pea-sized pieces of peridot, a prized gem. Carefully polished peridots were worn as jewelry by the Egyptian royal family and were buried with them when they died. Encased with mummies beneath the pyramids, the peridots of Zabargad lay hidden from the world until grave robbers broke into the tombs and carried them off. Eventually, the gems appeared in Europe, where jewelers set them into the crowns and scepters of European monarchs (►Fig. 5.1). Some of these peridots now glitter behind glass cases in museums, 3,000 years after first being pried free from the Earth, and perhaps 10 million years after first being formed by the bonding together of still more ancient atoms.

Peridot is one of an estimated 3,000 to 5,000 minerals that have been identified on Earth and fascinate collectors and geologists alike. Each different mineral has a name. Some names come from Latin, Greek, German, or English words describing a certain characteristic (for example, "albite" comes from the Latin word for "white," "orthoclase" comes from the German words meaning "splits at right angles," and olivine is olive-colored); some honor a person (sillimanite was named for Benjamin Silliman, a famous nineteenth-century mineralogist); some indicate the place where the mineral was first recognized (illite was first identified in rocks from Illinois); and some reflect a particular element in the mineral (chromite contains chromium). Some minerals have more than one name—for example, peridot is the gem-quality version of a common mineral named "olivine."

Although the vast majority of mineral types are rare, forming only under special conditions, many are quite common and are found in a variety of rock types at Earth's surface. You may want to note the names of the more common ones before proceeding (see Table 5.1), as these names will appear throughout the book. If

Sapphire gems are precious minerals found in many colors.

FIGURE 5.1 A royal scepter containing a variety of valuable jewels.

you have the chance to examine specimens of these minerals, you will find that with a little practice, you can recognize most of them on sight.

Why study minerals? Without exaggeration, we can say that minerals are the building blocks of our planet. To a geologist, almost any study of Earth materials depends on an understanding of minerals, for minerals make up the rocks and sediments that make up the Earth and its landscapes. But minerals are also important from a practical standpoint. **Industrial minerals** serve as the raw materials for manufacturing chemicals, concrete, and wallboard. **Ore minerals** are the source of valuable metals like copper and gold and provide energy resources like uranium (▶Fig. 5.2a, b). Certain forms of minerals, gems, delight the eye as jewelry. Unfortunately, not all minerals are beneficial; some pose environmental hazards. No wonder **mineralogy,**

TABLE 5.1 Some Mineral Names Worth Knowing

Name (Formula)	Description
Amphibole	Dark-colored; nonmetallic; a common rock-forming mineral.
Calcite ($CaCO_3$)	White, pink, or clear; has a milky luster; fizzes in acid; makes up chalk and limestone.
Chlorite	A dark-green mineral composed of thin flakes.
Clay	A group of mica-like minerals that occur only as very tiny flakes; makes mud.
Corundum (Al_2O_3)	Very hard; composed only of aluminum and oxygen; used as an abrasive; ruby.
Diamond (C)	The hardest mineral known; clear and glassy; used for jewelry and abrasives.
Dolomite ($CaMg[CO_3]_2$)	Similar to calcite, but less reactive with acid.
Feldspar	A group of light-colored hard minerals common in many rocks; accounts for about 60% of the Earth's crust.
Orthoclase ($KAlSi_3O_8$)	A common feldspar containing potassium; generally pinkish; also called K-feldspar.
Plagioclase	A common feldspar containing sodium or calcium; generally white, gray, or blue.
Galena (PbS)	Dark gray; occurs as metallic-looking cubic grains; a common type of lead ore.
Garnet	Dark brown to maroon; glassy; occurs in roundish, or dodecahedral, grains.
Graphite	Very soft, gray, and metallic; consists of carbon; used for pencil lead.
Gypsum	Clear and glassy; very soft; used for wallboard.
Halite (NaCl)	Clear or gray; occurs in cubic grains; also known as rock salt.
Hematite	Either earthy-colored or shiny bluish gray and metallic; a common type of iron ore.
Kyanite	Bluish; blade-shaped.
Magnetite (Fe_3O_4)	Dark gray to black; metallic-looking; very magnetic; also known as lodestone.
Mica	A group of minerals that occur as thin flakes or in books of thin sheets.
Biotite	A black mica.
Muscovite	A light-brown to clear mica.
Olivine ($[Mg,Fe]_2SiO_4$)	Olive green; glassy; occurs in sugar-like clusters; common in mantle rock.
Pyrite (FeS_2)	Golden bronze; occurs in metallic-looking cubic grains; also called "fool's gold."
Pyroxene	Dark-colored; nonmetallic; a common rock-forming mineral.
Quartz (SiO_2)	Glassy and hard; typically white, gray, or clear; found in beach sand.
Serpentine	A name used for many minerals that form as thin, thread-like needles (such as asbestos).
Talc	Very soft; used to make talcum powder.

FIGURE 5.2 (a) Museum specimen of malachite, a bright-green mineral containing copper (its formula is $Cu_2[CO_3][OH]_2$). Malachite is an ore mineral mined to produce copper, but because of its beauty, it is also used for jewelry. (b) Copper wire made by the processing of malachite and other ores of copper.

the study of minerals, is ever fascinating to professionals and amateurs alike.

The word "mineral" has a broader meaning in everyday English than it does in geology. For example, the first thing you ask when playing the game Twenty Questions is, "Is it animal, vegetable, or mineral?," "mineral" referring here simply to an inanimate object. Nutritionists talk about the "vitamins and minerals" in various types of foods—to them, a mineral is simply a metallic compound. In geology, however, a mineral is a special kind of substance with certain distinctive characteristics. In this chapter, we learn the geological definition of a mineral, then look at how minerals form and the main characteristics that allow us to identify them. Finally, we note the basic scheme that geologists use to classify minerals. This chapter assumes that you understand the fundamental concepts of matter and energy, especially the nature of atoms, molecules, and chemical bonds. If you are rusty on these topics, please review Appendix A. The basic definitions are summarized in Box 5.1, for convenience.

5.2 WHAT IS A MINERAL?

To geologists, a **mineral** is a homogeneous, naturally occurring, solid inorganic substance with a definable chemical composition and an internal structure characterized by an orderly arrangement of atoms, ions, or molecules in a lattice. Let's pull apart this mouthful of a definition and examine what its components actually mean.

• *Homogeneous:* Homogeneous materials are the same through and through—they cannot be physically broken into simpler components. When you smash a mineral specimen with a hammer, you get many tiny fragments of the same mineral. (Note that rocks are aggregates [combinations] of many separate mineral grains, or pieces. Fragments of a broken rock are not necessarily all the same mineral.)

• *Naturally occurring:* True minerals form by natural processes, not by the activity of a person. In recent decades, though, laboratory scientists have learned to manufacture materials that are essentially identical to naturally formed minerals. For example, some companies routinely manufacture diamonds by squeezing carbon under very high pressure. We sometimes call such materials **synthetic minerals,** to distinguish them from true minerals.

• *Solid:* A solid is a kind of matter that can maintain its shape indefinitely, and thus will not conform to the shape of its container. Therefore, liquids (like oil or water) and gases (like air) are not minerals.

• *Inorganic substance:* When we say that minerals are inorganic, we simply mean that they do not contain organic chemicals. (An **organic chemical** consists of carbon bonded to hydrogen along with varying amounts of oxygen, nitrogen, and other elements.) Thus, compounds like grain alcohol (CH_3CH_2OH) are not minerals, nor are more complex compounds like fat, protein, oil, or plastic. Note, however, that the presence of carbon alone does not disqualify a substance from being a mineral—pure carbon (graphite or diamond) and calcium carbonate ($CaCO_3$) *are* considered minerals.

• *Definable chemical composition:* This simply means that it is possible to write a chemical formula for a mineral (see Box 5.1). For example, the mineral quartz has the simple formula SiO_2—it contains the elements silicon and oxygen in the proportion of one silicon atom for every two oxygen atoms. Some formulas are more complicated: the formula for biotite is $K(Mg,Fe)_3(AlSi_3O_{10})(OH)_2$. While many minerals, like quartz, have only one composition, others have compositions that can vary slightly. In biotite, for example, iron (Fe) can substitute for magnesium (Mg)—which is why these elements are separated by a comma in the formula. Some samples of biotite contain more iron, while others contain more magnesium.

- *The orderly arrangement of atoms in a lattice:* Atoms, ions, or molecules in a mineral do not float about and rearrange into random configurations. Rather, they are fixed in a specific pattern that repeats at regular intervals. This orderly framework is called a **crystal lattice**—it's like the gridwork that makes up scaffolding (▶Fig. 5.3a,

b). Because minerals contain a crystal lattice, we say that they are **crystalline.**

With these definitions in mind, we can make an important distinction between a mineral and glass. Both minerals and glasses are solids, in that they can retain their shape indefinitely (see Appendix A). But a mineral is crystalline, while glass is not. While atoms, ions, or molecules in a min-

---

BOX 5.1

SCIENCE TOOLBOX

Some Basic Definitions from Chemistry

To describe minerals, we need to use several terms from chemistry (for a more in-depth discussion, see Appendix A).

- **Element:** a pure substance that cannot be separated into other elements.
- **Atom:** the smallest piece of an element that retains the characteristics of the element. An atom consists of a nucleus surrounded by a cloud of orbiting electrons; the nucleus is made up of protons and neutrons (except in hydrogen, whose nucleus contains only one proton and no neutrons). Electrons have a negative charge, protons have a positive charge, and neutrons have a neutral charge. An atom that has the same number of electrons as protons is said to be "neutral," in that it does not have an overall electrical charge.
- **Atomic number:** the number of protons in an atom of an element.
- **Atomic weight:** the approximate number of protons plus neutrons in an atom of an element.
- **Ion:** an atom that is not neutral. An ion that has an excess negative charge (because it has more electrons than protons) is an **anion,** while an ion that has an excess positive charge (because it has more protons than electrons) is a **cation.** We indicate the charge with a superscript. For example, Cl^- (chlorine) has a single excess electron; Fe^{2+} is missing two electrons.
- **Chemical bond:** an attractive force that holds two or more atoms together. **Covalent bonds** form when atoms share electrons. **Ionic bonds** form when a cation and anion (opposite charges) get close together and attract each other.
- **Molecule:** two or more atoms bonded together. The atoms may be of the same element or of different elements.

- **Compound:** a pure substance that can be subdivided into two or more elements. The smallest piece of a compound that retains the characteristics of the compound is a molecule.
- **Chemical:** a general name used for a pure substance (either an element or a compound).
- **Chemical formula:** a shorthand recipe that itemizes the various elements in a chemical and specifies their relative proportions. For example, the formula for water, H_2O, indicates that water consists of molecules in which two hydrogens bond to one oxygen.
- **Chemical reaction:** a process that involves the breaking or forming of chemical bonds. Chemical reactions can break molecules apart, or create new molecules.
- **Mixture:** a combination of two or more elements or compounds that can be separated without a chemical reaction. For example, cereal composed of bran flakes and raisins is a mixture—you can separate the raisins from the flakes without destroying either.
- **Solution:** a type of material in which one chemical (the solute) dissolves (becomes completely incorporated) in another (the solvent). In solutions, a dissolving compound may separate into ions during the process. For example, when salt (NaCl) dissolves in water, it separates into sodium (Na^+) and chlorine (Cl^-) ions. In a solution, atoms or molecules of the solvent surround atoms, ions, or molecules of the solute.
- **Precipitate:** (noun) a compound that forms when ions in solution join together to create a solid that settles out of the solution; (verb) the process of forming a precipitate. For example, when saltwater evaporates, solid salt crystals precipitate and settle to the bottom of the remaining water.

(a) (b)

(c)

(d)

FIGURE 5.3 (a) Internally, this quartz crystal contains an orderly arrangement of atoms. (b) This gridwork of scaffolding surrounding the Washington Monument in Washington, D.C., provides an analogy for the fixed arrangements of atoms in a mineral. (c) Disordered atoms, as occur in glass, do not define a regular pattern. (d) Ordered atoms like these are found in a mineral.

5.3 BEAUTY IN PATTERNS: CRYSTALS AND THEIR STRUCTURE

What's Inside a Crystal?

A **crystal** is a single, continuous (uninterrupted) piece of crystalline material bounded by flat surfaces (**crystal faces**) that formed naturally as the crystal lattice grew. The word comes from the Greek *krystallos,* meaning "ice." Crystals may have beautiful, regular shapes that look as though they came out of the pages of a geometry book. Back in the seventeenth century, a Danish monk named Nicholas Steno recognized that when two specimens of the same mineral were compared, the angle between any two adjacent crystal faces in one specimen was the same as the angle between the corresponding faces in the other specimen (▶Fig. 5.4a, b). For example, in the common mineral halite, or table (rock) salt, adjacent crystal faces intersect at a right angle, so that the crystal has the shape of a cube, regardless of its size. Not all crystals resemble cubes—some

FIGURE 5.4 For any given mineral, the angle between adjacent crystal faces is the same. (a) A small, well-formed crystal of quartz. The intersection between crystal faces makes an angle of 120°, as shown by the cross-section slice through the crystal. (b) A large, poorly formed crystal of quartz. Even though the dimensions of the faces differ from one another, the angle between the faces is still 120°, as measured by a goniometer (an instrument that measures angles).

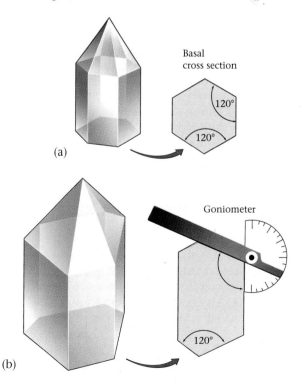

eral are ordered into a crystal lattice, like soldiers standing in formation, those in a glass are arranged in a semichaotic way, like a crowd of people at a party, in small clusters or chains that are neither oriented in the same way nor spaced at regular intervals (▶Fig. 5.3c, d). Note that the chemical compound silica (SiO_2) forms the mineral quartz when arranged in a crystalline lattice, but forms common window glass when arranged in a semichaotic way.

If you ever need to figure out whether a substance is a mineral or not, just check it against the criteria listed above. Is sugar a mineral? No—it's composed of organic chemicals (its chemical formula is $C_6H_{12}O_6$). Is table salt a mineral? Yes—it consists of a solid inorganic crystalline compound with the formula NaCl.

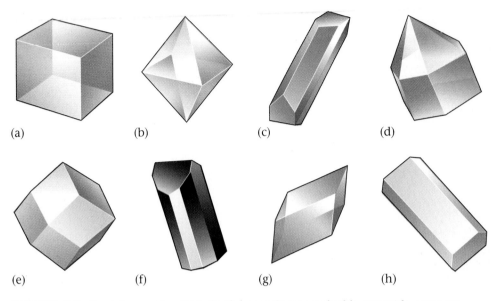

FIGURE 5.5 Crystals come in all kinds of shapes. Some are double pyramids, some are cubes, and some have blade shapes. Some crystals terminate at a point, and some terminate in a chisel-like wedge. (a) Halite, (b) diamond, (c) staurolite, (d) quartz, (e) garnet, (f) stibnite, (g) calcite, and (h) kyanite.

have brick- or plate-like shapes, some are trapezoids, pyramids, octahedrons, or hexagonal columns, and some resemble obelisks or needles. By studying the examples in ▶Figure 5.5, you'll notice that some crystals terminate at a point (d), while others have a knife-like edge (g).

Because of their regular geometric form, people have always considered crystals as something special, and many cultures have attributed magical powers to them. The shamans of legend often relied on talismans or amulets made of crystals, which supposedly brought power to their wearer or warded off evil spirits. Even in modern fantasy and science-fiction stories, crystals play a special role—in the TV series *Star Trek,* the starship *Enterprise* required a "dilithium crystal" to achieve "warp speed." (In reality, no such crystal exists.)

In recent years, it has become fashionable for people to wear crystals around their necks, suspend them over their heads at night, or place them in prominent places in their homes with the thought that crystals can somehow "channel" the "life force," or energy, of the universe into a person's soul. Many people have paid large fees to "crystal therapists" who claim to be able to cure ailments ranging from a stomach upset to schizophrenia. As far as the vast majority of geologists are concerned, such uses of crystals are pure bunk. A series of experiments has shown the proximity of crystals to have no demonstrable ef-

fect on health. Further, no documentation of magical powers has been reported in scientific journals, which require authors to justify conclusions with data. No unbiased tests have shown that crystal therapists have been more successful in healing illness than any other kind of adviser. For millennia, crystals have inspired awe because of the way they sparkle in light, but such behavior is simply a consequence of how crystal lattices interact with light, not an indicator of curing power.

While all reputable geologists scoff at the idea of crystals having magical powers, the scientific question of why crystals display the shapes they do has been a focus of study for over three centuries. In 1912, a German physicist named Max von Laue provided the key to this mystery when he showed that an X-ray beam passing through a crystal undergoes **diffraction,** meaning that it splits into many tiny beams that interfere with one another to create a pattern of spots on a projection screen (▶Fig. 5.6). Von Laue knew that diffraction occurs when electromagnetic radiation (for example, light or X-rays) passes through a material composed of regularly spaced obstacles, if the distance between the obstacles is comparable to the wavelength of the radiation. Thus, his observation of X-ray diffraction in crystals showed that crystals have an orderly internal arrange-

FIGURE 5.6 The pattern of dots on a screen that results when an X-ray beam passes through a crystal and is diffracted. X-rays interact this way with a crystal because of the crystal's orderly arrangement of atoms.

(a)

Sulfur

(b) Lead

FIGURE 5.7 (a) Repetition of a flower motif in wallpaper illustrates a regular pattern. (b) On the face of a crystal of galena (a type of lead ore), lead and sulfur atoms pack together in a regular array.

Cl⁻   Na⁺

(a)

(b)   (c)

FIGURE 5.8 (a) This diagram of a small piece of a cubic crystal of halite, called a ball-and-stick model, emphasizes the geometric arrangement of the chlorine (Cl^-) and sodium (Na^+) ions, with the chemical bonds represented by sticks. (b) Crystals have symmetry: one half of a halite crystal is a mirror image of the other half. (c) Snowflakes, crystals of ice, are symmetrical hexagons.

ment of atoms or ions (now an essential part of the definition of a mineral). This discovery also eventually allowed mineralogists to measure the spacing between atoms in crystals.

The orderly arrangement of atoms inside a crystal lattice provides one of nature's most spectacular examples of a pattern. As the pattern on a sheet of wallpaper may be defined by the regular spacing of, say, clumps of flowers, so the pattern in a crystal is defined by the regular spacing of atoms or ions (▶Fig. 5.7a, b). If the crystal contains more than one type of atom, the atoms alternate in a regular way.

The atoms or ions in a crystal are packed closely together. As an example, let's look at the insides of a crystal of halite (rock salt). Halite consists of ionically bonded sodium (Na^+) ions and chlorine (Cl^-) ions in equal proportion, and thus has the chemical formula NaCl. If you could see the ions in a crystal of salt, you would find that sodium and chlorine ions alternate in a cube-like grid (▶Fig. 5.8a). Since the ions are different sizes (Cl^- is much bigger than Na^+), this arrangement allows them to pack together as tightly as possible, with each smaller Cl^- ion surrounded by six larger Na^+ ions. Because the ions occur in a cube-like grid, crystals of halite have a cube-like form. The pattern of atoms or ions in a mineral displays **symmetry:** the shape of one part of a mineral is a mirror image of the shape of another part. If you

were to cut a halite crystal in half and place one half against a mirror, the crystal would look whole again (▶Fig. 5.8b, c). Crystals can be classified by their geometry.

Two different minerals that have the same chemical composition are known as **polymorphs.** For example, diamond and graphite are different minerals—they obviously look and behave nothing alike—but both consist exclusively of the element carbon. Diamond is the hardest substance known in nature, while graphite is so soft that you use it as the "lead" of your pencil; as you move a pencil across paper, tiny flakes of graphite peel off the pencil point and adhere to the paper. Why are these two mineral forms of carbon so different in properties? The explanation lies in their respective crystal lattice structures (▶Fig. 5.9a–d). Diamond has a three-dimensional arrangement of carbon atoms packed tightly together, with very strong bonds between them. It is difficult to break the atoms apart, and this is why diamond is so hard. By contrast, the carbon atoms in graphite comprise parallel sheets that are bonded to one another by weak bonds that can easily be broken.

The Formation and Destruction of Minerals

New mineral crystals can form in one of three ways. First, they can form by **the solidification of a melt,** meaning

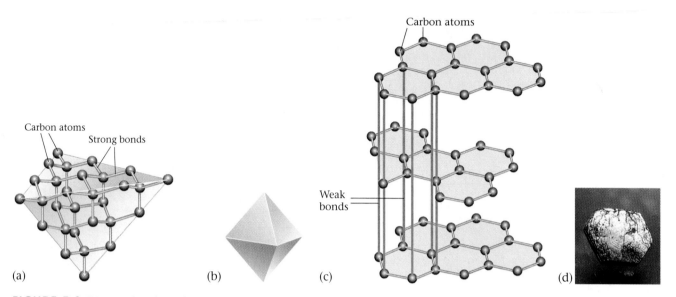

FIGURE 5.9 Diamond and graphite both consist of carbon (that is, they have the same chemical composition), but they have different crystal structures. Thus, they are called polymorphs. (a, b) In diamond, carbon atoms are strongly bonded in a 3-D grid. The strength of these bonds makes diamond very hard. (c, d) In graphite, the atoms are strongly bonded in 2-D sheets, but the sheets are bonded together only by weak bonds. The sheets of graphite can be easily peeled apart, as happens when you move a graphite pencil across paper and leave behind a trail of tiny graphite flakes.

the freezing of a liquid; for example, ice crystals, a type of mineral, are made by freezing water. Second, they can form by **precipitation from a solution,** meaning that atoms, molecules, or ions dissolved in water bond together and separate out of the water—salt crystals, for example, develop when you evaporate saltwater. And third, they can form by **solid-state diffusion,** the movement of atoms or ions through a solid to arrange into a new crystal lattice; this process takes place very slowly.

Regardless of the process, the first step in forming a crystal is the chance formation of a **seed,** or an extremely small crystal (▶Fig. 5.10a–c). Once the seed exists, other atoms in the surrounding material attach themselves to the face of the seed; here on the face, an electric charge pulls in and holds on (bonds) to atoms or ions in the surrounding material. As the crystal grows, crystal faces move outward from the center of the seed but maintain the same orientation. Thus, the youngest part of the crystal is always its outer edge (▶Fig. 5.11a, b).

In the case of crystals formed by the solidification of a melt, atoms begin to attach to the seed when the melt becomes so cool that thermal vibrations can no longer break

FIGURE 5.10 (a) New crystals nucleate (begin to form) in a water solution. They grow outward from the walls of the container. (b) At a later time, the crystals have grown larger. (c) On a crystal face, atoms in the solution are attracted to the surface and latch on.

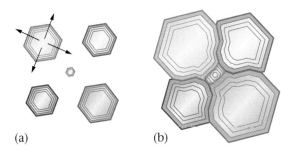

(a)   (b)

FIGURE 5.11 (a) Crystals grow outward from the central seed. (b) Crystals maintain their shape until they interfere with each other. When that happens, the crystal shapes can no longer be maintained.

apart the attraction between the seed and the atoms in the melt. Crystals formed by precipitation from a solution develop when the solution becomes **saturated,** meaning the number of dissolved atoms, ions, or molecules per unit volume of solution becomes so great that they can get close enough to one another to bond together. If a solution is not saturated, dissolved atoms, ions, or molecules are surrounded by solvent molecules and are shielded from the attractive forces of their neighbors. Sometimes crystals formed by precipitation from a solution grow from the walls of the solution's "container" (for example, a crack or pore in a rock). A spectacular example of this phenomenon is a **geode,** a roughly spherical cavity in rock in which crystals precipitate out of water solutions passing through the rock (▶Fig. 5.12a). In some cases, crystals form within the solution and then settle to the bottom.

Crystals created by solid-state diffusion represent the rearrangement of atoms or ions already within a rock, so that as a new crystal gradually forms, old ones gradually disappear. We will see in Chapter 8 how important this process is in the formation of metamorphic rock.

As crystals grow, they develop their particular **crystal shape,** based on the geometry of the lattice. The shape is defined by the relative dimensions of the crystal (needle-like, sheet-like, etc.) and the angles between crystal faces. If a mineral's growth is uninhibited so that it displays well-formed crystal faces, then it is a **euhedral crystal** (▶Fig. 5.12b). Typically, however, the growth of minerals is restricted in one or more directions, because existing crystals act as obstacles. In such cases, minerals grow to fill the space that is available, and their shape is controlled by the shape of the surroundings. Crystalline minerals without well-formed crystal faces are called **anhedral grains** (▶Fig. 5.13a, b). In the case of rocks created by the solidification of melts, many crystals grow at about the same time, competing with each other for space. As a consequence, these minerals grow into one another, forming anhedral grains that interlock like pieces of a jigsaw puzzle.

A mineral can be destroyed by melting, dissolving, or other chemical reaction. Melting involves heating a mineral to the **melting temperature,** the temperature at which thermal vibration of the atoms or ions in the lattice is sufficient to break the chemical bonds holding them to the lattice; the atoms or ions then separate, either individually or in small groups, to move around again freely. Different minerals have different melting temperatures, and the melting temperature of a given mineral changes with changing pressure.

Dissolving (dissolution) occurs when a mineral is immersed in a solvent (like water). Atoms or ions then separate from the crystal face and are surrounded by solvent molecules. Some minerals, such as salt, dissolve easily, but most do not dissolve much at all. Other chemical reactions that can destroy minerals happen when a mineral comes in contact with reactive materials. For example, reaction with oxygen oxidizes minerals that contain iron and turns them into rust.

FIGURE 5.12 (a) A geode, in which euhedral crystals grow from the wall into the center. (b) An enlargement of a euhedral crystal, showing that the surfaces are crystal faces.

(a)

(b)

(a)   **(b)**

FIGURE 5.13 A crystal growing in a confined space is anhedral, one whose surface is not composed of crystal faces. (a) A crystal stops growing when it meets the surfaces of other grains, and continues growing to fill in gaps. (b) The resulting mineral grain, if it were to be separated from other grains, would have an anhedral shape.

FIGURE 5.14 The range of colors of quartz, displayed by different crystals: milky, clear, and rose quartz.

5.4 HOW DO YOU TELL ONE MINERAL FROM ANOTHER?

We've seen that minerals have a definable chemical composition and a specific crystalline structure. These characteristics control a mineral's **physical properties**—familiar ones like color and hardness, as well as some less familiar ones, such as streak, luster, specific gravity, crystal form, crystal habit, and cleavage—which make each mineral type (there are over 3,000) uniquely different from the others. You can identify hand specimens of common minerals in the field or lab based on an analysis of their physical properties, a description of which follows. (See Appendix B for mineral-identification flow charts.)

- *Color:* **Color** results from the way a mineral interacts with light. Sunlight contains the whole spectrum of colors, where each color has a different wavelength. A mineral absorbs certain wavelengths, so the color you see when looking at it represents the wavelengths the mineral does not absorb. Certain minerals always have the same color (galena is always gray, for example), but many show a range of colors. For example, quartz can be clear, white, purple, gray (smoky), red (rose), or just about anything in between (▶Fig. 5.14). Purple quartz is known as amethyst (Fig. 5.12a). Color variations in a mineral reflect the presence of tiny amounts of impurities in the crystal. For example, a trace of iron may produce a reddish or purple tint, and manganese may produce a pinkish tint.

- *Streak:* The **streak** of a mineral refers to the color of a powder produced by pulverizing the mineral. You can obtain a streak by scraping the mineral against an unglazed ceramic plate (▶Fig. 5.15). The color of a

mineral powder tends to be less variable than the color of a whole crystal, and thus provides a fairly reliable clue to a mineral's identity. Calcite, for example, always yields a white streak even though pieces of calcite may be white, pink, or clear.

- *Luster:* **Luster** refers to the way a mineral surface scatters light. Geoscientists describe luster simply by comparing the appearance of the mineral with the appearance of a familiar substance, and thus use self-explanatory adjectives like "metallic" and "nonmetallic" (▶Fig. 5.16a, b). Types of nonmetallic luster may be further described, for example, as silky, glassy, satiny, resinous, pearly, or earthy.

- *Hardness:* **Hardness** is a measure of the relative ability of a mineral to resist scratching, and therefore represents the resistance of bonds in the lattice to being broken. The atoms or ions in crystals of a hard mineral are more strongly bonded than those in a soft mineral. Hard minerals can scratch soft minerals, but soft minerals cannot scratch hard ones. Diamond is the hardest

FIGURE 5.15 A streak plate, showing the red streak of hematite.

(a)

(b)

TABLE 5.2 Mohs Hardness Scale

Number	Mineral or Object Name
10	Diamond
9	Corundum (ruby)
8	Topaz
7	Quartz
6	K-feldspar
5.5	*Steel knife, glass*
5	Apatite
4	Fluorite
3.5	*Copper penny*
3	Calcite
2.5	*Fingernail*
2	Gypsum
1	Talc

FIGURE 5.16 (a) This specimen of pyrite looks like a piece of metal because of its shiny gleam; we call this metallic luster. (b) These specimens of feldspar have a nonmetallic luster. The white one on the left is plagioclase, and the pink one on the right is orthoclase (K-feldspar).

mineral known—it can scratch anything, which is why it is used to cut glass. In the early 1800s, a mineralogist named Friedrich Mohs listed some minerals in sequence of relative hardness; a mineral with a hardness of 5 can scratch all minerals with a hardness of 5 or less. This list, now called the **Mohs hardness scale,** helps in mineral identification. When you use the scale (Table 5.2), it might help to compare the hardness of a mineral with a common item like your fingernail, a penny, or a glass plate. Note that not all of the minerals in Table 5.2 are common or familiar.

- *Specific gravity:* **Specific gravity** represents the density of a mineral, as specified by the ratio between the weight of a volume of the mineral and the weight of an equal volume of water. For example, one cubic centimeter of quartz has a weight of 2.65 grams, while one cubic centimeter of water has a weight of 1.00 gram. Thus, the specific gravity of quartz is 2.65. People sometimes say that a joke, for instance, "went over like a lead balloon," because lead has a specific gravity of 11. In practice, you can develop a feel for specific gravity by hefting minerals

in your hands. Galena (lead ore) "feels" heavier for its size than quartz. Ore minerals, which contain a lot of a useful metal, typically have high specific gravities.

- *Crystal form:* **Crystal form** refers to the geometry of a euhedral crystal, one with natural crystal faces that grew unimpeded. The form is defined by the angular relation between crystal faces, which depends on the internal arrangement of atoms. Recall that the angle between adjacent crystal faces in a mineral is the same, even though the relative dimensions of faces may change from specimen to specimen.

- *Crystal habit:* Loosely defined, **crystal habit** refers to the general shape or character of a crystal or cluster of crystals that grew unimpeded. Geologists use adjectives like "needle-like," "fibrous," "cubic," "blade-like," "platy," "grape-like," or "prismatic" to describe habit (▶Fig. 5.17a–c). Some mineral types have more than one habit. The habit reflects, in part, the ease with which a mineral grows in a certain direction. For example, a needle-like crystal grows faster in one direction than in all others.

- *Cleavage:* A mineral's **cleavage** is the way it breaks. Minerals tend to break where the bonds holding the atoms together in the crystal are the weakest. When they break, a series of surfaces parallel to these bonds may form; the surfaces are called **cleavage planes** (▶Fig. 5.18a–f). Some minerals have one direction of cleavage (for example, mica, which splits into parallel sheets), some have two or three that intersect at a specific angle (like halite, which has three sets of cleavage planes that intersect at right angles), and others have none at all (like quartz, which breaks to form **conchoidal fractures,**

FIGURE 5.17 Crystal habit refers to the shape or character of the mineral. (a) Kyanite, which generally occurs in clusters of blades. (b) Prismatic crystals. (c) A spray of needle-like, or fibrous, crystals.

FIGURE 5.18 Mineral cleavage refers to the way a crystal breaks. Some crystals break in only one direction, some in two or three, and some in many. Others have no cleavage at all.
(a) Muscovite has one direction of cleavage and splits into thin sheets. (b) Pyroxene has two directions of cleavage at right angles. (c) Amphibole has two directions of cleavage, where one plane makes an angle of less than 90° with respect to the other two. (d) Halite has three directions of cleavage, all at right angles to each other. (e) Calcite has three directions of cleavage, one of which is inclined at an angle of less than 90°. (f) Calcite reacts with hydrochloric acid to produce carbon dioxide gas.

FIGURE 5.19 Minerals without cleavage break on random fractures. Some materials, like quartz and glass, break on conchoidal fractures, which have a curving, scoop shape.

smoothly curving, clamshell-shaped surfaces (▶Fig. 5.19). In the case of mica, atoms within mica sheets stick together by strong bonds, but the sheets themselves stick together only by weak bonds. It's easy to peel the sheets apart, because that action simply involves breaking weak bonds (Fig. 5.18a).

One final note about cleavage: because both cleavage planes and crystal faces reflect light, it's easy to confuse them. Cleavage forms by breaking, so you can recognize a cleavage plane if it is one of several parallel planes arranged like steps. There can be numerous parallel cleavage planes, but only one crystal face in a given orientation (▶Fig. 5.20a, b).

- *Special properties:* Some minerals have distinctive properties that readily distinguish them from other minerals. For example, calcite ($CaCO_3$) reacts with dilute hydrochloric acid (HCl) to produce carbon dioxide (CO_2) gas (Fig. 5.18f), graphite makes a gray mark on paper (it's the "lead" in pencils), magnetite attracts a magnet, halite tastes salty, and plagioclase has striations (thin parallel corrugations or stripes, on cleavage planes).

FIGURE 5.20 You can distinguish between cleavage planes and crystal faces because (a) cleavage planes are repeated, like a series of steps or terraces, while (b) a crystal face is a single surface. Note that there are no repetitions of the crystal face within a crystal.

(a)   (b)

5.5 ORGANIZING OUR KNOWLEDGE: MINERAL CLASSIFICATION

Though there are thousands of different minerals, they can be separated into a relatively small number of groups, or **mineral classes.** You may think, "Why bother?" but classification schemes are useful because they help organize information and streamline discussion. Biologists, for example, classify animals into groups based on how they feed their young and on the architecture of their skeletons, and botanists classify plants according to the way they reproduce and the shape of their leaves. A Swedish chemist, Baron Jöns Jacob Berzelius (1779–1848), analyzed the chemicals making up minerals and noted chemical similarities among many of them. Berzelius, along with his students, then established that most minerals can be classified by specifying the principal anion (negative ion) within the mineral. (Note that some anions consist of single atoms, while others consist of a group of atoms that act as a unit; see Box 5.1.) We now take a look at these mineral groups, focusing especially on silicates, the class that comprises most of the rock in the Earth.

The Mineral Classes

Mineralogists distinguish seven principal classes (and another five minor ones) of minerals, based on their chemical composition.

- *Silicates:* The fundamental component of **silicates** is the SiO_4^{4-} anionic group, a silicon atom surrounded by four oxygen atoms that are arranged to define the corners of a tetrahedron, a pyramid-like shape with four triangular faces. Mineralogists commonly refer to this anion as the **silicon-oxygen tetrahedron.** We can identify a huge variety of silicate minerals, differing from one another in the way the tetrahedra link and in the cations (positively charged ions) present in the mineral. Olivine, a common example, has the formula $(Mg,Fe)_2SiO_4$. Another well-known example, quartz (Fig. 5.14) has the formula SiO_2. We will learn more about silicates in the next section.

- *Oxides:* **Oxides** consist of metal cations bonded to oxygen anions. Typical oxide minerals include magnetite (Fe_3O_4) and hematite (Fe_2O_3). Because oxides commonly contain a relatively high proportion of metal atoms, they often serve as ore minerals.

- *Sulfides:* **Sulfides** consist of a metal cation bonded to sulfide (S^{2-}), the anion of sulfur. As with oxides, the metal forms a high proportion of the mineral, so sulfides are commonly considered ore minerals. Examples include galena (PbS) and pyrite (FeS_2; Fig. 5.16a). Many sulfide minerals have a metallic luster.

- *Sulfates:* **Sulfates** consist of a metal cation bonded to the SO_4^{2-} anionic group. Many sulfates form by precipitation out of water at or near the Earth's surface. An example is gypsum ($CaSO_4 \cdot 2H_2O$), in which water molecules bond to the calcium sulfate molecules. Pulverized gypsum mixed with water can be spread out in thin sheets that harden when they dry. Contractors use these sheets as wallboard ("sheetrock") in houses.

- *Halides:* The anion in a **halide** is a halogen ion (such as chlorine [Cl^-] or fluorine [F^-]), an element from the second column from the right in the periodic table (see Appendix A). Halite, or rock salt ($NaCl$; Fig. 5.18d), and fluorite (CaF_2), a source of fluoride, are common examples.

- *Carbonates:* In **carbonates,** the molecule CO_3^{2-} serves as the anion. Elements like calcium or magnesium bond to this anion. The two most common carbonates are calcite ($CaCO_3$; Fig. 5.18e) and dolomite ($CaMg[CO_3]_2$).

- *Native metals:* **Native metals** consist of pure masses of a single metal. Copper and gold, for example, may occur as native metals. A gold nugget is a mass of native gold that has been broken out of a rock.

Silicates: The Major Rock-Forming Minerals

In the rocks exposed at the surface of the continents, we can find minerals from a diversity of classes. But below a depth of around 5 km, on average, silicate minerals compose over 95% of the continental crust. Rocks making up oceanic crust and the Earth's mantle consist almost entirely of silicate minerals. Thus, silicate minerals are the most common on Earth.

As noted above, all silicate minerals consist of combinations of a fundamental building block called the silicon-oxygen tetrahedron. A silicon-oxygen tetrahedron is an anion that consists of a single silicon atom surrounded by four oxygen atoms (SiO_4^{4-}), arranged so that they define the points of a tetrahedron (a pyramid-like shape with four triangular faces; ▶Fig. 5.21a–c). Oxygen atoms are bigger than silicon atoms, so the silicon atom hides in the center of a tetrahedron, cozily fitting into the space between the oxygen atoms. Silica tetrahedra can link together, forming larger molecules, by sharing oxygen atoms.

Silicate minerals are divided into seven groups, of which five are described below. These are distinguished from each other based on how silicon-oxygen tetrahedra in the mineral link together. The number of links determines how many oxygen atoms are shared between tetrahedra and, therefore, the ratio of Si to O in the mineral. Each group includes several distinct minerals that differ from one another primarily in terms of the cations that are present.

- *Independent tetrahedra:* In this group, the tetrahedra are independent and do not share any oxygens (▶Fig. 5.22a). The mineral is held together by the attraction between the tetrahedra and positive ions (metal cations, such as iron or magnesium). This group includes olivines, a glassy green mineral that usually occurs in small, sugar-like crystals. Garnets are also members of this group.

- *Single chains:* In a single-chain silicate, the tetrahedra link to form a chain by sharing two oxygen atoms each (▶Fig. 5.22b). The most common of the many different types of single-chain silicates are pyroxenes, a group of black or

FIGURE 5.21 (a) The basic building block of a silicate mineral is the SiO_4^{4-} ion, the silicon-oxygen tetrahedron. In this compound, the oxygens occupy the corners of a tetrahedron, which can be portrayed in different ways. (b) In this ball model, the silicon atom is hidden. (c) A tetrahedron has been superimposed on this ball-and-stick model.

(a)

(b)

(c)

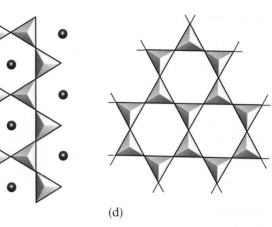

FIGURE 5.22 (a) Independent tetrahedra, as in olivine, share no oxygens. Positive ions hold them together. (b) Two single chains of tetrahedra, as in pyroxene, held together by positive ions. In each chain, a tetrahedron shares two oxygens. (c) A double chain of tetrahedra, as in amphibole. Here, two single chains link by sharing oxygens. Some tetrahedra share two oxygens, some share three. (d) A sheet of tetrahedra, as in mica. Each tetrahedron shares three oxygens. (e) A 3-D network (framework) of tetrahedra. Note that within the framework, each tetrahedron shares all four oxygens with its neighbors.

dark-green minerals that occur in elongate crystals with two mutually orthogonal cleavage directions (Fig. 5.18b).

- *Double chains:* In a double-chain silicate, the tetrahedra link to form a double chain by sharing two or three oxygens (▶Fig. 5.22c). Amphiboles are the most common type; these typically are black or dark-brown elongate crystals, with two cleavage directions. You can distinguish amphiboles from pyroxenes because the cleavage planes of amphiboles lie at about 60° to each other (Fig. 5.18c).

- *Sheet silicates:* The tetrahedra in this group all share three oxygens and therefore link to form two-dimensional sheets (▶Fig. 5.22d). Other ions and, in some cases, water molecules fit between the sheets. Because of their structure, sheet silicates have a single strong cleavage in one direction, and they occur in books of very thin sheets. In this group we find micas, a type of sheet silicate including muscovite (light-brown or clear mica; Fig. 5.18a) and biotite (black mica). Clay minerals are also sheet silicates and have a crystal structure similar to that of mica, but clay occurs only in extremely tiny flakes.

- *Framework silicates:* In a framework silicate, each tetrahedron shares all four oxygens with its neighbors, forming a 3-D structure (▶Fig. 5.22e). Examples include

feldspar and quartz. The two most common types of feldspar are plagioclase, which tends to be white, gray, or blue, and orthoclase (also called potassium feldspar, or K-feldspar), which tends to be pink (Fig. 5.16b). Feldspars typically contain a variety of different elements including aluminum (which substitutes for silicon in the tetrahedra), calcium, sodium, and potassium. Quartz, in contrast, contains only silicon and oxygen. As the ratio of silicon to oxygen in quartz is 1:2, the mineral has the familiar formula SiO_2.

5.6 SOMETHING PRECIOUS—GEMS!

Mystery and romance follow famous gems. Consider the stone now known as the Hope Diamond, recognized by name the world over. No one knows who first dug it out of the ground. Was it mined in the 1600s, or was it stolen off an ancient religious monument? What we do know is that in the 1600s, a French trader named Jean Baptiste Tavernier obtained a large (112.5 carats, where 1 carat ≈ 200 milligrams), rare blue diamond in India, perhaps from a Hindu statue, and carried it back to France. King Louis XIV bought the diamond and had it fashioned into a jewel of 68 carats.

FIGURE 5.23 The Hope Diamond, now on display in the Smithsonian Institution, Washington, D.C.

then became known as the Hope Diamond (▶Fig. 5.23). It changed hands several times until 1958, when the famous New York jeweler Harry Winston donated it to the Smithsonian Institution in Washington, D.C., where it now sits behind bulletproof glass in a heavily guarded display. Legend has it that whoever owns the stone suffers great misfortune.

What makes stones like the Hope Diamond so special that people risk life and fortune to obtain them? What is the difference between a gemstone, a gem, and any other mineral? A **gemstone** is a mineral that has special value because it is rare and people consider it beautiful. A **gem** is a cut and finished stone ready to be used in jewelry. Jewelers distinguish between **precious stones** (for example, diamond, ruby, sapphire, emerald), which are particularly rare and expensive, and **semiprecious stones** (topaz, tourmaline, aquamarine, garnet), which are less rare and less expensive. All the stones mentioned so far are transparent crystals, though most have some color (see Table 5.3). The category of semiprecious stones also includes opaque or translucent minerals such as lapis, malachite (Fig. 5.2a), and opal.

In everyday language, pearls and amber may also be considered gemstones. Unlike diamonds and garnets, which

This jewel vanished in 1762 during a burglary. Perhaps it was lost forever—perhaps not. In 1830, a 44.5-carat blue diamond mysteriously appeared on the jewel market for sale. Henry Hope, a British banker, purchased the stone, which

TABLE 5.3 Common Gems

Gem Name	Material/Formula	Comments
Amber	Fossil tree sap	Composed of organic chemicals; amber is not a mineral.
Amethyst	Quartz/SiO_2	The best examples precipitate from water in openings in igneous rocks; a deep-purple version of quartz.
Aquamarine	Beryl/$Be_3Al_2Si_6O_{18}$	A bluish version of emerald.
Diamond	Diamond/C	Brought to the surface from the mantle in igneous bodies called diamond pipes; may later be mixed in deposits of sediment.
Emerald	Beryl/$Be_3Al_2Si_6O_{18}$	Occurs in coarse igneous rocks (pegmatites) (see Chapter 6).
Garnet	Garnet/(e.g., $Mg_3Al_2[SiO_4]_3$)	A variety of different types (Ca, Fe, Mg, and Mn versions); occurs in metamorphic rocks (see Chapter 8).
Jade	Jadeite/$NaAlSi_2O_6$ Nephrite/$Ca_2(Mg,Fe)_5Si_8O_{22}(OH)_2$	Jade can be one of two minerals, jadeite (a pyroxene) or nephrite (an amphibole); both occur in metamorphic rocks.
Opal	Composed of microscopic spheres of hydrated silica packed together	Most opal comes from a single mining district in central Australia; occurs in bedrock that has reacted with water near the surface.
Pearl	Aragonite/$CaCO_3$	Formed by oysters, which secrete coatings around sand grains that are accidentally embedded in the soft parts of the organism. Cultured pearls are formed the same way, but the impurity is a spherical bead that is intentionally introduced.
Ruby	Corundum/Al_2O_3	The red color is due to chromium impurities; found in coarse igneous rocks called pegmatites and as a result of contact metamorphism (see Chapters 6 and 8).
Sapphire	Corundum/Al_2O_3	A blue version of ruby.
Topaz	$Al_2SiO_4(F,OH)_2$	Found in igneous rocks, and as a result of the reaction of rock with hot water.
Tourmaline	$Na(Mg,Fe)_3Al_6(BO_3)_3(Si_6O_{18})(OH,F)_4$	Forms in igneous and metamorphic rocks.
Turquoise	$CuAl_6(PO_4)_4(OH)_8 \cdot 4H_2O$	Found in copper-bearing rocks; a popular jewelry gem in the American Southwest.

form inorganically in rocks, pearls form in living oysters when the oyster extracts calcium and carbonate ions from water and precipitates them around an impurity, like a sand grain, embedded in its body. Most pearls used in jewelry today are "cultured" pearls, made by artificially introducing round sand grains into oysters in order to stimulate round pearl production. Amber is also formed by organic processes—it consists of fossilized tree sap. But because amber consists of organic compounds, it does not meet the definition of a mineral.

Rare means hard to find, and some gemstones are indeed hard to find. Most diamond localities, for example, occur in isolated regions of Congo, South Africa, Brazil, Canada, Russia, India, and Borneo (see Box 5.2). In some cases, it is not the mineral itself but rather the "gem-quality" versions of the mineral that are rare. For example, garnets are found in many rocks in such abundance that people use them as industrial abrasives. But most garnets are quite small and contain **inclusions** (specks of other minerals and/or bubbles) or fractures, so they are not particularly

Where Do Diamonds Come From?

BOX 5.2

THE REST OF THE STORY

As we saw earlier, diamond consists of the element carbon. Accumulations of carbon develop in a variety of ways: soot (pure carbon) results from burning plants at the surface of the Earth; coal (which consists mostly of carbon) forms from the remains of plants buried to depths of up to 15 km; and graphite develops from coal or other organic matter buried to still greater depths (15–70 km) in the crust during mountain building. Experiments demonstrate that the temperatures and pressures needed to form diamond are so extreme that, in nature, they generally occur only at depths of around 150 km below the Earth—that is, in the mantle. Under these conditions, the carbon atoms that were arranged in hexagonal sheets in graphite rearrange to form the much stronger and more compact structure of diamond. (Of note, engineers can duplicate these conditions in the laboratory; corporations manufacture several tons of diamonds a year.)

How does carbon get down into the mantle, where it transforms into diamond? Geologists speculate that the process of subduction provides the means. Carbon-containing rocks and sediments formed on oceanic lithosphere plates at the Earth's surface can be carried into the mantle at a convergent plate boundary. This carbon transforms into diamond, some of which becomes trapped in the lithospheric mantle beneath continents. But if diamonds form in the mantle, then how do they return to the surface? One possibility is that the process of rifting cracks the continental crust and causes a small part of the underlying lithospheric mantle to melt. Magma generated during this process rises to the surface, bringing the diamonds with it. Near the surface, the magma cools and solidifies to form a special kind of igneous rock called **kimberlite** (named for Kimberley, South Africa, where it was first found); some of the crystals embedded in this rock are diamonds. Kimberlite magma contains a lot of

dissolved gas and thus froths to the surface very rapidly. Kimberlite rock commonly occurs in carrot-shaped bodies 50–200 m across and at least 1 km deep, that are called **kimberlite pipes** (▶Fig. 5.24a).

Controversial measurements suggest that many of the diamonds that sparkle on engagement rings today were created when subduction carried carbon into the mantle 3.2 billion years ago. The diamonds sat at depths of 150 km in the Earth until two rifting events, one of which took place in the late Precambrian and the other during the late Mesozoic era, released them to the surface, like genies out of a bottle. The Mesozoic rifting event led to the breakup of Pangaea.

In places where diamonds occur in solid kimberlite, they can be obtained only by digging up the kimberlite and crushing it, to separate out the diamonds (▶Fig. 5.24b). But nature can also break diamonds free from the Earth. In places where kimberlite has been exposed at the ground surface for a long time, the rock chemically reacts with water and air (a process called weathering; see Chapter 7). These reactions cause most minerals in kimberlite to disintegrate, creating sediment that washes away in rivers. Diamonds are so strong that they remain as solid grains in river gravel. Thus, many diamonds have been obtained simply by separating them from recent or ancient river gravel.

Diamond-bearing kimberlite pipes are found in many places around the world, particularly where very old continental lithosphere exists, among them southern and central Africa, Siberia, northwestern Canada, India, Brazil, Borneo, Australia, and the U.S. Rocky Mountains. Rivers and glaciers, however, have transported diamond-bearing sediments great distances from their original sources. In fact, diamonds have even been found in farm fields of the Midwest. Not all natural diamonds are valuable: value depends on color and clarity. Diamonds that contain imperfections (cracks, or specks of other material) or are dark gray in color won't be used for jewelry. These stones, called industrial diamonds, are used instead as abrasives, for diamond powder is so hard (10 on the Mohs hardness scale) that it can be used to grind away any other substance.

Gem-quality diamonds come in a range of sizes. Jewelers measure diamond size in carats, where one **carat** equals 200 milligrams (0.2 grams)—one ounce equals 142 carats. (Note that a carat measures gemstone weight, while a **karat** specifies the purity of gold. Pure gold is 24 karat, while 18-karat gold is an alloy containing 18 parts of gold and 6 parts of other metals.) The largest diamond ever found, a stone called the Cullinan Diamond, was discovered in South Africa in 1905. It weighed 3,106 carats (621 grams) before being cut into nine large gems (the largest weighing 516 carats) as well as many smaller ones. By comparison, the diamond on a typical engagement ring weighs less than 1 carat. Diamonds are rare, but not as rare as their price suggests. A worldwide consortium of diamond producers stockpile the stones so as not to flood the market and drive the price down.

FIGURE 5.24 (a) A mine in a kimberlite pipe. (b) A raw diamond still imbedded in kimberlite.

(a)

(b)

beautiful. Gem-quality garnets, however—clean, clear, large unfractured crystals—are unusual. In some cases, gemstones are merely pretty and rare versions of more common minerals. For example, ruby is a special version of the common mineral corundum, and emerald is a special version of the common mineral beryl (▶Fig. 5.25). As for the beauty of a gemstone, this quality lies basically in its color and, in the case of transparent gems, its **fire**—the way the stone bends the light passing through it and disperses the light into a spectrum. Fire makes a diamond sparkle more than a similarly cut piece of glass.

As the table shows, gemstones are created in a huge variety of ways. Some solidify from a melt along with other minerals of igneous rock, some form by diffusion in a metamorphic rock, some precipitate out of a water solution in cracks, and some are a consequence of the chemical interaction of rock with water near the Earth's surface. Many gems come from **pegmatites,** particularly coarse-grained igneous rocks formed by the solidification of steamy melt.

Most gems when used in jewelry are "cut" stones. The smooth **facets** on a gem are ground and polished surfaces made with a faceting machine (▶Fig. 5.26a). Facets are not the natural crystal faces of the mineral, nor are they cleavage planes, though gem cutters sometimes make the facets parallel to cleavage directions and will try to break a large gemstone into smaller pieces by splitting it on a cleavage plane. A faceting machine consists of a doping arm, a device that holds a stone in a specific orientation, and a **lap,** a ro-

FIGURE 5.25 Beryl crystals in rock.

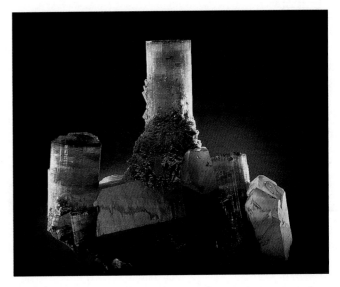

FIGURE 5.27 A spectacular museum specimen of a mineral cluster. The arrangement of colors and shapes is like abstract art.

tating disk covered with a wet paste of grinding powder and water. The gem cutter fixes a stone to the end of the doping arm and positions the arm so that it holds the stone against the moving lap. The movement of the lap grinds a facet. When the facet is complete, the gem cutter rotates the arm by a specific angle, lowers the stone, and grinds another facet. The geometry of the facets defines the cut of the stone. Different cuts have different names, like "brilliant," "French," "star," "pear," and "kite." Grinding facets is a lot of work—a typical engagement-ring diamond with a brilliant cut has fifty-seven facets (▶Fig. 5.26b)!

Some mineral specimens have special value simply because their geometry and color before cutting are beautiful. Prize specimens exhibit shapes and colors reminiscent of fine art and may sell for tens of thousands of dollars (▶Fig.

5.27). It's no wonder that mineral "hounds" risk their necks looking for a cluster of crystals protruding from the dripping roof of a collapsing mine or hidden in a crack near the smoking summit of a volcano.

CHAPTER SUMMARY

- Minerals are homogeneous, naturally occurring, solid inorganic substances with a definable chemical composition and an internal structure characterized by an orderly arrangement of atoms, ions, or molecules in a lattice.

- In the crystalline lattice of minerals, atoms occur in a specific pattern—one of nature's finest examples of ordering.

- Minerals can form by the solidification of a melt, precipitation from a water solution, or diffusion through a solid.

- There are over 3,000 different types of minerals, each with a name and distinctive physical properties (color, streak, luster, hardness, specific gravity, crystal form, crystal habit, and cleavage).

- The unique physical properties of a mineral reflect its chemical composition and crystal structure. By observing these physical properties, you can identify minerals.

- The most convenient way for classifying minerals is to group them based on their chemical composition. Mineral classes include: silicates, oxides, sulfides, sulfates, halides, carbonates, and native metals.

- The silicate minerals are the most common on Earth. The silicon-oxygen tetrahedron, a silicon atom surrounded

FIGURE 5.26 The shiny faces on gems in jewelry are made by a faceting machine. (a) In this faceting machine, the gem is held against the face of the spinning lap. (b) Top and side views show the many facets of a brilliant-cut diamond.

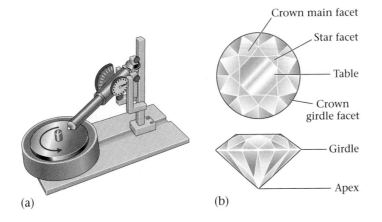

Crown main facet

Star facet

Table

Crown girdle facet

Girdle

Apex

(a)   (b)

by four oxygen atoms, is the fundamental building block of silicate minerals.

• There are several groups of silicate minerals, distinguished from one another by the ways in which the silicon-oxygen tetrahedra that comprise them are linked.

• Gems are minerals known for their beauty and rarity. The facets on cut stones used in jewelry are made by grinding and polishing the stone with a faceting machine.

KEY TERMS

anhedral grains (p. 117)
carat (p. 126)
carbonates (p. 122)
cleavage (p. 119)
conchoidal fracture (p. 119)
crystal (p. 113)
crystal faces (p. 113)
crystal form (p. 119)
crystal habit (p. 119)
crystal lattice (p. 112)
diffraction (p. 114)
euhedral crystal (p. 117)
facet (p. 126)
gem (p. 124)
gemstone (p. 124)
geode (p. 117)
halides (p. 122)
industrial minerals (p. 110)
kimberlite (p. 125)
luster (p. 118)
mineral (p. 111)

mineralogy (p. 110)
Mohs hardness scale (p. 119)
ore minerals (p. 110)
oxides (p. 121)
pegmatite (p. 126)
polymorph (p. 115)
precious stone (p. 124)
precipitation from a solution (p. 116)
saturated (p. 117)
seed (p. 116)
semiprecious stone (p. 124)
silicates (p. 121)
silicon-oxygen tetrahedron (p. 121)
solid-state diffusion (p. 116)
solidification of a melt (p. 115)
specific gravity (p. 119)
streak (p. 118)
sulfates (p. 121)
sulfides (p. 121)

REVIEW QUESTIONS

1. What is a mineral, as geologists understand the term? How is this definition different from the everyday usage of the word?

2. Why is glass not a mineral?

3. Salt is a mineral, but sugar is not. Why not? Is pepper a mineral?

4. Diamond and graphite have an identical chemical composition (pure carbon), yet differ radically in physical properties. Explain in terms of their crystal lattices.

5. In what way does the arrangement of atoms in a mineral define a pattern? How can X-rays be used to study these patterns?

6. Describe the three ways that mineral crystals can form.

7. Why do some minerals contain beautiful euhedral crystals, while others contain anhedral grains?

8. List the principal physical properties used to identify a mineral.

9. How can you determine the hardness of a mineral? What is the Mohs hardness scale?

10. How do you distinguish cleavage surfaces from crystal faces on a mineral?

11. What is the prime characteristic that geologists use to separate minerals into classes?

12. On what basis are silicate minerals further divided into distinct groups?

13. What is the relationship between the way in which silicon-oxygen tetrahedra bond in micas and the characteristic cleavage of micas (the way they split into sheets)?

14. How do sulfate minerals differ from sulfides?

15. Look at the periodic table of the elements in Appendix A. Note that carbon and silicon are in the same column (meaning that they both have ±4 charges and tend to form complex structures with covalent bonds). Why, then, are the most common minerals silicates?

16. Why are some minerals considered gems? How do you make the facets on a gem?

SUGGESTED READING

Berry, L. G., B. Mason, and R. V. Dietrich. 1983. *Mineralogy.* San Francisco: Freeman.

Cepeda, J. C. 1994. *Introduction to Minerals and Rocks.* New York: Macmillan.

Dietrich, R. V., and B. J. Skinner. 1979. *Rocks and Rock Minerals.* New York: Wiley.

———. 1990. *Gems, Granites, and Gravels: Knowing and Using Rocks and Minerals.* New York: Cambridge University Press.

Ernst, W. G. 1969. *Earth Materials.* Englewood Cliffs, N.J.: Prentice-Hall.

Hochleitner, R. 1994. *Minerals: Identifying, Classifying, and Collecting Them.* Hauppauge, N.Y.: Barron's Educational Series.

Klein, C., and C. S. Hurlbut, Jr. 1993. *Dana's Manual of Mineralogy.* 21st ed. New York: Wiley.

Zoltai, T., and J. H. Stout. 1984. *Mineralogy: Concepts and Principles.* Minneapolis: Burgess.

Rock Groups

In the Grand Canyon of Arizona, a river has sliced into the Earth, exposing a
great variety of rocks.

A.1 INTRODUCTION

During the 1849 gold rush in the Sierra Nevada Mountains
of California, only a few lucky individuals actually became
rich. The rest of the "forty-niners" either slunk home in
debt or took up less glamorous jobs in boomtowns like San
Francisco. Soon, the American West Coast was demanding
large quantities of manufactured goods from East Coast factories. Making the goods was no problem, but getting them
to California meant either a stormy voyage around the
southern tip of South America or a trek with stubborn mule
teams through the deserts of Nevada or Utah. The time was
ripe to build a railroad linking the East and West Coasts of
North America, and, with much fanfare, the Central Pacific

line decided to punch one right through the
peaks of the Sierras. In 1863, while the Civil
War raged elsewhere in the United States, the
company transported six thousand Chinese laborers across the Pacific in the squalor of un-
ventilated cargo holds, and set them to work
chipping ledges and blasting tunnels. Foremen
measured progress in terms of feet per day—if
they were lucky. Along the way, untold numbers of laborers died of frostbite, exhaustion,
mistimed blasts, landslides, or avalanches.

Through their efforts, the railroad laborers
certainly gained an intimate knowledge of how
rock feels and behaves—it's solid, heavy, and
hard! They also found that some rocks seemed
to break easily into layers, while others did not,
and some rocks were dark while others were
light. They realized, like anyone who looks
closely at rock exposures, that rocks are not just
gray, featureless masses, but rather come in a
great variety of colors and textures. Why are
there so many distinct types of rocks? The answer is simple: there are many different ways in
which rock can form, and many different materials out of which rocks can be made. Because
of the relationship between rock type and the
process of formation, *rocks provide a historical
record of geological events*. The next few chapters are devoted
to a discussion of rocks and a description of how rocks
form; this interlude serves as a general introduction. Here,
we learn what the term "rock" means to geologists, and
how to distinguish the three principal groups of rocks. We
also look at how geologists study rocks.

A.2 WHAT IS ROCK?

To geologists, **rock** is a coherent, naturally occurring solid,
consisting of an aggregate of minerals or a mass of glass.
Now let's take this definition apart.

- *Coherent:* A pile of unattached grains that can move around one another does not comprise a rock. A rock holds together, and it must be broken to separate into pieces. As a result of its coherence, rock can form cliffs or can be carved into sculptures.

- *Naturally occurring:* Geologists consider only naturally occurring materials to be rocks, so manufactured materials like concrete and brick do not qualify. (As a minor point, the term **stone** usually refers to rock used as construction material.)

- *An aggregate of minerals or a mass of glass:* The vast majority of rocks consist of an aggregate (a collection) of many mineral grains, or crystals, stuck or grown together. Technically, a single mineral crystal is simply a **mineral specimen,** not a rock, even if it is meters long. Some rocks contain only one kind of mineral, while others contain several different kinds. A few of the rock types that form at volcanoes (see Chapter 6) consist of glass, which may occur either as a homogeneous mass or as an accumulation of tiny shards (flakes).

What holds rock together? Grains in nonglassy rock stick together to form a coherent mass either because they are bonded by natural **cement,** mineral material that precipitates from water and fills the space between grains (▶Fig. A.1a–c), or because they interlock with one another like pieces in a jigsaw puzzle (▶Fig. A.2a–d). Rocks whose grains are stuck together by cement are called **clastic,** while rocks whose crystals interlock with one another are called **crystalline.**

A.3 ROCK OCCURRENCES

At the surface of the Earth, rock occurs either as broken chunks (pebbles, cobbles, or boulders; see Chapter 7) that have moved from their point of origin by falling down a slope or after being transported in ice, water, or wind; or as **bedrock,** which is still attached to the Earth's crust. Geologists refer to an exposure of bedrock as an **outcrop.** An outcrop may appear as a rounded knob out in a field, as

(a)

(b)

FIGURE A.1 (a) A hand specimen (fist-sized chunk) of sandstone. (b) A magnified image of the sandstone shows that it consists of round white sand grains, surrounded by cement. (c) This exploded image of the rock emphasizes how the cement surrounds the sand grains.

(c)

FIGURE A.2 (a) A hand specimen of granite, a rock formed when melt cools underground. (b) The texture of granite is different from that of sandstone. In granite, the grains interlock with one another, like pieces of a jigsaw puzzle. (c) An artist's sketch emphasizes the irregular shapes of grains and how they interlock. (d) This exploded image highlights the grain shapes.

a ledge along a cliff or ridge, on the face of a stream cut (where a river has cut down into bedrock), or along human-made road cuts and excavations (▶Fig. A.3a–d).

To people who live in cities or forests or on farmland, outcrops of bedrock may be unfamiliar, as they may be covered by vegetation, sand, mud, gravel, soil, water, asphalt, concrete, or buildings. Outcrops are particularly rare in regions like the midwestern United States, where, during the past million years, ice-age glaciers melted and left behind thick deposits of gravel and debris (see Chapter 22). These deposits completely buried preexisting valleys and hills, so today the bedrock surface lies as deep as 100 m below the ground. The depth of bedrock plays a key component in urban planning, because architects prefer to set the foundations of large buildings on bedrock rather than on loose sand or mud. Because of this preference, the skyscrapers of New York City rise in two clusters on the island of Manhattan, one at the south end and the other in the center, locations where bedrock lies close to the surface.

A.4 THE BASIS OF ROCK CLASSIFICATION

People have developed classification schemes for just about every group of materials on the planet—for insects, trees, airplanes, books, and so on. Why? Classification schemes help us organize information and remember significant details about materials, and they help us recognize similarities and differences between them. After struggling with the issue of rock classification for over a century, geologists finally realized that the most useful way to classify rocks is according to how a rock formed. We call this scheme a **genetic classification,** because it focuses on the genesis—the origin—of the rock.

In the contemporary genetic classification of rocks, geologists recognize three basic groups: (1) **igneous rocks,** which form by the freezing (solidification) of molten rock, or melt (▶Fig. A.4a); (2) **sedimentary rocks,** which form

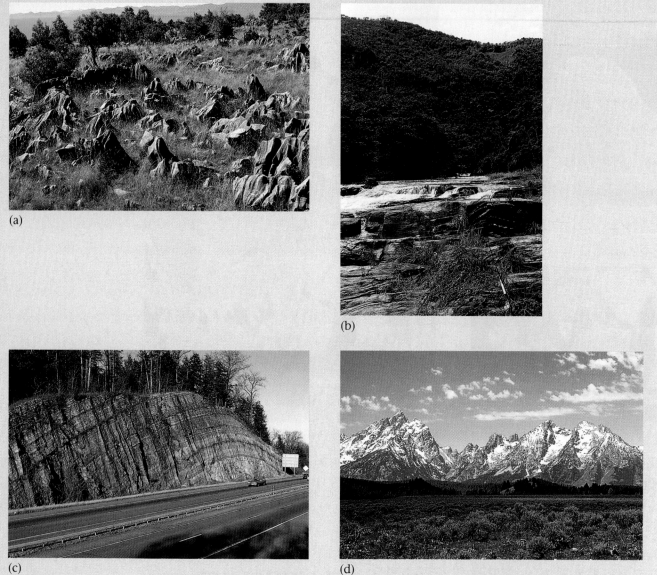

FIGURE A.3 (a) Outcrops (natural rock exposures) in a field. (b) A stream cut, meaning an outcrop that forms when a stream's flow removes overlying soil and vegetation. Note that dense forest covers most of the nearby hills, obscuring outcrops that may be exposed there. (c) Road cuts, like this one along a highway near Kingston, New York, are made by setting off dynamite placed at the bottom of drill holes. Note that the layers of rock exposed in this road cut are curved—such a bend is called a fold (see Chapter 11). (d) Mountain cliffs provide immense exposures of rock. These cliffs, in the Grand Teton Mountains of Wyoming, rise above a lowland in which bedrock has been covered by a layer of sediment, which hosts fields of sagebrush.

either by the cementing together of fragments (grains) broken off preexisting rocks or by the precipitation of mineral crystals out of water solutions at or near the Earth's surface (►Fig. A.4b); and (3) **metamorphic rocks,** which form when preexisting rocks change into new rocks in response to an increase in pressure and temperature (►Fig. A.4c). Metamorphic change occurs in the "solid state," which means that it does not require melting. Each of the three

groups contains many different individual rock types, distinguished from one another by physical characteristics such as

• *grain size:* The dimensions of individual **grains** (fragments of mineral crystals) in a rock may be measured in millimeters or centimeters. Some grains are so small that they can't be seen without a microscope, while others

(a)

(b)

(c)

FIGURE A.4 Examples of the major rock groups. (a) The volcano in the background erupted lava (molten rock) that flowed over the landscape and eventually froze, forming black basalt, an igneous rock. (b) Sand, deposited on a beach, eventually becomes buried to form layers of sandstone, a sedimentary rock, such as those exposed in the cliffs behind the beach. (c) When preexisting rocks become buried deeply during mountain building, the increase in temperature and pressure transforms them into metamorphic rocks. The lichen-covered outcrop in the foreground was once deeply buried beneath a mountain range, but was later exposed when glaciers and rivers stripped off the overlying rocks of the mountain.

are as big as a fist. Some grains are **equant,** meaning that they have the same dimensions in all directions, while some are **inequant,** meaning that the dimensions are not the same in all directions (▶Fig. A.5a–c). In some rocks, all the grains are the same size, while other rocks contain a variety of different-sized grains.

- *composition:* Not all rocks consist of the same kinds of minerals. Rock composition refers to the proportions of different minerals comprising the rock. The proportions of minerals, in turn, defines the proportions of chemicals comprising the rock.

- *texture:* This term refers to the arrangement of grains in a rock and the way grains connect to one another. The concept of rock texture will become easier to grasp as we look at different examples of rocks in the following chapters.

- *layering:* Some rock bodies appear to contain distinct layering, defined either by bands of different compositions or textures, or by the alignment of inequant grains so that they parallel one another. Different types of layering occur in different kinds of rocks. For example, the layering in sedimentary rocks is called **bedding,** while the layering in metamorphic rocks is called **metamorphic foliation** (▶Fig. A.6a, b).

Each individual rock type has a name, based on the dominant component making up the rock, on the region where the rock was first discovered or is particularly abundant, on a root word of Latin origin, or on a traditional name used by people in an area where the rock is found. All told, there are hundreds of different rock names, though in this book we learn only about thirty.

FIGURE A.5 (a) This boulder of metamorphic rock is an aggregate of mineral grains. (b) At high magnification, we can see that the rock consists of both equant and inequant grains. (c) Using this comparison chart, we can measure the size of the grains, in millimeters.

A.5 STUDYING ROCK

Outcrop Observations

The study of rocks begins by examining a rock in an outcrop. If the outcrop is big enough, such an examination will reveal relationships between the rock you're interested in and the rocks around it, and will allow you to detect layering. Geologists carefully record observations about an outcrop, then break off a **hand specimen,** a fist-sized piece, which they can examine more closely with a **hand lens,** or magnifying glass. Observation with a hand lens allows geologists to identify sand-sized or larger mineral grains, and may allow them to describe the texture of the rock.

Thin-Section Study

Geologists often must examine rock composition and texture in minute detail in order to identify a rock and develop a hypothesis for how it formed. To do this, they take a specimen back to the lab and make a very thin slice (about 3/100 mm thick, the thickness of a human hair), or **thin section,** of the rock (▶Fig. A.7a–c), which they examine with a **petrographic microscope** ("petro" comes from the Greek word for "rock"; geologists who specialize in the study of rocks are called **petrologists**). A petrographic microscope differs from an ordinary microscope in that it illuminates the thin section with **transmitted polarized light.** This means that the illuminating light beam first passes through a special filter that makes all the light waves in the beam vibrate in the same plane, and then passes up through the thin section; an observer can then look through the thin section as if it were a window.

When illuminated with transmitted polarized light, each type of mineral grain displays a unique suite of colors. The

FIGURE A.6 Layering in rocks. (a) Bedding in a sedimentary rock, in this case defined by alternating layers of sand and gravel. (b) Foliation in a metamorphic rock, in this case defined by alternating light and dark layers.

FIGURE A.7 (a) To prepare a thin section, a geologist cuts a brick-shaped chip out of rock using a rock saw (a rotating circular blade with a diamond-studded rim). (b) The chip is glued to a slide and then ground down until it's paper thin. (c) The thin section is then labeled and ready to examine.

specific color the observer sees depends on both the identity of the grain and on its orientation with respect to the waves of polarized light—a crystal lattice interferes with polarized light and allows only certain colors to pass through.

The brilliant colors and strange shapes in a thin section rival the beauty of an abstract painting or stained glass. By examining a thin section with a petrographic microscope, geologists can identify most of the minerals comprising the rock, and can describe the way in which the grains connect to one another (▶Fig. A.8a).

Note that to make a thin section, a geologist first uses a special saw with a spinning diamond-studded blade, a **rock saw,** to cut a small rectangular block, or chip, of rock. Rock saws work by grinding their way through rock—the hard diamonds em-

bedded in the saw blade scratch and pulverize minerals as the saw blade rubs against the rock. The geologist then cements the chip to a glass microscope slide with an epoxy adhesive (glue). By using a **lap,** a spinning plate coated with abrasive, the geologist can grind the chip down until only a thin slice, still cemented to the glass slide, remains, ready for examination with a petrographic microscope.

The concept of looking *through* rock may seem strange at first, because hand specimens of most rocks appear opaque. But people have known for centuries that many kinds of rock, if fashioned into thin slices, can be made transparent or at least translucent. In fact, medieval architects used thin slices of rock as windows in mosques.

High-Tech Analytical Equipment

Beginning in the 1950s, high-tech electronic instruments became available that allowed petrologists to examine rocks at an even finer scale than was permitted by the petrographic microscope. Modern research laboratories typically boast instruments like **electron microprobes,** which can focus a beam of electrons on a small part of a grain to create a signal that defines the chemical composition of the mineral (▶Fig. A.8b); **mass spectrometers,** which analyze the proportions of isotopes (see Appendix A) of a particular element contained in a rock; and **X-ray diffractometers,** which identify minerals by looking at the way X-ray beams pass through crystals in a rock. Such instruments, in conjunction with optical examination, can provide petrologists with highly detailed characterizations of rocks, which in turn help them understand how the rocks formed and where they came from. This information allows geologists to use the study of rocks as a basis for deciphering Earth history.

FIGURE A.8 (a) A rock photographed through a petrographic microscope. The colors are caused by the interaction of polarized light with the crystals. The long dimension of this photo is 2 mm. (b) An electron microprobe uses a beam of electrons to analyze the chemical composition of minerals.

CHAPTER 6

Up from the Inferno: Magma and Igneous Rocks

6.1 INTRODUCTION

Every now and then, an incandescent liquid—molten rock, or **melt**—begins to fountain from a crater (pit) or crack on the big island of Hawaii, for Hawaii is a **volcano,** a vent at which melt from inside the Earth spews out onto the planet's surface. Some of the melt, called **lava,** lies in pools around the vent, while the rest flows down the mountainside as a syrupy red-yellow stream called a **lava flow.** Near its source, the flow moves swiftly, cascading over escarpments at speeds of over 30 km per hour (▶Fig. 6.1a). At the base of the mountain, the lava flow slows, but advances nonetheless, engulfing any roads, houses, or vegetation in its path. At the edge of the flow, beleaguered plants incinerate in a burst of flames. As the flow cools, it slows down, and its surface darkens and crusts over, occasionally breaking to reveal the hot, sticky mass within (▶Fig. 6.1b). Finally, it stops moving entirely, and within a few days the once red-hot melt has become a hard, black solid through and through (▶Fig. 6.1c). A new **igneous rock,** made by the *freezing* of a melt, has formed. Considering the fiery heat of the lava from which igneous rocks develop, the name "igneous"—from the Latin *ignis,* meaning "fire"—makes sense.

It may seem strange to speak of freezing here, when most people think of freezing as the transformation of liquid water to solid ice on the surface of a lake when the temperature drops below 0°C (32°F). Nevertheless, the freezing of liquid rock to form solid igneous rock represents the same phenomenon, except that igneous rocks freeze at temperatures between 650°C and 1,100°C; to put such temperatures in perspective, remember that home ovens only attain a maximum temperature of 260°C (500°F).

Igneous rocks are the most abundant kind on Earth, for they comprise all of the mantle, all of the oceanic crust, and much of the continental crust. The oldest igneous rocks formed when the planet was still young, for after it coalesced out of

A river of molten rock (lava) weaves across a stark terrain of already solidified igneous rock. The volcano, the vent from which the lava has spilled out onto Earth's surface from the interior, can be seen in the distance.

(a)

(b)

(c)

FIGURE 6.1 (a) Lava fountains in this crater of a volcano on Hawaii, and a river of lava streams out of a gap in its side. As the lava moves rapidly away from the crater, it cools, and a black crust forms on the surface. (b) Farther down the mountain, the surface of the lava has completely crusted over with newborn rock, while the insides of the flow remain molten, allowing it to creep across the highway (in spite of the stop sign). Smoke comes from burning vegetation. (c) Eventually the flow cools through and through, and a new layer of basalt rock has formed. This rock is only a few weeks old.

planetesimals, the Earth became so hot that it began to melt. The heat came from four sources.

- As the Earth grew, gravity compressed, or squeezed together, the matter making up the planet; and when materials undergo compression, they heat up.

- Heavier materials, like iron alloy, sank toward the center of the planet; their potential energy transformed into kinetic energy, which converted to heat.

- Countless meteors, residual planetesimals pulled toward the Earth by gravity, slammed into the young planet. Their kinetic energy also converted to heat.

- The Earth contains radioactive elements that, when they decay, generate new heat (see Appendix A).

For a time, the surface of the planet may have turned into a sea of melt. But eventually, the Earth cooled suffi-ciently for its surface temperature to drop below the freez-ing point of rock. When this happened, a thin skin of ig-neous rock formed on its surface. Soon, the entire mantle froze, becoming a thick shell of solid, though still warm and plastic, igneous rock. Thus, igneous rocks were the first rocks to form on the Earth. But did they all originate early in Earth history? Clearly no. Igneous rocks continue to form today, for even though the crust and mantle are mostly solid, melt still can be generated at special locations inside the Earth.

Although some igneous rocks solidify at the surface dur-ing volcanic eruptions like those on Hawaii, a vastly greater volume result from solidification of melt underground, out of sight. Geologists refer to melt that exists below the Earth's surface as **magma,** and melt that has erupted from a volcano at the surface of the Earth as **lava.** Further, geolo-gists call rock made by the freezing of magma underground,

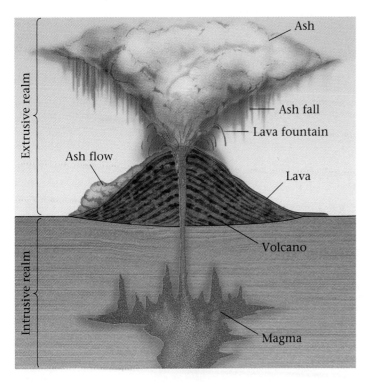

FIGURE 6.2 Extrusive igneous rocks, namely ash and lava, form above the Earth's surface, while intrusive rocks develop below. Melt that erupts from a volcano is lava, while underground melt is magma.

after it has pushed its way ("intruded") into preexisting rock of the crust, **intrusive igneous rock,** and rock that forms by the freezing of lava above ground, after it spills out, or "extrudes," onto the surface of the Earth and comes into contact with the atmosphere or ocean, **extrusive igneous rock** (▶Fig. 6.2). Extrusive igneous rock includes rock formed when lava flows solidify and **pyroclastic debris,** particles formed from lava that sprayed or exploded into the atmosphere. Pyroclastic debris includes fine particles of glass called **ash,** as well as larger pieces called **bombs** and **cinders.**

A great variety of igneous rocks exist on Earth. To understand why and how these rocks form, and why there are so many different kinds, we must first understand why magma forms, why it rises, why it sometimes erupts as lava, how it freezes in intrusive and extrusive environments, and how it transforms into rock. We then look at the scheme that geologists use to classify igneous rocks.

6.2 THE FORMATION OF MAGMA

As mentioned previously, the popular image that the solid crust of the Earth floats on a sea of molten rock is not correct. In reality, magma forms in the lithosphere and crust,

but only in special places where preexisting solid rock melts. Following are the conditions that lead to melting.

Melting as a result of a decrease in pressure (decompression). The Earth is quite hot inside, for the heat that developed within the planet during its birth radiates into space only very slowly, and radioactivity continues to generate new heat. At a depth of 4 km beneath typical continental crust, temperatures reach 100°C; at a depth of 35 km (that is, at the continental Moho), temperatures reach 500°–600°C; and at a depth of about 100–150 km (the base of the lithosphere), temperatures reach 1,280°C. The change in temperature with depth, the **geotherm,** can be expressed on a graph by a line. The geotherm in ▶Figure 6.3 indicates that beneath typical continental crust, temperatures comparable to those of lava (650°–1,100°C) generally occur in the upper mantle. But even though the upper mantle is very hot, its rock stays solid because it is also under high pressure from the weight of overlying rock. That is, thermal energy causes rock to melt because thermal vibration breaks apart the lattice of mineral crystals,

FIGURE 6.3 The graph plots the Earth's geotherm (solid line) and the melting curve for mantle rock (peridotite; dashed line). Note that the geothermal gradient (the rate of change in temperature), indicated by the geotherm, decreases with greater depths; if it were constant, the geotherm would be a straight line. A rock that starts at point A and moves to point B (that is, stays the same temperature but feels a decrease in pressure) will begin to melt, a process called decompression melting.

Mantle plume and
a hot-spot volcano

Subduction yields
a volcanic arc.

Melting occurs beneath
a mid-ocean ridge.

Melting occurs beneath
a continental rift.

Large volumes of magma erupt at a hot-spot volcano on
the oceanic crust, creating an oceanic island.

The Formation of Igneous Rocks

Molten rock, or melt, develops only in special locations in
the Earth: where a plume of hot mantle rock rises to the
base of the lithosphere (a volcano above such a plume is a
hot-spot volcano); in the asthenosphere above subducting
oceanic lithosphere at a convergent plate boundary (the
chain of volcanoes that results is a volcanic arc); in the
asthenosphere beneath a mid-ocean ridge, or divergent
plate boundary; and along a continental rift. While the
melt remains underground, it is called magma, but when
the melt spills out of a volcano, it is called lava.

When magma or lava cools, different minerals form in
sequence until the melt solidifies (freezes), and igneous rock
forms. The composition of a melt depends on its origin and
cooling history. For example, partial melting of the mantle
results in basaltic magma. Basaltic magma is very hot, so
when it rises into the continental crust, it can transfer heat
and cause partial melting of the crust, yielding rhyolitic
magma. Lava or magma that cools quickly tends to be fine-
grained or glassy, while lava or magma that cools slowly
tends to be coarse-grained.

Igneous rock that forms by the solidification of magma
underground is intrusive rock. Blob-like intrusions are
plutons, while sheet-like intrusions are dikes (if they cut
across preexisting layers, and sills if they intrude parallel to
preexisting layers; in rock containing no preexisting layers,
dikes are vertical and sills are horizontal. Intrusions that are
shaped like a blister are called laccoliths.

Volcanoes erupt both lava flows and pyroclastic debris
(ash and other fragmental material ejected explosively).
Igneous rock that forms by the extrusion of lava or the
accumulation of pyroclastic debris is called extrusive rock.

Less silica   More silica

Granite forms from the cooling of a silicic melt, such as in a
continental pluton, while basalt results from the cooling of
a mafic melt, as at an oceanic hot-spot volcano.

The Extrusive Environment

Dikes

Pyroclastic debris

Lava flow

Laccolith

Volcanic neck

Ring dikes

Country rock

Sills

Pluton

Magma chamber

The Intrusive Environment

If you examine granite with a microscope, you'll see that it consists of interlocking crystals of several minerals. We call this a crystalline texture.

The texture of an igneous rock depends on the cooling rate.

Fast-cooling

Obsidian (glassy)

Rhyolite (fine-grained)

Granite (coarse-grained)

Slow-cooling

Minerals in an igneous rock form in succession as the melt cools.

Hotter

Cooler

but pressure counters this effect by squeezing the atoms together, thereby preventing them from breaking free.

Because pressure prevents melting, a decrease in pressure can permit melting. Specifically, if the pressure affecting hot mantle rock decreases while the temperature remains unchanged, some of the minerals that comprise the mantle melt and a magma forms. This kind of melting, called **decompression melting** (see Box 6.1), occurs where hot mantle rock rises to shallower depths in the Earth, because pressure decreases toward the surface, and rock is such a good insulator that it doesn't lose much heat as it rises (▶Fig. 6.4a).

Melting as a result of the addition of volatiles. Magma also forms at locations where chemicals called volatiles mix with hot mantle rock. **Volatiles** are elements or compounds, such as water (H_2O) and carbon dioxide (CO_2), that evaporate easily and can exist in gaseous forms at the Earth's surface. When volatiles mix with hot rock, they help break the bonds that fix atoms to the surfaces of solid mineral crystals, so the addition of volatiles decreases the melting temperature—if you add volatiles to a solid, hot dry rock, the rock begins to melt (▶Fig. 6.4b). We can represent the effect of volatiles by looking at the contrast between the **melting curve** (the line defining the range of temperatures and pressures at which a rock melts) for wet basalt and that for dry basalt (▶Fig. 6.4c). Note that wet basalt (basalt containing volatiles) melts at much lower temperatures than dry (volatile-free) basalt.

Melting as a result of heat transfer from rising magma. When rock, or magma, from the mantle rises up into the crust, it brings heat with it. This heat flows into and raises the temperature of the surrounding crustal rock. In some cases, the rise in temperature may be sufficient to melt the crustal rock (Fig. 6.4a). Imagine injecting hot fudge into ice cream; the fudge transfers heat to the ice cream, raises its temperature, and causes it to melt. We call such melting **heat-transfer melting,** because it results from the transfer of heat from a hotter material to a cooler one. Since mantle-derived magmas are very hot (about 1,100°C) and rocks of the crust melt at temperatures of about 650°–850°C, when mantle-derived magma intrudes into the crust, it can raise the temperature of the surrounding crust enough to melt it.

6.3 WHAT IS MAGMA MADE OF?

All magmas contain silicon and oxygen, which bond to form the silicon-oxygen tetrahedron. But magmas also contain varying proportions of other elements like aluminum (Al), calcium (Ca), sodium (Na), potassium (K), iron (Fe), and magnesium (Mg). Because magma is a liquid, its atoms

---

BOX 6.1

THE REST OF THE STORY

Understanding Decompression Melting

Look again at Figure 6.3. The horizontal axis on the graph represents temperature, in degrees centigrade, and the vertical axis represents pressure. Since pressure in the Earth results from the weight of overlying rock, we can calibrate the vertical axis either in units of depth (km below the surface) or in units of pressure, such as bars. (Note: 1 bar approximately equals 1 atm, where 1 atm, or atmosphere, is the air pressure at sea level.)

The solid line on this graph is the **geotherm,** which defines the temperature as a function of depth in the Earth. Notice that the rate of increase in temperature, a quantity called the **geothermal gradient** (expressed in degrees per km), decreases with increasing depth. The dashed line represents the **melting curve** for mantle rock (peridotite). This line defines the conditions of pressure and temperature at which mantle rock begins to melt, as determined by laboratory measurements.

Values to the right of the dashed line correspond to possible conditions of pressure and temperature in which part of a mantle rock melts, whereas values to the left indicate pressures and temperatures for which the rock stays entirely solid.

To see what happens during decompression, imagine a mantle rock at a depth of 200 km (point A on the graph). According to the graph, the pressure ≈ 65,000 bars and the temperature ≈ 1,300°C at this depth. Now imagine that the rock moves closer to the Earth's surface without cooling, as in a rising mantle plume, and reaches point B, where the pressure is only about 35,000 bars. It has undergone decompression. Notice that point B lies on the melting curve, so the rock begins to melt—the thermal vibration of atoms in the rock, no longer countered by pressure, can cause the atoms to break free of crystals. Also notice that decompression melting takes place without adding new heat.

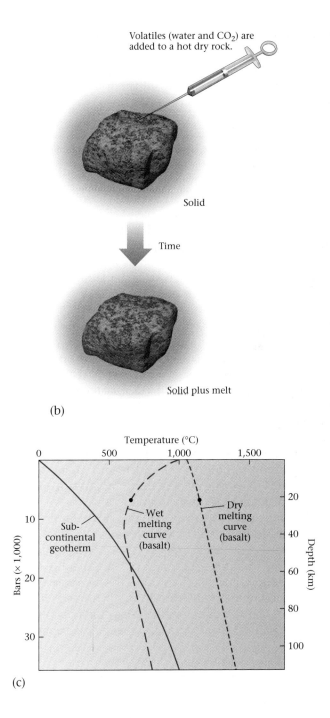

FIGURE 6.4 The three main causes of melting and magma formation in the Earth. (a) Decompression melting occurs when a hot rock rises to a shallow depth, where the pressure is less. Melting as a result of heat transfer happens when hot magma rises into rock that has a lower melting temperature. For example, hot basaltic magma rising from the mantle can make the surrounding intermediate-composition crust melt. (b) Melting as a result of the addition of volatiles occurs when compounds like water and carbon dioxide percolate into a solid hot rock. It's as if an "injection" of volatiles triggers the melting. (c) The addition of volatiles triggers melting because it decreases the melting temperature. For example, at a depth of 20 km, the melting temperature of wet basalt (basalt that contains volatiles) is about 500°C less than the melting temperature of dry basalt.

do not lie in an orderly crystalline lattice but are grouped instead in clusters or short chains, free to move with respect to one another.

Wet magmas include up to 15% dissolved volatiles such as water, carbon dioxide, nitrogen (N_2), hydrogen (H_2), and sulfur dioxide (SO_2). These volatiles come out of the Earth at volcanoes. Usually water constitutes about 50% of the gas erupted at a volcano, carbon dioxide about 20%. Thus, magma not only contains the elements that comprise solid minerals in rocks, it also contains molecules that become water or air.

The Major Types of Magma

Geologists distinguish between four types of magmas, based on the relative proportion of silica (SiO_2) to the sum of magnesium (Mg) and iron (Fe) oxides. **Silicic magmas** are particularly rich in silica (70%), with relatively little magnesium and iron. **Intermediate magmas** contain 55% silica. **Mafic magmas** are relatively rich in magnesium (hence the "ma") and iron (hence the "fic," from the Latin *ferrum*), with less silica (50%). **Ultramafic magmas** have even more magnesium and iron and even less silica (40%).

Typically, the temperature of a magma depends on its silica content: silicic magmas are the coolest (as low as 650°C), while ultramafic magmas are the hottest (over 1,100°C).

Why are there so many different types of magma? There are four main reasons.

Source rock composition. When you melt ice, you get water, and when you melt wax, you get liquid wax. There is no way to make water by melting wax. Clearly, the composition of a melt reflects the composition of the solid from which it was derived. Not all magmas form from the same source rock, so not all magmas have the same composition. Specifically, some magmas form in the upper mantle and some in the crust; magmas formed from crustal sources don't have the same composition as magmas formed from mantle sources, because the crust and mantle have different compositions to start with (see Chapter 2).

Partial melting. The melting of rock differs markedly from the melting of water or wax, in that water and wax contain only one kind of material, while most rocks consist of a variety of different minerals. Typically, when rock melts, only *some* of the minerals making up the rock melt, because each mineral has a different melting temperature. (To picture this, imagine you have a bowlful of chocolate chips mixed with plastic beads. If you stick the bowl in the oven at a low temperature, the chocolate melts but not the plastic. But in a very hot oven, both the chocolate and plastic melt.) In other words, magma generally forms by **partial melting,** not complete melting, of the source rock.

During partial melting, minerals with lower melting points transform into melt, while those with higher melting temperatures remain solid (▶Fig. 6.5a). The melt that forms moves away from the source rock before the entire rock can melt. Therefore, magma does not have the same composition as the original rock from which it formed. Silica-rich minerals (such as quartz or K-feldspar) melt at lower temperatures than do mafic minerals (olivine, pyroxene, or am-

(a)

(b)

FIGURE 6.5 (a) The concept of partial melting. Rock does not all melt at once; at lower temperatures, only part of the rock melts. The first melt tends to be more silicic than the later-formed melt, as the graph shows. When the rock first starts to melt, the molten rock films around still-solid grains. Grains that melt at lower temperatures melt first, while grains that melt at higher temperatures remain. With further melting, a crystal "mush" develops, with relict solid grains surrounded by melt. (b) The concept of magma contamination. Blocks of rock fall into a magma, melt, and become mixed with the magma. Also, wall rock begins to partially melt and contributes new magma to the rising magma column. (c) The concept of fractional crystallization. The highest-melting-temperature (mafic) minerals begin to crystallize. These early-formed minerals sink to the bottom of the magma body. Elements incorporated in these minerals, therefore, are extracted from the melt. The remaining melt becomes more silicic.

(c)

phibole), so magmas tend to be more silicic than the rock from which they are extracted. For example, the partial melting of ultramafic rock yields a mafic magma, and the partial melting of intermediate rock yields a silicic magma.

Contamination. As magma moves from its original source to the location where it finally solidifies, it incorporates components of rock that it passes through. This process is known as **magma contamination.** For example, when mafic magma derived by the partial melting of the mantle rises and intrudes into the continental crust, some makes it to the Earth's surface or nearly there, where it solidifies to form mafic igneous rock, but some stalls within the crust. As the stalled magma sits in the crust, it partially melts the surrounding crust. The magma that forms from the intermediate rock of the crust is silicic. This magma then mixes with mafic magma, making the latter less mafic. Contamination may also result when blocks of crust fall into the

magma and dissolve, like a sugar cube in water. This process of digesting blocks is called **assimilation** (▶Fig. 6.5b).

Fractional crystallization. Just as not all minerals in a magma melt at the same temperature, not all minerals freeze at the same temperature. In the case of a mafic magma, for example, mafic minerals solidify at a higher temperature than do silicic minerals. Thus, when a magma starts to cool, mafic minerals begin to form first. They may settle out of the magma, because they are denser than magma, before other minerals begin to form. As mafic minerals are rich in iron and magnesium, they take iron and magnesium out of the magma when they settle, so the remaining magma becomes more silicic (▶Fig. 6.5c). Because of this process, called **fractional crystallization,** a magma becomes progressively more silicic as it cools. (See Box 6.2 to learn the sequence in which minerals form during fractional crystallization.)

Bowen's Reaction Series

BOX 6.2

THE REST OF THE STORY

In the 1920s, Norman L. Bowen, a geologist at the Carnegie Institution in Washington and later at the University of Chicago, began a series of laboratory experiments designed to determine the sequence in which silicate minerals crystallized from a melt. Bowen first melted powdered mafic igneous rock in a sealed crucible by raising its temperature to about 1,280°C. Then he cooled the melt just enough to cause part of it to solidify, and "quenched" the remaining melt by submerging it quickly in mercury. **Quenching,** which means a sudden cooling to form a solid, caused any remaining liquid to turn into glass, a material without a crystalline structure. The early-formed crystals were trapped in the glass. Bowen identified mineral crystals formed before quenching by examining thin sections of the glass in which the crystals were embedded. Then he analyzed the composition of the remaining glass. After repeated experiments at different temperatures, Bowen found that different minerals crystallized in a specific sequence.

Olivine and calcium-rich plagioclase form first. Some of these crystals dissolve back into the melt after they form, but some settle out of the liquid and extract iron, magnesium, and calcium from the remaining melt, causing the remaining melt to become more silicic. As the magma continues to cool, eventually pyroxene, then amphibole, then biotite crystallize. All the while, plagioclase continues to form, but notably,

the plagioclase formed at lower temperatures contains more sodium and less calcium—the calcium had gone into making the early-formed crystals. At temperatures of between 650° and 850°C, only 10% melt remains, and silica-rich minerals—quartz, K-feldspar, and muscovite—crystallize.

Geoscientists now refer to the sequence in which different silicate minerals crystallize during the progressive cooling of a mafic melt as **Bowen's reaction series.** There are two tracks in the series. The **discontinuous reaction series** refers to the sequence olivine, pyroxene, amphibole, biotite, K-feldspar/muscovite/quartz: each step yields a different class of silicate mineral. The **continuous reaction series** refers to the progressive change from calcium-rich to sodium-rich plagioclase: the steps yield different versions of the same mineral (▶Fig. 6.6). In the discontinuous reaction series the first mineral to form is composed of isolated silicon-oxygen tetrahedra, the second contains single chains of tetrahedra, the third double chains of tetrahedra, the fourth 2-D sheets of tetrahedra, and the last 3-D network silicates.

Bowen's studies provided a remarkable demonstration of how laboratory experiments can help us understand processes that take place in locations (such as a deep magma chamber) that no human can visit directly. But Bowen paid a heavy price for his pioneering efforts. He used mercury to quench his melts because it can absorb heat rapidly. However, inhaling or touching mercury can cause brain damage. In fact, nineteenth-century hat makers who used mercury to make felt commonly incurred mental illness (remember the Mad Hatter from *Alice in Wonderland*?). In later life, Bowen unfortunately suffered from the effects of mercury poisoning.

FIGURE 6.6 Bowen's reaction series. Minerals that crystallize at higher temperatures are at the top of the series. On the left is the discontinuous reaction series, consisting of a succession of different minerals. On the right is the continuous reaction series, consisting of progressively changing plagioclase compositions. Rocks formed from minerals at the top of the series are mafic, while rocks made from minerals at the bottom of the series are silicic.

6.4 THE MOVEMENT OF MAGMA AND LAVA

Forces Driving the Rise of Magma

If magma stayed put once it formed, new igneous rocks would not develop in or on the crust. But it doesn't stay put; magma tends to move upward, away from where it formed—in some cases, it reaches the Earth's surface to create volcanoes. Magma rises slowly through the crust, because it has to work its way up through cracks and narrow conduits, and friction between the magma and the walls of the conduits slows it down. In a broad sense, the movement of magma from deep in the Earth up to the crust or even onto the surface transfers both matter and heat from the interior of the Earth upward. This movement is a key component of the Earth system, because it provides the raw material for new rocks and for the atmosphere and ocean. But why does magma rise?

For one thing, magma is less dense than surrounding rock, because rock expands as it melts and because magma tends to be more silicic than its source, so buoyancy drives magma upward through the solid rock just as it drives a wood block up through water. Of note, when volatile-rich magma rises to a shallower depth, the volatiles dissolved within it come out of solution and form bubbles, much as carbon dioxide makes bubbles in soda water when you pop the bottle cap off and release the pressure. The presence of gas bubbles within a magma further decreases its density and creates additional buoyancy force to drive the magma upward. (In some cases, the bubbles are preserved when the rock freezes and appear as open holes, called **vesicles,** in the rock.) The other reason magma rises is that the weight of overlying rock creates a pressure at depth that literally squeezes magma upward, much as mud squeezes up between your toes when you step into a puddle barefoot.

Melt Viscosity

The speed with which magmas or lavas move is affected by their viscosity—their resistance to flow. Magmas with low viscosity flow more easily than those with high viscosity, just as water flows more easily than molasses. Magma viscosity depends on temperature, volatile content, and silica content. Hotter magma is less viscous than cooler magma,

FIGURE 6.7 The behavior of erupting lava reflects its viscosity. (a) Viscous lava (silica-rich lava) forms a blob or mound-like dome at the volcano's vent. (b) Nonviscous lava (hot mafic lava) spreads out in a thin flow and can travel far from the vent; it can also fountain.

just as hot tar is less viscous than cool tar, because thermal energy breaks bonds and allows atoms to move more easily (remember that the Hawaiian lava slowed down as it cooled). Similarly, magmas or lavas containing more volatiles are less viscous than dry (volatile-free) magmas, because the volatile atoms also tend to break apart bonds. And magmas or lavas containing more silica are more viscous than those with less silica, because silicon-oxygen tetrahe-

dra tend to link together in the magma to create long chains that can't move past one another as easily as can smaller molecules. Thus, hotter mafic lavas have relatively low viscosity and flow in thin sheets over wide regions, while cooler silicic lavas are highly viscous and clump at the volcanic vent (▶Fig. 6.7a, b).

6.5 EXTRUSIVE VERSUS INTRUSIVE ENVIRONMENTS

Recall that there are two environments in which igneous rocks form. If magma erupts at the Earth's surface and freezes in contact with the atmosphere or the ocean, then the rock it forms is called **extrusive igneous rock.** The term implies that the melt was extruded from (it flowed or exploded out of) a vent in a volcano. In contrast, if magma freezes underground, the rock it forms is called **intrusive igneous rock,** implying that the magma pushed—intruded—into preexisting rock of the crust.

Extrusive Igneous Settings

Not all volcanic eruptions are the same, so not all extrusive rocks are the same (as will be discussed further in Chapter 9). Some volcanoes erupt streams of low-viscosity lava that run down the flanks of the volcano and then spread over the countryside. When this lava freezes, it forms a sheet of igneous rock also known as a **lava flow.** In contrast, some volcanoes erupt viscous masses of lava that pile into domes, and still others erupt explosively, sending clouds of volcanic

FIGURE 6.8 Types of volcanic extrusion. (a) Ash falls sprinkle down from clouds of ash blown high in the atmosphere. Ash flows behave like avalanches cascading down the side of the mountain. Lava flows are streams of liquid. (b) This lava flow is buried by layers of ash. Later erosion on the cliff face allows us to see the internal structure of the flow. Notice that it breaks into columns; the fractures bounding these columns are called columnar joints.

(a)

(b)

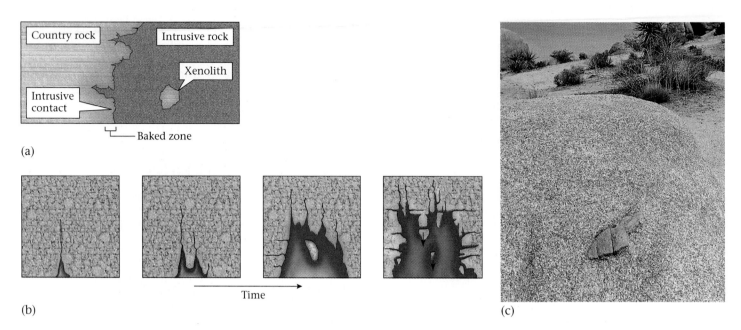

(a)

(b)

Time

(c)

FIGURE 6.9 How a magma rises. (a) An intrusive contact, showing the baked contact, blocks of country rock, fingers of the intrusion protruding into the country rock, and a xenolith. (b) A magma stoping into country rock, gradually breaking off and digesting blocks as it moves. (c) Xenoliths in a granite.

ash skyward or avalanches of ash—**ash flows**—that tumble down the sides of the volcano. **Volcanic ash** consists of tiny glass shards, formed when a fine spray of exploded lava freezes instantly upon contact with the atmosphere. The ash that billows into the sky (an **ash cloud**) cools and falls to Earth like snow, creating an **ash fall** that blankets the countryside. The ash flows remain hot, eventually settling and welding together (▶Fig. 6.8a, b). (See art on pp. 140–41.)

Which type of eruption occurs depends largely on a magma's composition and volatile content. As noted above,

silicic lavas tend to be viscous and form bulbous domes, while mafic lavas tend to have low viscosity and spread in broad, thin flows. Volatile-rich silicic lavas tend to erupt explosively and form thick ash deposits (see Chapter 9).

Intrusive Igneous Settings

Magma rises and intrudes into preexisting rock by slowly percolating upward between grains or by filling in cracks. The magma that doesn't make it to the surface freezes solid

FIGURE 6.10 (a) Dikes and sills are vertical or horizontal bands, respectively, on the face of an outcrop. (b) If we were to strip away the surrounding rock, dikes would look like walls, and sills would look like tabletops.

(a)

(b)

(a)

(c)

(b)

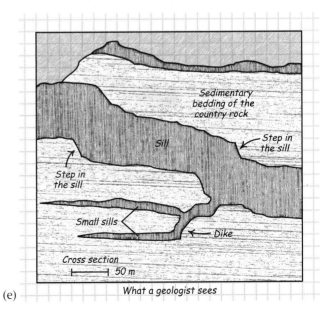

(d)

FIGURE 6.11 (a) A basalt dike looks like a black stripe painted on an outcrop of granite (here, in Arizona). But the dike actually intrudes, wall-like, into the outcrop. In this example, the dike happens to curve. (b) At this ancient volcano at Shiprock, New Mexico, ash and lava flows have eroded away, leaving a "volcanic neck" (the solid igneous rock that cooled in a magma chamber within the volcano). Large dikes radiated outward from the center, like spokes of a wheel. The softer rocks that once surrounded the dikes have eroded away, leaving a wall-like remnant of the dike exposed. (c) These Precambrian dikes exposed in the Canadian Shield formed when the region underwent stretching over a billion years ago; at that time, numerous cracks in the crust filled with magma. (d) This dark sill, exposed on a cliff in Antarctica, is basalt; the white rock is sandstone. (e) This geologist's sketch shows the cliff face as viewed face on. The sill appears to have a few steps in it, climbing to a higher bed. Thin sills connect to the main one at the base of the diagram, and the base of another sill forms the crest of the mountain.

Sedimentary bedding of the country rock

Sill

Step in the sill

Step in the sill

Small sills

Dike

Cross section

50 m

(e)   What a geologist sees

underground in contact with preexisting rock and becomes intrusive igneous rock. Geologists commonly refer to the preexisting rock into which magma intrudes as **country rock,** or **wall rock,** and the boundary between country rock and an intrusive igneous rock as an **intrusive contact** (▶Fig. 6.9a). If the country rock was cold to begin with, then heat from the intrusion "bakes" and alters it in a narrow band along an intrusive contact (see Chapter 8).

Where does the room for an intrusion come from? Magma can make some room for itself by surrounding and breaking off blocks of wall rock, which then sink into the magma. This process is called **stoping** (▶Fig. 6.9b). During stoping, some of the blocks melt entirely and mix with the magma, while others may remain solid blocks and become relicts surrounded by intrusive rock when the magma freezes. These relict blocks are called **xenoliths,** after the Greek word *xeno,* meaning "foreign" (▶Fig. 6.9c). Generally, however, magma fills space that opens as one crustal block moves away from another as a consequence of plate movement. During such movement, empty space never develops (it would collapse), because magma fills the space as it appears.

Geologists distinguish between different types of intrusions based on their shape. **Tabular intrusions,** or sheet intrusions, are planar and of roughly uniform thickness; they range in thickness from millimeters to tens of meters, and can be traced for meters to hundreds of kilometers. In places where tabular intrusions cut across rock that does not have layering, a nearly vertical, wall-like tabular intrusion is called a **dike,** while a nearly horizontal, tabletop-shaped tabular intrusion is a **sill.** In places where tabular intrusions enter rock that has layering (bedding or foliation), dikes are defined as intrusions that cut across layering, while sills are intrusions that are parallel to layering (▶Figs. 6.10a, b; 6.11a–e). Spectacular groups of dikes cut across the countryside of interior Canada, and a large sill, the Palisades Sill, makes up the cliff along the west bank of the Hudson River opposite New York City. Another sill forms the ledge on which Hadrian's Wall, which bisects Britain, was built. Some sills dome upward, creating a blister-shaped intrusion known as a **laccolith.**

Plutons are irregular or blob-shaped intrusions that range in size from tens of meters across to tens of kilometers across (▶Figs. 6.12a–c; 6.13a, b). The intrusion of numerous plutons in a region creates a vast composite body that may be several hundred kilometers long and 100 km wide; such immense masses of igneous rock are called **batholiths.** The rock making up the Sierra Nevada Mountains of California is the relict of a batholith created from plutons that intruded between 145 and 80 million years ago (▶Fig. 6.14a–d).

If intrusive igneous rocks form beneath the Earth's surface, why can we see them exposed today? The answer comes from studying the dynamic activity of the Earth. Over long periods of geologic time, mountain building,

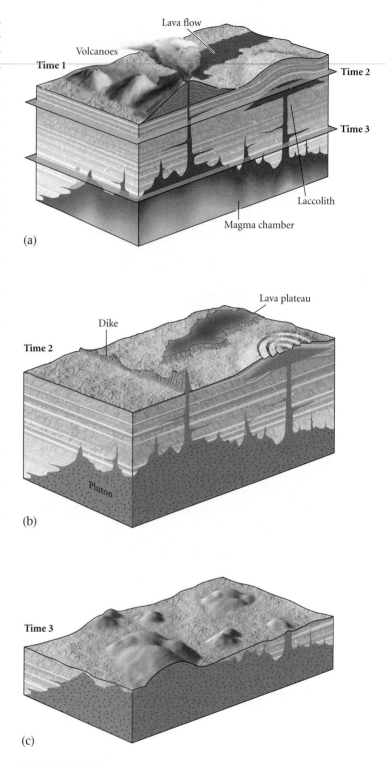

FIGURE 6.12 (a) While a volcano is active, a magma chamber exists underground, dikes, sills, and laccoliths intrude, and lava and ash erupt at the surface. (b) Later, the bulbous magma chamber freezes into a pluton. The soft parts of the volcano erode, leaving wall-like dikes and column-like volcanic necks. Hard lava flows create resistant plateaus. (c) With still more erosion, volcanic rocks and shallow intrusions are removed, and we see plutonic intrusive rocks.

(a)

(b)

What a geologist sees

FIGURE 6.13 (a) Torres del Paines, a spectacular group of mountains in southern Chile. The light rock is a granite pluton, and the dark rock is the remains of the country rock into which the pluton intruded. A screen of country rock (in the lower half) hides the front of the pluton. (b) A geologist's sketch, labeling the two major rock units.

FIGURE 6.14 (a) The batholiths of western North America today. (b) The geography of western North America about 100 million years ago, showing the position of the subduction zone responsible for the batholiths. Note that the west coast, south of Idaho, lay much farther east than it does today; a volcanic arc, associated with a convergent plate margin, existed there. (c) The Sierra Nevada Batholith as exposed today. The rounded, light-colored hills are all composed of granite-like intrusive igneous rock. (d) A geologist's sketch illustrates that today's land surface once lay several kilometers beneath a chain of volcanoes.

(a)

(b)

(c)

(d)

What a geologist imagines

driven by plate interactions, slowly uplifts huge masses of rock. Moving water, wind, and ice eventually strip away great thicknesses of overlying rock and exposes the intrusive rock that had formed below. For example, the Sierra Nevada Batholith that we now see at the ground surface developed several kilometers below a chain of volcanoes.

6.6 TRANSFORMING MAGMA INTO ROCK

Magma begins to turn into solid igneous rock when it cools below its freezing point. For this process to occur, the magma must move to a cooler environment. Because temperatures decrease toward the Earth's surface, magma automatically enters a cooler environment when it rises. The cooler environment may be cool country rock if the magma intrudes underground, or it may be the atmosphere or ocean if the magma extrudes as lava at the Earth's surface.

The rate at which a magma cools depends on how fast it is able to transfer heat into its surroundings. By analogy, if you pour hot coffee into a thermos bottle and seal it, the coffee stays hot for hours, but if you spill it onto a table, it cools quickly; because of insulation, the coffee in the thermos loses heat to the air outside only very slowly. Like the thermos bottle, rock acts as an insulator, in that it transports heat away from a magma very slowly, so magma cooled underground (in an intrusive environment) cools slowly. In contrast, lava that erupts at the ground surface, like coffee spilled on the table, cools quickly because it is surrounded by air or water, which conduct heat away quickly. In sum, extrusive rocks cool much more quickly than intrusive rocks.

Three factors control the cooling rate of magma that intrudes below the surface.

- *The depth of intrusion:* Intrusions deep in the crust cool more slowly than shallow intrusions, because warm country rock surrounds deep intrusions while cold country rock surrounds shallow intrusions, and warmer country rock slows the escape of heat (similarly, food stays warmer longer if you serve it on a hot plate rather than a cold one).

- *The shape and size of a magma body:* Heat escapes from magma at its surface, so the greater the surface area for a given volume of magma, the faster it cools. Thus, a pluton cools more slowly than a tabular intrusion with the same volume (because a tabular intrusion has a greater surface area across which heat can be lost), and a large pluton cools more slowly than a small pluton of the same shape (because the surface area per unit volume decreases as the volume increases). Similarly, droplets of

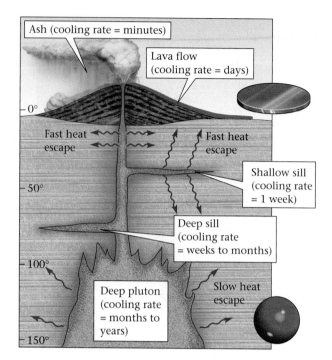

FIGURE 6.15 The cooling rate of an intrusion decreases with greater depth (because country rock is hotter at greater depth). The cooling rate also depends on the shape of the intrusion (a thin sheet cools faster than a sphere of the same volume) and on the size of the intrusion (a small intrusion cools faster than a large one). Thus, ash (formed when droplets of lava come in contact with air) cools fastest, followed by a thin sheet of lava, a shallow sill, a deep sill, and a deep pluton.

lava cool faster than a lava flow, and a thin flow of lava cools more quickly than a thick sheet (▶Fig. 6.15).

- *The presence of circulating groundwater:* Water passing through magma absorbs and carries away heat, much like the coolant that flows around an automobile engine.

6.7 IGNEOUS ROCK TEXTURES

If you ever examine a variety of igneous rock samples, you'll find that some consist of interlocking grains, some are made of solid glass, and some consist of individual fragments that appear to be packed or welded together. Further, you'll see that some rocks with interlocking crystals contain large grains, and some contain small grains. We refer to any igneous rock composed of mineral grains, regardless of size, as **a crystalline igneous rock.** In crystalline igneous rocks, mineral grains interlock with one another like pieces of a jigsaw puzzle, creating an **interlocking texture** (▶Fig. 6.16a). This texture occurs because once some grains have developed, they interfere with the growth of later-formed grains. In fact, the last grains to grow end up filling irregular

0 Mm 0.5   0 Mm 0.5

(a)   (b)

(c)   (d)   (e)

(f)   (g)   (h)

FIGURE 6.16 (a) This thin section of granite shows relatively large interlocking crystals. (b) Thin section of a glassy, fine-grained volcanic rock. The light-colored grains are crystals, and the dark matrix (the region between the crystals) is glass. (c) Rhyolitic welded tuff, (d) granite, (e) basalt, (f) gabbro, (g) pumice, (h) obsidian.

spaces. **Glassy igneous rocks** are made up entirely of glass, or of tiny crystals surrounded by a glass matrix (▶Fig. 6.16b). (A **matrix** consists of finer-grained material that surrounds larger grains.) Igneous rocks made from fragments that are packed or welded together are called **pyroclastic.**

The cooling rate determines whether an igneous rock consists of glass or of crystalline grains, and if it consists of grains, the cooling rate determines their size. If the melt cools so fast that the atoms within it don't have time to arrange into crystal lattices, the melt solidifies to form glass. (The process of forming natural igneous glass is identical to

the process for manufacturing window glass, except that window glass consists only of silica, while natural igneous glass contains other elements as well.) If the melt cools slightly more slowly, many mineral seeds form and grow, so that the resulting rock consists of numerous small grains and is **fine-grained.** If the melt cools very slowly, relatively few seeds form and grow, so the resulting rock consists of fewer but larger grains and is **coarse-grained.**

Not surprisingly, lava forms glass if it cools very quickly or is quenched by erupting into water or by spraying into the air; shallow sills and dikes as well as lava flows cool somewhat more slowly to create fine-grained igneous rocks; and plutons deep below the surface cool very slowly to form coarse-grained igneous rocks. Some melts cool in two stages: first the melt cools slowly and partially solidifies underground, and then it erupts and cools quickly and the remainder solidifies. Such rocks typically have two sizes of grains. The grains that form while the magma cools slowly tend to be large—these grains are called **phenocrysts.** Phenocrysts commonly consist of plagioclase, a mineral that begins to grow early during Bowen's reaction series (see Box 6.2). The grains that form when the rock cools quickly are small and collectively comprise the matrix.

Geologists use special terms for describing igneous rock grain size. We call coarse-grained igneous rocks **phaneritic,** fine-grained igneous rocks **aphanitic,** and glassy igneous rocks simply **glassy.** Igneous rocks with a texture characterized by phenocrysts distributed through a matrix are **porphyritic.**

One important igneous rock type, **pegmatite,** doesn't quite fit the grain size/cooling rate scheme just described. Pegmatite, a very coarse-grained igneous rock, contains crystals of up to tens of centimeters across and occurs in thin, dike-shaped intrusions called **pegmatite dikes.** Gold and many precious gemstones are found in pegmatites. Because pegmatite is found in dikes, which cool quickly, the coarseness of the rock may seem surprising, but pegmatites are coarse because they form from water-rich melts in which atoms can move around so fast that large crystals grow very rapidly.

All the igneous rocks discussed above form directly by the cooling of a melt. But some extrusive rocks consist of igneous debris—volcanic ash or fragments of preexisting volcanic rock—blown out of a volcano during an explosion. Again, rocks created from the accumulation of such debris are called **pyroclastic rocks,** from the Greek *pyro* ("fire") and *clastic* ("composed of clasts [grains] that are stuck together").

6.8 CLASSIFYING IGNEOUS ROCKS

Since melt comes in a variety of compositions and can freeze to form igneous rocks in many different environments both above and below the surface of the Earth, we observe a wide spectrum of igneous rock types, which can be classified according to their texture and composition (►Fig. 6.16c–h). Studying a rock's texture tells us about the rate at which it cooled and therefore the environment in which it formed, while studying its composition tells us about the original source of the magma and about the way in which the magma evolved before finally solidifying.

The classification scheme for the principal types of crystalline (nonglassy) igneous rocks is really quite simple. The different compositional classes are distinguished based on

silica content—ultramafic, mafic, intermediate, or silicic—while the different textural classes are distinguished based on whether or not the grains are large enough to be identified with the naked eye. The chart in ►Figure 6.17a gives the composition and texture of the most commonly used rock names. As a rough guide, the color of an igneous rock reflects its composition: mafic rocks tend to be black or dark gray, intermediate rocks tend to be lighter gray or greenish gray, while silicic rocks tend to be light tan to pink or maroon. Unfortunately, color can be a misleading basis for rock identification, so geologists use a petrographic microscope to confirm their identifications. Different types of porphyritic rocks, not listed in Figure 6.17a, are distinguished from one another according to the composition of their matrix. For example, andesite porphyry is an andesite containing phenocrysts.

Note that according to Figure 6.17a, rhyolite and granite have the same chemical composition, but differ in grain size. Which of these two rocks will develop from a melt of granitic composition depends on the cooling rate. A silicic lava that solidifies quickly at the Earth's surface or in a thin dike or sill turns into fine-grained rhyolite, but the same magma, if solidified slowly at depth in a pluton, turns into coarse-grained granite. A similar situation holds for mafic lavas—a mafic lava that cools quickly in a lava flow forms basalt, but a mafic magma that cools slowly forms gabbro.

Geologists also distinguish several different types of glassy igneous rocks, based on their texture (►Fig. 6.17b). Some are fragmental (they're composed of separate pieces stuck together), while others are not.

• A solid mass of volcanic glass is called **obsidian.** Obsidian, which tends to be black or brown, splits into sharp-edged pieces when hit with a hammer, for it develops conchoidal fractures. Preindustrial people worldwide used such pieces for arrowheads, scrapers, and knife blades. Obsidian generally develops from silicic lava, because this lava is so viscous that atoms within cannot migrate through it easily, and thus the magma solidifies before crystals can form.

• **Pumice,** a volcanic rock that is full of open pores, giving it the appearance of a sponge, forms by the quick cooling of frothy lava that resembles the foam head in a glass of beer. Such foam develops when large amounts of gas bubble out of a lava. In some cases, pumice contains so many air-filled pores that it can actually float on water like Styrofoam. Ground-up pumice makes the grainy abrasive that blue-jean companies use to "stonewash" jeans.

• **Scoria** resembles pumice, except that the air-filled pores constitute less than 50% of the rock.

• **Tuff** is a pyroclastic igneous rock composed of volcanic ash and/or fragmented lava and pumice. Tuff forms either

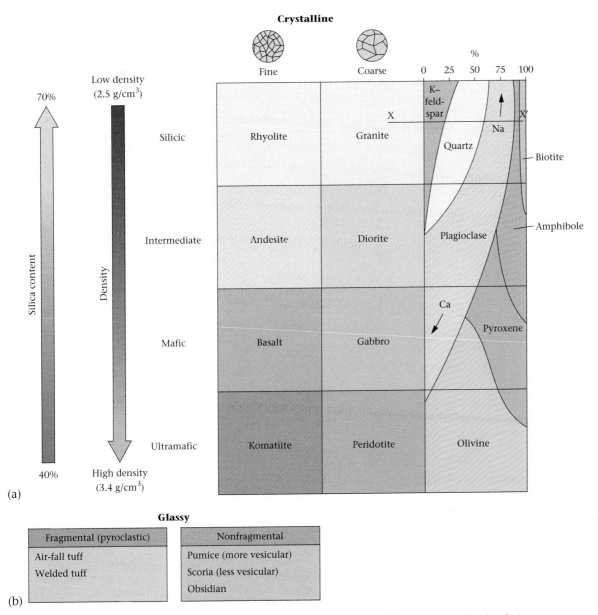

FIGURE 6.17 (a) Crystalline (nonglassy) igneous rocks are distinguished from one another by their grain size and composition. The right side of the chart shows the proportions of different minerals in the different rock types. To read this chart, draw a horizontal line next to a rock name; the minerals that the line crosses are the minerals found in that rock. (For example, granite [line X–X'] includes K-feldspar, quartz, plagioclase, amphibole, and biotite.) (b) The principal types of glassy (noncrystalline) igneous rocks can be separated into two categories, fragmental and nonfragmental.

from material that settles from the air (in an ash fall) and then is cemented together or sticks together, or from material that avalanches down the side of a volcano (an ash flow) and is still so hot when it settles that the glass shards weld together. Tuff that settles from air is called **air-fall tuff**, and tuff formed by the welding together of hot shards is called **welded tuff.**

6.9 WHERE DOES IGNEOUS ACTIVITY OCCUR, AND WHY?

If you look at a map showing the distribution of **igneous activity**—the formation, movement, and in some cases eruption of molten rock—around the world (▶Fig. 6.18),

▲▲▲ Convergent boundary ━━━ Ridge ──── Transform ▲ Volcanoes

FIGURE 6.18 The distribution of submarine and subaerial volcanoes worldwide. Note that volcanic activity occurs all along mid-ocean ridges, though most is submerged beneath the water and can't be seen. Most subaerial volcanoes lie in volcanic arcs bordering convergent plate boundaries. Others are found along continental rifts and at hot spots.

you'll see that igneous activity occurs in four settings: (1) in volcanic arcs bordering deep-ocean trenches, (2) at isolated hot spots, (3) within continental rifts, and (4) along mid-ocean ridges. As evident from this list, most igneous activity takes place at established or newly forming plate boundaries. (Hot-spot volcanoes, however, which erupt in the interiors of plates, violate this rule.) Most volcanic activity along mid-ocean ridges happens underwater, at **submarine volcanoes,** while most volcanic activity in rifts, along convergent margins, and at hot spots takes place under the air (above ground), at **subaerial volcanoes.**

The Formation of Igneous Rocks at Volcanic Arcs—the Product of Subduction

Most subaerial volcanoes on Earth occur in long, curving chains adjacent to the deep-ocean trenches that mark convergent plate boundaries. Some of these chains, such as the volcanoes of the Andes Mountains in South America, fringe the edge of a continent and are called **continental volcanic arcs;** others, such as the volcanoes of the Aleutian Islands in Alaska, form oceanic islands and are called **volcanic island arcs.** They are called arcs because in many locations the volcanic chain defines a curve on a map. Recall that at convergent plate boundaries, where volcanic

arcs are found, oceanic lithosphere subducts and sinks into the mantle. How does this process trigger volcanic activity?

The oceanic lithosphere that subducts at convergent boundaries is made up of oceanic crust and the underlying lithospheric mantle. Some of the minerals comprising the oceanic crust contain volatiles like water or carbon dioxide. When the subducting crust sinks into the mantle to a depth of about 150 km, it is heated by the surrounding mantle to such a degree that its volatile compounds bubble out and into the adjacent hot asthenosphere. The addition of these volatiles causes the peridotite (ultramafic rock) of the asthenosphere to partially melt and produce a basaltic magma. This magma either rises directly, to erupt as basaltic lava, or undergoes fractional crystallization before erupting, to create andesitic or even rhyolitic lava. Whether some of the crust on the downgoing slab also partially melts and contributes to the volcanic system is not known.

In continental volcanic arcs, not all the mantle-derived basaltic magma rises directly to the surface; some gets trapped at the base of the continental crust, and some in magma chambers deep in the crust. When this happens, heat transfers from the basaltic magma into the surrounding continental crust and causes partial melting of this crust. Because much of continental crust is mafic to inter-

FIGURE 6.19 Mt. Rushmore, a pluton of granite in the Black Hills of South Dakota, has been carved to display the colossal heads of four U.S. presidents (Washington, Jefferson, Lincoln, and Theodore Roosevelt). Sculptor Gutzon Borglum was able to carve these heads because of the uniformity and coherence of the pluton.

mediate in composition to start with, the resulting magmas are intermediate to silicic in composition. This magma rises, leaving the basalt behind, and cools higher in the crust to form granitic or dioritic plutons (▶Fig. 6.19).

Volcanic arcs form wherever subduction takes place. Subduction now occurs along 60% of the margin of the present Pacific Plate, so volcanic arcs border almost 60% of the margin of the Pacific Ocean; the Pacific rim is therefore commonly referred to as the "Ring of Fire." The volcanic arcs of the Ring of Fire include the Andes of South America, the Cascades of the northwestern United States, the Aleutian Islands of Alaska, the Kuril Islands off the east coast of Russia, Japan (including Mt. Fuji), and several smaller arcs in the southwestern Pacific, such as the Mariana arc. Other notable volcanic arcs, which do not lie along the Ring of Fire, include the Scotia arc off the southern tip of South America and the Caribbean arc at the eastern edge of the Caribbean Sea. Small arc segments also occur in the Mediterranean Sea.

The Formation of Igneous Rocks at Hot Spots—a Surprise of Nature

The volcanic islands of Hawaii lie in the middle of the Pacific Plate, far from any plate boundary. In fact, many of the fabled tropical paradise islands of the South Pacific are volcanoes or were volcanoes in the not-so-distant past; their native cultures revered the volcano as the giver, and taker, of life.

Hawaii and other South Pacific island volcanoes are **hot-spot volcanoes,** isolated volcanoes that are independent of plate-boundary interactions. About one hundred currently active hot-spot volcanoes are scattered around the world. (See Figs. 4.20–4.22.) Some, like Hawaii, erupt in the interior of oceanic plates away from convergent or divergent boundaries. Some, like Iceland, sit astride a divergent boundary (mid-ocean ridge). (Geoscientists associate Iceland with a hot spot because its volcanoes generate far more lava than normal mid-ocean ridge volcanoes—the extra basalt built a pile high enough to rise above the sea and form Iceland.) And some hot-spot volcanoes grow on continents; for example, a few currently erupt in Africa.

Perhaps the most famous continental hot-spot volcano, at least to North Americans, created the stunning landscape of Yellowstone National Park, in northwestern Wyoming and adjacent states. The "yellow stone" exposed in the park consists of sulfur- and iron-stained layers of volcanic ash. Though the last major eruption at the Yellowstone hot spot occurred about 600,000 years ago, heat from a magma chamber still heats the ground beneath the park. This heat boils underground water, which then bursts out of the ground in geysers.

As we learned in Chapter 4, most researchers now agree that hot-spot volcanoes form above column-shaped plumes of hot mantle rock that rise from the core-mantle boundary. These plumes probably originate because heat from Earth's core warms the base of the mantle so much that it becomes buoyant relative to the overlying mantle and begins to rise upward. Note that the rise of the plumes occurs by the slow flow of solid rocks—in other words, a plume does not consist of magma. But when the hot rock of a plume rises to the base of the lithosphere, decompression causes the rock (peridotite) of the plume to partially melt and generate a mafic magma. The mafic magma then rises through the lithosphere, pools in a magma chamber in the crust, and then eventually erupts at the surface, forming a volcano. In the case of hot spots created in the ocean, only mafic magma erupts, forming basalt. In the case of hot spots created in continents, some of the mafic magma erupts to form basalt, but some transfers heat to the continental crust, which then partially melts and creates silicic magmas that erupt to form rhyolite. Thus, continental hot-spot volcanoes may alternately yield basalt and rhyolite.

Remember that the position of a hot-spot plume remains relatively fixed with respect to the moving plates, so in the case of plate-interior hot spots, the drift of the plates causes a volcano that grew during one interval of time eventually to move off the hot spot. When this happens, the volcano dies, and a new volcano forms. As a consequence, active hot-spot volcanoes commonly occur at the end of a chain of dead volcanoes, and this chain is called a **hot-spot track.** Hawaii, for example, is at the end of the Hawaiian-Emperor chain of seamounts (the older dead volcanoes of the chain have sunk below sea level; see Fig. 4.22b). Similarly, Yellowstone Park lies at the end of a hot-spot track manifested by the Snake River Plain in Idaho.

The Formation of Igneous Rocks at Rifts—a Record of Continental Breakup

According to the theory of plate tectonics (Chapter 4), continental rifts are places where continental lithosphere is being stretched and thinned—successful rifting splits a continent in two and gives birth to a new mid-ocean ridge. As the continental lithosphere thins (before the continent splits), the weight of rock overlying the asthenosphere decreases, so pressure in the asthenosphere decreases and decompression melting takes place. The partial melting of asthenosphere yields basaltic magma, which rises into the crust. Some of this magma makes it to the surface, following the cracks that appear in the crust as a consequence of the stretching and breaking that accompany rifting, and erupts as basalt. Some of the magma gets trapped in the crust and transfers heat to the crust. The resulting partial melting of the crust yields silicic magmas that erupt as rhyolite. Thus, rift volcanoes, like continental hot-spot volcanoes, erupt two kinds of igneous magmas—basalt and rhyolite. Much of the rhyolite erupts in the form of ash.

The most famous active rift, the East African Rift, presently forming a 4,000-km-long gash in the crust of Africa, has produced numerous volcanoes, including Mt. Kilimanjaro. Recent rifting in North America has yielded the Basin and Range Province of Utah, Nevada, Arizona, and southeastern California. Though no currently erupting volcanoes exist in this region, the abundance of basalt flows and rhyolitic ash deposits testifies to the occurrence of igneous activity in the recent past. And geoscientists are now monitoring the Mono Lakes volcanic area, which lies along the western edge of the Basin and Range Rift, because of the possibility that volcanoes in this area may erupt in the near future.

In addition to active rifts (those currently opening), continents contain abundant examples of extinct rifts—places where a continent began to split but rifting stopped before an ocean formed. A narrow but deep rift almost split the United States in two about 1 billion years ago. This structure, known as the Mid-Continent Rift, which can be tracked from Lake Superior southward to Kansas, filled with over 15 km of basalt.

In some cases, mantle plumes underlie portions of rifts and cause hot spots to form along the rift. An unusually large quantity of magma is generated at the hot spot, because temperatures are higher in the underlying plume, so more partial melting occurs there than at other places along the rift. As a result, huge floods of basalt erupt at the hot spot as the rift grows. The particularly hot basaltic lava that erupts at such localities has such low viscosity that it can flow tens to hundreds of kilometers across the landscape. Geoscientists refer to such flows as **flood basalts.** Flood basalts make up the bedrock of the Columbia River Plateau in Oregon and Washington (►Fig. 6.20a), the Paraná

(a)

(b)

FIGURE 6.20 (a) Flood basalts underlie the Columbia River Plateau in Washington and Oregon, the dark area on this map. (b) Iguazu Falls, on the Brazil-Argentina border. The falls flow over the huge flood basalt sheet (the black rock) of the Parana Plateau. Flood basalt underlies all of the region in view.

Plateau in southeastern Brazil (►Fig. 6.20b), the Karoo region of southern Africa, and the Deccan region of southwestern India.

The Formation of Igneous Rocks at Mid-Ocean Ridges—Hidden Plate Formation

Most igneous rocks undoubtedly form at mid-ocean ridges, that is, along divergent plate boundaries. Think about it—the entire oceanic crust, a 7–10-km-thick layer of basalt and gabbro that covers 70% of the Earth's surface, forms at mid-ocean ridges. And this entire volume gets subducted only to be replaced by new crust, over a period of about 200 million years.

Igneous magmas form at mid-ocean ridges for much the same reason they do at hot spots and rifts. As sea-floor spreading occurs, and oceanic lithosphere plates drift away from the ridge, hot asthenosphere rises to fill the resulting

(a)

(b)

(c)

FIGURE 6.21 (a) The formation of pillow basalt. (b) This pillow basalt forms part of an ophiolite, a slice of sea floor that was pushed up onto the surface of a continent during mountain building. (c) A cross section through a single pillow shows the glassy rind, with a more crystalline center.

space. As this asthenosphere rises, it undergoes decompression, which leads to partial melting and the generation of basaltic magma. As noted earlier, this magma rises into the crust and pools in a magma chamber. Some cools slowly along the margins of the magma chamber to form massive

gabbro, while some intrudes upward to fill vertical cracks that appear as newly formed crust splits apart. Magma that cools in the cracks creates basalt dikes, and magma that makes it to the sea floor and extrudes as lava forms basalt flows. The basalt flows of the sea floor don't look like those that erupt on land, because the seawater cools the lava so fast that it can't flow very far before solidifying into a pillow-shaped blob with a glassy rind. Eventually, the pressure of the lava inside a pillow breaks the glassy rind, and another pillow extrudes. Thus, sea-floor basalt is made up of a pile of pillows, known by geologists as **pillow basalt** (▶Fig. 6.21a–c).

In this chapter, we've focused on the diversity of igneous rocks, and why and where they form. We see that extrusive rocks develop at volcanoes. There's a lot more to say about volcanoes—eruptions have the potential to cause great harm. We will look at volcanic eruptions in detail in Chapter 9.

CHAPTER SUMMARY

- Magma is liquid rock (melt) under the Earth's surface. Lava is melt that has erupted from a volcano at the Earth's surface.

- Magma forms when hot rock in the Earth melts. This process only occurs under certain circumstances—where the pressure decreases (decompression), where volatiles (such as water or carbon dioxide) are added to hot rock, and where heat is transferred by magma rising from the mantle into the crust.

- Magma comes in a range of compositions: silicic, intermediate, mafic, and ultramafic. Mafic magma is hotter than silicic magma. The composition of magma is determined in part by the original composition of the rock from which the magma formed and by the fact that magma forms by the partial melting of rock. Contamination (the addition of material to the magma from the surroundings) and fractional crystallization (the settling out of early-formed crystals) may change the composition of a magma once it has formed.

- During partial melting, only part of the source rock melts to create magma. Magma tends to be more silicic than the rock from which it was extracted, because silica-rich minerals tend to melt first.

- Magma rises from the depth because of its buoyancy and because the pressure caused by the weight of overlying rock squeezes magma upward.

- Magma viscosity (its resistance to flow) depends on its composition. Silicic magma is more viscous than mafic magma.

- Geologists distinguish between two types of igneous rocks. Extrusive igneous rocks form from lava that erupts out of a volcano and freezes in contact with air or the ocean. Intrusive igneous rocks develop from magma that freezes inside the Earth.

• Lava may solidify to form flows or domes, or it may be exploded into the air to form ash.

• Intrusive igneous rocks form when magma injects into preexisting rock (country rock) below Earth's surface. They can be tabular, sheet-like, or blob-shaped. Blob-shaped intrusions are called plutons. Sheet-like intrusions that cut across layering in country rock are dikes, and sheet-like intrusions that form parallel to layering in country rock are sills. Huge intrusions, made up of many plutons, are known as batholiths.

• The rate at which intrusive magma cools depends on the depth at which it intrudes, the size and shape of the magma body, and whether circulating groundwater is present. The cooling rate is reflected in the grain size of an igneous rock. Instantly cooled melt produces glass, less quickly cooled melt produces fine-grained rock, and slowly cooled melt produces coarse-grained rock.

• Crystalline (nonglassy) igneous rocks are classified according to texture and composition. (For example, granite and rhyolite are both silicic rocks, but granite is coarse-grained, while rhyolite is fine-grained.) Glassy igneous rocks are classified based on texture (a solid mass is obsidian, while ash that has cemented or welded together is a tuff).

• The origin of igneous rocks can readily be understood in the context of plate tectonics. Magma forms at continental or island volcanic arcs along convergent margins, mostly because of the addition of volatiles to the asthenosphere above the subducting slab. Igneous rocks form at hot spots owing to the decompression melting of a rising mantle plume. (Where mantle-derived mafic magma rises into continental crust, it can partially melt the crust and generate silicic magma.) Igneous rocks form at rifts as a result of decompression melting of the asthenosphere below the thinning lithosphere. Igneous rocks form along mid-ocean ridges because of decompression melting of the rising asthenosphere.

KEY TERMS

aphanitic (p. 154)
ash cloud (p. 148)
ash fall (p. 148)
ash flow (p. 148)
assimilation (p. 145)
batholith (p. 150)
Bowen's reaction series (p. 145)
continental volcanic arc (p. 156)
country (wall) rock (p. 150)
decompression melting (p. 142)
dike (p. 150)
extrusive igneous rock (p. 139)
flood basalt (p. 158)
fractional crystallization (p. 145)
geotherm (p. 139)
glassy (p. 154)
heat-transfer melting (p. 142)
hot-spot track (p. 157)
hot-spot volcanoes (p. 157)
igneous rock (p. 137)
interlocking texture (p. 152)
intermediate magmas (p. 143)
intrusive contact (p. 150)
intrusive igneous rock (p. 139)
laccolith (p. 150)
lava (p. 137)
lava flow (p. 137)
mafic magmas (p. 143)
magma (p. 138)
magma contamination (p. 145)
matrix (p. 153)
melt (p. 137)
melting curve (p. 142)
obsidian (p. 154)
partial melting (p. 144)
pegmatite (p. 154)
phaneritic (p. 154)
phenocrysts (p. 154)
pillow basalt (p. 159)
pluton (p. 150)
porphyritic (p. 154)
pumice (p. 154)
pyroclastic debris (p. 139)
pyroclastic rocks (p. 153)
quenching (p. 145)
scoria (p. 154)
silicic magmas (p. 143)
sill (p. 150)
stoping (p. 150)
subaerial volcano (p. 156)
submarine volcano (p. 156)
tabular intrusion (p. 150)
tuff (p. 154)
ultramafic magmas (p. 143)
volatiles (p. 142)
volcanic ash (p. 148)
volcanic island arc (p. 156)
xenoliths (p. 150)

REVIEW QUESTIONS

1. How is the process of freezing magma similar to that of freezing water? How is it different?

2. How did the first igneous rocks on the planet form?

3. How did Hutton show that igneous rocks came from a melt and not from water?

4. Describe the three processes that are responsible for the formation of magmas.

5. Why are there so many different types of magmas?

6. Why do magmas rise to the surface?

7. What factors control the viscosity of a melt?

8. What two criteria would you use to determine if an igneous rock intruded while hot, or simply eroded and became covered by sedimentary rock?

9. What factors control the rate of cooling of a magma within the crust?

10. How does grain size reflect the rate of cooling of a magma?

11. What does the mixture of grain sizes in a porphyritic igneous rock indicate about its cooling history?

12. How does pumice differ from scoria?

13. Describe the way magmas are produced in subduction zones.

14. How are hot-spot volcanoes created?

15. Describe how magmas are produced at continental rifts.

16. Why does melting take place beneath the axis of a mid-ocean ridge?

SUGGESTED READING

Barker, D. S. 1983. *Igneous Rocks.* Englewood Cliffs, N.J.: Prentice-Hall.

Best, M. G. 1982. *Igneous and Metamorphic Petrology.* San Francisco: Freeman.

Cepeda, J. C. 1994. *Introduction to Minerals and Rocks.* New York: Macmillan.

Dietrich, R. V., and B. J. Skinner. 1979. *Rocks and Rock Minerals.* New York: Wiley.

———. 1990. *Gems, Granites, and Gravels: Knowing and Using Rocks and Minerals.* New York: Cambridge University Press.

Hess, P. C. 1989. *Origin of Igneous Rocks.* Cambridge, Mass.: Harvard University Press.

Hughes, C. J. 1982. *Igneous Petrology.* New York: Elsevier.

Hyndman, D. W. 1985. *Petrology of Igneous and Metamorphic Rocks.* New York: McGraw-Hill.

Maaloe, S. 1985. *Principles of Igneous Petrology.* New York: Springer Verlag.

Middlemost, E. A. K. 1985. *Magma and Magmatic Rocks.* London: Longman.

Nichols, J. W., and J. K. Russell. 1990. *Modern Methods of Igneous Petrology: Understanding Magmatic Processes.* Washington, D.C.: Mineralogical Society of America.

Philpotts, A. R. 1990. *Igneous and Metamorphic Petrology.* Englewood Cliffs, N.J.: Prentice-Hall.

CHAPTER 7

A Surface Veneer: Sediments and Sedimentary Rocks

7.1 INTRODUCTION

In the 1950s, Egypt decided to build the Aswan High Dam to trap the water of the Nile River in a huge reservoir. Designers wanted to anchor the dam on solid bedrock for stability, so they hired geologists to find where bedrock lay close to the ground surface. The geologists drilled numerous holes adjacent to the Nile, but to their surprise, no matter where they drilled, they penetrated thick layers of loose sand and mud—in fact, in some places, bedrock lay 1.5 km (4,900 feet) below the surface. Apparently, the home of the pharaohs lies over a huge canyon, about the size of the Grand Canyon in Arizona; but unlike the Grand Canyon, the Nile "canyon" has completely filled with sand and mud, and its floor lies far below sea level (▶Fig. 7.1). The existence of the Nile canyon baffled geologists, because rivers stop flowing when they reach the sea and cannot carve out canyons below sea level. Could the surface of the Mediterranean Sea once have lain 1.5 km below its present level?

The mystery remained unsolved until the summer of 1970, when geologists studying the floor of the Mediterranean stumbled on a major clue. When they drilled holes into the sea floor, they expected to find the shells of plankton that had settled to the sea floor like snow, but instead found a 2-km-thick layer of halite and gypsum, minerals that precipitate out of seawater only when seawater dries up. By calculating the amount of salt held in the Mediterranean, the researchers realized that to yield the layer of halite and gypsum they had observed, the entire Mediterranean would have had to dry up completely and become a desert several times, with the sea refilling between each drying event. Now the formation of the Nile canyon could be explained. When the sea dried up, its floor lay 1.5 km below the present sea level, so the Nile River cut down to meet it. When the Mediterranean refilled with water, this river canyon flooded and filled with sand and gravel up to present-day sea level.

But what could cause the Mediterranean to dry up in the first place? Plate tectonics provides an answer. Only 10% of the water in the Mediterranean enters the

The colorful layers of sedimentary rock, exposed by erosion in the walls of Bryce Canyon, in Utah, were originally deposited in streams, lakes, and floodplains.

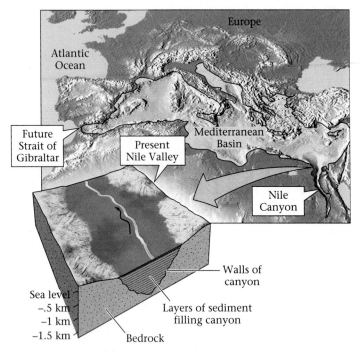

FIGURE 7.1 The present Nile River overlies a much deeper canyon, now filled with layers of sand and mud. The bottom of the canyon lies 1.5 km below sea level.

sea from rivers, and since the sea lies in a hot, dry region, ten times that amount of water evaporates from its surface each year. Thus, most of the water in the Mediterranean enters through the Strait of Gibraltar from the Atlantic Ocean—stop this flow, and the Mediterranean Sea disappears. About 6 million years ago, a geological event took place that did just that. When the northward-drifting African Plate collided with the European Plate, the strait closed. Later, when global sea level dropped a little as it sometimes does, water stopped flowing from the Atlantic over this natural dam, and the Mediterranean evaporated. All the salt that had

been dissolved in its water accumulated as a solid deposit of halite and gypsum on the floor of the resulting basin. Eventually, the sea level rose, and **erosion,** the wearing away and removal of rock and debris, breached the Gibraltar dam, and a gigantic flood rushed from the Atlantic into the Mediterranean, filling the basin again. This process was repeated some two dozen times. Then about 5.5 million years ago, the Mediterranean filled to its present level, and sand, mud, and gravel accumulated in the Nile canyon.

Geologists refer to the kinds of deposits just described—sand, mud, gravel, halite and gypsum, shell fragments—as sediment. **Sediment,** in general, consists of loose rock fragments, grains (small pieces) of minerals broken down from once-intact rock, mineral crystals that precipitate directly out of water, and shells (formed when organisms extract ions from water). Much of what we know about the history of the Earth (including the amazing story of the Mediterranean Sea) comes from studying sediments—not only those that remain **unconsolidated** (loose and not connected), but also those that have transformed into a type of rock called sedimentary rock. Formally defined, **sedimentary rock** is rock that formed at or near the surface of the Earth by the precipitation of minerals from water solutions, by the growth of skeletal material in organisms, or by the cementing together of shell fragments or loose grains derived from preexisting rock. Layers of sediment or sedimentary rock are like the pages of a book, recording tales of ancient events and ancient environments on the ever-changing face of the Earth.

Sediments and sedimentary rocks only occur in the upper part of the crust—in effect, they form a surface veneer, or **cover,** on older igneous and metamorphic rocks, which make up the **basement** of the crust. This veneer ranges from nonexistent, in places where igneous and metamorphic rocks crop out, to almost 20 km thick beneath continental shelves (▶Fig. 7.2). While sediments and sedimentary rocks cover more than 80% of the Earth's surface,

FIGURE 7.2 Near the bottom of the Grand Canyon, we can see the boundary between the sedimentary veneer, or cover (here, a succession of horizontal layers), and the older basement (here, the steep cliff of dark metamorphic rock that goes down to the river). A geologist's sketch emphasizes the contact, or boundary, between cover and basement.

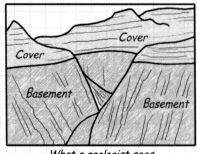

What a geologist sees

they constitute less than 1% of the Earth's mass. Nevertheless, they represent a uniquely important rock type, for they contain the bulk of our energy resources, as we'll see in Chapter 14.

In this chapter, we first learn about weathering and erosion, the process by which sediment forms from fresh bedrock. Some sediment becomes soil, an important global resource, whereas some transforms into new sedimentary rock—we'll see how each develops. We then look at the various kinds of sedimentary rock, and examine the different geological environments in which sedimentary rocks form.

7.2 WEATHERING: THE FORMATION OF SEDIMENT

The Mountains Crumble

If you ever have the chance to hike or drive through granitic mountains, like the Sierra Nevadas of California or the Coast Mountains of Canada, you may notice that in some outcrops the granite surface looks hard and smooth and contains shining crystals of feldspar, biotite, and quartz, while in other outcrops the granite surface looks grainy and rough—feldspar crystals appear dull, biotite flakes have spots of rust, and the rock may peel apart like an onion. Why are these two types of outcrop different? The first type exposes **fresh rock,** rock whose mineral grains have their original composition and shape, while the second type exposes **weathered rock,** rock that has reacted with air and/or water at or near the Earth's surface (▶Fig. 7.3).

Weathering refers to the processes that break up and corrode solid rock, eventually transforming it into sediment. All mountains and other features on the Earth's surface sooner or later crumble away because of weathering. Geologists distinguish between two types of weathering: physical weathering and chemical weathering. Just as a plumber can unclog a drain by using physical force (with a plumber's snake) or by causing a chemical reaction (with a dose of liquid drain opener), nature attacks rocks in two ways.

Physical Weathering

Physical weathering, sometimes referred to as **mechanical weathering,** breaks intact rock into smaller grains or chunks, collectively called **detritus.** Each size range has a name (measurements are grain diameters):

- *boulders*   more than 256 millimeters (mm)
- *cobbles*   between 64 mm and 256 mm

FIGURE 7.3 This outcrop shows the contrast between fresh and weathered granite. The rock below the notebook is fresh—the outcrop face is a fairly smooth fracture. The rock above the notebook is weathered—the outcrop face is crumbly, breaking into grains that have fallen and collected on the ledge.

- *pebbles*   between 2 mm and 64 mm
- *sand*   between 1/16 mm and 2 mm
- *silt*   between 1/256 mm and 1/16 mm
- *mud*   less than 1/256 mm

We refer to boulders, cobbles, and pebbles as coarse sediment, sand as medium sediment, and silt and mud as fine sediment. Many different phenomena contribute to physical weathering.

Jointing. Rocks buried deep in the Earth's crust endure enormous pressure (owing to the weight of overlying rock, the **overburden**) and high temperatures. If deeply buried rocks rise toward the Earth's surface, as happens when erosion removes overlying rock, the pressure squeezing the rocks decreases, and they cool. This causes the rocks to change shape slightly, as a rubber ball will if you squeeze it and then let go. Unlike a rubber ball, however, when rock changes shape near the surface of the Earth, it cracks and breaks into pieces (▶Fig. 7.4a, b). Such naturally formed cracks in rocks are known as **joints.**

Almost all rock outcrops contain joints, some of which are fairly planar, some curving, and some irregular. The spacing between adjacent joints varies from less than a centimeter to tens of meters. Joints can break rock into large or small rectangular blocks, onion-like sheets, irregular chunks, or pillar-like columns (see Chapter 11). Typically, large granite plutons split into onion-like sheets along joints that lie parallel to the mountain face. This process is

Downward pressure

Sedimentary
rock layers

Exfoliation joints

Joint-bounded
blocks

Time

Vertical joints

Bedding

Granite pluton

(a)

(b)

FIGURE 7.4 (a) The weight of overburden creates pressure on rocks at depth. Removal of the overburden by erosion allows once-deep rocks to be exposed at the Earth's surface. (b) The exposure of once-deep rocks causes them to crack. Different rock types crack in different ways. Here, the granite pluton develops exfoliation joints as well as vertical joints, while the sedimentary rock layers develop mostly vertical joints. Joint-bounded blocks break off the outcrop.

called **exfoliation** (▶Fig. 7.5a). Sedimentary rock layers, in contrast, tend to develop vertical joints, and thus break into rectangular blocks (▶Fig. 7.5b). But regardless of their orientation, joints gradually transform once-intact bedrock into a jumble of blocks. Eventually, these blocks fall from the outcrop at which they formed. After a while, they may col-

lect in an apron of rock rubble at the base of a slope—such aprons are called **talus** (▶Fig. 7.5c). The blocks may start their journey when washed away during a storm, when shaken loose by an earthquake or by the wind, or when undermined by the removal of an underlying layer. They may also be wedged free by ice and roots, as we now see.

(a)

(b)

Joint

FIGURE 7.5 (a) Exfoliation joints in the Sierra Nevadas. (b) Vertical joints in sedimentary rock (Brazil). (c) Talus, an apron-shaped pile of fragmental rock, has accumulated at the base of these cliffs near Mt. Snowdon, in Wales.

(c)

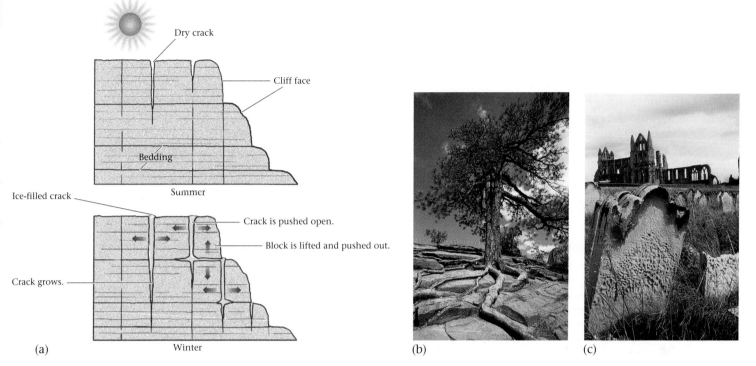

FIGURE 7.6 Examples of processes contributing to physical weathering. (a) During the summer, cracks are closed. During the winter, water in the cracks freezes and forces rocks apart. Ice can even lift blocks up. (b) The roots of this old pine tree in Zion National Park, Utah, originally grew in exfoliation joints. Eventually, the roots pried the rock above the joints free. Thus, the roots are now exposed. (c) These gravestones, near the ruin of a medieval abbey on the seacoast near Whitby, England, absorbed salt from the sea spray. Salt wedging has resulted in honeycomb weathering.

Frost wedging. Freezing water bursts pipes and shatters bottles, because water expands when it freezes and pushes the walls of the container apart. The same phenomenon happens in rock. When the water trapped in a joint freezes, it forces the joint open and may cause the joint to grow. Such **frost wedging** helps break blocks free from intact bedrock (▶Fig. 7.6a). Of course, frost wedging is most common where water periodically freezes and thaws, as occurs in temperate climates or at high altitudes in mountains.

Root wedging. Have you ever noticed how the roots of an old tree can break up a sidewalk? Even though the wood of roots doesn't seem very strong, as roots expand, they apply pressure to their surroundings. Tree roots that grow into joints can push joints open (▶Fig. 7.6b). Even the roots of small plants, fungi, and lichen get into the act, by splitting open small cracks and pores.

Salt wedging. In arid climates, dissolved salt in groundwater crystallizes and grows in open pore spaces in rocks, and pushes apart the surrounding grains. This process, called **salt wedging,** so weakens the rock that when exposed to wind and rain, the rock disintegrates into separate grains

(▶Fig. 7.6c). The same phenomenon happens in coastal areas, where salt spray percolates into surface rock and then dries.

Salt wedging contributed to the destruction of hieroglyphics on the face of Cleopatra's Needle, in New York City's Central Park. The "needle" is a red granite obelisk that was carved in southern Egypt in about 1500 B.C.E. Caesar Augustus moved it, along with a twin (which now stands in London), to Alexandria around 14 B.C.E. During the nineteenth century, the needle was taken down and stored in a basement, where it absorbed salty groundwater. Growing salt crystals began to break apart the rock's surface, so when it was finally erected in Central Park in 1880, the once-sharp hieroglyphics had been blunted. The moist air of New York caused further growth of salt crystals, and further destruction of the hieroglyphics, particularly on the windward side of the needle—these hieroglyphics are now unreadable. Thus, inscriptions that had lasted over 3,000 years in the desert air of Egypt vanished in a century when subjected to salt wedging.

Thermal expansion. When the heat of an intense forest fire bakes a rock, the outer layer of the rock expands. On cooling, the layer contracts. This change creates forces in the rock sufficient to make the outer part of the rock **spall,** or break off in sheet-like pieces.

Animal attack. Animal life also contributes to physical weathering: burrowing creatures, from earthworms to gophers, push open cracks and move rock fragments. And in the past century, humans have become perhaps the most energetic agent of physical weathering on the planet. When we excavate quarries, foundations, mines, or roadbeds by digging and blasting, we shatter and displace rock that might otherwise have remained intact for millions of years more.

Chemical Weathering

Up to now we've taken the "plumber's snake approach" to breaking up rock; now let's look at the "liquid drain opener approach." **Chemical weathering** refers to the chemical reactions that alter or destroy minerals when rock comes in contact with water solutions or air. Because these reactions proceed more quickly in warm, wet conditions, chemical weathering takes place much faster in the tropics than it does in deserts, on high mountain peaks, or near the poles. Chemical weathering in warm, wet climates can produce a layer of rotten rock, called **saprolite,** over 100 m thick. Common reactions involved in chemical weathering include the following.

Dissolution. Chemical weathering during which minerals dissolve into water is called **dissolution.** Dissolution primarily affects salts and carbonate minerals, but even quartz dissolves slightly (▶Fig. 7.7a, b). Some minerals, like halite, can dissolve rapidly in pure rainwater. But some, like cal-

FIGURE 7.7 Weathering by dissolution. (a) A salt crystal consists of ions that can be attracted by polar water molecules. (b) Eventually, water molecules pluck sodium and chlorine ions off the face of the crystal, surround them, and carry them away. (c) Dissolution enlarges joints on the surface of a limestone outcrop and dissolves away sharp edges. In this example from Ireland, wildflowers find a home in the troughs that have formed by dissolution. (d) The chemical weathering of pyrite. Once shiny and metallic, these cubic crystals are now oxidized and dull.

(a)

— Water molecule

— Liquid water

— Salt crystal

(b)

— Dissolved ion

— Pitted crystal face

(c)

(d)

cite, dissolve rapidly only when the water is **acidic,** meaning that it contains an excess of hydrogen ions (H^+) (recall from Chapter 5 that you can identify calcite by its reaction with acid to produce gas bubbles). Acidic water develops quite commonly at the surface of the Earth, because as it falls, rainwater dissolves carbon dioxide gas in the atmosphere to make carbonic acid ($H^+ + HCO_3^-$). Because of the solubility of calcite, limestone and marble (two types of rock composed of calcite) dissolve in wet climates, widening joints and leading to the formation of caverns (►Fig. 7.7c; see Chapter 19).

Hydrolysis. During **hydrolysis,** water chemically reacts with minerals and breaks them down (*lysis* means "loosen" in Greek). Hydrolysis works faster in slightly acidic water. The following formula shows what happens when carbonic acid reacts with potassium (K) feldspar, a common mineral in granite:

$$2KAlSi_3O_8 \quad + \quad 2H^+ + 2HCO_3^- \quad + \quad H_2O \quad \rightarrow$$
potassium feldspar   carbonic acid   water

$$Al_2Si_2O_5(OH)_4 \quad + \quad 2K^+ \quad + \quad 2HCO_3^- \quad + \quad 4SiO_2.$$
kaolinite clay   potassium ions   bicarbonate ions   dissolved silica

Note that some of the atoms making up feldspar become ions that dissolve in water, but some rearrange to form a completely different mineral, kaolinite, a type of clay. In other words, chemical weathering of feldspar produces clay. In the weathered granite described earlier, the dullness of the feldspar indicates the beginning of this reaction. Hydrolysis reactions break down not only feldspars, but other silicate minerals as well—amphiboles, pyroxenes, micas, and olivines all react slowly and transform into clay. Quartz undergoes hydrolysis so slowly that it is a residual product of weathering in most climates. Only in the tropics is there enough rainfall to hydrolyze quartz and produce dissolved silica.

Oxidation. Chemists refer to a reaction during which an element loses electrons as an **oxidation reaction,** because commonly such a loss takes place when elements combine with oxygen. The oxidation, or rusting, of iron, a familiar process, involves two reactions:

$$4Fe^{2+} \quad + \quad 3O_2 \quad \rightarrow \quad 2(Fe^{3+})_2O_3,$$
ferric iron   oxygen   hematite

and

$$4Fe^{2+} \quad + \quad O_2 \quad + 2H_2O \rightarrow 4(Fe^{3+})O(OH).$$
ferric iron   oxygen   water   goethite

In this way, oxidation reactions in rocks transform iron-bearing minerals (such as biotite and pyrite) into a rusty-brown mixture of various iron-oxide and iron-hydroxide minerals, like hematite and goethite (►Fig. 7.7d).

Hydration. **Hydration,** the absorption of water into the crystal structure of minerals, causes some minerals, such as biotite and a type of clay called smectite, to expand.

The relative stability of minerals during chemical weathering. Not all minerals undergo chemical weathering at the same rates. Some weather in a matter of months or years, while others remain unweathered for millions of years. In humid climates, for example, halite and calcite weather faster than most silicate minerals. Of the silicate minerals, those that crystallize at the highest temperatures (such as olivine and pyroxene) are generally less stable under the cool temperatures of the Earth's surface than those that crystallize at lower temperatures (like quartz). The difference depends partly on crystal structure and partly on chemical composition. Specifically, minerals with fewer linkages between silicon-oxygen tetrahedra tend to have weaker structures and thus weather faster than do minerals with stronger structures. And minerals containing ions like iron, magnesium, sodium, potassium, and aluminum tend to weather faster than minerals without them. Thus, quartz (pure SiO_2), which consists of a 3-D network with strong bonds in all directions, is very stable. When you chemically weather a granite (which contains quartz, mica, and feldspar), after enough time everything but quartz has weathered to clay. Beaches typically consist of quartz sand, because quartz is the only mineral left after the other minerals turn to clay and wash away.

Chemical weathering produced by organisms. Until fairly recently, geoscientists tended to think of chemical weathering as strictly an inorganic chemical reaction, occurring entirely independently of life forms. But it is now clear that organisms play a major role in the chemical-weathering process. For example, the roots of plants, fungi, and lichens secrete organic acids that help dissolve minerals in rocks; these organisms extract nutrients from the minerals, and thus undermine the structure of mineral crystals. Amazingly, microbes—nearly invisible bacteria—literally eat minerals for lunch. Bacteria can metabolize an incredible

Fewer cracks,
less surface area

More cracks,
more surface area

Surface area = 6 m^2 Surface area = 12 m^2 Surface area = 60 m^2

FIGURE 7.8 The surface area per unit volume of a block increases every time you break the block into more pieces. For example, a 1 m^2 block has a surface area of 6 m^2 and a volume of 1 m^3. Divide the block into 8 pieces, and the surface area increases to 12 m^2, but the volume stays the same. Divide the block into 1,000 pieces, and the surface area increases to 60 m^2. The rate of chemical weathering increases as the surface area increases, because the weathering reactions occur at the surface—an increase in surface area provides more places for the reactions to take place. (To picture this, recall how fast granular sugar dissolves as compared with a solid cube of sugar.)

range of compounds, depending on the environment they are living in. They pluck off compounds from minerals, and use the energy from the compound's chemical bonds to supply their own life force. Mineral-eating bacteria live at depths of up to a few kilometers in the Earth's crust; below these depths, temperatures are too high for them to survive. If microbes can live off the minerals below the surface of the Earth, can they do so beneath the surface of Mars? Future missions to Mars may provide the answer. (Notably, some researchers claim to have found fossils of microbes in a sample of a meteorite that reached Earth after being knocked off the surface of Mars. But this proposal remains controversial.)

Physical and Chemical Weathering Working in Concert

Although we've looked at the processes of chemical and physical weathering separately here, in the real world they happen together, aiding each other in disintegrating rock to form sediment.

Physical weathering speeds up chemical weathering. Chemical-weathering reactions take place at the surface of a material, so the overall rate at which chemical weathering occurs depends on the ratio of surface area to volume—the greater the surface area, the faster the volume as a whole can chemically weather. When jointing (physical weathering) breaks a large block of rock into smaller pieces, the surface area increases, so chemical weathering happens faster (▶Fig. 7.8).

Similarly, chemical weathering speeds up physical weathering, because chemical weathering—by dissolving away grains or cements that hold a rock together, by transforming hard minerals (like feldspar) into soft minerals (like clay), or by causing minerals to absorb water and expand—makes the rock weaker, so it can disintegrate more easily (▶Fig. 7.9a–d). If you take a block of fresh granite and drop it on the ground, it will most likely stay intact, but if you drop a block of rotten granite on the ground, it will crumble into a pile of sand and clay.

Notably, both kinds of weathering happen faster at edges, and even faster at the corners of broken blocks. This is because weathering attacks a flat face from only one direction, an edge from two directions, and a corner from three directions. Thus, with time, edges become blunt and corners become rounded (▶Fig. 7.10a). In rocks like granite, which do not contain layering that can affect weathering rates, rectangular blocks transform into a spheroidal shape (▶Fig. 7.10b, c).

FIGURE 7.9 Chemical weathering aids physical weathering by weakening the attachments between grains; hold up a complete jigsaw puzzle, and it stays together, but take out some pieces (or make them soft), and the puzzle falls apart. (a) This rock is solid. (b) Susceptible minerals have started to weather. (c) The rock crumbles. (d) Weaker minerals break up or react to form clay and wash away.

Under a given set of environmental conditions, not all rock types weather at the same rate. Some develop more joints than others, and some contain more reactive minerals than others. When different rocks in an outcrop undergo weathering at different rates, we say that the outcrop has undergone **differential weathering.** Because of differential weathering, cliffs composed of a variety of rock layers take on a stair-step or sawtooth-like shape (▶Fig. 7.10d). Weak layers may weather away beneath a more resistant layer, creating an overhang. Similarly, the rate at which the land surface weathers depends on the rock type, so valleys tend to develop over weak rocks, while strong rocks hold up hills.

You can easily see the consequences of differential weathering if you walk through a graveyard. The inscriptions on some headstones are sharp and clear, while those on other stones have become blunted or have even disappeared. That's because the minerals in these different stones have different resistances to weathering. Granite, an igneous rock with a high quartz content, retains inscriptions the longest (though, as illustrated by Cleopatra's Needle, they can be destroyed by salt wedging). Inscriptions on slate, a metamorphic rock composed of recrystallized clay minerals (see Chapter 8), can last for a long time too. But marble, a metamorphic rock composed of calcite, dissolves away rapidly in acidic rain. And some sandstone, held together by calcite cement, crumbles when the cement dissolves. So if you want your headstone to remain readable for generations, choose granite.

FIGURE 7.10 (a) Weather attacks more vigorously at edges and most vigorously at corners, resulting in a rounded block. (b) Over time, solid bedrock becomes a pile of rounded boulders. (c) Rounded blocks of granite in Joshua Tree National Monument, California. (d) Sawtooth shape of an outcrop of weathered sedimentary rock. Weak shale layers are softer than sandstone layers, so the sandstone layers stick out relative to the shale.

7.3 SOIL: DERIVED FROM SEDIMENT AT THE EARTH'S SURFACE

Once sediment forms, it can either (1) evolve in-place at the Earth's surface and change into soil, (2) be carried elsewhere, or (3) become buried by other sediment and transform into sedimentary rock. In this section, we examine the first fate—soil formation. "Soil" is a word that comes up in any discussion of agriculture, forestry, ranching, or home gardening. What makes soil different from just any old sediment? **Soil** consists of sediment that has undergone changes at the surface of the Earth, including reaction with rainwater and the addition of organic material, so that it can support the growth of plant life. Soil is one component of **regolith** (from the Greek *rhegos,* which means "cover"), which refers to any kind of unconsolidated debris that covers bedrock—by "unconsolidated," we mean loose or unconnected. Besides soil, regolith includes accumulations of rock fragments, sand, and gravel that have not evolved into soil.

Three processes taking place at or just below the surface of the Earth contribute to soil formation. First, animals, plants, microbes, and fungi interact with sediment, absorbing nutrients and leaving behind their waste and remains. Second, rainwater enters sediments and percolates downward. Closer to the surface, in a region called the **zone of leaching,** this water dissolves ions and picks up very fine clay; as the water moves downward, it carries the ions and clay with it. Farther down, in the **zone of accumulation,** new minerals precipitate out of the water, and the water leaves behind its load of fine clay. (To picture this phenomenon, imagine making coffee or tea in a leaky cup containing bread crumbs. As you pour water onto the coffee grounds, the water percolates down and *leaches* [absorbs] the coffee flavor. The coffee-saturated water accumulates temporarily in the cup, where it is absorbed by the bread crumbs, so even when the coffee leaks away, some stays attached to the bread crumbs.) Third, burrowing organisms like ants, worms, and gophers churn the soil, so its fabric becomes different from that of the original sediment, and

FIGURE 7.11 During the formation of soil, the downward percolation of water creates a zone of leaching and a zone of accumulation. (a) The same process happens when you pour hot water through coffee grounds or tea leaves into a leaky mug containing bread crumbs (elements in the coffee or tea dissolve in the water and are carried down and collect in the bread crumbs; the water eventually leaks from the mug). (b) In soil, the percolating water carries ions and clay downward. Soil formation also involves the metabolism of microbes and fungi and the addition of organic matter at the surface and underground.

organic material from the ground surface gets mixed in (▶Fig. 7.11a, b).

Because of the way soil forms, it typically contains distinct zones, known as **horizons,** arranged in a vertical sequence called a **soil profile** (▶Fig. 7.12a). Let's look at an idealized soil profile, from top to bottom, using a soil formed in a temperate climate as our example. The highest horizon is the **O-horizon** (the prefix stands for "organic"), so called because it consists almost entirely of humus, partially decayed organic matter. Below the O-horizon, we find the **A₁-horizon,** in which humus has decayed further and has mixed together with mineral grains (clay, silt, and sand). The O-horizon and the A₁-horizon together comprise the nutrient-rich **topsoil,** usually dark in color, which is the portion of soil that farmers till for planting crops—topsoil contains diverse life, such as bacteria, fungi, insects, and rodents. The A₁-horizon grades downward into the **A₂-horizon** (also referred to as the **E-horizon;** E stands for "eluviation," the washing down of fine components), which tends to be lighter-colored and contain less organic material than the overlying topsoil.

Water percolates downward through the O- and A-horizons and carries dissolved ions and clay into the **B-horizon,** or **subsoil,** where the ions precipitate to form new minerals (such as iron oxides), and the clay collects. Thus, the O- and A-horizons represent the zone of leaching, while the B-horizon represents the zone of accumulation. Commonly, the B-horizon is somewhat red, because of the presence of iron-oxide minerals and the lack of organic matter.

Finally, at the base of a soil profile, we find the **C-horizon,** which consists of material derived from the substrate (base) that's been chemically weathered and broken apart. If the substrate consists of bedrock, the C-horizon grades downward into unweathered bedrock; if the substrate consists of sediment, the C-horizon grades down into unweathered sediment.

As farmers, foresters, and ranchers well know, the soil in one locality can differ greatly from the soil in another, in both composition and thickness. And crops that grow well in one type of soil may wither and die in another. Such diversity exists because the makeup of a soil depends on several factors (▶Fig. 7.12b–d).

(a)

(b)

(c)

(d)

FIGURE 7.12 The process of soil formation results in distinctive soil profiles. (a) In this soil exposed on a cliff face, the dark layer (horizon) at the top is the organic rich layer. Because of the redistribution of elements, the different horizons have different colors. The thickness of a soil at a given latitude depends on (b) the composition of the substrate (base), because less resistant underlying rocks weather more deeply; (c) the steepness of a slope, because soil washes or slides off steep slopes; and (d) the duration of soil formation; young soil is thinner than an old soil.

- *Substrate:* Some soils form on basalt, some on granite, some on ash, some on recently deposited sediment. These different substrates consist of different materials, so the soils formed on them end up with different chemical compositions. For example, a soil formed on basalt tends to be richer in iron than a soil formed on granite, because basalt is mafic, while granite is silicic. Also, soils tend to develop faster on unconsolidated material (ash or sediment) than on bedrock.

- *Climate:* Rainfall quantities determine how much leaching takes place. Large amounts of rainfall accelerate chemical weathering and leach most of the soluble minerals. In regions with small amounts of rainfall, chemical weathering rates are slow, and soils can retain unweathered minerals and soluble components.

- *Slope steepness:* A thick soil can accumulate under flat-lying land, but on a steep slope regolith may wash away before it can evolve into a soil. Thus, all other factors being equal, soil thickness increases as the slope angle decreases.

- *Duration of soil formation:* Because soil formation is an evolutionary process, a young soil tends to be thinner and less evolved than an old soil.

- *Vegetation type:* Different kinds of plants extract or add different nutrients and quantities of organic matter to a soil. Also, some plants have deeper root systems than others, and more effectively prevent soil from washing away.

Soil scientists recognize many different classes of soil. For example, a temperate-climate soil formed on granite is called a **pedalfer soil;** this type of soil is characterized by well-defined soil horizons and an organic-rich A-horizon (▶Fig. 7.13a). **Pedocal soils,** which are formed in desert climates, tend to be thin. The A-horizon in a pedocal soil contains unweathered minerals, rock fragments, and a relatively high concentration of soluble minerals like calcite, but very little organic matter. Calcite in a pedocal soil collects in the B-horizon and cements the soil together, creating a solid mass called **caliche** (▶Fig. 7.13b). In a **laterite soil,** which is formed on iron-rich rock in a

FIGURE 7.13 (a) In a pedalfer soil formed in a temperate climate, because of the moderate amount of rainfall, materials leached from the A-horizon can accumulate in the B-horizon. In this example, the C-horizon consists of weathered granite. (b) Because of low rainfall, a thin pedocal soil in a desert has only a thin A-horizon. Soluble minerals, specifically calcite, that would be washed away in a temperate climate can accumulate in the B-horizon, creating caliche. In this example, the C-horizon consists of weathered limestone. (c) In a thick tropical laterite soil, so much water percolates down from the heavy rainfall that all reactive minerals dissolve or break down and get carried away. This leaves only a residue of iron oxide and/or aluminum oxide (these are very stable). There is no real zone of accumulation, but at depth, iron-rich clays collect. Here, the C-horizon is weathered metamorphic rock. (d) Soil thickness varies with latitude, because of variations in temperature, rainfall, and vegetation.

tropical environment, an abundance of percolating water leaches just about all mineral components, leaving only a dark-red mass of insoluble iron and/or aluminum oxide; this soil has no B-horizon (▶Fig. 7.13c). (If a laterite soil contains abundant aluminum, we call it **bauxite.**) Considering the lushness of rain forests, you might expect thick humus in a laterite soil, but in fact, organic matter in a tropical climate decays so rapidly that such soils generally contain hardly any humus at all. Rain-forest trees can grow permanently in a laterite soil, but crop plants use its nutrients so fast that farm fields created by clear-cutting a rain forest typically become infertile in a year or two, after which even the rain-forest trees cannot return. The character and thickness of a soil vary with latitude, because of differences in climate and vegetation (▶Fig. 7.13d).

Clearly, soils take time to form, so those capable of supporting agriculture or forests should be considered a natural resource worthy of protection. Unfortunately, practices such as agriculture, overgrazing, and clear-cutting, which remove the cover of vegetation that protects soil, have led to **soil erosion,** the removal of soil by wind or runoff. When this happens, heavy rainfall washes soil into rivers (▶Fig. 7.14), and wind storms rip the soil away and suspend it in the air as a dark cloud. During the 1930s, a succession of droughts killed off so much vegetation in the American high plains that wind stripped the land of soil and caused devastating dust storms.

FIGURE 7.14 After a heavy rain, soil erosion has taken place on this farm field. Unprotected by vegetation, the rain has carved gullies, carrying away soil in the process.

7.4 INTRODUCING SEDIMENTARY ROCKS

So far, we've learned that weathering attacks rock and breaks it down into sediment, which includes fragments of the original rock, grains of durable minerals, and new grains formed by chemical reactions. In addition to these solid materials, weathering yields dissolved ions. What happens next? Solid grains get picked up and moved to a new location by wind, water, or ice until they eventually settle out. Then they either become incorporated in soil or become buried deeply. Dissolved ions are either carried to the sea in streams or transported underground in groundwater. Eventually, the ions precipitate as new minerals in the sea or as new minerals filling open spaces between grains underground. New sedimentary rocks develop from buried sediment that gets stuck or cemented together by precipitating minerals, or from minerals that have precipitated out of open water, with or without the help of organisms.

Clearly, sedimentary rocks can form from a variety of materials in a variety of settings. Thus, there are many different kinds of sedimentary rock. For most discussions, geologists divide sedimentary rocks into four major classes, based on their mode of origin. (1) **Clastic,** or **detrital, sedimentary rocks** consist of cemented-together detritus (solid fragments and grains) derived from preexisting rocks ("clastic" comes from the Greek *klastos,* meaning "broken"). (2) **Biochemical sedimentary rocks** are made up of the shells of organisms. (3) **Organic sedimentary rocks** consist of carbon-rich relicts of plants. And (4) **chemical sedimentary rocks** are made up of minerals that precipitate directly from water solutions. In some situations, it is also useful to distinguish among different kinds of sedimentary rocks based on their composition. **Siliceous rocks** contain quartz, **argillaceous rocks** contain clay, and **carbonate rocks** contain calcite or dolomite.

7.5 CLASTIC SEDIMENTARY ROCKS

Formation

Nine hundred years ago, a thriving community of Native Americans inhabited the high plateau of Mesa Verde, Colorado. In the hollows beneath huge overhanging ledges, they built multistory stone-block buildings that have survived to this day. Clearly, the blocks are solid and durable—they are, after all, rock. But if you were to rub your thumb along one, it would feel gritty, and small grains of quartz would break free and roll under your thumb, for the block consists of quartz sand grains cemented together. Geologists call such rock a sandstone.

Sandstone is an example of **clastic,** or **detrital,** sedimentary rock, rock created from solid grains (**clasts,** or detritus) stuck together to form a solid mass. The grains can consist of individual minerals (grains of quartz or flakes of clay) or fragments of rock (pebbles of fresh granite). The loose grains of sediment transform into clastic sedimentary rock by the following five steps (▶Fig. 7.15a, b).

- *Weathering (detritus formation):* Detritus forms by weathering either directly on an outcrop or later, when grains in regolith break down as they are moving in wind, water, or ice.

- *Erosion:* Once formed, sediment either falls off an outcrop and becomes incorporated in regolith or gets picked up by moving wind, water, or ice. **Erosion** refers to the combination of processes that loosen rock or regolith, separate it from its substrate, and carry it away.

- *Transportation:* Wind, water, or ice transports the sediment some distance. The ability of a medium to carry sediment depends on its viscosity and velocity. Solid ice can carry sediment of any size, regardless of how slowly the ice moves. Very fast moving, turbulent water can transport coarse fragments (cobbles and small boulders), moderately fast moving water can carry only sand and gravel, and slowly moving water carries only silt and mud. Strong winds can move sand and dust, while gentle breezes carry only dust. Thus, during transportation, sediment can be **sorted** by size.

- *Deposition:* **Deposition** is the process by which sediment settles out of the transporting medium. When the ice of a glacier melts, its sediment load settles on the ground. Sediment settles out of wind or moving water when these fluids slow, because as the velocity decreases, the fluid no longer has the ability to move sediment. For example, when wind dies down, the sand and dust that it carried settles out.

- *Lithification (compaction and cementation):* Geologists refer to the transformation of loose sediment into solid rock as **lithification.** The formation of clastic sedimentary rocks first requires the burial of the sediment by more sediment.

FIGURE 7.15 (a) The basic steps during the development of a sedimentary rock: weathering → erosion → transportation → deposition → lithification. (b) As sediment moves from its source to the site of deposition, it becomes finer-grained.

When the sediment has been buried deep in the ground, pressure caused by the overburden (overlying rock) squeezes out water and air that had been trapped between clasts, and clasts press together tightly. We refer to this phase as **compaction** (▶Fig. 7.16). Mud, a mixture of clay and water, compacts (that is, decreases in volume) by 50–80% when buried. Sand, on the other hand, compacts by only 10–20%. Compacted sediment may then be bound together to make coherent sedimentary rock by the process of **cementation. Cement,** consisting of minerals (commonly quartz or calcite) that precipitate from groundwater, partially or completely fills the spaces between clasts and attaches each grain to its neighbor.

Classification

Let's look again at the fate of a rock fragment, and its component mineral grains, as it gradually moves from a cliff face in the mountains via a river to its ultimate resting place on a seashore (Fig. 7.15). We'll see that different kinds of sediment develop along its route. And each of these different types of sediment, when buried and lithified, produces a different kind of sedimentary rock.

Imagine that a large block of granite breaks off a cliff, falls, and slams into a block already at the bottom. The impact shatters the granite block into a pile of angular fragments, with sharp edges and corners. If these fragments were to be cemented together, the resulting rock would be **breccia.** Later, a hurricane passes through, and floodwaters pick up the smaller fragments and carry them into a turbulent river, where they bang into each other and into the bed of the stream, a process that shatters them into still smaller pieces and breaks off their sharp edges. The angular clasts become rounded clasts, with smooth edges and corners, ranging in size from pebbles to cobbles or small boulders, and in shape from spherical to elliptical or pancake-shaped. Eventually, after the flood, the pebbles and cobbles, along with sand, come to rest, forming a bar of gravel; the burial and lithification of these rounded clasts produce **conglomerate.** (Some conglomerates consist entirely of similar-sized clasts, while some consist of large clasts, like cobbles, surrounded by small clasts, like sand. In such cases, we refer to the smaller clasts as the **matrix** of the rock.)

If the gravel stays put for a long time, it undergoes chemical weathering: biotite in the clasts hydrate and expand, and some feldspar grains hydrolyze and transform into clay. As a consequence, the cobbles and pebbles begin to break apart into individual mineral grains, a mixture of quartz, weathered biotite, and remaining feldspar, and new clay. The next storm that comes along causes the river to flood, and these sediments start to wash downstream; as the grains rub against each other, they become rounder. The clay is so fine that it's washed away entirely, but when the river flow slows down again, the sand-sized sediment settles

FIGURE 7.16 The process of lithification. As sediment is buried, it becomes compacted (expelling the water between the grains), and the grains pack tightly together. Groundwater passing through the rock precipitates ions to form mineral cements that bind the grains together. If there is clay in the rock, weak chemical bonds may cause the clay grains to stick to one another.

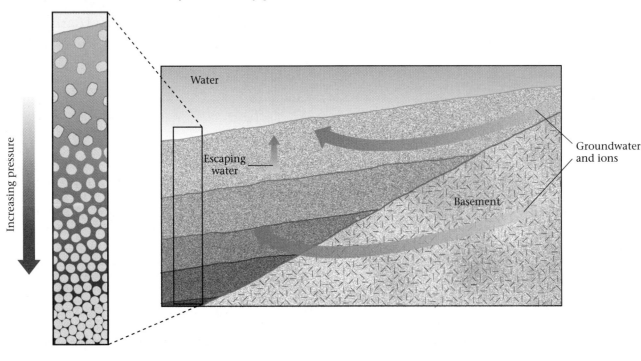

out to form sand bars. In these bars, we find a mixture of quartz and some feldspar grains—this sediment, when buried and lithified, becomes **arkose.** Whatever feldspar grains remain in the sand bar continue to weather into clay, so gradually during successive storms that transport the sediment farther downstream, the sand loses feldspar and ends up being composed entirely of quartz grains—this sediment, when buried and lithified, becomes quartz **sandstone.** Still farther downstream, in slower-moving sections of the river and at the coast where the river reaches the sea and slows to a halt, silt (very fine sand) and mud (a mixture of clay and water) accumulate. The silt lithifies into **siltstone,** and the mud lithifies into **shale** (if it breaks into thin sheets) or **mudstone** (if it does not split into sheets).

Breccia, conglomerate, arkose, sandstone, siltstone, shale, and mudstone are the most common types of sedimentary rocks. But there's one additional type that does not form in rivers—**graywacke,** a rock that consists of a range of grain sizes, from mud to coarse sand or pebbles, all mixed together. The finer clasts, in which the larger ones are suspended, comprise the matrix of the rock. Typically, some of the coarser clasts are not single mineral grains, but are actually chips or fragments of rocks, called **lithic clasts.** The sediment comprising graywacke typically collects in submarine avalanches on the margins of continents.

From the discussion so far, we see that three factors come into play when classifying clastic sedimentary rocks: composition (mineral content), grain size, and grain shape. In sequence from coarsest to finest, the basic types of clastic sedimentary rock are (▶Fig. 7.17a–d):

- *very coarse*   breccia (angular fragments)
  conglomerate (rounded clasts)
- *coarse*   arkose (sand-sized quartz plus feldspar grains)
  sandstone (almost pure quartz sand; other compositions are possible)
  graywacke (a mixture of different clast sizes; contains lithic clasts)
- *fine*   siltstone (fine sand)
- *very fine*   shale (clay; splits into thin sheets)
  mudstone (clay; does not split into thin sheets)

We have observed that different sizes of sediment collect at different places along the river—that's because moving fluids (water or air) sort sediment by size, gradually winnowing out finer-grained sediments and carrying them downstream, leaving behind coarse sediments. Geologists use the term **sorting** to refer to the range of clast sizes in a collection of sediment (▶Fig. 7.18a). Samples in which all clasts have the same size are well sorted, while those in which a variety of grain sizes occur are poorly sorted

(graywacke, with sand and lithic clasts embedded in a mud matrix, is an example of a poorly sorted rock).

Our narrative also illustrates that clasts can be distinguished from each other by grain shape (▶Fig. 7.18b). Geologists describe shape in two ways. **Sphericity** measures the degree to which a clast approaches the shape of a sphere; clasts with low sphericity, for example, may be shaped like blades or bricks. **Angularity** measures the degree to which grains have sharp or rounded edges and corners. Basically, all these terms simply provide a shorthand means for describing a rock. If a geologist says, "I saw an outcrop containing a poorly sorted conglomerate composed of sand- to cobble-sized subrounded clasts with moderate sphericity," a clear image of the rock should immediately come to mind. How does this description compare with the sample in Figure 7.17a?

Finally, you may have noted from our narrative that as sediment moves downstream, grains overall become smaller and rounder, and grains composed of minerals that are susceptible to chemical weathering progressively break down and disappear. Geologists use the term **sediment maturity** to refer to the degree to which a sediment has evolved from being just a crushed-up version of the original rock to a sediment that has lost its easily weathered components and has become well sorted and rounded. Thus, we can say that an immature sandstone is one that contains large angular clasts, is poorly sorted, and contains easily weathered minerals, while a mature sandstone is one that contains only small, rounded grains of resistant minerals (▶Fig. 7.18c).

7.6 BIOCHEMICAL AND ORGANIC SEDIMENTARY ROCKS: BYPRODUCTS OF LIFE

The Earth system involves interactions between living organisms and the physical planet. Already in this chapter, we've seen how living organisms play a role in chemically weathering rocks by secreting chemicals or by physically forcing open cracks. Here, we learn how some living organisms play a role in creating the materials comprising sedimentary rocks.

Numerous organisms have developed the ability to extract dissolved ions from seawater to make solid shells. Some organisms construct their shells out of calcium (Ca^{2+}) and carbonate (CO_3^{2-}) ions, which they merge to make the minerals calcite ($CaCO_3$) or its polymorph, aragonite, while other organisms make their shells out of dissolved silica. When the organisms die, the solid material in their shells turns into sediment that eventually becomes incorporated in the class of sedimentary rocks called

FIGURE 7.17 Sediments and the rocks that form from them.
(a) Gravel, like this beach deposit, becomes conglomerate.
(b) Desert sand forms sandstone, like this ledge at Mesa Verde,
Colorado. Centuries ago, Native Americans built a village, from
sandstone blocks, under the ledge. (c) Mud forms shale. This
shale bed (with a coin for scale) has split into thin sheets.
(d) This electron photomicrograph shows the clay flakes in a
shale. Each flake is about 0.001 mm across.

FIGURE 7.18 (a) In a poorly sorted sediment, there is a great variety of different grain sizes, while in a well-sorted sediment, all the grains are the same size. (b) A grain with high sphericity (top row) has roughly the same length in all directions, while one with low sphericity is elongate or flattened (bottom row). Sphericity is independent of roundness, which refers to whether the grain has sharp edges or not. Grains on the left are more angular than grains on the right. (c) As sediments are transported progressively farther, weatherable sediments like feldspar break down and convert to clay, which washes away, so the proportion of sediment consisting of resistant minerals like quartz increases. Further, the physical bouncing and grinding that accompanies transport progressively rounds the quartz grains and sorts them.

biochemical sedimentary rocks. Thus, biochemical sedimentary rocks consist of shells and shell debris—the soft flesh of the organisms that constructed the shells either rots away or transforms into oil (see Chapter 14). Plants, such as trees, ferns, grass, and moss, also yield materials that can be incorporated in sedimentary rocks, but since plants do not have shells, the rocks formed from plant debris consist mostly of organic chemicals and pure carbon derived from organic chemicals. We therefore call rocks made of altered plant material **organic rocks.**

Limestone (Biochemical Type)

A snorkeler gliding above the Great Barrier Reef of Australia, a buildup of marine life in shallow seawater, sees an incredibly diverse community of coral and algae, around which creatures such as clams, oysters, snails (gastropods), and lampshells (brachiopods) live, and above which plankton float (▶Fig. 7.19a). Though they all look so different from one another, many of these organisms share an important characteristic: they make solid shells of calcite (or its polymorph, aragonite). When the organisms die, their skeletons either stay in place, as is the case with reef builders like coral, or are transported to another location, where they eventually settle out. During transport, shells may break up into small fragments. Rocks formed from the calcite or aragonite skeletons of organisms are the biochemical version of **limestone** (▶Fig. 7.19b, c). Note that geologists use the name "limestone" for any rock composed of calcite (and/or aragonite), regardless of origin; later, we'll meet chemical limestone, a type that precipitates directly out of water.

There are a great variety of limestones, differing from one another according to the material from which they formed. Some retain the internal structure of coral colonies, some consist of large, angular shell fragments, some are made up of rounded grains of calcite that rolled about in the surf, some consist of lime mud (tiny grains of calcite formed by the breakup of larger grains, or tiny shards of calcite that were embedded in the tissue of algae), and some form from plankton shells (particularly from microscopic animals called **foraminifera,** or **forams** for short). Three common types are **fossiliferous limestone,** consisting of identifiable shells and shell fragments, **micrite,** consisting of lime mud, and **chalk,** consisting of plankton shells.

Typically, ancient limestone, as exposed on the high cliffs of the Canadian Rockies, is a massive light-gray to dark-bluish-gray rock that breaks into chunky blocks—it doesn't look much like a pile of shell fragments. That's because with time, the calcite and aragonite in limestone tend to **recrystallize,** meaning that the ions composing them rearrange into new crystals: all aragonite transforms into calcite, a more stable mineral, and smaller crystals of calcite merge to make larger ones.

Chert (Biochemical Type)

If you walk beneath the north end of the Golden Gate Bridge in San Francisco, you will find outcrops of reddish, almost porcelain-like rock occurring in 3–15-cm-thick layers, one on top of another in a sequence with a total thickness of hundreds of meters (▶Fig. 7.20a). Hit it with a hammer, and the rock cracks, almost like glass, creating smooth, spoon-shaped (conchoidal) fractures. Geologists call this rock biochemical **chert;** it's made from **cryptocrystalline quartz** (*crypto* is Greek for "hidden"), quartz grains that are too small to be seen without the extreme magnification of an electron microscope. The chert beneath the Golden Gate Bridge formed from the silica shells of plankton (particularly of microscopic animals called radiolaria), which accumulated to form an ooze, or gel, at the floor of the deep ocean. Gradually, after deep burial, this ooze solidified to make chert.

By the way, how did this chert end up beneath the Golden Gate Bridge? About 145 million years ago, the western United States was a convergent plate boundary. As the Pacific Ocean floor slipped beneath North America during subduction, the chert was scraped off the downgoing plate,

FIGURE 7.19 (a) In this modern coral reef, corals produce shells of calcite or aragonite. If buried and preserved, these become limestone. (b) A quarry face in Vermont shows the typical gray color of limestone. The thinly laminated layers are lime mud; the white mounds are relicts of small reefs. (c) This particular specimen of fossiliferous limestone consists entirely of small fossil shells and shell fragments. Not all fossiliferous limestones contain such a high proportion of fossils.

(a)   (b)   (c)

FIGURE 7.20 (a) This bedded chert, which crops out near the foundation of the Golden Gate Bridge in San Francisco, developed on the deep sea floor by the deposition of radiolaria, a type of plankton that secretes silica shells. Later, it was scraped off the sea floor by subduction and became part of an accretionary prism. The bends in the layers, called folds (see Chapter 11), formed when the layers were squeezed and wrinkled as they were scraped off the sea floor. (b) Coal is deposited in layers just like other kinds of sedimentary rocks. Here, we see a coal seam (a miner's term for a coal layer) between layers of sandstone and shale.

like snow in front of a plow, and became incorporated in an accretionary prism that grew between the deep-ocean trench and the Sierra Nevada volcanic arc. The foundation of the Golden Gate Bridge stands on the remnants of this accretionary prism.

Organic Rocks: Coal and Oil Shale

The Industrial Revolution of the nineteenth century, which transformed the world's economy from an agricultural to an industrial base, depended on power provided by steam engines. After decimating forests to provide fuel for these engines, industrialists turned to coal. **Coal** is a black, combustible rock consisting of over 50% carbon, and so differs markedly from the other sedimentary rocks discussed so far—the carbon of coal occurs as pure carbon or as an element in organic chemicals, not in minerals. Still, we consider coal a sedimentary rock because it is made up of detritus (of plants) deposited in layers (▶Fig. 7.20b). We'll look more at coal formation in Chapter 14. Here, we simply need to know that the organic chemicals making coal come from the remains of plant material that died and accumulated on the floor of a forest or swamp. The remains were buried deeply, and the heat and pressure at depth compacted the plant material and drove off volatiles (hydrogen, water, carbon dioxide, ammonia), leaving a concentration of carbon.

Organic material, in the form of chemicals derived from fats and proteins that comprised the flesh of plankton, commonly mix with mud and may be incorporated in shale. The organic material, which tends to color shale black, gradually transforms into oil. When the percentage of organic matter exceeds 50%, we call the rock **oil shale.**

7.7 CHEMICAL SEDIMENTARY ROCKS

The colorful terraces, or mounds, around the vents of hot-water springs, the immense layers of salt that underlie the floor of the Mediterranean Sea, the smooth, sharp point of an ancient arrowhead—these materials all have something in common. They all consist of rock formed by the precipitation of minerals out of water solutions at the surface of the Earth. We call such rocks **chemical sedimentary rocks.** They differ from clastic rocks in that they do not contain clasts of preexisting rocks and can become solid rock without undergoing burial, compaction, or cementation. And they differ from biochemical rocks in that the minerals comprising them precipitate, at least in part, without the life activity of organisms. Chemical sedimentary rocks typically have a crystalline texture, partly formed during the original precipitation and partly the result of later recrystallization.

Evaporites: The Products of Saltwater Evaporation

In 1965, two daredevil drivers in jet-powered cars battled to be the first to break the land-speed record of 600 mph. On November 7, Art Arfons, in the "Green Monster," peaked at 576.127 mph, but eight days later Craig Breedlove, driving the "Spirit of America," touched the 600.601 mph mark. Traveling at such speeds, a driver must maintain an absolutely straight course; any turn will catapult the vehicle out of control, because its tires simply can't grip the ground. So the race course must be extremely long and flat. Not many places meet that criterion—the Bonneville Salt Flats, near the Great Salt Lake of central Utah, do.

How did this vast salt plain come into existence? Streams bringing water from Utah's Wasatch Mountains into the Salt Lake basin, like all streams, carry trace amounts of dissolved ions (provided to the water by chemical weathering). Most lakes have an outlet, so the water in them constantly flushes out and the ion concentration stays low. But the Great Salt Lake has no such outlet, so water escapes from the lake only by evaporating. Evaporation removes just the water; dissolved ions stay behind, so over time, the lake water has become a concentrated solution of dissolved ions—in other words, very salty (▶Fig. 7.21a). In the past, when the region had a wetter climate, the Great Salt Lake was larger and covered the region of the Bonneville Salt Flats. Along its shores, water dried up and salt precipitated. When the Great Salt Lake shrank to its present dimension,

the vast extent of the Bonneville Salt Flats was left high and dry, and covered with salt (▶Fig. 7.21b). Such salt precipitation occurs wherever there is saturated saltwater—along desert lakes with no outlet (for example, the Dead Sea, between Israel and Jordan) and along margins of restricted seas (the Persian Gulf, where seawater evaporates faster than new seawater can replace it). For thick deposits of salt to form, large volumes of water must evaporate (▶Fig. 7.21c, d). This may happen when plate tectonic movements temporarily cut off arms of the sea (as we saw in the case of the Mediterranean Sea) or during continental rifting, when seawater first begins to spill into the rift valley.

Because salt deposits form as a consequence of evaporation, geologists refer to them as **evaporites.** The specific type of salt comprising an evaporite depends on the amount of evaporation. When 80% of the water evaporates, gypsum forms; and when 90% of the water

FIGURE 7.21 (a) In lakes with no outlet, the tiny amount of salt brought in by "fresh" water streams stays behind as the water evaporates. Along the margins of the lake, salts precipitate. If the whole lake evaporates, a flat surface of salt forms. (b) Recently deposited evaporites along the margin of a salt lake. (c) Salt precipitation can also occur along the margins of a restricted marine basin, if saltwater evaporates faster than it can be resupplied. The entire restricted sea may dry up if it is cut off from the ocean. (d) Thick layers of salt accumulate in a rift and later are buried deeply. The salt then recrystallizes. Here, salt layers are being mined.

evaporates, halite precipitates. If seawater were to evaporate entirely, the resulting evaporite would consist of 80% halite and 13% gypsum; the remainder consists of other salts and carbonates.

Travertine (Chemical Limestone)

Around some hot springs, places where hot-water solutions spill out of the Earth (see Chapter 19), terraces of chemical limestone accumulate (▶Fig. 7.22a). Such buildups develop because when the hot water reaches the ground surface, it cools and degasses (meaning that dissolved carbonic acid bubbles off as carbon dioxide gas), and thus can dissolve less calcite; as a result, the calcite precipitates to form rock. Calcite precipitation similarly occurs on the walls of caves, where groundwater seeps out, building intricate growth forms (▶Fig. 7.22b). Geologists refer to the rock making up the buildups on cave walls and around hot springs as **travertine.** Since travertine consists of calcite, we consider it a type of limestone.

Though geologists traditionally view travertine as a chemical precipitate, new observations suggest that complex microbial communities inhabit groundwater and may be involved in the precipitation of calcite to create travertine. Regardless of its exact origin, travertine makes a popular building stone, because if you slice into it, you find a beautiful succession of layers, representing seasonal episodes of precipitation.

Dolostone: Replacing Calcite with Dolomite (a Type of Diagenesis)

Dolostone differs from limestone in that it contains the mineral **dolomite** ($CaMg[CO_3]_2$). Most dolomite forms by a chemical reaction between solid calcite and magnesium-bearing groundwater. Much of the dolostone you may find in an outcrop actually originated as limestone but later recrystallized so that dolomite replaced the calcite. This recrystallization—which may take place beneath lagoons along a shore soon after the limestone formed, or a long time later, after the limestone has been buried—may destroy fossils in the rock. Geologists classify dolostone as a chemical rather than a biochemical sedimentary rock because dolomite crystals form by chemical reaction, not directly by the activity of a living organism. The formation of dolostone is a type of **diagenesis,** the recrystallization of already formed sedimentary rock as a result of pressure, increased temperature, and reaction with water. The process of diagenesis occurs at temperatures of less than about 250°C, too cool for metamorphism to take place.

Chert (Replacement Type)

A tribe of Native Americans, the Onondaga, once inhabited the eastern part of New York State. In this region, outcrops of limestone contained layers of a black chert (▶Fig. 7.22c). Because of the way it breaks, artisans could fashion sharp-edged tools (arrowheads and scrapers) from this chert, so

FIGURE 7.22 (a) A travertine buildup at Mammoth Hot Springs. (b) A travertine buildup on the wall of a cave. (c) Replacement chert occurring as nodules in a limestone.

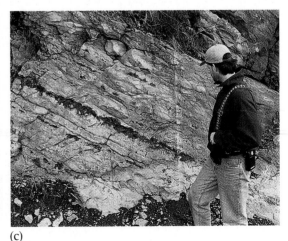

(a)   (b)   (c)

the Onondaga collected it for their own toolmaking industry and for use in trade with other tribes. Unlike the deep-sea (biochemical) chert described earlier, the chert collected by the Onondaga formed when cryptocrystalline quartz gradually replaced calcite crystals within a body of limestone after the limestone was deposited; geologists thus call it **replacement chert.**

Chert comes in many colors (black, white, red, brown, green, gray), depending on the impurities it contains. Black chert, or **flint,** made the tools of the Onondaga. Red chert, or **jasper,** which like all chert takes on a nice polish, makes beautiful jewelry. **Petrified wood** is chert that's made when ash from a volcanic eruption buries a forest. The ash contains silica, which dissolves in groundwater that passes into trees. Dissolved silica precipitates as cryptocrystalline quartz within wood, gradually replacing the wood's cellulose. The chert retains the shape of the wood and even its growth rings. Some chert, known as **agate,** grows in concentric rings inside hollows in a rock and ends up with a striped appearance, caused by variations in the content of impurities while precipitation took place.

7.8 SEDIMENTARY STRUCTURES

In the photo of a stark ledge in Figure 7.10d, note the distinct lines across its face. In 3-D, we see that these lines are the traces of individual surfaces, separating the rock into sheets. In fact, sedimentary rocks in general contain distinctive layering. The layers themselves may have a characteristic internal arrangement of grains or distinctive markings on their surface. We use the term **sedimentary structure** for the layering of sedimentary rocks, surface features on layers formed during deposition, and the arrangement of grains in beds. Now let's see how these structures provide important clues to the environment in which the sediment was deposited.

Bedding and Stratification

Geologists have developed a jargon for sedimentary structures. A layer of sediment or sedimentary rock with a recognizable top and bottom is called a **bed,** or **stratum;** the boundary between two beds is a **bedding plane;** several beds together comprise **strata;** and the overall arrangement of sediment into a sequence of beds is its **bedding,** or **stratification.** From the word "strata," we derive other words, like **stratigrapher** (a geologist who specializes in studying strata) and **stratigraphy** (the study of strata).

When you examine a sequence of strata in a region with good exposure, the bedding generally stands out clearly. Beds appear as bands across a cliff face (Fig. 7.10d). Typically, a contrast in rock type distinguishes one bed from adjacent beds. For example, a sequence of strata may contain a bed of sandstone overlain by a bed of shale, overlain by a bed of limestone. Each bed has a definable thickness (from a couple of centimeters to tens of meters) and some marked contrast (composition, color, grain size) that separates it from its neighbors. But in many examples, adjacent beds all appear to have the same composition. In such cases, bedding may be defined by subtle changes in grain size, by surfaces that represent interruptions in deposition, or by joints that have formed parallel to the flat surfaces of grains pressed together during compaction.

Why does bedding form? To find the answer, we need to think about how sediment is deposited. Changes in the source of sediment, climate, or water depth control the type of sediment deposited at a location at a given time. For example, on a normal day a slow-moving river may carry only silt, which is deposited at the mouth of the river. During a flood, the river carries sand and pebbles, so a layer of gravel forms over the silt layer. Then when the flooding stops, more silt buries the gravel. If these sediments become lithified and exposed for you to see, they appear as alternating beds of siltstone and sandy conglomerate (▶Fig. 7.23a–d).

FIGURE 7.23 Bedding forms as a result of changes in the environment. (a) During a normal river flow, a layer of silt is deposited. (b) During a flood, turbulent water brings in a layer of coarse sand and gravel. (c) When the river returns to normal, another layer of silt is deposited. (d) Later, after lithification, uplift, and exposure, a geologist sees these layers as beds on an outcrop.

(a)   Basement (substrate)   Silt

(b)   Gravel

(c)   Gravel   Silt

(d)   Conglomerate   Siltstone / Bed / Bedding plane

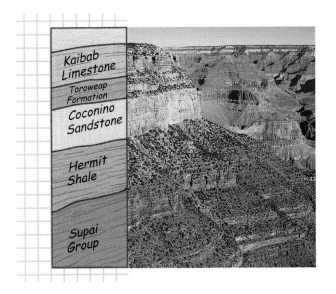

FIGURE 7.24 A particularly thick bed, a sequence of beds of the same composition, or a sequence of beds of alternating rock types can be called a stratigraphic formation, if the sequence is distinctive enough to be traced across the countryside. In this photo of the Grand Canyon, we can see five formations. Formations that consist primarily of one rock type may take the rock-type name (for example, Kaibab Limestone), but formations containing more than one rock type may just be called "formation." The Supai "Group" is a group because it consists of several related formations, which are too thin to show here. Formations and groups are examples of stratigraphic units. Note that each formation consists of many beds, and that beds range greatly in thickness. The boundaries between units are called contacts.

In the river example, changes that yield beds of different composition or character last for only a short time. But during geologic time, major, long-term changes in a depositional environment can also take place. If a sequence of strata is distinctive enough to be traced across a fairly large region, geologists call it a stratigraphic **formation** (▶Fig. 7.24). For example, a region may contain a succession of alternating sandstone and shale beds deposited by rivers, overlain by beds of marine limestone deposited later when the region was submerged by the sea. A stratigrapher might identify the sequence of sandstone and shale beds as one formation and the sequence of limestone beds as another. Formations are often named after the locality where they were first found and studied. For example, the Schoharie Formation was recognized and described from exposures near Schoharie, New York.

Special Kinds of Bedding

Two distinctive types of bedding form under special conditions. **Cross beds,** thin bands that tilt at an angle to the boundary of the overall bedding in a sequence of strata, develop during the deposition of sediment in a current. For example, sediment deposited in a river may build a layer in the downstream direction, at the end of which is a step. When new sediment carried by the river reaches this step, it slips down the face of the step and settles. The slope of this face defines the cross bed. Similarly, when blowing sand builds into sand dunes in a desert, the sand tumbles up the windward side of the dune (the side from which the wind is blowing), then avalanches down the leeward side (the side toward which the wind is blowing) and settles in quieter air. The sloping leeward face becomes the cross bed (▶Fig. 7.25a–c).

During a storm or earthquake, loose sediment on a slope starts to slip downslope. The sediment mixes with water to create a murky cloud that is denser than clear water, and thus flows downslope like an underwater avalanche. We call this moving underwater cloud a **turbidity current.** As the turbidity current slows, sediment settles out of the water, with the coarsest sediment settling first, because large grains sink faster. As a result, a **graded bed** of sediment forms, with coarser sediment (sand) at the base and finer sediment (mud) at the top (▶Fig. 7.26a, b).

Bed-Surface Markings: Further Clues to Past Environments

A number of features appear on the surface of a bed as a consequence of events that happen during deposition or soon after, while the sediment layer remains soft. These **bed-surface markings** include the following.

- *Ripple marks:* As a current of wind or water flows over a sediment layer, the sediment builds into elongated ridges, called **ripple marks,** that are spaced at a roughly uniform distance from each other. Ripple marks lie perpendicular to the current (▶Fig. 7.27a, b).

- *Mud cracks:* If a mud layer dries up after deposition, it cracks into roughly hexagonal plates that typically curl up at their edges. We refer to the openings between the plates as **mud cracks.** Later, these fill with sediment and can be preserved (▶Fig. 7.27c, d).

- *Fossils:* **Fossils** are relics of past life. Some fossils are shell imprints or footprints on a bedding surface (Fig. 7.19c; see Interlude D).

The presence of bed-surface markings tells us a lot about the environment of deposition. Clearly, if we see ripple marks, we know that the sediment was deposited in a current. If we see mud cracks, we know the sediment was exposed to air periodically, so it could not have been deposited on the floor of a deep ocean. Fossils can pinpoint the environment of deposition, because different environments harbor different organisms. For example, the presence of a land-plant fossil in a shale means that the shale was deposited in a lake or river, not in the deep sea.

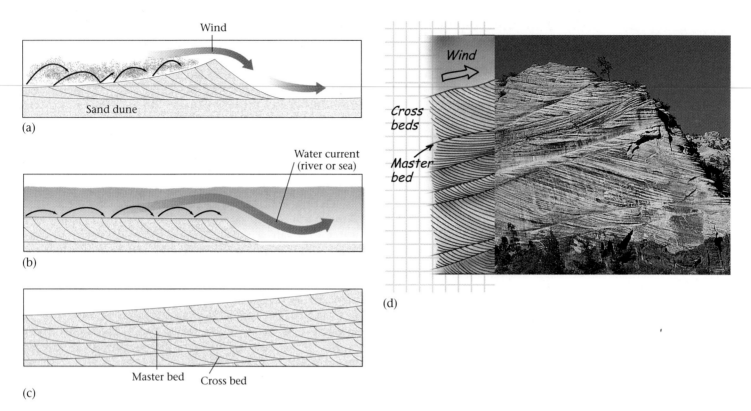

FIGURE 7.25 (a) Cross beds formed on the leeward face of a sand dune by wind-transported sediment. (b) Cross beds formed on the downstream side of river-deposited sediment. (c) Successive layers, or master beds, of cross-bedded strata. (d) On this cliff face of sandstone in Zion National Park, we see remnants of ancient sand dunes. Cross beds indicate the wind direction during deposition.

7.9 SEDIMENTARY ENVIRONMENTS

Geologists refer to the conditions in which sediment was deposited as the **sedimentary environment.** Examples include beach environments, glacial environments, and river environments. To identify these environments, geologists, like detectives, look for such clues as grain size, composition, sorting, and roundness of clasts, which can tell us how far the sediment has traveled from its source and whether it was deposited from the wind, from a fast-moving current, or from a stagnant body of water. Clues like fossil content and sedimentary structures can tell us whether the sediments were deposited subaerially, just off the coast, or in the deep sea. Geologists refer to the set of sediments and sedimentary structures indicative of a particular environment as a **sedimentary facies.** For example, a beach facies gives us the set of sediments and sedimentary structures found on a beach. Different facies can occur in adjacent locations at a given time—strata containing a reef facies may accumulate just offshore of strata comprising a beach facies.

Now let's look at some examples of different sedimentary environments and the facies that form in them by imagining that we are taking a journey from the mountains to the sea, examining sediments as we go. We begin with **terrestrial sedimentary environments,** those formed on dry land, and end with **marine sedimentary environments,** those formed along coasts and under the waters of the ocean. (See art on pp. 192–93.)

Terrestrial (Nonmarine) Sedimentary Environments

Glacial environments. We begin high in the mountains, where it's so cold that more snow collects in the winter than melts away, so glaciers—rivers of ice—develop and slowly flow downslope. Because ice is a solid, it can move sediment of any size. So as a glacier moves down a valley in the mountains, it carries along *all* the sediment that falls on its surface from adjacent cliffs or gets plucked from the ground at its base. At the end of the glacier, where the ice finally melts away, it drops its load and makes a pile of **glacial till** (►Fig. 7.28a). Till is unsorted and unstratified—it contains clasts ranging from clay size to boulder size all mixed together, with large clasts distributed through a matrix of fine clasts.

Shore line

Sediment breaks loose and avalanches down canyon.

Sea level

Submarine canyon

Turbidity current, a cloud of debris, fans out and settles.

Cross section

Submarine fan

Turbulent

Settling

Graded bed

(a)

Bed 2

Bed 1

Younger

Clay, silt

Fine sand

Older

Coarse sand

FIGURE 7.26 (a) An earthquake or storm triggers an underwater avalanche (turbidity current), which mixes sediment of different sizes together. When the current slows, the larger grains settle faster, gradually creating a graded bed. (b) In this example of a graded bed, pebbles lie at the bottom of the bed, and silt at the top.

Top (fine)

Base (coarse)

(b)

FIGURE 7.27 (a) Modern ripples forming in the silt on a beach. (b) Ancient ripples preserved in a layer of quartzite that's over 800 million years old. (c) Mud cracks in dried red mud, from Utah, as viewed from the top. Note how the edges of the mud cracks curl up. (d) Ancient mud cracks preserved in ancient mudstone (about 280 million years old), here viewed from the side (in cross section). Note the upward curl on the side of the mud crack.

Thus, in a sequence of strata, a layer of unsorted sediment that was deposited before it was extensively weathered may be the record of an ancient episode of glaciation.

Mountain stream environments. As we walk down beyond the end of the glacier, we enter a realm where turbulent streams rush downslope in mountain valleys. This fast-moving water has the power to carry large clasts; in fact, during floods, boulders and cobbles tumble down the stream bed, occasionally bouncing into each other and breaking. Between floods, the largest clasts settle out to form thick gravel and boulder beds, while the slower water of the stream carries finer sediments like sand and mud away (▶Fig. 7.28b). Clasts in this mountain stream environment have not had the opportunity to weather substantially, and thus still contain their original

minerals. A mountain stream facies would include coarse conglomerate.

Mountain front environments. Our journey now takes us to the mountain front, where the steep, fast-moving stream abruptly empties out onto a flat plain. In arid regions, where there is insufficient water for the stream to flow continuously, the stream deposits its load of sediment right at the mountain front, creating a large, wedge-shaped apron called an **alluvial fan** (▶Fig. 7.28c). Deposition takes place here because when the stream pours from a canyon mouth and spreads out over the plain, friction with the ground and the decrease in slope cause the water to slow down, and slow-moving water does not have the power to move coarse sediment. Because we are still so close to the mountains—the source of the sediment—the sand still contains feldspar

(a)   (b)   (c)

FIGURE 7.28 (a) Glacial till deposited at the end of a melting glacier in New Zealand. (b) Coarse gravel deposited by a flooding mountain stream. (c) An alluvial fan. Note the road for scale.

grains, for these have not yet broken up and have not yet weathered into clay. Alluvial-fan sediments, when later buried and transformed into sedimentary rock, become arkose and conglomerate. During lithification, the sediment may oxidize in the air, so that the iron it contains rusts to bright red (iron oxide). Geologists commonly call red, oxidized sediment layers **redbeds.**

Sand-dune environments. In deserts, relatively few plants grow, so the ground lies exposed to the wind. The strongest winds can transport both sand and dust, while gentler breezes carry away only the dust. As a result, **sand dunes,** large piles of well-sorted sand, accumulate in deserts (Fig. 7.17b). Because the sand tumbles and bounces about during wind storms, the grains become frosted (in the same way that rubbing glass with sandpaper makes the glass frosted) and well rounded. Typically, wind blows sand up and over the crest of a dune, and the sand settles on the steeper leeward side, leading to the development of large cross beds. Thus, thick layers of well-sorted sandstone, in which we see large (meters-high) cross beds, are relicts of a sand-dune facies (Fig. 7.25d).

Lake environments. From the dry regions, we continue our journey into a humid realm, where water remains at the surface throughout the year. Some of this water collects in lakes, relatively quiet water that is unable to move coarse sediment; any coarse sediment brought into the lake by a stream settles out along the shore. Only fine clay makes it out into the center of the lake, where it eventually settles to form mud on the lake bed. Thus, lake sediments (also called **lacustrine sediments**) typically consist of laminated shale (▶Fig. 7.29a).

River environments. The lake drains into a stream that carries water onward toward the sea. As we follow the stream, it merges with tributaries to become a large, slow-moving river, winding back and forth across a plain. Slow-moving rivers transport sand, silt, and mud. The coarser sediments tumble along the bed in the river's channel, while the finer sediments drift along, giving the water in which it is suspended a muddy color (▶Fig. 7.29b). This fine sediment settles out along the banks of the river, or on the **floodplain,** the flat region on either side of the river that is covered with water only during floods. Since river sediment is deposited in a current, the sediment surface develops ripple marks, and the sediment layers internally have small cross beds. On the floodplain, mud layers dry out between floods, leading to the formation of mud cracks.

Because the river has transported sediment for a great distance, the minerals comprising its sediment have undergone chemical weathering. As a result, very little feldspar remains—most of it has changed into clay. Thus, river sediments lithify to form sandstone, siltstone, and shale. Typically, channels of coarser sediment (sandstone) are surrounded by layers of fine-grained floodplain deposits; in cross

Glacial environment

Estuary

Beach

Bar

Continental shelf

Coastal
erosion

Turbidity
current

Submarine fan

Deep-sea current

Layers of sedimentary
rock accumulate.

Mountain building folds
the rock layers.

Forming an unconformity

The mountains are eroded; the
folded layers are submerged.

New sedimentary layers
accumulate.

Forming Sedimentary Rocks

Categories of sedimentary rocks include clastic sedimentary rocks, chemical sedimentary rocks (formed from the precipitation of minerals out of water), and biochemical sedimentary rocks (formed from the shells of organisms). Clastic sedimentary rocks develop when grains (clasts) break off preexisting rock by weathering and erosion and are transported to a new location by

wind, water, or ice; the grains are deposited to create sediment layers, which are then cemented together. We distinguish among types of clastic sedimentary rocks based on grain size.

The character of a sedimentary rock depends on the composition of the sediment and on the environment in which it accumulated. For example, glaciers carry sediment of all sizes, so

Lake environment

Desert environment

Saline lake

Fluvial environment

Sand dunes

Coastal environment

Coastal swamp

Reef

Delta

Shale

Siltstone

Sandstone

Conglomerate

Fossiliferous limestone

Unconformity

they leave deposits of poorly sorted (different-sized) till; streams deposit coarser grains in their channels and finer ones on floodplains; a river slows down at its mouth and deposits an immense pile of silt in a delta. Fossiliferous limestone develops on coral reefs. In desert environments, sand accumulates into dunes, and evaporates precipitate in saline lakes. Offshore, submarine canyons channel avalanches of sediment, or turbidity currents, out to the deep-sea floor.

Sedimentary rocks tell the history of the Earth. For example, the layering, or bedding, of sedimentary rocks is initially horizontal. So where we see layers bent or folded, we can conclude that the layers were deformed during mountain building. Where horizontal layers overlie folded layers, we have an unconformity: for a time, sediment was not deposited, and/or older rocks were eroded away.

FIGURE 7.29 (a) Finely laminated lake-bed shales in the Green River Formation, Wyoming. (b) The character of river sediment varies with distance from the source. In the steep channel, the turbulent river can carry boulders and cobbles. As the river slows, it can only carry sand and gravel. And as the river winds across the floodplain, it carries sand, silt, and mud. The coarser sediment is deposited in the river channel, the finer sediment on the floodplain. (c) This exposure shows the lens-like shape of an ancient gravel-filled river channel in cross section. (d) A geologist's sketch emphasizes the channel shape.

section, the channel has a lens-like shape (►Fig. 7.29c, d). Since river sediments (also called **fluvial sediments,** from the Latin word for "river") commonly become exposed to air or to oxygen-rich water, the iron they contain turns into red iron oxide (rust), so deposits of river strata that we see in the stratigraphic record can be classified as redbeds.

Marine Sedimentary Environments

Delta deposits. After following the river downstream for a long distance, we reach its mouth, where it empties into the sea. Here, the river builds a **delta** of sediment out into the ocean. Deltas get their name from their shape: some of them, like the Nile Delta on the north coast of Egypt, look like triangles on a map and thus resemble the Greek letter *delta* (Δ) (►Fig. 7.30a). River water stops flowing when it en-

ters the sea, and below wave depth the water is so quiet that clay and silt from the river settle out. As the delta builds out into the sea, the river flows over it, and the older portions get buried by swampy floodplain mud. Channels transport sediment across the swamp to the front of the delta, where it spills down into deeper water. In a simple model of a small delta, we see a distinctive geometry. The horizontal beds of swampy mudstone layers, known as the **topset beds,** build out over silts and clays that accumulate as **foreset beds** on the slope. The foreset beds, in turn, slope downward into fine horizontal mudstone layers, the **bottomset beds,** deposited in deep water at the foot of the delta (►Fig. 7.30b). Large marine deltas are more complex, because the sea level changes as they are deposited. Nevertheless, if you find a thick succession of silt and mud in an outcrop, you might be looking at the foreset and bottomset beds of a delta.

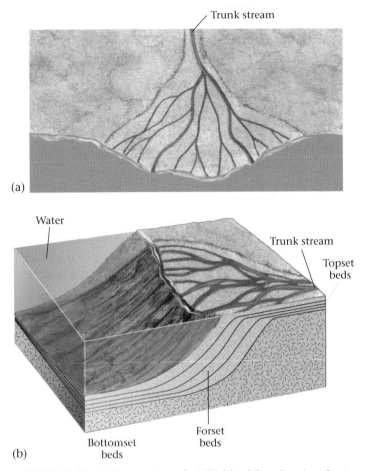

Trunk stream

(a)

Water

Trunk stream

Topset beds

Forset beds

Bottomset beds

(b)

FIGURE 7.30 (a) A map view of a Nile-like delta, showing the Δ shape. Note that the main river, called the trunk stream, divides into many smaller streams that fan out (b) Topset beds in a delta build out over foreset beds, which build out over bottomset beds as the delta grows.

Coastal beach sands. Now we leave the delta and wander along the coast. Oceanic currents transport sand along the coastline, and if the sea level rises, successive layers of sand are deposited. The sand washes back and forth in the surf, so it becomes well sorted (waves winnow out mud and silt) and well rounded, and because of the back-and-forth movement of ocean water over the sand, the sand surface becomes rippled. Thus, if you find a facies of well-sorted, medium-grained sandstone containing ripple marks, you may be looking at the remnants of a beach environment.

Shallow-marine clastic deposits. From the beach, we proceed offshore. Finer sediment gets washed out to sea by the waves and comes to rest in quieter water below the waves, offshore. As the water here may be only meters to a few tens of meters deep, geologists refer to this depositional setting as a shallow-marine environment. Clastic sediments that

accumulate in this environment tend to be fine-grained, well-sorted, well-rounded silt, and are inhabited by a great variety of organisms like mollusks and worms. Such sediment does not contain the mud cracks, ripple marks, or terrestrial plant and animal fossils found in sediment of similar grain size in river floodplains. Thus, if you see smooth beds of marine-fossil-rich siltstone and mudstone, you may be looking at sediment deposited offshore of a beach.

Shallow-water carbonate environments. In shallow-marine settings where relatively little clastic sediment (sand and mud) enters the water, and where the water is fairly warm, clear, and full of nutrients, most sediment is made up of the shells of organisms. These shells consist of carbonate minerals (calcite and aragonite), so we call such environments carbonate environments. To reach one, we would have to journey to a coastline far from a river's mouth, in a tropical climate. The margins of tropical islands, away from the clastic debris of land, provide ideal carbonate environments (▶Fig. 7.31a). If we bend over and examine the white sand of an island's beach, we will find that it consists of tiny rounded shells and shell fragments. It may also include **ooids,** snowball-like spheres of calcite formed as sand grains roll back and forth in water saturated in carbonate. In carbonate environments, the nature of sediment depends on the water depth. Beaches collect sand composed of shell fragments, lagoons (quiet water) are sites where lime mud accumulates, and reefs consist of coral and coral debris. Shells of mobile creatures (like snails and scallops) may fall in any of these places. Farther offshore of a reef, we can find a sloping apron of reef fragments (▶Fig. 7.31b).

Shallow-water carbonate environments transform into sequences of fossiliferous limestone and micrite. If you see ancient limestone exposed on a cliff in the snowy Canadian Rockies today, it means that when the sediment making the rock was deposited, the region was a tropical paradise. How could this be? Because of plate tectonics, Canada once occupied an equatorial latitude, then drifted to its present location.

Deep-marine deposits. We conclude our journey by sailing offshore. Along the transition between coastal regions and the deep ocean, turbidity currents deposit **turbidites,** graded beds of graywacke (Fig. 7.26). Typically, marine turbidites accumulate in a fan-shaped deposit known as a **submarine fan** at the base of a submarine canyon (a valley cut into the continental shelf). Farther offshore, in the deep-ocean realm, only fine clay and plankton provide a source for sediment. The clay eventually settles out onto the deep sea floor, forming deposits of finely laminated mudstones, and plankton shells settle to form chalk (from calcite shells; ▶Fig. 7.32a, b) or chert (from siliceous shells).

FIGURE 7.31 (a) A coral reef and adjacent lagoon surrounding an island in the South Pacific. (b) The different carbonate environments associated with a reef.

Thus, deposits of mudstone, chalk, or bedded chert indicate a deep-marine source.

You Can Be a Sedimentary Detective!

Now you should be able to look at most sedimentary rocks in quarries, road cuts, and cliffs and take a pretty good guess as to what ancient environments they represent. From now on, when you see limestone in a quarry, it's not just a limestone, it's the Bahamas or the Great Barrier Reef. And a sandstone cliff—think ancient dune or beach. Coarse conglomerates should scream "alluvial fan!" or "mountain stream!" and a shale should bring to mind a floodplain, a lake bed, or the floor of the deep sea. Every sequence of strata has a story to tell.

7.10 SEDIMENTARY BASINS

The sedimentary veneer on Earth's surface varies greatly in thickness. Where igneous and metamorphic rocks crop out at the surface, the veneer does not exist, but on the margins of continents, where rivers dump their sedimentary load in vast deltas, a wedge of sediment up to 20 km thick may develop. Thick accumulations of sediment form only in regions where the surface of the Earth's lithosphere sinks relative to sea level, thereby creating a depression that fills with a thick accumulation of sediment. We use the term **subsidence** to refer to the process by which the surface of the lithosphere sinks, and the term **sedimentary basin** for the depression that fills with sediment. What geologic conditions lead to subsidence and the development of

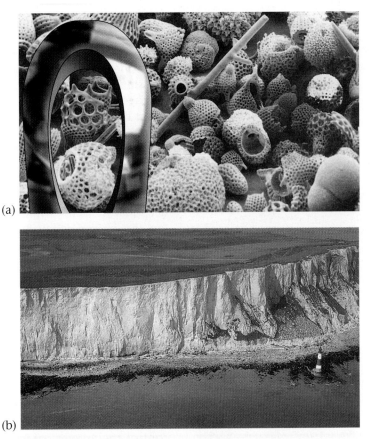

FIGURE 7.32 (a) These plankton shells, which make up deep-marine sediment, are so small that they could pass through the eye of a needle. (b) The chalk cliffs of Dover, England. These were originally deposited on the sea floor and later uplifted.

sedimentary basins? The theory of plate tectonics provides some answers.

Types of Basins in the Context of Plate Tectonics Theory

The surface of the lithosphere can subside (sink) because the underlying asthenosphere is relatively soft and able to flow, thereby creating room for the sinking lithosphere. (If the asthenosphere were completely rigid, subsidence could not occur.) But what conditions cause the lithosphere to sink? The lithosphere can sink when a heavy load is placed on its surface, just as the surface of a raft sinks when you climb up onto it. Of course, the weight of a person or even a city cannot make the lithosphere subside; but huge weights, like mountain belts or volcanoes, can. The surface of the lithosphere can also sink when the lithosphere stretches (a detailed explanation of this kind of subsidence can be found in advanced geology books).

Geologists have identified many different types of sedimentary basins that result from subsidence, based on the geologic setting in which they occur.

- **Rift basins:** These form over continental rifts where the surface of the lithosphere becomes lower, relative to its surroundings, as a consequence of stretching and thinning. We can use a stick of chewing gum to help visualize this process: if you pull on either end of the stick, the gum stretches and the thin center sinks.

- **Passive-margin basins:** These form along the edges of continents, where the lithosphere stretches during rifting. After rifting ceases, the stretched lithosphere gradually cools and becomes denser and heavier, and, as a result, sinks. The sediment in the passive-margin basins can reach a thickness of 15 to 20 km.

- **Intracontinental basins:** These develop in the interiors of continents, probably over an unsuccessful rift, where crustal stretching began but didn't get very far before it stopped. Illinois and Michigan are each underlain by an intracontinental basin in which up to 7 km of sediment has accumulated.

Floodplain
Swamp
Shore
Redbeds
Peat
Coal

Shore

Transgression
Shore
Shore migrates inland.

Regression
Maximum limit of transgression
Shore migrates seaward.
Shore
Uplift and erosion

Redbeds
Coal
Sandstone
Shale
Sandstone
Coal
Redbeds

FIGURE 7.33 The concept of transgression and regression. As the sea level rises and the shore migrates inland, coastal sedimentary environments overlap terrestrial environments. Eventually, deeper-water environments overlap shallower ones. Thus, a regionally extensive layer does not all form at the same time. During regression, the sea level falls, and the shore moves seaward.

• **Foreland basins:** These form on the continent side of a mountain belt, because as the mountain belt grows, large slices of rock are pushed out onto the surface of the continent along large faults. The weight of these slices pushes down on the surface of the lithosphere, creating a wedge-shaped depression adjacent to the mountain range that fills with sediment.

Different groups of sedimentary facies form in different sedimentary basins. For example, rift basins typically fill with alluvial-fan facies, fluvial facies, and lacustrine facies. In contrast, a passive-margin basin consists of shallow- to deep-marine facies, while a foreland basin contains alluvial-fan, fluvial, and deltaic facies.

Transgression and Regression

Sea-level changes control the succession of sediments that we see in a sedimentary basin. At times during Earth history, the sea level has risen by as much as a few hundred meters, creating shallow seas that covered the interiors of continents; there have also been times when the sea level has fallen by a few hundred meters, exposing even the continental shelves to air. Sea-level changes may be due to a number of factors, including climate changes, which control the amount of ice stored in polar ice caps (when ice caps grow, more of Earth's surface water is stored in ice, and the sea level drops; when ice caps melt, water returns to the ocean and the sea level rises), and changes in the volume of mid-ocean ridges, which determine the amount of room in ocean basins for water (because ridges occupy volume, when they grow, they displace water, so the sea level rises).

When the sea level rises, the coast migrates inland—we call this process **transgression.** As the coast migrates, the sandy beach migrates with it, and the site of the former beach gets buried by deeper-water sediment. Thus, as transgression occurs, an extensive layer of beach sand eventually forms. This layer may look like a blanket of sand that was deposited all at once, but in fact the sand deposited at one location differs in age from the sand deposited at another location. When the sea level sinks, the coast migrates seaward—we call this process **regression** (▶Fig. 7.33). Typically, the record of a regression will not be preserved as well, because as the sea level drops, areas that had been sites of deposition become exposed to erosion. A succession of strata deposited during a cycle of transgression and regression is called a **sedimentary sequence.** In the time since the end of the Precambrian (that is, during the past 545 million years), six major sequences, representing six major transgressions and regressions, have formed in the interior of North America.

CHAPTER SUMMARY

• Sediment consists of detritus (mineral grains and rock fragments derived from preexisting rock), mineral crystals that precipitate directly out of water, and shells (formed when organisms extract ions from water).

• Rocks at the surface of the Earth undergo physical and chemical weathering. During physical weathering, intact rock breaks into detritus (mineral grains and rock fragments). Processes like jointing and frost wedging aid physical weathering. During chemical weathering rocks react with water and air. Chemical weathering includes reactions like dissolution, hydrolysis, and oxidation. It can produce new minerals like clay, and ions in solution.

• The covering of loose rock fragments, sand, gravel, and soil at the Earth's surface is regolith. Soil differs from other types of regolith in that it has been changed by activities of organisms, by downward-percolating rainwater, and by the mixing in of organic matter. Downward-percolating water redistributes ions and clay.

• Geologists recognize four major classes of sedimentary rocks. Clastic (detrital) rocks form from cemented-together detritus (grains or rock fragments). Biochemical rocks develop from the shells of organisms. Organic rocks consist of plant debris, or of altered plankton remains. Chemical rocks precipitate directly from water.

• Weathering, erosion, transportation, deposition, and lithification lead to the development of clastic rocks. We distinguish different types of clastic rocks based on mineral content, grain size, and grain shape.

• Limestone, which consists of calcite or aragonite, forms either biochemically from the shells of organisms, or chemically by precipitation from water. Dolostone, which consists of dolomite, forms when water that contains magnesium reacts with limestone.

• Chert forms either biochemically, from the silica shells of radiolaria (a type of plankton) or chemically, when quartz replaces other minerals in a preexisting rock. Coal (which is over 50% carbon) forms from plant detritus deposited in layers.

• The precipitation of salts, when saltwater evaporates, produces deposits called evaporites.

• Sedimentary rocks occur in layers called beds. The term "strata" refers to a succession of beds. Sedimentary structures include bedding (layering) of sedimentary rocks, surface features on beds formed during deposition, and the arrangement of grains in beds.

• Cross beds are inclined layers within a thicker bed, formed when sediment is deposited in a current. In graded beds, which settle out of undersea avalanches called turbid-ity currents, the grain size of clasts progressively decreases from the base to the top of the bed. Ripple marks, mud cracks, and fossils develop on the surface of beds.

• Glaciers, mountain streams and fronts, sand dunes, lakes, rivers, deltas, beaches, shallow seas, and deep seas each accumulate a different sedimentary facies, a set of sedimentary rocks and structures. Thus, by studying sedimentary rocks, we can reconstruct the characteristics of past environments.

• Thick piles of sedimentary rocks accumulate in sedimentary basins, regions where the lithosphere sinks, creating a depression at the Earth's surface. Basins form where the lithosphere has been loaded by a weight or where it has undergone stretching.

• Sea level changes with time. Transgressions occur when the sea level rises and the coastline migrates inland. Regressions occur when the sea level falls and the coastline migrates seaward.

KEY TERMS

A_1-, A_2-horizon (p. 173)
alluvial fan (p. 190)
angularity (p. 179)
arkose (p. 179)
B-horizon (subsoil) (p. 173)
bauxite (p. 176)
bed (p. 186)
biochemical sedimentary
  rocks (p. 176)
breccia (p. 178)
C-horizon (p. 173)
caliche (p. 175)
carbonate rock (p. 176)
cementation (p. 178)
chalk (p. 182)
chemical sedimentary rocks
  (p. 176)
chemical weathering (p. 168)
chert (p. 182)
clastic (detrital) sedimentary
  rocks (p. 176)
clasts (p. 177)
coal (p. 183)
compaction (p. 178)
conglomerate (p. 178)
cross bed (p. 187)
cryptocrystalline quartz (p. 182)
delta (p. 194)
deposition (p. 177)

diagenesis (p. 185)
differential weathering (p. 171)
dissolution (p. 168)
dolomite (p. 185)
dolostone (p. 185)
erosion (p. 164)
evaporite (p. 184)
exfoliation (p. 166)
fluvial (p. 194)
foreland basins (p. 198)
formation (p. 187)
fossiliferous limestone (p. 182)
frost wedging (p. 167)
glacial till (p. 188)
graded bed (p. 187)
graywacke (p. 179)
hydration (p. 169)
hydrolysis (p. 169)
intracontinental basins
  (p. 197)
joints (p. 165)
lacustrine (p. 191)
laterite soil (p. 175)
limestone (p. 182)
lithification (p. 177)
micrite (p. 182)
mud cracks (p. 187)
mudstone (p. 179)
O-horizon (p. 173)

REVIEW QUESTIONS

1. In most places, it would be impossible to empty an ocean basin. Explain the peculiar circumstances (plate tectonics, climate, oceanographic conditions, and the balance between rivers, ocean, and evaporation) that allowed the Mediterranean to dry up.

2. How does physical weathering differ from chemical weathering?

3. Describe the processes that produce joints in rocks.

4. Feldspars are among the most common minerals in igneous rocks, but they are relatively rare in sediments. Why are they more susceptible to weathering? What chemical reactions are responsible for their breakdown? What common sedimentary minerals are produced from all the chemicals released by weathered feldspar?

5. What kind of minerals tend to weather more quickly?

6. Describe the different horizons in a typical soil profile.

7. What factors affect the nature of soils in different regions?

8. Describe how a sedimentary rock is formed from its unweathered parent rock.

9. Clastic and chemical sedimentary rocks are both made of material that has been transported. How are they different?

10. Is a rock made of shell fragments a clastic or a chemical sedimentary rock, or both? Explain your reasoning.

11. Describe how grain size and shape, sorting, sphericity, and angularity change as sediments move downstream.

12. Describe the two different kinds of chert. How are they similar? How are they different?

13. What kinds of conditions are required for the formation of evaporites?

14. What minerals precipitate out of seawater first? next? last? What does this suggest when geologists find huge volumes of pure gypsum in the Earth's crust?

15. How is dolostone different from limestone?

16. Describe how cross beds form. How can you read the current direction from cross beds?

17. Describe how a turbidity current forms and moves. How does it produce graded bedding?

18. Compare the geometry and typical sediments of an alluvial fan with a typical river environment and with a deep-marine deposit.

19. Why don't sediments accumulate everywhere? What kind of tectonic conditions are required to create basins?

SUGGESTED READING

Birkeland, P. W. 1984. *Soils and Geomorphology*. New York: Oxford University Press.

Blatt, H., G. Middleton, and R. Murray. 1980. *Origin of Sedimentary Rocks*. Englewood Cliffs, N.J.: Prentice-Hall.

Boggs, S. 1992. *Petrology of Sedimentary Rocks*. New York: Macmillan.

Collinson, J. D., and D. B. Thompson. 1989. *Sedimentary Structures*. London: Allen and Unwin.

Julien, P. Y. 1995. *Erosion and Sedimentation*. New York: Cambridge University Press.

Martini, I. P., and W. Chesworth, eds. 1992. *Weathering, Soils, and Paleosols*. New York: Elsevier.

Newton, C. R., and L. F. Laporte. 1989. *Ancient Environments*. 3rd ed. Englewood Cliffs, N.J.: Prentice-Hall.

Ollier, C., and C. Pain. 1996. *Regolith, Soils, and Land Forms*. New York: Freeman.

Prothero, D. R., and F. Schwab. 1996. *Sedimentary Geology*. New York: Freeman.

Reading, H. G., ed. 1986. *Sedimentary Environments and Facies*. Boston: Blackwell.

Retallack, G. J. 1990. *Soils of the Past.* Boston: Hyman.

Robinson, D. A., and R. B. G. Williams, eds. 1994. *Rock Weathering and Landform Evolution.* New York: Wiley.

Selley, R. C. 1982. *An Introduction to Sedimentology.* New York: Academic Press.

Siever, R. 1988. *Sand.* New York: Scientific American Library.

Wild, A. 1993. *Soils and the Environment: An Introduction.* New York: Cambridge University Press.

CHAPTER 8

Change in the Solid State: Metamorphic Rocks

8.1 INTRODUCTION

In the early years of geological study, researchers struggling to develop an understanding of how rocks formed faced a difficult puzzle. They could explain the origin of sedimentary rocks fairly easily, for they could see direct similarities between modern sediment and older sedimentary rock. Likewise, they could explain the origin of igneous rocks, because they could observe lava cooling on the flanks of a volcano, and could imagine magma cooling underneath. But some rocks seemed to defy explanation.

For example, marble, the favorite stone of sculptors, seems to have the same mineral composition as fossiliferous limestone (both rocks consist of calcite), but marble does not contain fossils and exhibits a coarse, interlocking crystalline texture that resembles the texture of granite (▶Fig. 8.1a, b). As another example, slate resembles shale but is much harder and splits easily into thin sheets, making it an excellent choice for roofing shingles. On close examination, geologists recognized that the planes on which slate splits are not parallel to bedding; how did these planes form? And geologists exploring in mountain ranges like the Alps found extensive areas underlain by hard rocks containing bizarre swirls of alternating dark and light layers, as well as exotic minerals they had not seen in sedimentary or igneous rocks (▶Fig. 8.2). How did these "unusual" rocks come to be?

The first clue to this puzzle came when geologists traced belts of these unusual rocks into regions where they also found rocks with familiar sedimentary or igneous characteristics. In many cases, they found that the transition between the unusual and the familiar rocks was gradual. This observation led to the conclusion that the unusual rocks formed when the familiar rocks underwent some sort of change or transformation. So geologists began to refer to the unusual rocks as "metamorphic rocks," from the Greek words *meta* (meaning "beyond" or "change") and *morphe* (meaning "form"), and the process by which rocks underwent a change as "metamorphism."

An outcrop of Precambrian metamorphic rock, exposed in the Wasatch Mountains, in Utah. The foliation formed during metamorphism.

(a)   (b)

FIGURE 8.1 (a) This thin section of an Ordovician-age limestone, a sedimentary rock, shows small fossil shells distributed in a matrix of lime mud. Most minerals comprising this rock consist of calcite. (b) This thin section of marble, a metamorphic rock, shows how a rock's fabric changes during metamorphism. Atoms have completely rearranged to form new grains that interlock with one another. Fossils are no longer recognizable. This photo is taken with polarized light; the color of an individual grain depends on its orientation with respect to the light. In nonpolarized light, all grains would have the same color.

Why does metamorphism occur? Geologists found that some examples of metamorphic rocks occur in a belt surrounding a large pluton of igneous rock. From this observation, they deduced that heat from the pluton somehow "cooked" the surrounding rocks, and thus that metamorphism could be caused by a change in temperature. Geologists also found that metamorphic rocks crop out in mountain ranges, and speculated that the process of mountain building somehow caused metamorphism, but they didn't really understand why. The story of metamorphism became clearer only in the twentieth century, when geologists learned to simulate metamorphism in the laboratory and realized that it involves complex chemical reactions that replace old minerals with new ones and/or cause mineral grains to change their shape and orientation. Based on this work, we now know that metamorphic rocks can form in response to changes in temperature, changes in pressure, and the squashing and shearing of rock. And when the theory of plate tectonics became established, geologists finally developed explanations for why metamorphic rocks appear where they do.

In this chapter, we first learn about the concept of metamorphism and the basic properties of metamorphic rocks. Then we look at the physical processes that lead to the formation of metamorphic rocks, and the classification scheme that geologists use to distinguish the different types. Finally, we discover the locations in the crust in which metamorphic rocks form and the reasons, in light of plate tectonics theory, that rocks end up in such locations.

FIGURE 8.2 Some metamorphic rocks, called gneiss, contain alternating bands of light and dark minerals, which sometimes display interesting swirls.

8.2 WHAT IS A METAMORPHIC ROCK?

Formally defined, a **metamorphic rock** is a rock that changed from its original form into a new one, without first becoming a melt or a sediment. Note that this definition implies that metamorphic rocks result from a change *in the solid state.* (If a rock changes by first melting, then the rock that forms when it cools is a new igneous rock, and if a rock changes by first breaking down into sediment, then the rock that forms when the sediment lithifies is a new sedimentary rock.) As metamorphosis is the amazing process by which a caterpillar transforms into a butterfly, **metamorphism** is the equally amazing process by which one kind of rock transforms into a different kind of rock. (Geologists do not use the term "metamorphism" to mean the chemical weathering that happens when rocks react with the atmosphere or water at the surface of the Earth, nor do we use the term to refer to the process of diagenesis or to the process by which sediment transforms into sedimentary rock.)

A metamorphic rock can be as different from its **protolith,** the original rock from which it formed, as a butterfly is from a caterpillar. For example, the metamorphism of a red shale (a sedimentary rock composed of clay, very fine quartz, and hematite; ▶Fig. 8.3a) may yield a sparkling coarse-grained metamorphic rock consisting of alternating dark and light bands speckled with brilliant purple crystals of garnet (▶Fig. 8.3b). Metamorphism can thus replace a rock's original group of minerals with a new group, and can alter its texture (the size, shape, and arrangement of grains).

If someone were to put a rock on a table in front of you, how would you know whether it was metamorphic? Two features characterize a metamorphic rock, and allow geologists to distinguish them from igneous or sedimentary rocks. As illustrated by the example above, metamorphic rocks possess either a **metamorphic mineral assemblage** (a group of minerals that form in a rock as a result of metamorphism), a **metamorphic foliation** (parallel surfaces or layers that develop in a rock as a result of metamorphism), or both. As you read on, the nature of these two characteristics will become clear.

8.3 CAUSES AND CONSEQUENCES OF METAMORPHISM

Caterpillars undergo metamorphosis because of hormonal changes in their bodies as they age. Rocks undergo metamorphism when acted on by one or more **agents of metamorphism:** heat, hot groundwater, pressure, and differential stress.

Heating and Recrystallization

When you bake cake batter (a mixture of flour, sugar, eggs, and water), chemical reactions take place to form a new material, cake. Similarly, when you heat a rock, its ingredients (a mixture of minerals) react to create a new material, metamorphic rock. Why does this happen? Think about what happens in a rock as it gets warmer. Heat causes the atoms making up mineral crystals to vibrate more rapidly (see Appendix A), and the vibration stretches and bends the chemical bonds that lock the atoms together to make solid crystals. Every now and then, this thermal vibration breaks a bond, and when this happens, an atom detaches from a crystal, moves a short distance, and then reattaches in another space. The process repeats often enough that, over time, atoms can rearrange entirely to form a set of new crystals whose structure can more easily withstand the higher heat (▶Fig. 8.4; see Box 8.1). We call the formation of new crystals from old **recrystallization,** since the change occurs in already crystalline material.

Recrystallization takes place more rapidly at higher temperatures than at lower, because bonds vibrate faster and break more frequently as the temperature increases. Cold (less

FIGURE 8.3 (a) Hand specimen of red shale, consisting of clay flakes, quartz, and iron oxide (hematite). (b) Hand specimen of gneiss containing bright-purple garnets. The shale was the protolith of this gneiss.

(a)   (b)

No crystal is perfect—there are holes (vacancies) in the lattice. These give atoms spaces in which to move.

Atoms can migrate and rearrange, without destroying the integrity of the solid.

The dark atoms are now closer together than they were earlier.

FIGURE 8.4 Here we see how recrystallization (by solid-state diffusion) takes place at an atomic scale. The dark atoms rearrange as atoms shuffle around, even though the whole block stays solid. Note that at the end, the spacing between the dark atoms has changed.

than 250°C) rock near the Earth's surface can remain unchanged for billions of years, while at elevated temperatures recrystallization occurs rapidly enough for mineral assemblages to change in a relatively short geologic time (less than a million years). Under particularly high temperatures, the atoms migrate, so that layers of different composition develop in a rock: light-colored mineral layers alternate with dark ones, a phenomenon known as **compositional banding.**

We saw earlier that red shale consists of a mixture of quartz, clay, and hematite; together, these minerals provide atoms of silicon, oxygen, iron, magnesium, calcium, aluminum, and potassium, elements that can be rearranged to form a great variety of minerals, including mica and garnet. In other words, protoliths consisting of many different

elements recrystallize to form metamorphic rocks with many different minerals.

In contrast, the metamorphism of a protolith made of only one simple mineral may yield a metamorphic rock composed of the same mineral. For example, metamorphism transforms pure quartz sandstone into **quartzite,** a metamorphic rock also composed of quartz (►Fig. 8.5a, b), and metamorphism changes pure calcite limestone into **marble,** which also consists only of calcite (Fig. 8.1). During the formation of quartzite or marble, recrystallization does occur, and atoms do rearrange to form new crystals, but the new crystals have the same composition as the original crystals of the protolith. Note that the new crystals may contain atoms that came from several different sedimentary grains or from the rock's cement. Also, during recrystallization, pore space (open space between grains) disappears, and large crystals may grow, using atoms that had once been distributed among many small crystals.

Heating a rock may also trigger changes in its chemical makeup. Specifically, some minerals contain molecules of –OH (hydroxyl) or –CO$_3$ (carbonate). When heated to high temperatures during metamorphism, these molecules detach and migrate out of the rock in the form of water and carbon dioxide. In other words, high heat drives off water, carbon dioxide, and other volatile compounds.

BOX 8.1

SCIENCE TOOLBOX

Diffusion

Chemists call the process by which atoms migrate through a material **diffusion.** Diffusion occurs in gases, liquids, and solids. You experience the consequences of gaseous diffusion every time you smell an odor: when a person wearing perfume or cologne enters the room, the random motions of the molecules of scent allow them to gradually migrate through the air and reach your nose. If you've ever dripped a drop of ink in water, you've seen liquid diffusion: with time, the molecules of ink migrate through the liquid until they are evenly distributed throughout. Gaseous and liquid diffusion take place much more rapidly than diffusion

through solids, because the atoms of gases and liquids can move more easily and there is plenty of room for atoms to go when they move. In a solid, atoms diffuse slowly, because most of the time they are locked into place by chemical bonds, and there is not much room for them to travel. In fact, the only reason that diffusion in solids can take place at all is because crystals contain imperfections, little spaces or holes in the lattice called **vacancies.** For example, in Figure 8.4, the dark atoms can change position only because the light atoms are able to rearrange and make room for them.

Sandstone

Quartzite

Recrystallization

(a) Cement   Crack

(b)   Crack

FIGURE 8.5 The transition of quartz sandstone to quartzite. (a) In a sandstone, you can see rounded sand grains glued together by quartz cement. If the rock cracks, the crack follows the surface of the grains. (b) In a quartzite, the grains have grown. Each grain contains SiO_2 (quartz), derived from both the sand grains and cement. If the rock cracks, the crack ignores the grain boundaries.

Reaction with Hot Groundwater: Metasomatism and Veining

Hydrothermal (hot-water) solutions passing through a rock may pick up some dissolved ions and drop off others, just like a bus on its route through a city, and thus can help change the chemical composition of a rock during metamorphism. The process by which a rock's overall chemical composition changes because of a reaction with hot water is called **metasomatism** (the word *soma* in Greek means "body").

The dissolved ions carried away by hydrothermal solutions may eventually precipitate elsewhere in the cracks that cut through cooler rock. The resulting seams of precipitated minerals are called **veins** (▶Fig. 8.6). Many veins in metamorphic rock consist of quartz (which contains silica), for silica tends to be left over after metamorphic reactions and enters hydrothermal solutions relatively easily.

Pressure and Its Consequences

As you swim underwater in a swimming pool, water squeezes against you equally from all sides—in other words, your body feels **pressure.** If pressure becomes large enough, it can cause a material to collapse inward; for example, if you pulled an air-filled balloon down to a depth of 10 m in a lake, the balloon would become significantly smaller. Pressure has the same effect on minerals. At low pressures, minerals with relatively large spaces between their atoms can survive. Take these minerals down several kilometers in the crust, where pressures are high, and the atoms rearrange

FIGURE 8.6 These milky white quartz veins cutting across a metamorphic rock form by the precipitation of silica from hot water in cracks.

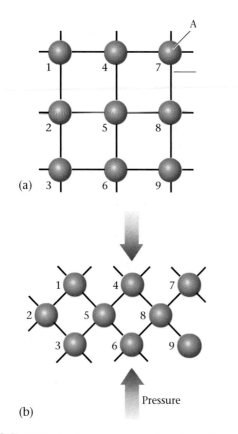

FIGURE 8.7 (a) Under low pressure, minerals can have open lattices in which atoms are widely spaced. (b) High pressure can cause atoms to rearrange and move closer together, to form minerals that have a tight lattice and occupy less volume.

so that they pack into a smaller space (►Fig. 8.7a, b). This rearrangement contributes to the formation of new minerals.

Changing Pressure and Temperature Together: Mineral Stability

So far, we've looked at changes in pressure and temperature as separate phenomena. But in the Earth, a rock may be subjected to simultaneous changes in pressure and temperature. For example, the progressive burial of a rock causes an increase in both temperature and pressure. Geologists use the term **metamorphic conditions** to refer to the range of temperature and pressure at which metamorphic reactions take place. Temperature tends to be more important in determining if metamorphism occurs; only at temperatures greater than 250°–300°C do atoms vibrate fast enough to break their bonds and migrate.

By simulating metamorphism in the laboratory, geologists found that a particular mineral remains **stable,** or unchanging, only under a limited range of temperature and pressure conditions. If the temperature or pressure in the

rock changes, then the mineral becomes **unstable,** and the atoms composing it redistribute into a new mineral. We can illustrate the concept of stability by studying the behavior of the compound Al_2SiO_5 (aluminum silicate) as portrayed on a graph that plots temperature and pressure (►Fig. 8.8). This compound can exist in the form of three different minerals. From the graph, you can see that at lower pressures and temperatures, it exists as andalusite. An increase in pressure causes andalusite to become unstable and kyanite to form. At a high pressure, an increase in temperature causes kyanite to transform into sillimanite. Geologists studying metamorphic rocks search hard to find andalusite, kyanite, and sillimanite, because their presence can define the pressure and temperature at which a rock formed.

Differential Stress: The Formation of Preferred Mineral Orientation

Imagine that you have just built a house of cards, and, being in a destructive mood, you step on it so the cards collapse. They collapse because the downward push you apply with your foot is greater than the pushes in other directions. If a material is squeezed (or stretched) unequally from different sides, we say that it is subjected to a **differential stress,** meaning that the push or pull in one direction differs in magnitude from the push or pull in another direction (►Fig. 8.9a). We distinguish two components of differential stress.

FIGURE 8.8 The "stability fields" for three metamorphic minerals (kyanite, andalusite, and sillimanite) that are polymorphs of Al_2SiO_5 (aluminum silicate). If an aluminum-rich rock is taken down to a pressure of 2 kbar at a temperature of 450°C (point X), various atoms from preexisting minerals of the protolith combine to form andalusite. If the pressure increases so that the rock feels 5 kbar at the same temperature (point Y), then the andalusite recrystallizes to form kyanite. If the temperature increases to 650°C while the rock stays at 5 kbar (point Z), then the kyanite recrystallizes to form sillimanite.

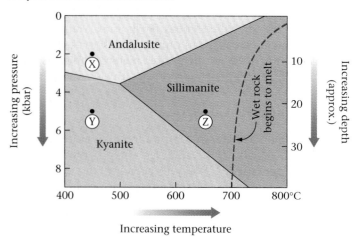

- *Normal stress:* **Normal stress** pushes or pulls perpendicular to a surface. We call a push **compression** and a pull **tension.** Compression squashes a material (for example, if you compress a sphere, it flattens into a pancake; ▶Fig. 8.9b), while tension stretches a material (if you apply tension to a sphere, it stretches into a cigar shape).

- *Shear stress:* **Shear stress** moves one part of a material sideways relative to another. If, for example, you place a deck of cards on a table, then set your hand on top of the deck and move your hand parallel to the table, you shear the deck and spread the cards across the table (▶Fig. 8.9c).

When rocks are subjected to differential stress during metamorphism, they can change shape without breaking. For example, a piece of rock that is squeezed or sheared may slowly become flatter. As it changes shape, the internal

Spheroidal (equant)

Elongate

Platy

Inequant

FIGURE 8.10 Some of the basic shapes in nature. A spheroidal grain (for example, a sand grain) is equant, while elongate and platy grains are inequant.

FIGURE 8.9 The concept of differential stress. (a) A house of cards feels only air pressure as you start to bring your foot down. But when you step on the cards, they feel a differential stress, which flattens the house. (b) A normal stress applied to a ball of dough flattens the ball into a pancake. (c) A shear stress smears a pack of cards out parallel to the table.

(a)

(b)

(c)

texture of a rock—the arrangement and orientation of its mineral grains—also changes, sometimes resulting in the development of **preferred mineral orientation.** By this, we mean that **platy** (pancake-shaped) grains lie parallel to one another, and/or **elongate** (cigar-shaped) grains align in the same direction. Both platy and elongate grains are **inequant,** meaning that the length of a grain is different in different directions (▶Fig. 8.10). The preferred orientation of minerals in a rock makes the rock appear layered; this layering is a type of metamorphic foliation.

Preferred orientation can develop in four ways. (1) At relatively high temperatures, **ductile (plastic) deformation** takes place: mineral grains behave like soft plastic, and when compressed or sheared, they become flattened or elongate (▶Fig. 8.11a). (2) In wet rocks, **pressure solution** takes place, during which the compressed faces of mineral grains dissolve, producing ions that then precipitate elsewhere on faces where compression is less (▶Fig. 8.11b). (3) Rigid, inequant grains distributed through a soft matrix may undergo **grain rotation**—grains simply rotate into parallelism as the rock changes shape, much like logs scattered in a flowing river align with the current (▶Fig. 8.11c). (4) Finally, inequant grains may *grow* with a preferred orientation, because the direction

(a)

(b)

or

Plastic change   Original grains   Pressure solution

— Sites of dissolution

— New mineral growth

Random fabric

(Before compression)   (After compression)

(c)

— Direction of preferred orientation

Compression

2 mm

— Direction of preferred orientation

(d)

FIGURE 8.11 Squeezing or shearing a rock under metamorphic conditions can result in preferred mineral orientation, in four ways. (a) At higher temperatures in dry rock, the grains can plastically (ductilely) change shape (by the rearrangement of atoms inside). (b) The grains can undergo pressure solution, during which they dissolve on the side where the stress is greatest; the dissolved ions migrate through a water film and precipitate on the side where the stress is least. Commonly, the dissolved side becomes pitted. (c) Inequant grains distributed through a soft matrix may rotate into parallelism as the rock changes shape in response to compression. (d) Inequant grains can grow with a preferred orientation.

in which a crystal grows is affected by differential stress (▶Fig. 8.11d).

Recap: The Formation of Metamorphic Rock

Before proceeding, let's briefly review how metamorphism leads to the development of metamorphic mineral assemblages and metamorphic foliation. If a rock is subjected to a change in temperature or pressure, it undergoes recrystallization to form new mineral crystals. If the protolith (original rock) contained a variety of elements, the new minerals can comprise a new metamorphic mineral assemblage that differs radically from the original mineral assemblage. If the protolith consists of just quartz or calcite, it recrystallizes into quartz or calcite, respectively.

Recrystallization can take place simply by rearranging atoms into new crystals without changing the overall elemental composition of the rock. But at high temperatures, water and carbon dioxide (volatile elements) may be driven out of the rock, and if hydrothermal solutions pass through the rock, they may drop off or remove ions. Thus, under certain circumstances, the composition of a rock can change during metamorphism. At particularly high temperatures, recrystallization leads to the separation of light-colored minerals from dark-colored minerals, resulting in compositional banding.

If, during metamorphism, the rock is subjected to compression or shear, inequant minerals may form and/or may align parallel to one another, to create preferred orientation. Metamorphic rocks in which preferred orientation and/or compositional banding have developed display metamorphic foliation.

8.4 TYPES OF METAMORPHIC ROCKS

Geologists separate metamorphic rocks into two fundamental classes, foliated rocks and nonfoliated rocks. Each class contains several rock types, distinguished from each other by their composition, grain size, and, in the case of foliated rocks, the nature of their foliation.

Foliated Metamorphic Rocks

Let's now look at **foliation** in more detail. "Foliation" comes from the Latin word (*folium*) for "leaf." Geologists use the word to refer to the repetition of any surfaces or layers in a metamorphic rock; some layers are indeed as thin as a leaf, but some may be over a meter thick. Foliation gives some metamorphic rocks a striped or streaked appearance in an outcrop, and makes others capable of splitting into thin sheets. A metamorphic rock has foliation either be-

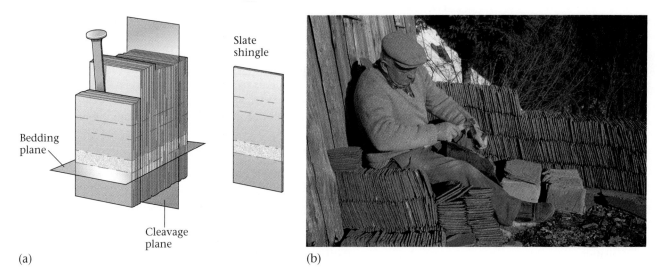

(a)   (b)

FIGURE 8.12 (a) A block of rock with slaty cleavage splits along cleavage planes into thin sheets. Note that the bedding (the thin, relict sandy layer) and cleavage in this example are not parallel. Originally, the slate was shale, which had bedding parallel to the sandy layer. (b) Slate easily splits into thin sheets, which have been used as shingles on roofs for a millennium.

cause its mineral crystals are aligned parallel to one another (defining preferred mineral orientation) or because the rock developed compositional banding, the alternation of light and dark layers.

Foliated metamorphic rocks can be distinguished from one another according to their composition, grain size, and the nature of their foliation.

• *Slate:* **Slate,** the finest-grained type of foliated metamorphic rock, forms by the metamorphism of shale (a sedimentary rock consisting of clay) under relatively low pressures and temperatures. The foliation, or **slaty cleavage,** in slate reflects the preferred orientation of clay (▶Fig. 8.12a). Slate tends to break on slaty cleavage planes, and roofers use the rock to shingle a roof because it easily splits into thin, impermeable sheets (▶Fig. 8.12b).

Slaty cleavage develops as a consequence of the compression and shearing of shale. Typically, the cleavage planes form perpendicular to the direction in which the rock squashed the most, and thus do not necessarily parallel the bedding of the shale protolith. For example, if a sequence of horizontal shale beds is compressed in the direction parallel to the bedding (end-on), then slaty cleavage forms perpendicular to the bedding (▶Fig. 8.13a). Folds may develop in association with slaty cleavage (▶Fig. 8.13b). The formation of slaty cleavage involves pressure solution and recrystallization. Also, as the slate forms, some relict grains physically rotate into the plane of foliation.

• *Metasandstone and metaconglomerate:* Shale is commonly found in a sequence of beds that includes sandstone and conglomerate. Under the metamorphic conditions that transform shale to slate, sandstone and conglomerate also metamorphose slightly, thereby creating a foliation defined by pancake- or cigar-shaped quartz grains or conglomerate clasts (▶Fig. 8.14). The flattening of quartz grains or conglomerate clasts occurs by plastic deformation and pressure solution. Geologists refer to sand-

FIGURE 8.13 (a) The compression of a bed end-on will create slaty cleavage at an angle perpendicular to the bedding. (b) Commonly, the rock folds (bends into curves) at the same time cleavage forms. The dashed lines indicate the original shape of the rock that was deformed.

(a)   (b)

FIGURE 8.14 In a deformed conglomerate, the pebbles and cobbles are squashed into pancake-like shapes that align with one another to define a foliation.

stones and conglomerates that contain a metamorphic foliation as **metasandstone** and **metaconglomerate,** respectively.

• *Phyllite:* The root of the word comes from the Greek word (*phyllon*) for "leaf"—the same root as for "phyllo," the dough used to make flaky Greek pastry. **Phyllite** is a fine-grained metamorphic rock with a foliation caused by the preferred orientation of fine-grained mica (muscovite) and chlorite. The fine-grained mica gives phyllite a silk-like sheen, known as **phyllitic luster** (▶Fig. 8.15a). Phyllite forms by the heating of slate at a temperature high enough to cause clay to recrystallize into mica. If the recrystallization process happens while the rock is being sheared or squashed, the new mica grains grow roughly perpendicular to the direction in which the rock flattens.

• *Schist:* **Schist** is a medium- to coarse-grained metamorphic rock that possesses a type of foliation, called **schistosity,** defined by the preferred orientation of large mica flakes (▶Fig. 8.15b). Again, the parallelism of the flakes develops because of shearing and squashing during metamorphism. Schist forms at a higher temperature than phyllite and differs from phyllite in that the mica grains, formed by recrystallization, are larger. Typically, schist also contains other minerals like quartz, feldspar, and garnet. The specific name given to a sample of a schist indicates the major minerals in the rock (for example, a quartz-mica schist is a rock that has schistosity and is composed predominantly of quartz and mica). Schist can form from shale but also from a great variety of other protoliths, as long as the protolith contains the appropriate elements to make mica. The chemical composition of the protolith determines the mineralogy of the specific schist that results.

• *Amphibolite:* Mafic rocks like basalt, which contain relatively little silica and an abundance of iron and/or magnesium, can't transform into a mixture of quartz and mica when metamorphosed—they don't have the right mix of chemicals to yield these minerals. Rather, such mafic rocks transform into **amphibolite,** a dark-colored metamorphic rock containing amphibole, plagioclase, and garnet.

• *Gneiss:* **Gneiss** is a compositionally banded metamorphic rock, typically composed of alternating dark and light layers that range in thickness from millimeters to meters. The compositional banding gives gneiss a striped appearance (see Fig. 8.2). Light layers contain light-colored minerals like quartz, feldspar, and muscovite (also called white mica), whereas the dark layers contain dark-colored minerals like amphibole, pyroxene, and biotite (dark

FIGURE 8.15 (a) The sheen in this phyllite comes from the reflection of light off the tiny mica flakes that comprise the rock. Folding (bending) of the layer has wrinkled the foliation. The tiny folds that result, called crenulations, make the rock look like the bellows on an accordion. (b) A hand specimen of schist.

(a)

(b)

mica). If a gneiss contains mica, the mica-rich layers also display schistosity. But not all gneiss contains mica, because at very high temperatures mica reacts to form other minerals and thus disappears.

We've already seen one way in which compositional banding forms: during recrystallization at very high temperatures, the atoms that make light minerals separate from the atoms that make dark minerals. Compositional banding may also take place when the compositional contrasts between beds in a sedimentary protolith are retained during metamorphism—a rock made of alternating beds of sandstone and shale may metamorphose into a gneiss consisting of alternating beds of quartzite and mica schist. Finally, compositional banding may develop when layers of igneous rock intrude like thin sills between layers of the protolith.

- *Migmatite:* At even higher temperatures than those needed to form gneiss, a rock begins to partially melt. Remember, when this happens, only the lower-melting-temperature minerals become liquid, while other minerals stay solid. In some cases, the melt does not move very far before freezing again. As a result, layers of new igneous rock grade into layers of relict gneiss, making a mixture of igneous and metamorphic rock we call **migmatite** (▶Fig. 8.16).

Nonfoliated Metamorphic Rocks

A **nonfoliated metamorphic rock** contains minerals that recrystallized during metamorphism, but it has no foliation.

- *Hornfels:* Some rock undergoes metamorphism simply because of a change in temperature resulting from a nearby igneous intrusion, without being subjected to a differential stress. The resulting rock is called **hornfels.** Because hornfels forms in the absence of differential stress, it contains randomly oriented crystals and thus has no foliation. The specific mineral content of a hornfels depends on the composition of the protolith and on the temperature of metamorphism.

- *Quartzite:* **Quartzite** forms by the metamorphism of quartz sandstone. During the metamorphism, the pre-existing quartz grains recrystallize, creating new, generally larger grains. In the process, the distinction between cement and grains disappears, open pore space disappears, and the grains become interlocking. In fact, recrystallization makes the grains of a quartzite effectively weld together, so that when quartzite cracks, the resulting fracture cuts across grain boundaries—in contrast, fractures in sandstone curve *around* grains (Fig. 8.5). Quartzite tends to look glassier than sandstone, and blocks of quartzite do not exhibit the grainy, sandpaper-like surface charac-

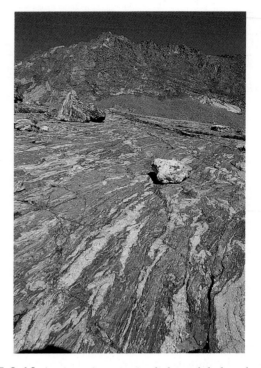

FIGURE 8.16 A migmatite contains light and dark rocks that swirl together like vanilla and chocolate batter in a marble cake. The light rock solidified from a melt that formed at temperatures between metamorphic and igneous conditions. The dark rock, whose minerals have high melting temperatures, thus did not melt; however, it was very soft at the time of metamorphism.

teristic of sandstone. Depending on the impurities contained in the quartz, quartzite can vary in color from white to gray, purple, or green. Most quartzite is considered a nonfoliated rock because it does not commonly show pronounced foliation, owing to the lack of mica or compositional layering. In some cases, however, quartzite has undergone deformation in response to differential stress, and *has* developed foliation, whose planes are defined by the parallel alignment of pancake-shaped quartz grains. We call such rock **foliated quartzite.**

- *Marble:* The metamorphism of limestone yields **marble.** During the formation of marble, the calcite composing the protolith recrystallizes; as a consequence, original sedimentary features like fossil shells can no longer be recognized, pore space disappears, and the distinction between grains and cement disappears. Marble typically consists of a fairly uniform mass of interlocking calcite crystals. Sculptors favor marble because the rock is relatively soft and easy to carve, and its uniform texture gives it the cohesiveness and homogeneity needed to fashion large, smooth, highly detailed sculptures. Marble comes in a variety of colors—white, pink, green, and black—depending on the impurities present. Michelangelo, the

(a)   (b)

FIGURE 8.17 (a) The marble in this unfinished sculpture by Michelangelo is fairly soft and easy to carve, but does not crumble. (b) Various impurities (such as iron and carbon) create the color banding of decorative marble, as illustrated by the beautiful stripes in this seventeenth-century column.

great artist of the Italian Renaissance, sought blocks of creamy white marble from the quarries in the Italian Alps, from which he created his masterpieces (▶Fig. 8.17a). Like quartzite, most marble is considered a nonfoliated rock since it commonly contains equant grains (and therefore no preferred mineral orientation) and no mica. But in many examples, impurities like iron oxide or graphite may create beautiful color banding in marble, making it also a prized stone for building facades, floors, and table-tops (▶Fig. 8.17b). Such **foliated marble** develops because marble is a relatively weak rock under metamorphic conditions; it can flow like soft plastic, though very slowly. This flow smears out different-colored portions of marble into contorted planes.

8.5 GRADES OF METAMORPHISM

Metamorphism, as we know, occurs in response to changes in temperature and pressure—but how big do these changes have to be? ▶Figure 8.18 gives a rough sense of the range of conditions leading to metamorphism. At low pressures and temperatures, found in the upper few kilometers of the Earth's crust and away from a volcanic region, only lithification takes place, while at very high temperatures, near the ground surface in volcanic regions or otherwise only at

FIGURE 8.18 The conditions of pressure and temperature under which metamorphism occurs. At low pressures and temperatures, only lithification takes place. At progressively higher pressures and temperatures, a rock passes from low to medium to high grade. If the temperature increases but the pressure stays low, we call the metamorphism "thermal." If the rock contains water, it begins to melt at the upper limit of high-grade metamorphism, but if it does not contain water, melting doesn't begin until higher temperatures.

great depth in the crust, melting takes place. The conditions for metamorphism lie in between.

The graph of Figure 8.18 also emphasizes that not all metamorphism occurs under the same conditions of pressure and temperature. Generally speaking, rocks that metamorphose under relatively low temperatures (less than 320°C) are called **low-grade metamorphic rocks,** and rocks that metamorphose under relatively high temperatures (over 500°C) are called **high-grade metamorphic rocks. Intermediate-grade metamorphic rocks** form under temperatures between these two extremes. (Note that temperature is the main factor that determines grade.) Different grades of metamorphism yield different groups of metamorphic minerals and different kinds of foliation. For example, high-grade rocks tend to be coarser-grained. Also, they contain "drier" minerals (minerals that contain less water bound to minerals) than do low-grade rocks; this is because water and other volatile elements of compounds are released during higher-grade metamorphic reactions.

An Example: The Metamorphism of Shale to Progressively Higher Grade

To better understand the concept of metamorphic grade, let's look at what happens to shale during metamorphism to progressively higher grade (►Fig. 8.19). The protolith, shale, consists of clay minerals and some fine-grained quartz, along with water (H_2O) molecules occurring both as a liquid in the pore space and as molecules wedged between layers of atoms in the clay flakes. Under low-grade conditions, shale transforms into slate, which involves the expulsion of the water from pore spaces and from between layers in clay flakes; pressure solution; and recrystallization. All these processes produce tightly packed, slightly larger, and better-formed clay flakes that have a strong preferred orientation. The change in the rock fabric explains the physical characteristics of slate—the hardness of slate compared with shale reflects the tighter bonding between clay flakes and the lack of water, while the ability of slate to split into thin sheets reflects the strong preferred orientation of the clay flakes.

As the metamorphic grade increases (as the temperature and pressure rise), the clay minerals recrystallize to form fine-grained muscovite. As a result, the rock transforms into phyllite; the characteristic silky sheen of phyllite comes from the parallelism of translucent muscovite flakes, for these reflect light. Under intermediate-grade conditions, muscovite crystals grow larger, and metamorphic reactions generate brand-new minerals like quartz and garnet, so the rock becomes a schist. Muscovite has the complicated-looking chemical formula $K_2Al_4[Si_6Al_2O_{20}](OH)_4$; notice that it includes OH, a compound containing the two elements, oxygen and hydrogen, found in water. During metamorphic reactions under high-grade conditions, the OH separates from minerals and becomes volatile water that escapes from

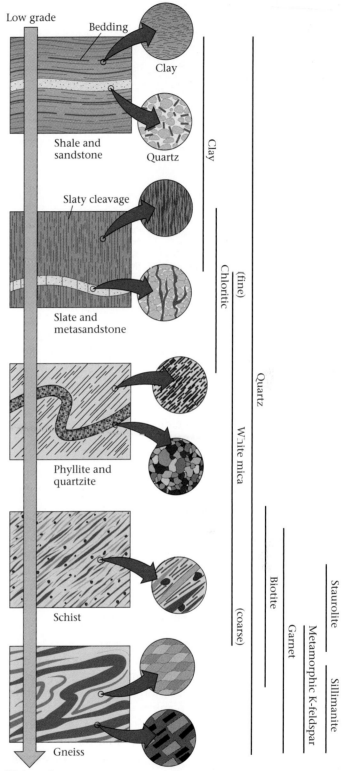

FIGURE 8.19 When shale progressively metamorphoses from low grade to high grade, it first becomes slate, then phyllite, then schist, then gneiss. In many cases, gneiss and schist can form under the same conditions. The side graph shows the stability range of various minerals.

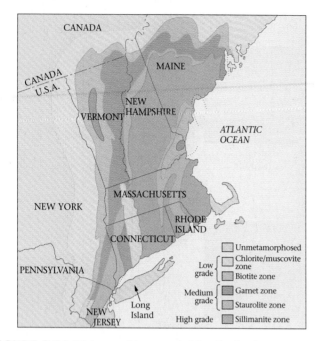

FIGURE 8.20 Metamorphic zones in New England. Isograds, defined by the first appearance of an index mineral, separate the zones. Within each zone, rocks contain the specified index mineral.

the rock. Thus, under high-grade conditions, yet another assemblage of minerals forms, including feldspar and quartz, which contain no water in their structure. Usually this process is accompanied by the formation of gneissic layering. If the temperature goes any higher, the rock begins to melt, yielding magma, and a migmatite may develop. Note that all the different metamorphic rocks (slate, phyllite, schist, gneiss, migmatite) just described formed from the same protolith—the differences between them reflect differences in mineral content and the nature of the foliation.

Index Minerals and Metamorphic Zones

Some of the minerals that form in metamorphic rocks are also found in igneous and sedimentary rocks. For example, mica occurs in both igneous and metamorphic rocks, and quartz in igneous, sedimentary, and metamorphic rocks. But metamorphic rocks may also contain minerals (such as sillimanite, kyanite, and andalusite) that form only under metamorphic conditions. Geologists have discovered that certain minerals serve as good indicators of metamorphic grade, for they appear only when the temperature and pressure reach a certain value. We call these **index minerals.** When mapping the countryside, geologists separate grades of rock by drawing lines representing the location where a particular index mineral first appears. On one side of the line, rocks contain the index mineral, and on the other side, they do not. These lines

are called **isograds** (from the Greek *iso,* meaning "equal") because rocks at all points along the line are taken to be at the same metamorphic grade. **Metamorphic zones** are the regions between two isograds; zones are named after the index mineral that occurs within them (▶Fig. 8.20). As an example, let's consider once again what happens to shale during metamorphism: progressively higher temperatures first produce chlorite, then metamorphic biotite, then metamorphic garnet. So on a map, we can draw the chlorite, biotite, and garnet isograds where these rocks appear.

Metamorphism that occurs as temperatures and pressures are increasing is called **prograde metamorphism.** The progressive transformation of shale to phyllite to schist to gneiss is an example of prograde metamorphism, during which water separates out of the minerals and leaves the rock. But metamorphism can also take place when temperatures and pressures decrease; this is known as **retrograde metamorphism.** In order for retrograde metamorphism to occur, water must be added back to the rock. Thus, retrograde metamorphism does not happen unless hot-water solutions pass through the rock to provide the water molecules; it does not automatically happen even when a high-grade metamorphic rock is uplifted from great depth to the surface of the Earth, because recrystallization under cold and dry conditions proceeds too slowly to change a rock significantly, even after billions of years (see Box 8.1). It is for that reason that we are able to see high-grade rocks formed early during Earth history exposed today at the surface.

We can represent the concept of prograde and retrograde metamorphism on the pressure-temperature graph (▶Fig. 8.21). As a rock gets buried progressively deeper and heats up, it follows the prograde path on the graph. As the rock moves back toward the Earth's surface when uplift and

FIGURE 8.21 As a rock experiences increased heating and pressure, it undergoes prograde metamorphism. As the temperature and pressure decrease, the rock undergoes retrograde metamorphism, if water can be added back.

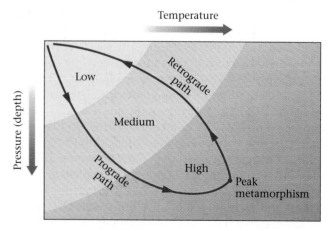

Metamorphic Facies

BOX 8.2

THE REST OF THE STORY

Geologists gradually came to realize that the identification of one mineral alone cannot completely define the pressure and temperature at which a metamorphic rock formed, because a single mineral can exist under a broad range of metamorphic conditions. Eventually, researchers discovered that they could more precisely determine the conditions under which a rock underwent metamorphism by identifying an assemblage, or group, of several minerals. But they also found that the mineral assemblage formed under a given range of temperature and pressure depends on the original composition of the rock—metamorphosed basalt does not contain the same assemblage of minerals as metamorphosed shale (though it metamorphoses under the same conditions). Geologists proposed the term **metamorphic facies** to refer to assemblages, or groups, of metamorphic minerals formed under a specific range of pressures and temperatures. There are seven major facies: **zeolite, hornfels, greenschist, amphibolite, blueschist, eclogite,** and **granulite.** Several different metamorphic rock types can exist in the same facies; though all the rock types in a facies form under the same range of pressure and temperature, each has a different composition and therefore a different metamorphic mineral assemblage.

We can represent the conditions under which metamorphic facies formed on another graph showing pressure and temperature (▶Fig. 8.22). Each area on the graph, labeled with a facies name, represents the range of temperatures and pressures in which mineral assemblages characteristic of that particular facies form. For example, a rock subjected to the pressure and temperature at point A (4 kbar and 350°C) develops a mineral assemblage characteristic of the greenschist facies. We can also plot different geothermal gradients (lines representing the change in temperature with depth) on the graph. In places where there is volcanic activity, temperatures can become so high near the Earth's

surface that hornfels forms under very small pressure. Beneath mountain ranges, the geothermal gradient passes through the zeolite, greenschist, amphibolite, and granulite facies. The geothermal gradient in an accretionary prism created at a subduction zone passes into the blueschist facies, because temperatures in the prism, composed of cold sediment scraped off the sea floor, remain relatively cool, even at high pressures.

"Grade" and "facies" are related terms, in that they both distinguish among rocks formed under different metamorphic conditions, but geologists use "grade" to give an approximate sense of metamorphic conditions, and "facies" to indicate a more specific temperature and pressure range in which a metamorphic rock developed. To define the facies, we must usually carefully examine a thin section of a rock under a microscope. In general, zeolite and lower-greenschist facies rocks are considered low grade, upper-greenschist facies through lower-amphibolite facies intermediate grade, and upper-amphibolite to granulite facies rocks high grade. ("Lower" and "higher" simply mean cooler and warmer conditions, respectively.)

The names for the different facies are based on a distinctive feature found in the first rock to be associated with a facies. For example, the "green" in greenschist comes from the presence of chlorite, a green mineral found in metamorphic basalt. The name "amphibolite" was coined because mafic rocks of this facies contain metamorphic amphibole. And the "blue" in blueschist comes from the presence of a bluish mineral called glaucophane.

FIGURE 8.22 The common metamorphic facies. Note that most granulite facies rocks form at pressure-temperature conditions to the right of the wet melting curve; such rocks develop only if they are dry. The dotted lines indicate various geothermal gradients (the change in temperature with depth) found on Earth. ("PP" stands for "prehnite-pumpellyite," two unusual metamorphic minerals that form under the pressure and temperature conditions indicated on the graph.)

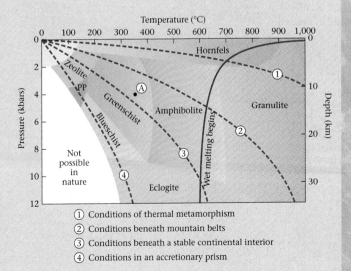

① Conditions of thermal metamorphism
② Conditions beneath mountain belts
③ Conditions beneath a stable continental interior
④ Conditions in an accretionary prism

Environments of Metamorphism

Metamorphism at a Convergent Margin

Contact metamorphism

Blueschist formation in an accretionary prism

Blueschist

Foliation resulting from deformation

Squashing

Shearing

Increasing temperature

Increasing pressure

Low grade

Slate

Schist

Increasing metamorphic grade

Gneiss

High grade

Blueschist

Hornfels formation

Mylonite in a shear zone

Regional Metamorphism in an Orogenic Belt

Unmetamorphosed shale

Metamorphic rocks form when a preexisting rock (a protolith) undergoes changes in texture and/or mineral content in the solid state, in response to changes in temperature, pressure, or differential stress. Some metamorphic rocks are unfoliated (they do not have metamorphic layering), while others are foliated (they do have metamorphic layering). Foliation results when rock is squashed or sheared during metamorphism, causing minerals to grow or rotate into parallelism with each other. Regional metamorphism occurs during mountain building, when a region is buried deeply, and during subduction, when sea-floor sediment is carried to the base of an accretionary prism. Contact metamorphism takes place around an igneous intrusion, or pluton, caused by the heat released by the pluton.

Geologists distinguish metamorphic rocks according to the type of foliation and the mineral assemblage a rock contains. Hornfels is unfoliated and forms as a result of contact metamorphism. Mylonite develops when shearing creates a foliation but not necessarily a change in types of minerals. Slate, which forms from shale, contains slaty cleavage; clay flakes are typically aligned at an angle to the bedding. Schist contains coarse grains of mica, aligned parallel to each other. Gneiss has compositional banding. Migmatite has chaotic layering, formed when part of the rock melts. Quartzite is composed only of quartz (it is metamorphosed sandstone), while marble is composed only of calcite (it is metamorphosed limestone). Quartzite and marble are usually unfoliated.

The types of minerals and foliation in a metamorphic rock indicate the rock's grade. Higher-grade rocks, like gneiss, form at higher temperatures and pressures, while lower-grade rocks, like schist, form at lower pressures and temperatures. Blueschist is an unusual metamorphic rock that develops under relatively high pressures but relatively low temperatures—the environment of an accretionary prism.

Migmatite

Gneiss

Compositional bands

Schist

Schistosity

Slate

Relict bedding

Slaty cleavage

erosion strip away rock from above, it follows the retrograde path (but remember, retrograde reactions don't occur without the addition of hot water).

8.6 ENVIRONMENTS OF METAMORPHISM

At this point, you should have a clear sense of what happens during metamorphism and what kinds of rocks develop as a consequence. You should also understand that the type of metamorphic rock that forms depends not only on the original composition of the rock, but also on the conditions of pressure, temperature, water content, and differential stress under which the rock was metamorphosed. These conditions depend in turn on the **environment of metamorphism**—the geologic setting in which metamorphism occurs. Different environments yield different conditions, and therefore different types of metamorphic rocks.

Metamorphism next to Plutons (Thermal, or Contact, Metamorphism)

As magma rises into the shallower, and therefore cooler, crust, it brings with it a large amount of heat. Imagine that a large volume of hot magma intrudes as a pluton into sedimentary rock (shale, limestone, and sandstone) at a depth of about 2 km below the Earth's surface. The magma initially has a temperature of about 900°C, while the sedimentary rock, before the intrusion, has a temperature of about 50°C. Heat slowly transfers from the pluton to the surrounding sedimentary (country) rock. The heat provided by the pluton cannot melt the country rock, but it can heat the country rock sufficiently to cause recrystallization. Further, hot water expelled from the pluton may react with the surrounding rock. Depending on the size and shape of the pluton, the amount of heat transferred into the country rock can metamorphose rock for a distance of tens to hundreds of meters away from the boundary of the pluton. This region is called the **metamorphic aureole.** Larger intrusions provide more heat and thus create larger aureoles.

Metamorphism caused by heat conducted into country rock from an igneous intrusion is called either **thermal metamorphism** (see Box 8.3), to emphasize that it results from heating without compression, shearing, or an increase in pressure, or **contact metamorphism,** since it occurs in contact with an intrusion (►Fig. 8.23). Because metamorphism in a contact aureole takes place without compression or shearing, aureoles typically contain hornfels, a nonfoliated metamorphic rock.

Where does contact metamorphism occur? Anywhere that the intrusion of plutons occurs. In the context of plate tectonics theory, plutons intrude at convergent plate boundaries, in rifts, and during the mountain building that takes place when continents collide. In the case of collisional mountain building, only those plutons that intrude after the squashing and shearing has stopped yield easily recognizable contact aureoles; the aureoles surrounding plutons that intrude during the squashing and shearing tend to be too distorted to identify.

Metamorphism as a Result of Shearing in Fault Zones (Dynamic Metamorphism)

Faults are fractures or zones on which one piece of crust slides past another. Near the Earth's surface (in the upper 10–15 km), this movement shatters the rock, breaking it into angular fragments and ultimately crushing it to make powder; as we'll see in Chapter 10, the breaking can cause an earthquake. But at greater depths, depending on the geothermal gradient, rock in the crust is so warm that it is too

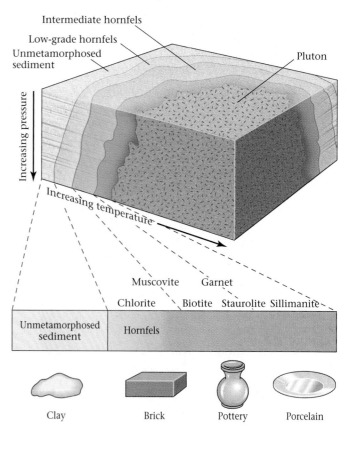

FIGURE 8.23 In a metamorphic aureole, the highest-grade thermally metamorphosed rocks directly border the pluton. The grade decreases away from the pluton. The gradation is analogous to the gradation from clay to porcelain.

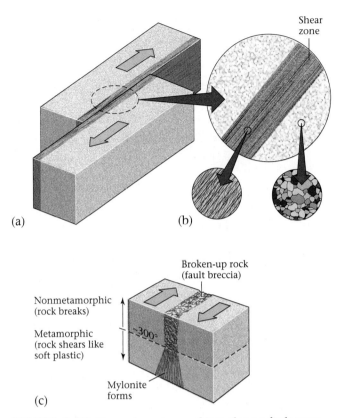

(a)

(b)

Shear zone

Broken-up rock (fault breccia)

Nonmetamorphic (rock breaks)

Metamorphic (rock shears like soft plastic)

~300°

Mylonite forms

(c)

FIGURE 8.24 Dynamic metamorphism along a fault zone. (a) Note the band of sheared rock on either side of the slip surface. (b) The rock outside the shear zone has a different texture from the rock inside. (c) The block formed in (a) must have developed at a depth where metamorphic conditions exist, so that mylonite forms; otherwise, it would break up during movement.

soft and plastic to break. Thus, when the shearing movement takes place, the rock deforms like soft plastic and smears out like taffy. During this process, the minerals in the rock recrystallize. We call this process **dynamic metamorphism,** because it occurs as a consequence of shearing alone, with no change in temperature or pressure. The resulting rock, a **mylonite,** has a pronounced foliation that roughly parallels the fault (▶Fig. 8.24a–c).

Faulting takes place at all plate boundaries (and sometimes in plate interiors as well), and dynamic metamorphism takes place anywhere that faulting occurs under warm conditions. For example, if you were to drill a hole down to a depth of 15 km or more along the San Andreas Fault, the major transform plate boundary that runs along the coast of California, you would likely find a several-kilometer-wide belt of mylonite created by the shearing of the Pacific Plate past the North American Plate. Because of the geothermal gradient, rock at these depths deforms at temperatures in the range of 300° to 350°C—that is, metamorphic temperatures. Notably, because the shearing that produces mylonite takes place without cracking or

breaking the rock, mylonite can form without generating earthquakes.

Metamorphism Beneath Mountains (Dynamothermal or Regional Metamorphism)

In mountain ranges like the Andes, which build at a convergent boundary, large bodies of magma rise from depth to shallower regions of the crust. The heat that the magma transports with it causes thermal metamorphism in surrounding rocks. But because mountain building occurs in response to differential stress (compression and shear stresses develop as plates push together and shear past one another), some of these rocks may also develop a foliation. Since this metamorphism involves not only heat but also shearing and squashing, we call it **dynamothermal metamorphism.** Typically, such metamorphism affects a large region, so many geologists also call it **regional metamorphism.**

Dynamothermal metamorphism also develops during continental collisions (▶Fig. 8.25a, b). To see why, imagine a layer of shale exposed in an outcrop at the surface of the Earth, at a temperature of 20°C and a pressure of 1 atm (caused by the weight of the overlying atmosphere). What happens to the shale if it somehow gets buried by 15 km of

FIGURE 8.25 (a) Metamorphism occurs where there is plutonic activity along a convergent boundary. Some may be thermal, but because of compression and shearing along convergent boundaries, some may be dynamothermal. (b) The sedimentary rock that lay at the top of a passive margin (point A) gets carried to great depth in a continental collision that leads to mountain building. As a result, it undergoes dynamothermal metamorphism. A broad region beneath a collision will lie in the field of metamorphism, and with continued shearing, rocks that metamorphose at depth will be squeezed up to the top of the mountain range.

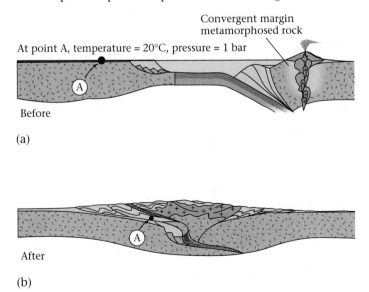

Convergent margin metamorphosed rock

At point A, temperature = 20°C, pressure = 1 bar

A

Before

(a)

A

After

(b)

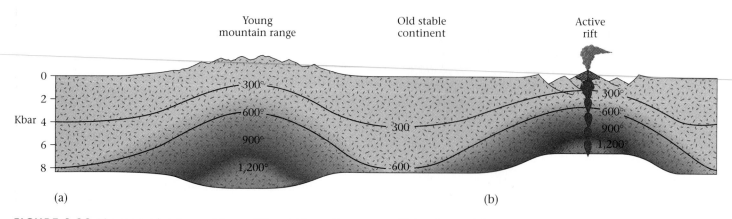

(a)   (b)

FIGURE 8.26 The range of metamorphic conditions (a) beneath a young collisional mountain range (such as the one shown in Figure 8.25) and (b) beneath an active rift. The solid lines are isotherms: they connect locations where the temperature has the same value (along the 300°C isotherm, the temperature is 300°C). Note that isotherms rise to shallower depths beneath mountain ranges because plutons carry heat with them up to shallower depths. The crust is fairly cool beneath old, stable continents.

other rock? Because of the geothermal gradient and the weight of the overlying rock, the temperature in the shale rises to about 400°C and the pressure rises to about 3.5 kbar (3,500 atm). Under these conditions, the minerals comprising the shale become unstable and recrystallize to form new metamorphic minerals. Note that because of the change in burial depth, the rock experiences both temperature *and* pressure change (▶Fig. 8.26a).

Now you are probably asking, how do rocks travel from the surface of the Earth to a depth of 15 km or even 60 km,

BOX 8.3

THE HUMAN ANGLE

Pottery Making—an Analogue for Thermal Metamorphism

A crumbling brick in the wall of an adobe house, an earthenware pot, a stoneware bowl, and a translucent porcelain teacup may all be formed from the same lump of soft clay, scooped from the surface of the Earth and shaped by human hands. This pliable and slimy muck is a mixture of water and very fine clay minerals, formed during the chemical weathering of rock. Fine potters' clay for making white china consists mainly of a particular clay mineral called kaolinite ($Al_2Si_2O_5[OH]_4 \cdot 2H_2O$), named after the locality in China (called Kauling, meaning "high ridge") where it was originally discovered.

People in arid climates make adobe bricks simply by pressing damp clay into a mold, which they then dry in the sun. Such bricks can be used for construction *only* in arid climates, because if it rains heavily, the bricks will rehydrate and turn back into sticky muck—drying clay in the sun does not change the structure of the clay minerals. In more humid climates, brickmakers fire the clay at high temperatures in a kiln. This process makes the bricks harder and impervious to water.

Potters use the same process of firing in a kiln when making vases and plates. Firing makes the vessels so durable that liquids cannot pass through them. In fact, fired clay jugs, called amphorae, that were used for storing wine and olive oil have been found intact in sunken Greek and Phoenician ships that have rested on the floor of the Mediterranean Sea for thousands of years.

Clearly, the firing of a clay pot fundamentally and permanently changes clay in a way that makes it physically different (Fig. 8.23). In other words, firing causes a thermal metamorphic change in the mineral assemblage that composes pottery. The extent of the transformation depends on the kiln temperature. Potters usually fire earthenware at about 1,100°C and stoneware (which is harder than a knife or fork) at about 1,250°C. To produce porcelain—fine china—the clay is partially melted at even higher temperatures. Just as it begins to melt, the potter cools it quickly (quenches it). Quenching of the melt creates glass, which gives porcelain its translucent, vitreous (glassy) appearance.

and then return to the surface? Once again, the answer comes from plate tectonics theory. We learned in Chapter 4 that in collision zones, mountain ranges develop as blocks of continental crust squeeze together. When this happens, large faults are created on which one piece of crust slides over another (see Chapter 11). During this movement, sedimentary rocks once at the Earth's surface (on the block that has been overridden) not only feel progressively greater temperatures and pressures, but also feel differential stress that causes shearing and squashing, and thus undergo dynamothermal metamorphism. Because of the changes in temperature and pressure, new metamorphic mineral assemblages grow, and because of the shearing and squashing, the rocks develop foliation. Thus, in such regions, you will find slate, phyllite, schist, and gneiss.

Metamorphosed rocks formed at depth beneath a mountain range return to the surface for several reasons. First, continued shearing and squashing can squeeze the rocks up from depth, or can stretch and thin the near-surface portion of the mountain range above the rocks. Second, rivers and glaciers at the surface erode the mountain range away, and the deeper parts of the crust move upward. The overall process of returning deeply buried rocks to the surface is called **exhumation.** Because of exhumation, dynamothermal metamorphic rocks become exposed at the surface. The different metamorphic zones we now see in a mountain range correspond, simplistically, to different depths of metamorphism—high-grade rocks (granulite or amphibolite facies rocks; see Box 8.2) were taken to great depth, while low-grade rocks (greenschist facies rocks) were taken to moderate depth.

Metamorphism as a Result of Rifting and Transform Faulting

During continental rifting, igneous magma rises beneath the axis of a rift as continental crust stretches and pulls apart. As a consequence, the temperature in the crust beneath rifts becomes quite high, and thermal metamorphism takes place (▶Fig. 8.26b). Hot-water solutions associated with these igneous intrusions may cause the retrograde metamorphism of preexisting metamorphic rocks. Along faults in the rift, dynamic metamorphism creates mylonite. Igneous activity rarely occurs along transform faulting, but when shearing takes place at depths greater than about 15 km, dynamic metamorphism again produces mylonite.

Metamorphism at Mid-Ocean Ridges

Hot magma rises beneath the axis of mid-ocean ridges. When cold seawater sinks into the crust, gaining access through cracks and faults, it heats up as it passes through or near the magma. The resulting hot water then rises along

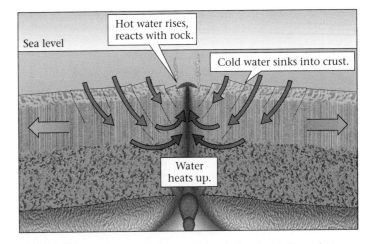

FIGURE 8.27 Along a mid-ocean ridge, the circulation of seawater through the oceanic crust causes retrograde metamorphism.

the ridge axis, creating the black smokers discussed in Chapter 4. As this water rises through the crust, it causes a retrograde metamorphism of the ocean-floor basalt (▶Fig. 8.27). This process leads to the growth of chlorite (green mica) in the basalt, giving the rock an overall greenish hue. Once the metamorphosed basalt is formed, sea-floor spreading carries it away from the ridge. Most sea-floor basalt has undergone this kind of retrograde metamorphism.

Metamorphism in Subduction Zones (the Blueschist Puzzle)

Blueschists are rare rocks that contain an unusual, blue-colored type of amphibole called glaucophane. Laboratory experiments showed that blueschist requires very high pressures but relatively low temperatures to form. These conditions do not typically occur in the crust—usually, at the high pressures needed to generate blueschist, temperatures are also high, so greenschist or amphibolite facies rock would be created instead. Where on Earth do high pressures and relatively low temperatures develop?

Plate tectonics theory provides the answer to the blueschist puzzle. Researchers realized that blueschist only occurs in accretionary prisms next to subduction zones (▶Fig. 8.28). When they calculated the temperature and pressure conditions found within an accretionary prism, they realized that because oceanic lithosphere is generally quite cool when it subducts, relatively little heat rises from the oceanic lithosphere into the base of the accretionary prism. As the accretionary prism grows, it becomes very thick, and at its thickest point the sediment at the base may be at a depth of almost 20 km. At such depths, pressures are high, but because so little heat rises

At point C, temperature = 600, pressure = 7,000 atm (amphibolite conditions)

At point B, temperature = 300, pressure = 7,000 bars (blueschist conditions)

FIGURE 8.28 Because oceanic lithosphere is cool when it subducts beneath accretionary prisms, the isotherms drop to great depths; rock at the base of the accretionary prism feels high pressures but relatively low temperatures. Under these conditions, blueschist forms.

from the underlying oceanic lithosphere, temperatures remain low, yielding the appropriate conditions for blueschist.

8.7 WHERE DO YOU FIND METAMORPHIC ROCKS?

If you want to study metamorphic rocks, you may want to go hiking in a mountain range for a start. The formation and then erosion of mountain belts first metamorphose rocks and then bring metamorphic rocks from depth back to the Earth's surface. Thus, in the interior of a mountain range, you will see towering cliffs of gneiss and schist. The process of uplift and exposure can happen relatively quickly, geologically speaking; some metamorphic rocks now visible in young mountains formed only a few million years ago. Where ancient mountain ranges once existed, we can find belts of metamorphic rocks cropping out at the ground surface even if the high peaks of the range have long since eroded away.

Even more metamorphic rocks crop out in older regions of continents, which geologists refer to as **shields.** Here, extensive areas of Precambrian rock (more than 545 million years old) make up the ground surface, because overlying younger rock has eroded away (▶Fig. 8.29a). In

fact, a substantial part of a shield consists of so-called high-grade gneiss terranes (areas where gneiss is prevalent), consisting of banded gneiss and migmatite. Some of these high-grade gneiss terranes contain the oldest rocks found on Earth—over 3.9 billion years. In North America, the shield encompasses about half the area of Canada (▶Fig. 8.29b). Extensive shields also occur in South America, northern Europe, Africa, India, and Siberia.

In the United States, a veneer of Paleozoic and Mesozoic sedimentary rocks continues to cover most of the Precambrian metamorphic rocks, so you can see these metamorphic rocks only where they were uplifted and exposed in young mountains (such as the Rocky Mountains), and in places like the Grand Canyon or the Black Canyon of the Gunnison River, where rivers sliced deep enough down in the Earth to expose Precambrian rocks (▶Fig. 8.29c). In such places, you can see the contact, or boundary, between the Precambrian metamorphic rocks and the overlying sedimentary rocks. Effectively, the metamorphic rocks formed the **basement** on which the **cover** of younger sedimentary rocks was deposited. For example, in the Grand Canyon, the 1.8-billion-year-old Vishnu Schist, the dark cliffs bordering the Colorado River, makes up the basement for the overlying cover of Paleozoic sedimentary rocks.

CHAPTER SUMMARY

• Metamorphism refers to changes in a rock that result in the formation of a metamorphic mineral assemblage and/or a metamorphic foliation, without the rock's first melting or becoming sediment. The new rock that results from these changes is a metamorphic rock.

• Metamorphic mineral assemblages form when the original minerals in a protolith become unstable owing to pressure and temperature changes, and recrystallization rearranges atoms into new mineral crystals. If hot-water solutions bring in or remove atoms, we say that metasomatism has occurred.

• Metamorphic foliation can be defined either by compositional banding or by a preferred mineral orientation (aligned inequant crystals). Compositional banding forms when light minerals separate from dark minerals during recrystallization. Preferred mineral orientation develops where differential stress causes the squashing and shearing of a rock, so that its inequant grains align parallel with each other.

• Geologists separate metamorphic rocks into two classes, foliated rocks and nonfoliated rocks, depending on whether

FIGURE 8.29 (a) The distribution of shield areas (exposed Precambrian metamorphic and igneous rock) on the Earth. (b) The Canadian Shield as viewed from the air. (c) The walls of the Black Canyon of the Gunnison River, in Colorado, display high-grade metamorphic rocks. These rocks underlie the Rocky Mountains of Colorado. The stripes are pegmatite dikes.

the rock contains foliation (preferred mineral orientation or compositional banding).

• The class of foliated rocks includes slate, metasandstone and metaconglomerate, phyllite, schist, amphibolite, gneiss, and migmatite; the class of nonfoliated rocks includes hornfels and some kinds of quartzite and marble.

• Metamorphic conditions are the temperatures and pressures necessary to cause metamorphism. Temperature is the more important factor. Rocks formed under relatively low temperatures are known as low-grade rocks, while those formed under high temperatures are known as high-grade rocks. Intermediate-grade rocks develop under intermediate conditions. Slate and phyllite are low grade, most schist is intermediate grade, and most gneiss and all migmatite are high grade.

• Geologists track the distribution of different grades of rock by looking for index minerals, which indicate the temperature and pressure at which a rock formed. We then can map out metamorphic zones, regions in which an index mineral occurs, and isograds, the boundary lines between zones.

• A metamorphic facies is a group of metamorphic rocks that develop under a specified range of temperature and pressure conditions (more precisely defined than those indicated by grade). Facies are based on the occurrence of an assemblage of minerals; the assemblage in a given rock depends on the composition of the protolith as well as the metamorphic conditions.

• Thermal metamorphism (also called contact metamorphism) occurs in an aureole surrounding an igneous intrusion. Because there is no shearing involved, nonfoliated rocks form in contact aureoles. Dynamically metamorphosed rocks form along faults, where rocks are only sheared, under metamorphic conditions. Dynamothermal metamorphism (also called regional metamorphism) results when rocks are buried deeply during mountain building. Because such metamorphism involves shearing and squashing as well as heat, rocks develop foliation.

• Metamorphism occurs because of plate interactions: the process of mountain building in either convergent or collisional zones causes dynamothermal metamorphism; shearing along plate boundaries causes dynamic metamorphism; and igneous plutons in rifts cause thermal metamorphism. The circulation of hot water causes a retrograde metamorphism of oceanic crust at mid-ocean ridges. Unusual metamorphic rocks called blueschists form at the base of accretionary prisms, where pressures are high but temperatures are low.

• We find extensive areas of metamorphic rocks in mountain ranges. Vast regions of continents known as shields expose ancient (Precambrian) metamorphic rocks.

KEY TERMS

amphibolite (p. 212)
basement (p. 224)
blueschist (p. 225)
compositional banding (p. 206)
compression (p. 209)
differential stress (p. 208)
diffusion (p. 206)
ductile (plastic) deformation (p. 209)
dynamic metamorphism (p. 221)
dynamothermal (regional) metamorphism (p. 221)
elongate (p. 209)
foliation (p. 210)
gneiss (p. 212)
hornfels (p. 213)
index minerals (p. 216)
isograds (p. 216)
marble (p. 206)
metaconglomerate (p. 212)
metamorphic aureole (p. 220)
metamorphic conditions (p. 208)
metamorphic facies (p. 217)
metamorphic grade (p. 214)
metamorphic mineral assemblage (p. 205)

metamorphic rock (p. 204)
metamorphic zone (p. 216)
metasandstone (p. 212)
metasomatism (p. 207)
migmatite (p. 213)
phyllite (p. 212)
platy (p. 209)
preferred orientation (p. 209)
pressure solution (p. 209)
prograde metamorphism (p. 216)
protolith (p. 205)
quartzite (p. 206)
recrystallization (p. 205)
retrograde metamorphism (p. 216)
schist (p. 212)
schistosity (p. 212)
shear stress (p. 209)
shield (p. 224)
slate (p. 211)
slaty cleavage (p. 211)
tension (p. 209)
thermal (contact) metamorphism (p. 220)

REVIEW QUESTIONS

1. How are metamorphic rocks different from igneous and sedimentary rocks?

2. What two features characterize most metamorphic rocks?

3. How do heat, hot groundwater, and pressure change a rock?

4. How does differential stress determine mineral orientation?

5. What is foliation?

6. How is a slate different from a phyllite? How does a phyllite differ from a schist?

7. Why are quartzites and marbles usually nonfoliated?

8. What is a metamorphic grade? How is it recognized in a rock?

9. How does prograde metamorphism differ from retrograde metamorphism?

10. Describe the geologic settings where thermal, dynamic, and dynamothermal metamorphism take place.

11. How does plate tectonics explain the peculiar combination of low-temperature but high-pressure minerals found in a blueschist?

The Rock Cycle

The three rock types of the Earth system. (a) Igneous rock, here formed by the cooling of lava at a volcano. (b) Sedimentary rock, here eroding to form sediment. (c) Metamorphic rock, here exposed in a mountain belt. Over time, materials composing one rock type may be incorporated in another.

(a)   (b)   (c)

B.1 INTRODUCTION

"Stable as a rock." This familiar expression implies that rock is immutable, unchanging over time—it isn't. In the time frame of Earth's history, a span of over 4.5 billion years, atoms making up one rock type may be rearranged or moved elsewhere, eventually becoming part of another rock type. Later, the atoms may move again to form a third rock type, and so on. Geologists call the progressive transformation of Earth materials from one rock type to another the **rock cycle** (▶Fig. B.1), one of many examples of cycles acting in or on the Earth. We focus on the rock cycle here be-

227

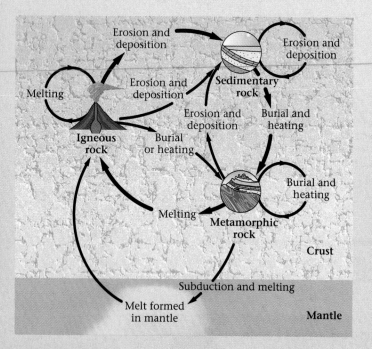

FIGURE B.1 The stages of the rock cycle, with its various alternative pathways. Note that new material comes from the mantle, and some old material returns to the mantle.

cause it illustrates the relationships between the three rock types described in the previous three chapters.

A **cycle** (from the Greek word for "circle" or "wheel") is a series of interrelated events or steps that occur in succession and can be repeated, perhaps indefinitely. During **temporal cycles,** such as the phases of the Moon or the seasons of the year, events happen according to a timetable, but the materials involved do not necessarily change. The rock cycle, in contrast, is an example of a geologic **mass-transfer cycle,** one that involves the transfer or movement of materials (mass) to different parts of the Earth system. (The hydrologic cycle, which we will learn about in Interlude E, is another type of mass-transfer cycle.)

There are many paths around or through the rock cycle. For example, igneous rock may weather and erode to produce sediment, which lithifies to form sedimentary rock. The new sedimentary rock may become buried and form metamorphic rock, which could partially melt to create magma. This magma later solidifies to form new igneous rock. We can symbolize this path as igneous → sedimentary → metamorphic → igneous; that is, a complete loop around the rock cycle. But alternatively, the metamorphic rock could be uplifted and eroded to form new sediment and ultimately new sedimentary rock without melting, taking a shortcut path through the cycle that we can symbolize as igneous → sedimentary → metamorphic → sedimentary. Likewise, the igneous rock could be metamorphosed di-

rectly, without first turning to sediment. This metamorphic rock could again be turned into sedimentary rock, defining another shortcut path: igneous → metamorphic → sedimentary. To get a clearer sense of how the rock cycle works, we'll look at one particular example.

B.2 A CASE STUDY

Material first enters the rock cycle when basaltic magma rises from the mantle. Suppose the magma erupts and forms basalt (an igneous rock) at a continental hot-spot volcano (▶Fig. B.2a). Movement of the magma brought atoms from the mantle up to the Earth's surface, where they arranged to form various mineral crystals (such as plagioclase, pyroxene, and olivine). Interaction with wind, rain, and vegetation gradually weathers the basalt, physically breaking it into smaller fragments and chemically altering it to create clay. As rainwater washes over the newly formed clay, it carries the clay away and transports it downstream—if you've ever seen a brown-colored river, you've seen clay en route to a site of deposition. Eventually the river reaches the sea, where the water slows down and the clay settles out.

Let's imagine, for this example, that the clay settles out along the margin of continent X and forms a deposit of mud. Gradually, through time, the mud becomes progressively buried by still younger sediment. Water in the mud escapes, and the clay flakes pack tightly together, resulting in a new sedimentary rock, shale. The shale resides comfortably 5 or 6 km deep along the continental margin for millions of years, until the adjacent oceanic plate is consumed at a convergent boundary and a neighboring continent, Y, collides with X. The shale gets buried very deeply when the edge of the encroaching continent pushes over it. As mountains are built, the shale that had once been 5 or 6 km below the surface now ends up 15–20 km below the surface, and under the pressure and temperature conditions present at this depth, it metamorphoses into schist (▶Fig. B.2b).

The story's not over. Once mountain building stops, erosion gradually grinds away the mountain range, and the schist is exposed at the ground surface. Some of the schist formed during collision transforms directly into sediment, which is carried off and deposited elsewhere to form new sedimentary rock—this material takes a shortcut through the rock cycle. But some schist remains preserved below the surface. Eventually, continental rifting takes place at the site of the former mountain range, and the crust containing the schist begins to split apart. When this happens, new mafic magma develops in the underlying mantle and rises into the crust, where it transfers enough heat to the schist to partially melt it and generate a new silicic magma. This silicic magma rises to the surface of the crust and freezes to create rhyolite, a new igneous rock (▶Fig. B.2c). In terms of the

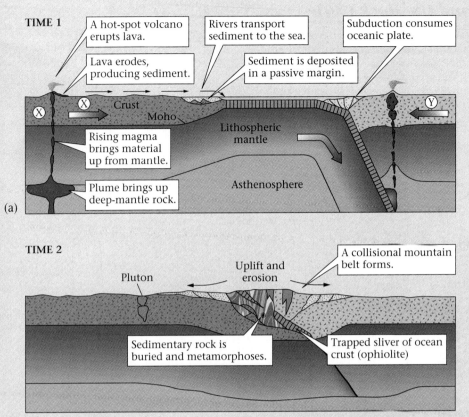

TIME 1

A hot-spot volcano erupts lava.

Rivers transport sediment to the sea.

Subduction consumes oceanic plate.

Lava erodes, producing sediment.

Sediment is deposited in a passive margin.

Crust

Moho

Lithospheric mantle

Rising magma brings material up from mantle.

Asthenosphere

Plume brings up deep-mantle rock.

(a)

TIME 2

Pluton

Uplift and erosion

A collisional mountain belt forms.

Sedimentary rock is buried and metamorphoses.

Trapped sliver of ocean crust (ophiolite)

(b) ▨ Metamorphic rock   ☐ Sediment eroded from mountains

TIME 3

Metamorphic rock is pre-served in a mountain belt.

Mountains erode away.

Sediment is deposited on continent Y.

(c)

TIME 4

Rifting occurs, and the crust stretches and breaks.

Rift-related volcanoes erupt.

Partial melting occurs in the asthenosphere as it rises.

Rock is partially melted by heat transferred into the crust.

(d)

FIGURE B.2 (a) At the beginning of the rock cycle (time 1), atoms, originally comprising peridotite in the mantle, rise in a mantle plume. The peridotite partially melts at the base of the lithosphere, and the atoms become part of a basaltic magma that rises through the lithosphere of continent X and erupts at a volcano. At this time, the atoms become part of a lava flow; that is, an igneous rock. Weathering breaks the lava down, and the resulting clay is transported to a passive-margin basin. After the clay is buried, the atoms become part of a shale—a sedimentary rock. Note that the ocean floor to the east of the passive-margin basin is being consumed beneath continent Y. (b) At time 2, continents X and Y collide, and the shale is buried deeply beneath the resulting mountain range (at the dot). Now the atoms become part of a schist—a metamorphic rock. (c) At time 3, the mountain range erodes away and the schist rises but does not reach the surface. (d) At time 4, rifting begins to split the continents apart, and igneous activity occurs again. At this time, the atoms of the schist become part of a new melt, which eventually freezes to form a rhyolite, another igneous rock.

Rock-Forming Environments and the Rock Cycle

There are many different environments in which rocks form. Igneous rocks develop where melt rises from depth and cools. Intrusive igneous rocks form where magma cools underground, extrusive igneous rocks form where lava and ash erupt at the surface.

Weathering and erosion break up existing rock and produce sediment. Different kinds of sediments develop in different places, reflecting both the composition of the source and the setting in which the sediment is deposited. We distinguish among sediment that accumulates in alluvial fans, desert dunes, river channels and floodplains, deltas, coral reefs, coastlines, the continental shelf, the deep sea, and the toe of a glacier. When this sediment eventually gets buried and undergoes lithification, new sedimentary rocks form.

Drainage networks collect surface water that can transport sediment to the ocean.

Sand dunes form from grains carried by the wind.

In a desert environment, rock weathers and fragments. Debris falls in landslides.

Flash floods carry sediment out of canyons to form an alluvial fan.

Volcanic eruptions emit lava and ash, which form new igneous rock at Earth's surface.

Sedimentary rocks make a cover on the surface of continents.

The crust and lithospheric mantle stretch and thin in a rift.

Magma rises from the mantle. Heat from this magma causes contact metamorphism.

Deep levels of continents consist of ancient metamorphic and igneous rocks. This is the basement of the continents.

Continental margins slowly sink and are buried by new sediment.

km

0

10

20

30

40

50

60

70

80

90

100

Partial melting occurs in the asthenosphere to produce new magma.

Glaciers erode rock and can transport sediment of all sizes.

In a region of continental collision, rocks that were near the surface are deeply buried and metamorphosed.

In humid climates, thick soils develop.

Magma that cools and solidifies underground forms igneous intrusions.

Along coastal plains, rivers meander. Sediment collects in the channel and floodplain.

Reefs grow from calcite-secreting organisms. These will eventually turn into limestone.

Where a river enters the sea, sediment settles out to form a delta.

Many different kinds of sediment accumulate along coastlines, building out a continental shelf.

Underwater avalanches carry a cloud of sediment that settles to form a submarine fan.

Fine clay and plankton shells settle on the oceanic crust.

The oceanic crust consists of igneous rocks formed at a mid-ocean ridge.

    Under certain conditions, preexisting rocks can undergo change in the solid state—metamorphism—which produces metamorphic rocks. Contact metamorphism is due to heat released by an intrusion of magma. Regional metamorphism occurs where tectonic processes cause rocks from the surface to be buried very deeply.
    Because the Earth is dynamic, environments change through time. Tectonic processes cause new igneous rocks to form. When exposed at the surface, these rocks weather to make sediment. The slow sinking of some regions creates sedimentary basins in which sediment accumulates and new sedimentary rocks form. Later, these rocks may be buried deeply and metamorphosed. Uplift as a result of mountain building exposes the rocks to the surface, where they may once again be transformed into sediment. This progressive transformation is called the rock cycle.

rock cycle, we're back at the beginning, having once again made igneous rock (▶Fig. B.2d).

The route around the rock cycle in this case study is not the only one possible. For example, instead of being buried and metamorphosed, the shale could have been uplifted, eroded, and transformed directly into a different sedimentary rock. Alternatively, the igneous volcanic rock that began our rock cycle could have been buried or heated and thereby transformed directly into metamorphic rock without ever having gone through the sedimentary stage. There is no one path through the rock cycle. What happens to a rock at any given stage reflects the changing environmental conditions at its location, and these conditions depend on many geologic variables.

Note that atoms, as they pass through the rock cycle, do not always stay within the same mineral. In our example, a silicon atom in a pyroxene crystal of the basalt may become part of a clay crystal in the shale, part of a muscovite crystal in the schist, and part of a feldspar crystal in the rhyolite. Further, a rock doesn't necessarily retain the same composition through all stages in the cycle. For example, the rhyolite magma formed by a partial melting of the schist did not have the same composition as the schist, because partial melting yields a magma that is more silicic than the original rock.

B.3 RATES OF MOVEMENT THROUGH THE ROCK CYCLE

We saw that not all atoms pass through the rock cycle in the same way. Similarly, not all atoms pass through the rock cycle at the same rate, and for that reason we find rocks of many different ages at the surface of the Earth. Some rocks remain in one form for less than a few million years, while others stay the same for most of Earth history. The location where a rock happens to have developed governs, to some extent, how fast it passes through the rock cycle.

Rocks comprising the **cratons,** the stable interiors of continents, have remained unchanged for billions of years. For example, North America's craton includes the Canadian Shield, where rocks as old as 3.9 billion years have been found. In contrast, rocks in the Appalachian Mountains have passed through stages of the rock cycle many times in the past few billion years, because the eastern margin of North America has been subjected to multiple events of basin formation, mountain building, and rifting since the shield to the west developed. A silicon atom with an East Coast address (such as eastern Massachusetts or Nova Scotia) may have been part of a granite 2.7 billion years ago when the region was a volcanic arc, part of a gneiss 1.8 billion years ago when the region underwent a continental collision, part of a shale 1.3 billion years ago when the region

was a passive margin, part of a schist 1.0 billion years ago when the region was once again in a collision, part of a sandstone 0.5 billion years ago when the region once more became a passive margin, part of a quartzite 0.25 billion years ago when the region yet again underwent a collision, and may be part of a beach sand today, as streams carry the erosional products of the Appalachians to the Atlantic coast.

Studies of the past two decades suggest that most of the rock now making up the Earth's continental crust contains atoms that were extracted from the mantle over 2.5 billion years ago. Yet we see rocks of many different ages in the continents today. That is because nature recyles these atoms again and again, similar to the way people recycle the metal of old cars to make new ones. And just as the number of late-model cars on the road today exceeds the number of vintage cars, younger rocks are more common than ancient rocks. At the surface of continents, sedimentary rocks created during the last several hundred million years are the most widespread type, whereas rocks recording the early history of the Earth are quite rare. But even though most continental crustal rocks are recycled, some new ones continue to be freshly extracted from the mantle each year, adding to the continent at volcanic arcs or hot spots.

Do the atoms in continental rocks ever get a chance to start the rock cycle all over, by returning to the mantle? Yes. Some sediment that erodes off a continent ends up in deep-ocean trenches, and some of this is dragged back into the mantle by subduction. In fact, recent research suggests that metamorphic and igneous rocks at the base of the continental crust may be ground away and carried back down into the mantle by the grinding action of the downgoing slab at subduction zones.

Our tour of the rock cycle has focused on continental rocks. What about the oceans? Oceanic crust consists of igneous rock (basalt and gabbro) overlain by sediment. Because a layer of water blankets the crust, erosion does not affect it, so oceanic crustal rock does not follow the path into the sedimentary loop of the rock cycle. But sooner or later, oceanic crust subducts. When this happens, the rock of the crust first undergoes metamorphism, for as it sinks, it is subjected to progressively higher temperatures and pressures. And eventually, some of the rock may melt and become new magma, which then rises at a volcanic arc and thus enters the continental rock cycle.

B.4 WHAT DRIVES THE ROCK CYCLE?

The rock cycle occurs because beneath and on the Earth's surface, two systems interact and battle one another. Inside the Earth, the planet's internal heat and gravitational field drive plate movements. Plate interactions cause the uplift of mountain ranges, a process that exposes rock to weathering, erosion, and sediment production. Plate interactions also generate the geological settings in which metamorphism occurs, where rock melts to provide magma, and where sedimentary basins develop.

At the surface of the Earth, the gases released by volcanism collect to form the ocean and atmosphere. Heat (from the sun) and gravity drive convection in the atmosphere and oceans, leading to wind, rain, ice, and currents—the agents of weathering and erosion. Weathering and erosion grind away at the surface of the Earth and send material into the sedimentary loop of the cycle. In sum, solar heat, Earth's internal heat, and gravity all play a role in driving the rock cycle, by keeping the mantle, atmosphere, and oceans in constant motion.

Tectonic Activity of a Dynamic Planet

*E*arlier in this book, we learned that the map of the Earth constantly changes (although very slowly) because of plate tectonics, and that as plates move relative to each other, they interact along their boundaries. We saw that geologists recognize three types of boundaries: divergent (spreading mid-ocean ridges), convergent (subduction zones), and transform—and each results in a unique suite of geologic features.

We now explore some of the most dramatic consequences of tectonic activity: volcanoes (Chapter 9), earthquakes (Chapter 10), and mountains (Chapter 11). We not only learn how volcanoes and earthquakes result from plate tectonics, we also see how they are natural hazards with a profound impact on society. Geologists have invested much effort in trying to understand how to predict when volcanic eruptions and earthquakes will take place and how to prevent loss of life if they do. In our discussion of mountains, we also review geologic structures—the bends, breaks, and sliding surfaces that develop in rocks in response to stress. And to fill in our picture of the Earth's interior, Interlude C introduces discoveries that were made by studying how the energy produced by earthquakes passes through the Earth.

C H A P T E R   9

The Wrath of Vulcan: Volcanic Eruptions

Glowing waves rise and flow, burning all life on their way, and freeze into black, crusty rock which adds to the height of the mountain and builds land, thereby adding another day to the geological past. . . . I became a geologist forever by seeing with my own eyes: the earth is alive!

—H. CLOOS (1947), on seeing an eruption of Mt. Vesuvius

9.1 INTRODUCTION

Every few hundred years, one of the hills on Vulcano, an island in the Mediterranean Sea off the west coast of Italy, rumbles and spews out molten rock, glassy cinders, and dense "smoke" (actually a mixture of various gases, like water and sulfur dioxide; fine ash, composed of tiny glass shards; and aerosols, tiny droplets of water or acid). Ancient Romans thought that such **eruptions** happened when Vulcan, the god of fire, fueled his forges beneath the island to manufacture weapons for the other gods. Geological study suggests, instead, that the eruptions take place when hot magma, formed by melting inside the Earth, rises through the crust. Though the Roman myth of Vulcano lost believers long ago, the island's name evolved into the English word **volcano,** which geologists use to designate either an erupting vent or fissure through which molten rock reaches the Earth's surface, or a mountain built from the products of eruption.

On the main peninsula of Italy, not far from Vulcano, another volcano, Mt. Vesuvius, towers more than 1 km over the nearby Bay of Naples (▶Fig. 9.1a, b). Two thousand years ago, Pompeii, a prosperous Roman resort and trading town of 20,000 inhabitants, sprawled at the foot of Vesuvius. Then one morning in 79 C.E., earthquakes signaled the mountain's awakening. At 1:00 P.M., a dark mottled cloud boiled up above Mt. Vesuvius's summit to a height of 27 km. As lightning sparked in its crown, the cloud drifted over Pompeii, turning day into night. Blocks and

---

During a volcanic eruption, molten rock rises from inside the Earth and flows out as lava or blasts into the air as ash.

237

FIGURE 9.1 (a) Pompeii, once buried by 6 m of volcanic debris from Mt. Vesuvius, was excavated by archaeologists in the late nineteenth century. Vesuvius rises in the distance. (b) The streets of Pompeii were paved with soft stone. Over time, chariots carved tracks in the stone. (c) A plaster cast of an unfortunate inhabitant of Pompeii, found buried by ash in the corner of a room, where the person crouched for protection. The flesh rotted away, leaving only an open hole that could be filled by plaster.

pellets of rock fell like hail, while fine ash and choking fumes enveloped the town. Frantic people rushed to escape, but for many it was too late. As the growing weight of volcanic debris began to crush buildings, an avalanche of ash swept over Pompeii, and by the next day the town had vanished beneath a 6-m-thick gray-black blanket. Pompeii was protected so well by its covering that when archaeologists excavated the town 1,800 years later, they found an amazingly complete record of Roman daily life. Interestingly, during their work, archaeologists discovered open spaces in the debris. Out of curiosity researchers filled the spaces with plaster, and realized that the spaces were fossil casts of Pompeii's unfortunate inhabitants, their bodies twisted in agony or huddled in despair (▶Fig. 9.1c).

Clearly, volcanoes are unpredictable and dangerous. Volcanic activity can build a towering, snow-crested mountain or can blast one apart. It can provide the fertile soil that enables a civilization to thrive, or it can snuff out a civilization in a matter of minutes. Because of the diversity of volcanic activity and its consequences, this chapter sets out

ambitious goals. We first review the products of volcanic eruptions and the basic characteristics of volcanoes. Then we look again at the different kinds of volcanic eruptions on Earth, in terms of the geologic settings in which they occur as defined by the theory of plate tectonics. We'll learn that volcanoes are *not* randomly distributed around the globe—their positions reflect the locations of plate boundaries, rifts, and hot spots. Finally, we examine the hazards posed by volcanoes and note efforts by geoscientists to predict eruptions and help minimize the damage they cause. We'll see that eruptions may affect the climate and civilization.

9.2 THE PRODUCTS OF VOLCANIC ERUPTIONS

The drama of a volcanic eruption transfers materials from inside the Earth to our planet's surface. Products of an eruption come in three forms. (1) **Lava flows** are sheets or mounds of lava that flowed onto the ground surface or sea floor in molten form and then solidified. (2) **Pyroclastic debris** is fragmented material that was thrown out of a volcano and landed on the ground or sea floor in solid form. Pyroclastic debris includes both brand-new clasts that form when ejected drops or blobs of molten lava freeze in air or water, and clasts created by the fragmenting of preexisting volcanic rock. (3) **Volcanic gas** consists of elements or compounds that bubble out of magma or lava in gaseous form. Some of this gas may condense, when it enters the cooler air above the volcano, to make tiny liquid droplets called **aerosols,** which remain suspended in the air for a long time. The "smoke" we see above a volcano is a mixture of fine pyroclastic debris, gas, and aerosols.

Lava and Lava Flows

Lavas come in a variety of compositions, and geologists describe these compositions by specifying the silica (SiO_2) content relative to the iron oxide and magnesium oxide ($FeO + MgO$) content. Lavas high in silica are called silicic (or rhyolitic), lavas with an intermediate silica content are called intermediate (or andesitic), while lavas low in silica are called mafic (or basaltic). Rocks formed from lava come in a variety of textures, depending on their composition and also on their cooling rate. All these rocks are fine-grained, but silicic lavas tend to be glassier than mafic lavas, and lavas that cool quickly tend to be glassier than lavas that cool slowly. In fact, rapidly cooled, silicic lavas form **obsidian,** massive volcanic glass.

The viscosity, or resistance to flow, of lava also depends on its composition and temperature. The less silica in a lava, the less its viscosity (because silica molecules link up in chains, or polymers, that hold lava together), so the greater the ease with which it can flow. Temperature and gas content also affect viscosity: the hotter the lava, the less its viscosity, because heat tends to break the bonds linking chains of silica molecules together; and the more gas-rich a lava, the less its viscosity. A sheet or mound of volcanic rock that forms by extrusion and flow of lava is called a **lava flow.** Depending on their viscosity, different kinds of lava produce different-shaped flows (▶Fig. 9.2a–d).

Basaltic lava flows. Basaltic (mafic) lavas have low viscosity and can travel easily in relatively fast-moving streams that generate relatively thin and long lava flows. In fact, typical basaltic lava can flow for tens of kilometers from a vent, while particularly hot basaltic lava can flow over 500 km from a vent (▶Fig. 9.3a).

The surface texture of a basaltic lava flow when it finally freezes reflects the timing of freezing relative to its movement. Flows with warm, pasty surfaces wrinkle into smooth, glassy, rope-like ridges; geologists have adopted the Hawaiian word **pahoehoe** (pronounced pa-hoy-hoy) for such flows

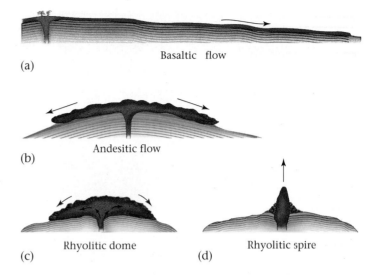

(a) Basaltic flow

(b) Andesitic flow

(c) Rhyolitic dome

(d) Rhyolitic spire

FIGURE 9.2 The character of a lava flow depends on the viscosity of the lava. Hotter lavas are less viscous than cooler lavas, and mafic lavas are less viscous than silicic lavas. (a) A basaltic lava flow is very fluid-like and can travel a great distance, forming a thin sheet. (b) An andesitic flow is too viscous to travel far, and tends to break up as it flows. (c) Rhyolitic lava is so viscous that it piles up at the vent in the shape of a dome. (d) In some cases, rhyolitic lava is too viscous to flow at all, and rises out of the vent as a columnar plug.

(▶Fig. 9.3b). Flows that continue to move after their surface freezes break up into a jumble of sharp, angular fragments, creating a rubbly flow also called by its Hawaiian name, **aa** (pronounced ah-ah) (▶Fig. 9.3c). Footpaths made by people living in volcanic regions follow the smooth surface of pahoehoe flows rather than the rough, foot-slashing surface of aa flows.

Even after the surface of a large lava flow has solidified, the inner part may continue to stream downslope. In some cases, the stream localizes in a natural tunnel below the surface of the otherwise solid flow. If lava drains out of the tunnel, an empty space known as a **lava tube** develops (▶Fig. 9.3d). Eventually, during the final stages of cooling, the interior of a thick lava flow contracts and fractures into roughly hexagonal columns, a type of fracturing called **columnar jointing** (▶Fig. 9.4a, b).

Basaltic flows that erupt underwater look different from those that erupt on land, because the lava cools so much more quickly. In fact, water removes heat thirty times faster than air does (that's why you soon get hypothermia when immersed in cold water). Because of rapid cooling, submarine basaltic lava forms a glass-encrusted blob, or **pillow,** upon freezing. The rind of a pillow momentarily stops the flow's advance, but within minutes the pressure of the lava squeezing into the pillow breaks the rind, and a new blob of lava squirts out, perhaps moving 0.5–2 m before itself freezing into a pillow. The process repeats until the lava pillows

(a)

(b)

(c)

(d)

FIGURE 9.3 Characteristics of basaltic (mafic) lava flows. (a) A basaltic lava flow on Hawaii has such low viscosity that it can flow a great distance, creating a thin sheet that here covers a highway. (b) A pahoehoe lava flow at Craters of the Moon National Park, in Idaho, displays the characteristic smooth, ropy surface. (c) An aa flow nearby, in contrast, has a rough and rubbly surface. (d) Lava travels to the end of a flow through a tunnel called a lava tube. Some tubes drain, as did this one in Hawaii.

FIGURE 9.4 (a) Lava flows contract when they cool, and some crack to form columnar joints. These columns crop out in Yellowstone Park. (b) In cross section, the columns are somewhat hexagonal.

(a)

(b)

freeze through and through, and the lava at the vent pushes up and makes another layer of pillows above the first (see Fig. 6.21). As a result, a mound of pillow lava develops. **Hyaloclastite** ("broken glass rock"), debris consisting of small glassy fragments of volcanic rock, breaks off the rinds of pillows and fills the spaces between pillows.

Andesitic lava flows. Because of its higher silica content and thus its greater viscosity, andesitic lava cannot flow as easily as basaltic lava. When erupted, andesite lava first forms a large mound above the vent. This mound then advances slowly down the volcano's flank at only about 1–5 m a day, in a lumpy flow with a bulbous snout. The length of the flow depends on the viscosity of the specific flow (which depends in turn on how hot and gas-rich it is) and on the steepness of the slope. Typically, even the hottest andesite flows are less than 10 km long. Because the lava moves so slowly, the outside of the flow has time to solidify; so as it moves, the surface breaks up into angular blocks, and the whole flow looks like a jumble of rubble. In places where submarine eruptions yield andesitic lava, a cone of splintery, glassy volcanic fragments (hyaloclastite) builds up around the vent.

Rhyolitic lava flows. Rhyolitic lava is the most viscous of all lavas because it is the most silicic and the coolest. Therefore, it tends to accumulate either in a dome-like mass, called a **lava dome,** above the vent or in short and bulbous flows rarely more than 1–2 km long. Sometimes rhyolitic lava freezes while still in the vent, and then pushes upward as a column-like spire up to 100 m above the vent. Rhyolitic flows, where they do form, have broken and blocky surfaces, because the rind of the flow shatters as the inner part fills with lava and expands. Submarine rhyolitic flows also yield hyaloclastite.

Pyroclastic Debris

Pyroclastic debris comes in a great variety of sizes (▶Fig. 9.5a). During some eruptions, clots of lava fountain into the air and freeze into glassy rock before landing (▶Fig. 9.5b). The resulting fragments include **ash** (powder sized), **lapilli,** or **cinders** (marble-to-plum-sized; ▶Fig. 9.5c), and **blocks** (basketball-to-refrigerator-sized). If blocks become streamlined as they travel through the air, they are called **volcanic bombs** (▶Fig. 9.5d). An accumulation of blocks or bombs is called **volcanic agglomerate.** While falling, some basaltic droplets mold into teardrop-shaped glassy beads, known in Hawaii as **Pelé's tears** after the Hawaiian goddess of volcanoes, while others stretch out into long strands like melted pizza cheese, to become **Pelé's hair.** In andesitic and rhyolitic eruptions, fragments ejected from a volcano include **pumice,** sponge-like volcanic glass filled with gas bubbles formed by the ejection of froth in the lava.

Much of the debris covering Pompeii, for example, consists of pumice lapilli. The explosion of lava into the air creates a fine mist that cools instantly to make tiny flakes or shards of glass (less than 2 mm in diameter) called **volcanic ash** (▶Fig. 9.5e).

Unconsolidated accumulations of pyroclastic grains (ash or lapilli) comprise **tephra,** which when cemented becomes **tuff.** Some ash settles gently from the air, like falling snow, to form **air-fall tuff.** But some ash mixes with air to make a fast-moving avalanche called a **pyroclastic flow** (also called a **nuée ardente,** French for "glowing cloud") (▶Fig. 9.6a). The deposits of pyroclastic flows when solidified into rock are known as **ignimbrite.** In particularly thick pyroclastic flows, the base of the flow remains so hot that the glass shards comprising the ash weld together, creating a **welded tuff.**

Perhaps the most famous pyroclastic flow erupted from Mt. Pelée, a volcano on the otherwise quiet tropical West Indies island of Martinique. In April 1902, a small eruption shed fine white air-fall ash over the town of St. Pierre, at the foot of the volcano. A mayoral election was in process, and the opposition candidate called for an evacuation, while the incumbent mayor urged people to stay and remain calm. As a spire of rhyolite began to push upward out of Mt. Pelée's summit, like a cork slowly working out of a champagne bottle, the mayor continued to request residents to stay. His plea cost the town's population their lives. Early one morning, the "cork" popped, and like the froth that streams down the side of a champagne bottle, a pyroclastic flow swept down Pelée's flank. The cloud of burning ash, blocks, gas, and debris, at a temperature of 200°–450°C, rode a cushion of air and may have reached a velocity of over 300 km per hour before it slammed into St. Pierre two minutes later. One breath of the super-hot ash meant instant death, and within moments 28,000 people lay dead of asphyxiation or incineration. Buildings toppled into a chaos of rubble and twisted metal, stockpiles of rum barrels exploded and sent flaming liquid into the streets, and ships at anchor in the harbor capsized. Only two people survived; one was a prisoner named Louis-Auguste Sylbaris who, though burned, was protected from the full cataclysm by the stout walls of his underground cell.

Similar eruptions began in the late 1990s on the island of Montserrat, another volcano in the eastern Caribbean (▶Fig. 9.6b, c). Numerous pyroclastic flows have buried once-lush fields and forests and destroyed the towns, but fortunately the island's 12,000 inhabitants have been evacuated from the danger zone.

Not all of an eruption's debris forms directly from molten rock. Volcanoes also blast out chunks of preexisting rock, broken out of the crust below the volcano or from the volcano itself. These fragments may mix together with tephra. Also, some eruptions generate landslides that tumble or slide down the side of the volcano. Volcanic landslide blocks can measure

FIGURE 9.5 (a) Some volcanoes erupt large quantities of pyroclastic debris. Some ash may rise all the way to the stratosphere, while some falls back to earth, growing progressively finer farther away from the volcano. Ash may also cascade down the side of a volcano as a pyroclastic flow (also called a nuée ardente). (b) An eruption of Stromboli, in Italy, sends clots of lava into the air. (c) Small clots of lava form lapilli, or cinders. (d) Large clots of lava form volcanic bombs. (e) A 1997 eruption of the Soufrière Hills volcano, in the West Indies, sends rolling clouds of ash skyward.

(a)

(c)

(b)

FIGURE 9.6 (a) A pyroclastic flow rushes down the side of a volcano in Japan. (b) Pyroclastic flows have left a swath of devastation on the flank of the Montserrat volcano, in the Caribbean. The ash has made a delta along the shore. (c) Some of the ash erupted from Montserrat has blanketed the town of Plymouth.

up to half a kilometer across. When ash and debris mix with water, either in rivers or from rain or melting snow and ice on the flank of a volcano, the result is a thick watery mixture, or slurry, called a **lahar.** Lahars flow down river channels or valleys at speeds of up to 50 km per hour, until they slow and settle out to develop thick, conglomerate-like layers composed of a chaotic mixture of pebbles and cobbles of volcanic rock suspended in mud (►Fig. 9.7).

Unwelded pyroclastic debris around a volcano is not very strong, and thus has a tendency to slip down the volcano's slope long after it originally accumulated. The debris moves either as a wet, chaotic mass called a debris flow or as sediment carried by flowing water. The moving water sorts the debris, so that when it finally settles at a site of deposition, it may be separated into layers with different grain sizes. These layers have the same array of sedimentary structures as any sedimentary rock, but they differ from other sedimentary rocks in that they consist entirely of volcanic material.

Volcanic Gas

Most magma contains dissolved gases, including water, carbon dioxide, sulfur dioxide, and hydrogen sulfide (H_2O, CO_2, SO_2, H_2S). In fact, up to 9% of a magma's composition may consist of gaseous components (generally, lavas with more silica contain a greater proportion of gas). Because of the sulfur in volcanic gas, the smoke cloud above a volcano typically smells like rotten eggs. The SO_2 dissolves in water droplets to create an aerosol of corrosive sulfuric acid.

FIGURE 9.7 A lahar flowed down a river on the flanks of Mt. Saint Helens following the volcano's 1980 eruption, destroying homes and property along the banks of the river.

Volcanic gases come out of solution when the magma approaches the Earth's surface, just as bubbles come out of solution in a soda or in champagne when you pop the bottle top off.

In low-viscosity magma (basalt), gas bubbles can rise faster than the magma moves, and thus most reach the surface of the magma and enter the atmosphere. Nevertheless, some bubbles freeze into the lava to create holes called **vesicles** (►Fig. 9.8). In high-viscosity magmas (andesite and rhyolite), the gas has trouble escaping because bubbles can't push through the sticky lava. As these magmas approach the Earth's surface, and pressure decreases, the gas expands so that in some cases bubbles may account for as much as 50–75% of the volume of the magma.

FIGURE 9.8 Vesicles are the holes made by gas bubbles trapped in a freezing lava. This boulder of basalt, from Sunset Crater National Monument, in Arizona, contains vesicles of various sizes.

9.3 THE ARCHITECTURE AND SHAPE OF VOLCANOES

As we saw in Chapter 6, melting in the upper mantle and lower crust produces magma, which rises into the upper crust. Typically, this magma accumulates underground in a **magma chamber,** an open space or a zone of highly fractured rock that can contain a large quantity of magma. Some of the magma freezes in the magma chamber and transforms into intrusive igneous rock, but some rises through a conduit to the Earth's surface and erupts to form a volcano. In some volcanoes, the conduit has the shape of a long vertical pipe (known as a **chimney**), while in others the conduit is a long crack (called a **fissure**) (►Fig. 9.9a–c).

With time, the solid products of eruption (lava and/or pyroclastic debris) accumulate around a chimney to form a mound or cone (or on both sides of a fissure to create a pair of ridges). At the top of the mound, a circular depression called a **crater** (shaped like a bowl, up to 500 m across and 200 m deep) develops, either during eruption as material ac-

FIGURE 9.9 (a) Some volcanoes erupt out of a circular vent above a tube-shaped chimney. (b) Other volcanoes erupt out of a long crack, called a fissure, and produce a curtain of lava. (c) A small fissure eruption along the flank of a Hawaiian volcano. Note the alignment of small, steaming vents along the crack.

(a)   (b)

(c)

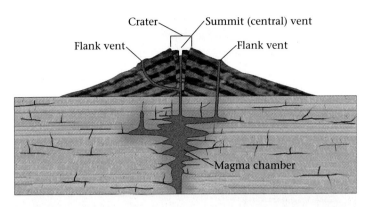

FIGURE 9.10 The plumbing beneath a volcano can be complex. A central vent may lie directly above the magma chamber, but some of the lava may erupt at flank vents.

cumulates around the summit vent or just after eruption as the summit collapses into the drained chimney. Eruptions that happen in the summit crater are **summit eruptions.** In some volcanoes, a secondary chimney or fissure breaks through along the sides, or flanks, of the volcano, causing a **flank eruption** (▶Fig. 9.10).

After major eruptions, the center of the volcano may collapse into the large, drained magma chamber below, creating a **caldera,** a big circular depression (up to thousands of meters across and up to several hundred meters deep) with steep walls and a fairly flat floor (▶Fig. 9.11a–c). If an eruption begins anew, a new mound or cone (called a **resurgent dome**) will grow within a caldera (▶Fig. 9.11d). Note that calderas differ from craters in terms of size, shape, and mode of formation.

Geologists distinguish between three different shapes of subaerial volcano. **Shield volcanoes,** so named because

FIGURE 9.11 (a) During an eruption, the magma chamber beneath a volcano is inflated with magma. (b) If the eruption drains the magma chamber, the volcano collapses downward to form a circular depression called a caldera. Later eruptions may cover the caldera with new lava. (c) A moderate-sized caldera has formed at the summit of Mt. Kilimanjaro, in the East African Rift. (d) A resurgent dome of rhyolite.

(a)

(b)

(c)

(d)

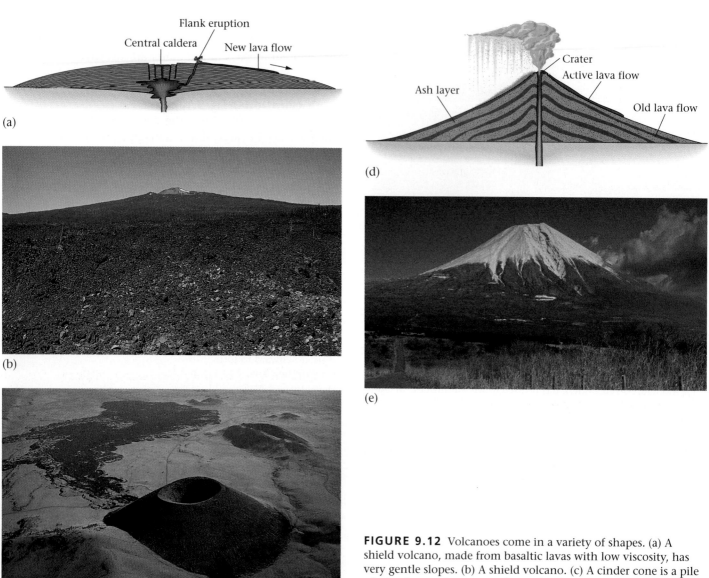

FIGURE 9.12 Volcanoes come in a variety of shapes. (a) A shield volcano, made from basaltic lavas with low viscosity, has very gentle slopes. (b) A shield volcano. (c) A cinder cone is a pile of ash whose sides assume the angle of repose. (d) A composite volcano consists of alternating ash and lava. (e) Mt. Fuji, in Japan, is a composite volcano.

they resemble a soldier's shield lying on the ground, are broad, gentle domes (►Fig. 9.12a, b). Shields form either from low-viscosity basaltic lava or from large pyroclastic sheets. **Cinder cones** consist of cone-shaped piles of tephra (►Fig. 9.12c). The slope of the cone approaches the **angle of repose** of tephra, meaning the steepest slope that the pile can attain without collapsing from the pull of gravity (about 40°, like a sandpile). Typically, cinder cones are symmetrical and have deep craters at their summit. **Stratovolcanoes,** also called **composite volcanoes,** are large and cone-shaped, and consist of alternating layers of lava and tephra (►Fig. 9.12d, e). Their shape, exemplified by Japan's Mt. Fuji, supplies the classic image most people have of a volcano, although this shape is commonly disrupted by explosions or landslides.

Volcanoes come in a great range of sizes (►Fig. 9.13). Shield volcanoes tend to be the largest, followed by composite volcanoes. Cinder cones tend to be relatively small and are often found on the surface of larger volcanoes. Submarine volcanoes don't fit any of these categories, because they usually grow as irregularly shaped mounds, modified by huge landslides along their margins.

FIGURE 9.13 Volcanoes come in different sizes. Large shield volcanoes, like Hawaii, are many times larger than cinder cones.

9.4 ERUPTIVE STYLES: WILL IT FLOW, OR WILL IT BLOW?

We saw that the eruption of Mt. Vesuvius in 79 C.E. generated lots of tephra but not much lava. In contrast, a 1990 eruption of Kilauea in Hawaii produced lava lakes and rivers of molten rock that cascaded down the mountain's flanks. An eruption in Brazil during the Cretaceous period created a flood of lava that spread over several thousand square kilometers. The 1980 eruption of Mt. Saint Helens in Oregon climaxed with a tremendous explosion. And an explosion at Crater Lake, in Oregon, about 6,850 years ago spewed 50 cubic km of pyroclastic debris over the western United States.

Clearly, different volcanoes erupt in different ways; in fact, successive eruptions from the same volcano may differ from one another. Geologists refer to the character of an eruption as the **eruptive style,** and make the following distinctions.

- **_Effusive eruptions:_** These eruptions produce mainly lava flows (►Fig. 9.14a). Most yield low-viscosity basaltic lavas, which can stream tens to hundreds of kilometers.

FIGURE 9.14 Contrasting eruptive styles. (a) This effusive eruption on Hawaii, though relatively quiet, has produced a large lake of molten lava. The surface of the lake has frozen to form a black crust, but convection within the lake cracks the crust and allows us to see the red molten rock below. (b) This eruption of Surtsey, off the coast of Iceland, is a phreatomagmatic eruption, caused by steam explosions that result when seawater enters the magma chamber.

(a)

(b)

Volcano

Volcanic eruptions are a sight to behold, and in some cases a hazard to fear. Beneath a volcano, magma formed in the upper mantle or the lower crust rises to fill a magma chamber near the Earth's surface. When the pressure in this magma chamber becomes great enough, magma is forced upward through a chimney, or crack, to the ground surface, and erupts.

Once molten rock has erupted at the surface, it is called lava. Some lava flows down the side of the volcano to make a lava flow. Lava flows eventually cool, forming solid rock. In some cases, lava spatters or fountains out of the volcanic vent in little blobs or drops that cool quickly in the air to create fragmental igneous rock called tephra, or cinders. Larger blobs ejected by a volcano become volcanic bombs, which attain a streamlined shape as they fall. Cinders may accumulate in a cone-shaped pile called a cinder cone.

Relative sizes of volcanoes

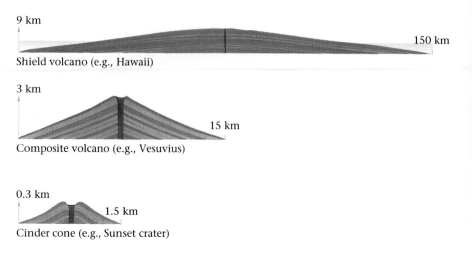

9 km

150 km

Shield volcano (e.g., Hawaii)

3 km

15 km

Composite volcano (e.g., Vesuvius)

0.3 km

1.5 km

Cinder cone (e.g., Sunset crater)

Caldera formation (for example, Crater Lake, Oregon)

Volcano starts to erupt.

Full magma chamber

Ash and debris

Main explosive eruption

Magma chamber empties.

Newly formed caldera

Collapsed blocks

Empty magma chamber

New volcanic cone grows.

Lake fills caldera.

Side vent

Eroded cone

Lava cone

Mud flows (older)

Sills

Lava flow

Dikes

Cinder cones

Lava pavement (cracked/broken)

Chimney

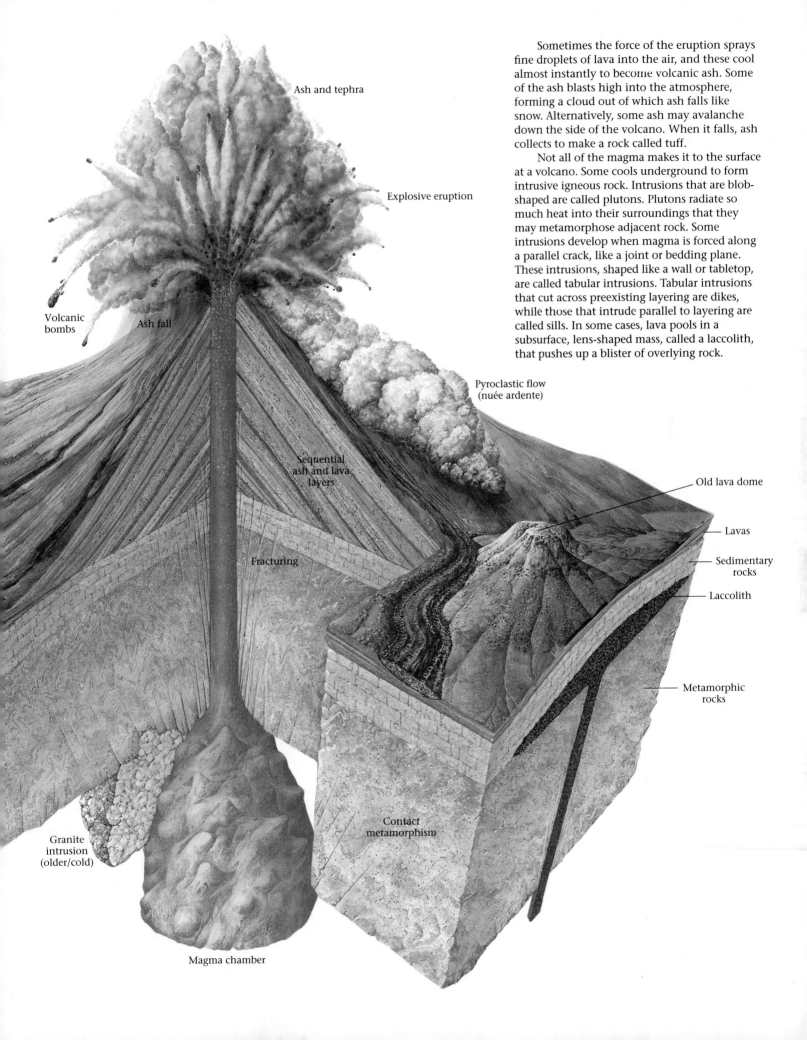

Ash and tephra

Explosive eruption

Volcanic
bombs

Ash fall

Sequential
ash and lava
layers

Pyroclastic flow
(nuée ardente)

Old lava dome

Lavas

Fracturing

Sedimentary
rocks

Laccolith

Metamorphic
rocks

Granite
intrusion
(older/cold)

Contact
metamorphism

Magma chamber

Sometimes the force of the eruption sprays
fine droplets of lava into the air, and these cool
almost instantly to become volcanic ash. Some
of the ash blasts high into the atmosphere,
forming a cloud out of which ash falls like
snow. Alternatively, some ash may avalanche
down the side of the volcano. When it falls, ash
collects to make a rock called tuff.

Not all of the magma makes it to the surface
at a volcano. Some cools underground to form
intrusive igneous rock. Intrusions that are blob-
shaped are called plutons. Plutons radiate so
much heat into their surroundings that they
may metamorphose adjacent rock. Some
intrusions develop when magma is forced along
a parallel crack, like a joint or bedding plane.
These intrusions, shaped like a wall or tabletop,
are called tabular intrusions. Tabular intrusions
that cut across preexisting layering are dikes,
while those that intrude parallel to layering are
called sills. In some cases, lava pools in a
subsurface, lens-shaped mass, called a laccolith,
that pushes up a blister of overlying rock.

In some effusive eruptions, lava lakes develop around the vent, whereas in others, lava sprays up in fountains that produce a cinder cone around the vent. To understand why fountaining occurs, watch the droplets of liquid ejected into the air above a frothing glass of soda. The bursting bubbles of gas eject liquid into the air. It's the rise of gas that propels lava upward in fountains.

• **Explosive (pyroclastic) eruptions:** These eruptions produce clouds and avalanches of pyroclastic debris (Fig. 9.5e). Pyroclastic eruptions happen when gas expands in the rising magma but cannot escape. Eventually, the pressure becomes so great that it blasts the lava, along with previously solidified volcanic rock, out of the volcano. The process is similar to the way the rapidly

---

BOX 9.1

THE
HUMAN
ANGLE

Volcanic Explosions to Remember

Explosions of volcanic-arc volcanoes generate enduring images of destruction. Let's look at two notable historic cases (see ▶Fig. 9.15).

Mt. Saint Helens, a snow-crested stratovolcano in the Cascade mountain chain, had not erupted since 1857. However, geologic evidence suggested that the mountain had a violent past, punctuated by many explosive eruptions. On March 20, 1980, an earthquake announced that the volcano was awakening once again. A week later, a crater 80 m in diameter burst open at the summit and began emitting gas and pyro-

clastic debris. Geologists who set up monitoring stations to observe the volcano noted that its north side was beginning to bulge markedly, suggesting that the volcano was filling with magma and that the magma was making the volcano expand like a balloon. Their concern that an eruption was imminent led local authorities to evacuate people in the area.

The climactic eruption came suddenly. At 8:32 A.M. on May 18, geologist David Johnston, monitoring the volcano from a distance of 10 km, shouted over his two-way radio, "Vancouver, Vancouver, this is it!" An earthquake had triggered a huge landslide that caused 3 cubic km of the volcano's weakened north side to slide away. The sudden landslide released pressure on the magma in the volcano, causing a sudden and violent expansion of gases that blasted through the side of the volcano (▶Fig. 9.16a–c). Rock, steam, and ash screamed north at the speed of sound and flattened a forest and everything in it over an area of 600 square km (▶Fig. 9.16d, e). Tragically, Johnston, along with sixty others, vanished forever. Water-saturated ash flooded river valleys, carrying away everything in their path. Seconds after the sideways blast, a vertical column convected about 540 million tons of ash (about 1 cubic km) 25 km into the sky, where the jet stream carried it away so that it was able to circle the globe. In towns near the volcano, a blizzard of ash choked roads and buried fields. Measurable quantities of ash settled over an area of 60,000 square km. When the eruption was over, the once cone-like peak of Mt. Saint Helens had disappeared—the summit now lay 440 m lower, and the once snow-covered mountain was a gray mound with a large gouge in one side.

An even greater explosion happened nearly a hundred years earlier. Krakatau, a volcano in the sea between Indonesia and Sumatra, where the Indian Ocean floor subducts beneath Southeast Asia, had grown to become a 9-km-long island. Then, on May

FIGURE 9.15 The graph shows the relative amounts of pyroclastic debris (in cubic km) ejected during major historic eruptions of the past two centuries. Notice that the 1815 Tambora eruption was over five times bigger than the 1883 Krakatoa eruption, which in turn was over five times larger than the 1980 Mt. Saint Helens eruption.

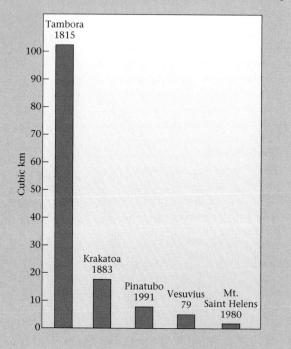

20, 1883, the island began to erupt with a series of large explosions, yielding ash that settled as far as 500 km away. Smaller explosions continued through June and July, and steam and ash rose from the island, forming a huge black cloud that rained ash into the surrounding straits. Ships sailing by couldn't see where they were going, and their crews had to shovel ash off the decks.

The climax came at 10 A.M. on August 27, perhaps when the volcano cracked and the magma chamber flooded with seawater. The resulting blast, the largest explosion in recorded history (five thousand times greater than the Hiroshima atomic bomb explosion), could be heard as far as 4,800 km away, and subaudible sound waves traveled around the globe seven times. Giant waves pushed out by the explosion slammed into coastal towns, killing over 36,000 people. Near the volcano, a layer of pumice up to 40 m thick fell from the sky. When the air finally cleared, Krakatau was gone, replaced by a submarine caldera some 300 m deep. All told, the eruption shot 20 cubic km of rock into the sky. Some ash reached as high as 80 km into the sky. Because of this ash, the world was treated to spectacular sunsets for the next few years.

FIGURE 9.16 (a) Before Mt. Saint Helens exploded in 1980, the magma chamber inside was empty. (b) As magma rose from below, the chamber inflated with molten rock. The pressure caused rock comprising the volcano to crack, and created a large bulge on the north side of the volcano. (c) The weakened north flank suddenly slipped, releasing the pressure on the magma chamber. The sudden decrease in pressure caused dissolved gases in the magma to expand and blast laterally out of the volcano. (d) The actual explosion. (e) The neighboring forest, flattened by rock, steam, and ash.

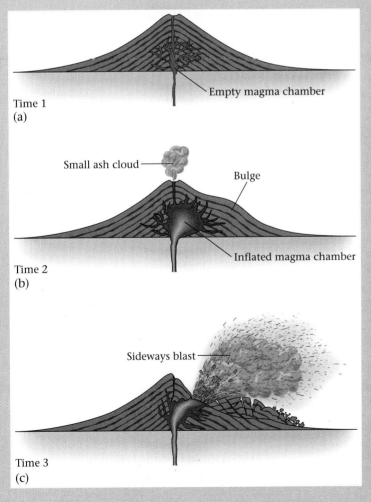

Empty magma chamber

Time 1
(a)

Small ash cloud

Bulge

Inflated magma chamber

Time 2
(b)

Sideways blast

Time 3
(c)

(d)

(e)

expanding gas accompanying the explosion of gunpowder in a cartridge shoots a bullet out of a gun.

In some cases, an explosive eruption blasts the volcano apart and leaves behind a large caldera. Such explosions, awesome in their power and catastrophic in their consequences, eject cubic kilometers of igneous particles upward at initial speeds of up to 90 m per second. Convection in the cloud can carry ash up through the entire troposphere and into the stratosphere (30–45 km above the Earth's surface). The resulting plume of debris resembles the mushroom cloud above a nuclear explosion. Coarse-grained ash and lapilli settle from the cloud close to the volcano, while finer ash settles farther away.

Some explosive eruptions take place when water gains access to the magma chamber and suddenly transforms into steam—the steam pressure blasts the magma apart and energetically expels debris. Geologists refer to pyroclastic eruptions involving the introduction of water to the magma chamber as **phreatomagmatic eruptions** (Fig. 9.14b).

Notably, the type of volcano (shield, cinder cone, or composite) depends on its eruptive style. Volcanoes that have only effusive eruptions become shield volcanoes, those that generate small pyroclastic eruptions yield cinder cones, and those that alternate between effusive and large pyroclastic eruptions become composite volcanoes. Large explosions yield calderas and blanket the surrounding countryside with tapering sheets of ignimbrite. Why are there such contrasts in eruptive style, and therefore in volcano shape? Eruptive style depends on the viscosity and gas pressure of the magma in the volcano, which in turn depend on the composition and temperature of the magma and on the environment (subaerial or submarine) in which the eruption occurs. Let's look at these controls in detail.

- *The effect of viscosity on eruptive style:* Low-viscosity (basaltic) lava flows out of a volcano easily, while high-viscosity (andesitic and rhyolitic) lava can clog up a volcano's plumbing. Thus, effusive basaltic eruptions produce shield volcanoes.

- *The effect of gas pressure on eruptive style:* The injection of magma into the magma chamber and conduit generates an outward "push" or pressure inside the volcano. The presence of gas within the magma increases this pressure, because gas expands greatly as it rises toward the Earth's surface. In runny (basaltic) magma, gas bubbles can rise to the surface of the magma and pop, causing the lava to fountain into the sky; a small cinder cone of bombs and cinders results. In a viscous (andesite or rhyolite) magma, however, the gas bubbles cannot escape and thus move with the magma toward the Earth's surface. As pressure on the magma from overlying rock decreases, the gas bubbles expand and create a tremendous outward pressure. Eventually, the gas pressure shatters the partially solidified

magma and sends a large cloud of pyroclastic debris into the sky or down the flank of the volcano, causing a pyroclastic eruption. Thus, large pyroclastic eruptions typically occur only when the lava is intermediate to silicic in composition.

- *The effect of the environment on eruptive style:* The ability of a lava to flow depends on where it erupts. Lava flowing on dry land cools more slowly than lava erupting beneath water. Thus, a basaltic lava that could flow easily down the flank of a subaerial volcano will pile up in a mound of pillows around the vent of a submarine volcano.

Traditionally, geologists have classified volcanoes according to their eruptive style, each style named after a well-known example (Strombolian, Vulcanian, etc.). But now it is perhaps more meaningful to relate eruptive styles to the geologic setting in which the volcano forms, in the context of plate tectonics theory. In some cases, different kinds of volcanoes form in the same plate setting (see Fig. 6.18 and art on pp. 248–49).

9.5 HOT-SPOT ERUPTIONS

Hot-spot volcanoes develop above finger-like plumes of hot mantle that rise from near the core-mantle boundary. When the plume, composed of hot, plastic, but still solid rock, reaches the base of the lithosphere, it partially melts, generating quantities of basaltic magma that rise and fuel a volcano at the Earth's surface above. Most of the over one hundred hot-spot volcanoes currently on our planet speckle plate interiors, but some lie along divergent plate boundaries. We'll now look at two well-known examples of hot-spot volcanism.

Oceanic Hot-Spot Volcanoes (Hawaii)

When a mantle plume rises beneath oceanic lithosphere, basaltic magma erupts at the surface of the sea floor and forms a submarine volcano. At first, such submarine eruptions yield an irregular mound of pillow lava. With time, the volcano grows up above the sea surface and becomes an island. But when the volcano emerges from the sea, the basalt lava that erupts no longer freezes so quickly, and thus flows as a thin sheet over a great distance. Thousands of thin basalt flows pile up, layer upon layer, to build a broad, dome-shaped shield volcano with gentle (less than 1.5 degrees) slopes (Fig. 9.12a, b). Note that such shield volcanoes develop their distinctive shape because the low-viscosity, hot basaltic lava that comprises them spreads out like pancake syrup and cannot build up into a cone-shaped mound. As the volcano grows, submarine portions of it slip seaward, creating large slumps. (These may trigger huge waves.)

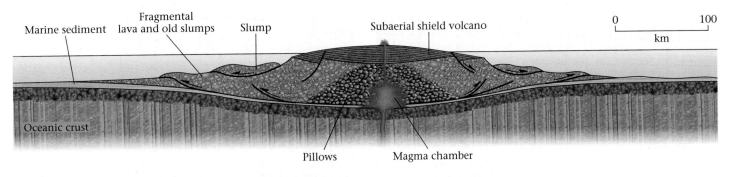

FIGURE 9.17 The inside of an oceanic hot-spot volcano is a mound of pillow basalt built on the surface of the oceanic crust. When the mound emerges above sea level, a shield volcano forms on top. Volcanic debris accumulates along the margin of the volcano. The weak material occasionally slumps seaward on sliding surfaces (indicated with arrows).

Eventually, the volcano becomes so heavy that it pushes down the surface of the sea floor. Thus, in cross section, hot-spot volcanoes are quite complex (▶Fig. 9.17).

The big island of Hawaii, one of the largest oceanic hot-spot volcanoes on Earth today, currently consists of four shield volcanoes, each built around a different vent. The island now towers over 9 km above the adjacent ocean floor (about 4.2 km above sea level), the greatest relief from base to top of any mountain on Earth (by comparison, Mt. Everest rises 8.85 km above the plains of India). Calderas up to 3 km wide have formed at the summit, and basaltic lava has extruded from both chimneys and fissures. During some eruptions, the lava fountains into the air, so mounds of tephra gather around the vents. So much lava may erupt in a short time that it accumulates locally in **lava lakes** (Fig. 9.14a). The lakes gradually drain to feed fast-moving (30 km per hour) streams of low-viscosity lava that cascade down the flanks of the volcano. These lava streams, which may extend tens of kilometers, undulate in smooth curves where they flow over bumps or dips, and bank as they careen around curves. When a lava flow reaches shallower slopes, it begins to flow more slowly and cool. Lava continues to reach the toe of the flow via large lava tubes (tunnels beneath the solidified lava surface), some of which are 10 m in diameter. In fact, some Hawaiian lava tubes carry lava all the way to the sea, where the glowing molten rock drips into the water and instantly disappears in a cloud of steam.

Because of the continuing northwestward movement of the Pacific Plate, Hawaii will eventually move off the mantle plume, and when it does, it will cease erupting. Already, a new submarine volcano, Loihi, has begun erupting to the southeast of Hawaii, but Loihi's peak still lies 1 km below sea level. Recall that the other islands of the Hawaiian chain, and of the Emperor seamount chain to the northwest, comprise a **hot-spot track,** a line of now-dead volcanoes transported off the mantle plume by the movement of the Pacific Plate.

Continental Hot-Spot Volcanoes (Yellowstone National Park)

North America has also been drifting westward above a large mantle plume for millions of years. Today, the plume lies beneath Yellowstone National Park, yielding fascinating landforms, rock deposits, and geysers. The track of this hot spot underlies the Snake River Plain, an elongate belt covered by volcanic rocks (▶Fig. 9.18a, b).

Eruptions at the Yellowstone hot spot differ from those on Hawaii in an important way: unlike Hawaii, the Yellowstone hot spot erupts both basaltic lava and rhyolitic pyroclastic debris. This happens because the basaltic magma rising from the top of the mantle plume must pass through thirty or so kilometers of continental crust before it reaches the surface. Some of the magma makes it all the way through and erupts as basalt, building small shield volcanoes, but some stalls in the continental crust and partially melts the crust, to produce silicic magma. This silicic magma has produced cataclysmic rhyolitic eruptions, creating thick tuffs that now crop out as yellow and red rocks in the canyon of the Yellowstone River (▶Fig. 9.18c).

About 620,000 years ago, an immense pyroclastic flow, as well as a cloud of ash, blasted out of the Yellowstone region during an explosion many times larger than that of Mt. Saint Helens. Close to the eruption, ignimbrites up to tens of meters thick formed, and ash from the giant cloud sifted down over the United States as far east as the Mississippi River. The eruption left a huge caldera, almost 100 km across, which now dominates the landscape of Yellowstone. Think of it—less than a million years ago, a blister of Earth 100 km across burst, sending a piece of North America skyward in an event ten or twenty times more powerful than the explosion of Krakatau. Geological studies indicate that the Yellowstone caldera is only one of at least a dozen calderas that underlie the Snake River Plain.

(a)

(c)

(b)

FIGURE 9.18 (a) Volcanic rocks from hot spots formed in two places in the western United States. First, the Columbia River basalt plateau erupted about 17 million years ago (m.y. = million years ago), when the plume rose beneath a rift. Then, either a new plume or the lower part of the same plume rose beneath northern Nevada. As North America drifted to the west-southwest, eruptions yielded the calderas along the Snake River Plain; the numbers indicate the age of the calderas (millions of years). The hot spot now lies underneath Yellowstone National Park. (b) Numerous flows of basalt piled one on top of the other in the Snake River Plain. The Snake River has cut a canyon through these basalts. (c) The "yellow stone" of Yellowstone National Park consists of silicic tuffs. The Yellowstone River has been able to cut a deep canyon through these tuffs, because they are relatively soft.

In sum, the activity of a continental hot spot like Yellowstone yields effusive, pyroclastic, and explosive eruptions resulting in a variety of volcanic shapes, ranging from shield volcanoes to calderas. Which type of volcanism happens at a given time depends on the type of magma being erupted.

Flood-Basalt Eruptions

When a plume lies beneath a continental rift (where a continent is stretching and breaking apart), particularly voluminous amounts of lava erupt along fissures in the rift. This happens because the very hot asthenosphere of the plume undergoes a greater amount of partial melting than does the somewhat cooler asthenosphere that normally underlies rifts. Recall that magma forms beneath rifts because of decompression (see Chapter 6). The decompression of normal asthenosphere leads to 4–6% partial melting, while the decompression of the plume's asthenosphere may cause up to 10% partial melting. Thus, there is more basaltic magma above a mantle plume. When rift faults crack the lithosphere, this basalt rushes to the surface and, because of its especially high temperature and low viscosity, spreads out in sheets over vast areas. Geologists refer to these sheets as **flood basalt.**

About 15 million years ago, rifting above a plume created the region that now comprises the Columbia River Plateau of Washington and Oregon (Fig. 9.18a). Eruptions

FIGURE 9.19 The flood basalts of western India, known as the "Deccan traps," are exposed in a canyon near the village of Ajanta. Between about 100 B.C.E. and 700 C.E., Buddhists carved a series of monasteries and meeting halls into the solid basalt. These are decorated by huge statues, carved in place, as well as spectacular frescoes, painted on cow-dung plaster.

yielded sheets of basalt up to 30 m thick that flowed as far as 550 km from the source. Gradually, layer upon layer erupted, creating a pile of basalt up to 500 m thick over a region of 220,000 square km. Even larger flood-basalt provinces formed elsewhere in the world, notably the Deccan Plateau of India (▶Fig. 9.19), the Paraná Basin of Brazil, and the Karroo Plateau of South Africa. The Paraná and Karroo flood basalts developed above a hot spot along the rift that separated South America from Africa about 100 million years ago.

9.6 ERUPTIONS ALONG MID-OCEAN RIDGES

Most volcanoes worldwide occur along the mid-ocean ridge system, the plate boundaries at which new sea-floor crust forms. In fact, mid-ocean ridge volcanism takes place on the upper couple of kilometers of all oceanic crust, and thus produces the covering for 70% of our planet's surface. We don't generally see this volcanic activity, however, because the ocean hides most of it beneath a blanket of water.

Mid-ocean ridge volcanoes, which develop along fissures parallel to the ridge axis, are not all continuously active. Each one turns on and off in a time scale measured in tens to hundreds of years. They erupt basalt, which, because it's underwater, forms pillow-lava mounds. Water that heats up as it circulates through the crust near the magma chamber bursts out of hydrothermal (hot-water) vents (see Chapter 4). Some of these vents are called black smokers, because they contain dissolved sulfide minerals that precipitate to form tiny dark mineral crystals as soon as the hot water mixes with seawater and cools. Near some hydrothermal vents, researchers in the submersible *Alvin* have observed a thriving ecosystem populated by giant clams, meter-long worms, shrimp, and various other creatures in a food chain that starts with bacteria that thrive by eating the sulfide minerals.

Iceland—a Hot Spot on a Ridge

Iceland is one of the few places on Earth where mid-ocean ridge volcanism built a mound of basalt that protrudes above the sea. The island formed where a mantle plume lies beneath the Mid-Atlantic Ridge—the presence of this plume means that far more magma erupted here than beneath other mid-ocean ridges. (The extra heat provided by the plume causes more partial melting, which produces more magma.) Because Iceland straddles a divergent plate boundary, it is being stretched apart, with faults forming as a consequence. Indeed, the central part of the island is a narrow rift, in which the youngest volcanic rocks of the island are found (▶Fig. 9.20a, b).

Iceland's volcanoes appear along the rift faults, as the faults slip to accommodate divergence between the European and North American Plates. Faulting cracks the crust and so provides a conduit to a magma chamber. Thus, eruptions on Iceland tend to be **fissure eruptions,** yielding either curtains of lava that are many kilometers long or linear chains of small cinder cones (Fig. 9.9b).

Not all volcanic activity on Iceland occurs subaerially. Some eruptions take place under glaciers. During 1996, for example, an eruption at the base of a 600-m-thick glacier melted the ice and produced a column of steam that rose several kilometers into the air. Meltwater accumulated under the ice for six days, until it burst through the edge of the glacier and became a flood that lasted two days and destroyed roads, bridges, and telephone lines.

Some of Iceland's volcanic activity occurs under the sea. Continuing eruptions off the coast yielded the island of Surtsey, whose birth was first signaled by huge quantities of steam bubbling up from the ocean. Eventually, steam pressure explosively ejected ash as much as 5 km into the atmosphere. Surtsey finally emerged from the sea in a phreatic eruption on November 14, 1963, building up a cone of ash and lapilli that rose almost 200 m above sea level in just three months (Fig. 9.14b). Waves could easily have eroded the cinder cone away, but the island became permanent when it erupted lava that flowed down and encased its flanks like armor.

(a)

(b)

FIGURE 9.20 (a) Iceland consists of volcanic rocks that erupted from a hot spot along the Mid-Atlantic Ridge. Because the island straddles a divergent boundary, it gradually stretches, leading to the formation of faults. The central part of the island is an irregular northeast-trending rift, where we find the youngest rocks of the island. (b) The surface of Iceland has dropped down along the faults that bound the central rift.

9.7 ERUPTIONS ALONG CONVERGENT BOUNDARIES

Most of the subaerial volcanoes on Earth lie along convergent plate boundaries (subduction zones). The volcanoes form when volatile (gaseous) compounds like water and car-bon dioxide boil off the subducting plate and rise into the overlying hot mantle, generating magma which then rises through the lithosphere and erupts. Some volcanoes start out as submarine volcanoes and later grow into **volcanic island arcs** (such as the Aleutians of Alaska), while others grow on continental crust, building **continental volcanic arcs** (such as the Cascade mountain chain of Washington and Oregon). Typically, individual volcanoes in volcanic arcs lie about 50–100 km apart. Subduction zones border over 60% of the Pacific Ocean, creating a 20,000-km-long chain of volcanoes known as the **"Ring of Fire."**

Many different kinds of magma form at volcanic arcs. Basalt may erupt if the magma rises directly from the mantle; andesite may erupt if the magma has time to undergo fractional crystallization or is contaminated by elements from the crust; and rhyolite may erupt if hot magma rising from the mantle transfers heat into the continental crust and causes the crust to partially melt. As a result, subduction zone volcanoes sometimes have effusive eruptions and sometimes pyroclastic eruptions—and occasionally they explode. Such eruptions yield both composite volcanoes, like the elegant symmetrical cone of Mt. Fuji (Fig. 9.12e), and the blasted-apart remnants of composite volcanoes, like the gray hulk of Mt. Saint Helens.

Because the volcanoes of island arcs initially erupt underwater, their foundation consists of volcanic material that froze in contact with water, or of debris that was redeposited underwater. The layers that comprise such a volcano include pillow basalts, hyaloclastites, submarine debris flows, and even turbidites (graded sedimentary beds; see Chapter 7) composed of volcanic clasts.

9.8 ERUPTIONS IN CONTINENTAL RIFTS

The rifting of continental crust yields a wide array of different types of volcanoes, because (as in the case of continental hot spots) the magma that feeds these volcanoes comes both from the partial melting of the mantle, which yields basalt, and from the partial melting of the crust, which yields rhyolite. Rifts host basaltic fissure eruptions, in which curtains of lava fountain up or linear chains of cinder cones develop. They also host explosive silicic volcanoes, and in some places composite volcanoes.

Rift volcanoes are active today in the East African Rift (Fig. 9.11c). During the past 25 million years, rift volcanoes were active in the Basin and Range Province of Nevada, Utah, and Arizona. Geologists have found abundant basaltic flows and ignimbrite sheets in the province, representing the products of different types of volcanoes that were active during the rifting. About 1 billion years ago, a narrow but deep rift formed in the middle of the United States and filled with over 15 km of basalt; this Mid-Continent Rift runs from the tip of Lake Superior to central Kansas.

9.9 VOLCANOES IN THE LANDSCAPE

Why do volcanoes look the way they do? First of all, the shape of a volcano depends on whether it has been erupting recently or ceased erupting long ago. For erupting volcanoes, the shape (shield, stratovolcano, or cinder cone) depends primarily on the eruptive style. If a volcano erupts from a single vent for a long period of time, it builds up a single high mountain, but if it erupts from several vents, perhaps aligned along a fissure, it will consist of overlapping cones. The cataclysmic explosion of a volcano instantly destroys its form and leaves a caldera surrounded by a circular or horseshoe-shaped ring. In cases where a caldera develops beneath the sea, the high parts of the rim rise above the sea to create islands.

At an erupting volcano, the process of construction happens faster than the process of erosion, and in the context of geologic time, volcanoes can develop very rapidly.

FIGURE 9.21 Devil's Tower, now a national monument in Wyoming, rising about 260 m above the surrounding land surface, formed in the magma chamber of a volcano about 40 million years ago. Huge columnar joints, 2.5 m wide at the base, developed when the magma cooled about half a kilometer below the ground at that time. Subsequently, erosion stripped away the softer tuff and flows that once composed the volcano surrounding the magma chamber—the harder rock of the magma chamber is now all that remains. In Native American legend, the ribbed surface of Devil's Tower represents the claw marks of a giant bear, trying to reach a woman seeking refuge on its summit.

Hawaii, for example, is less than a few million years old. In southern Mexico, the volcano Paricutín began to spatter out of a cornfield on February 20, 1943. Its eruption continued for nine years, and by the end 2 cubic km of tephra had piled up, creating a cone almost half a kilometer high.

Once a volcano stops erupting, erosion attacks. The rate at which a volcano is destroyed depends on whether it's composed of pyroclastic debris or lava. Cinder cones and ash piles can wash away quickly. For example, in the summer of 1831, a cinder cone grew 60 m above the surface of the Mediterranean. As soon as the island appeared, Italy, Britain, and Spain laid claim to it, and shortly the island had at least seven different names. But the volcano stopped erupting, and within six months it was gone, fortunately before a battle for its ownership had begun. In contrast, composite or shield volcanoes, which have been armor-plated by lava flows, can withstand the attack of water and ice for quite some time.

In the end, however, erosion wins out, and you can tell an old volcano that has not erupted for a long time from a volcano that has erupted recently by the extent to which river or glacial valleys have carved into its flanks. In some cases, the softer exterior of a volcano completely erodes away, leaving behind the plug of harder frozen magma that once lay just beneath the volcano and the network of dikes that radiate from this plug. You can see good examples of such landforms at Shiprock, New Mexico (Fig. 6.11b), and at Devil's Tower, Wyoming (▶Fig. 9.21). Given enough time (millions to tens of millions of years, depending on the uplift and erosion rates), erosion will strip away all remnants of a volcanic chain and expose the intrusive rocks that formed at depth in the crust beneath.

9.10 BEWARE: VOLCANOES ARE HAZARDS!

Not all volcanic activity is bad. Over time, volcanic activity has played a major role in making the Earth a habitable planet. Eruptions and underlying igneous intrusions have produced the rock making up the Earth's crust, and gases emitted by volcanoes provided the raw materials from which the atmosphere and oceans formed. The black smokers surrounding vents along mid-ocean ridges may have served as a birthplace for life, and volcanic islands in the oceans have hosted isolated populations whose evolution adds to the diversity of life on the planet. Volcanic activity continues to bring nutrients from Earth's interior to the surface, and provides fertile soils that allow for plant growth. And in more recent time, people have exploited the mineral and energy resources generated by volcanic eruptions.

Nevertheless, like earthquakes, volcanoes are natural hazards that carry the potential to cause great destruction to humanity, in both the short term and long term. According to one estimate, volcanic eruptions in the last two thousand years have caused about a quarter-million deaths—much fewer than those caused by earthquakes, but nevertheless a sizable number. Considering the rapid expansion of cities that has taken place in the past half century, far more people live in dangerous proximity to volcanoes today than ever before, so if anything, the hazard posed by volcanoes has gotten worse—imagine if a Krakatau-like explosion were to occur next to a major city today. Let's now look at the different kinds of threats posed by volcanic eruptions.

Threat of lava flows. When you think of an eruption, perhaps the first threat that comes to mind is the lava, and indeed on many occasions lava has overwhelmed towns. Basaltic lava from effusive eruptions is the greatest threat, because it can flow quickly and spread over a broad area. In Hawaii, recent lava flows have buried roads, housing developments, and cars (Fig. 9.3a). In one place, basalt almost completely submerged a parked (and empty) school bus (►Fig. 9.22a). Usually people have time to get out of the way of such flows, but not necessarily with their possessions. All they can do is watch helplessly from a distance as an advancing flow engulfs their home (►Fig. 9.22b). Before the lava even touches it, the building may burst into flames from the intense heat. Similarly, forests, orchards, and sugarcane fields are burned and buried, their verdure replaced by blackness.

Threat of ash falls, pumice, and lapilli falls. During a pyroclastic eruption, large quantities of ash erupt into the air, later to fall back to Earth. Close to a volcano, pumice and lapilli tumble out of the sky, smashing through or crushing roofs of nearby buildings (for this reason, Japanese citizens living near volcanoes keep hard hats handy), and can accumulate into a blanket up to several meters thick. We saw earlier that the eruption of Mt. Vesuvius covered Pompeii with lapilli. Winds can carry fine ash over a broad region. In the Philippines, for example, a typhoon spread heavy air-fall ash from the 1991 eruption of Mt. Pinatubo so that it covered a 4,000-square-km area (►Fig. 9.22c). Such ash buries crops, and may spread toxic chemicals that poison the soil. Ash insidiously infiltrates machinery and quickly wears out moving parts.

Fine ash from an eruption can also present a dangerous hazard to airplanes. Ash clouds rise so fast that they may be at airplane heights (11 km) long before the volcanic eruption has been reported, especially if the eruption occurs in a remote locality; and at high elevations, the ash cloud may be too dilute for a pilot to see. Like a sandblaster, the sharp, angular ash abrades turbine blades, greatly reducing engine efficiency, and the ash, along with sulfuric acid formed from the volcanic gas, scores windows and damages the fuselage. Also, when heated inside a jet engine, the ash melts, creating a liquid that sprays around the turbine and freezes; the resulting glassy coating blocks the air flow and shuts down the engine.

For example, in 1982, a British Airways 747 flew through the ash cloud over a volcano on Java. Corrosion turned the windshield opaque, and ingested ash caused all four engines to fail. For thirteen minutes, the plane glided earthward, dropping from 11.5 km (37,000 feet) to 3.7 km (12,000 feet) above the black ocean below. As passengers assumed a brace position for ditching at sea, the pilots tried repeatedly to restart the engines. Suddenly, in the oxygen-rich air of lower elevations, the engines roared back to life. The plane swooped back into the sky and headed for an emergency landing in Jakarta, where, without functioning instruments and with an opaque windshield, the pilot brought his 263 passengers and crew in for a safe landing. To land, he had to squint out an open side window, with only his toes touching the controls. During the month after the 1991 eruption of Mt. Pinatubo, fourteen jets flew through the resulting ash cloud, and of these, nine had to make emergency landings because of engine failure. Later that year, pilots and scientists met in Seattle to discuss rapid ways of alerting flights that an ash cloud had erupted.

Threat of pyroclastic flows. Pyroclastic flows race down the flanks of a volcano at speeds of 100–300 km per hour (►Fig. 9.22d). The largest can travel tens to hundreds of kilometers. The volume of ash contained in such glowing avalanches is not necessarily great—St. Pierre on Martinique was covered only by a thin layer of dust after the avalanche from Mt. Pelée had passed (see Chapter 6)—but the cloud can be so hot and poisonous that it means instant death to anyone caught in its path, and because it moves so fast, the force of its impact can flatten buildings and forests (Fig. 9.6a–c). Huge pyroclastic flows, erupting hundreds of cubic kilometers of ash and pumice, have not occurred in historic time—were one to happen now, its consequences would be dire.

Threat of the blast. Most exploding volcanoes direct their fury upward. But some, like Mt. Saint Helens, explode sideways. The forcefully ejected gas and ash, like the blast of a bomb, flattens everything in its path. In the case of Mt. Saint Helens, the region around the volcano had been a beautiful pine forest; but after the eruption, the once-towering trees were stripped of bark and needles and lay scattered over the hill slopes like matchsticks (Fig. 9.16e).

Threat of landslides. Eruptions commonly trigger large landslides along the volcano's flanks. The debris, composed of ash and solidified lava that erupted earlier, can move quite fast (250 km per hour) and far. During the eruption of Mt. Saint Helens, 8 billion tons of debris took off down the mountainside, careened over a 360-m-high ridge, and channeled down a river valley, until the last of it finally came to rest over 20 km from the volcano.

(a)

(b)

(c)

(d)

(e)

FIGURE 9.22 (a) This empty school bus was engulfed by a basalt flow in Hawaii. (b) When lava at over 1,000°C comes close to a house, the house erupts in flame. (c) A blizzard of ash falling from the eruption of Mt. Pinatubo, in the Philippines, blankets a nearby town in ghostly white. (d) A pyroclastic flow rushes toward fleeing firefighters in Japan, during the eruption of Mt. Unzen. (e) A devastating lahar buried the town of Armero, Colombia.

Threat of lahars. When volcanic ash and other debris mix with water, the result is a slurry that resembles freshly mixed concrete. This slurry, known as a lahar, can flow downslope at speeds of over 50 km per hour. Because lahars are denser and more viscous than water, they pack more force than flowing water and can literally carry away everything in their path. The lahars of Mt. Saint Helens traveled more than 40 km from the volcano, following existing drainages. When they had passed, they left a gray and barren wake of mud, boulders, broken bridges, and crumpled houses, as if a giant knife had scraped the landscape. Widespread lahars also swept down the flanks of Mt. Pinatubo in 1991, the water provided by the typhoonal and monsoonal rains.

Lahars may develop in regions where snow and ice cover an erupting volcano, for the eruption melts the snow and ice, thereby creating an instant supply of water. Perhaps the most destructive lahar of recent times accompanied the eruption of snow-crested Nevado del Ruiz in Colombia on the night of November 13, 1985. The lahar surged at 30 km per hour down a valley of the Rio Lagunillas like a 40-m-high wave, hitting the sleeping town of Armero, 60 km from the volcano. When it had passed, 90% of the buildings in the town were gone, replaced by a 5-m-thick layer of mud (▶Fig. 9.22e), which now entombs the bodies of 25,000 people and 15,000 animals.

Threat of earthquakes. Earthquakes accompany almost all major volcanic eruptions, for the movement of magma breaks rocks underground and causes subterranean explosions in the magma chamber. Such earthquakes may trigger landslides on the volcano's flanks, and can cause buildings to collapse and dams to rupture, even before the eruption itself begins.

Threat of tsunamis (giant waves). Where explosive eruptions occur in the sea, the blast and the underwater collapse of a caldera generate huge sea waves, tens of meters (in rare cases, over 100 meters) high. Most of the 36,000 deaths attributed to the 1883 eruption of Krakatau were due not to ash or lava, but rather to tsunamis that slammed into nearby coastal towns.

Threat of gas. We have already seen that volcanoes erupt not only solid material, but also large quantities of gases like water, carbon dioxide, sulfur dioxide, and hydrogen sulfide into the atmosphere, with notable consequences. Usually the gas eruption accompanies the lava and ash eruption, with the gas contributing only a minor part of the calamity. But occasionally gas erupts alone and snuffs out life in its path without causing any other damage. Such an event occurred in 1986 near Lake Nyos in Cameroon, western Africa.

Lake Nyos is a small but deep lake filling the crater of an active hot-spot volcano in Cameroon. Though only

FIGURE 9.23 Cattle near Lake Nyos, Camaroon, fell where they stood, victims of a cloud of carbon dioxide.

1 km across, it reaches a depth of over 200 m. Because of its depth, the cool bottom water of the lake does not mix with warm surface water, and for many years the bottom water remained separate from the surface water. During this time, carbon dioxide gas slowly bubbled out of cracks in the floor of the crater and dissolved in the cool bottom water. Apparently, by August 21, 1986, the bottom water had become supersaturated in carbon dioxide. On that day, perhaps triggered by an earthquake or a storm, the lake burped, expelling huge bubbles of carbon dioxide (together comprising 1 cubic km of gas). Because it is denser than air, this invisible gas flowed down the flank of the volcano and spread out over the countryside for about 23 km, before dispersing. While not toxic, carbon dioxide cannot provide oxygen for metabolism or oxidation (for this reason, it is the principal component of dry fire extinguishers). When the gas cloud engulfed the village of Nyos, it quietly put out the cooking fires and suffocated the sleeping inhabitants, most of whom died where they lay. The next morning, the landscape looked exactly as it had the day before, except that it was littered with the bodies of 1,742 people and about 6,000 head of cattle (▶Fig. 9.23).

9.11 PROTECTION FROM VULCAN'S WRATH

Active, Dormant, and Extinct Volcanoes

In the geological record, volcanoes come and go. For example, while a particular convergent plate boundary exists, a volcanic arc exists; but if subduction ceases, the volcanoes in the arc die and erode away. But even when alive, individual volcanoes erupt only intermittently. In fact, the average

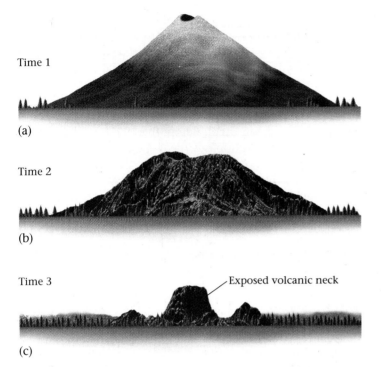

Time 1

(a)

Time 2

(b)

Time 3

Exposed volcanic neck

(c)

FIGURE 9.24 (a) The shape of an active volcano is defined by the surface of the most recent lava flow or ash fall. Little erosion affects the surface. (b) A dormant volcano has been around long enough for the surface to be modified by erosion. In humid climates, these volcanoes have gullies carved into their flanks and may be partially covered with forest. (c) An extinct volcano has been so deeply eroded that only the neck of the volcano may remain.

time (the **recurrence interval**) between successive eruptions ranges from a few years to a few centuries, and in some cases millennia.

Geologists refer to volcanoes that have erupted within the past few centuries as **active volcanoes,** and distinguish them from **dormant volcanoes,** which have not erupted for hundreds to thousands of years but do have the potential to erupt again in the future. Volcanoes that were active in the past but have shut off entirely and will not erupt in the future are called **extinct volcanoes.** As examples, geologists consider Hawaii's Kilauea to be active, for it currently erupts and has erupted frequently during recorded history. Similarly, Pompeii's Mt. Vesuvius, which has erupted over fifty times in the last two millennia, most recently in 1944, is an active volcano even though it currently emits no cloud of gas and ash. In contrast, Mt. Rainier in the Cascades last erupted centuries to millennia ago, but since subduction continues along the western edge of Oregon and Washington, the volcano could erupt in the future, and so it is considered dormant. Devil's Tower, in Wyoming, is the remnant of a volcano that was active millions of years ago but is now extinct, for the geologic cause

for volcanism in the area no longer exists, and thus the volcano will never erupt again.

Clearly, when we know whether a volcano is active, dormant, or extinct, we have a basis for determining the hazard that the volcano represents. Active volcanoes are a threat, because they will likely erupt in the near future. Dormant volcanoes may be a threat, but they might not come alive until far in the future. (However, many so-called dormant volcanoes have switched to the active category.) Extinct volcanoes are no threat at all. Commonly, the amount of erosion that affects a volcano provides a key to its classification. Active volcanoes, which display eruptive landforms (such as the shape of a cone), typically have a coating of recently erupted lava or pyroclastic debris that has barely been eroded, and may be free of vegetation or weathering. Dormant volcanoes display weathered surfaces, have been dissected by erosion, and may be covered by lush vegetation, but nevertheless they still look like volcanoes. Extinct volcanoes may have been eroded so much that they no longer look like volcanoes (▶Fig. 9.24a–c).

Predicting Eruptions

Little can be done to predict an eruption at a given volcano beyond a few months or years, except to define the recurrence interval. But short-term (weeks to months) predictions of impending volcanic activity, unlike short-term predictions of earthquakes, *are* actually feasible. Some (not all) volcanoes send out distinct warning signals announcing that an eruption may take place very soon, for as magma squeezes into the magma chamber, it causes a number of changes that geologists can measure.

- *Changes in heat flow:* The presence of hot magma increases the local **heat flow,** the amount of heat passing through rock. When you pour hot coffee into a mug, for example, within a short time the sides of the mug feel warm, because heat flows from the coffee to the surface of the mug. Similarly, when hot magma fills a magma chamber, the heat flows through the volcano to its surface. In some cases, the increase in the heat flow melts snow or ice on the volcano, triggering floods and lahars even before an eruption occurs.

- *Changes in shape:* As magma fills the magma chamber inside a volcano, it pushes outward and can cause the surface of the volcano to bulge; the same effect happens when you blow into a balloon. For example, two months before Mt. Saint Helens exploded, the north flank of the volcano began to bulge outward, up to 1.5 m per day (Fig. 9.16b). Eventually, the surface of the volcano pushed out by over 100 m.

- *Earthquake activity:* When a magma chamber fills, rocks surrounding the chamber crack, and blocks slip with

FIGURE 9.25 Hot springs indicate hot magma is below.

respect to one another. Such cracking and shifting causes earthquakes. In addition, the formation and bursting of gas bubbles in the magma chamber leads to explosions, which also register as earthquakes. Thus, in the days or weeks preceding an earthquake, the region between 1 and 7 km beneath a volcano becomes seismically active, generating small- and moderate-sized earthquakes with increasing frequency.

• *Increases in gas emission and steam:* Even though magma remains below the surface, gases bubbling out of the magma, or steam formed by the heating of groundwater by the volcano, percolate upward through cracks in the Earth and rise from the volcanic vent. So an increase in the volume of gas emission, or of new hot springs, indicates that magma has entered the ground below (▶Fig. 9.25).

Because geologists can determine when magma has moved into the magma chamber of a volcano, government agencies now send monitoring teams to a volcano at the first sign of activity. These teams set up instruments to record earthquakes, measure the heat flow, determine changes in the volcano's shape, and analyze emissions. In some cases, the monitoring comes to naught because the magma freezes in the magma chamber without ever erupting. But in other cases, the work becomes dangerous, and over the years several members of monitoring teams have been killed by the eruptions they were trying to predict. Such tragedies happen because while monitoring can yield a prediction that an eruption is imminent, it usually cannot pinpoint the exact time or eruptive style.

Controlling Volcanic Hazards

Danger assessment maps. Let's say that a given volcano has the potential to erupt in the near future. What can we do to prevent the loss of life and property? Since we can't prevent

the eruption, the first and most effective precaution is to define the regions that can be directly affected by the eruption—to compile a **volcanic-danger assessment map** (▶Fig. 9.26). These maps delineate areas that lie in the path of lava flows, lahars, debris flows, or pyroclastic flows.

River valleys initiating on the flanks of a volcano are particularly dangerous areas, because lahars and debris flows channel down them. Before the 1991 eruption of Mt. Pinatubo in the Philippines, geologists had defined areas potentially in the path of pyroclastic flows, and had predicted which river valleys were likely hosts for lahars. While the predicted pyroclastic-flow paths proved to be accurate, the region actually affected by lahars was much greater. Nevertheless, many lives were saved by evacuating people in areas thought to be under threat. If danger assessment maps had been prepared before communities and farms had

FIGURE 9.26 A danger-assessment map for the Mt. Rainer area, in Washington (courtesy of the U.S. Geological Survey). The different colors on the map indicate different kinds of hazards. Note that lahars can travel long distances down river valleys—some may threaten the city of Tacoma.

been built in the first place, the loss of life and property might have been less.

Evacuation. Unfortunately, because of the uncertainty of prediction, the decision about whether or not to evacuate is a hard one. More than once, people have fled a smoking volcano only to observe the eruption die out without any catastrophe, because the magma froze in the magma chamber without reaching the surface. In 1976, debate over the need for an evacuation around a volcano on Guadeloupe, in the French West Indies, was fierce. Eventually, the population of a threatened town *was* evacuated, but as months passed, the volcano did not erupt. Instead, tempers did, and anger at the cost of the evacuation translated into lawsuits. In the case of Mt. Saint Helens, hundreds of lives were saved by timely evacuation, but in the case of Mt. Pelée, thousands of lives were lost because warning signs were ignored.

Diverting flows. In traditional cultures, people believe that gods or goddesses control volcanic eruptions, so when a volcano rumbles, they provide offerings to appease the deity. More recently, people have used direct force to change the direction of a flow or even to stop it. For example, during a seventeenth-century eruption of Mt. Etna, a lively volcano on the Italian island of Sicily, basaltic lava formed a glowing orange river that began to spill down the side of the mountain. When the flow approached the town of Catania, 16 km from the summit, fifty townspeople protected by wet cowhides boldly hacked through the chilled side of the flow to create an opening through which the

lava could exit. They hoped thereby to cut off the supply of lava feeding the end of the flow, near their homes. Their strategy worked, and the flow began to ooze through the new hole in its side. But unfortunately, the diverted flow began to move toward the neighboring town of Paterno. Five hundred men of Paterno then chased away the Catanians so that the hole would not be kept open, and eventually the flow swallowed part of Catania.

More recently, people have used high explosives to blast breaches in the flanks of flows, and have built dams and channels to divert flows. Inhabitants of Iceland used a particularly creative approach in 1973 to stop a flow before it overran a town: they sprayed cold seawater onto the flow to freeze it in its tracks (▶Fig. 9.27a, b). The flow did stop short of the town, but whether this was a consequence of the cold shower it received remains unknown.

9.12 VOLCANOES AND CLIMATE

In 1783, Benjamin Franklin was living in France as the United States' minister. He found Europe at the time to be particularly cool and hazy. Franklin, who was a scientist as well as a statesman, learned that in June of that year there had been a huge volcanic eruption along a fissure in Iceland. In addition to lava, the eruption produced large quantities of ash, sulfur dioxide, and carbon dioxide, which spread over Europe. Aerosols from the eruption produced the haze that so impressed Franklin. He wondered whether the haze

FIGURE 9.27 (a) In 1973, a volcano erupted next to the seaport of Eldfell, Iceland. Curtains of lava lit the night sky. (b) As a basalt flow encroached on the town, firefighters used forty-three pumps to dump over 6 million cubic m of seawater on the lava to freeze it and stop the flow.

(a)

(b)

caused the unusual coolness of the climate by preventing sunlight from reaching the Earth, and reported this idea at a meeting in 1789, perhaps the first scientist ever to do so.

Franklin's idea seemed to be confirmed in 1815, when Mt. Tambora in Indonesia exploded. Tambora's explosion ejected over 100 cubic km of ash and pumice into the air (compared with 1 cubic km for Mt. Saint Helens). So much ash filled the world's sky that stars dimmed by a full magnitude. Temperatures dipped so low in the Northern Hemisphere that 1816 became known as "the year without a summer." The unusual weather of that year left a permanent impact on Western culture. Memories of fabulous sunsets and the hazy glow of the sky inspired the luminous and atmospheric quality that made the landscape paintings of the English artist Joseph M. W. Turner so famous (▶Fig. 9.28). Two English writers also documented the phenomenon. Lord Byron's 1816 poem "Darkness" contains the gloomy lines "The bright Sun was extinguish'd, and the stars / Did wander darkling in the eternal space . . . Morn came and went—and came, and brought no day"; and two years later, Mary Shelley, trapped in her house by bad weather, wrote *Frankenstein*, with its numerous scenes of gloom and doom.

Geoscientists have witnessed other examples of eruption-triggered coolness more recently. In the months following the 1883 eruption of Krakatau, the 1982 eruption of El Chichón in southern Mexico, and the 1991 eruption of Pinatubo, global temperatures noticeably dipped. Classical literature provides more evidence of the volcanic impact on climate. For example, Plutarch wrote around 100 C.E., "Among events of divine ordering there was . . . after Caesar's murder . . . the obscuration of the Sun's rays. For

FIGURE 9.28 The glowing sunset depicted in this 1840 painting by the English artist Joseph M. W. Turner was typical in the years following the 1815 eruption of Tambora in Indonesia.

during all the year its orb rose pale and without radiance . . . and the fruits, imperfect and half ripe, withered away." Similar conditions appear to have occurred in China the same year, as described in records from the Han dynasty, and may be a consequence of volcanic eruption.

To further determine the effect of volcanic activity on climate, geologists have studied ice cores extracted from the glaciers of Greenland and Antarctica. Increases in sulfuric acid in the ice indicate the past presence of high concentrations of volcanic aerosols in the atmosphere, for these aerosols dissolved in snow. In a number of cases, years in which ice contains acid correspond to years during which the thinness of tree rings elsewhere indicates a cool growing season.

How can a volcanic eruption create these cooling effects? When a large explosive eruption takes place, fine dust and aerosols enter the stratosphere. It takes only about two weeks for the ash and aerosols to circle the planet. Particles stay suspended in the stratosphere for many months to years, because they are above the weather and do not get washed away by rainfall. The haze they produce causes cooler average temperatures, because it absorbs incoming visible solar radiation during the day but does not absorb the infrared radiation that rises from the Earth's surface at night. A Krakatau-scale eruption can lead to a drop in global average temperature of about 0.3° to 1°C, and, according to some calculations, a series of large eruptions over a short period of time could cause a global average temperature drop of 6°C.

Notably, the observed effect of volcanic eruptions on the climate provides a model with which to predict the consequences of a nuclear war. Researchers have speculated that so much dust and gas would be blown into the sky in the mushroom clouds of nuclear explosions (which strongly resemble ash clouds over volcanic eruptions) that a "nuclear winter" would ensue, one that could disrupt the global food chain and cause many species to go extinct.

9.13 VOLCANOES AND CIVILIZATION

Volcanoes and people have lived in close association since the first human-like ancestors walked the Earth 3 million years ago. In fact, one of the earliest relics of human ancestors consists of footprints fossilized in a volcanic ash layer in the East African Rift. Since volcanic ash contains abundant nutrients (potassium, magnesium, sulfur, calcium, and phosphorous) that make crops prosper, people tend to populate volcanic regions. It's amazing how soon after an eruption a volcanic soil in a humid climate sprouts a cloak of green. Only twenty years after the eruption of Mt. Saint Helens, new plants covered much of the affected area.

But, as we have seen, volcanic eruptions also pose a hazard to life. Eruptions may even lead to the demise of civilizations. The history of the Minoan people, who inhabited several islands in the eastern Mediterranean during the Bronze Age, illustrates this possibility. Beginning around 3000 B.C.E., the Minoans built elaborate cities and prospered. Then their civilization waned and disappeared (▶Fig. 9.29a). Geologists have discovered that the disappearance of the Minoans corresponded with a series of eruptions of the Santorini volcano, whose remnants now comprise Thera, one of the islands of Greece. After a huge eruption, the center of the volcano sank into the sea, leaving only a steep-walled caldera (▶Fig. 9.29b). Archaeologists speculate that pyroclastic debris from the eruptions periodically darkened the sky, burying Minoan settlements and destroying crops. In addition, related earthquakes crumbled homes, and large tsunamis generated by the eruptions washed away Minoan seaports. Perhaps the Minoans took these calamities as a sign of the gods' displeasure, became demoralized,

FIGURE 9.29 (a) Archaeologists have uncovered Minoan cities in the eastern Mediterranean, remnants of a culture that disappeared before the rise of classical Greece. (b) The cataclysmic eruption of the Santorini volcano in about 1645 B.C.E. may have contributed to the demise of the Minoan culture. All that is left of Santorini is a huge caldera, whose rim still lies above sea level, forming the island of Thera.

(a)

(b)

and left the region. Eventually the eruptions stopped, and the Mycenaeans moved in. Their culture evolved into that of Classical Greece.

The Minoans, though, were not completely forgotten. Plato, in his *Dialogues*, refers to a lost city, home of an advanced civilization that bore many similarities to that of the Minoans. According to Plato, this city, which he named Atlantis, disappeared beneath the waves of the sea. Perhaps this legend came from the true history of the Minoans, as modified by Egyptian scholars who passed it on to Plato. The idea remains highly controversial, but if correct suggests that the perpetually popular legend of the "Lost City of Atlantis" may have its roots in real volcanic eruptions.

Numerous cultures living along the Pacific Ring of Fire have evolved religious practices that are based on volcanic activity—not surprisingly, considering the awesome might of a volcanic eruption in comparison with the power of humans. In some cultures, this reverence took the form of sacrifice: people were cast into the volcanic crater to appease a god or goddess, in hopes of preventing an eruption that could destroy villages and bury food supplies. In traditional Hawaiian culture, Pelé, goddess of the volcano, created all the major landforms of the Hawaiian Islands. She gouged out the craters that top the volcano mountains, her fits and moods bring about the eruptions, and her tears are the smooth, glassy lapilli ejected from the lava fountains.

Even local details of the landscape figure in Hawaiian legends. According to one, when a group of warriors challenged Pelé, the goddess transformed them into rock. The "petrified soldiers" actually formed when a lava flow engulfed a forest. The lava froze and created a carapace of rock (a tree mold) around each tree, just as liquid wax freezes to make a carapace around your finger when you dip it in. If the lava flow drains away before it entirely freezes, the tree molds remain standing high, like sentinels. The trees themselves burn up.

9.14 VOLCANOES ON OTHER PLANETS

We conclude this chapter by looking beyond the Earth, for our planet is not the only one in the solar system to have hosted volcanic eruptions. We can see the effects of volcanic activity on our nearest neighbor, the Moon, just by looking up on a clear night. The broad darker areas of the Moon, the **maria** (after the Latin word for "sea"), consist of flood basalts that erupted over 3 billion years ago. Geologists propose that the flood basalts formed when huge meteors collided with the Moon, blasting out giant craters. The craters decreased the pressure in the Moon's mantle by so much that the mantle partially melted, generating hot and fluid basaltic magma that rose to the surface and filled the craters.

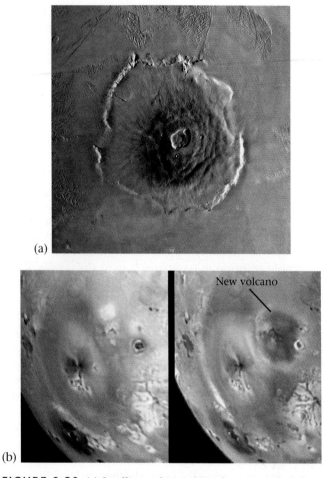

(a)

(b)

New volcano

FIGURE 9.30 (a) Satellites orbiting Mars have provided this digital image of Olympus Mons, an immense shield volcano. Notice the caldera at the summit. The escarpment surrounding the volcano may indicate that water once covered the base of the volcano and eroded it away. (b) A satellite image caught a volcano on Io, one of the moons of Jupiter, in the act of erupting. The bluish bubble is a cloud of erupting gas.

On Venus, about 22,000 volcanic edifices have been identified. Some of these even have familiar caldera structures at their crests, and some may be currently active. Though no volcanoes currently erupt on Mars, the planet's surface displays a record of a spectacular volcanic past. The largest known mountain in the solar system, Olympus Mons (▶Fig. 9.30a), is an extinct shield volcano. The base of Olympus Mons is 600 km across, and its peak rises 25 km above the surrounding plains, making it three times as high as Mt. Everest. A caldera 65 km in diameter developed on its crest.

Active volcanism currently occurs on Io, one of the many moons of Jupiter. Recently, space imagery caught a huge volcano on Io in the act of spraying a semicircular cloud of gas into space (▶Fig. 9.30b). Nine such volcanoes have since been observed on the satellite. But the nature of Io's volcanism remains mysterious. Since radioactivity and leftover heat from Io's formation cannot provide enough energy to drive such volcanism, researchers have proposed that the volcanic activity is due to tidal power: Io moves around Jupiter in an elliptical orbit, and Jupiter is a huge planet, so tidal forces caused by variations in Jupiter's gravitational pull alternately stretch and then squeeze Io, creating sufficient friction to keep its mantle hot enough to melt sulfur.

CHAPTER SUMMARY

• Volcanoes are vents at which molten rock (lava), pyroclastic debris (ash, pumice, and fragments of volcanic rock), gas, and aerosols erupt at the Earth's surface. A hill or mountain created from the products of an eruption is also called a volcano.

• The characteristics of a lava flow depend on its viscosity, which in turn depends on its temperature and composition. Silcic (rhyolitic) lavas tend to be more viscous than mafic (basaltic) lavas.

• Basaltic lavas can flow great distances. Pahoehoe flows have smooth, ropy surfaces, while aa flows have rough, rubbly surfaces. In some cases, columnar joints form in a lava flow when it cools.

• Andesitic and rhyolitic lava flows tend to pile into mounds at the vent.

• Pyroclastic debris includes powder-sized ash, marble-sized lapilli, and basketball-to-refrigerator-sized blocks. Frothy lava freezes to form sponge-like pumice, and cemented ash makes up tuff.

• Fast-moving pyroclastic flows (glowing avalanches) solidify into a rock called ignimbrite.

• Volcanic eruptions may emit many kinds of gas. Gas bubbles frozen into rock are vesicles.

• Eruptions may occur at a volcano's summit or from fissures on its flanks. The summit of an erupting volcano may collapse to form a bowl-shaped depression called a caldera.

• A volcano's shape depends on the type of eruption. Shield volcanoes are broad, gentle domes. Cinder cones are steep-sided, symmetrical hills composed of tephra. Composite volcanoes can become quite large, and consist of alternating layers of pyroclastic debris and lava.

• The type of eruption depends on the lava's viscosity and gas content. Effusive eruptions produce only flows of lava, while explosive eruptions produce clouds and flows of pyroclastic debris.

• Volcanoes form in a variety of plate-tectonic settings: oceanic hot spots, continental hot spots, mid-ocean ridges, subduction zones, and rifts. Different kinds of volcanoes develop in these settings.

- Volcanic eruptions pose many hazards: lava flows overrun roads and towns, ash falls blanket the landscape, pyroclastic flows incinerate towns and fields, landslides and lahars bury valleys, earthquakes topple structures and rupture dams, tsunamis wash away coastal towns, and invisible gases suffocate nearby people and animals.

- Geologists distinguish between active, dormant, and extinct volcanoes on the basis of the likelihood that the volcano will erupt.

- Eruptions can be predicted through changes in heat flow, changes in shape of the volcano, earthquake activity, and the emission of gas and steam.

- We can minimize the consequences of an eruption by avoiding construction in danger zones (such as the path of lahars) and by drawing up evacuation plans. In a few cases, it may be possible to divert flows.

- Volcanic gases and ash, erupted into the stratosphere, may keep the Earth from receiving solar radiation and thus may affect climate. Eruptions also bring nutrients from inside the Earth to the surface, and eruptive products may evolve into fertile soils.

- The largest known volcano in the solar system, Olympus Mons, towers over the surface of Mars. Satellites have documented eruptions on Io, a moon of Jupiter.

KEY TERMS

aa (p. 239)
caldera (p. 245)
chimney (p. 244)
cinder cone (p. 246)
cinders (p. 241)
columnar jointing (p. 239)
continental volcanic arc
  (p. 256)
crater (p. 244)
effusive eruption (p. 247)
eruption (p. 237)
explosive (pyroclastic) eruption
  (p. 250)
fissure eruption (p. 255)
flank eruption (p. 245)
flood basalt (p. 257)
heat flow (p. 261)
hot-spot track (p. 253)
hyaloclastite (p. 241)
ignimbrite (p. 241)
lahar (p. 243)
lapilli (p. 241)
lava dome (p. 241)
lava flow (p. 238)
lava lakes (p. 253)
lava tube (p. 239)

magma chamber (p. 244)
nuée ardente (p. 241)
obsidian (p. 239)
pahoehoe (p. 239)
Pelé's hair (p. 241)
Pelé's tears (p. 241)
phreatomagmatic eruption
  (p. 252)
pillow lava (p. 239)
pumice (p. 241)
pyroclastic debris (p. 239)
pyroclastic flow (p. 241)
recurrence interval (p. 261)
resurgent dome (p. 245)
Ring of Fire (p. 256)
shield volcano (p. 245)
stratovolcano (composite
  volcano) (p. 246)
summit eruption (p. 245)
tephra (p. 241)
tuff (p. 241)
vesicles (p. 244)
volcanic ash (p. 241)
volcanic bombs (p. 241)
volcanic gas (p. 239)
volcanic island arc (p. 256)

REVIEW QUESTIONS

1. Describe the three different kinds of material that can erupt from a volcano.

2. How does a lava tube form?

3. How does a pillow lava form?

4. Describe the differences between a pyroclastic flow and a lahar.

5. How is a crater different from a caldera?

6. Describe the differences between shield volcanoes, stratovolcanoes, and cinder cones. How are these differences explained by the composition of their lavas and other factors?

7. Why do some volcanic eruptions consist mostly of lava flows, while others are explosive and have no flow?

8. Explain how viscosity, gas pressure, and the environment affect the eruptive style of a volcano.

9. Describe the activity in the mantle that leads to hot-spot eruptions.

10. How do continental rift eruptions form flood basalts?

11. How are black smokers formed, and where are they found?

12. What is the "Ring of Fire"?

13. Contrast an island volcanic arc with a continental volcanic arc.

14. How are volcanoes sometimes beneficial?

15. List some of the major volcanic hazards.

16. How do scientists predict volcanic eruptions?

SUGGESTED READING

Chester, D. 1993. *Volcanoes and Society.* London: Edward Arnold.

Decker, R. W., and B. B. Decker. 1991. *Mountains of Fire.* Cambridge: Cambridge University Press.

Francis, P. 1993. *Volcanoes: A Planetary Perspective.* Oxford: Clarendon Press.

Grove, N. 1992. *Crucibles of Creation. National Geographic* 182: 5–41.

Macdonald, G. A. 1972. *Volcanoes.* Englewood Cliffs, N.J.: Prentice Hall.

Scarth, A. 1994. *Volcanoes: An Introduction.* College Station: Texas A & M University Press.

Vitaliano, D. B. 1973. *Legends of the Earth.* Bloomington: Indiana University Press.

A Violent Pulse: Earthquakes

10.1 INTRODUCTION

As the morning of January 17, 1994, approached, residents of Northridge, a sub-urb near Los Angeles, slept peacefully in anticipation of the Martin Luther King Day holiday. But beneath the quiet landscape, a disaster was in the making. For many years, the imperceptibly slow movement of the Pacific Plate relative to the North American Plate had been bending the rocks of the California crust. But like a stick that you flex with your hands, rock can bend only so far before it snaps (▶Fig. 10.1a, b). Under the San Fernando Valley, the "snap" happened at 4:31 A.M., 10 km down, sending shock waves racing through the crust at an average speed of 11,000 km (7,000 miles) per hour, ten times the speed of sound.

When the first shock wave reached the Earth's surface, it pushed the ground up with a violent jolt. Sleepers bounced off their beds, homes slipped off their foundations, and freeway bridges disconnected from their supports. As more and more shocks arrived, the ground bucked up and down and side to side, causing walls to sway and cave in, roofs to fall, and rail lines to buckle (▶Fig. 10.2). Early risers brewing coffee in their kitchens tumbled to the floor, under attack by dishes and cans catapulting out of cupboards. Trains careened off their tracks, and steep hill slopes bordering the coast gave way, dumping heaps of rock, mud, and bro-ken houses onto the beach below. Ruptured gas lines fed fires that had been ig-nited by sparking wires and water heaters in the rubble. Then, forty seconds after it started, the motion stopped, and the shouts and sirens of rescuers replaced the crash and clatter of breaking masonry and glass. A major **earthquake**—an episode of ground shaking—had occurred.

Earthquakes have affected the Earth since the formation of its solid crust al-most 4 billion years ago, and have afflicted human civilization since the con-struction of the first village. Earthquakes, most of which ultimately result from the movement of lithosphere plates, punctuate each step in the growth of moun-tains, the drift of continents, and the opening and closing of ocean basins. And, perhaps of more relevance to us, earthquakes have directly caused the deaths of over 3.5 million people during the past two millennia (see Table 10.1). Ground

An oblique air view, generated by a computer, of the San Francisco region. The red lines are faults on which earthquakes have occurred and will in the future. The complex topography of this region results from fault movement.

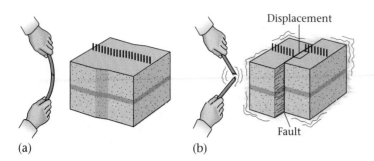

(a)   (b)

FIGURE 10.1 Most earthquakes happen when rock in the ground first bends slightly and then suddenly snaps and breaks, like a stick you flex in your hands. (a) Before an earthquake, the crust bends (the amount of bending is greatly exaggerated here). (b) When the crust breaks, sliding suddenly occurs on a fault, generating vibrations.

FIGURE 10.2 In the 1994 Northridge, California, earthquake, this building facade tore free of its supports and collapsed.

shaking, giant waves, landslides, and fires during earthquakes turn cities to rubble. Some earthquakes may even have changed the course of civilization.

What does an earthquake feel like? When you're in one, time seems to stand still, so even though most earthquakes take less than a minute, they seem much longer. Because of the lurching, bouncing, and swaying of the ground and buildings, people become disoriented, panicked, and even seasick. A witness to the 1923 earthquake in Tokyo observed that "large boulders and large pine trees were tossed about like peas in a boiling pot."[1] Some people recall hearing a dull rumbling or a series of dull thumps, as well as crashing and clanging. Earthquake vibrations shake dust into the air, creating a fine, fog-like mist.

Earthquakes are a fact of life on planet Earth: primarily because of plate movements and the forces they generate, almost 1 million detectable earthquakes happen every year. Fortunately, most cause no damage or casualties, because either they are too small or they occur in unpopulated areas. But a few hundred earthquakes per year rattle the ground sufficiently to damage buildings and injure their occupants, and every five to twenty years, on average, a great earthquake triggers a horrific calamity. What geologic phenomena generate earthquakes? Why do earthquakes take place where they do? How do they cause damage? Can we predict when earthquakes will happen, or even prevent them from happening? These questions have puzzled **seismologists** (from *seismos*, Greek for "shock" or "earthquake"), geoscientists who study earthquakes, for decades. In this chapter, we learn some of the answers, answers that can help us to cope with living in earthquake country.

---

[1]John Gribbin, *This Shaking Earth: Earthquakes, Volcanoes, and Their Impact on Our World* (New York: Putnam, 1978), p. 13.

10.2 FAULTS AND THE GENERATION OF EARTHQUAKES

Ancient cultures offered a variety of explanations for **seismicity** (earthquake activity), most of which involved the action or mood of a giant animal or god: earthquakes happened when a giant catfish or turtle holding up the Earth moved, for example. Scientific study suggests that seismicity can occur for several reasons, including

• the sudden formation of a new **fault** (a fracture on which sliding occurs);

• sudden slip on an existing fault;

• a sudden change in the arrangement of atoms in the minerals of rock;

• the cracking of a volcano when it fills with magma;

• the explosion of a volcano;

• giant landslides;

• a meteorite impact; or

• underground nuclear-bomb tests.

As we learned in Chapter 2, the place in the Earth where rock ruptures and slips, or the place where an explosion occurs, is the **focus** (or **hypocenter**) of the earthquake. Energy radiates from this focus. The point on the surface of the Earth that lies directly above the focus is the **epicenter** (▶Fig. 10.3a–c). The formation and movement of faults cause the vast majority of destructive earthquakes, so typically the focus of an earthquake lies on a fault plane (the surface of the fault). Thus, we'll begin our investigation with a look at how faults develop and why their movement generates earthquakes.

TABLE 10.1 Notable Earthquakes

Year	Location	Number of Deaths
1999	Calaraca/Armenia, Colombia	2,000
1995	Kobe, Japan	5,100
1994	Northridge, California	51
1989	Loma Prieta, California	65
1988	Armenia	24,000
1985	Mexico City	9,500
1983	Turkey	1,300
1978	Iran	15,000
1976	T'ang-shan, China	250,000
1976	Caldiran, Turkey	8,000
1976	Guatemala	23,000
1972	Nicaragua	12,000
1971	Los Angeles	50
1970	Peru	20,000
1968	Iran	12,000
1964	Anchorage, Alaska	131
1963	Skopje, Yugoslavia	1,000
1962	Iran	12,000
1960	Agadir, Morocco	12,000
1960	Southern Chile	6,000
1949	Khait, USSR	12,000
1939	Erzincan, Turkey	40,000
1939	Chile	30,000
1935	Quetta, Pakistan	60,000
1932	Kansu, China	70,000
1923	Tokyo	143,000
1920	Kansu, China	180,000
1915	Avezzano, Italy	30,000
1908	Messina, Italy	160,000
1906	San Francisco	500
1898	Japan	22,000
1886	Charleston, South Carolina	60
1866	Peru and Ecuador	25,000
1811–12	New Madrid, Missouri (3 events)	few
1783	Calabria, Italy	50,000
1755	Lisbon, Portugal	30,000
1556	Shen-shu, China	830,000

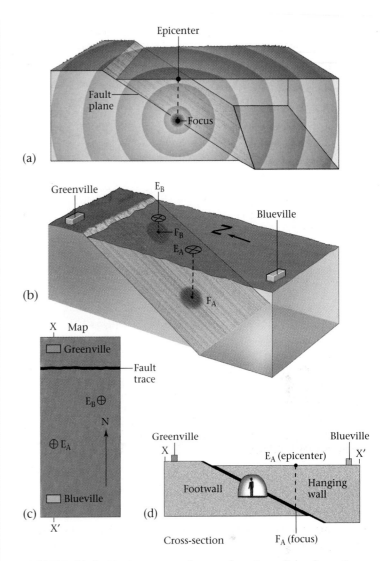

FIGURE 10.3 (a) The energy of an earthquake radiates from the focus, the place underground where rock has suddenly broken. The point on the ground surface directly above the focus is the epicenter. (b) Different locations on this tilted fault plane can slip at different times. Each slip event causes an earthquake. F_A is the focus (and E_A is the epicenter) of earthquake A, and F_B is the focus (E_B is the epicenter) of earthquake B. (c) We can represent the location of epicenters on a map. The intersection of the fault with the ground surface is the fault trace. (d) The hanging wall is the portion of the fault above the fault plane, while the footwall is the portion below the fault plane. Note that Blueville is on the hanging-wall block, while Greenville is on the footwall block.

Faults in the Crust

Faults are fractures on which slip or sliding occurs. They can be pictured as planes that cut through the crust. Some faults are vertical, but most slope at an angle. Nineteenth-century miners who encountered faults in mine tunnels referred to the rock mass above a sloping fault plane as the **hanging wall,** because it hung over their head, and the rock mass below the fault plane as the **footwall,** because it lay beneath their feet (▶Fig. 10.3d). The miners described the direction in which rock masses slipped on a fault by specifying the direction that the hanging wall moved in relation to the footwall, and we still use these terms today. When the hanging wall slips down the slope of the fault, it's

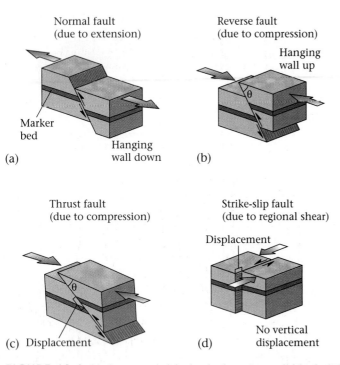

Normal fault
(due to extension)

Reverse fault
(due to compression)

Hanging
wall up

Marker
bed

(a)

Hanging
wall down

(b)

Thrust fault
(due to compression)

Strike-slip fault
(due to regional shear)

Displacement

θ

(c) Displacement

(d)

No vertical
displacement

FIGURE 10.4 (a) On a normal fault, the hanging-wall block slides down the slope of the fault plane, relative to the footwall block. (b) A reverse fault is a steeply sloping fault on which the hanging-wall block slides up, relative to the footwall block. (c) A thrust fault is a gently sloping fault on which the hanging-wall block slides up, relative to the footwall block. (d) On a strike-slip fault, one block slides laterally past the other, and there is no up or down motion.

a **normal fault,** and when the hanging wall slips up the slope, it's a **reverse fault** if steep and a **thrust fault** if shallowly sloping (▶Fig. 10.4a–c). **Strike-slip faults** have near-vertical planes on which slip occurs parallel to an imaginary horizontal line, called a **strike line,** on the fault plane—no up or down motion takes place here (▶Fig. 10.4d).

Normal faults form in response to stretching of the crust, reverse or thrust faults develop in response to squeezing and shortening of the crust, and strike-slip faults form where one block of crust slides past another laterally. By measuring the distance between the two ends of a distinctive sedimentary bed or igneous dike that's been offset by a fault, geologists define the **displacement,** the amount of slip, on the fault (▶Fig. 10.5a, b). During a single earthquake, displacement ranges from a few centimeters for a small earthquake to ten meters for a large one. If repeated earthquakes happen on the same fault over a long time (thousands to millions of years), cumulative displacements can range from a few centimeters on small faults to hundreds of kilometers on large faults. In fact, over 1,000 km of displacement may occur on some oceanic transform faults.

Faults are found almost everywhere—but don't panic! Not all of them are likely to be the source of earthquakes in the future. Faults that have moved recently or are likely to move in the near future are called **active faults** (and if they generate earthquakes, they are **earthquake faults**); faults that last moved in the distant past and probably

FIGURE 10.5 (a) This wooden fence was built across the San Andreas Fault. During the 1906 San Francisco earthquake, slip on the fault broke and offset the fence; the displacement of the fence indicates that the fault is strike-slip, as we see no evidence of up or down motion. The rancher quickly connected the two ends of the fence so no cattle could escape. (b) The amount the fence was offset indicates the displacement on the fault.

(a)

(b)

Displacement

Fault
trace

What a geologist sees

Before erosion

(a)

Normal fault

After erosion

(b)

(c)

Reverse fault

(d)

(e)

Blind fault

(f)

(g)

FIGURE 10.6 (a) Where an active fault (here, a normal fault) intersects the ground surface, a fault scarp forms; after time, it will erode away. (b) An eroded fault scarp for a normal fault. (c) A fault scarp for a reverse fault. (d) An eroded fault scarp for a reverse fault. Note that after erosion, we may not be able to distinguish between a normal and reverse fault from the shape of the ground surface alone. (e) A blind fault does not intersect the ground surface. (f) After erosion, a fold may be exposed over a blind fault. If erosion continues, eventually the fault plane itself will be exposed. (g) The trace of the strike-slip fault that ruptured the ground surface during the Hector Mine earthquake in the southern California desert. Note the cracks in the ground and the small ridges and depressions.

won't move again in the near future (but are still recognizably faults because of the displacement across them) are called **inactive faults.** Some faults have been inactive for billions of years, while others have reactivated on and off during Earth's history.

The intersection between a fault and the ground surface is a line we call the **fault trace,** or **fault line** (Fig. 10.3a–c). In places where an active normal or reverse fault intersects the ground, one side of the fault moves vertically with respect to the other side, creating a small step called a **fault scarp** (▶Fig. 10.6a–d). Active strike-slip faults tend to form narrow bands of low ridges and narrow depressions, because they break up the ground when they move (▶Fig. 10.6g). Not all active faults, though, intersect the ground surface. Those that don't are called **blind faults** (▶Fig. 10.6e, f). Blind faults worry seismologists because they pose a risk of earthquakes in unexpected places. Many fault traces that we see on maps represent inactive faults. The

portion of the fault that we see may once have been far below the surface of the Earth, becoming visible only because of erosion.

The Formation of Faults

As we learned in Chapter 8, **stress** is the push, pull, or shear that a material feels when subjected to a force. In a drawing, we can represent stress by an arrow showing the direction in which the stress acts. Objects can change shape in response to the application of a stress; this change in shape is referred to as a **strain.** (We will learn more about stress and strain in Chapter 11.)

What does stress have to do with fault formation? Stress causes faulting. To see why, imagine that you grip each side of a brick-shaped rock with a clamp (▶Fig. 10.7a–c). Now, suppose you apply a stress to the rock by moving one side upward and the other side downward. As soon as the move-

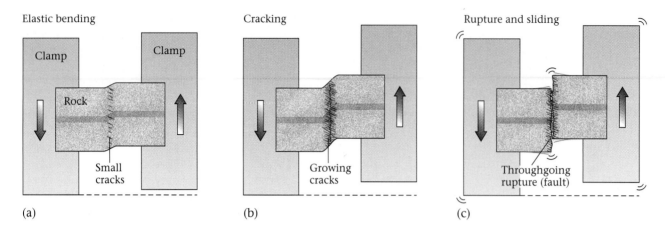

FIGURE 10.7 The stages in the development of a fault can be illustrated by breaking a rock block gripped at each end by a clamp. (a) As one clamp moves relative to the other (indicated by arrows), the rock feels a stress and begins to bend elastically. (b) If bending continues, the rock begins to crack, and then the cracks grow and start to connect. (c) When the cracks connect sufficiently to make a throughgoing fracture, the rock ruptures into two pieces. The instant one piece slides past another, the rupture becomes a fault.

ment begins, the rock begins to change shape (a line traced across the middle of the brick bends into an S-like curve), but it doesn't break, and if you were to remove the stress at this stage, the rock would return to its original shape, just as a stretched rubber band returns to its original shape when you let go (▶Fig. 10.8). A change in the shape of an object that disappears when stress is removed is called an **elastic strain.** If you apply a larger stress, so that the sides of the rock shift further, the rock starts to crack. First, a series of small cracks form, but as movement continues, the cracks grow and connect to one another to create a fracture that cuts across the entire block of rock. The instant this thor-

FIGURE 10.8 "Strain" refers to a change in the shape of an object. When you pull on a rubber band, it changes shape by becoming longer. The strain developed here is elastic, because when you let go, the rubber band returns to its original shape.

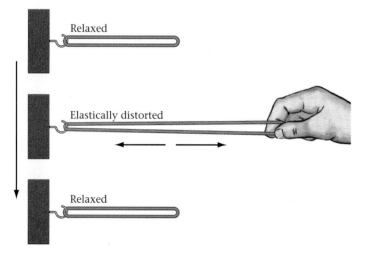

oughgoing fracture forms, the rock on one side of the fracture slides past the rock on the other side, and the fracture becomes a fault. And as soon as the fault forms, the once-bent line across the middle separates into two segments that no longer align with each other.

Friction and Stick-Slip

If you slide a book across a tabletop, it eventually slows down and stops because of friction. Similarly, once a fault forms and rock starts to slip, it doesn't slip forever, because of friction. **Friction,** the resistance to sliding on a surface, regulates movement in our Universe. Friction occurs because, in reality, no surface can be perfectly smooth—rather, all surfaces contain little bumps or depressions, called **asperities.** As one surface moves against another, the bumps on one surface snag on the bumps of the opposing surface, acting like little anchors that slow down and eventually stop the movement (▶Fig. 10.9a, b).

Though friction quickly stops slip on a fault, plate movements in earthquake-prone areas cause stress acting across an existing fault to gradually build up again. Eventually, the magnitude of stress becomes so great that friction can no longer prevent movement, and the instant this happens, the fault slips once again. Notably, the stress required to overcome friction and reactivate an existing fault tends to be less than the stress that's needed to fracture intact rock and form a new fault; in effect, once a fault has formed, it's like a permanent scar that is weaker than the surrounding crust. Thus, existing faults may reactivate many times. In sum, between faulting events, stress builds up on a fault. When slip suddenly occurs, the stress drops and sliding takes place. Friction quickly brings sliding to a halt. Then

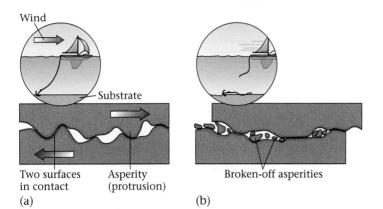

Wind

Substrate

Two surfaces in contact

Asperity (protrusion)

(a)

Broken-off asperities

(b)

FIGURE 10.9 At a microscopic scale, real surfaces have protrusions and indentations, also known as asperities. (a) Before movement, the protrusions in rock lock together, causing friction that prevents sliding. Like a boat, one block is "anchored" to the other. (b) Like a boat whose anchor cable snaps, when the protrusions break off, sliding can take place.

stress builds up until another sliding event occurs. Geologists refer to such start-stop movement on a fault as **stick-slip behavior** (▶Fig. 10.10a–f).

How Faulting Generates Earthquakes

How does the formation of a fault generate an earthquake? The moment a fault slips, the rock around the fault feels a sudden push or pull. Like a hammer blow, this movement sends a sudden pulse of energy, a "shock," into the sur-

rounding rock. An instant after the first shock, the rock vibrates back and forth for a while before coming to a complete stop, just like a stretched spring that has been released (▶Fig. 10.11a, b). Each back-and-forth motion sends an additional shock into surrounding rock. The series of shocks generated by the sudden slip on the fault and the subsequent vibration create the shaking we feel as an earthquake. The bigger the amount of slip and the greater the amount of rock that moves, the greater the vibrations and, therefore, the larger the earthquake.

A major earthquake may be preceded by smaller ones, called **foreshocks,** which possibly represent the development of the smaller cracks that will eventually link up to form a major rupture. Smaller earthquakes that follow a major earthquake, called **aftershocks,** may occur for two or three days. The largest aftershock tends to be ten times smaller than the main shock; most are much smaller. Aftershocks happen because the movement of rock during the main earthquake creates new stresses, which may be large enough to reactivate small portions of the main fault or to activate small, nearby faults.

The Amount of Slip on Faults

How much of a fault slips during an earthquake? The answer depends on the size of the earthquake: the larger the earthquake, the larger the slipped area. For example, the major earthquake that hit San Francisco in 1906 ruptured a 430-km-long (measured parallel to the Earth's surface) by 15-km-high (measured perpendicular to the Earth's surface)

FIGURE 10.10 The concept of stick-slip behavior can be illustrated by means of a heavy block with a spring attached. (a) The block starts at rest. (b) We pull on the spring and the spring stretches, but friction keeps the block in place. Stretching the spring builds up stress. (c) Suddenly, the block slides along the tabletop (analogous to the slip on a fault), and the spring relaxes (analogous to a drop in stress). (d) We pull on the spring again, and the stress builds up once more. (e) The block slides again, and the stress drops. (f) The graph shows how stress gradually builds up and then suddenly drops during stick-slip behavior.

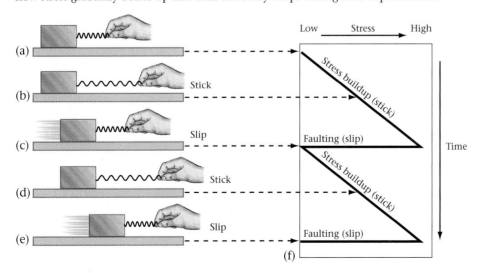

Faulting in the Crust

Normal fault
(a result of stretching of the crust)

Faceted spurs

Uplifted land

Fault scarp

Hanging wall

Footwall

Fault-line scarp

Faults are fractures along which one block of crust slides past another block. Sometimes movement takes place slowly and smoothly, without earthquakes, but other times the movement is sudden, and rocks break as a consequence. The sudden breaking of rock sends shock waves, called seismic waves, through the crust, creating vibrations at the Earth's surface—an earthquake.

Geologists recognize three types of faults. If the hanging-wall block (the rock above a fault plane) slides down the fault's slope relative to the footwall block (the rock below the fault plane), the fault is a normal fault. (Normal faults form where the crust is being stretched apart, as in a continental rift.) If the hanging-wall block is being pushed up the slope of the fault relative to the footwall block, then the fault is a reverse fault. (Reverse faults develop where the crust is being compressed or squashed, as in a collisional mountain belt.) If one block of rock slides past another and there is no up or down motion, the fault is a strike-slip fault. Strike-slip fault planes tend to be nearly vertical.

If a fault displaces the ground surface, it creates a ledge called a fault scarp. Sometimes we can identify the trace (or line) of a fault on the land surface because the rock of the hanging wall has a different resistance to erosion than the rock of the footwall; a ledge formed from this line is a fault-line scarp. Where fault scarps cut a system of rivers and valleys, the ridges are truncated to make triangular facets. Strike-slip faults may offset ridges, streams, and orchards sideways. If there is a slight extension along the fault, the land surface sinks, and a sag pond develops.

An Earthquake!

Catastrophic damage

A new fault surface

Fractured rock adjacent to the fault

Seismic waves

Focus of earthquake

**Reverse fault
(a result of shortening of the crust)**

Strike-slip fault (one block of crust slides laterally past another)

Offset rows of trees in an orchard

Offset stream

Sag pond

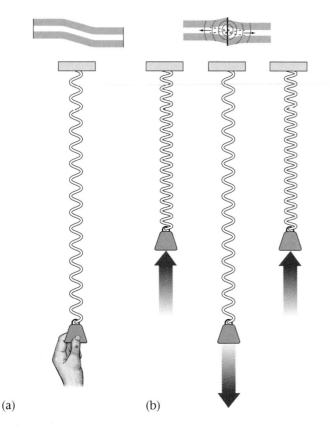

(a)   (b)

FIGURE 10.11 Rock vibrates when it breaks, like a spring that bounces up and down when released. (a) The elastic straining of the rock is comparable to stretching a spring with a weight on the end. (b) The sudden breaking of rock and slip on a fault is comparable to the up-and-down motion of a spring.

segment of the San Andreas Fault. The fault slip (the amount of sliding along the fault) died out at the ends of the segment, but toward the middle it reached 7 m, in a strike-slip sense. Slip on a thrust fault caused the 1964 "Good Friday" earthquake in southern Alaska: at depth in the Earth, slip reached a maximum of 12 m, and at the Earth's surface, the faulting uplifted the ground over a 500,000-square-km area by as much as 2 m. Smaller earthquakes, like the one that hit Northridge in 1994, resulted in about 0.5 m slip on a break that was about 5 km long and 5 km wide. The smallest-felt earthquakes (which rattle the dishes but not much more) represent displacements measured in millimeters to centimeters.

While the cumulative movement on a fault during a human life span may not amount to much, over geologic time the cumulative movement becomes significant. For example, if earthquakes occurring on a reverse fault cause 1 cm of uplift over ten years on average, the fault's movement will yield 1 km of uplift after 1 million years. Thus, earthquakes mark the incremental movements that create mountains. To take another example, movement on the San

Andreas Fault in California averages around 6 cm per year. As a result, Los Angeles, which is to the west of the fault, will move northward by 6,000 km in 100 million years.

Deformation Without Earthquakes

When a material subjected to stress cracks and fractures, we say that it has undergone **brittle deformation.** Thus, earthquake faulting, like a glass plate shattering on the floor, is a type of brittle deformation. Generally, rock must be fairly cool and must be stressed fairly quickly to behave in a brittle way. If a rock is warm and undergoes stress slowly, or is particularly weak, it can bend and flow without breaking. We call such behavior **ductile deformation** (▶Fig. 10.12a, b). If the glass plate were heated to a high temperature, it could be bent in a ductile manner, like chewing gum. Ductile deformation does not cause earthquakes.

Because the temperature of the Earth increases with depth, most brittle deformation and, therefore, earthquake faulting in continental crust occurs in the upper 15–20 km of the crust. At greater depths, shear and movement can take place, but it is accomplished by means of ductile deformation. In cold oceanic plates, earthquake faulting can happen in the lithospheric mantle as well.

In some cases, movement on faults in the upper 15–20 km of the crust takes place slowly and steadily, without generating earthquakes. When movement on a fault happens without generating earthquakes, we call the movement **fault creep.** Seismologists do not completely understand fault creep, but speculate that it occurs in particularly weak

FIGURE 10.12 The contrast between brittle and ductile deformation. (a) A plate when dropped cracks brittlely and breaks into fragments. (b) A wax candle placed in a warm oven ductilely changes shape without breaking.

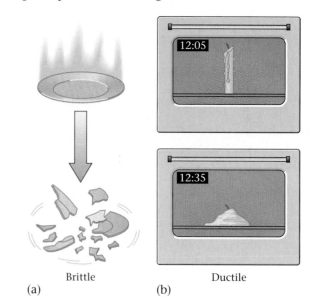

       Brittle   Ductile
(a)   (b)

rock, which can change shape without breaking or can slip smoothly without creating shock waves.

10.3 SEISMIC WAVES

How does the energy emitted at the focus of an earthquake travel to the surface or even pass through the entire Earth? Like other kinds of energy, earthquake energy travels in the form of waves (see Box 1.3). We call these waves **seismic waves** (or **earthquake waves**). You feel them if you touch one end of a brick and tap the other end with a hammer—the energy of the hammer blows travels to your fingertip in the form of waves. Friction absorbs energy as a wave passes through a material, so the amount of energy carried by seismic waves decreases the farther they travel from the focus: people near the epicenter of a large earthquake may be thrown off their feet, but those 100 km away barely feel it.

Seismologists distinguish between different types of seismic waves based on where and how the waves move. **Body waves** pass through the interior of the Earth (that is, within the body of the Earth), while **surface waves** travel along the Earth's surface. Waves in which particles of material move back and forth parallel to the direction in which the wave itself moves are called **compressional waves.** As a compressional wave passes, the material first compresses (or squeezes) together, then dilates (or expands). To see this kind of motion in action, push on the end of a spring and watch as the little pulse of compression moves along the length of the spring. Waves in which particles of material move back and forth perpendicular to the direction in

TABLE 10.2 Velocities of P-waves Through Different Materials

Material	P-Wave Velocity (km per second)
Seawater	1.5
Unconsolidated sediment	2.5
Limestone	4.0
Granite	5.5
Peridotite	8.0

which the wave itself moves are called **shear waves.** To see shear-wave motion, wiggle the end of a rope up and down and watch how the up-and-down motion travels along the rope. With these concepts in mind, we can define four basic types of seismic waves (▶Fig. 10.13a–f):

- **P-waves** (P stands for "primary") are compressional body waves.
- **S-waves** (S stands for "secondary") are shear body waves.
- **R-waves** (R stands for Rayleigh, the name of a physicist) are surface waves that cause the ground to ripple up and down, like water waves in a pond.
- **L-waves** (L stands for Love, the name of a seismologist) are surface waves that cause the ground to ripple back and forth, creating a snake-like movement.

P-waves travel the fastest through rock and thus arrive first. S-waves travel more slowly, at about 60% of the speed of P-waves, so they arrive later. Surface waves (R- and L-waves) are the slowest of all, traveling at about 55% of the speed of

Moonquakes

When the *Apollo* astronauts landed on the Moon in the 1960s and 1970s, they left instruments behind that would measure moonquakes, shaking events on the Moon. The instruments found that moonquakes happen far less often than earthquakes (only about 3,000 a year) and are very small. Geologists were not surprised, because plate movement does not occur on the Moon, so there's no volcanism, rifting, subduction, or collision to generate the forces that cause earthquakes. The Moon has a diameter of about 1,738 km. Of this, the outer 1,000 km constitutes the lithosphere, which we know because instruments have detected moonquakes with foci as deep as 1,000 km. Thus, rigid lithosphere accounts for almost 60% of the Moon's di-

ameter (as opposed to only about 1.7% of Earth's), and this ultra-thick lithosphere is simply too strong to break into plates in response to the flow in the Moon's thin asthenosphere.

But if plates don't move on the Moon, then what causes moonquakes? Undoubtedly some are caused by the impacts of meteorites, which hit the Moon's surface like a large hammer. Most moonquakes, though, seem to occur when the Moon reaches its closest distance to the Earth, as it travels along its elliptical orbit, suggesting that gravitational attraction between the Earth and the Moon creates tidal-like motions that crack the Moon's lithosphere or make the faults in the Moon slip.

FIGURE 10.13 (a) Compressional waves (like P-waves) can be generated by pushing on the end of a spring. The pulse of energy compresses in sequence down the length of spring. Note that the back-and-forth motion of the coils occurs in the same direction the wave travels. (b) The wavelength of P-waves is defined by the distance between successive pulses of compression. (c) Shear waves (like S-waves) resemble the waves in a rope. Note that the back-and-forth motion occurs in a direction perpendicular to the direction the wave travels. (d) The wavelength of S-waves is defined as the distance between successive peaks (or troughs). (e) Rayleigh waves make the surface of the ground go up and down, like ripples on the surface of a pond. (f) Love waves cause the surface of the ground to shear sideways, like a moving snake.

S-waves. And all these earthquake waves travel at different speeds in different materials (see Table 10.2).

10.4 MEASURING AND LOCATING EARTHQUAKES

Most news reports about earthquakes provide information on the "size" and "location" of an earthquake. What does this information mean, and how do we obtain it? What's the difference between a great earthquake and a minor one? How do seismologists locate an epicenter? Understanding how a seismograph works and how to read the information it provides will allow you to answer these questions.

Seismographs and the Record of an Earthquake

When, in 1889, a German physicist realized that a pendulum in his lab had moved in reaction to a deadly earthquake that had occurred in Japan (see Chapter 2), his observation confirmed speculations that earthquake energy can pass through the planet. On reading of this discovery, other researchers saw a way to construct an instrument, called a **seismograph** (or **seismometer**), that can systematically record the ground motion from an earthquake happening anywhere on Earth. Seismologists now use two basic configurations of seismographs, one for measuring vertical (up-and-down) ground motion and the other for measuring horizontal (back-and-forth) ground motion (▶Fig. 10.14a, b). Typically, the instruments are placed on bedrock in sheltered areas, away from traffic noise (▶Fig. 10.14c).

A mechanical **vertical-motion seismograph** consists of a heavy weight (like a pendulum) suspended from a spring (▶Fig. 10.15a–c). The spring connects to a sturdy frame that has been bolted to the ground. A pen extends sideways from the weight and touches a vertical revolving cylinder of paper that has been connected to the seismograph frame. When an earthquake wave arrives and causes the ground surface to move up and down, it makes the seismograph frame also move up and down. The weight, however, because of its **inertia** (the tendency of an object at rest to remain at rest), remains fixed in space. As a consequence, the revolving paper roll moves up and down under the pen, which traces out the waveforms representing the up-and-down movement. Note that if the paper cylinder did not revolve, the pen would move back and forth in the same place on the paper. A mechanical **horizontal seismograph** works on the same principle, except that the paper cylinder is horizontal and the weight is suspended from a wire. Back-and-forth movement of this seismograph causes the pen to trace out waveforms (Fig. 10.14b). In sum, the key to a seismograph is the presence of a weight that stays fixed in space while everything else moves around it.

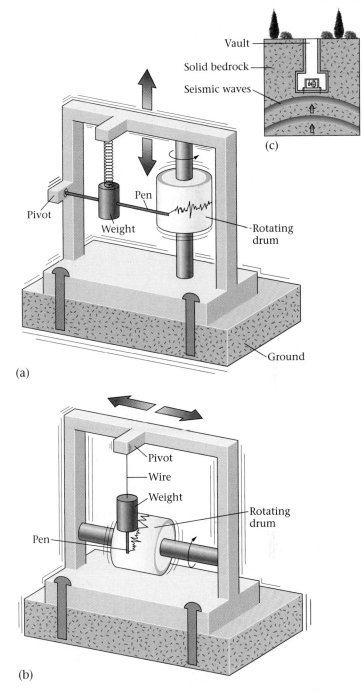

FIGURE 10.14 (a) A vertical-motion seismograph records up-and-down ground motion. (b) A horizontal-motion seismograph records back-and-forth ground motion. (c) Seismographs are bolted to the bedrock in a protected shelter or vault.

The wave train traced by the pen on a seismograph provides a record of the earthquake called a **seismogram** (▶Fig. 10.15d, e). In order to be able to determine when a particular earthquake wave arrives, the record displays lines representing time. At first glance, a typical seismogram looks like a messy squiggle of lines, but to a seismologist it

FIGURE 10.15 How a seismograph works (here, a vertical-motion seismograph). (a) Before an earthquake, the needle traces a straight line. (b) During an earthquake, when the ground and the frame of a seismograph go down, the weight stays in place because of inertia, so the needle rises relative to the paper roll. (c) When the ground and the seismograph frame rise, the needle goes down. (d) On a real seismograph record, one revolution of the paper cylinder corresponds to an hour; a single paper roll holds the record for a whole day. This unrolled paper cylinder contains the record of a large earthquake and a couple of smaller ones. The vertical rules represent time. The little squiggles that are not part of the earthquake represent noise (wind, traffic, construction). (e) This close-up of the record (seismogram) for a single earthquake shows the signals generated by different kinds of seismic waves.

contains a wealth of information. The horizontal axis represents time, and the vertical axis represents the amplitude (the size) of the seismic waves. The instant at which an earthquake wave appears is the **arrival time** of the wave. The first squiggles on the record represent P-waves, because

P-waves travel the fastest; next come the S-waves, and finally the surface waves (Rayleigh and Love). Typically, the first surface waves to arrive have the largest amplitude. With time, the surface waves become smaller and then disappear.

FIGURE 10.16 In a modern electronic seismograph, the weight has been replaced by a heavy magnet. Movement of the wire coil inside the magnet, in response to ground motion, generates an electric current. The voltage of the current represents the size of the motion.

Seismologists the world over have agreed to certain standards for measuring earthquakes, and they calibrate their seismographs to a precise time signal so that they can compare seismograms from different parts of the world. Today, seismologists work with digital records produced by modern electronic seismographs. In these instruments, the simple weight has been replaced by a heavy cylindrical magnet surrounding a coil of wire (▶Fig. 10.16). The coil connects to a solid frame anchored to the ground. During an earthquake, the magnet stays in place while the coil and the frame move. This movement generates an electrical current, whose voltage indicates the amount of movement. A magnetic tape keeps a record of the movement. Modern seismographs are so sensitive that they can record ground movements of a millionth of a millimeter (only ten times the diameter of an atom)—movements that we can't sense. Seismologists can now access data from over two thousand stations comprising a **worldwide seismic network.**

Since the dawn of the atomic era, seismograms have been of more than just academic interest, for they can record underground nuclear explosions as well as natural earthquakes. During the Cold War, government-funded seismic studies on both sides of the Iron Curtain monitored the progress of the arms race by analyzing the tests of the other side's weapons. Today, monitoring efforts continue to

determine whether nations are complying with the Nuclear Test-Ban Treaty, or if nonacknowledged nuclear powers have developed weapons. Seismograms of nuclear explosions differ from those of natural earthquakes because nuclear explosions release all their energy in a quick pulse, so the record shows the sudden arrival of large waves, which then quickly die out. Natural earthquakes, in contrast, release energy over a slightly longer period, and are accompanied by foreshocks and aftershocks.

On a seismogram recorded at a great distance from the earthquake epicenter (such as the one shown in Fig. 10.15d), the earthquake waves arrive over a long period of time—up to an hour—even though at the epicenter the earthquake lasted only less than a minute. This phenomenon happens because different waves travel at different velocities, so the farther the waves travel, the greater the distance between them. As an analogy, consider a three-way automobile race. The first car travels at 100 mph, the second at 80 mph, and the third at 60 mph (▶Fig. 10.17a). At the starting gate (representing the earthquake's focus), the cars line up next to each other. But a half hour into the race, the fastest car (P-wave) is 10 miles ahead of the second car (S-wave) and 20 miles ahead of the slowest car (R- and L-waves), and an hour into the race, the fastest car is 20 miles ahead of the second car and 40 miles ahead of the slowest. P- and S-waves travel along curved paths through the Earth (see Interlude C), and surface waves move along the ground at a slower rate (▶Fig. 10.17b).

FIGURE 10.17 (a) Different seismic waves travel at different velocities, like cars racing at different speeds. (b) Thus, different waves arrive at different times at seismograph stations. P-waves arrive first, then S-waves, and finally surface (R- and L-) waves. The difference in velocity explains the seismograph shown in Figure 10.15e.

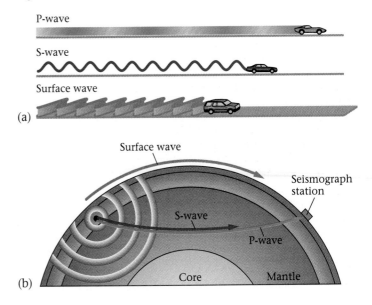

Finding the Epicenter

If an earthquake happens in or near a populated area, seismologists can approximate the location of the epicenter by noting where there's the most damage. But how do we find the epicenters of earthquakes that occur in uninhabited areas or in the ocean lithosphere, thousands of kilometers from land?

The difference in velocity, and therefore arrival time, of the different kinds of earthquake waves provides the key. Seismologists use the time delay between the arrivals at a seismograph station of the P-wave and the S-wave for the calculation. The delay between P-wave and S-wave arrival times increases as the distance from the epicenter increases (▶Fig. 10.18a). We can represent this time delay on a graph called a **travel-time curve,** which plots the time since the

earthquake began on the vertical axis and the distance to the epicenter on the horizontal axis. For example, the graph in ▶Figure 10.18b shows that after five minutes a P-wave has traveled about 2,600 km, while the S-wave has traveled only about 1,200 km.

To use a travel-time curve to determine the distance of an epicenter, start by measuring the time difference between the P- and S-waves on your seismograph. Then draw a line segment on a piece of paper to represent this amount of time, at the scale used for the vertical axis of the graph. Move the line segment back and forth until one end lies on the P-wave curve and the other end lies on the S-wave curve. You have now determined the distance at which the time difference between the two waves equals the time difference you observed. Extend the line down to the horizontal axis, and simply read off the distance to the epicenter.

(a)

(b)

FIGURE 10.18 (a) The greater the distance between the epicenter and the seismograph station, the greater the time delay between the P-wave and S-wave arrival times. In this example, station 1 is closest to the epicenter, and station 3 is farthest. Note that the P-wave arrives later at station 3 than at station 1, and that the time interval between P- and S-wave arrivals is greater at station 3 than at station 1. Arrivals at station 2 are in-between. (b) We can represent the contrasting arrival times of P-waves and S-waves on a travel-time graph. (c) If an earthquake epicenter lies 2,000 km from station 1, we draw a circle with a radius of 2,000 km around the station. Following the same procedure for stations 2 and 3, we can locate the epicenter: it lies at the intersection point of the three circles.

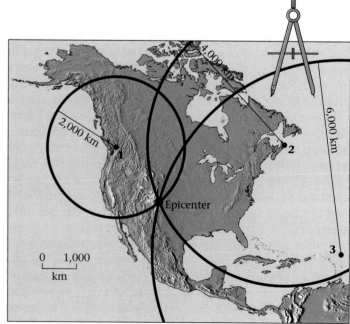

(c)

The analysis of one seismogram tells you only the distance between the epicenter and the seismograph station; it does not tell you in which direction the epicenter lies. To determine the map position of the epicenter, we use a method called **triangulation,** plotting the distance between the epicenter and three stations. For example, say you know that the epicenter lies 2,000 km from station 1, 4,000 km from station 2, and 6,000 km from station 3. On a map, draw a circle around each station, whose radius is the distance between the station and the epicenter, at the scale of the map. The epicenter lies at the intersection of the three circles, for this is the only point at which the epicenter has the appropriate measured distance from all three stations (▶Fig. 10.18c).

Defining the Size of an Earthquake

Some earthquakes shake the ground violently and cause extensive damage, while others can barely be felt. Seismologists have developed means to define size in a uniform way, so that they can make systematic comparisons between earthquakes. Three different scales for measuring earthquake size are used today.

Mercalli intensity scale. In 1902, the Italian scientist Giuseppe Mercalli developed the first widely used scale for characterizing earthquake size. This scale, called the **Mercalli intensity scale,** defines the intensity of an earthquake by the amount of damage it causes—that is, by its destructiveness. We denote different Mercalli intensities (M) with Roman numerals, as shown in Table 10.3. Note that the specification of the intensity of an earthquake depends on a subjective assessment of the damage, not on a particular measurement with an instrument. Note also that the Mercalli intensity varies with distance from the epicenter, because earthquake energy dies out as the waves travel farther through the Earth. Seismologists draw lines, called contours, around the epicenter delimiting zones in which the earthquake has a specific intensity (▶Fig. 10.19). A major earthquake has a large intensity value over the epicenter; also, its intensity contours cover a wider area. Maps showing the regional variation in intensity expected for earthquakes in a given location provide useful guidelines for urban planners trying to specify building codes.

Richter magnitude scale. Most people have heard of the **Richter magnitude scale,** developed by the American seismologist Charles Richter in 1935, because the popular press usually uses this term when describing earthquakes. The magnitude of an earthquake on the Richter scale specifies the amplitude of the largest **ground motion** generated by the earthquake at a seismograph station 100 km from the epicenter. To determine the ground motion, Richter simply used a ruler to measure the amplitude (the

TABLE 10.3 Mercalli Intensity Scale

M	*Destructiveness (Perceptions of the Extent of Damage)*
I	Detected only by instruments; causes no damage.
II	Felt by a few stationary people, especially in upper floors of buildings; suspended objects, like lamps, may swing.
III	Felt indoors; standing automobiles sway on their suspensions; it seems as though a heavy truck is passing.
IV	Shaking awakens some sleepers; dishes and windows rattle.
V	Most people awaken; some dishes and windows break, unstable objects tip over; trees and poles sway.
VI	Shaking frightens some people; plaster walls crack, heavy furniture moves slightly, and a few chimneys crack, but overall little damage occurs.
VII	Most people are frightened and run outside; a lot of plaster cracks, windows break, some chimneys topple, and unstable furniture overturns; poorly built buildings sustain considerable damage.
VIII	Many chimneys and factory smokestacks topple; heavy furniture overturns; substantial buildings sustain some damage, and poorly built buildings suffer severe damage.
IX	Frame buildings separate from their foundations; most buildings sustain damage, and some buildings collapse; the ground cracks, underground pipes break, and rails bend; some landslides occur.
X	Most masonry structures and some well-built wooden structures are destroyed; the ground severely cracks in places; many landslides occur along steep slopes; some bridges collapse; some sediment liquefies; concrete dams may crack; facades on many buildings collapse; railways and roads suffer severe damage.
XI	Few masonry buildings remain standing; many bridges collapse; broad fissures form in the ground; most pipelines break; severe liquefaction of sediment occurs; some dams collapse; facades on most buildings collapse or are severely damaged.
XII	Earthquake waves cause visible undulations of the ground surface; objects are thrown up off the ground; there is complete destruction of buildings and bridges of all types.

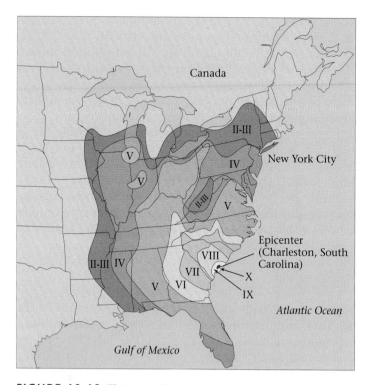

FIGURE 10.19 This map shows the contours of Mercalli intensity for the 1886 Charleston, South Carolina, earthquake. Note that near the epicenter, ground shaking reached M = X, and in New York City, ground shaking reached M = II to III.

need to use a chart that takes into account the distance from the epicenter to determine the magnitude (▶Fig. 10.20). We specify magnitude (using a lowercase m) with an Arabic number.

Very small earthquakes (where m is less than 2) can be recorded only by seismographs near the epicenter. Earthquakes with magnitudes greater than m = 3 or 4 can be recorded globally. The largest recorded earthquakes have a magnitude of about 8.9, the smallest a magnitude of about −2. There are no specific boundaries to the Richter scale, but it really can't measure earthquakes larger than about 8.9 because the needle then goes off the scale on a seismogram. To illustrate what the magnitudes mean, Table 10.4 indicates the type of damage associated with different-magnitude earthquakes.

Because earthquakes come in such a great range of sizes, Richter made his scale logarithmic. That means that a magnitude 8 earthquake results in ground motion that is actually ten times greater than that of a magnitude 7 earthquake and a thousand times greater than that of a magnitude 5 earthquake. To make life easier, seismologists often use familiar adjectives to describe the size of an earthquake, namely: "great" (8.0 and larger), "major" (7.0–7.9), "strong" (6.0–6.9), "moderate" (5.0–5.9), "light" (4.0–4.9), and "minor" (less than 3.9). Note that small earthquakes happen far more frequently than do large ones (▶Fig. 10.21a).

As stated earlier, earthquakes release energy. For example, seismologists estimate that a magnitude 8 earthquake releases an amount of energy equivalent to about 0.1% of the annual energy consumption in the United States. Notably, an increase in Richter magnitude by one integer represents approximately a thirty-three-fold increase in energy. Thus, a magnitude 8 earthquake releases about 1 million times more energy than a magnitude 4 earthquake (▶Fig. 10.21b). That

distance from top to bottom) of the largest deflection on a seismogram; typically, the largest deflection represents one of the surface waves. Because wave size depends on the distance from the epicenter, and most seismograph stations do not happen to lie 100 km from the epicenter, seismologists

TABLE 10.4 Richter Magnitude Scale

m	Effects	Average Number per Year	Approximate Energy Equivalent
8.9	Absolute devastation (M = XII)	0.03	m = 8.9 ≈ 3 billion tons of TNT
8.0–8.8	Nearly total destruction (M = XI)	0.1	m = 8.0 ≈ H-bomb (100 million tons of TNT)
7.4–7.9	Great damage	4	
7.0–7.3	Serious damage	15	
6.2–6.9	Moderate to serious damage	100	
5.5–6.1	Slight to moderate damage	500	m = 6.0 ≈ atomic bomb (100,000 tons of TNT)
4.9–5.4	Felt by all; slight damage	1,400	
4.3–4.8	Felt by many (M = V)	4,800	
3.5–4.2	Felt by some; recorded globally	30,000	m = 4.0 ≈ 100 tons of TNT
2.0–3.4	Not felt but recorded at a distance	800,000	m = 2.0 ≈ lightning bolt
less than 2.0	Not felt; recorded only locally	millions	m = 1.0 ≈ like a 2-ton truck driving by

Based on data from the U.S. Geological Survey.

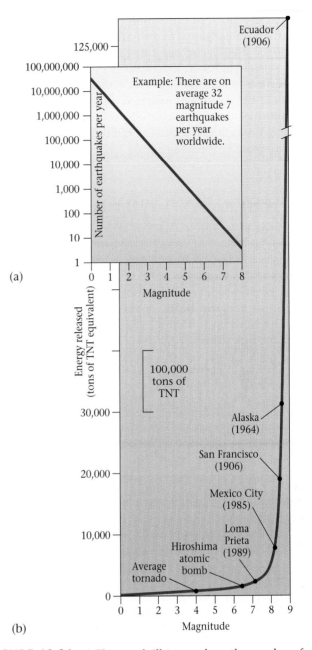

FIGURE 10.20 To calculate the Richter magnitude from a seismogram, first measure the S-minus-P time to determine the distance to the epicenter, then measure the height, or amplitude (in mm), of the largest wave recorded by the seismograph. Draw a line from the point on column A representing the S-minus-P time to the point on column C representing the wave amplitude, and read the Richter magnitude (m) off column B. Note that if the earthquake were much closer, then the same amplitude in the seismogram would yield a smaller-magnitude earthquake. We must take the distance to the epicenter into account because seismic waves grow smaller in amplitude with increasing distance from the epicenter.

FIGURE 10.21 (a) This graph illustrates how the number of earthquakes of a given magnitude decreases with increasing magnitude. According to the graph, there are over 1 million m = 1 earthquakes per year, but only about 10,000 m = 4 earthquakes. These numbers differ from those given in Table 10.4, because both sets of numbers are just approximations. (b) This graph emphasizes that the energy released by an earthquake increases dramatically with magnitude.

means that one magnitude 8.9 earthquake releases almost as much energy as all other earthquakes in a given year combined. Though earthquake energy seems immense to us because of the damage it can do, all earthquake energy together actually represents less than 0.01% of the heat energy released by the Earth to space in a given year, or about 3% of humanity's annual energy consumption.

Seismic-moment magnitude scale. Though we usually read of the Richter magnitude in the media, seismologists today actually prefer to use a different scale, the **seismic-moment magnitude scale,** for describing earthquake size. Seismologists define the seismic moment by the equation

seismic moment = (the amount of slip) × (length of rupture)
× (depth of rupture) × (rock strength).

Bigger earthquakes rupture more rock and by a greater amount. Bigger earthquakes also occur in stronger rock, because larger stresses can build up in stronger rock before it breaks. Seismic-moment magnitude values differ somewhat from Richter magnitude values. For example, the 1964 earthquake of Alaska was assigned a Richter magnitude of 8.4 while its moment magnitude was 9.2, and the 1906 California quake had a Richter magnitude of 8.3 and would have had a moment magnitude of 7.9.

10.5 WHERE AND WHY DO EARTHQUAKES OCCUR?

Earthquakes do not take place everywhere on the globe. By plotting the distribution of earthquake epicenters on a map, seismologists find that most, but not all, earthquakes occur in fairly narrow **seismic belts,** or **seismic zones.** Many seismic belts define plate boundaries, so we refer to earthquakes within them as **plate-boundary earthquakes,** and the ones that occur away from plate boundaries as **intraplate earthquakes** (the prefix "intra" means "within") (▶Fig. 10.22). Notably, 80% of the earthquake energy released on Earth comes from plate-boundary earthquakes in the belts surrounding the Pacific Ocean. Most of the remainder comes from earthquakes in the collision zone on the north side of the African and Indo-Australian Plates. Though numerous earthquakes happen along mid-ocean ridges and within plates, together they account for relatively little energy.

Seismologists distinguish three classes of earthquakes according to how deep their focus is: **shallow-focus earthquakes** occur in the top 20 km of the Earth, **intermediate-focus earthquakes** take place between 20 and 300 km, and **deep-focus earthquakes** occur down to a depth of about 670 km. Earthquakes cannot happen deep in the Earth, because rock is too ductile, meaning that it flows instead of breaks when subjected to stress. Shallow-focus

FIGURE 10.22 A map of epicenters shows that most earthquakes occur in distinct belts that define plate boundaries, and most take place in the highly populated Pacific Rim region. Epicenters that do not lie along plate boundaries are the result of rifting (in, for example, the western United States and eastern Africa) or collision (along the Alpine-Himalayan collision zones), or are in intraplate settings (the eastern United States and Canada, central Australia, the interior of the Indian Ocean Plate). Intermediate- and deep-focus earthquakes occur only along convergent plate boundaries, except for a few that happen in collisional zones.

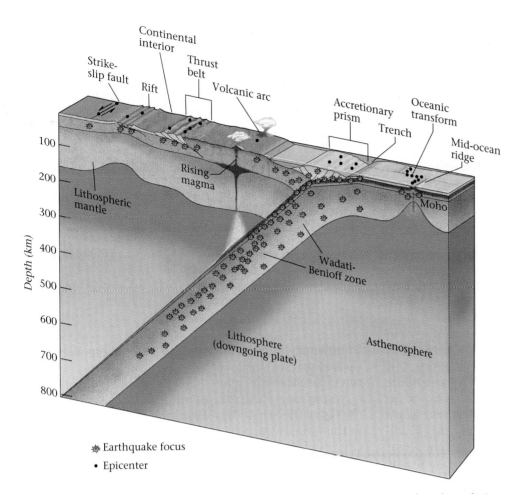

FIGURE 10.23 Earthquakes along oceanic divergent boundaries and transform boundaries are all shallow-focus. Shallow-focus earthquakes also occur in the overriding plate of a convergent boundary. Earthquakes in the downgoing plate of a convergent boundary define the Wadati-Benioff zone and include intermediate- and deep-focus earthquakes.

earthquakes cause by far the most damage, because the waves do not lose much energy before vibrating the surface. In this section, we look at the characteristics of various types of earthquakes and learn why they take place where they do. As we will see, different kinds of earthquakes occur at different kinds of plate boundaries.

Earthquakes at Plate Boundaries

The majority of earthquakes happen at plate boundaries, because the relative motion between plates takes place primarily by slip at their boundaries. Away from plate boundaries, lithosphere plates are fairly rigid, too strong to break in response to the forces present, and much less movement takes place. We find different kinds of faulting at different types of plate boundaries.

Divergent plate boundary seismicity. At divergent plate boundaries (mid-ocean ridges), two oceanic plates form and move apart. Divergent boundaries are segmented, and

spreading segments are linked by transform faults. Therefore, two kinds of faults develop at divergent boundaries. Along spreading segments, newly formed crust at or near the mid-ocean ridge stretches and ruptures, generating normal faults, whereas along the transform faults that link spreading segments, strike-slip faults occur (▶Fig. 10.23). Seismicity along mid-ocean ridges takes place at shallow depths (less than 10 km), but because most ridges are out in the ocean, far away from civilization, people rarely feel these earthquakes, and they don't cause much damage. Only in a few localities (like Iceland) that lie astride a mid-ocean ridge can divergent plate boundary earthquakes be of concern.

Convergent plate boundary seismicity. Convergent plate boundaries are complicated regions at which several different kinds of earthquake take place. Shallow-focus earthquakes occur in both the subducting plate and the overriding plate. Specifically, as the downgoing plate begins to subduct, it bends and shears along the base of the overriding plate. The bending stretches the downgoing plate and generates small normal faults in the plate. Of more importance, large thrust faults develop along the contact between the downgoing and overriding plates. The shear between the two plates generates the great earthquakes of convergent boundaries. Finally, the push applied by the downgoing plate compresses, or squeezes, the overriding plate and causes shallow-focus reverse faulting in the overriding plate.

In contrast to other types of plate boundaries, convergent plate boundaries also host intermediate- and deep-focus earthquakes. These occur in the downgoing slab as it sinks into the mantle, defining the sloping band of seismicity called the **Wadati-Benioff zone,** after the seismologists who first recognized it (Fig. 10.23). Intermediate- and deep-focus earthquakes happen in response to stresses caused by shear between the downgoing plate and the mantle, and by the "slab pull" of the deeper part of the plate on the shallower part.

Considering the depth at which they occur, these earthquakes present a problem: shouldn't the Earth be too warm to undergo brittle deformation at such depths?

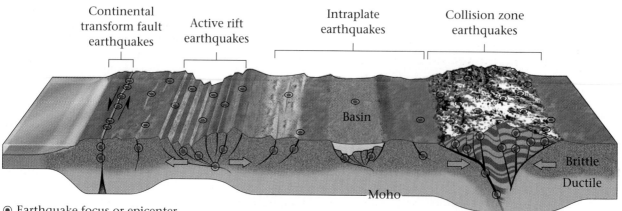

⊙ Earthquake focus or epicenter

FIGURE 10.24 Earthquakes occur in continental transform faults (such as the San Andreas), in continental rifts (the East African Rift, the Basin and Range Province), in intraplate settings (usually by reactivating old faults), and in collision zones (mountain ranges). Continental earthquakes mostly happen in the brittle crust, at depths of less than about 15 km.

Seismologists suggest that intermediate-focus earthquakes can happen because the downgoing plate takes a long time to heat up, and the rock within it remains cool and brittle enough to break even down to about 300 km. Deep-focus earthquakes, however, remain a mystery. Some researchers suggest that they occur when minerals in the rock suddenly "collapse" as a result of great pressure, and the atoms rearrange into new minerals that take up less space—the sudden change in volume would generate a shock. The fate of subducting lithosphere below a depth of 670 km remains unclear. Recent evidence suggests that some plate material accumulates at 670 km, while some sinks still deeper but no longer generates earthquakes, because all minerals have completed the transition to more compact forms, and the rock is too warm to break.

Earthquakes in southern Alaska, Japan, the west coast of South America, Mexico, and other places along the Pacific Rim are examples of convergent boundary earthquakes. One such event happened on Good Friday of 1964, near Anchorage, Alaska. The m = 8.4 (seismic-moment magnitude = 9.2) earthquake—equivalent in energy to a megaton H-bomb—wrecked substantial parts of towns all along the coast. In 1995, the m = 7.2 earthquake of Kobe, Japan, similarly caused immense devastation (buildings collapsed, and a large elevated expressway tipped over) and killed over 5,000 people.

Transform plate boundary seismicity. At transform plate boundaries, where one plate slides past another without the creation or consumption of oceanic lithosphere, most faulting results in strike-slip motion. The majority of transform faults in the world link segments of oceanic ridges (Fig. 10.23), but a few, such as the San Andreas Fault of California and the Alpine Fault of New Zealand, cut across

continental lithosphere (▶Fig. 10.24). All transform-fault earthquakes are shallow-focus, so the larger ones on land can cause disaster.

As an example of a transform-fault earthquake, consider the slip of the San Andreas Fault near San Francisco in 1906. In the wake of the gold rush, San Francisco was a booming city with broad streets and numerous large buildings. But it was built on the transform boundary along which the Pacific Plate moves north at an average of 6 cm per year relative to North America. Because of the stick-slip behavior of the fault, this movement doesn't occur smoothly but happens rather in little jerks, each of which causes an earthquake. At 5:12 A.M. on April 18, the fault slipped by as much as 6 m, and earthquake waves slammed into the city. In the words of one witness, "The whole street was undulating, as if the waves of the ocean were coming toward me."[2] Buildings swayed and banged together, laundry lines stretched and snapped, church bells rang, and then towers, facades, and houses toppled. Two main shocks hit the city, the first lasting about forty seconds and the other arriving ten seconds later and lasting about twenty-five seconds—based on the damage, seismologists estimate that the largest shock would have registered as 8.3 on the Richter scale (and 7.9 on the seismic-moment magnitude scale). Most building collapse took place downtown, while buildings on surrounding hills remained standing. Fire followed soon after (perhaps started by overturned cooking stoves), consuming huge areas of the city, for most buildings were made of wood. In the end, about five hundred people died, and a quarter of a million were left homeless. The mayor ordered out troops and commissioned vigilantes who sometimes

___
[2]Gribbin, *This Shaking Earth*, p. 89.

brutally controlled looting by refugees. Remarkably, the city was largely rebuilt in about three years.

The San Francisco earthquake has not been the only one to strike along the San Andreas and nearby related faults. Over a dozen major earthquakes have happened on these faults during the past two centuries, including the 1857 (m = 8.7) earthquake just east of Los Angeles, and the 1989 (m = 7.1) Loma Prieta earthquake, which occurred 100 km south of San Francisco but nevertheless shut down a World Series game and caused the collapse of a double-decker freeway, flattening cars in the resulting concrete sandwich. Notably, the recent magnitude 6.7 earthquake at Northridge, described earlier, occurred not on a strike-slip fault, but rather on a blind thrust fault formed just south of a bend in the San Andreas Fault. The geometry of this bend has created the local buildup of stress.

Earthquakes at Continental Rifts and Collisions

Continental rifts. The stretching of continental crust at continental rifts generates normal faults (Fig. 10.24). Active rifts today include the East African Rift, the Basin and Range Province (mostly in Nevada, Utah, and Arizona), and the Rio Grande Rift (in New Mexico). In all these places, shallow-focus earthquakes occur, similar in nature to the earthquakes at mid-ocean ridges. But in contrast to mid-ocean ridges, these seismic zones can be located under populated areas, like Salt Lake City, and thus cause major damage.

Though a great earthquake hasn't occurred near Salt Lake City since the city became large, geologic evidence demonstrates that significant earthquakes *have* happened there, for faults cut across and displace glacial moraines (sediment piles deposited by glaciers) at the foot of the nearby Wasatch Mountains. The moraines accumulated during the last ice age, perhaps as recently as 12,000 years ago, so the faulting developed more recently than that.

Collision zones. Zones where two continents collide after the oceanic lithosphere that once separated them has been completely subducted yield great mountain ranges like the Alpine-Himalayan chain in South Asia. Though a variety of earthquakes happen in these zones, the most common result from movement on thrust faults, which formed when the crust was compressed, or squeezed (Fig. 10.24).

The complex collision between Africa and Europe generated the stress that triggered an earthquake near Lisbon, Portugal, on All Saints' Day (November 1), 1755. According to an eyewitness account,

The first alarm was a rumbling noise that sounded like exceptionally heavy traffic. Then there was a brief pause and a devastating shock followed, lasting over two minutes, that brought down roofs, walls, facades of churches, palaces and houses and shops in a dreadful deafening roar of destruction. Close on this came a third trembling to complete the disaster, and then a dark cloud of suffocating dust settled fog-like on the ruins of the city.

About fifteen minutes later, fires began to spread and consume the rubble, and then the waters of the harbor "rocked and rose menacingly, and then poured in three great towering waves over its banks."[3] Aftershocks lasted for days. Innumerable artworks by great Renaissance masters were destroyed, along with a library that housed all the records of Portuguese exploration.

The effects of the Lisbon earthquake went beyond the immediate destruction. Voltaire (1694–1778), the great French writer, immortalized the earthquake in his novel *Candide*. How, pondered Voltaire, could this world be the best of all possible worlds if such a calamity could happen and ruin good people? The earthquake, and Voltaire's writing about it, brought to an end the "era of optimism" and ushered in the "era of realism" in Europe.

Intraplate Earthquakes

Intraplate earthquakes occur in the interiors of plates and are not associated with plate boundaries, active rifts, or collision zones (Fig. 10.24). They account for only about 5% of the earthquake energy released in a year. Almost all have a shallow focus and thus represent faulting in the crust. In many cases, geologists have demonstrated that they occur along ancient fault zones, in some cases dating back to the Precambrian, that appear to have remained weak since their formation.

Seismologists are still trying to understand the causes of intraplate earthquakes. Most favor the idea that force applied to the boundary of the plate can cause the interiors of plates to break at preexisting zones of weakness. Alternatively, the activity may be due to forces resulting from shear between the lithosphere and the underlying asthenosphere, or when the upper crust readjusts to loads from glaciers or sedimentary piles.

Because the crust in an intraplate setting tends to be composed of stronger rock, overall, than crust along a plate boundary, it tends to transmit seismic waves more efficiently than does plate-boundary crust. As an analogy, if you were to bang a hammer on a block of solid rock and on a block of styrofoam, you would feel the impact on the other side of the rock, but the impact would be absorbed by the styrofoam. Thus, intraplate earthquakes can cause damage far from the epicenter.

---

[3]James R. Newman, "The Lisbon Earthquake," in Frank H. T. Rhodes and Richard O. Stone, eds., *Language of the Earth* (New York: Pergamon Press, 1981), p. 59.

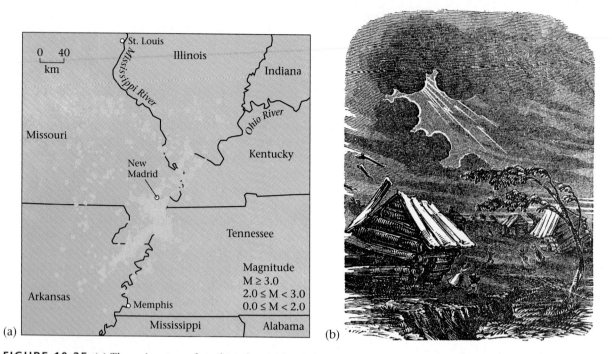

FIGURE 10.25 (a) The epicenters of earthquakes in the New Madrid, Missouri, area, recorded by modern instruments. The earthquakes, most of which are minor, appear to occur along two sets of ancient, weak faults that break in response to stress in the interior of North America. The New Madrid zone is not a plate boundary. (b) The 1811 New Madrid earthquake destroyed cabins and disrupted the Mississippi River.

In North America, intraplate earthquakes occur in distinct clusters, most notably in the vicinity of New Madrid, Missouri; Charleston, South Carolina; eastern Tennessee; Montreal (Quebec); and the Adirondack Mountains (New York). A magnitude 6.5–7.5 earthquake occurred near Charleston in 1886, ringing church bells up and down the coast and vibrating buildings as far away as Chicago (Fig. 10.19). In Charleston itself, over 90% of the buildings were damaged, and sixty people died. In 1811, the region of New Madrid, which lies near the Mississippi River in southernmost Missouri, was inhabited by a small population of Native Americans and an even smaller population of European descent (▶Fig. 10.25a, b). In the winter of that year, an estimated m = 8.5 earthquake, the first of a series of three, struck the region. In the words of a settler,

> The Mississippi first seemed to recede from its banks, and its waters gathered up like a mountain, leaving for a moment many boats on the bare sand. . . . Then, rising 15 to 20 feet perpendicularly . . . the banks overflowed with a retrograde current rapid as a torrent. The boats were now torn from their moorings . . . the river took with it whole groves of young cottonwood trees.[4]

---
[4]Gribbin, *This Shaking Earth,* p. 15.

The quake changed the course of the Mississippi and caused the formation of Reelfoot Lake. It would have toppled a city, had one been there. Such earthquakes result from slip on faults in a Precambrian rift that underlies the Mississippi Valley. Both St. Louis, Missouri, and Memphis, Tennessee, were built close to the epicenter, so earthquakes in the area could be disastrous if they occurred today.

Induced Seismicity

Most earthquakes happen independently of human activity. But the timing of some earthquakes relative to human-caused events suggests that in certain cases, people *can* influence seismicity. **Induced seismicity,** or seismic events caused by people, generally occurs in response to changes in groundwater pressure. In the Earth, the pressure of water can slightly push apart the opposing surfaces of faults and thereby decrease the friction that resists motion on them. So when people increase groundwater pressure, a fault may suddenly slip.

Seismologists observed such a relationship near Denver, Colorado, when engineers pumped waste water from a military installation down a well; as soon as the pumping began, earthquakes started. Induced seismicity is a particular danger when people build dams and create large reservoirs over faults. Faulting generally breaks up rock, making it more erodible, so, not surprisingly, deep valleys commonly form over large faults. When a reservoir fills over a fault, water seeps down into the fault and, under the pressure caused by the water column above, triggers earthquakes. A big earthquake could destroy the dam, causing calamity downstream.

10.6 DAMAGE FROM EARTHQUAKES

An area ravaged by a major earthquake is a heartbreaking sight. The terror and sorrow etched on the faces of survivors mirrors the inconceivable destruction. This destruction comes as a result of many processes.

Ground Shaking and Displacement

As earthquake body waves arrive at the surface of the Earth, or surface waves pass along the ground, the ground moves. For great earthquakes, this movement can have an amplitude of as much as 1 m, but for moderate earthquakes, motions fall in the range of a few centimeters or less. The type of ground motion reflects the type of earthquake waves passing through an area. Some waves make the surface buck up and down, some make it squash back and forth, some create a rolling motion like ocean waves, while some cause lateral back-and-forth shifting (▶Fig. 10.26a, b). Interference among the different kinds of waves causes motion, like the choppiness on the surface of a pond, to be anything but regular. Feeling the ground move is a terrifying and disorienting experience that throws off your sense of balance, and if the motions are large enough, they may knock you down or even throw you in the air. Ground accelerations caused by moderate earthquakes are in the range of 10–20% of g (where g is the acceleration resulting from gravity), and during a great earthquake accelerations may approach 1 g.

The damage caused by ground motion depends not only on proximity to the focus (ground shaking dies off with distance from the focus, as the earthquake waves die down), but also on the nature of the substrate (the material underlying the ground surface). For example, vibrations in softer sediment have a lower frequency than do vibrations in hard rock. Lower-frequency vibrations cause buildings to sway more than higher-frequency vibrations do, and thus can cause more damage.

People can withstand a few g's of acceleration, so if you're out in an open field during an earthquake, ground motion alone won't kill you—you may be knocked off your feet and bounced around a bit, but your body is too flexible to break. Buildings aren't so lucky (▶Fig. 10.27a–g). When earthquake waves pass, buildings sway, twist back

(a)

(b)

FIGURE 10.26 (a) Rayleigh waves make the ground undulate in a rolling motion, like the sea surface. (b) Love waves cause the ground to undulate laterally.

and forth, or lurch up and down, depending on the type of wave motion. As a result, connectors between the frame and facade of a building may separate, so the facade crashes to the ground. The flexing of walls shatters windows and makes roofs collapse. Floors may rise up and slam down on the columns that support them. This motion can crush the columns, so some buildings collapse with their floors piling on top of one another like pancakes in a stack (▶Fig. 10.28a). The majority of earthquake-related deaths and injuries result when people are hit by flying debris or are crushed beneath falling walls or roofs. Aftershocks worsen the problem, because they may be the final straw that breaks already weakened buildings, trapping rescuers.

Roads, rail lines, and bridges suffer the same fate as buildings: ground motion breaks and buckles them (▶Fig.

FIGURE 10.27 Shaking causes (a) a concrete-slab building (or bridge) to disconnect and collapse, (b) a building's facade to fall off, (c) a poorly supported bridge to topple, (d) a bridge span to disconnect and collapse, (e) neighboring buildings to collide and shatter (floors inside a tall building may collapse), (f) a concrete-block, brick, or adobe building to crack apart and collapse, (g) a steep cliff to collapse, carrying buildings with it.

10.28b–e). In the case of bridges, the span may rise off its supports, then slam down and buckle them so the bridge collapses, or the span may slip off the supports. For unlucky drivers on the bridge, the road literally disappears in front of them.

Ground motion can cause water in lakes, bays, reservoirs, and pools to slosh back and forth, in some cases thousands of kilometers from the epicenter. The water's rhythmic movement, known as a **seiche,** can build up waves almost 10 m high and can last for hours. Seiches capsize small boats and flood shoreline homes. And if they occur in reservoirs, seiches may wash over and weaken retaining dams, creating a potential for failure.

If the ruptured portion of a fault intersects the ground surface, it displaces the ground surface. Strike-slip faults cause lateral displacements, while normal or reverse faults cause vertical displacements and fault scarps. If a building, fence, road, pipeline, or rail line straddles a fault trace, slippage on the fault can crack the structure and separate it into two pieces. Note that, in contrast to what happens in movies, the ground doesn't actually open up as a result of fault displacement; faulting makes one piece of crust slide past another, but the two pieces remain in contact.

Landslides and Avalanches

The shaking of an earthquake can cause ground on steep slopes or ground underlain by weak sediment to give way. This movement results in **landslides,** the tumbling and flow of soil and rock downslope. A landslide destroys everything that it carries—trees, buildings, roads—and most of what's in its path, and may completely bury roads and buildings at the base of the slope. Similarly, avalanches of snow may be triggered by earthquakes and, like landslides, cause havoc in their path.

Landslides commonly occur along the coast of California, for movement on faults has rapidly uplifted this coastline in the past few million years, resulting in the development of steep cliffs. When earthquakes take place, the cliffs collapse, often carrying expensive homes down to the beach below (▶Fig. 10.29a). Such events lead to the misperception that "California will someday fall into the sea." While small portions of the coastline do collapse, the state as a whole remains firmly attached to the continent, despite what Hollywood scriptwriters say.

Sediment Liquefaction

Certain types of sediment contain abundant clay. As we learned in Chapter 5, clay consists of tiny flakes of mica-like minerals. When clay-rich sediment has been slightly compacted, electric charges on the surface of these flakes stick the flakes together, trapping the water in pores be-

(a)

(b)

(c)

(d)

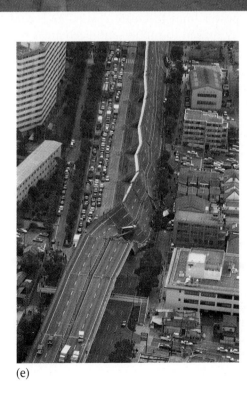

(e)

FIGURE 10.28 (a) During a 1999 earthquake in Turkey, this concrete building collapsed when the supports gave way and piled onto one another like pancakes. (b) Reinforced concrete bridge supports were crushed during the 1994 Northridge earthquake. (c) New apartment buildings are commonly placed on columns, making room for parking below. When the columns collapse, the building crushes parked cars. (d) Ground shaking during the Northridge earthquake toppled highway bridges from their supports. (e) An elevated bridge, balanced on a single row of columns, simply tipped over during the 1995 Kobe, Japan, earthquake.

(a)   (b)

FIGURE 10.29 (a) A landslide along the California coast, triggered by the Northridge earthquake, carried part of a luxury home down to the beach below. (b) The foundation of this apartment in Hsingchung, Taiwan, liquefied during a 1999 earthquake, and the building tipped over, intact.

tween them. However, when the ground shakes during an earthquake, the motion unsticks the clay flakes from one another, and the sediment becomes a slurry of clay and water. We call this process **liquefaction.** If liquefaction occurs in sediment that lies beneath the foundation of a building, the building will topple over (▶Fig. 10.29b). And if liquefaction occurs in sediment beneath a slope, the ground may give way and slip downslope.

Such an event happened during the 1964 Alaska earthquake, when the sediment beneath a housing development in the coastal suburb of Turnagain Heights liquefied (▶Fig. 10.30a–c). The land broke into a series of blocks that slid seaward on the liquefied sediment, carrying with it dozens of houses that transformed into a jumble of splintered wood and shattered windows. Throughout the city of Anchorage, buildings sank into the ground by as much as 3 m. During the 1989 Loma Prieta earthquake in the San Francisco area, the liquefaction of landfill (places where dirt had been piled into coastal marsh to create new land) caused buildings in the Marina district to be damaged more than those of other areas; landfill tends to liquefy.

In some cases, the liquefaction of sand layers below the ground surface makes the sand erupt through holes or cracks in overlying clay, producing small mounds of sand called **sand volcanoes,** or **sand blows.** Liquefaction may also cause bedding in unconsolidated sequences of sediment to break up and become contorted (▶Fig. 10.31a, b).

Fire

The shaking during an earthquake can make lamps, stoves, or candles with open flames tip over, and it may break wires or topple power lines to create sparks. As a consequence, areas already turned to rubble, and even areas not so badly damaged, may be consumed by fire. Ruptured gas pipelines and oil tanks feed the flames, sending columns of fire erupting skyward (▶Fig. 10.32a, b). Firefighters might not even be able to reach the fires, because the doors to the firehouse won't open or rubble blocks the streets; moreover, firefighters may find themselves without water, for ground shaking can damage water lines.

If the fire gets a good start, it can become an unstoppable inferno. Most of the destruction of the 1906 San Francisco earthquake, in fact, resulted from a fire that ignited after the shaking had stopped. For three days, the blaze spread through the city until firefighters contained it by blasting a firebreak. But by then, five hundred blocks of structures had turned to ash, causing twenty times as much financial loss as the shaking itself. When a large earthquake hit Tokyo in September 1923, fires set by cooking stoves spread quickly through the wood-and-paper buildings, creating an inferno that heated the air above the city. The hot air rose like a balloon, and when cool air rushed in, creating wind gusts of over 100 mph, the wind stoked the blaze and incinerated 120,000 people.

(a) Wet clay layer

Clay
Bond
Water

Stationary clay flakes
stuck together

Earthquake !

(b) Slipping on clay layer   Sliding surface

Moving slurry

(c)

FIGURE 10.30 The Turnagain Heights (Alaska) disaster. (a) Before the earthquake, wet clay packed together in a subsurface layer of compacted but wet mud. (b) Ground shaking caused liquefaction of the wet clay layer in the sediment beneath a housing development. As a consequence, the land slumped and slid seaward on an array of sliding surfaces, carrying the houses with it. The ground broke into many slices. (c) The housing development after the slide.

(a)

(b)

FIGURE 10.31 (a) Ground shaking may cause wet sand buried below the surface to squirt up between cracks in overlying compacted mud. The result is a row of sand volcanoes. Generally, these are very small, but those formed after the 1811 New Madrid (Missouri) quake were tens of meters across. (b) Thin layers of sand interlayered with mud may get disrupted, folded, and broken up as a result of ground shaking.

FIGURE 10.32 (a) Fire may consume large areas of toppled buildings after an earthquake. (b) Broken gas mains erupt fountains of flame.

Tsunamis

If fault slip displaces the sea-floor surface, it may generate a wave in the overlying water. Specifically, if part of the sea floor drops as a result of normal faulting, the water level above suddenly sinks, and water rushes in from the sides to fill the space; if the sea floor rises as a result of reverse or thrust faulting, the water level suddenly rises, and water rushes off the uplifted area (▶Fig. 10.33a–f). In either case, the sudden movement of the sea surface creates waves that flow away from the epicenter. If this movement affects a large area, the amount of water that moves can be immense, even though the vertical change in the elevation of the sea floor may be slight. Similar waves may also be caused by landslides that dump rock into adjacent bodies of water, by volcanic explosions, and by submarine slumps (when large masses along the margins of continents or islands suddenly slip downslope).

Geologists refer to such waves as **tsunamis,** a Japanese word meaning "harbor waves," because the waves gained notoriety by destroying harbors. Until recently, the popular press incorrectly referred to them as "tidal waves"—true tidal waves are much smaller, and are the result of the transition between low and high tide. Geologists also use the term "tsunami" for a giant water wave caused by a landslide or volcanic explosion.

A small tsunami, generated by a small sea-floor displacement, may affect only nearby coastlines. But tsunamis generated by great earthquakes (m = 8 or larger) move much bigger areas of sea floor and therefore much more water. These waves can cross the entire ocean—at velocities of up to 800 km per hour, the speed of a jet plane. In the open ocean, where the sea reaches depths of over 4 km, such a tsunami may be only about 30 cm high, comparable to the vertical displacement of the sea floor, so a ship could cross one without even noticing. But the wave may be 200 km across, as measured perpendicular to the crest, and can carry more than 50,000 tons of water per kilometer. As it approaches the shore, friction with the sea floor slows it down, but the water builds up to form a monster wave up to

FIGURE 10.33 (a) Before a tsunami forms. (b) A normal fault creates a void, and water rushes to fill it. Water mounds up over the fault. (c) The resulting waves move toward shore, where they build into a huge breaker. A wave also races out to sea. (d) A similar process happens in response to a reverse or thrust fault. (e) This time, the rising sea floor shoves up the water surface. (f) Large breakers develop.

30 m high. And if the wave funnels into a narrow harbor or bay, it may reach heights of 70 m.

Tsunamis that travel across the Pacific strike distant shores hours after the earthquakes that cause them have struck. For example, a tsunami generated during a 1960 earthquake near the southern tip of South America hit Hawaii seven hours later and Japan twenty-two hours later. When a tsunami approaches the shore, water first recedes far below the low-tide level. Unsuspecting inhabitants of coastal villages may run out onto the exposed seashore to collect stranded fish. Then the wall of water rushes in, swallowing people along the beach. As it moves inland, it sweeps ships onto shore, crushes buildings, and floods low-lying land. Eventually the water recedes, leaving ships hundreds of meters inland and carrying people, animals, and debris out to sea (▶Fig. 10.34a–c). Because tsunamis occur in trains, the first wave may be followed by others at intervals of several minutes to a few hours. For example, five

tsunamis generated by the 1964 Alaska quake struck Crescent City, California. The first wave arrived at low tide and had little effect, but the last came in at high tide and drowned twelve people.

Predicting tsunamis can save thousands of lives. In Hawaii, a tsunami warning station has been established that keeps track of Pacific Rim earthquakes and uses data from tidal gauges to determine if a tsunami has formed. If evidence suggests that a tsunami has set off across the Pacific, the station radios alarms, thus giving people hours of warning.

Disease

Once the ground shaking and fires have stopped, disease may still threaten lives in an earthquake-damaged region. Earthquakes cut water and sewer lines, destroying clean water supplies and exposing the public to bacteria, and they

FIGURE 10.34 (a) Hawaiians flee in panic from a tsunami in the process of flooding Hilo. (b) From a safe distance, a photographer recorded this tsunami washing up onto the shore of Hawaii. (c) In the aftermath of a tsunami that hit Kodiak, Alaska, after the 1964 earthquake, ships rest amid the rubble of a coastal town.

cut transportation lines, preventing food and medicine from reaching the city. The bodies of victims may begin to rot, exposing survivors to additional contagions. The severity of such problems depends on the ability of emergency services to cope. During the January 1999 m = 6.3 earthquake in Colombia, over 2,000 people died owing to the collapse of unreinforced buildings, and tens of thousands were left homeless. Cut off from aid by impassable mountain roads, survivors faced the threat of cholera and starvation until an airlift could bring in supplies.

10.7 PREDICTING "THE BIG ONE"

We have seen that large earthquakes occurring near population centers bring catastrophes. Needless to say, many lives could be saved if only it were possible to know exactly when and where an earthquake will happen, so people could evacuate dangerous buildings, turn off gas and electricity, and, with more warning, build stronger structures.

Can seismologists predict earthquakes? The answer depends on the time frame of the prediction. With our present understanding of the distribution of seismic zones and the frequency at which earthquakes occur, we can make **long-term predictions** (of decades to centuries). For example, with some certainty, we can say that a major earthquake will rattle California during the next century, and that a major earthquake will not strike in central Canada during the next ten years. But despite extensive research, seismologists cannot make accurate **short-term predictions** (of hours to weeks). Thus, we cannot say that an earthquake will happen in Montreal at 2:43 P.M., January 17, 2002.

In this section, we look at the scientific basis of both long- and short-term predictions, and consider the consequences of a prediction. Seismologists refer to studies leading to predictions as **seismic-risk,** or **seismic-hazard, assessment.** Based on this work, they produce maps that assign levels of seismic risk to different regions.

Long-Term Predictions

When making a long-term prediction, we use the word "probability," because a prediction only gives the likelihood of an event. For example, a seismologist may say, "The probability of a major earthquake occurring in the next twenty years in southern California is twenty percent." This sentence implies that there's a one-in-five chance of the earthquake happening during a twenty-year period. Urban planners and civil engineers can use long-term predictions to help create building codes for a region—stronger buildings make sense for regions with greater seismic risk. They may also use predictions to determine whether or not to build vulnerable structures like nuclear power plants, hospitals, or dams in seismic areas.

Seismologists base long-term earthquake predictions on two pieces of information: the identification of seismic zones and the **recurrence interval** (the average time between successive events). To identify a seismic zone, seismologists produce a map showing the epicenters of earthquakes that have happened during a set period of time (say, twenty years). Clusters or belts of epicenters define the seismic zone. The basic premise of long-term earthquake prediction can be stated as follows: a region in which there have been many earthquakes in the past will likely experience more earthquakes in the near future. Seismic zones, therefore, are regions of greater seismic risk. This doesn't mean that a disastrous earthquake can't happen far from a seismic zone—they can and do—but the risk is less. It's like saying that a person who smokes has a higher risk of lung cancer than a person who doesn't smoke. That doesn't mean that nonsmokers don't get lung cancer, only that they are less likely to.

Epicenter maps can only be produced using data from the past fifty years or so, because before that time seismologists did not have sufficient seismographs to locate epicenters accurately. Fortunately, geological mapping can contribute information about faulting back through longer periods of time, because seismic zones by definition contain active faults. Thus, geologists can assess seismic risk by examining landforms for evidence of recent faulting. For example, they can look for young fault scarps, determine whether recent sedimentary deposits have been cut by faults, and decide whether faulting has offset stream channels or created small ponds or ridges (▶Fig. 10.35a–c).

Because a great earthquake releases far more energy than many small earthquakes combined, it is more dangerous. Thus, to determine seismic risk, seismologists must find the recurrence interval for great earthquakes in a seismic zone, which can be calculated by averaging the time between several such large earthquakes that happened in the past. Since the historical record does not provide information far enough back in time, we must also study geologic evidence for great earthquakes: for example, we can examine sedimentary strata near a fault to find layers of sand volcanoes and disrupted bedding in the stratigraphic record. Each layer, whose age can be determined by using radiocarbon dating of plant fragments, records the time of an

earthquake. We can obtain additional information by dating offset soil horizons now buried beneath other sediment (▶Fig. 10.35d).

As an example, imagine that disrupted layers formed 260, 820, 1,200, 2,100, and 2,300 years ago. By calculating the number of years between successive events and taking the average, we can say that the recurrence interval between events is 510 years, so the probability of an earthquake happening in any given year is 1 in 510, or about 0.2%. (See caption for a different example.) Note that a recurrence interval does not specify the exact number of years between events, only the average number.

Another way for determining the recurrence interval comes from simply estimating how long it would take, along a given plate boundary, for the average relative movement between the two plates to produce a large enough stress to generate a sizable slip. For example, if great earthquakes along a plate boundary result in 3 m of slip, and movement along the boundary occurs at 3 cm per year on average, then you might expect about 100 years to pass between major earthquakes. With such information, seismologists are able to produce regional maps illustrating seismic risk (▶Fig. 10.36a, b).

Seismologists suspect that places where an active fault has not moved for a number of years, referred to as **seismic gaps,** may be particularly dangerous (▶Fig. 10.36c). In a seismic gap, either the fault must be moving by nonseismic fault creep, or the stress must be building up to be released by a major earthquake. Many seismic gaps have been identified along the seismic belts bordering the Pacific Ocean—these areas are viewed with concern.

Short-Term Predictions

Short-term prediction, which could lead to such precautions as evacuating dangerous buildings, shutting off gas and electricity, and readying emergency services, is not reliable and may never be. Chinese seismologists claimed to have predicted that a major earthquake would happen near Haicheng during the first week in February 1975, and authorities evacuated the city before a major quake did strike on February 4. But no one predicted the T'ang-shan earthquake of 1976, which killed over a quarter-million people, the largest death toll of any earthquake in the twentieth century. This failure hints that the earlier prediction may have been a coincidence. More recently, seismologists predicted that an earthquake would take place on the San Andreas Fault in Parkfield, California, in mid-November 1993, because an earthquake has occurred there every twenty to twenty-two years since 1857. But despite an intense effort to monitor activity on the fault, the prediction proved false. In fact, many seismologists feel that seismicity is a somewhat random process that can't be accurately predicted.

FIGURE 10.35 The effects of recent fault movement on the land surface. (a) A normal fault truncates ridges, creating triangular facets. (b) A strike-slip fault offsets stream channels and ridges that cross it. (c) A strike-slip fault may bend along its length. Some bends restrain motion and therefore feel compression, while some bends do not, and thus open up during fault motion. Where there is compression, you find pressure ridges or transverse thrust belts. Where there is extension, you find sag ponds or pull-apart basins. Inset map shows the orientation of the restraining bend and the releasing bend, relative to the slip direction on the fault. (d) Here, earthquake events are represented by disrupted bedding (1), an offset ancient soil horizon, or paleosol (2), a layer of sand volcanoes (3), and a bent tree caused by the formation of a fault scarp (4). We can date these events from the wood fragments in the strata. If the disrupted bedding formed in 100 B.C.E., the offset paleosol in 800 C.E., the sand volcanoes in 1340, and the tilted tree in 1950, then the time gaps between earthquakes are 900, 540, and 610 years. The recurrence interval is 683 years (900 + 540 + 610 divided by 3), with a large uncertainty.

FIGURE 10.36 (a) This map projects expected levels of ground shaking from earthquakes in the United States during the next fifty years. The most dangerous areas include those that lie near seismic zones in which there have been recent great earthquakes. (b) This map shows the probability (in percent) of a strong-to-great earthquake occurring along segments of the San Andreas Fault during the next thirty years. When an earthquake takes place here, only part of the fault slips. Not all segments of the fault have the same likelihood of producing a large earthquake, because some segments have slipped more recently than others; seismologists assume that stress is lower across segments that have slipped more recently. (c) Earthquakes may be more likely to occur in the seismic gaps around the Pacific in the near future.

Nevertheless, there are clues to imminent earthquakes, and seismologists have been working hard to understand them. The first clue comes from the detection of foreshocks (▶Fig. 10.37). A swarm (a cluster of events during a short period of time) of foreshocks may indicate the cracking that precedes a major rupture along a fault zone. In this regard, foreshocks are analogous to the cracking noises you hear just before a tree limb breaks off and falls down. But foreshocks do not always occur, and even if they do, they usually can be identified only in hindsight, because they may be indistinguishable from other small earthquakes.

Another possible data source for short-term predictions comes from the precise laser surveying of the ground. Before an earthquake, a region of crust may undergo distortion, either in response to the buildup of elastic strain in the rock, or because of the development of small, open cracks

FIGURE 10.37 A sudden increase in the number of small earthquakes along a fault segment may be a possible precursor to a large earthquake along that segment; that is, the small earthquakes may be foreshocks.

that cause the crust to increase in volume. The land surface may bulge or sink, or straight lines on the ground may become bent. The detection of such movements by laser surveying can hint at an upcoming earthquake.

Other techniques that have been explored but have not yet proven to be precursors of earthquakes include the following.

- *Changes in the water level in wells:* The squeezing or stretching of rock or the development of cracks may change the volume of pore space (open spaces within rock), causing groundwater levels to rise or fall.

- *The detection of gas in wells:* Gases such as radon or helium may be released from pore spaces as rock begins to crack before an earthquake. Most investigators remain skeptical of this event as a precursor.

- *The detection of electrical conductivity in rock:* Cracking in rock may change its ability to conduct a current. Such changes conceivably can be measured, but here too, most investigators remain skeptical.

- *The detection of unusual animal behavior:* Folklore suggests that animals may sense an impending earthquake,

perhaps because of the transmission of cracking noises through rock. There have been reports of dogs howling, horses stampeding, fish jumping, and snakes crawling out of the ground before earthquakes, but this idea has not yet been scientifically demonstrated.

As long as short-term predictions remain questionable, emergency service planners must ask: "What if a prediction is wrong?" Should schools and offices be shut because of a prediction? Should millions be spent to evacuate people? Should a city be deserted, allowing for the possibility of looters? Should the public be notified, or should only officials be notified, creating a potential for rumor? If the prediction proves wrong, can seismologists be sued? No one really knows the answers to these questions.

10.8 EARTHQUAKE ENGINEERING AND ZONING

If we can't avoid earthquakes, can we prepare for them? A glance at Table 10.1 illustrates that the loss of human life from earthquakes varies widely. The loss depends on a number of factors, most notably the proximity of an epicenter to a population center, the depth of the focus, the style of construction in the epicentral region, whether or not the earthquake occurred in a region of steep slopes or along the coast, whether buildings lay on solid bedrock or on a weak substrate, whether the earthquake happened when people were outside or inside, and whether the government was able to provide emergency services promptly.

For example, the 1988 earthquake in Armenia was not much bigger than a 1971 earthquake in southern California, but it caused almost five hundred times as many deaths (24,000 versus 50). The difference in death toll reflects differences in the style and quality of construction, and the characteristics of the substrate. The unreinforced concrete-slab buildings and masonry houses of Armenia collapsed, whereas the structures in California had by and large been erected according to building codes that take into account stresses caused by earthquakes. Most flexed and twisted but did not fall down and crush people. The terrible 1976 earthquake in T'ang-shan, China, was such a calamity because the ground beneath the epicenter had been weakened by coal mining and collapsed, and because buildings were so poorly constructed. Mexico City's 1985 earthquake proved disastrous because the city lies over a sedimentary basin whose bowl-like shape focuses seismic energy, and whose content vibrated with the same frequency as five-to-fifteen-story buildings. Because of the similarity in frequency, the buildings swayed more than they might have otherwise, just as a child in a swing goes higher if the parent pushes at just the right moment. As a consequence, the buildings col-

FIGURE 10.38 How to prevent damage and injury during an earthquake. (a) By wrapping a bridge's support columns in cable (preventing buckling of the columns) and bolting the span to the columns (preventing the span from separating from the columns), the bridge will not collapse as easily. (b) Buildings will be stronger if they are wider at the base and if cross beams are added inside. (c) Placing buildings on rollers (or shock absorbers) will lessen the severity of the vibrations. (d) Build on a solid substrate. Structures built on soft sediment will sway more, and their foundations may sink into the ground; those built on hard bedrock sway through a smaller amplitude and don't sink.

lapsed. During the 1989 Loma Prieta, California, quake, portions of Route 880 in Oakland that were built on weak mud collapsed, while portions built on bedrock or gravel remained standing, for weak mud amplifies ground motion and tends to vibrate at a more damaging frequency.

The differences in the destructiveness of earthquakes demonstrate that we can mitigate or diminish their consequences by taking sensible precautions. Clearly, **earthquake engineering,** the design of buildings that can withstand shaking, and **earthquake zoning,** the determination of where land is stable and where it might collapse, can help save lives and property.

In regions prone to large earthquakes, buildings should be constructed so they are able to withstand vibrations without collapsing (▶Fig. 10.38a–d). They should be somewhat flexible so that ground motions can't crack them, and supports should be strong enough to maintain loads far in excess of the loads caused by the static (nonmoving) weight of the building. Bridge support columns should also be constructed with earthquakes in mind. By wrapping steel cables around the columns, they become many times stronger, and bolting the bridge spans to the top of a column prevents the spans from bouncing off. Concrete-block buildings, unreinforced concrete, and unreinforced brick buildings crack and tumble under conditions where wood-frame, steel-girder, or reinforced concrete buildings remain standing. Traditional heavy, brittle tile roofs shatter and bury the inhabitants inside, while sheet-metal or asphalt shingle roofs do not. Ancient Japanese temples have stood for centuries through many earthquakes, because the lightweight walls of the buildings do not touch the ground, so earthquake energy only transmits into support columns, which can easily sway. Loose decorative stone and huge open-span roofs do not fare well when vibrated, and should be avoided.

Similarly, developers should avoid construction on land underlain by weak, wet mud that could liquefy or vibrate at dangerous frequencies. They should not build on top of, on, or at the base of steep escarpments, and they should avoid locating large population centers downstream of dams (which could fail, causing a flood). And they should avoid constructing vulnerable buildings directly over active faults, whose movement could crack and destroy the structure. Cities in seismic zones need to draw up emergency plans to deal with disaster. Communication centers should be located in safe localities, and strategies for providing supplies under circumstances where roads may be impassable need to be implemented.

Finally, individuals should learn to protect themselves during an earthquake. In your home, keep emergency supplies, bolt bookshelves to walls, install locking latches on cabinets, know how to shut off the gas and electricity, and know where to go to find family members. Schools and offices should have earthquake-preparedness drills (▶Fig. 10.39a, b). If an earthquake strikes, stay outdoors and away from buildings if you can. If you're inside, stand near a wall or in a doorway near the center of the building, or crouch under a heavy table. And if you're on the road, stay away from bridges. As long as plates continue to move, earthquakes will continue to shake. But we can learn to live with them.

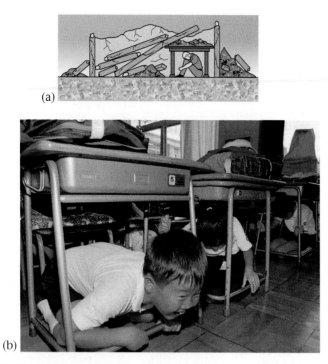

FIGURE 10.39 (a) If an earthquake strikes, take cover under a sturdy table near a wall. (b) Japanese schoolchildren practice earthquake preparedness during an earthquake drill.

CHAPTER SUMMARY

• Earthquakes are episodes of ground shaking, caused when earthquake waves reach the ground surface. Earthquake activity is called seismicity.

• Most earthquakes happen when rock breaks during faulting. A fault is a fracture on which sliding occurs. The place where rock breaks and earthquake energy is released is called the focus, and the point on the ground directly above the focus is the epicenter.

• We can distinguish between normal, reverse, thrust, and strike-slip faults, based on the relative motion of rock across the fault. The amount of movement is called the displacement.

• Active faults are faults on which movement is currently taking place. Inactive faults ceased being active long ago, but can still be recognized because of the displacement across them. Displacement on active faults that intersect the ground surface may yield a fault scarp. The intersection of the fault with the ground is the fault trace. Faults that do not intersect the ground are blind faults.

• During fault formation, rock elastically strains, then cracks form. Eventually, the cracks link to form a through-going rupture on which sliding occurs. When this happens, the elastically strained rock breaks and vibrates, and this generates an earthquake.

• Slip may occur on a fault more than once. Friction resists sliding until stress acting on the fault gets large enough. Thus, faults exhibit stick-slip behavior, in that they move in sudden increments.

• Earthquake faulting is a type of brittle deformation. Earthquakes in the continental crust can only happen in the brittle, upper part of the crust. At depth, where rocks become ductile, earthquakes don't occur.

• Earthquake energy travels in the form of seismic waves. Body waves, which pass through the interior of the Earth, include P-waves (compressional waves) and S-waves (shear waves). Surface waves, which pass along the surface of the Earth, include Rayleigh waves and Love waves.

• We can detect earthquake waves using a seismograph. A weight inside this instrument stays fixed in position, while a pen attached to the weight traces out seismic waves on a paper cylinder attached to the frame, which moves with the Earth.

• Seismograms demonstrate that different earthquake waves arrive at different times, because they travel at different velocities. Using the difference between P-wave and S-wave arrival times, graphed on a travel-time curve, seismologists can determine the distance from a seismograph station to an earthquake epicenter and can then pinpoint the epicenter location.

• The Mercalli intensity scale is based on documenting the damage caused by an earthquake. The Richter magnitude scale measures the size of the largest recorded earthquake wave on a seismogram. The seismic-moment magnitude scale takes into account the amount of slip, the length and depth of the rupture, and the strength of the ruptured rock.

• The largest earthquakes have a Richter magnitude of about 8.9. A magnitude 8 earthquake yields about ten times as much ground motion as a magnitude 7 earthquake, and releases about thirty-three times as much energy. Great earthquakes have a magnitude larger than about 8.0.

• Most earthquakes occur in seismic belts, or zones, of which the majority lie along plate boundaries. Intraplate earthquakes, which happen in the interior of plates, are relatively infrequent, but can be large, as happened in New Madrid, Missouri.

• Different kinds of earthquakes happen at different kinds of plate boundaries. Shallow-focus earthquakes associated with normal faults occur at divergent plate boundaries and in rifts. Earthquakes associated with thrust and reverse faulting occur at convergent and collisional boundaries. At convergent plate boundaries, we also observe intermediate- and deep-focus earthquakes, which define the Wadati-

Benioff zone. Shallow-focus strike-slip earthquakes occur along transform boundaries.

• Earthquakes can be induced at some locations by injecting water into the ground.

• Earthquake damage results from ground shaking (which can topple buildings), landslides (set loose by vibration), sediment liquefaction (the transformation of compacted clay into a muddy slurry), fire, and tsunamis (giant waves).

• Seismologists can predict that earthquakes are more likely in seismic zones than elsewhere, and can determine the recurrence interval (the average time between successive events) for great earthquakes by studying geologic features such as the distribution of sand volcanoes and disrupted bedding in a sequence of recent strata. But it may never be possible to pinpoint the exact time and place at which an earthquake will take place.

• Earthquakes may be more likely at seismic gaps (places where no recent earthquakes have happened) along seismic belts, presumably because stress is building in these locations.

• Earthquake hazards can be reduced with better construction practices and zoning, and by knowing what to do during an earthquake.

KEY TERMS

aftershock (p. 275)
blind fault (p. 273)
body waves (p. 279)
brittle deformation (p. 278)
compressional waves (p. 279)
displacement (p. 272)
ductile deformation (p. 278)
elastic strain (p. 274)
epicenter (p. 270)
fault (p. 270)
fault creep (p. 278)
fault scarp (p. 273)
fault trace (p. 273)
focus (hypocenter) (p. 270)
footwall (p. 271)
foreshock (p. 275)
friction (p. 274)
hanging wall (p. 271)
induced seismicity (p. 292)
intraplate earthquakes (p. 288)
liquefaction (p. 296)
Love waves (p. 279)
Mercalli intensity scale (p. 285)

normal fault (p. 272)
P-waves (p. 279)
plate-boundary earthquakes (p. 288)
Rayleigh waves (p. 279)
recurrence interval (p. 301)
reverse fault (p. 272)
Richter magnitude scale (p. 285)
S-waves (p. 279)
sand volcanoes (p. 296)
seiche (p. 294)
seismic belts (zones) (p. 288)
seismic gaps (p. 301)
seismicity (p. 270)
seismic-moment magnitude scale (p. 287)
seismic-risk assessment (p. 300)
seismic waves (p. 279)
seismograph (p. 281)
seismologist (p. 270)
shear waves (p. 279)
stick-slip behavior (p. 275)
strain (p. 273)

stress (p. 273)
strike line (p. 272)
strike-slip fault (p. 272)
surface waves (p. 279)

thrust fault (p. 272)
travel-time curve (p. 284)
tsunami (p. 298)
Wadati-Benioff zone (p. 289)

REVIEW QUESTIONS

1. Compare normal, reverse, and strike-slip faults.

2. Why are blind faults even more worrisome than faults with known surface breaks?

3. What is the difference between stress and strain? What are the three kinds of stress?

4. Describe elastic strain.

5. Compare brittle and ductile deformation.

6. Describe the motions of the four types of seismic waves. Which are body waves, and which are surface waves? What are their relative velocities?

7. How do the velocities of P-waves change as they pass through different materials? Which materials have the highest seismic velocity? Which have the slowest?

8. Explain how the vertical and horizontal components of an earthquake are detected on a seismograph.

9. Why was the worldwide seismic network of military as well as scientific value?

10. Compare the Mercalli, Richter, and seismic-moment scales in terms of what they measure.

11. How does seismicity on mid-ocean ridges compare with seismicity at convergent or transform boundaries?

12. What is the Wadati-Benioff zone, and why was it important in understanding plate tectonics?

13. Describe the different types of damage caused by earthquakes.

14. What is a seiche, and how does it move?

15. Explain how liquefaction occurs in an earthquake, and how it can cause damage.

16. How are long-term and short-term earthquake predictions made?

17. Why is it difficult to make accurate short-term predictions?

18. What types of structure are most prone to collapse in an earthquake? What types are most resistant to collapse?

19. What should you do when you feel an earthquake starting?

Seeing Inside the Earth

C.1 INTRODUCTION

We live on the Earth's skin, and can see light years into space just by looking up. But when we look down—that's another story! The opaqueness of rock hides the Earth's interior, so that descending even into the deepest mine allows us to glimpse only the outer 0.03% of the planet. Fortunately, as discussed in Chapter 2, nineteenth-century geologists realized that measurements of the Earth's mass and shape provide indirect clues to the mystery of what's inside. From these clues, they speculated that the Earth is not homogeneous but rather consists of three concentric layers: a crust (of low density), a mantle (of intermediate density), and a core (of high density). Further study showed that the crust beneath continents differs from the crust beneath oceans. Continental crust consists of a variety of silicic-, intermediate-, and mafic-composition rock, while oceanic crust consists almost entirely of mafic-composition rock (▶Fig. C.1). However, the determination of the depths of the boundaries between Earth's layers and the division of the layers into sublayers with distinct properties were not possible until the twentieth century, when studies of seis-

mic waves were made. Only by studying the speed and direction in which seismic waves travel through the Earth can seismologists "see" details of the planet's internal layers.

In this interlude, we look at the behavior of seismic waves as they pass through our planet, and learn how this behavior characterizes Earth's interior. We begin by reviewing a few key points about seismic waves, then move on to the general phenomena of wave reflection and refraction. Finally, we witness the discoveries of the different layer boundaries in the Earth. This interlude, incorporating the information about earthquakes and seismic waves provided in Chapter 10, completes the journey to the center of the Earth that we began in Chapter 2.

C.2 THE MOVEMENT OF SEISMIC WAVES THROUGH THE EARTH

Recall that a sudden rupture or the frictional slip of rock on a fault produces seismic waves. These waves move outward from the point of rupture, the earthquake focus, in all directions at once. A single earthquake produces many kinds of waves, distinguished from one another by where and how they move. Notably, surface waves (R-waves and L-waves) propagate along the planet's surface, while body waves (P-waves and S-waves) pass through the interior. P-waves are compressional, and resemble the waves generated when you push a spring back and forth in a direction parallel to the length of the spring. S-waves are shear, and resemble the waves generated when you wiggle a rope back and forth perpendicular to the length of the rope (see Fig. 10.13).

The boundary between the rock through which a wave has passed and the rock through which it has not yet passed is called a **wave front.** A wave front expands outward from the focus like a growing bubble. Since an earthquake generates many waves, we can represent one in a drawing by a series of concentric wave fronts. The changing position of an imaginary point on a wave front as the front moves through rock is called a **seismic ray.** Note that seismic rays are perpendicular to wave fronts, so that each point on the

FIGURE C.1 The nineteenth-century three-layer image of the Earth, showing the crust, mantle, and core. The inset shows the contrast between continental crust (silicic, intermediate, and mafic rock) and oceanic crust (mafic rock).

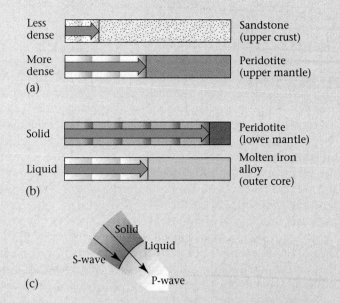

FIGURE C.2 An earthquake sends out waves in all directions. Seismic rays are perpendicular to the wave fronts.

FIGURE C.3 (a) Seismic waves travel at different velocities in different rock types. For example, they travel faster in peridotite than in sandstone. (b) Seismic waves travel faster in solid peridotite than in a liquid like molten iron alloy. (c) Both P-waves and S-waves can travel through a solid, but only P-waves can travel through a liquid.

wave front follows a slightly different ray (►Fig. C.2). The time it takes for a wave to travel from the focus to a seismograph station along a given ray is the **travel time** along that ray.

The ability of a seismic wave to travel through a certain material and the velocity at which it travels depend on the character of the material. Factors such as density, **rigidity** (how stiff or elastic-like a material is), and **compressibility** (how much a material's volume changes in response to squashing) all affect seismic-wave movement. Following are three observations to keep in mind.

- Seismic waves travel at different velocities in different rock types (►Fig. C.3a). For example, P-waves travel at 8 km per second in peridotite, but at only 3.5 km per second in low-density sandstone. Therefore, waves accelerate or slow down if they pass from one rock into another. P-waves in rock travel about ten to twenty-five times faster than sound waves in air. But even at this rate, they take about twenty minutes to pass entirely through the Earth along a diameter.

- In general, seismic waves travel faster in a solid than in a liquid. For example, they travel more slowly in magma than in solid rock, and more slowly in molten iron alloy than in solid peridotite (►Fig. C.3b).

- Both P-waves and S-waves can travel through a solid, but only P-waves can travel through a liquid (►Fig. C.3c). That's because liquids, which have no rigidity, can transmit compressional waves but not shear waves. If you shear a liquid, the moving liquid simply flows past the adjacent liquid and thus has no effect on it. For example, if you push down on the water surface in a pool, you send a pulse of compression (a P-wave) to the bottom of the pool. Now move (shear) your hand sideways through

the water. The water in front of your hand simply slides or flows past the water deeper down: your shearing motion has no effect on the water at the bottom of the pool (►Fig. C.4a, b).

FIGURE C.4 (a) Pushing down on a liquid creates a compressive pulse (P-wave) that can travel through a liquid. (b) Shearing your hand through water does not generate a shear wave; the moving water simply flows past the water deeper down.

C.3 THE REFLECTION AND REFRACTION OF WAVE ENERGY

Shine a flashlight into a container of water so that the light ray hits the boundary (or interface) between water and air at an angle. Some of the ray bounces off the water surface and heads back up into the air, while some enters the water (▶Fig. C.5a). The light ray that enters the water bends at the air-water boundary, so that the angle between the ray and the boundary in the air is different from the angle between the ray and the boundary in the water. Physicists refer to the light ray that bounces off the air-water boundary and heads back into the air as the **reflected ray,** and the ray that bends at the boundary as the **refracted ray.** The phenomenon of bouncing off is **reflection,** and the phenomenon of bending is **refraction.** Wave reflection and refraction take place at the interface between two materials, if the wave travels at different velocities in the two materials.

The amount and direction of refraction at a boundary depend on the contrast in wave velocity across the boundary and on the angle at which a wave hits the interface. As a rule, if waves enter a material through which they will travel more slowly, they bend away from the interface, while if the waves enter a material through which they will travel faster, they bend toward the interface. For example, the light beam in ▶Figure C.5b bends down when hitting the air-water boundary, because light travels more slowly in water. If the beam were to pass from a material in which it travels slowly into one in which it travels more rapidly, it would bend up (▶Fig. C.5c).

Like light, seismic energy travels in the form of waves, so seismic waves, like light beams, reflect and/or refract when reaching the interface between two rock layers if the waves travel at different velocities in the two layers. For example, imagine a layer of sandstone overlying a layer of peridotite. Seismic velocities in sandstone are faster than in peridotite, so as seismic waves reach the boundary, some reflect while some refract.

C.4 DISCOVERING THE CRUST-MANTLE BOUNDARY

In 1909, Andrija Mohorovičić, a Croatian seismologist, noted that P-waves arriving at seismograph stations less than 200 km from the epicenter traveled at an average

FIGURE C.5 (a) A beam of light, when it reaches the boundary between water and air, partly reflects and partly refracts. The refracted beam bends down as it enters the water. (b) A wave that enters a slower medium bends away from the boundary (like light reaching water from air). (c) A wave that enters a faster medium bends toward the boundary.

(a)

(b)

(c)

FIGURE C.6 (a) At a nearby seismograph station, seismic waves traveling through the crust reach the seismograph first. Seismic rays refract at the Moho, the crust-mantle boundary. (b) At a distant station, seismic waves traveling through the mantle reach the seismograph first, which means that seismic waves travel at a faster velocity in the mantle than in the crust. The Moho lies at a depth of 35–40 km beneath continental interiors.

C.5 DEFINING THE STRUCTURE OF THE MANTLE

After studying materials erupted from volcanoes, geologists have concluded that the entire mantle has roughly the chemical composition of the ultramafic igneous rock called peridotite. If the density, rigidity, and compressibility of peridotite were exactly the same at all depths, seismic velocities would be the same everywhere in the mantle, and seismic rays would be straight lines. But by studying travel times, seismologists have determined that seismic waves do not travel at the same velocity with depth. Let's now look at variations in seismic velocity that depend on mantle depth, and consider how these variations affect the shape of seismic rays.

Between about 100 and 200 km deep in the mantle beneath oceanic lithosphere, seismic velocities are slower than in the overlying lithospheric mantle (►Fig. C.7). In this **low-velocity zone,** the prevailing temperature and pressure conditions cause peridotite to partially melt, by up to 2%. The melt, a liquid, coats solid grains and fills voids between grains. Because seismic waves travel more slowly through liquids than through solids, the coatings of melt slow seismic waves down. In the context of plate tectonics theory, the low-velocity zone is the weak layer on which

speed of 6 km per second, while P-waves arriving at seismographs more than 200 km from the epicenter traveled at an average speed of 8 km per second. To explain this observation, he suggested that P-waves reaching nearby seismographs followed a shallow path through the crust of the Earth, in which they traveled more slowly, while P-waves reaching distant seismographs followed a deeper path through the mantle, in which they traveled more rapidly.

To understand Mohorovičić's proposal, examine ►Figure C.6a, which shows seismic P-waves, depicted as rays, generated by an earthquake in the crust. Seismic ray A, the shallower wave, travels through the crust directly to a seismograph. Seismic ray B, the deeper wave, heads downward, refracts at the crust-mantle boundary, curves through the mantle, refracts again at the boundary, and then proceeds through the crust up to the seismograph. At stations less than 200 km from the epicenter, ray A arrives first, because it has a shorter distance to travel. But at stations more than 200 km from the epicenter (►Fig. C.6b), ray B arrives first, even though it has farther to go, because it travels faster for part of its length. The distance at which the deeper waves overtake the shallower waves requires the crust-mantle boundary to be at a depth of about 40 km. As we learned in Chapter 2, this boundary is now called the **Moho,** in honor of Mohorovičić.

FIGURE C.7 The velocity of P-waves in the mantle changes with depth. Note the low-velocity zone between 100 and 200 km, and the sudden jumps in velocity in the transition zone between 400 and 670 km.

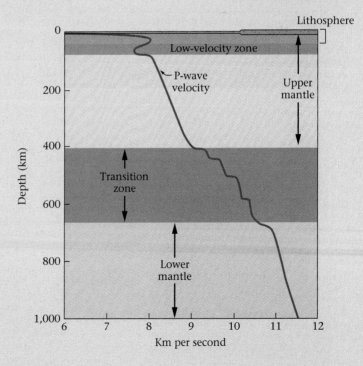

oceanic lithosphere plates move. Below the low-velocity zone, the mantle does not contain melt. Geologists do not find a well-developed low-velocity zone beneath continents; continental lithosphere extends down to a depth of 150 or 200 km.

Below about 200 km, seismic-wave velocities everywhere in the mantle increase with depth. Seismologists interpret this increase to mean that mantle peridotite becomes progressively denser and more rigid with depth. This proposal makes sense, considering that the weight of overlying rock increases with depth, and as pressure increases, the atoms making up rock squeeze together more tightly. Because of refraction, the progressive increase in seismic velocity with depth causes seismic rays to curve in the mantle. To understand the shape of a curved ray, look at ▶Figure C.8a, which represents a portion of the mantle by a series of layers, each with a slightly greater velocity than the layer above. Every time a seismic ray crosses the boundary between adjacent layers, it refracts a little toward the boundary. After the ray has crossed several layers, it has bent so much that it begins to head back up toward the top of the stack. Now if we replace the stack of distinct layers with a single layer in which velocity increases with depth at a constant rate, the wave follows a smoothly curving path (▶Fig. C.8b, c).

When depicting the entire mantle below a depth of 150 km, seismologists tend to portray seismic waves as smooth curves, as in Figure C.8b, for simplicity. But between 400 km and 670 km deep (see Fig. C.7), seismic velocity increases in a series of abrupt steps, so the stack of layers in Figure C.8a is actually a more realistic image. Experiments suggest that these **seismic-velocity discontinuities** occur at depths where pressure causes atoms in a mineral to rearrange and pack together more tightly, thereby changing the rock's density, compressibility, and rigidity. Each seismic-velocity discontinuity corresponds to a different rearrangement of atoms. For example, at 400 km, the mineral olivine, a major constituent of peridotite, becomes unstable and collapses to form a different mineral called magnesium spinel. Magnesium spinel has the same chemical composition as olivine but a different internal arrangement of atoms. Because of these seismic-velocity discontinuities, as we learned in Chapter 2 but now can see more clearly, seismologists subdivide the mantle into the **upper mantle** (above 400 km), **transition zone** (between 400 and 670 km), and **lower mantle** (below 670 km) (see Fig. 2.12).

FIGURE C.8 (a) In a stack of layers in which seismic waves travel at different velocities (fastest in the lowest layer), a seismic ray gradually bends around and heads back to the surface. The curve consists of several distinct segments. (b) If the mantle's density increased gradually with depth, the ray would define a smooth curve. (c) Since the velocity of seismic waves increases with depth, wave fronts are oblong and seismic rays curve.

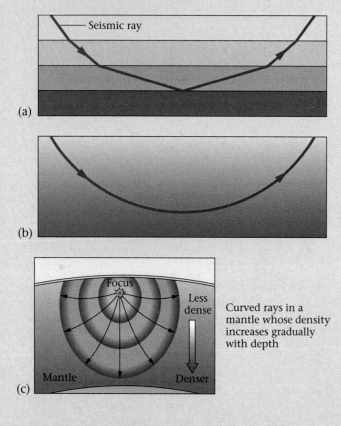

C.6 THE CORE-MANTLE BOUNDARY (THE P-WAVE SHADOW ZONE)

During the first decade of the twentieth century, seismologists installed seismographs at many stations around the world, expecting to be able to record waves produced by a large earthquake anywhere on Earth. In 1914, one of these seismologists, Beno Gutenberg, discovered that P-waves from an earthquake do not arrive at seismographs lying in a band between 103° and 143° from the earthquake epicenter, as measured along the circumference of the Earth. This band is now called the **P-wave shadow zone** (▶Fig. C.9). If the density of the Earth increased gradually with depth all the way to the center, the shadow zone would not exist, because rays passing into the interior would curve up and reach every point on the surface. Thus, the presence of a shadow zone means that deep in the Earth a major interface exists where seismic waves abruptly refract down (implying that the velocity of seismic waves suddenly decreases). This interface, now called the **core-mantle boundary,** lies at a depth of about 2,900 km.

To see why the P-wave shadow zone exists, follow the two seismic rays labeled A and B in Figure C.9. Ray A curves smoothly in the mantle (we are ignoring seismic-velocity discontinuities in the mantle) and passes just above the core-mantle boundary before returning to the surface. It reaches the surface 103° from the epicenter. In contrast, ray B just

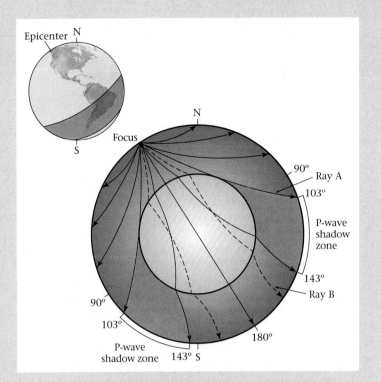

FIGURE C.9 P-waves do not arrive in the interval between 103° and 143° from an earthquake's epicenter, defining the P-wave shadow zone. The wave that arrives at 103° passed just above the core-mantle boundary. The next wave bent down so far that it arrives at about 160°. And the wave that arrives at 143° bent only slightly as it passed through the core. The inset shows the shadow zone on a globe. Note that P-waves bend down at the core-mantle boundary because velocities are slower in the core than in the mantle (the core is less rigid).

penetrates the boundary and refracts down into the core. Ray B then curves through the core and refracts again when it crosses back into the mantle. As a consequence, ray B intersects the surface at more than 143° from the epicenter.

The downward bending of seismic waves when they pass from the mantle down into the core indicates that seismic velocities in the core are slower than in the mantle. Thus, even though the core is deeper and denser than the mantle, it must be less rigid. Based on density calculations and on the study of meteorites thought to be fragments of another planet's interior, seismologists concluded that the core consists of iron alloy, which is less rigid than peridotite.

C.7 THE NATURE OF THE CORE (THE S-WAVE SHADOW ZONE)

While the study of P-waves defined the core-mantle boundary, the study of S-waves gave seismologists insight into the physical character of the core. Seismologists found that S-waves do not arrive at stations located between 103° and

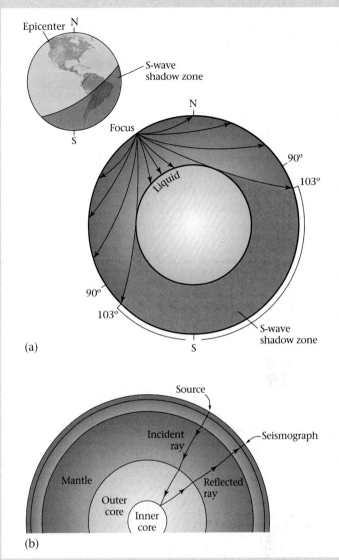

FIGURE C.10 (a) The S-wave shadow zone covers about a third of the globe, and exists because shear waves cannot pass through the liquid outer core. (b) The solid inner core was detected when seismologists observed that some seismic waves generated by nuclear explosions reflected off a boundary within the core.

180° from the epicenter (a band called the **S-wave shadow zone**). This means that S-waves cannot pass through the core at all—otherwise, an S-wave headed straight down through the Earth would appear on the other side. Remember that S-waves are shear waves, which by their nature can only travel through solids. Thus, the fact that S-waves do not pass through the core means that the core, or at least part of it, consists of liquid (▶Fig. C.10a).

At first, seismologists thought that the entire core was liquid iron alloy. But in 1936, a Danish seismologist, Inge Lehmann, discovered that P-waves passing through the core reflected off a boundary within the core. She then proposed that the core is made up of two parts: an **outer core** consisting of liquid iron alloy and an **inner core** consisting of solid iron alloy. Lehmann's work defined the existence of the inner core but could not locate the depth at which the inner core/outer core interface occurs. This depth was located by measuring the exact time it took for seismic waves generated by nuclear explosions to penetrate the Earth, bounce off the inner core/outer core boundary, and return to the surface (▶Fig. C.10b). The measurements showed that the inner core/outer core boundary occurs at a depth of about 5,155 km. Recent studies of seismic waves passing through the inner core have shown that the inner core rotates slightly faster than the rest of the Earth—it makes an extra revolution about once every twenty-five years.

C.8 FINE-TUNING OUR IMAGE OF THE EARTH'S LAYERS

Seismic Tomography

In recent years, sophisticated computers have been able to use global seismic data to create a three-dimensional image of seismic-wave velocities within the Earth. This type of analysis, known as **seismic tomography,** resembles the method employed by a medical CAT-scan machine ("CAT" stands for "computer-aided tomography"). Tomography (the word comes from the Greek word for "slice") allows us

to see slices through a human body or the Earth. In seismic tomography studies, researchers compare travel times for seismic waves that follow different paths through the Earth, and they distinguish regions in which waves travel unexpectedly fast from regions where waves travel unexpectedly slowly. Overall, tomographic studies show that a layered image of the Earth like that depicted in Figure C.8a, while reasonable as a rough approximation, is not accurate in detail (▶Fig. C.11a, b). Rather, it appears that there can be different seismic velocities at a given depth in the Earth.

For example, seismologists have discovered distinct zones of lower-than-expected velocities and distinct zones of higher-than-expected velocities at the same depth in the mantle. These velocity contrasts indicate that there are blotchy variations in density or rigidity within the mantle. Seismologists suggest that these variations in turn reflect temperature variations—hotter rocks are less rigid than cooler rocks. The distribution of hotter and cooler zones in the mantle may be an indication that convection occurs in the mantle—the hotter zones are rising, and the cooler zones are sinking. Tomographic studies in the last few years have become detailed enough to provide images of subducted lithosphere "graveyards" at depth.

Seismic-Reflection Profiling

Seismic techniques are also letting us fine-tune our image of the crust. During the past half century, geologists have found that by using dynamite, by banging large weights against the

(a)

(b)

FIGURE C.11 Tomographic images of the mantle show regions of faster and slower velocities. (a) Redder colors indicate regions where seismic waves travel more slowly than expected (the mantle is warmer and less rigid), while bluer colors are regions where seismic waves travel faster than expected (the mantle is cooler and more rigid). In the slowest regions, waves travel 1.5% more slowly, while in the fastest regions, waves travel 1.5% faster than expected. (b) Close-up of the mantle beneath North America. The East area may be the remnant of a cold subducted slab.

(a)

(b)

(c)

Oceanus 270 line 32

(d)

(e)

FIGURE C.12 (a) Trucks collecting data for a seismic-reflection profile. (b) Analyzing data with a computer. (c) A seismic-reflection profile. The colored stripes are layers of strata. (d) A ship collecting seismic data at sea. (e) This image shows layers of subsurface strata in 3-D. Computers can expose different cross-section and map-view slices of the image. From such data, important features like faults (indicated by colored surfaces) can be located.

Earth's surface, or by releasing bursts of compressed air into the water, they can create artificial seismic waves that propagate down into the Earth and reflect off the boundaries between different layers of rock in the crust. By recording the time at which these reflected waves return to the surface, geologists effectively create a cross-sectional view of the crust called a **seismic-reflection profile** (▶Fig. C.12a–e). This image defines the depths at which layers of strata occur and reveals the presence of subsurface folds (bends in layers) and faults. Oil companies produce thousands of seismic-reflection profiles a year, despite their high cost, because they allow geologists to identify likely locations for oil and gas deposits underground. Research geologists at universities have used the technique to obtain images of the Moho.

In the past decade, computers have become so fast that geologists can now pro-

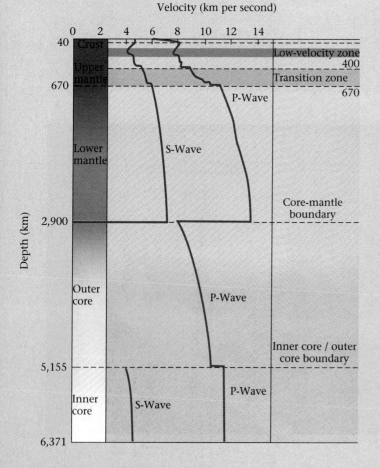

FIGURE C.14 The modern view of a complex and dynamic Earth interior. Note the convecting cells, the mantle plumes, and the subducted-plate graveyards.

FIGURE C.13 The velocity-versus-depth profile for the Earth.

duce three-dimensional seismic-reflection images of the crust. These provide so much detail that geologists can trace out a ribbon of sand representing the channel of an ancient stream even where the sand lies buried kilometers below the Earth's surface.

C.9 AN INTEGRATED VIEW OF THE EARTH

By the time of World War II, seismologists had identified the crust-mantle boundary (the Moho) and the core-mantle boundary, and had recognized that the core is divided into two parts—in other words, the prevailing image of the Earth's insides was an onion-like sequence of concentric zones. After World War II, seismologists set to the task of refining this image, a task made easier by the Cold War: because of the need to detect nuclear explosions, the Western nuclear powers built a vast array of precise seismograph stations scattered around the world. Through painstaking effort, seismologists eventually developed a graph, known as a **velocity-versus-depth curve,** that shows the depths at which seismic velocity suddenly changes (▶Fig. C.13).

Work in more recent years has gone beyond defining layers. Geologists now can see everything from mantle convection cells to subducted slabs to layers of strata—and are beginning to detect the mantle plumes that feed hot-spot volcanoes. As a result, a complicated image of the Earth's interior has evolved (▶Fig. C.14).

CHAPTER 11

Crags, Cracks, and Crumples: Crustal Deformation and Mountain Building

Innumerable peaks, black and sharp, rose grandly into the dark blue sky, their bases set in solid white, their sides streaked and splashed with snow, like ocean rocks with foam. . . . [Mountains] are nature's poems carved on tables of stone. . . . How quickly these old monuments excite and hold the imagination!

—JOHN MUIR, from *WILDERNESS ESSAYS*

11.1 INTRODUCTION

Geographers call the peak of Mt. Everest "the top of the world," for this mountain, which lies in the Himalayas of South Asia, rises higher than any other on Earth (▶Fig. 11.1). The cluster of flags at Mt. Everest's summit flap 8.85 km (29,029 feet) above sea level—almost the cruising height of modern jets. No one can live very long at the top, for the air there is too thin to breathe. In fact, even after spending weeks acclimatizing to high-altitude conditions at a base camp a couple of kilometers below the summit, most climbers need to use bottled oxygen during their summit attempt. In 1953, British explorer Sir Edmund Hillary and Tenzing Norgay, of Nepal, managed to reach the summit. By 1999, about 750 more people had also succeeded—but 150 died trying. So many casualties happen because a successful climb depends not just on the skill of the climber, but also on the path of the **jet stream,** a 200-km-per-hour current of air that flows at high elevations (see Chapter 20). If the jet stream crosses the summit, it engulfs climbers in heat-robbing winds that can freeze a person's face, hands, and feet even if they're swaddled in high-tech clothing.

Some of the world's most beautiful scenery can be found in mountainous regions. The Swiss Alps, in which this idyllic meadow lies nestled, rose when Italy collided with Europe, and they were carved by glaciers during the last ice age.

319

FIGURE 11.1 Mt. Everest (the large peak in the center) and the surrounding Himalayas, as viewed from the space shuttle *Atlantis*.

Mountains draw nonclimbers as well, for everyone loves a vista of snow-crested peaks. Their stark cliffs, clear air, meadows, forests, streams, and glaciers provide a refuge from the mundane. For millennia, mountain beauty has inspired the work of artists and poets, and in some cultures mountains served as a home to the gods. Geologists feel a special fascination with mountains, for they provide one of the most obvious indications of dynamic activity on Earth. To make a mountain, Earth forces lift cubic kilometers of crustal rock skyward against the pull of gravity. This uplift then provides the fodder for erosion, which, over time, grinds away at a mountain to make sediment, and in the process sculpts jagged mountain topography.

The process of forming a mountain not only uplifts the surface of the crust, but also imparts many changes

FIGURE 11.2 Digital map of world topography, showing the locations of major mountain ranges.

to the crust. It can cause rocks to undergo **deformation,** a process by which they squash, stretch, bend, or break in response to squeezing, stretching, or shearing. Deformation produces **geologic structures,** including **joints** (cracks), **faults** (fractures on which one body of rock slides past another), **folds** (bent or wrinkled layers), and **foliation** (layering as a consequence of the alignment of mineral grains). Mountain building can also cause rocks to undergo metamorphism or melting. In this chapter, we learn about the phenomena that happen during mountain building—deformation, igneous activity, sedimentation, metamorphism, uplift, and erosion—and discover why they occur, in the context of plate tectonics theory.

11.2 MOUNTAIN BELTS AND THE CONCEPT OF OROGENY

With the exception of hot-spot volcanoes, mountains do not occur in isolation, but rather appear in linear ranges variously called **mountain belts, orogenic belts,** or **orogens** (from the Greek words *oros,* meaning "mountain," and *genesis,* meaning "formation"). Geographers define about a dozen major orogenic belts and numerous smaller ones worldwide (▶Fig. 11.2). Some large orogens contain smaller ranges within. For example, the Sierra Nevada and Rocky Mountains lie within the North American Cordillera (system of mountain ranges).

At any given location, a mountain-building event, or **orogeny,** has a limited lifetime. The process of uplift begins, lasts for tens of millions of years, and then ceases. After an orogeny ceases, erosion can bevel the land surface almost back to sea level in less than about 50 million years. Thus, the mountain ranges we see today are comparatively young, meaning that most of Earth's present topography didn't exist before the Cretaceous period, and that orogeny is an ongoing process. Notably, a given region may be subjected to several different orogenies during geologic time. Geologists can identify ancient orogens by studying the rock record, for even long after erosion has eliminated its peaks, a belt of deformed (contorted or broken) and metamorphosed rocks marks the trace of an ancient orogen.

Why do orogens form? Scientific attempts to answer this question date back to the birth of geology, over two centuries ago. But workable explanations of the origin and distribution of mountains appeared only with the discovery of plate tectonics theory: orogens develop because of subduction at convergent plate boundaries, rifting, continental collisions, and motion on certain continental transform faults.

11.3 ROCK DEFORMATION IN THE EARTH'S CRUST

Deformation and Strain

As noted above, orogeny causes deformation—bending, breaking, squashing, stretching, or shearing—which yields geologic structures. To get a visual sense of deformation, let's compare a road cut along a highway in the central Great Plains of North America, a region that has not undergone orogeny, with a cliff face in the Alps of Switzerland, a region that *has* (▶Fig. 11.3a, b).

(a)

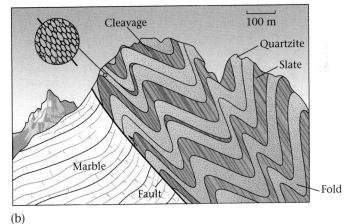

(b)

FIGURE 11.3 (a) In this road cut exposing beds of Paleozoic strata along a highway, as might occur in the interior (nonorogenic) region of North America, beds of shale alternate with beds of sandstone. The beds lie flat, though they have been cut by some vertical joints. Inset: An enlargement of the sandstone, showing its texture. The undeformed sandstone has spherical grains. (b) In this outcrop on a large cliff face as might occur in an orogenic belt like the Alps, note the folded layers of quartzite and slate and the fault. Inset: Grains of sand in the quartzite have become somewhat flattened, and are aligned parallel to one another. The slate has slaty cleavage.

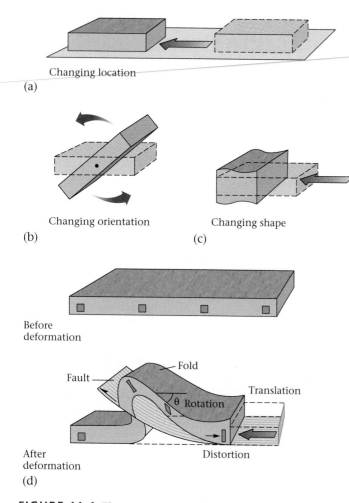

FIGURE 11.4 The components of deformation. (a) A block of rock changes location when it moves from one place to another. (b) It changes orientation when it tilts or rotates around an axis. (c) It changes shape when its dimensions change, or once planar surfaces become curved. (d) Folds and faults represent deformation, because they involve changes in location (sliding has occurred on a fault, for example), orientation (a layer has tilted to form a fold), and shape (the squares in the undeformed layer have become rectangles or parallelograms in the deformed layer).

The Great Plains road cut, which lies at an elevation of only about 100 m above sea level, exposes nearly horizontal beds of sandstone and shale—these beds have the same orientation that they had when first deposited. Notably, sand grains in sandstone have a nearly spherical shape (the same shape they had when deposited), and clay flakes in the shale lie roughly parallel to the bedding, because of compaction. This outcrop contains no geologic structures, except for a few joints. Its rock is undeformed.

In the Alpine cliff face, exposed at an elevation of 3 km, rocks look very different. Here, we find layers of quartzite and slate (the metamorphic equivalent of sandstone and shale) in contorted beds whose shapes resemble the wrinkles in a rug

that has been pushed across the floor. These structures are folds. Quartz (sand) grains in quartzite are not spheres, but resemble flattened eggs, and the clay flakes in slate are aligned parallel to one another and tilt at a steep angle to the bedding. In fact, the rock splits on planes that parallel the flattened sand grains and clay flakes, and thus cut across the bedding at a steep angle. Finally, if we try tracing the quartzite and slate layers along the outcrop face, we find that they abruptly terminate at a sloping surface marked by shattered rock. This surface is a fault. Thick beds of marble lie on the other side of this surface. The beds in the Alpine cliff have been deformed, and as a result the cliff exposes a variety of geologic structures. Beds no longer have the same shape and position that they had when first formed, and the shape

FIGURE 11.5 (a) Undeformed, flat-lying beds of sediment in Badlands National Monument, South Dakota. (b) Tilted beds of strata in Arizona. The tilting is a manifestation of deformation. (c) Folded layers of quartzite and schist in Australia. The folding is also a manifestation of deformation.

and orientation of grains has changed. In sum, deformation includes one or more of the following (▶Fig. 11.4a–d): (1) a change in location, (2) a change in orientation, (3) a change in shape. Deformation can be fairly obvious when observed in an outcrop (▶Fig. 11.5a–c).

Geologists use the term **strain** to refer to the change in shape that deformation brings. We distinguish between differ-

FIGURE 11.7 You can simulate shear strain by moving a deck of cards so that each card slides a little with respect to the one below. Note how a circle drawn on the side of the deck changes shape to become an ellipse, and that the angle between the bottom of the deck and the back side of the deck has changed from a right angle into an acute angle.

ent kinds of strain depending on how the rock changes shape. If a layer of rock becomes longer, it has undergone **stretching,** but if the layer becomes shorter, it has undergone **shortening** (▶Fig. 11.6a–c). If a change in shape involves the movement of one part of a rock body past another, the result is called **shear strain** (▶Figs. 11.6d; 11.7).

Kinds of Deformation: Brittle and Ductile Behavior

We saw in Chapter 10 that rocks temporarily change shape when subjected to stress, developing what we call an elastic strain, and then change back when the stress that caused the strain is removed. But rocks can also develop a permanent strain, in two fundamentally different ways. As mentioned in Chapter 10, during **brittle deformation,** a material breaks into two or more pieces, like a plate shattering on the floor, while during **ductile deformation,** a material changes shape without breaking, like a ball of dough squeezed beneath a book (▶Fig. 11.8a–d). Joints and faults are brittle structures, while folds and foliations are ductile structures. What actually happens in a rock during the different kinds of deformation? Recall that rocks are solids in which chemical bonds, like little springs, link atoms together. During brittle deformation, many bonds break at once so that rocks can no longer hold together, while during ductile deformation, some bonds break but new ones quickly form, so that rocks do not separate into pieces as they change shape.

Why do rocks inside the Earth sometimes deform brittlely and sometimes ductilely? The behavior of a rock depends on

FIGURE 11.6 Different kinds of strain. (a) First, an unstrained cube and an unstrained fossil shell (brachiopod). (b) Stretching changes the cube into a brick whose long dimension parallels the direction of stretching, and it makes the brachiopod longer. (c) Shortening changes the cube into a brick whose long dimension lies perpendicular to the shortening direction, and it makes the brachiopod taller. (d) Shear strain tilts the cube over and transforms it into a parallelogram, and it changes the angular relationships in the brachiopod shell.

FIGURE 11.8 (a) Brittle deformation occurs when you drop a plate and it shatters. (b) Ductile deformation takes place when you squash a soft ball of dough beneath a book and the dough flattens into a pancake without breaking. (c) Cracks (joints) in an outcrop result from brittle deformation. (d) Folds, like these in the marble of a quarry wall, form without breaking a rock, and thus represent ductile deformation.

- *Temperature:* Warm rocks tend to deform ductilely, while cold rocks tend to deform brittlely. To see this contrast, try an experiment with a candle. Chill a candle in a freezer, then press its middle against the edge of a table; the candle will brittlely snap in two. But if you first warm the candle in the sun, it will bend slowly and ductilely without breaking when pressed against the table.

- *Pressure:* Under great pressures deep in the Earth, rock behaves more ductilely than it does under low pressures near the surface. Pressure effectively prevents rock from separating into fragments.

- *Deformation rate:* A sudden change in shape causes brittle deformation, while a slow change in shape causes ductile deformation. For example, if you hit a thin marble bench with a hammer, it shatters, but if you leave it alone for a century, it gradually sags without breaking.

- *Composition:* Some rock types are softer than others; for example, halite (rock salt) can deform ductilely under conditions in which granite behaves brittlely.

Considering that pressure and temperature both increase with depth, geologists find that in typical continental crust rocks behave brittlely above about 10–15 km, while they behave ductilely below; we call this depth the **brittle-ductile transition.** Earthquakes in continental crust happen only above this depth.

In some cases, both brittle and ductile structures occur in the same outcrop. For example, in our Alpine cliff, you can see both faulting (a brittle structure) and folding (a ductile structure). Such an occurrence may seem like a paradox at first. But the juxtaposition of styles happens simply because of changes in the deformation rate during orogeny. Slow deformation yielded the folds, while a pulse of rapid deformation caused the fault to suddenly form.

Force, Stress, and the Causes of Deformation

Science museums and national parks commonly provide displays that describe mountain building. In such displays, you may read a caption that says something like, "The mountains you are looking at were caused by forces deep within the Earth!" What does this mean? Isaac Newton defined force by the equation **force** = mass × acceleration. This means that a force, when applied to a mass, causes the mass to change its velocity (accelerate or decelerate). For example, if you apply a force to a stationary ball, the ball starts rolling, and when a car drives into a wall, the wall exerts a force on the car that stops it from moving (decelerates it) and crumples the bumper. Plate interactions, rifting, continental collisions, and gravity similarly apply forces to rocks in the crust, and these forces cause the rocks to move (though very slowly). Such a force deforms rocks: it changes their position, reorients them, and strains them.

The consequences of applying a force depend on the area over which the force acts. So in practice, geologists use the word "stress" instead of "force" when talking about the cause of deformation. We define **stress** as the force applied per unit area in a material (that is, stress = force/area). To understand the difference between force and stress, imagine you are standing on a single aluminum soda can. The force applied by the weight of your body acts over a small area and thus creates a large stress,

(a)   (b)

FIGURE 11.9 (a) When you stand on a single can, you apply enough force to the can to crush it, for the can feels a large stress. (b) When you stand on a board resting on 100 cans, you apply the same force to the board, but now it is spread out over 100 cans. Therefore, each can feels only a small stress and does not crumple.

enough to crush the can. But if you place a hundred soda cans next to each other in a cluster, set a board over the top, and then stand on the board, the cans hold you up (▶Fig. 11.9a, b). In this case, the weight of your body applies the same downward force as before, because of gravity, but this time the force has been distributed over a broad area so each can feels only a small stress—too small to crush it.

Different kinds of stress occur in rock bodies. As we learned in Chapter 10, **compression** develops when a rock is squeezed, **tension** occurs when a rock is pulled apart, and **shear stress** develops when one side of a rock body moves sideways past the other side (▶Fig. 11.10a–d). Note that "stress" and "strain" have different meanings to geologists (though we tend to use them interchangeably in everyday English): stress refers to the amount of force per unit area of a rock, while strain refers to the change in shape of a rock. Thus, stress *causes* strain. Specifically, compression causes shortening, tension leads to stretching, and shear stress creates shear strain. With our knowledge of stress and strain, we can now look at the nature and origin of various classes of geologic structures.

11.4 JOINTS: NATURAL CRACKS IN ROCKS

If you look through the photographs of rock outcrops in this book, you'll notice thin black lines that cross the rock face. These lines represent traces of natural cracks, places where the rock broke and separated into two pieces during brittle deformation. Geologists refer to such natural cracks as **joints.** Rock bodies do not slide past each other on

joints. (Since joints are roughly planar structures, we define their orientation by their strike and dip; see Box 11.1.)

Joints develop in response to tensional stress in brittle rock: a rock splits open because it has been pulled slightly apart. They may form for a variety of geologic reasons. For example, some joints form when a rock cools and contracts, because contraction makes one part of a rock pull away from the adjacent part. Others develop when rock layers once at depth feel a decrease in compression as overlying rock erodes away, and thus change shape slightly. And still others form when rock layers bend.

Rock bodies may contain two categories of joints. **Systematic joints** are long planar cracks that occur fairly regularly through a rock body, while **nonsystematic joints** are short cracks that come in a range of orientations and are randomly spaced. A group of systematic joints comprise a **joint set,** a spectacular example of which can be seen in Arches National Park, in Utah (▶Fig. 11.12a). The sandstone in the park breaks into long parallel ridges bounded by joints. In sedimentary rocks, systematic joints typically are vertical planes (▶Fig. 11.12b).

If groundwater seeps through joints for a long period of time, minerals like quartz or calcite can precipitate out of the groundwater and fill the joint. Such mineral-filled

FIGURE 11.10 We represent the direction and magnitude of stress acting on each face of an object by arrows; the lengths of the arrows represent the magnitude of the stress. (a) Pressure occurs when an object feels the same stress on all sides. (b) Compression takes place when you squeeze an object. (c) Tension is created when you pull on the opposite ends of an object. (d) Shear stress occurs when you slide one surface of an object relative to the other surface (we depict the shear direction with half arrows).

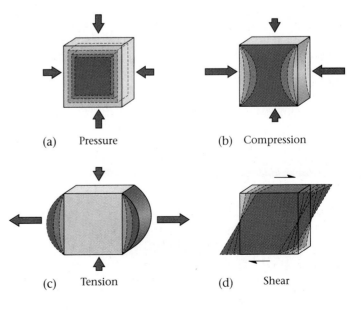

(a)   Pressure   (b)   Compression

(c)   Tension   (d)   Shear

BOX 11.1

THE REST
OF THE
STORY

Describing the Orientation of Structures

When discussing geologic structures, it's important to be able to communicate information about their orientation. Does a fault exposed in an outcrop at the edge of town extend beneath the nuclear power plant 3 km to the north, or does it extend beneath the hospital 2 km to the east? If we knew the fault's orientation, we might be able to answer this question. To describe the orientation of a geologic structure, geologists picture the structure as a simple geometric shape, then specify the angles that the shape makes with respect to a horizontal plane (a flat surface parallel to sea level), a vertical plane (a flat surface perpendicular to sea level), and the north direction (a line of longitude).

Let's start by observing planar structures like faults, beds, and joints. We call these structures planar because they resemble the geometric shape known as a plane. A planar structure's orientation can be specified by its strike and dip (▶Fig. 11.11a, b). The **strike** is the angle between an imaginary horizontal line (the **strike line**) on the plane and the direction to true north. We measure the strike with a magnetic compass. The **dip** is the angle of the plane's slope (more precisely, the angle between a horizontal plane and the **dip line,** an imaginary line parallel to the steepest slope on the plane, as measured in a vertical plane perpendicular to the strike). We measure the dip angle with an **inclinometer,** a type of protractor that measures slope angles. A horizontal plane has a dip of 0°, and a vertical plane has a dip of 90°. We represent strike and dip on a geologic map using the symbol shown in ▶Figure 11.11b.

Another kind of structure resembles a line rather than a plane; these linear structures include needle-shaped mineral crystals and scratches on a rock surface. Geologists specify the orientation of lines by giving their plunge and bearing (▶Fig. 11.11c). The **plunge** is the angle between the line and an imaginary horizontal plane, as measured with an inclinometer in the vertical plane that contains the line. A horizontal line has a plunge of 0°, and a vertical line has a plunge of 90°. The **bearing** is the compass heading of the line (more precisely, the angle between the projection of the line on the horizontal plane and the direction to true north).

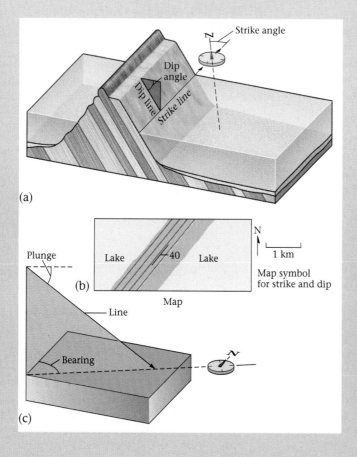

(a)

(b)

Map

(c)

FIGURE 11.11 (a) We use strike and dip to measure the orientation of planar structures like these tilted beds. The strike is the compass angle between the strike line (an imaginary horizontal line on the plane) and true north. The dip is the angle between the strike line and the dip line (an imaginary line parallel to the steepest slope on the plane). Note that the strike line and the dip line are perpendicular to each other. (b) On a map, the line segment represents the strike direction, while the tick on the segment represents the dip direction. The number indicates the dip angle as measured in degrees. (c) To specify the orientation of a line, we use plunge and bearing. The plunge is the angle between the line and an imaginary horizontal plane, while the bearing is the compass orientation of the line.

(a)

(b)

(c)

FIGURE 11.12 (a) This bedding plane in sandstone of Arches National Park, Utah, contains many systematic joints. (b) These vertical joints exposed on a cliff face near Ithaca, New York, run from the surface down into bedrock. (c) The veins in this outcrop, composed of milky white quartz, fill fractures in gray shale.

joints are called **veins** and look like white stripes cutting across a body of rock (▶Fig. 11.12c). Some veins contain small quantities of valuable metals, like gold.

Geological engineers, people who study the geologic setting of construction sites, pay close attention to jointing when recommending where to put roads, dams, and buildings. Water flows much more easily through joints than it does through solid rock, so it would be a bad investment to situate a water reservoir over rock with closely spaced joints—the water would leak down into the joints. Also, building a road on a steep cliff composed of jointed rock could be risky, for joint-bounded blocks separate easily from bedrock, and the cliff might collapse.

11.5 FAULTS: FRACTURES ON WHICH SLIDING HAS OCCURRED

After the San Francisco earthquake of 1906, geologists found a rupture that ripped across the landscape near the city. Where this rupture crossed orchards, it offset rows of trees, and where it crossed a fence, it broke the fence in two and moved the west side of the fence north by about 2 m (Fig. 10.5a). The rupture represented the trace of the San Andreas Fault. As we have seen, a **fault** is a fracture on which sliding occurs (another brittle structure); rock on one side of a fault moves along the fault surface relative to rock on the other side, creating a shear strain (▶Fig. 11.13a, b). Slip events, or **faulting,**

FIGURE 11.13 (a) The San Andreas Fault displacing rows of trees in an orchard. (b) An outcrop in the Rocky Mountains of Colorado, showing a fault offsetting strata in cross section.

(a)

(b)

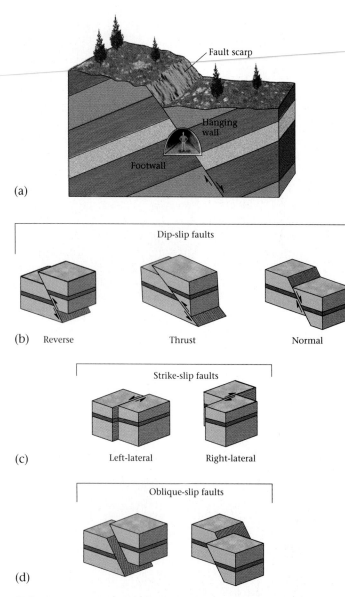

FIGURE 11.14 (a) A hanging-wall block and footwall block, relative to a sloping fault surface. The fault scarp is an exposure of the fault at the ground surface. (b) Three types of dip-slip faults, on which sliding parallels the dip line. (c) Two types of strike-slip faults, on which sliding parallels the strike line. (d) Two types of oblique-slip faults, on which sliding takes place diagonally along the surface.

generate earthquakes. Faults, like joints, are planar structures, so we represent their orientation by strike and dip.

Faults riddle the Earth's crust. Some are currently active (sliding has been occurring on them in recent geologic time), but most are inactive (sliding on them ceased millions of years ago). Some faults, like the San Andreas, intersect the ground surface and thus displace the ground when they move. Others accommodate the sliding of rocks in the crust at depth, and remain invisible at the surface unless later exposed by erosion.

Geologists study faults not only because the movement on some faults causes earthquakes, but also because they juxtapose bodies of rock that did not originally lie adjacent to each other and thus complicate the arrangement of rocks at the Earth's surface. (For example, in our Alpine cliff face, movement on a fault placed quartzite and slate beds against marble beds.) We must understand these rearrangements in order to make geologic maps and to predict where resources lie underground.

Fault Classification

Over the years, geologists have developed terminology to classify faults and describe movement on them. The fault plane can be vertical, horizontal, or at some angle in between, and we can describe its orientation by a strike and dip measurement. As we learned in Chapter 10, in the case of nonvertical faults (those that slope at an angle), we define the **hanging-wall block** as the rock above the fault plane, and the **footwall block** as the rock below the fault plane (▶Fig. 11.14a). If you stand in a tunnel along a fault plane, the hanging-wall block looms over your head, and the footwall block lies under your feet. We can distinguish several types of faults.

- *Dip-slip vs. strike-slip vs. oblique-slip faults:* On **dip-slip faults,** sliding occurs up or down the slope of the fault (therefore, up or down the dip); on **strike-slip faults,** one block slides past another (therefore, parallel to the strike line; and on **oblique-slip faults,** sliding occurs diagonally along the fault plane (▶Fig. 11.14b–d).

- *Types of dip-slip faults:* We subdivide dip-slip faults into two kinds, depending on which way the hanging-wall block moves relative to the footwall block. On **thrust faults** and **reverse faults,** the hanging-wall block moves up the slope of the fault. Thrust faults differ from reverse faults only in terms of the fault-plane's slope (or dip)—thrust faults have a slope (or dip) of less than about 35°, while reverse faults have a slope of greater than 35°. On **normal faults,** the hanging-wall block moves down the slope of the fault. "Normal" and "reverse" are relics of nineteenth-century miners' jargon. Normal faults were more common in the mines then, but globally, normal faults aren't any more common or typical than reverse faults.

- *Types of strike-slip faults:* Geologists distinguish between two types of strike-slip faults, based on the relative movement of one side of the fault with respect to the other. If you stand facing the fault, you can say that it is a **left-lateral strike-slip fault** if slip moved the block on the far side to your left, and that it is a **right-lateral strike-slip fault** if the block on the far side moved to your right. Note that strike-slip faults commonly have a vertical dip, so we cannot define the hanging-wall or footwall block.

Recognizing Faults

How do you recognize a fault when you see one? The most obvious criterion is the appearance of **displacement,** or **offset,** meaning the amount of movement across a fault plane (Fig. 11.13b; ▶Fig. 11.15a, b). Displacement disrupts the layers in rocks, so that layers on one side of a fault are not continuous with layers on the other side; in fact, the rock unit on one side may be entirely different from the rock unit on the other side. In our Alpine cliff example, we can spot the fault as the plane where quartzite and slate beds are juxtaposed against limestone beds. Typically, thrust or reverse faults cutting sedimentary beds place older beds on younger ones, while normal faults place younger beds on older. In some cases, layers of rock cut by a fault undergo folding during or just before slip; the resulting folds are informally called **drag folds.**

Faults may also leave their mark on the landscape. Those that intersect the ground surface while they are active can displace natural landscape features (such as stream valleys or glacial moraines) and human-made features (such as highways, fences, or rows of trees in orchards). Uplift of the hanging-wall block relative to the footwall block (on a dip-slip or oblique-slip fault) will make a small step on the ground surface; this step is called a **fault scarp** (▶Fig. 11.15c). And because faults tend to break up rock, the fault may be eroded; if this happens, the fault trace (the line of intersection between the fault and the ground surface) will be marked by a linear valley. Even if a fault did not intersect the ground surface while active, it may influence the landscape later in geologic history. For example, if the fault movement juxtaposed strong rock with weak rock, long-term erosion will produce a scarp on the landscape, marking the trace of the fault. Because of faults' influence on landscapes, geologists can identify them from air photos or satellite images.

Fault surfaces and their borders typically look different from bedding planes. For example, faulting under brittle conditions may crush or break adjacent rock. If this shattered rock consists of visible angular fragments, then it is called **fault breccia** (▶Fig. 11.15d), but if it consists of a fine powder, then it is called **fault gouge.** Some fault sur-

FIGURE 11.15 (a) A thrust fault, on which a distinct layer has been offset. (b) A geologist's sketch emphasizes the offset. (c) A fault scarp formed after an earthquake in Nevada. (d) This fault breccia along a fault consists of broken-up rock. (e) Slip lineations on a fault surface. (f) Mylonite formed in a shear zone. Note the fine grain size of the rock.

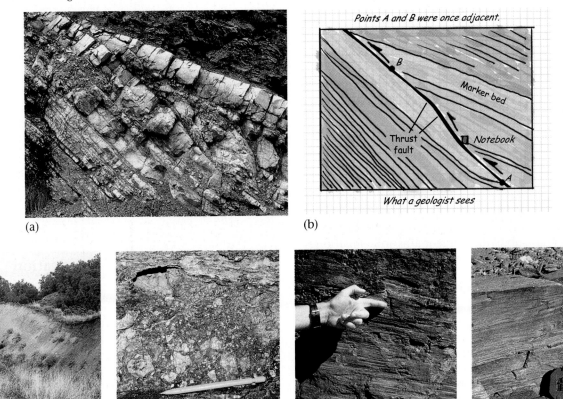

(a)

(b)

(c)   (d)   (e)   (f)

FIGURE 11.16 This large thrust fault puts older rock (Precambrian) over younger rock (Mesozoic). Erosion has removed much of the hanging-wall block, but a small remnant still lies to the east of the mountains. On the geologic map of the region, the triangular barbs point to the hanging-wall block.

faces are polished and grooved by the movement of the hanging wall past the footwall. Polished fault surfaces are called **slickensides,** and linear grooves on fault surfaces are **slip lineations** (▶Fig. 11.15e). Slip lineations, which indicate the direction of slip, can form when bumps protruding from one wall of the fault plow a furrow into the opposing wall as the rock moves. We specify the orientation of a slip lineation by giving its plunge and bearing (see Box 11.1).

The shear on some faults takes place at depth in the crust, under ductile conditions. Where this happens, rock does not break up into breccia or gouge along the fault zone, but rather shears ductilely to form a band of fine-grained foliated rock called **mylonite** (▶Fig. 11.15f). The fine grain size of mylonite results not from brittle fracturing during shear, but rather from a type of metamorphic recrystallization that subdivides large grains into small ones. Geologists refer to faults in which movement occurred ductilely as **shear zones.**

Faults on Geologic Maps

Faults can intersect the ground surface, either because they broke and offset the ground when active or because they were exposed by erosion. In either case, geologists represent the fault trace on a map by a heavy line. On steeply dipping

dip-slip faults, we indicate the upthrown block with a U and the downthrown block with a D. We portray a thrust fault by a line with triangular barbs on it—the barbs point to the hanging-wall block (▶Fig. 11.16). And we show strike-slip faults with two half arrows whose orientation indicates the direction of shear.

Fault Systems and the Resulting Strain

Numerous related faults are often found in groups called **fault systems.** Individual faults in a system may merge at depth with a nearly horizontal **detachment fault.** The faulting in a thrust-fault system shortens the crust; slip makes slices of the crust overlap like shingles (▶Fig. 11.17a). Faulting in normal-fault systems, on the other hand, stretches the crust (▶Fig. 11.17b). In this system, elongate basins separate high blocks of rock. Some individual basins may be bounded on each side by a normal fault, with each fault dipping toward the basin. Such a fault-bounded basin is called a **graben;** the high block between two graben is a **horst** (▶Fig. 11.18a, b). Typically, normal faults grow shallower with depth and merge with a detachment fault. As the hanging-wall block above such a curved fault slides down the fault, a triangular basin called a **half-graben** develops between the fault and the top surface of the rotated block.

11.6 FOLDS: CURVING ROCK LAYERS

Imagine a carpet lying flat on the floor. Push on one end of the carpet, and it will wrinkle or contort into a series of wave-like curves. Layers of rock can do the same: planar layers can be contorted into curves, which geologists call **folds.** In other words, a fold consists of curving rock layers (a kind of ductile deformation). The layers can be beds of sedimentary rock, flows of extrusive igneous rock, or banding or schistosity of metamorphic rock. Folds may also be defined by curving dikes, sills, or veins.

Fold Terminology and Classification

Not all folds look the same—some look like an arch, some like a trough, and some have other shapes. To describe these shapes, we first label the parts of a fold. The **hinge** refers to the portion of the fold where curvature is greatest, and the **limbs** are the sides of the fold that show less curvature. The **axial plane** is an imaginary surface that encompasses the hinges of successive layers. With these terms in hand, we can now describe types of folds.

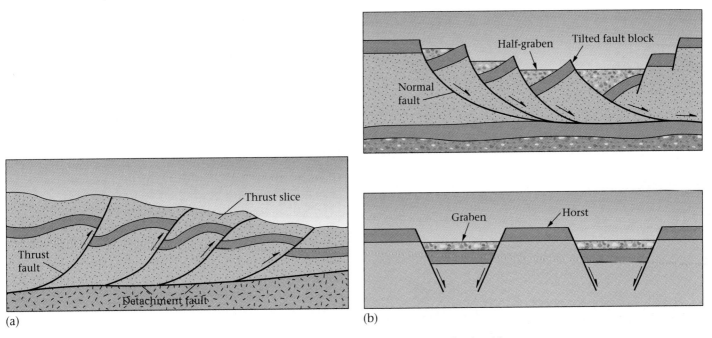

(a)

(b)

FIGURE 11.17 (a) In this thrust-fault system, several related thrust faults merge at depth with a detachment fault. Note that displacement on the thrusts shortens the layers of rock above the detachment. (b) A normal-fault system consists of several related normal faults. Note that the faults merge with a detachment fault at depth and that displacement on them stretches the layers of rock above the detachment. The rock layers have tilted as a consequence of slip on the faults. The basins bounded by faults are grabens, and the block between two grabens is a horst; the wedge-shaped depressions bounded by a fault on one side and the tilted rock layer on the other are half-grabens.

FIGURE 11.18 (a) Horsts and grabens exposed in a quarry wall in Brazil. (b) A geologist's sketch of the quarry wall, indicating the positions of the faults.

(a)

(b)

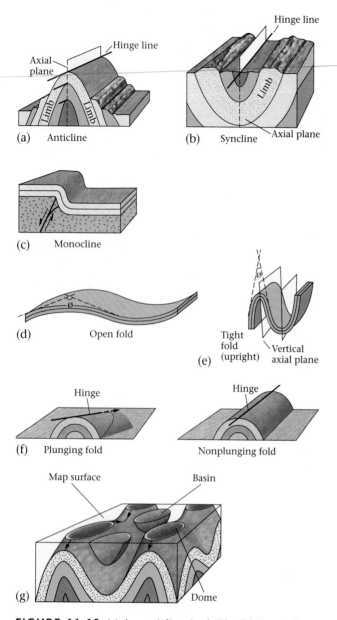

FIGURE 11.19 (a) An anticline (arch-like fold) and (b) a syncline (trough-like fold), showing the hinge, limbs, and axial plane. (c) A monocline. Note how it resembles a stair step. (d) An open fold has a large angle, ø, between the limbs. (e) A tight fold has a small angle, ø, between the limbs. (f) If the hinge of a fold is tilted, it's a plunging fold; if the hinge is horizontal, it's a nonplunging fold. (g) Domes (dome-shaped folds) and basins (bowl-shaped folds).

• *Anticlines, synclines, and monoclines:* Folds that have an arch-like shape in which the limbs dip away from the hinge are called **anticlines,** while folds with a trough-like shape in which the limbs dip toward the hinge are called **synclines** (▶Fig. 11.19a, b). A **monocline** has the shape of a carpet draped over a stairstep (▶Fig. 11.19c).

• *Open and tight folds:* If the angle between the limbs is large, then the fold is an **open fold,** but if the angle between the limbs is small, then the fold is a **tight fold** (▶Fig. 11.19d, e).

• *Nonplunging and plunging folds:* If the hinge is horizontal, the fold is called a **nonplunging fold,** but if the hinge is tilted, the fold is called a **plunging fold** (▶Fig. 11.19f).

• *Domes and basins:* A fold with the shape of an overturned bowl is called a **dome,** while a fold shaped like a right-side-up bowl is called a **basin** (▶Fig. 11.19g). Some small domes and basins are found at outcrops, but others measure hundreds of kilometers across.

Now see if you can use this terminology to identify the various folds shown in ▶Figure 11.20a–d.

Folds on Geologic Maps

You can recognize folds on a geologic map by the pattern of rock units. For example, a nonplunging anticline involving sedimentary beds appears as a series of parallel stripes, with the oldest layer in the center and progressively younger layers away from the center; the stripes are symmetrically positioned around the hinge. In a nonplunging syncline involving sedimentary beds, the youngest rocks crop out in the center and the oldest at the margins (▶Fig. 11.21a, b). Layers in a plunging fold have a U-shape on the map surface (▶Fig. 11.21c). We can represent the hinge of the fold by a heavy line bordered by outward-pointing arrows for an anticline and inward-pointing arrows for a syncline. Domes and basins both show circular outcrop patterns that look like a bull's-eye—the oldest layer occurs in the center of a dome, while the youngest layer is located in the center of a basin (▶Fig. 11.21d, e).

Forming Folds

Folds develop in two principal ways. During **flexing,** a stack of layers bends, and slip occurs between the layers (▶Fig. 11.22a). The same phenomenon happens when you bend a deck of cards—to accommodate the change in shape, the cards slide with respect to each other. **Flow folds** form when the rock overall is so soft that it behaves like weak plastic; these folds develop simply because different parts of the rock body move at different rates (▶Fig. 11.22b). (You can see the same process happen when scum or bubbles on the surface of a slow-moving river twist into complex curves.)

FIGURE 11.20 (a) An open anticline, exposed in a road cut near Kingston, New York. (b) An open syncline, exposed in a road cut near Sideling Hill, in Maryland. (c) A tight fold exposed along the coast of Brazil. (d) A train of folds exposed in sea cliffs in eastern Ireland. Note that the axial planes of these folds are not vertical.

FIGURE 11.21 (a) After erosion, folds involving alternating strong and weak rock layers may control the positions of valleys and ridges on the land surface. Commonly, valleys form over synclines and in the hinge area of anticlines. (b) On the map surface of a nonplunging anticline and syncline, the same units appear on either side of the hinge. (c) In the map view of a plunging anticline, the units curve around the hinge. (d) A dome. (e) A basin.

FIGURE 11.22 (a) This fold developed by flexing (the layers slid past each other like cards in a bending deck). (b) This fold developed by flowing. The lines of dots represent the boundaries of a layer. During flow, the lines move at somewhat different velocities, so the layer becomes folded.

Folds develop for a variety of reasons. Some layers wrinkle up, or **buckle,** in response to end-on compression (▶Fig. 11.23a). Others form where shear stress gradually moves one part of a layer up and over another part (▶Fig. 11.23b). Still others develop where rock layers move up and over bends in a fault and must curve to conform with the fault's shape (▶Fig. 11.23c). Finally, some folds form when a block of basement moves and bends the overlying sedimentary layers (▶Fig. 11.23d).

FIGURE 11.23 Different causes of folding. (a) If a layer becomes shortened along its length, it buckles (wrinkles up like a rug). (b) If a layer is sheared, it gradually bends over on itself to form a fold. (c) When layers move up and over step-shaped faults, they must bend into folds. (d) Faulting at depth may fold a layer closer to the ground surface. The folded layers drape over the uplifted fault block.

11.7 TECTONIC FOLIATION IN ROCKS

In an undeformed sandstone, the grains of quartz are roughly spherical, and in an undeformed shale, clay flakes press together into the plane of bedding so that shales tend to split parallel to the bedding. During ductile deformation, however, internal changes take place in a rock that gradually modify the original shape and arrangement of grains. For example, quartz grains may transform into cigar shapes, elongate ribbons, or tiny pancakes, and clay flakes may recrystallize or reorient so that they lie at an angle to the bedding. Overall, deformation causes inequant grains to align parallel with one another, thereby generating metamorphic foliation in the rock. We refer to layering created by the alignment of deformed and/or reoriented grains as **tectonic foliation** (▶Fig. 11.24a). Note that tectonic foliation can be oriented at a large angle to the original bedding.

We were introduced to foliations such as slaty cleavage, schistosity, and gneissic layering in Chapter 8. Here we add to the story by noting that they form in response to flattening and shearing in ductilely deforming rocks—in other words, foliations indicate that the rock has developed a strain. In rocks with slaty cleavage, the cleavage planes lie perpendicular to the direction of the shortening strain, because shortening flattens the grains. Thus, the cleavage may be parallel to the axial plane of folds that developed in association with the cleavage. In schists and gneisses, the foliation commonly lies parallel to or at a slight angle to the direction of shear, because shear smears grains out into the plane of shearing (▶Fig. 11.24b–f).

11.8 ROCK FORMATION DURING OROGENY

Many kinds of rock would never have formed were it not for orogeny. To start with, orogeny leads to various kinds of igneous activity (▶Fig. 11.25). This is because orogens form along convergent plate boundaries, accompany rifting, or result from continental collision—all environments where igneous activity takes place. Specifically, at convergent plate boundaries, subduction causes a volcanic arc to form, creating intrusive and extrusive igneous rocks. In rifts, the thinning of the lithosphere decompresses the upper mantle and generates magmas that seep to the surface along normal faults; these faults either inject into the crust as dikes or sills, or extrude onto the surface in lava flows. And during continental collisions, the last gasps of subduction may generate magmas that intrude into the collision zone.

FIGURE 11.24 The development of tectonic foliation in rock. (a) Compression of a layer causes it to shorten in one direction and lengthen in the other. Quartz grains flatten, and clay grains reorient. As a result, the rock develops cleavage. (b) Slaty cleavage oriented parallel to the axial plane of a fold and perpendicular to the direction of shortening. (c) A stream cut showing axial-planar cleavage. (d) A geologist's sketch of the stream cut. (e) Schistosity oriented at a low angle to the direction of shear. Note how large grains are all parallel to each other—they rotated during shear. (f) A geologist's sketch of the outcrop in (e), showing shear movement.

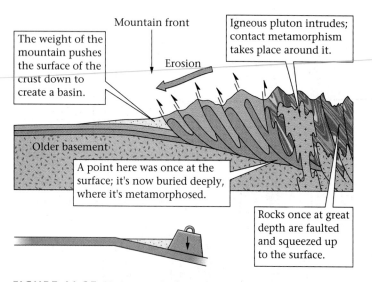

FIGURE 11.25 Various rocks form during orogeny. Sediment eroded from the orogen fills a sedimentary basin next to the mountain, rocks buried deeply in the orogen become squeezed, sheared, and heated to form metamorphic rocks, and igneous rocks intrude from below.

The weathering and erosion of mountains generate vast quantities of sediment that tumble down slopes or are carried away by glaciers or mountain streams. Sediment eroded from the mountains eventually moves to the **mountain front,** the boundary between the mountain range and the plains, where it accumulates in alluvial fans. The weight of orogens pushes down on the surface of the crust, creating a deep sedimentary basin at the mountain front.

Both contact metamorphism and regional metamorphism happen in association with orogeny. Contact metamorphic aureoles (see Chapter 8) border the igneous plutons of orogens, where country rock gets heated up. And regional metamorphism occurs because rock once near the surface is buried deeply during orogeny, when thrust faulting places one part of the crust over another. Once buried, the weight of the orogen squeezes the rock at depth. In addition, the deeply buried rock heats up, because the temperature of the crust increases with depth. These regional changes in environment, along with the shearing that accompanies fault movement, lead not only to metamorphic recrystallization, but also to the development of metamorphic foliation.

11.9 UPLIFT AND THE FORMATION OF MOUNTAIN TOPOGRAPHY

Leonardo da Vinci, the Renaissance artist and scientist, enjoyed walking in the mountains, sketching rock ledges and examining the rocks he found there. In the process, he discovered marine shells (fossils) in limestone beds cropping out a kilometer above sea level, and suggested that the rock containing the fossils had risen from below sea level up to its present elevation. Contemporary geologists agree with Leonardo, and now refer to the process by which the surface of the Earth moves vertically from a lower to a higher elevation as **uplift.** Mountain building requires substantial uplift of the Earth's surface (▶Fig. 11.26).

What kinds of distances are we talking about when referring to uplift in mountain ranges? As noted earlier, Mt. Everest rises 8.85 km above sea level. Although this distance may seem monumental—that's equal to 5,000 people standing one on top of the other—it represents only about 0.06% of the Earth's diameter. In fact, if the Earth were shrunk to the size of a billiard ball, its surface (mountains and all) would feel smoother than that of an actual billiard ball. Note that, in general, the individual peaks that you see in a mountain range represent only a fraction of its total height, for the plain at the base of the mountains may be significantly higher than sea level. Nevertheless, mountain heights are spectacular, and in this section we look at why uplift occurs, how erosion carves rugged landscapes into uplifted crust, and why mountains can't get much higher than Everest.

Crustal Roots and Mountain Heights

In the mid-1800s, Sir George Everest undertook a survey of India, which at the time formed part of the British Empire. Sir Everest, aside from contributing his name to Earth's highest mountain, discovered that the mass of the Himalaya Mountains was great enough to deflect the plumb bob (a

FIGURE 11.26 There are over 2 km of vertical relief between the base of Wyoming's Grand Teton Mountains and the peak. And the base already lies at a high elevation. The exposed rocks once lay 12 km or more beneath the surface of the Earth. Clearly, mountain building involves large vertical displacements of the surface of the crust.

lead weight at the end of a string) he was using to determine the vertical direction when setting up surveying equipment. But surprisingly, the amount of deflection caused by the mountains was actually much *less* than it should have been, given their size. Why? A nineteenth-century British scientist, George Airy, keeping in mind that Earth's crustal rocks are less dense than its mantle rocks, came up with an explanation. Airy suggested that the crust of the Earth is thicker beneath the Himalayas than elsewhere, and thus that a low-density **crustal root** protrudes downward into the dense mantle beneath the range. A mountain range with a low-density crustal root has less mass overall than a mountain range underlain by dense mantle, and so exerts less pull on a plumb bob.

Work in the twentieth century has confirmed that collisional and convergent-boundary mountain belts do sit above crustal roots. Whereas typical continental crust has a thickness of about 35–40 km (measured from the surface to the Moho), the crust beneath some mountain belts may reach a thickness of 50–70 km (about double its normal thickness) (▶Fig. 11.27a). Mountain building shortens the crust horizontally and thickens it vertically. Crustal roots are important because without their buoyancy, mountain ranges would not be so high. To see why, try an experiment in a bathtub.

Prepare four 6-cm-thick wooden blocks, two 8-cm-thick blocks, and one 12-cm-thick block. Line the blocks up in a row, with the thinner (shorter) blocks on the outside, in a tub full of water (▶Fig. 11.27b). You'll find that the tops of the thinner blocks lie at a lower elevation than the tops of the thicker blocks. That's because when you place a block of wood in water, it experiences a **buoyancy force** that pushes the block up until, as stated in **Archimedes' principle,** the mass of the water displaced by the block equals the mass of the whole block (see Box 4.1). Since wood (0.8 gm/cm^3) is less dense than water (1.0 gm/cm^3), a certain volume of water has the same mass as a larger volume of wood. Thus, wood floats and juts out of the water (20%, or 1.2 cm, of the 6-cm-thick blocks protrudes, while 2.4 cm of the 12-cm-thick block protrude).

The row of floating blocks resembles the cross section of a mountain range, with the highest part of the range lying over the thickest crustal root. Mountain belts are like icebergs, with only a small part of their mass above sea level and the bulk below. The extra thickness of continental crust beneath mountain ranges is the reason for their high elevations.

Keep in mind that it's not just the continental crust that floats on the mantle. Rather, it's the continental lithosphere as a whole that floats on the asthenosphere. The crust acts like a buoy that holds the lithosphere up; the bigger the buoy (the thicker the crust), the higher the lithosphere floats.

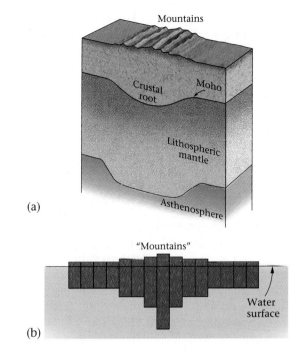

(a)

(b)

FIGURE 11.27 (a) Mountain belts have crustal roots, meaning that where the land surface rises to a higher elevation, the crust underneath is thicker. (b) In general, the lithosphere obeys Archimedes' principle of buoyancy. The surface of a thick buoyant block rises to a higher elevation than the surface of a thinner buoyant block. Also, the base of a thick block extends down to greater depths.

The Concept of Isostacy

The condition that exists when the buoyancy force pushing lithosphere up equals the gravitational force pulling lithosphere down is called **isostacy,** or **isostatic equilibrium.** In most places, isostatic equilibrium exists at the surface of the crust, so that the surface elevation of the crust reflects the level at which the lithosphere naturally floats. If a geologic event happens that changes the density or thickness of the lithosphere, then the surface of the crust slowly rises or falls to reestablish isostatic equilibrium, a process called **isostatic compensation.**

To picture the process of isostatic compensation, let's return to the bathtub. If you lay a 4-cm-thick block of wood on top of a 6-cm-thick block, the base of the block sinks to compensate for the load until isostacy has been reestablished. Because wood is less dense than water, the surface of the resulting 10-cm-thick block will lie at a higher elevation than did the surface of the 6-cm-thick block. If you lift the 4-cm-block off, the surface of the 6-cm-block rises. We see this same phenomenon happen at the surface of the Earth. For example, when glaciers grow during ice ages, their weight depresses the surface of the crust, and when they melt away after an ice age, the surface of the crust rises. Such isostatic compensation on Earth takes a long time, for asthenosphere must flow out of the way when lithosphere sinks and must flow back under the lithosphere when lithosphere rises, and asthenosphere can flow only very slowly because it is highly viscous.

The process of mountain uplift is effectively a manifestation of isostatic compensation. That is, uplift associated with collision or convergence occurs because rock deformation substantially thickens the crust, creating a crustal root. The thicker crust rises and floats higher, as the base of the crust sinks (like the blocks of wood).

Sculpting Mountains by Erosion

The image we most often associate with a mountain range is that of rugged topography with spire-like peaks, knife-edge ridges, precipitous cliffs, and deep valleys. This type of landscape is a consequence of erosion by ice and water, which, over time, sculpts uplifted land. The specific style of topography found within a mountain range depends on the climate and on various geologic factors.

The climate determines whether glaciers or rivers erode a mountain. If conditions in the mountains become cool enough, glaciers form and, as they flow, carve pointed peaks and steep-sided valleys, as we will see in Chapter 22. But in many ranges, we can see glacially carved landscapes even though we can't see the glaciers. Such landscapes formed during the last ice age and stopped forming when the ice melted away, less than 14,000 years ago (▶Fig. 11.28a). At lower elevations or in warmer climates, mountain landscapes reflect the consequences of river erosion (▶Fig. 11.28b). Soil

formation and vegetation growth may blunt escarpments, creating rounded hills. In desert regions, mountains are typically covered with rubble and expose steep rock escarpments, for thick soil does not develop, and rock debris falling off cliffs litters their slopes. Geologic factors such as bedrock composition also affect topography, for resistant rock units (quartzite or granite, for example) typically stand up as high ridges, whereas weak rock units (like shale) tend to erode.

In some cases, the geometry of geologic structures affects the shape of mountains. Mountain ridges formed by tilted layers of rock typically have an asymmetric shape, with a steep face on one side (where a cliff cuts across a layer) and a gentle slope on the other side (parallel to the layering, where the top layer becomes the surface of a hill). Such an asymmetric ridge is a **cuesta** (▶Fig. 11.29). Cuestas to the east of Paris helped France defend itself against Germany during World War I; the gentle slopes dipped toward Paris, while the steep cliffs faced Germany. The German army was trapped at the base of the cliffs.

What Goes Up Must Come Down: Uplift Has Its Limits

Mountains much higher than Mt. Everest cannot exist on the Earth, for two reasons. First, erosion attacks mountains as soon as they rise, and in some cases it can tear mountains down as fast as uplift occurs. Second, rock does not have infinite strength. As mountains rise, the weight of overlying rock presses down on the rock that is now buried at depth in the crust. Further, the rock buried at depth gradually grows

FIGURE 11.28 Evidence of erosion in mountain ranges. (a) Glacially carved peaks in Switzerland; (b) a deep, river-cut gorge in the Himalayas.

(a)

(b)

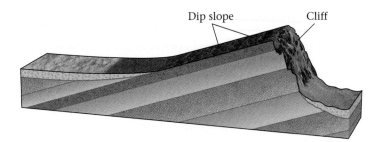

FIGURE 11.29 A cuesta is an asymmetric ridge underlain by dipping strata. Note the contrast between the shallow dip slope and the cliff face.

warmer and softer because of heat rising from Earth's interior. Eventually, the weight of the mountain range causes warm, soft rock at depth to flow slowly. Effectively, the mountains begin to collapse under their own weight and spread laterally like soft cheese that has been left out in the summer sun (►Fig. 11.30a, b). Geologists call this process **orogenic collapse.** Together, erosion and orogenic collapse expose metamorphic and plutonic rocks that were once deep in the crust, a process called **exhumation.**

A mountain range can exist only if the rate of uplift exceeds the rate of erosion. Once the rate of erosion is faster, the mountain range gradually erodes away, perhaps eventually down to sea level. As each kilometer is eroded from the top of the range, the base of the range rises by about one-third of a kilometer, to maintain isostacy. Thus, with time, the excess crustal thickness in the orogen is slowly removed.

FIGURE 11.30 (a) Soft cheese spreads out sideways as it warms up. (b) Similarly, mountain belts spread out sideways once they reach a certain thickness. The ductile crust at depth flows, while the upper (brittle) crust is broken by normal faults.

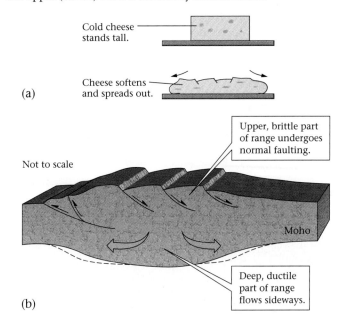

11.10 CAUSES OF MOUNTAIN BUILDING

Before plate tectonics theory was established, geologists were just plain confused about how mountains formed. For example, some geologists thought that mountains rose in response to the cooling of the Earth; they argued that as it cools, the Earth shrinks, so its crust wrinkles like the skin of an old apple. But this idea had to be discarded when the discovery of natural radioactivity led to the realization that the Earth is not cooling rapidly, for radioactive decay produces heat. Plate tectonics theory provided a new approach to interpreting mountains: as noted earlier, mountains form in response to convergent boundary/continental margin interactions, continental collisions, and rifting. Below, we look at these different settings and the types of mountains and geologic structures that develop in each one.

Mountains Related to Convergence

Along the margins of continental convergent plate boundaries, where oceanic lithosphere subducts beneath a continent, a continental volcanic arc forms and compression between the two plates causes a mountain range to rise (►Fig. 11.31a). Some convergent-margin orogens last as long as 200 million years. During this time, offshore island volcanic arcs, oceanic plateaus, and small fragments of continental crust may drift into the convergent margin. These blocks are too buoyant to subduct, so they collide with the convergent margin and **accrete,** or attach, to the continent. Geologists refer to such blocks before they attach as **exotic terranes** and once they have attached as **accreted terranes.** Once an exotic terrane accretes to a continent, a new subduction zone typically forms on the ocean side of the terrane, and a new volcanic arc develops on top of the terrane. In some convergent-margin orogens, numerous exotic terranes attach over time, making the continent grow laterally (to the side). The Western half of the North American Cordillera consists of accreted terranes (►Fig. 11.31b). Orogens that grow laterally by the attachment of exotic terranes have come to be known as **accretionary orogens.**

If plate movements push the continent tightly against the subduction zone, compression on the continent side of the volcanic arc generates a **fold-thrust belt.** In such belts, a thrust system has developed above a detachment (horizontal) fault, and folds appear as strata are pushed up the fault. Convergence along the west coast of South America, for example, has generated a thrust system on the east side of the Andes. A similar thrust system formed on the east side of the North American Cordillera during the Mesozoic and Cenozoic eras.

(a)

(b)

FIGURE 11.31 (a) In a convergent-margin orogen, volcanic arcs form, and there may be compression. Where this occurs, a large mountain range develops. Island arcs and small continental blocks may collide with the convergent margin and accrete to the orogen. (b) Much of the western portion of the North American Cordillera consists of accreted terranes.

Mountains Related to Continental Collision

Once the oceanic lithosphere between two continents completely subducts, the continents themselves collide with one another. Continental collision results in the creation of large mountain ranges such as the present-day Himalayas or the Alps (►Fig. 11.32) and the Paleozoic Appalachian Mountains. The final stage in the growth of the Appalachians happened when Africa and North America collided.

During collision, intense compression generates fold-thrust belts on the margins of the orogen. In the interior of the orogen, where one continent overrides the edge of the other, high-grade metamorphism occurs, accompanied by flow folds and the formation of tectonic foliation. During this process, the crust below the orogen thickens to as much as twice its normal thickness. Gradually, rocks squeeze upward in the hanging walls of large thrust faults and later become exposed by exhumation. Finally, as noted earlier, rock at depth in the orogen heats up and becomes so weak that the mountain belt may collapse and spread out sideways. The broad Tibet Plateau may have formed when crust, thickened during the collision of India with Asia, spread to the northeast.

Mountains Related to Continental Rifting

Continental rifts are places where continents are splitting in two. When rifts first form, there is generally significant uplift, and this uplift contributes to creating mountainous topography (►Fig. 11.33). Uplift occurs, in part, because as the lithosphere thins, hot asthenosphere rises, making the remaining lithosphere less dense. Because the lithosphere is less dense, it becomes more buoyant and thus rises to reestablish isostatic equilibrium.

As heating in a rift takes place, stretching causes normal faulting in the brittle crust above, creating a normal-fault system. Movement on the normal faults drops down blocks of

FIGURE 11.32 In a collisional orogen, two continents collide. The compression that results from the collision shortens and thickens the continental crust so that a large mountain range develops. Fold-thrust belts form along the margins of the orogen.

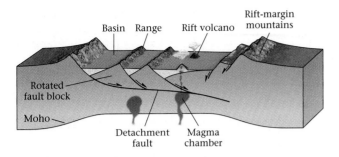

FIGURE 11.33 When the crust stretches in a continental rift, rift-related mountains form, as do normal faults. Displacement on the faults leads to the tilting of crustal blocks and the formation of half-grabens. The half-grabens fill with sediment eroded from the adjacent mountains. Later, the exposed ends on the tilted blocks create long, narrow ranges, and the half-grabens become flat basins. In the United States, the region containing such a structure is called the Basin and Range Province (located in Nevada, Utah, and Arizona).

crust, creating deep basins separated by narrow, elongate mountain ranges that contain tilted rocks. These ranges are sometimes called **fault-block mountains.** In addition, the rising asthenosphere beneath the rift partially melts, generating magmas that rise to form volcanoes within the rift. Today, the East African Rift clearly shows the configuration of rift-related mountains and volcanoes. And in North America, rifting yielded the broad Basin and Range Province of Utah, Nevada, and Arizona. If you drive across the province from east to west, you'll pass over two dozen fault-block mountain ranges, separated from each other by sediment-filled basins.

11.11 CRATONS AND THE DEFORMATION WITHIN THEM

A **craton** consists of crust that has not been affected by orogeny for at least 1 billion years. We can further divide cratons into two provinces: **shields,** in which Precambrian metamorphic and igneous rocks crop out at the ground sur-

FIGURE 11.34 Topographical map of the United States.

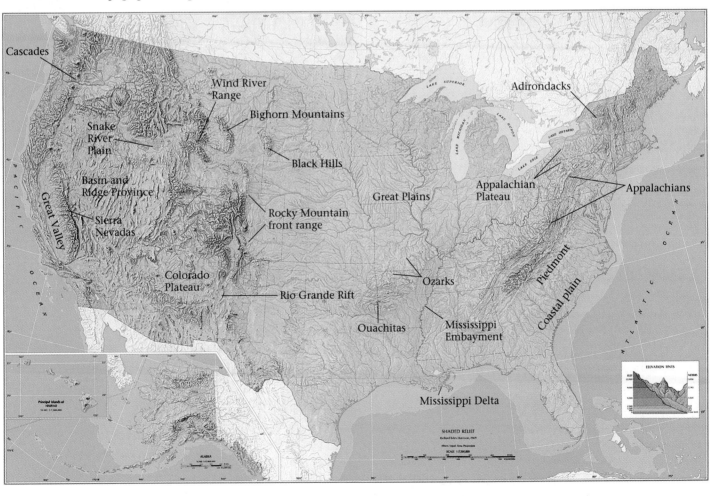

The Collision of India with Asia

The Himalaya Mountains and other important highlands of southern Asia, are a consequence of the collision of India, a small but very old and strong block of continental lithosphere, with Asia about 55 million years ago. At the time of the collision, the southern margin of Asia consisted of several smaller crustal blocks that had become stitched together by recent collisions, and thus was composed of younger, warmer, and softer lithosphere. Since then, the strong lithosphere of India has continued to push slowly into the soft lithosphere of Asia.

N

Ganges Plain

Himalaya Mountains

Small, north-south-trending rifts

Kathmandu

Suture between Indian and Asian Plates

Continental crust

Continental lithosphere

Lithospheric mantle

Indian Plate

Mt. Everest (Sagarmatha)

Normal fault

Thrust faults

Karakoram
Range

Tien Shan
Mountains

Tarim
Basin

Kunlun Mountains

Faults accommodating
a component of
strike-slip motion

Qaidam
Basin

Qilian
Mountains

Asian Plate

Region of thin
lithospheric mantle

Lithospheric mantle
sinking into asthenosphere

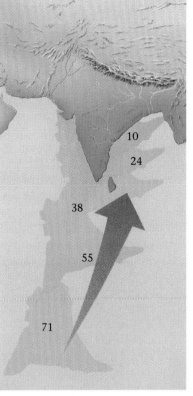

10
24
38
55
71

The collision of India with Asia has uplifted the Himalayas and Tibet. Portions of China and southeast Asia have slipped to the east to "escape" the collision. Faults in central Asia have become active, causing the uplift of ranges like the Tien Shan, as compressive forces build up.

The development of large thrust faults has uplifted the curving Himalayan chain where India begins to be overridden by Asia. Why the broad plateau of Tibet has risen remains something of a mystery. In part, the uplift may be a consequence of the thickening of the crust as it is squashed horizontally; continental crust is relatively weak, and so may spread laterally (like soft cheese in the sun), leading to the formation of normal faults and small rifts in the upper crust and a plastic-like flow in the deep crust. The uplift may also be due to the heating of the region when slabs of the underlying lithospheric mantle drop off and sink, to be replaced by hot asthenosphere.

As India has pushed into Asia, it may have squeezed blocks of China and Southeast Asia sideways, toward the east; this motion is accommodated by slip on strike-slip faults. The collision may also have caused reverse faults in the interior of Asia to become active, uplifting a succession of small mountain ranges, like the Tien Shan.

Tien Shan
Mountains

Tibet Plateau

Himalaya
Mountains

face, and the **cratonic platform,** where a relatively thin layer of Phanerozoic sediment covers the Precambrian rocks (▶Fig. 11.34; see Fig. 13.9).

In shield areas, we find intensively deformed metamorphic rocks—abundant examples of shear zones, flow folds, and tectonic foliation. That's because the cratons formed back in the Precambrian as a result of a succession of orogenies. Recent studies of the Canadian Shield, which occupies much of the eastern two-thirds of Canada, for example, reveal the traces of Himalaya-like collision zones, Andean-like convergent boundaries, and East African–like rifts, all formed more than 1 billion years ago (some over 3 billion years ago). These orogens are so old that erosion has worn away the original topography, in the process exposing deep crustal rocks at the Earth's surface. Also, because orogeny happened so long ago, the crust of the shield area has become quite cool and therefore very strong.

In the cratonic platform, we can't see the Precambrian rocks and structures, except where exposed by deep erosion. Younger strata do display deformation features, but in contrast to the deformation of orogens, cratonic-platform deformation is subtle and localized. In fact, the cratonic platform of the U.S. Midwest region includes two classes of structures: regional basins and domes, and local zones of folds and faults. Their formation yields at most small hills, not high mountains.

Regional basins and **regional domes** are broad areas that gradually sank or rose, respectively (▶Fig. 11.35a, b). Consider a slice of the upper crust running across Missouri and Illinois. In Missouri, strata arch around a broad dome, the Ozark Dome, whose diameter is 400 km. Individual sedimentary layers thin toward the top of the dome, because the dome was a high area while the sediment accumulated. Erosion during more recent geologic history has produced the characteristic bull's-eye pattern of a dome, with the oldest rocks (Precambrian granite) exposed near the center. In the Illinois Basin, strata appear to warp downward into a huge bowl that is also about 400 km across. Strata get thicker toward the center, indicating that the floor of the basin sank while sediment accumulated. The Illinois Basin also has a bull's-eye shape, but here the youngest strata are in the center. Geologists refer to the broad vertical movements that generate huge, but gentle, mid-continent domes and basins as **epeirogeny.**

Folds and faults are hard to find in the cratonic platform, because most have been buried by younger strata. But subsurface studies indicate that faults do occur at depth. Monoclines, step-shaped folds, develop over these faults; the folding appears to be a consequence of the fault movement. Most of these zones were likely active when major orogenies happened along the continental margin. This relation suggests that the orogenies created enough stress in the mid-continent to cause faults to move, but not enough to generate large mountains or to create foliation.

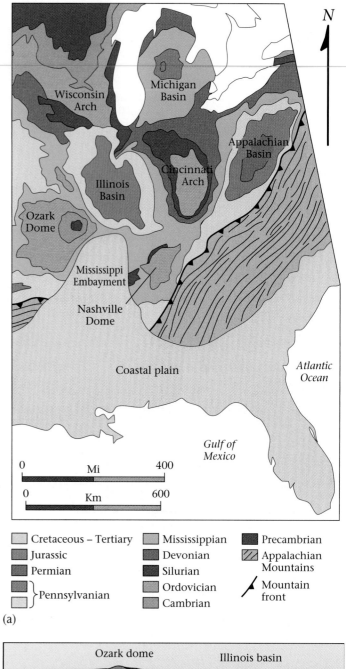

Cretaceous – Tertiary	Mississippian	Precambrian
Jurassic	Devonian	Appalachian Mountains
Permian	Silurian	
} Pennsylvanian	Ordovician	Mountain front
	Cambrian	

(a)

(b)

FIGURE 11.35 (a) Geologic map of the mid-continent region of the United States, showing the basins and domes and the faults that cut the region. (b) This cross section illustrates the geometry of the regional basins and domes. Note how layers of strata thin toward the crest of the Ozarks Dome and thicken toward the center of the Illinois Basin.

Grenville orogeny

Post-Grenville rifting

Paleozoic passive margin

Taconic orogeny

Convergence of Avalonia   Avalonia

Acadian orogeny   Africa

Future Valley and Ridge   Alleghenian orogeny

Mesozoic rifting

Present day

FIGURE 11.36 These idealized stages show the tectonic evolution of the Appalachian Mountains. Note that while mountains do form during rifting events, geologists traditionally assign names only to the collisional or convergent events.

11.12 LIFE STORY OF A MOUNTAIN RANGE: AN EXAMPLE

Perhaps the easiest way to bring together all the information in this chapter is to look at the life story of one particular mountain range—let's take the Appalachian Mountains of North America as our example. Geologists have constructed the range's life story by studying its structures, by determining the ages of igneous and metamorphic rocks, and by searching for strata formed from sediment eroded from the range.

About 1 billion years ago, the Appalachian region was involved in a massive collision with another continent (▶Fig. 11.36). This event, called the **Grenville orogeny,** yielded a belt of deformed and metamorphosed rocks that underlie the eastern fifth of the continent. For a while, the Appalachian region lay in the middle of a supercontinent. But this supercontinent rifted apart around 600 million years ago. Eventually, new ocean formed to the east, and the former rifted margin of eastern North America cooled, sank, and evolved into a passive-margin sedimentary basin. From 600 to about 420 million years ago, this basin filled with a thick sequence of sediment.

Between around 420 and 370 million years ago, two collisions took place between North America and exotic terranes. During the first convergent event, called the **Taconic orogeny,** a volcanic island arc collided with eastern North America, and during the second convergent event, the **Acadian orogeny,** continental crustal slivers accreted to the continent. The accretion of these terranes deformed the sediment that had accumulated in the passive-margin basin, and made the continent grow eastward. Then, 270 million years ago, Africa collided with North America. This event, the **Alleghenian orogeny,** yielded a huge mountain range resembling the present-day Himalayas and created a wide fold-thrust belt along the mountains' western margin. Eroded folds of this belt make up the topography of the present Valley and Ridge Province in Pennsylvania (▶Fig. 11.37). When the Alleghenian orogeny ceased, the Appalachian region once again lay in the interior of a supercontinent (Pangaea), where it remained until about 180 million years ago. At that time, rifting split the region open again, creating the Atlantic Ocean. As you can see from this example, major ranges like the Appalachians incorporate the products of multiple orogenies.

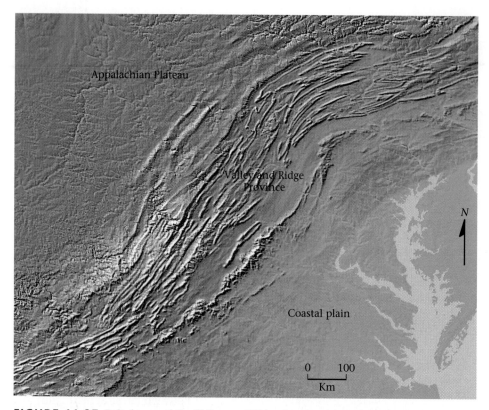

FIGURE 11.37 Relief map of the Valley and Ridge Province, Pennsylvania to Virginia. The ridges, which outline the shapes of plunging folds in the fold-thrust belt, are composed of resistant sandstone beds.

11.13 MEASURING MOUNTAIN BUILDING IN PROGRESS

Mountains are not just "old monuments," as John Muir mused. The rumblings of earthquakes and the eruptions of volcanoes attest to present-day movements in mountains. Geologists can measure the rates of these movements through field studies and satellite technology. For example, geologists can determine where coastal areas have been rising relative to the sea level by locating ancient beaches that now lie high above the water. And they can tell where the land surface has risen relative to a river by identifying places where a river has recently carved a new valley down into sediments that it had previously deposited. In addition, geologists now can use the satellite **global positioning system** (GPS) to measure rates of uplift and horizontal shortening in orogens. With present technology, we can "see" India pressing into Asia at a rate of a couple of centimeters per year, and we can "watch" as mountains along this collisional boundary rise by a couple of millimeters per year.

CHAPTER SUMMARY

• Mountains occur in linear ranges called mountain belts, orogenic belts, or orogens. An orogen forms during an orogeny, or mountain-building event. Orogenies, which last for millions of years, are a consequence of continental collision, subduction at a convergent plate boundary, or rifting. Even after erosion, we can see in linear belts of metamorphic and igneous rock the roots of collisional or convergent-margin orogens.

• Mountain building causes rocks to bend, break, squash, stretch, and shear. Because of such deformation, rocks change their location, orientation, and shape.

• During brittle deformation, rocks crack and break into two or more pieces. During ductile deformation, rocks change shape without breaking. Brittle deformation occurs at relatively low pressures and temperatures, such as those found at shallow depths in the crust (above 10–15 km). Ductile deformation takes place at relatively high temperatures and pressures, deeper in the crust.

- Rocks undergo three kinds of stress: compression (squeezed), tension (pulled apart), and shear (when one side of a rock moves sideways past the other).

- Strain refers to the way rocks change shape when subjected to a stress. Compression causes shortening, tension causes stretching, and shear stress leads to shear strain.

- Deformation results in the development of geologic structures. Brittle structures include joints and faults, while ductile structures include folds and foliation.

- Structures can be visualized as geometric lines or planes. We can define the orientation of a plane by giving its strike and dip and the orientation of a line by giving its plunge and bearing.

- Joints are natural cracks in rock, formed in response to tension under brittle conditions. Some joints develop when rock cools and contracts, others form when erosion decreases the pressure on rocks buried at depth. Systematic joints are long planar cracks that are spaced fairly regularly through a rock body.

- Veins develop when minerals precipitate out of water passing through joints.

- Faults are fractures on which there has been shearing. In the case of nonvertical faults, the rock above the fault plane is the hanging-wall block, and the rock below the fault plane is the footwall block.

- On normal faults, the hanging-wall block slides down the surface; on reverse faults, the hanging-wall block slides up the surface; on strike-slip faults, rock on one side of the fault slides horizontally past the other; and on oblique-slip faults, rock slides diagonally across the surface.

- Faults can be recognized by the presence of broken rock (breccia) or fine powder (gouge). Scratches or grooves on fault surfaces are called slip lineations.

- Several faults together make up a fault system. Thrust-fault systems shorten the crust, while normal-fault systems stretch the crust.

- Folds are curved layers of rock. Anticlines are arch-like folds, synclines are trough-like, monoclines resemble the shape of a carpet draped over a stair step, basins are shaped like a bowl, and domes are shaped like an overturned bowl.

- Tectonic foliation forms when grains are flattened or rotated so that they align parallel with one another, or when new platy grains grow parallel to one another.

- The process of orogeny may yield new igneous, metamorphic, and sedimentary rocks.

- Large mountain ranges are underlain by relatively buoyant roots, which contribute to their height. The height of such mountains is controlled by isostasy, the condition that exists when the buoyancy force pushing lithosphere up equals the gravitational force pushing it down.

- Once uplifted, mountains are sculpted by the erosive forces of glaciers and rivers. Also, when the crust thickens during mountain building, the lower part eventually becomes warm and weak and begins to flow, leading to orogenic collapse.

- Mountain belts formed by convergent margin tectonism may incorporate accreted terranes, blocks of buoyant crust left over when intervening ocean crust has been subducted.

- Continental collision, which resulted in the Alps, Himalayas, and Appalachians, tends to create metamorphic rocks and tectonic foliation. Fold-thrust belts form on the continental edge of collisional and convergent-margin orogens.

- Tilted blocks of crust in rifts become narrow, elongate mountain ranges.

- Cratons are the old, relatively stable parts of continental crust. They include shields, where Precambrian rocks are exposed at the surface, and platforms, where the Precambrian rocks are buried by a thin layer of sedimentary rock. Broad regional domes and basins form in platform areas because of epeirogeny.

- The rocks comprising the Appalachian range were subjected to convergent and collisional orogeny, beginning around 1 billion years ago.

- With modern satellite technology, it is now possible to measure the slow movements of mountains.

KEY TERMS

accreted terrane (p. 339)
anticline (p. 332)
basin (p. 332)
bearing (p. 326)
brittle deformation (p. 323)
compression (p. 325)
craton (p. 341)
cuesta (p. 338)
deformation (p. 321)
detachment fault (p. 330)
dip (p. 326)
dip line (p. 326)
dip-slip fault (p. 328)
dome (p. 332)
drag fold (p. 329)
ductile deformation (p. 323)
epeirogeny (p. 344)
exhumation (p. 339)
exotic terrane (p. 339)
fault (p. 327)
fault-block mountains (p. 341)

fault breccia (p. 329)
fault gouge (p. 329)
fault system (p. 330)
fold (p. 330)
fold-thrust belt (p. 339)
foliation (p. 321)
force (p. 324)
graben (p. 330)
half-graben (p. 330)
hinge (p. 330)
horst (p. 330)
isostasy (isostatic equilibrium) (p. 337)
joint (p. 325)
limbs (of fold) (p. 330)
monocline (p. 332)
mountain front (p. 336)
mylonite (p. 330)
nonplunging fold (p. 332)
normal fault (p. 328)
oblique-slip fault (p. 328)

offset (p. 329)
open fold (p. 332)
orogenic belts (orogens)
   (p. 321)
orogenic collapse (p. 339)
orogeny (p. 321)
plunge (p. 326)
plunging fold (p. 332)
reverse fault (p. 328)
shear strain (p. 323)
shear stress (p. 325)
shear zone (p. 330)
shield (p. 341)
slickensides (p. 330)

slip lineations (p. 330)
strain (stretching, shortening)
   (p. 323)
stress (p. 324)
strike (p. 326)
strike line (p. 326)
strike-slip fault (p. 328)
syncline (p. 332)
tectonic foliation (p. 334)
tension (p. 325)
thrust fault (p. 328)
tight fold (p. 332)
uplift (p. 336)
veins (p. 327)

REVIEW QUESTIONS

1. What are the changes that rocks undergo in an orogenic belt like the Alps?

2. What is the difference between brittle and ductile deformation?

3. What factors influence whether a rock will behave in brittle or ductile fashion?

4. How are stress and strain different?

5. How is a fault different from a joint?

6. Compare the motion of normal, reverse, and strike-slip faults.

7. How do you recognize faults in the field?

8. Describe the differences between an anticline, a syncline, and a monocline.

9. Explain how certain kinds of igneous, sedimentary, and metamorphic rocks are formed during orogeny.

10. Describe the principle of isostasy.

11. What happens to the isostatic equilibrium of a mountain range as it is eroded away?

12. What happens to a mountain range when its uplift rate slows down?

13. Discuss the processes by which mountain belts are formed in convergent margins, in continental collisions, and in continental rifts.

14. How are the structures of a craton different from a typical orogenic belt?

SUGGESTED READING

Condie, K. 1997. *Plate Tectonics and Crustal Evolution.* 4th ed. Woburn, Mass.: Butterworth-Heinemann.

Davis, G. H., and S. J. Reynolds. 1996. *Structural Geology of Rocks and Regions.* New York: Wiley.

Hancock, P. L., ed. 1994. *Continental Deformation.* Oxford, Eng.: Pergamon Press.

Howell, D. G. 1989. *Tectonics of Suspect Terranes: Mountain Building and Continental Growth.* London: Chapman and Hall.

Hsu, K. J. 1982. *Mountain Building Processes.* New York: Academic Press.

Miyashiro, A., K. Aki, and A. M. C. Sengör. 1982. *Orogeny.* New York: Wiley.

Moores, E. M., and R. J. Twiss. 1995. *Tectonics.* New York: Freeman.

Park, R. G. 1988. *Geologic Structures and Moving Plates.* New York: Chapman and Hall.

Van der Pluijm, B. A., and S. Marshak. 1997. *Earth Structure: An Introduction to Structural Geology and Tectonics.* Dubuque, Iowa: WCB/McGraw-Hill.

History Before History

*P*erhaps the most important contribution that the science of geology has made to our understanding of Earth is the demonstration that our planet existed long, long before humans took their first footsteps. In Part IV, we peer back into this history. First we look at fossils, remnants of ancient life that allow us to correlate life's evolution with that of Earth. Then, in Chapter 12, we learn how geologists have addressed the issue of "deep time"—geologic time, or the time since Earth's beginnings—first by determining the relative ages of geologic features (whether one feature is older or younger than another), and then by learning how to calculate the numerical ages (the ages in years) by measuring the ratios of radioactive elements to their daughter products in minerals. With the background provided in Chapter 12, we're ready for Chapter 13's brief synopsis of Earth history, from the birth of the planet to the present and beyond. We see how plate tectonics has redistributed continents and built mountains, and how the sea level has risen and fallen and the climate has changed.

Every rock exposure has a tale to tell about Earth history. On the coast of France at Etretat, the layers of chalk record a time before the cliffs rose and the entire region in view was submerged by the sea.

Memories of Past Life: Fossils and Evolution

D.1 THE DISCOVERY OF FOSSILS

Rocks throughout the world contain shapes that closely resemble animal shells or bones, or plant stems or leaves (▶Fig. D.1). How do these shapes form? People initially assumed that they were the handiwork of supernatural beings, placed in rock as pranks. Then, about 450 B.C.E., the Greek historian Herodotus suggested that imprints of shells in rocks were the remains of sea creatures that lived when the ocean covered what is now dry land. Beginning around 300 B.C.E., however, students of Aristotle abandoned this hypothesis and argued that lifelike shapes had actually grown within rocks and had nothing to do with life; this concept of inorganic growth became dogma through the medieval millennium.

At the dawn of the Renaissance, lifelike shapes, along with any other items dug up from the Earth (such as mineral and rock specimens), came to be known as fossils (from the Latin word *fossilis*, which means "dug up"), and their origin once again became an issue for debate. Leonardo da Vinci favored Herodotus's idea, but Georgius Agricola (born Georg Bauer), in his 1546 *De natura fossilium* (considered the first geology textbook), maintained the medieval idea that the shapes grew in place. Some authors suggested another alternative, that fossil shells formed when eggs washed down cracks and grew inside rocks.

In the early 1600s, the term **fossil** assumed its present definition: the remnant or trace of an ancient living organism that has been preserved in rock or sediment. Fossils include preserved tracks (like footprints and worm burrows), bones that have been transformed into rock, insects encased in amber, and the imprints of shells. The modern era of fossil study began in 1669, when Nicholaus Steno suggested that fossil-containing rocks originated as loose sediment incorporating the remains of organisms; when the sediment hardened into rock, the remains became fossils. During the next two centuries, geologists (mostly amateurs who studied rocks as an avocation) described thousands of fossils and established museum collections (▶Fig. D.2a, b). By 1800, geologists realized that in all but the most recent layers of strata, these specimens came from species that no longer existed on Earth. Further, they found that different fossils occur in different layers of strata within a sequence of sedimentary rock. A certain fossil species present in the lower layers may be absent in the higher layers. The level at which the species disappears represents the time it went extinct. Once a species disappears, it never reappears higher in the sequence. As we will see in Chapter 12, this concept, now called the principle of fossil succession, implies that extinction is forever.

The nineteenth century saw **paleontology,** the study of fossils, ripen into a science. (Note that paleontology differs from archaeology: the former focuses on the remains of

FIGURE D.1 This bedding surface in limestone contains fossils of organisms that lived about 420 million years ago. These particular species no longer exist on Earth.

(a)

(b)

FIGURE D.2 (a) A drawer of labeled fossils. Paleontologists from around the world study such collections to help identify unknown specimens. (b) Robert Hooke, a British scientist who developed the microscope in order to study cells, published these sketches of fossil shells (ammonites) in 1703. They are among the first such sketches to be published.

ancient organisms, while the latter focuses on the remains, or artifacts, of human culture.) William Smith, an engineer supervising canal construction in England during the 1830s, showed that because of fossil succession, the group of fossils found in a particular layer of strata can be used as a basis for determining the age of that layer relative to other layers. Thus, fossils became an indispensable tool for studying geologic time and the evolution of life. As this subject is the focus of the next two chapters, we learn here what fossils are and how they develop.

D.2 FOSSILIZATION

What Kinds of Rocks Contain Fossils?

Most fossils are found in sediments or sedimentary rocks, for they form when organisms die and become buried by sediment, or when organisms travel over or through sediment and leave their mark. Rocks created from sediments

deposited under anoxic (oxygen-free) conditions in quiet water (such as lake beds or lagoons) preserve particularly fine specimens. Rocks made from sediments deposited in high-energy environments, on the other hand—where strong currents tumble shells and bones and break them up—contain at best only small fragments of fossils mixed with other clastic grains.

Some fossils are found in volcanic ash deposits, for ash settles just like sediment and can bury an organism or preserve its footprint. Fossils can survive very low-grade metamorphism, but not the mineral transformations and shearing that accompany the creation of intermediate or high-grade metamorphic rocks. Similarly, fossils do not occur in intrusive igneous rocks that crystallize directly from a melt, for the organisms from which fossils form can't live in this intrusive environment (it's too hot). Occasionally, however, lava flows preserve the shapes of trees, because the lava surrounds a tree and freezes before the tree has completely incinerated—the space containing a little charcoal that remains is, strictly speaking, a fossil.

Forming a Fossil

Paleontologists refer to the process of forming a fossil as **fossilization.** To see how a typical fossil develops in sedimentary rock, let's follow the fate of an old dinosaur as it searches for food along a riverbank (▶Fig. D.3). On a scalding summer day, the poor hungry dinosaur, plodding through the muddy ground, succumbs to the heat and collapses dead into the mud. Scavengers over the coming days strip the skeleton of meat and scatter the bones among the dinosaur footprints. But before the bones have time to react with the atmosphere and turn into dust, a storm causes the river to flood and bury the bones, along with the footprints, under a layer of silt, which then protects the bones and footprints from weathering away. Silt from succeeding floods buries the bones and prints still deeper. Later, the sea level rises and buries the fluvial sediment beneath a thick sequence of marine sediment, whose weight squashes and flattens the bones somewhat.

Eventually, the sediment containing the bones turns to rock (siltstone). The footprints remain outlined by the contact between the siltstone and mud, while the bones reside within the siltstone. Minerals precipitating from the groundwater passing through the siltstone gradually replace some of the chemicals comprising the bones, until the bones themselves have become rock. The buried bones and footprints are now fossils. One hundred million years later, the region of the dinosaur's grave is uplifted and eroded. Part of a fossil bone protrudes from a rock outcrop. A lucky fossil collector observes the fragment and starts excavating, gradually un-

FIGURE D.3 How a dinosaur eventually becomes a fossil.

covering enough of the bones to reconstruct the beast's skeleton, and finds the shapes of its footprints. The dinosaur rises again, but this time in a museum.

Similar tales can be told for fossil seashells buried by sediment settling in the sea, for insects trapped in hardened tree sap (amber), and for mammoths drowned in the muck of a tar pit. In all cases, fossilization involves the burial and preservation of an organism or the trace (like a footprint) of an organism. Once buried, the organism may be altered to varying degrees by pressure and chemical interaction with groundwater.

The Many Different Kinds of Fossils

Perhaps when you think of a fossil, you picture a dinosaur bone or a seashell. In fact, paleontologists distinguish many different kinds of fossils, based on the specific way in which the organism was fossilized. These include the following.

- *Frozen or dried bodies:* In a few environments, whole bodies of organisms may be preserved. Most of these fossils are fairly young, by geological standards—their ages can be measured in thousands, not millions, of years. Examples include woolly mammoths that became incorporated in the permafrost (permanently frozen ground) of Siberia and have stayed frozen since their death. When one such mammoth was excavated after being buried for 100,000 years, it still had edible meat on its bones and retained a nearly complete coat of fur (▶Fig. D.4a). In desert climates, organisms become desiccated (dried out) like Egyptian mummies and can last for a long time. Paleontologists have found the desiccated bodies of extinct giant ground sloths buried in the debris of desert caves.

- *Fossils preserved in amber or tar:* Insects landing on the bark of trees may become trapped in the sticky sap or resin the trees produce. This golden syrup envelops the insects and over time hardens into **amber,** the semiprecious "stone" used for jewelry. Amber can preserve insects, as well as other delicate items such as feathers, for 40 million years or more (▶Fig. D.4b).

  Tar similarly serves as a preservative. In isolated regions where oil has seeped to the surface, the more volatile components of the oil evaporate away and bacteria degrade what remains, leaving behind a sticky residue. At one such locality, the La Brea Tar Pits in Los Angeles, tar accumulated in a swampy area. While drinking at the swamp, large numbers of animals became mired in the tar and sank into it. Their bones have been remarkably well preserved, though their flesh and fur has largely vanished.

- *Preserved or replaced bones, teeth, and shells:* Bones (the internal skeletons of vertebrate animals) and shells (the external skeletons of invertebrate animals) consist of durable minerals, which may survive in rock. Eventually, however, the minerals recrystallize or are replaced by new minerals that precipitate from groundwater solutions (▶Fig. D.4c). But even when this happens, the shape of the bone or shell remains in the rock.

- *Petrified (permineralized) organisms:* **Petrified** literally means "turned to stone." Geologists use the term (or the more technical **permineralization**) to refer to a process by which plant material becomes transformed into rock. Once a plant has been buried, the cellulose comprising the plant's cell walls can survive. Groundwater passing through the rock or sediment containing the plant carries dissolved silica. Microcrystalline quartz (or chert) precipitates inside the cell walls, and the plant gradually transforms into a solid mass of chert. The fine details of the plant structure can still be visible in the chert. Petrified wood, admired because of the colorful growth rings it displays, is solid chert formed by the permineralization of trees (▶Fig. D.4d). It typically forms when a volcanic eruption rapidly buries a forest in siliceous (silica-containing) ash. Some trees even become buried while still in growth position.

- *Molds and casts:* As sediment compacts around a shell, it conforms to the shape of the shell. If the shell later disappears, either because of mechanical weathering or because it dissolves in acidic groundwater, a cavity called a **mold** remains (▶Fig. D.4e). (Sculptors use the same term to refer to the shape into which they pour bronze or plaster.) A mold preserves the delicate shape of the shell's surface; it looks like an indentation on a rock bed. The sediment that had filled the shell also preserves the shell's shape; this **cast** protrudes from the surface of the adjacent bed.

- *Carbonized impressions:* Impressions are simply flattened molds created when soft or semisoft organisms (leaves, insects, shell-less invertebrates, sponges, feathers, jellyfish) get pressed between layers of sediment. Chemical reactions eventually remove the organic chemicals that composed the organism, leaving only a thin film of carbon on the surface of the impression (▶Fig. D.4f). Fossil plant leaves appear as carbonized impressions in a lighter-colored rock.

- *Trace fossils:* These include footprints, feeding traces, burrows, and dung that organisms leave behind in sediment (▶Fig. D.4g, h). Dinosaurs, for example, have left remarkable footprints in shale, and also have left their gizzard stones (rocks that were originally used in their digestive track to crush food).

The discussion of fossilization so far has dealt primarily with **macrofossils,** fossils large enough to be seen with the naked eye. But some rocks and sediments also contain abundant **microfossils,** which can be seen only with a mi-

(a)

(b)

(c)

(d)

(e)

(f)

FIGURE D.4 (a) A frozen mammoth, found in the permafrost (permanently frozen ground) of Siberia. (b) A piece of amber containing a fossil insect. (c) Fossil dinosaur bones exposed on a tilted bed of sandstone in Dinosaur National Monument, Utah. (d) Petrified wood from the Petrified National Forest, Arizona. Petrified wood is much harder than the surrounding tuff and thus remains after the tuff has eroded away. (e) Molds and casts of shells. (f) The carbonized impression of fern fronds. (g) Dinosaur footprints in mudstone. (h) Worm burrows on a block of siltstone.

(g)

(h)

croscope or even an electron microscope (►Fig. D.5). Microfossils include remnants of plankton, algae, cyanobacteria, and pollen. Some deep-sea sediments consist almost entirely of microfossils derived from plankton. Pollen proves to be a particularly valuable microfossil for studying ancient climates (paleoclimates), as pollen can be used to identify the types of plants that live in a certain area, and the plants that grow depend on climate.

Fossil Preservation

Not all living organisms become fossils when they die. In fact, only a small percentage do, for it takes special circumstances—namely, the following four—to create a fossil.

• *Death in an anoxic (oxygen-free) environment:* A dead squirrel by the side of the road won't become a fossil. As

FIGURE D.5 Fossil plankton from deep-marine sediment. Because of their small size (a fraction of a millimeter), these specimens are considered microfossils.

time passes, birds, dogs, or other scavengers may come along and eat the carcass. And if that doesn't happen, maggots, bacteria, and fungi will infest the carcass and gradually digest it. Flesh that has not been eaten or does not rot reacts with oxygen in the atmosphere and transforms into carbon dioxide gas. The remaining skeleton weathers in air and turns to dust. Thus, before the dead squirrel can become incorporated in sediment, it has vanished. In order for fossilization to occur, it helps for a carcass to settle into an oxygen-free environment, where oxidation reactions don't happen, where scavenging organisms aren't as abundant, and where bacterial metabolism takes place very slowly. In such environments, found at the bottom of stagnant lakes, in the deep ocean, or in tar or amber, the organism won't rot away before it has a chance to be buried and preserved.

- *Rapid burial:* If an organism dies in a depositional environment where sediment accumulates rapidly, it may be buried before it has time to rot, oxidize, or be eaten even if oxygen is present. For example, if a storm suddenly buries an oyster bed with a thick layer of silt, the oysters die and become part of the sedimentary rock derived from the sediment; similarly, if a river suddenly floods and buries a dinosaur carcass in silt, the carcass will become fossilized.

- *The presence of hard parts:* Organisms without durable shells or skeletons, collectively called "hard parts," commonly won't be fossilized, as soft flesh decays long before hard parts do under most depositional conditions. For this reason, paleontologists know much more about the fossil record of bivalves (a class of organisms, including clams and oysters, with strong shells) than they

do about the fossil record of jellyfish (which have no shells) or spiders (which have very fragile shells).

- *Lack of diagenesis or metamorphism:* Processes such as diagenesis (changes that occur in a sedimentary rock as a result of deep burial and subsequent reaction with groundwater) or metamorphism (changes in a rock's recrystallization and fabric as a result of shearing and changes in pressure and temperature) may destroy fossils that had been present. For example, groundwater may dissolve fossils away, and metamorphism may distort fossils or stimulate the growth of new minerals that obscure the fossils' form.

Paleontologists, by carefully studying modern organisms, have been able to provide rough estimates of the **preservation potential** of organisms, meaning the likelihood that an organism will be buried and eventually transformed into a fossil. For example, in a typical modern-day shallow-marine environment, such as the mud-and-sand sea floor close to a beach, about 30% of the organisms have sturdy shells and thus a high preservation potential, 40% have fragile shells and a low preservation potential, and the remaining 30% have no hard parts at all and are not likely to be fossilized except in special circumstances. Of the 30% with sturdy shells, though, few happen to die in a depositional setting where they actually *can* become fossilized.

Extraordinary Fossils: A Special Window to the Past

Though only hard parts survive in most fossilization environments, paleontologists have discovered a few special locations where rock contains relics of soft parts as well; such fossils are known as **extraordinary fossils.** We've already seen how extraordinary fossils such as insects and even feathers can be preserved in amber, and how complete skeletons have been found in tar pits. Another environment in which extraordinary fossils are found is the anoxic floor of lakes or lagoons or the deep ocean. Here, organic-rich mud accumulates, oxidation cannot occur, and flesh does not rot before burial. Carcasses of animals that settle into the mud gradually become fossils, but because they were buried before the destruction of their soft parts, fossil impressions of their soft parts surround the fossils of their bones.

A small quarry near Messel, Germany, for example, has revealed extraordinary fossils of 49-million-year-old mammals, birds, fish, and amphibians who died in a shallow-water lake (▶Fig. D.6a). Bird fossils from the quarry include the delicate imprints of feathers, bat fossils come complete with impressions of ears and wing flaps, and mammal fossils have an aura of carbonized fur. Exposures of the Solenhofen Limestone, an approximately 150-million-year-old rock derived from lime mud deposited in a stagnant lagoon, contain extraordinary fossils of *Archaeopteryx,* one of

FIGURE D.6 Extraordinary fossils: (a) a mammal from Messel, Germany; (b) *Archaeopteryx* from the Solenhofen Limestone; (c) reproduction of a soft-bodied creature from the Burgess Shale, in Canada.

In a few cases, extraordinary fossils include actual tissue, a discovery that has led to a research race to find the oldest preserved DNA. (**DNA,** short for deoxyribonucleic acid, is the complex molecule, shaped like a double helix, that contains the code that guides the growth and development of an organism. Individual components of this code are called **genes,** and the study of genes and how they transmit information is called **genetics.**) Paleontologists have isolated small segments of dinosaur DNA—certainly not enough to clone dinosaurs, as suggested in the popular movie *Jurassic Park,* but enough to suggest that it may someday be possible to study the genetics of extinct species.

D.3 CLASSIFYING LIFE

The classification of fossils follows the same principles used for classifying living organisms, so before we learn how paleontologists classify fossils, we must first examine how biologists classify life forms. The principles of classification were first proposed in the eighteenth century by Carolus Linnaeus, a Swedish biologist. The study of such relationships is now referred to as **taxonomy.** Linnaeus's scheme has a hierarchy of divisions. First, all life is divided into kingdoms, and each kingdom consists of one or more phyla. A phylum, in turn, consists of several classes; a class, of several orders; an order, of several families; a family, of several genera; and a genus, of one or more species (▶Fig. D.7). Kingdoms, then, are the broadest category and species the narrowest. Where needed, biologists and paleontologists separate out subphyla and subclasses. Biologists now recognize six kingdoms (▶Fig. D.8):

- **Archaeobacteria** ("old bacteria," found in extreme environments like hot springs; also called "Archaea");
- **Eubacteria** ("true bacteria," including the kind that cause infections);
- **Protista** (various unicellular and simple multicellular organisms; these include algae [such as diatoms] and forams, two of the major plankton types in the oceans);
- **Fungi** (such as mushrooms and yeasts);
- **Plantae** (trees, grasses, ferns, etc.); and
- **Animalia** (such as sponges, corals, snails, dinosaurs, ants, and people).

In some classifications, biologists emphasize that the last four kingdoms share enough common characteristics to be lumped together as a group called "Eukarya," and thus consider life to be divided into three "domains": Archaeobacteria, Eubacteria, and Eukarya.

Figure D.7 shows the subdivisions below the level of kingdom that apply to humans. Humans, together with monkeys, apes, and lemurs, are primates because they have

the earliest birds (▶Fig. D.6b). And exposures of the Burgess Shale in the Canadian Rockies of British Columbia have yielded a plentitude of fossils that show what shell-less invertebrates who inhabited the deep-sea floor about 510 million years ago looked like (▶Fig. D.6c). The Burgess Shale fauna (animal life) is so strange—for example, it includes organisms with circular jaws—that it has been hard to determine how these organisms are related to present-day ones.

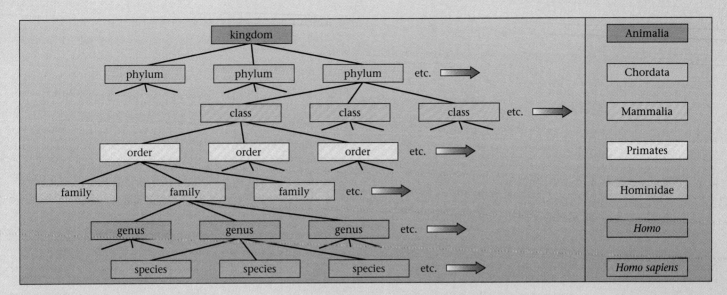

FIGURE D.7 The taxonomic subdivisions. The right-hand column shows how the names apply to human beings.

opposable thumbs and similar facial characteristics. Primates, together with dogs, cats, horses, cows, elephants, whales, and many other animals, are mammals because they have fur and nurse their young with milk. Mammals, along with reptiles, birds, and fish, are chordates because they have an internal skeleton with a spinal canal. And chordates, along with clams, insects, lobsters, and worms, are an-

FIGURE D.8 The basic kingdoms of life on Earth. Archaeobacteria and Eubacteria are both prokaryotes, because they don't have nuclei, while all other life forms are eukaryotes, because they consist of cells with a nucleus.

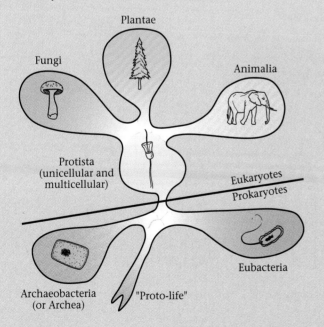

imals because they share a similar cell structure. Note that genus and species names are given in Latin and italicized.

D.4 CLASSIFYING FOSSILS

As you'll recall from Chapter 5, mineralogists recognize thousands of different types of minerals. A given mineral has a specific crystal structure and a chemical composition that varies within a definable range. Not all specimens of a given mineral look the same—they can differ in terms of color, shape, size, and crystal form. In contrast, paleontologists traditionally distinguish different fossil species from one another according to **morphology** (the form or shape) alone. A fossil clam is a fossil clam because it looks like one. In some cases, the characteristics that distinguish one fossil species from another may be pretty subtle (for example, the number of ridges on the surface of a shell, or the relative length of different leg bones), but even beginners can distinguish the major groups of fossils from one another on sight. Though presently morphology provides the only basis for fossil identification, paleontologists may someday be able to determine relationships between organisms by specifying the percentages of shared protein sequences comprising their DNA.

When it came to naming fossils, paleontologists began by comparing fossil forms with known organisms. They looked at the nature of the skeleton (was it internal or external?), the symmetry of the organism (was it bilaterally symmetric like a mammal, or did it have fivefold symmetry like a starfish?), the design of the shell (in the case of invertebrates), and the design of the jaws or feet (in the case of vertebrates). For example, a fossil organism with a spiral shell

FIGURE D.9 Examples of the diversity of gastropods (snails). Note that although all these shells have a spiral shape, they differ in detail from each other.

that does not contain internal chambers is classified a member of the class Gastropoda (the snails) within the phylum Mollusca (▶Fig. D.9). Gastropod species with similar-shaped shells fall within the same genus, and the species themselves may differ from one another in the way their shells coil and in the design of the shell's opening.

Not all fossil species resemble living families of organisms. In such cases, the comparisons must take place at the level of orders or even higher. For example, a group of extinct organisms called trilobites have no close living relatives. But they were clearly segmented invertebrate animals, and as such resemble arthropods like insects and crustaceans. Thus, they are considered to be an extinct member of the phylum Arthropoda.

There's nothing magical about classifying fossils; major characteristics at the level of class or higher are fairly easy to distinguish just from the way they look. For instance, skeletons of birds (class Aves) are hard to mistake for skeletons of mammals (class Mammalia), and snail shells (class Gastropoda) are hard to mistake for clam shells (class Pelecypoda). ▶Figure D.10 shows examples of some of the major classes and subphyla of invertebrate fossils. With this chart, you should be able to identify many of the fossils you'll find in a common bed of limestone. Particularly common invertebrate fossils include the following.

• **Trilobites:** These have a segmented shell that is divided across its width into three parts. They are a type of arthropod.

• **Gastropods** (snails): These have a spiral shell that does not contain internal chambers.

• **Bivalves** (clams and oysters): These have a shell that can be divided into two similar halves.

• **Brachiopods** (lamp shells): These have a shell that can be divided into two unequal halves. The shells have ridges radiating out from the hinge.

• **Bryozoans:** These resemble a screen-like grid of cells. Each cell is the shell of a single animal. The animals lived together in a colony.

• **Crinoids** (sea lilies): These organisms look like a flower but actually are animals. Their stalks consist of numerous circular plates stacked on top of each other.

• **Graptolites:** These look like tiny carbon saw blades in a rock. They are remnants of colonial animals that floated in the sea.

• **Ammonites:** These have a spiral or straight shell that contains internal chambers and has a ridged surface. These organisms were squid-like.

• **Corals:** These include colonial organisms that form distinctive mounds or columns. Paleozoic examples are solitary.

FIGURE D.10 Common types of invertebrate fossils.

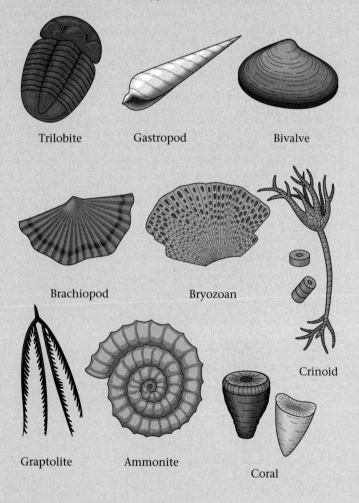

Trilobite   Gastropod   Bivalve

Brachiopod   Bryozoan   Crinoid

Graptolite   Ammonite   Coral

D.5 THE FOSSIL RECORD

A Brief History of Life

As we will see in Chapter 12, the fossil record defines the long-term evolution of life on planet Earth. Bacteria fossils are found in rocks as old as about 3.5 billion years, indicating that these organisms are the earliest forms of life on Earth. Researchers once thought that bacteria somehow developed by chemical reactions in concentrated "soups" that formed when seawater trapped in shallow pools evaporated. Growing evidence suggests, however, that they may instead have developed in warm groundwater beneath the Earth's surface or at hot springs. An alternative view is that they were brought to Earth by an asteroid or comet.

For the first billion years or so of life history, bacteria were the only type of life on Earth. Then, about 2.5 Ga (billion years ago), organisms of the protist kingdom first appeared. Early multicellular organisms, shell-less invertebrates of the animal kingdom, and fungi of the plant kingdom came into existence perhaps 1.0–1.5 billion years ago. Within each kingdom, life radiated (divided) into different phyla, and within each phylum, life radiated into different classes. The great variety of shelly invertebrate classes appeared a little over 500 million years ago, and organisms that persist today appeared at different times since then: first came invertebrates, then fish, land plants, amphibians, reptiles, mammals, and finally birds. Paleontologists represent their ideas about which groups radiated from which ancestors on a chart called the "tree of life," or, more formally, the **phylogenetic tree** (▶Fig. D.11). Study of the DNA of different organisms is beginning to allow researchers to understand the relationship between molecular processes and evolutionary change.

Is the Fossil Record Complete?

By some estimates, more than 250,000 species of fossils have been collected and identified to date, by thousands of geologists working on all continents during the past two centuries. These fossils define the framework of life evolution on planet Earth. But the record is not complete—every intermediate step in the evolution of every organism cannot be accounted for by known fossils. Considering that there may be as many as 5 million species living on Earth today (not counting bacteria), over the billions of years that life has existed there may have been 5 billion to 50 billion species. Clearly, known fossils represent at most a tiny percentage of these species. Why is the record so incomplete?

First, despite all the fossil-collecting efforts of the past two centuries, paleontologists have not even come close to sampling every square centimeter of sedimentary rock exposed on Earth. Just as biologists have not yet identified

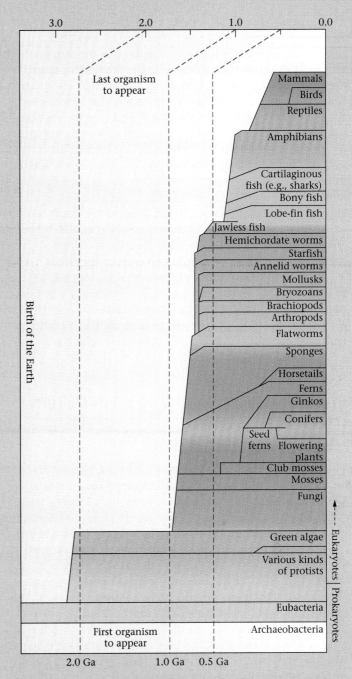

FIGURE D.11 A simplified tree of life for organisms on Earth, showing when different branches of the tree appeared.

every living species of insect, paleontologists have not yet identified every species of fossil. New species and even genera of fossils continue to be discovered every year.

Second, not all organisms are represented in the rock record, because not all organisms have a high preservation potential. As noted earlier, fossilization occurs only under special conditions, and thus only a minuscule fraction of the organisms that have lived on Earth become fossilized. There may be few, if any, fossils of a vast number of extinct species, so we have no way of knowing that they ever existed.

Finally, as we will learn in Chapter 12, the sequence of sedimentary strata that exists on Earth does not account for every minute of time since the formation of our planet. Sediments only accumulate in environments where conditions are appropriate for deposition and not for erosion—sediments do not accumulate, for example, on the dry great plains or on mountain peaks, but do accumulate in the sea and in the floodplains and deltas of rivers. (In fact, on mountain peaks, erosion actively cuts down into rock and removes it.) Because Earth's climate changes through time and because the sea level rises and falls, certain locations on continents are sometimes sites of deposition and sometimes aren't, and on occasion are sites of erosion. Therefore, the sequence of strata records only part of geologic time.

In sum, a rock sequence provides an incomplete record of Earth history, organisms have a low probability of being preserved, and paleontologists have found only a small percentage of the fossils preserved in rock. So the incompleteness of the fossil record should come as no surprise.

D.6 EVOLUTION AND EXTINCTION

Earth's Inhabitants Through Time

The succession of fossils we find in the rock record can best be explained by the **theory of evolution,** proposed by Charles Darwin in the mid–nineteenth century, which states that species evolve, or change, over time. As a young man, Darwin was unable to settle on a career but had developed a strong interest in natural history, so he jumped at the opportunity to accompany the *H.M.S. Beagle* as the ship's naturalist on its around-the-world surveying cruise. During the five years of the cruise, Darwin amassed immense specimen collections from South America, Australia, and Africa. His observations during these years, and the influence of Charles Lyell's 1830 textbook *Principles of Geology,* which argued that the Earth had a long history, led Darwin to develop his concept of evolution.

But only when Alfred Russell Wallace, a naturalist working in Indonesia, wrote to Darwin (in 1858) outlining almost identical ideas did Darwin publish the concept in a book called *On the Origin of Species by Means of Natural Selection.* Here he observed that characteristics of domestic animal breeds could change by selective breeding, and speculated that similar processes took place without the intervention of breeders. Specifically, since populations do not grow exponentially, they must be limited by competition for scarce resources in the environment. In the case of domesticated plants and animals, breeders choose characteristics they want (such as large size), even if these characteristics might not make the organism better able to survive. In nature, only

the fittest organisms survive to pass on their characteristics to the next generation, a process called **natural selection.** Over long periods of Earth history, natural selection would have caused species to evolve. Also through natural selection, populations gradually adapt to take better advantage of ecological niches.

Darwin and Wallace based their ideas on observations of whole organisms. Studies during the late nineteenth and early twentieth centuries, beginning with the work of an Austrian monk named Gregor Mendel, showed how genetic mutations could lead to new traits that could then be passed on to offspring, and this provided a mechanism for evolution. Traits that make an organism less likely to survive are not passed on, because the organism dies before it has offspring. With the discovery of DNA in 1953, biologists began to understand the molecular nature of genes and mutations and thus of evolution. Though the overall concept of evolution has attained the status of theory because it so successfully explains a vast suite of observations, developing a detailed understanding of how it operates will require further research.

Paleontologists disagree about the rate at which evolution takes place. Traditionally, it was assumed that evolution happened at a constant, slow rate—this concept is called **gradualism.** More recently, however, researchers have suggested that evolution takes place in fits and starts: evolution occurs very slowly for quite a while (the species are in equilibrium), and then during a relatively short period it takes place very rapidly. This concept is called **punctuated equilibrium.** Factors that could cause sudden pulses of evolution include (1) a sudden mass extinction event, during which many organisms disappear, leaving ecological niches open for new species to colonize; (2) a sudden change in the Earth's climate that puts stress on organisms—organisms that evolve to survive the new stress survive, while others go extinct; and (3) the sudden formation of new environments, as may happen when rifting splits apart a continent and generates a new ocean with new coastlines.

Extinction: When Species Vanish

Extinction occurs when the last members of a species die, so there are no parents to pass on their genetic traits to offspring. These days, we take for granted that species go extinct, because a great number have, unfortunately, vanished from the Earth during human history. Before the 1770s, however, few geologists thought that extinction occurred; they thought that fossils that didn't resemble known species must have living relatives somewhere on Earth. Considering that large parts of the Earth remained unexplored, this idea wasn't so far-fetched. But by the end of the eighteenth century, it became clear that numerous fossil organisms did not have modern-day counterparts. The bones of mastodons and woolly mammoths, for example, were too different from those of elephants to be of the same species, and the animals were too big to hide.

What causes extinction? Initially, paleontologists assumed that Noah's flood (recounted in the Old Testament) was the cause. But as the fossil record began to accumulate, and the principle of fossil succession was documented, they realized that not all extinction could have happened in a single event. Some authors proposed that extinction was the consequence of a series of global catastrophic floods, but continued geologic research demonstrated that such floods have not taken place.

Twentieth-century studies have concluded that many different phenomena can contribute to extinction. Some extinctions may happen suddenly, when all members of a species die off in a short time, while others may occur over longer periods, when the replacement rate of population simply becomes lower than the mortality rate. By examining the number of species on Earth through time (that is, by studying variations in the **diversity of life,** or **biodiversity**), paleontologists have found that there is a varying rate of extinction. Generally, the rate is fairly slow, but on occasion a **mass extinction event** occurs, during which a large number of species worldwide disappear. At least five major mass extinction events have happened during the past half billion years (▶Fig. D.12). These events define the boundaries between some of the major intervals into which geologists divide time. For example, a major extinction event marks the end of the Cretaceous period, 65 million years ago. During this event, dinosaur species (with the ex-

FIGURE D.13 Painting of an asteroid impact in Yucatán that possibly eliminated most if not all dinosaur species.

ception of their modified descendants, the birds) vanished, along with many marine invertebrate species.

Following are some of the geological factors that may cause extinction.

- *Global climate change:* At times, the Earth's mean temperature has been significantly colder than today's, while at other times it has been much hotter. These shifts affect ocean currents and sea level, and may trigger ice ages or droughts. Because of a change in climate, an individual species may lose its habitat, and if it cannot adapt to the new habitat or migrate to stay with its old one, the species will disappear.

- *Tectonic activity:* Tectonic activity causes mountain building in narrow belts, the gradual vertical movement of the crust over broad regions, changes in sea-floor spreading rates, and changes in the amount of volcanism. These phenomena can make the sea level rise or fall, or can bring about changes in elevation, thus modifying the distribution and area of habitats. Species that cannot adapt to a new habitat die off.

- *Asteroid impact:* Many geologists have concluded that impacts of large meteorites or asteroids with the Earth have been catastrophic for life (▶Fig. D.13). An impact would send so much dust and debris into the atmosphere that it would blot out the Sun and plunge the Earth into darkness and cold. Such a change, though relatively short-lived, could interrupt the food chain and send many species into extinction.

- *The appearance of a predator or competitor:* Some extinctions may happen simply because a new predator appears on the scene. Some researchers suggest that this phenomenon explains the mass extinction that occurred during the past 20,000 years, when a vast number of large mammal species vanished from North America. The timing of these extinctions appears to coincide with the appearance of the first humans on the continent. If a more efficient competitor appears, the competitor steals an ecological niche from the weaker species, whose members can't obtain enough food and thus die out.

FIGURE D.12 This graph illustrates the variation in diversity of life with time. Steep dips in the curve mean that the number of species on Earth suddenly decreased substantially. The largest dips represent major mass extinctions.

CHAPTER 12

Deep Time: How Old Is Old?

If the Eiffel Tower were now representing the world's age, the skin of paint on the pinnacle-knob at its summit would represent man's share of that age; and anybody would perceive that that skin was what the tower was built for. I reckon they would, I dunno.

—MARK TWAIN (1835–1910)

12.1 INTRODUCTION

In May of 1869, a one-armed Civil War veteran named John Wesley Powell set out with a team of nine geologists and scouts to explore the previously unmapped expanse of the Grand Canyon, the greatest gorge on Earth. Though Powell and his companions battled fearsome rapids and the pangs of starvation, most managed to emerge from the mouth of the canyon three months later (▶Fig. 12.1). During their voyage, seemingly insurmountable walls of rock both imprisoned and amazed the explorers, and led them to pose important questions about the Earth and its history, thoughts that even casual tourists to the canyon ponder today: Did the Colorado River sculpt this marvel, and if so, how long did it take? When did the rocks making up the walls of the canyon form? Was there a time before the colorful layers accumulated? These questions pertain to **geologic time,** the span of time since Earth's formation. Powell realized that, like the pages in a book, rock layers of the Grand Canyon contain a record of Earth history.

In this chapter, we first learn how geologists developed the concept of geologic time and thus a frame of reference for describing the ages of rocks, fossils, structures, and landscapes. Then we look at the tools geologists use to determine the age of the Earth and its features. With the concept of geologic time in hand, a hike down a trail through the Grand Canyon becomes a trip into the distant past, into what author John McPhee calls *Deep Time.* The geological discovery that the planet's history extends billions of years into the past changed humanity's perception of the Universe as profoundly as did the astronomical discovery that the limit of space extends billions of light years beyond the edge of our solar system.

How do geologists determine the age of rocks? This cliff face in Missouri shows two rock units. The sedimentary layers on the left were deposited on the granite to the right. From this relation, we can determine the relative age of the two units—the granite is older. But to determine the age of the granite in years, its numerical age, we must date it radiometrically.

365

FIGURE 12.1 Woodcut illustration of the "noonday rest in Marble Canyon," from J. W. Powell, *The Exploration of the Colorado River and Its Canyons* (1895). The explorers have just entered a quiet stretch of the river: "We pass many side canyons today that are dark, gloomy passages back into the heart of the rocks. . . . It rains again this afternoon . . . until great streams are formed. Although the walls of the canyon are chiefly limestone, the adjacent country is of red sandstone; and now the waters, loaded with these sands, come down in rivers of bright red mud, leaping over the walls in innumerable cascades. It is plain now how these walls are polished."

12.2 TIME: A HUMAN OBSESSION

When you plan your daily schedule, you have to know not only where you need to be, but when you need to be there. Because time assumes such significance in human consciousness, we have developed elaborate tools to measure it and formal scales to record it. We use a **second** as the basic unit of time measurement. What exactly is a second? From 1900 to 1968, we defined the second as 1/31,556,925.947 of the year 1900, but now we define it as the time it takes for the magnetic field of a cesium atom to flip polarity 9,192,631,770 times, as measured by an atomic clock. We sum sixty seconds into one minute, sixty minutes into one hour, and twenty-four hours into one **day,** about the time it takes for Earth to spin once on its axis.

In the preindustrial era, each locality kept its own time, setting noon as the moment when the Sun reached the highest point in the sky. But with the advent of train travel and telegraphs, people needed to calibrate schedules from place to place. So in 1883, countries around the globe agreed to divide the world into 15°-wide bands of longitude called **time zones**—in each time zone, all clocks keep the same **standard time.** The times in each zone are set in relation to **Greenwich Mean Time (GMT),** the time at the astronomical observatory in Greenwich, England.

12.3 THE CONCEPT OF GEOLOGIC TIME

The Birth of Geologic Time

Most people develop a sense for the duration of time by remembering when events took place in terms of human lifetimes. So if an average generation spans twenty years, then recorded history began about two to four hundred generations ago. Many cultures viewed geologic time to be about the same duration as historical time. For example, in 1654, Irish archbishop James Ussher added up the generations of patriarchs described in the Judeo-Christian Old Testament and concluded that the Earth came into being on October 23, 4004 B.C.E. Scholars at the time assumed that the Earth formed just as we see it, and has changed only by floods, volcanic eruptions, storms, earthquakes, and human actions.

Beginning in the Renaissance, however, scientists studying geological features began to speculate that geological time might far exceed historical time. The discovery of seashell-like shapes in rocks triggered this revolution in thinking. Although most people dismissed these shapes, which came to be known as fossils (from the Latin word *fossilis,* for "dug up"), as a coincidence or a supernatural trick, some concluded that they were the relics of ancient sea creatures trapped in the rocks. But even though Leonardo da Vinci had proposed the organic origin of fossils as early as 1500, the idea would not gain acceptance until the work of Nicolaus Steno (1638–1686) became known, over a century and a half later.

Steno, born Niels Stenson in Denmark, served as court physician to the Grand Duke of Tuscany in Florence, Italy. During walks in the nearby Apennine Mountains, Steno frequently came upon fossils and wondered how they ended up in solid rock hundreds of meters above sea level. One day, local fishermen gave Steno a shark's head, which he dissected out of curiosity. Inside its mouth, he found rows of distinctive triangular teeth, identical to so-called tongue stones, fossils thought to be the petrified tongues of dragons (▶Fig. 12.2). Steno became convinced that these fossils, and

FIGURE 12.2 A fossilized shark tooth. Before Steno explained the origin of these fossils, they were known as dragons' tongues.

by implication others as well, did not form supernaturally but were the relicts of ancient life, and introduced this idea in his 1669 book *Forerunner to a Dissertation on a Solid Naturally Occurring Within a Solid.* The title may seem peculiar until you think about the puzzle Steno was trying to solve: how did a solid object (a fossil) get into another solid object (a sedimentary rock)? Steno concluded that the rock must have once been loose sand that incorporated the teeth and shells while soft and only later hardened into rock. His discovery implied that geologic features developed not all at once, but rather by a series of events that took a long time, longer than human history.

The next major step in the evolution of the idea that geologic time exceeded historical time came from the work of a Scottish doctor and gentleman farmer named James Hutton (1726–1797). Hutton lived during the Age of Enlightenment, when philosophers like Voltaire, Kant, Hume, and Locke encouraged people to cast aside the constraints of dogma and think for themselves, and the writings of Sir Isaac Newton (who proposed that phenomena obeyed physical laws) made people look to natural, not supernatural, processes to explain the Universe. As Hutton wandered around Scotland, observing and thinking about the rocky crags he saw there, he eventually came up with an idea that provided the foundation of geology and led to his title: "father of geology." This idea, the **principle of uniformitarianism,** states simply that physical processes we observe today also operated in the past and were responsible for the formation of geologic features we see in present-day outcrops; more concisely, the present is the key to the past. If this principle is correct, Hutton reasoned, the Earth must be much older than Ussher's calcu-

lations allowed, for observed geologic processes work very slowly. In his 1785 book *Theory of the Earth with Proofs and Illustrations,* he mused on the issue of geologic time, and suggested that "there is no vestige of a beginning, no prospect of an end."

Relative versus Numerical Age

In the early nineteenth century, geologists struggled to develop ways to divide and describe geologic time. Like historians, geologists want to establish both the sequence of events that created an array of geologic features (such as rocks, structures, and landscapes) and the exact dates on which the events happened. We specify the age of one feature with respect to another as its **relative age** and the age of a feature given in years as the **numerical age** (or, in older literature, the **absolute age**) (▶Fig. 12.3a, b). Geologists developed ways of defining relative age long before they did so for numerical age, so we will look at relative-age determination first.

FIGURE 12.3 The difference between relative and numerical age. (a) The relative ages of selected wars in the twentieth century. (b) The numerical ages of these same wars. Clearly, this chart provides more information, for it displays the duration of events and indicates the amount of time between events.

RELATIVE AGE	NUMERICAL AGE
Gulf War (Persian Gulf)	2000
	Gulf War
Falkland War (Argentina-England)	
	Falkland War — 1980
6-day War (Arab-Israeli)	
	Vietnam War
Vietnam War	6-day War
	Bay of Pigs — 1960
Bay of Pigs Invasion (Cuba-U.S.)	
	Korean War
Korean War	
	World War II — 1940
World War II	
	Spanish Civil War
Spanish Civil War	
	1920
World War I	
	World War I
Russian Revolution	
	Russian Revolution
Russo-Japanese War	Russo-Japanese War
	Boer War — 1900
Boer War (South Africa)	

(a)   (b)

(a)

(b)

(c)

(d)

(e)

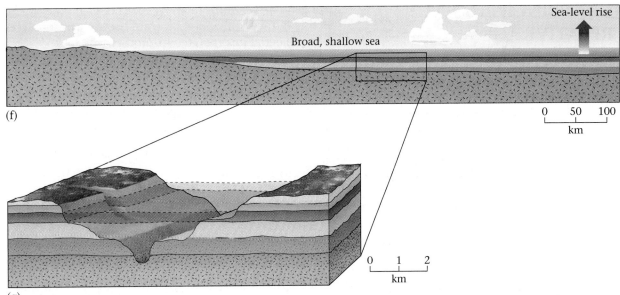

Sea-level rise

Broad, shallow sea

(f)

0   50   100
km

0   1   2
km

(g)

Time 1
(h)

Dike

Time 2

Inclusion (xenolith)

Sill

Inclusion (pebble)

Flow

(i)

Baked contact

Pluton

(j)

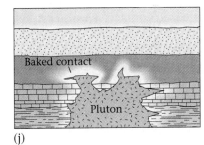

FIGURE 12.4 (a, b) Uniformitarianism: Geologists assume that the physical processes that created these mud cracks in a tidal flat in (a) also formed these mud cracks preserved on the surface of a Paleozoic siltstone bed in (b). Note that the mud cracks in (b) were inside solid rock and are visible now only because the overlying bed has been removed by erosion. (c) Superposition: If you fill a glass cylinder with different colors of sand, the oldest sand (white) must be on the bottom. (d, e) Original horizontality: When sediment is originally deposited, like this silt on the tidal flat near Mont St. Michel, France, it forms a flat layer. If layers do not become deformed, they can remain horizontal for hundreds of millions of years, as illustrated by these sandstone beds in Kings Canyon, Australia. (f, g) Original continuity: The sediments on the floor of this broad shallow sea (f), which formed when the sea level rose and flooded the continent, accumulated in nearly horizontal layers, continuous over a broad region. We assume that the layers of sediment were originally continuous across the canyon before the canyon developed (g). (h) Cross-cutting relations: The sedimentary beds existed first; then they were cut by igneous magma rising to a volcano, so today we see a dike cutting across the bedding. Since the dike does the cutting, it must be younger. (i) Inclusions: In the left-hand example, a sill intrudes between a limestone layer and a sandstone. The sill incorporates fragments (inclusions) of limestone and sandstone, so it must be younger. In the right-hand example, the igneous rock is a lava flow that existed before the sandstone was deposited; the sandstone contains pebbles (inclusions) of lava. (j) Baked contacts: The intrusion of the pluton creates a metamorphic aureole (baked contact) in the surrounding rock, so the pluton must be younger. Note that the pluton also crosscuts bedding, confirming this interpretation.

12.4 PHYSICAL PRINCIPLES FOR DEFINING RELATIVE AGE

Nineteenth-century geologists such as Charles Lyell recast the ideas of Steno and Hutton into formal, usable geological principles. These principles, defined below, continue to provide the basic framework within which geologists read the record of Earth history.

- **The principle of uniformitarianism:** As noted earlier, the present is the key to the past: physical processes we observe operating today also operated in the past at comparable rates (▶Fig. 12.4a, b).

- **The principle of superposition:** In a sequence of sedimentary rock layers, each layer must be younger than the one below, for a layer of sediment cannot accumulate unless there is already a substrate on which it can collect. Thus, the layer at the bottom of a sequence is the oldest, and the layer at the top is the youngest (▶Fig. 12.4c).

- **The principle of original horizontality:** Sediments on Earth settle out of a fluid in a gravitational field. Typically, the surfaces on which sediments accumulate (such as a floodplain or the bed of a lake or sea) are fairly horizontal. If sediments were deposited on a steep slope, they would likely slide to the bottom before lithification, and so would not be preserved as sedimentary layers. Therefore, layers of sediment when originally deposited are fairly horizontal. With this principle in mind, we know that when we see folds and tilted beds, we are seeing the consequences of deformation that postdates deposition (▶Fig. 12.4d, e).

- **The principle of original continuity:** Sediments generally accumulate in continuous sheets. If today you find a sedimentary layer cut by a canyon, then you can assume that the layer once spanned the canyon but was later eroded by the river that formed the canyon (▶Fig. 12.4f, g).

- **The principle of cross-cutting relations:** If one geologic feature cuts across another, the feature that has been cut is older. For example, if an igneous dike cuts across a sequence of sedimentary beds, the beds must be older than the dike (▶Fig. 12.4h). If a fault cuts across and displaces layers of sedimentary rock, then the fault must be younger than the layers. But if a layer of sediment buries a fault, the sediment must be younger than the fault.

- **The principle of inclusions:** If an igneous intrusion contains xenoliths (fragments) of another rock, the fragments must be older than the intrusion. If a layer of sediment deposited on an igneous layer includes pebbles of the igneous rock, then the sedimentary layer must be younger. The xenoliths in an igneous body and the pebbles in the sedimentary layer are **inclusions,** or pieces of one material incorporated in another. The rock containing the inclusion must be younger than the inclusion (▶Fig. 12.4i).

- **The principle of baked contacts:** An igneous intrusion "bakes" (metamorphoses) surrounding rocks. The rock that has been baked must be older than the intrusion (▶Fig. 12.4j).

Now let's use these principles to determine the relative ages of features shown in ▶Figure 12.5. In so doing, we develop a **geologic history** of the region, defining the relative ages of events that took place there.

First, note what kinds of rocks and structures the figure contains. Most of the rocks comprise a sequence of folded sedimentary beds, but we also see a granite pluton, a basalt dike, and a fault. Let's start our analysis by looking at the sedimentary sequence. By the principle of superposition, we know that the oldest layer, the limestone labeled 1, occurs at the bottom and that progressively younger beds lie above the limestone. We can confirm that layer 1 predates layer 2 by applying the principle of inclusions, because layer 2 contains pebbles (inclusions) of layer 1. As the same holds true for the other layers, the sedimentary beds from oldest to youngest are 1, 2, 3, 4, 5, 6, 7. (There are other rocks below layer 1, but we do not see them in our cross section.) And considering the principle of original horizontality, we conclude that the layers were folded sometime after deposition.

Now let's look at the relationships between the igneous rocks and the sedimentary rocks. The granite pluton cuts across the folded sedimentary rocks, so the intrusion of the pluton occurred after the deposition of the sedimentary beds and after they were folded. The layer of igneous rock that parallels the sedimentary beds could be either a sill that intruded between the sedimentary layers or a flow that spread out over the sandstone and solidified before the shale was deposited. From the principle of inclusions, we deduce that the layer is a sill, because it contains xenoliths of both the underlying sandstone and the overlying shale. Since the sill is folded, it intruded before folding took place. By applying the principle of baked contacts, we also can tell that the sill intruded before the pluton did, because the baked zones (metamorphic aureole) surrounding the pluton affected the sill. The dike cuts across both the pluton and the sill, as well as all the sedimentary layers, and thus formed later.

Finally, let's consider the fault and the land surface. Because the fault offsets the granite pluton and the sedimentary beds, by the principle of cross-cutting relations the fault must be younger than those rocks. But the fault itself has been cut by the dike, and so must be older than

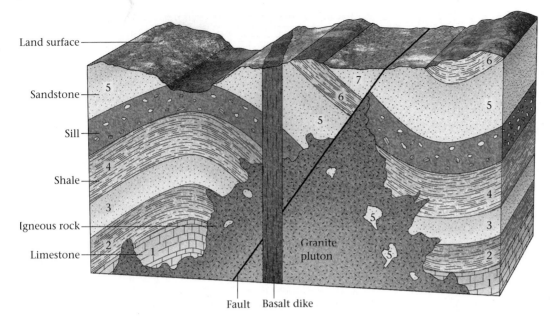

FIGURE 12.5 Geologic principles allow us to interpret the sequence of events leading to the development of the features shown here. Beds 1–7 were deposited first. Intrusion of the sill came next, followed by folding, intrusion of the granite, faulting, intrusion of the dike, and erosion to yield the present land surface.

the dike. The present land surface erodes all rock units and the fault, and thus must be younger. We can now propose the following geologic history for this region: (a) deposition of the sedimentary sequence, in order from layers 1 to 7; (b) intrusion of the sill; (c) folding of the sedimentary layers and the sill; (d) intrusion of the granite pluton; (e) faulting; (f) intrusion of the dike; (g) formation of the land surface.

12.5 ADDING FOSSILS TO THE STORY: FOSSIL SUCCESSION

As England entered the Industrial Revolution in the late eighteenth and early nineteenth centuries, new factories demanded coal to fire their steam engines. During the 1830s, the government decided to build a network of canals to transport coal and iron, and hired an engineer named William Smith (1769–1839) to survey the excavations. These excavations provided fresh exposures of bedrock, which previously had been covered by vegetation. Smith quickly learned to recognize distinctive layers of sedimentary rock and to identify the **fossil assemblage** (the group of fossil species) that they contained (▶Fig. 12.6). He also realized that a particular fossil species can be found only in a limited interval of strata, and not above or below this interval. Thus, once a fossil species dis-

appears at a horizon in a sequence of strata, it never reappears higher in the sequence. In other words, extinction is forever. Smith's observation has been repeated at millions of locations around the world, and has been codified as the **principle of fossil succession.**

To see how this principle works, examine ▶Figure 12.7, which depicts a sequence of strata. Bed 1 at the base contains fossil species A, bed 2 contains fossil species A and B, bed 3 contains B and C, bed 4 contains C, and so on. From these data, we can define the **range** of specific fossils in the sequence. Note that the sequence contains a definable succession of fossils (A, B, C, D, E, F), that the range in which a particular species occurs may overlap with the range of other species, and that once a species vanishes, it does not reappear higher in the section.

Because of the principle of fossil succession, we can define the relative ages of strata by looking at fossils. For example, if we find a bed containing fossil A, we can say that the bed is older than a bed containing, say, fossil F. Geologists have now identified and determined the relative ages of so many fossil species that by examining the fossils

FIGURE 12.6 A bedding surface containing many fossils.

FIGURE 12.7 The principle of fossil succession. Note that each species has only a limited range in a succession of strata, and ranges of different fossils may overlap. Widespread fossils with a short range are index fossils. The inset illustrates how overlapping fossil ranges can be used to limit the age range of a given bed.

of rock exposed there consisted of alternating beds of gray sandstone and shale, while another consisted of red sandstone and conglomerate (▶Fig. 12.8a, b). The beds of gray sandstone and shale were nearly vertical, while the beds of red sandstone and conglomerate were roughly horizontal. Further, the horizontal layers seemed to lie across the truncated ends of the vertical layers, like a handkerchief lying across a row of books. We can imagine that as Hutton examined this odd geometric relationship, the tide came in and deposited a new layer of sand on top of the rocky shore; and with the principle of uniformitarianism in mind, he suddenly realized the significance of what he saw. Clearly, the gray sandstone/shale sequence had been deposited, then tilted, and then truncated by erosion before the red sandstone/conglomerate beds were deposited. Just as rock outcrops were being covered by sand along the shoreline as Hutton watched, at some time in the past the gray sandstone/shale sequence formed the rocky shore, and the red sand and conglomerate were deposited above.

FIGURE 12.8 (a) The Siccar Point unconformity in Scotland. (b) A geological interpretation of the unconformity.

alone, we can determine the relative age of a bed at one locality with respect to a bed at another locality.

12.6 UNCONFORMITIES: GAPS IN THE RECORD

James Hutton often strolled along the coast of Scotland because the shore cliffs provided good exposures of rock, stripped of soil and shrubbery. He was particularly puzzled by an outcrop along the shore at Siccar Point. One sequence

Hutton deduced that the surface between the gray and red rock sequences represented a long interval of time during which new strata were not deposited at Siccar Point and/or older strata may have been eroded away. We now call such surfaces, representing a period of nondeposition and/or erosion, **unconformities.** The interval of time between deposition of the youngest rock below an unconformity and deposition of the oldest rock above is called a **hiatus.** Essentially, an unconformity forms wherever the land surface does not receive and accumulate sediment. For example, dry land, except where sand dunes or alluvial fans are growing, will be an unconformity in the future. Geologists recognize three kinds of unconformity:

- *Angular unconformity:* Rocks below an **angular unconformity** were tilted or folded before the unconformity developed (▶Fig. 12.9a). Thus, the unconformity cuts across the layers below, which have a different orientation from the layers above. (We can see an angular unconformity in the outcrop at Siccar Point.) Angular unconformities form where rocks were squashed, uplifted, and eroded during mountain building, and then later became submerged during a rise in sea level.

- *Nonconformity:* A **nonconformity** is a type of unconformity at which sedimentary rocks overlie intrusive igneous rocks and/or metamorphic rocks (▶Fig. 12.9b). The igneous or metamorphic rocks must have cooled, been uplifted, and been exposed by erosion to form the substrate on which new sedimentary rocks were deposited. At a nonconformity, you typically find pebbles of the igneous or metamorphic rock in the lowermost bed of the sedimentary sequence.

- *Disconformity:* Imagine that a sequence of sedimentary beds has been deposited beneath a shallow sea. Then the sea level drops, and the recently deposited beds are exposed for some time. During this time, no new sediment accumulates, and some of the preexisting sediment gets eroded away. Later, the sea level rises, and a new sequence of sediment accumulates over the old. The boundary between the two sequences is a **disconformity** (▶Fig. 12.9c). Even though the beds above and below the disconformity are parallel, the contact between them represents an interruption in deposition. Disconformities may be hard to recognize unless you notice evidence of erosion (such as stream channels) or soil formation at the disconformity surface, or can identify a distinct gap in the succession of fossils.

The succession of strata at a particular location provides a record of Earth history there. But because of unconformities, the record preserved in the rock layers is incomplete (▶Fig. 12.9d). It's as if geologic history is being chronicled by a tape recorder that turns on only intermittently—when it's on (times of deposition), the rock record accumulates,

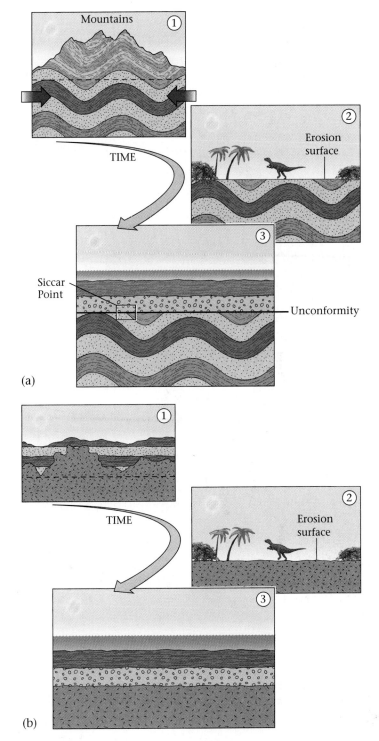

FIGURE 12.9 (a) The stages during the development of an angular unconformity. (b) The stages during the development of a nonconformity.

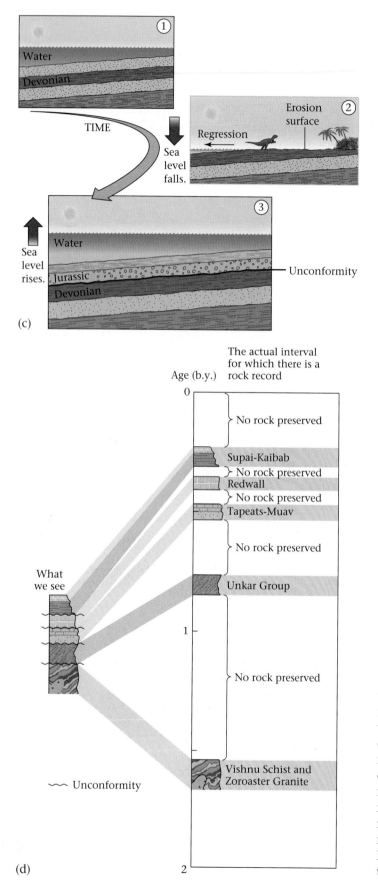

(c)

(d)

FIGURE 12.9 (continued) (c) The stages during the development of a disconformity. Note that regardless of the details, an unconformity represents a surface of erosion and/or a period of nondeposition. (d) Because of unconformities, the stack of strata exposed in the Grand Canyon represents only bits and pieces of geologic history. If the strata are projected on a numerical time scale, you can see that large intervals of time are not accounted for.

but when it's off (times of nondeposition and/or erosion), an unconformity develops. Because of unconformities, no single location on Earth contains a complete record of Earth history.

12.7 STRATIGRAPHIC FORMATIONS AND THEIR CORRELATION

The traveler on the brink looks from afar and is overwhelmed with the sublimity of massive forms; the traveler among the gorges stands in the presence of awful mysteries.
—JOHN WESLEY POWELL (1834–1902), describing the Grand Canyon

Geologists summarize information about the sequence of strata at a location by drawing a **stratigraphic column.** Typically, we draw columns to scale, so that the relative thicknesses of layers portrayed on the column represent the thicknesses of layers in the outcrop. Then, for ease of reference, geologists divide the sequence of strata represented on a column into **stratigraphic formations** (**formations** for short), recognizable layers of a specific rock type or group of rock types deposited during a specific time interval, that can be traced over a fairly broad region. The boundary surface between two formations is a type of geologic **contact.**

Let's see how the concept of a stratigraphic formation applies to the Grand Canyon. The walls of the canyon look striped, because they expose a variety of rock types that differ in color and in resistance to erosion. Geologists identify the major contrasts and use them as a basis to divide the strata into formations, each of which may consist of many beds (▶Fig. 12.10a–c). Note that some formations include a single rock type, while others include interlayered beds of two or more rock types. Also, note that not all formations have the same thickness. Typically, geologists name a formation after a locality where it was first identified. If the formation consists of only one rock type, we may incorporate that rock type in the name (for example, Kaibab Limestone), but if the formation contains more than one rock type, we use the word "formation" in the

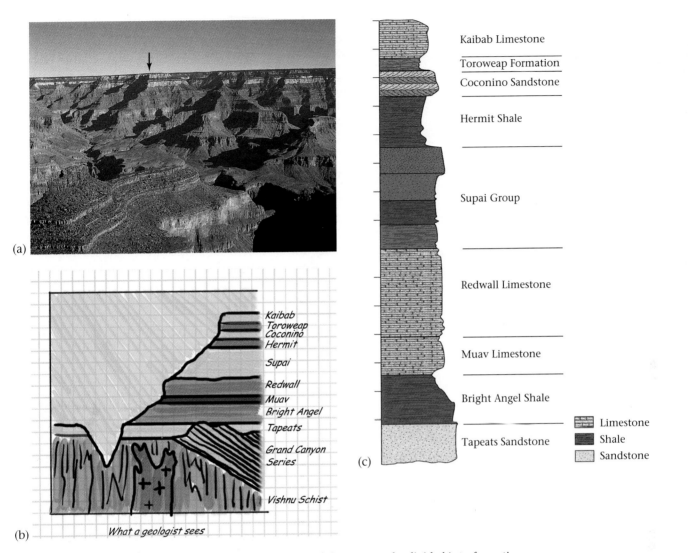

FIGURE 12.10 (a, b) The succession of rocks in the Grand Canyon can be divided into formations, based on notable changes in rock type and changes in fossil assemblages. (c) The sequence of strata in the Grand Canyon can be represented on a stratigraphic column. The vertical scale gives relative thicknesses. The right-hand edge of the column represents resistance to erosion (for example, Coconino Sandstone is more resistant than Hermit Shale).

name (Toroweap Formation; note that both words are capitalized). Several related formations in a succession may be lumped together as a **group.**

While he was excavating canals in England, William Smith discovered that formations cropping out at one locality resembled formations cropping out at another, in that their beds looked similar and contained similar fossil assemblages. In other words, Smith was able to define the age relationship between the strata at one locality and the strata at another, a process called **correlation.**

How does correlation work? Typically, geologists correlate formations between *nearby* regions based on similarities in rock type. We call this method of correlation **lithologic correlation** (▶Fig. 12.11a–c). For example, the sequence of strata on the south rim of the Grand Canyon clearly correlates with the sequence on the north rim, because they contain the same rock types in the same order. In some cases, a sequence contains a **key bed,** a particularly unique bed that provides a definitive basis for correlation. But to correlate units over *broad* areas, we must rely on fossils to define the relative ages of sedimentary units. We call this method **fossil correlation.** Geologists must rely on fossil correlation for studies of broad areas because sources of sediments and depositional environments may change from one location to another. The beds deposited at one location during a given time interval may look quite different from the beds deposited at another location during the same time interval.

(a)

|← 1 km →|

(b)

(c)

A   B   C

- Oswaldo Formation
- Franklin Ash
- Emma Shale
- Rufus Limestone
- David Sandstone
- Metamorphic basement
- Unconformity

FIGURE 12.11 The principle of lithologic correlation. (a) These three outcrops of rock are a couple of kilometers from each other. (b) The stratigraphic sections at each location are somewhat different, but the columns can be correlated with one another by matching rock types. Note that the Franklin Ash is a key bed, a distinctive layer that can be correlated with certainty. And the Rufus Limestone "pinches out" (thins and disappears along its length), so the contact between the Emma Shale and the David Sandstone in column C is an unconformity. (c) Geologists reconstruct a sedimentary basin using correlation. The eastward thinning of sedimentary layers suggests that the basin tapered to the east.

Now let's look at an example of fossil correlation by tracing the individual formations exposed in the Grand Canyon into the mountains just north of Las Vegas, 150 km to the west (▶Fig. 12.12a, b). Near Las Vegas, we find a sequence of sedimentary rocks that includes a limestone formation called the Monte Cristo Limestone. The Monte Cristo Limestone contains the same fossils as the Redwall Limestone of the Grand Canyon, but is much thicker. Because the formations contain the same fossils, we conclude that they were deposited during the same time interval, and thus we say that they correlate with one another. Not only are the units thicker in the Las Vegas area than in the Grand Canyon area, there are more of them. In the Grand Canyon, the Temple Butte Limestone lies directly be-

neath the Redwall. But the unit near Las Vegas that contains the same fossils as the Temple Butte lies 1,000 m below the Monte Cristo Limestone. In fact, the stratigraphic column at Las Vegas includes many formations that do not correlate with any units at the Grand Canyon. This discovery indicates that during part of the time when thick sediments were deposited near Las Vegas, none accumulated near the Grand Canyon. Thus, the contact between the Grand Canyon's Redwall Limestone and Temple Butte Limestone is an unconformity.

The contrast between the stratigraphic columns of Las Vegas and the Grand Canyon represents a contrast in depositional setting. The Las Vegas column accumulated on the thinner crust of a passive-margin basin that sank (subsided) rapidly and remained submerged below the sea almost continuously. The Grand Canyon column, on the other hand, accumulated on the thicker crust of a craton that periodically emerged above sea level.

By correlating strata at many locations, William Smith realized that he could trace individual formations of strata

FIGURE 12.12 (a) Based on their fossil content, the Redwall Limestone of the Grand Canyon correlates with the Monte Cristo Limestone near Las Vegas. But the fossils in the Muav Limestone, which lies directly beneath the Redwall at the Grand Canyon, correlate with those of a unit 1,000 m below the Monte Cristo Limestone. The sequence of strata between the two limestones at Las Vegas is not represented by any rock at the Grand Canyon; thus, the Redwall-Muav contact is an unconformity. (b) In the Paleozoic era, a sedimentary basin thinned radically between Las Vegas and the Grand Canyon. Thus, the sequence of strata deposited at the Grand Canyon was thinner and less complete than the section deposited at Las Vegas.

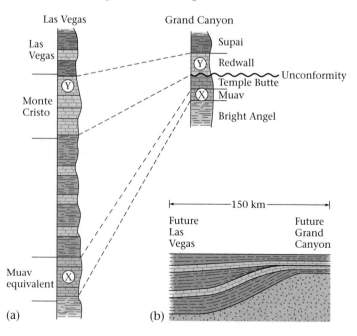

Las Vegas   Grand Canyon

- Las Vegas
- Monte Cristo
- Muav equivalent

- Supai
- Redwall
- Unconformity
- Temple Butte
- Muav
- Bright Angel

|← 150 km →|

Future Las Vegas   Future Grand Canyon

(a)   (b)

(a)

Block diagram

(b) Map

Explanation

(unconformity)

5 km

FIGURE 12.13 (a) The block diagram depicts an angular unconformity with horizontal strata above and folded strata below. (b) The map shows what this ground surface would look like if viewed from above. On the map, the contacts have been projected onto a flat piece of paper.

over fairly broad regions. In 1815, he plotted the distribution of formations and created the first modern **geologic map,** which portrays the spatial distribution of rock units at the Earth's surface. As ▶Figure 12.13 shows, we can use the information provided by a geologic map to identify geologic structures.

12.8 THE GEOLOGIC COLUMN

As stated earlier, no one locality on Earth provides a complete record of our planet's history, because stratigraphic columns can contain unconformities. But by correlating rocks from locality to locality at millions of places around the world, geologists have pieced together a *composite stratigraphic column,* called the **geologic column,** that represents just about the entirety of Earth history (▶Fig. 12.14a–c). The column is divided into segments, each of which represents a specific interval of time. The largest sub-

divisions break Earth history into the Hadean, Archean, Proterozoic, and Phanerozoic **eons.** (The first three comprise the **Precambrian.**) The suffix "zoic" means "life," so "Phanerozoic" means "visible life," and "Proterozoic" means "beginning life" (a bit misleading, because fossil evidence shows that the earliest life, bacteria, appeared during the Archean eon). In Phanerozoic time, organisms with hard parts (shells and, later, skeletons) became widespread, so there are abundant fossils from this eon, whereas in Precambrian time, there were either no fossils or only small organisms with no shells; Precambrian fossils are rare and hard to see.

The Phanerozoic eon is also subdivided into **eras.** In order from oldest to youngest, they are the Paleozoic ("ancient life"), Mesozoic ("middle life"), and Cenozoic ("recent life") eras. We can further divide each era into **periods** and each period into **epochs.**

Where do the names of the periods come from? They refer either to localities where a fairly complete stratigraphic column representing that time interval was identified (for example, rocks representing the Devonian period crop out near Devon, England) or to a characteristic of the time (rocks from the Carboniferous period contain a lot of coal), or they come from Latin roots. Some of the names may be familiar to you ("Jurassic," for example), but many may be new. The terminology was not set up in a planned fashion that would make it easy to learn; instead, it grew haphazardly in the years between 1760 and 1845 as geologists began to refine their understanding of geologic history and fossil succession.

Note that the way geologists divide time into named intervals resembles the way historians divide human history. For example, intervals of Chinese history have names like "Han dynasty" and "Chin dynasty," while intervals of British history are called "Elizabethan," "Victorian," or "Edwardian," each named after the reigning monarch of the time. Geologists know that Jurassic rocks are younger than Permian rocks, just as historians know that Victorian furniture is younger than Elizabethan furniture.

Fossils and the Geologic Column

The succession of fossils in the geologic column describes the course of life's evolution through Earth history. In fact, geologists now define the age range of fossil species in terms of the periods and epochs of the geologic column, and can use fossils alone to determine the relative age of a sedimentary rock. Notably, when Charles Darwin sailed on the *Beagle* in 1831, he carried with him a copy of Charles Lyell's 1830 *Principles of Geology,* which detailed Hutton's uniformitarian view of the Earth, the principle of fossil succession, and William Smith's concepts of stratigraphy. Reading this book primed Darwin to develop the theory of evolution by natural selection (as described in Interlude D).

Limestone: reef in warm seas

Cross-bedded sandstone:
sand dunes in a desert

Fault scarp:
a consequence
of recent faulting

Present-day erosion surface

Gypsum beds: a playa lake in a desert

Granite:
an intrusion
of silicic
magma at
depth

Basalt dike:
a result of
igneous activity

Trilobite

Cephalopod

**Fossils for
dating**

Metamorphic
aureole

Brachiopod

The Record in Rocks: Reconstructing Geologic History

Ignimbrite (welded tuff): an explosive volcanic eruption

Limestone: reef in warm seas

Redbeds: sand and mud deposited in a river channel and bordering floodplain

Basalt lava: flows from a volcano

Radiometric dating

Daughter isotope

Feldspar crystal

Parent isotope

Conglomerate: debris eroded from a cliff

Redbeds: sand and mud deposited by distributaries of a delta plain

Unconformity

Conglomerate: deposits of a pebble beach

Gneiss: metamorphism at depth beneath a mountain belt

When geologists examine a sequence of rocks exposed on a cliff, they see a record of Earth history that can be interpreted by applying the basic principles of geology, searching for fossils, and using radiometric dating. In this canyon, we see evidence for many geologic events—the layers of sediment (and the sedimentary structures they contain), the igneous intrusions, and the geologic structures tell us about past climates and past tectonic activity.

The insets show the way the region looked in the past, based on the record in the rocks. For example, the presence of gneiss at the base of the canyon indicates that at one time the region was mountain belt, for the protoliths of the gneiss were buried deeply. Unconformities indicate that the region underwent uplift and erosion. Sedimentary successions record transgressions and regressions of the sea, igneous rocks are evidence of volcanic and intrusive activity, and faults indicate deformation.

Clearly, the land surface portrayed in this painting was sometimes a river floodplain or a delta (indicated by red beds), sometimes a shallow sea (limestone), and sometimes a desert dune field (cross-bedded sandstones). And at several times in the past, volcanic activity occurred in the region. We can gain insight into the age of the sedimentary rocks by studying the fossils they contain, and the age of the igneous and metamorphic rocks by using radiometric dating methods.

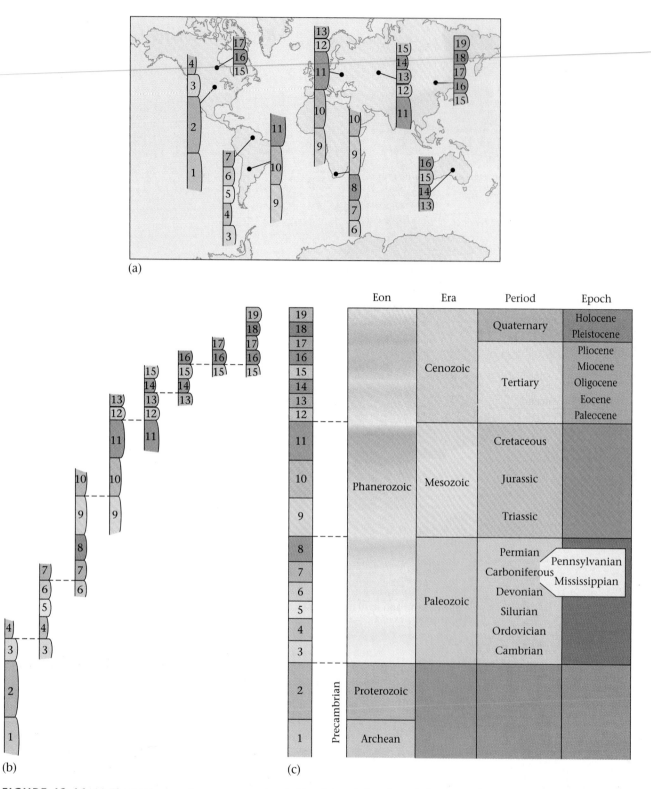

FIGURE 12.14 (a) The geologic column was constructed by determining the relative ages of stratigraphic columns from around the world. (b) By correlation, these columns were stacked one on top of the other. (c) The column was first built without knowledge of numerical ages, so while we can depict the sequence of periods, we cannot indicate their relative duration.

FIGURE 12.15 Life evolution in the context of the geologic column. Some of the names here will not be introduced until later in the chapter, or in Chapter 13.

By studying the succession of fossils in strata, paleontologists have been able to lay out the framework of life's evolution. Simple bacteria appeared in the Archean eon, while complex invertebrates did not evolve until the late Proterozoic. The appearance of invertebrates with shells defines the Precambrian-Cambrian boundary, while the first vertebrates, fish, appeared in the Ordovician period. Before the Silurian period, the land surface was barren of multicellular life, but in Silurian time, land plants spread over the landscape, followed in the Devonian by the first amphibians. Though reptiles evolved during the Pennsylvanian period, the first dinosaurs did not pound across the land until the Triassic. Dinosaurs continued to inhabit the Earth until their sudden extinction at the end of the Cretaceous period. For this reason, we refer to the Mesozoic era as the "Age of Dinosaurs." Small mammals also appeared in the Triassic period, but the **diversification** (development of many different species) of mammals to fill a wide range of ecological

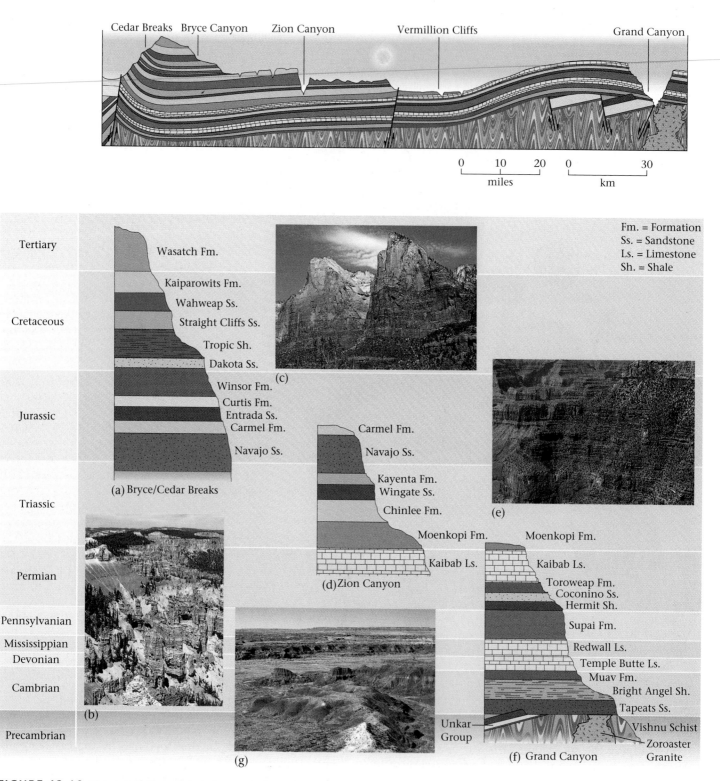

Cedar Breaks   Bryce Canyon   Zion Canyon   Vermillion Cliffs   Grand Canyon

0   10   20   0   30
miles   km

Tertiary

Wasatch Fm.

Cretaceous

Kaiparowits Fm.
Wahweap Ss.
Straight Cliffs Ss.
Tropic Sh.
Dakota Ss.

(c)

Fm. = Formation
Ss. = Sandstone
Ls. = Limestone
Sh. = Shale

Jurassic

Winsor Fm.
Curtis Fm.
Entrada Ss.
Carmel Fm.
Navajo Ss.

Carmel Fm.
Navajo Ss.

(a) Bryce/Cedar Breaks

Triassic

Kayenta Fm.
Wingate Ss.
Chinlee Fm.

Moenkopi Fm.

(e)

Moenkopi Fm.

Kaibab Ls.

(d) Zion Canyon

Kaibab Ls.

Permian

Toroweap Fm.
Coconino Ss.
Hermit Sh.

Pennsylvanian

Supai Fm.

Mississippian
Devonian

Redwall Ls.
Temple Butte Ls.
Muav Fm.

Cambrian

Bright Angel Sh.
Tapeats Ss.

(b)

Precambrian

Unkar
Group

Vishnu Schist

(g)

(f) Grand Canyon

Zoroaster
Granite

FIGURE 12.16 The correlation of strata between the various national parks of Arizona and Utah: (a, b) Bryce Canyon/Cedar Breaks, (c, d) Zion Canyon, (e, f) the Grand Canyon, (g) the Painted Desert. The inset at the top shows a cross section of the region.

niches did not happen until the beginning of the Cenozoic era, so we now call the Cenozoic the "Age of Mammals." Birds also appeared during the Age of Dinosaurs (specifically, at the beginning of the Cretaceous period), but underwent great diversification in the Cenozoic era (▶Fig. 12.15).

By correlating strata, geologists have concluded that the oldest strata in the national parks of the Colorado Plateau (in northern Arizona and southern Utah) are located in the Grand Canyon. These layers were once overlain in succession by the rocks now exposed in the Painted Desert, the Petrified Forest, Zion Canyon, Bryce Canyon, and Cedar Breaks (▶Fig. 12.16a–g). Walking through these parks is thus like walking through time—each layer gives an indication of the climate and topography of the past. For example, when the Precambrian metamorphic and igneous rocks exposed in the inner gorge of the Grand Canyon first formed, the region was a high mountain range, perhaps as dramatic as the Himalayas today. When the fossiliferous beds of the Kaibab Limestone at the rim of the canyon first developed, the region was not a high, dusty plateau but rather a Bahama-like carbonate reef and platform, bathed in a warm, shallow sea. And when the rocks making up the towering red cliffs of the Navajo Sandstone in Zion Canyon were deposited, the region was a Sahara-like desert, blanketed with huge sand dunes.

Clearly, the Earth changes significantly through time. And by looking at the fossils in these rocks, we find creatures that vanished long ago from the planet, showing that life has changed too. Still, spectacular as the stratigraphic record of the Colorado Plateau may be, it is not complete, for it contains many unconformities.

12.9 NUMERICAL AGE AND THE RADIOMETRIC CLOCK

By saying that World War II was fought after World War I, a historian provides the relative ages of the two events. By saying that World War II began in 1939, while World War I began in 1914, a historian automatically defines the relative ages of the events, but also provides their numerical ages by placing them on a scaled time line (calibrated in years) of human history. Historians know the numerical ages of events because they can consult a written record keyed to the calendar. Geologists since the days of Hutton could determine relative ages of geologic events and could construct the geologic column, but unlike historians, they had no way to specify numerical ages; they could not define a time line for Earth history or determine the duration of events. This situation changed with the discovery of radioactivity in the twentieth century.

Simply put, radioactive elements decay at a constant rate that can be measured in the lab and can be specified in years. Around 1950, geologists developed techniques for using measurements of radioactive elements to calculate the ages of rocks. We call the science of dating geologic events in years **radiometric dating** (or **geochronology**). But before looking at this technique, we must first learn about radioactive decay.

Radioactive Decay and the Concept of a Half-Life

All atoms of a given element have the same number of protons in their nucleus—we call this number the **atomic number** (see Appendix A). However, not all atoms have the same number of neutrons in their nucleus. Therefore, not all atoms of a given element have the same **atomic weight**—the number of protons plus neutrons. Different versions of an element, called **isotopes** of the element, have the same atomic number but a different atomic weight. For example, all uranium atoms have 92 protons, but the uranium-238 isotope (abbreviated ^{238}U) has an atomic weight of 238 and thus has 146 neutrons, while the ^{235}U isotope has an atomic weight of 235 and thus has 143 neutrons.

Some isotopes of an element are stable, meaning that they last essentially forever. **Radioactive isotopes** are unstable: after a given time, they undergo a change called **radioactive decay,** which converts them into a different element. Radioactive decay can take place by a variety of reactions. In these reactions, the isotope that undergoes decay is the **parent isotope,** while the decay product is the **daughter isotope.** Some kinds of decay reactions create daughter isotopes with the same atomic weight as the parent, while others result in a daughter isotope with a different atomic weight. For example, rubidium-87 (^{87}Rb) is a radioactive isotope that decays to strontium-87 (^{87}Sr), and potassium-40 (^{40}K) is a radioactive isotope that decays to argon-40 (^{40}Ar), but uranium-238 (^{238}U) decays to form lead-206 (^{206}Pb).

Physicists cannot specify how long an individual radioactive isotope will survive before it decays, but they *can* measure how long it takes for half of a group of an element's isotopes to decay. This time is called the **half-life** of the element. ▶Figure 12.17 can help you visualize the concept of a half-life. Imagine a crystal containing 16 radioactive parent isotopes. After one half-life, 8 isotopes have decayed, so the crystal now contains 8 parent and 8 daughter isotopes. After a second half-life, 4 of the remaining parent isotopes have decayed, so the crystal contains 4 parent and 12 daughter isotopes. And after a third half-life, 2 more parent isotopes have decayed, so the crystal contains 2 parent and 14 daughter isotopes. Note that we cannot predict which specific isotopes decay at which time, only that during a half-life, half the parent isotopes decay to form daughter isotopes, and that the overall rate of decay decreases with time because there are fewer parent isotopes left.

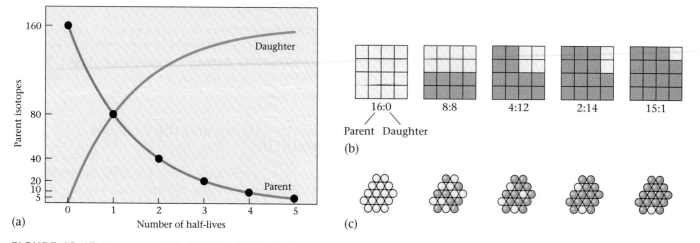

FIGURE 12.17 The concept of a half-life. (a) Graph showing the decrease in the number of parent isotopes as time passes; the curve is "exponential," meaning that the rate of change in the parent-daughter ratio decreases with time. (b) The ratio of parent-to-daughter isotopes changes with the passage of each successive half-life. (c) A cluster of isotopes undergoing decay. There's no way to predict which parent will decay in a given time interval.

Radiometric Dating Technique

Like the ticktock of a clock, radioactive decay proceeds at a known rate and thus provides a basis for telling time. As we can measure daily time in terms of revolutions of a clock's hands around the dial, so we can measure geologic time with the "radiometric clock." Because an element's half-life is a constant, the ratio between parent and daughter isotopes changes at a known rate (quickly at first, more slowly later on). Thus, by knowing the rate of decay, we can calculate the age of a mineral by measuring the ratio of parent to daughter isotopes in the mineral.

How do geologists actually obtain a radiometric date? First, we must find the right kind of elements to work with. Although there are many different pairs of parent and daughter isotopes among the known elements, only a few have long enough half-lives and occur in sufficient abundance to be useful for radiometric dating. Particularly useful elements include ^{238}U, which decays to ^{206}Pb; ^{235}U, which decays to ^{207}Pb; ^{40}K, which decays to ^{40}Ar; ^{87}Rb, which decays to ^{87}Sr; and samarium-147 (^{147}Sm), which decays to neodymium-143 (^{143}Nd) (see Table 12.1). Each radioactive element has its own half-life.

Second, we must identify the right kind of minerals to work with. Not all minerals contain radioactive elements, but fortunately some common minerals do. For example, feldspar, mica, and hornblende contain potassium and rubidium,

zircon contains uranium, and garnet contains samarium. Once we have identified appropriate minerals containing appropriate elements, we can set to work. Radiometric dating consists of the following steps.

- *Collecting the rocks:* Geologists collect fresh (unweathered) rocks for dating. The chemical reactions that happen during weathering may allow daughter isotopes to leak out of minerals, in which case a date from the rock has no valid meaning.

- *Separating the minerals:* The fresh rocks are crushed, and the appropriate minerals are separated from the debris.

- *Extracting parent and daughter isotopes:* To separate out the parent and daughter isotopes from minerals, geologists either dissolve the minerals in acid or evaporate them

TABLE 12.1 Isotopes Used in the Radiometric Dating of Rocks

Parent → Daughter	Half-Life (years)	Minerals in Which the Isotopes Occur
$^{147}Sm \rightarrow {}^{143}Nd$	106 billion	Garnets, micas
$^{87}Rb \rightarrow {}^{87}Sr$	47 billion	Potassium-bearing minerals (mica, feldspar, hornblende)
$^{238}U \rightarrow {}^{206}Pb$	4.5 billion	Uranium-bearing minerals (zircon, uraninite)
$^{40}K \rightarrow {}^{40}Ar$	1.3 billion	Potassium-bearing minerals (mica, feldspar, hornblende)
$^{235}U \rightarrow {}^{207}Pb$	713 million	Uranium-bearing minerals (zircon, uraninite)

(a)

(b)

FIGURE 12.18 (a) A lab used in preparing samples for radiometric dating. The air must be exceedingly clean so that stray parent or daughter isotopes don't contaminate the samples. (b) The heart of this mass spectrograph, used for radiometric dating, is a large magnet.

with a laser. This work must take place in a very clean lab, to avoid contaminating samples with parent or daughter isotopes from the atmosphere (▶Fig. 12.18a).

• *Analyzing the parent-daughter ratio:* Geologists pass the dissolved or evaporated minerals through a mass spectrometer, a complex instrument that uses a magnet to separate isotopes from one another according to their respective weight, and then measures the ratio of parent to daughter isotopes (▶Fig. 12.18b).

At the end of the laboratory process, geologists can define the ratio of parent to daughter isotopes in a mineral, and from this ratio calculate the age of the mineral. Needless to say, the description of the procedure here has been simplified—in reality, obtaining a radiometric date is time consuming and expensive and requires complex calculations. Generally, when they report radiometric dates, geologists also report the **uncertainty** of the measurement. Uncertainty, which defines the range of values in which the true measurement lies, arises because no instrument can count atoms perfectly. Uncertainties for radiometric dates may be on the order of 1% or less. For example, a date may be reported as 200 ± (plus or minus) 2 million years. The older the rock, the larger the uncertainty.

What Does a Radiometric Date Mean?

At high temperatures, isotopes in a crystal lattice vibrate so rapidly that chemical bonds can break and reattach relatively easily. As a consequence, parent and daughter isotopes escape from or move into crystals, so parent-daughter ratios are meaningless. Because radiometric dating is based on the parent-daughter ratio, the radiometric clock starts only when crystals become cool enough for both parent and daughter isotopes to be locked into the lattice. The temperature below which isotopes are no longer free to move is called the **blocking temperature** of a mineral—the blocking temperature is typically hundreds of degrees cooler than the melting temperature of a mineral (▶Fig. 12.19a, b). Not all

FIGURE 12.19 (a) In this example, above the blocking temperature, daughter isotopes escape, so that the radiometric clock cannot work. (b) Below the blocking temperature, both parent and daughter isotopes remain locked in the crystal.

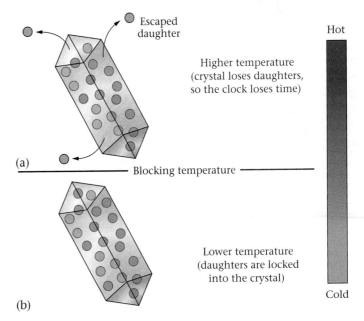

(a)

(b)

minerals have the same blocking temperature; for example, the blocking temperature of hornblende (an amphibole) is much higher than that of biotite (a mica). When we specify a radiometric date for a rock, we are defining the time at which minerals in the rock cooled below their respective blocking temperatures.

With the concept of blocking temperature in mind, we can interpret the meaning of radiometric dates. In the case of igneous rocks, radiometric dating tells you when a magma or lava cooled to form a solid, cool igneous rock. In the case of metamorphic rocks, a radiometric date tells you when a rock cooled from the high temperature of metamorphism down to a low temperature. If a rock cools quickly (as when a lava flow freezes), then all minerals are roughly the same age, but if a rock cools slowly (as when a pluton cools slowly at depth in the Earth), minerals with high blocking temperatures give older ages than minerals with low blocking temperatures.

Can we radiometrically date a sedimentary rock directly? No. If we date the minerals in a sedimentary rock, we determine only when the minerals comprising the sedimentary rock first crystallized as part of an igneous or metamorphic rock, not the time the minerals were deposited as sediment nor the time the sediment lithified to form a sedimentary rock. For example, if we date the feldspar grains contained in a granite pebble in a conglomerate, we're dating the time the granite cooled below feldspar's blocking temperature, not the time the pebble was deposited by a stream.

Other Methods of Determining Numerical Age

Counting rings in trees or layers in sediment. The changes in seasons affect a wide variety of phenomena, including the following.

- *The growth rate of trees:* Trees grow seasonally, with rapid growth during the spring and no growth during the winter.

- *The organic productivity of lakes and seas:* During the winter, less light reaches the Earth, water temperatures decrease, and the supply of nutrients decreases.

- *The sediment supply carried by rivers:* During the rainy season, the volume and velocity of water in rivers increase, and thus rivers carry more sediment.

- *The growth rate of chemically precipitated sedimentary rocks (such as travertine):* Factors that control precipitation rates, such as the rate of groundwater flow and temperature, change seasonally in some locations.

- *The growth rate of shell-secreting organisms:* Organisms grow and produce new shell material on a seasonal basis.

- *The proportion of snow, relative to dust, that falls in cold climates:* During the snowy season, more snow falls relative to dust than during the dry season; annual snowfall includes a dusty layer and a clean layer.

BOX 12.1

THE REST OF THE STORY

Carbon-14 Dating

Many people who have heard of **carbon-14 (^{14}C) dating** assume that it can be used to define the numerical age of rocks. But this is not the case. Rather, ^{14}C dating tells us the ages of organic material—such as wood, cotton fibers, charcoal, flesh, bones, and shells—that contain carbon originally extracted from the atmosphere by photosynthesis in plants. ^{14}C, a radioactive isotope of carbon, forms naturally in the atmosphere when cosmic rays (charged particles from space) bombard atmospheric nitrogen-14 (^{14}N) atoms. When plants consume carbon dioxide during photosynthesis, or when animals consume plants, they ingest a tiny amount of ^{14}C along with ^{12}C, the more common isotope of carbon. After an organism dies and can no longer exchange carbon with the atmosphere, the ^{14}C

in its body begins to decay back to ^{14}N again. Thus, the ratio of ^{14}C to ^{12}C changes at a rate determined by the half-life of ^{14}C.

We can use ^{14}C dating to determine the age of prehistoric fire pits or of organic debris in sediment. ^{14}C has a short half-life—only 5,730 years. Thus, the method cannot be used to date anything older than about 70,000 years, for after that time essentially no ^{14}C remains in the material. But this range makes it a useful tool for geologists studying sediments of the last ice age and for archaeologists studying ancient cultures or prehistoric peoples. Since rocks do not contain organic carbon, and may be significantly older than 70,000 years, we cannot date rocks with ^{14}C.

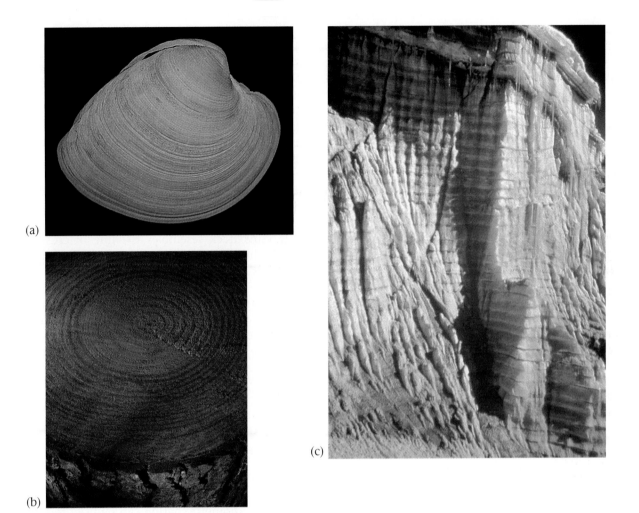

FIGURE 12.20 (a) A shell with growth rings; (b) tree rings; (c) rhythmic layers in the ice of a glacier.

As a consequence of these seasonal changes, **growth rings** develop in trees, travertine deposits, and shelly organisms, and **rhythmic layering** develops in sedimentary accumulations and glacier ice (▶Fig. 12.20a–c). By counting rings or layers, geologists can determine how long an organism survived, or how long a sedimentary accumulation took to form. If rings or layers have developed right up to the present, we can count backward and determine how long ago they began to form.

Dendrochronologists, scientists who study and date tree rings, have found that not only do tree rings count time, they preserve a record of the climate: in warm, rainy years, trees grow much faster than they do in drought years. Thus, by studying tree rings, dendrochronologists can provide a dated record of the climate in prehistoric times. The oldest living trees, bristlecone pines, are almost 4,000 years old. By correlating the older rings of these trees with rings in logs preserved in sediment accumulations, dendrochro-

nologists have extended the tree-ring record back for many thousands of years.

Similarly, geologists have found that glacier ice preserves a valuable record of the past—ice cores drilled through the Greenland ice cap, for example, contain a continuous record of glacial accumulation back through 160,000 years. Further, geologists have found that the ratio of different oxygen isotopes in the water molecules making up the ice reflects the global temperature at the time the snow fell to create the ice, and thus provides another clue to past climate. These data help geologists to date ice-age events.

Magnetostratigraphy. As discussed in Chapter 3, the polarity of Earth's magnetic field flips every now and then through geological time. For igneous rocks less than about 4.5 million years old, the uncertainty of radiometric dating methods is smaller than the length of a polarity chron.

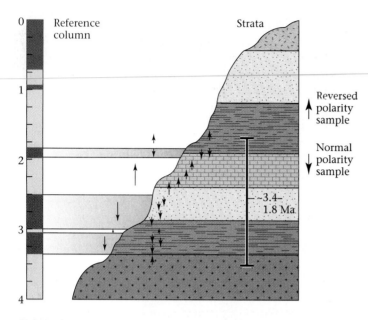

FIGURE 12.21 Magnetostratigraphy compares the sequence of reversals in strata with the sequence of reversals in a global reference column to determine the age of the strata.

FIGURE 12.22 (a) Photomicrograph of fission tracks in a crystal. Though only a few atoms wide, the tracks are visible because the sample has been treated with a solvent that preferentially dissolves the crystal along the tracks. (b) How a fission track is formed.

Thus, geologists have determined when reversals took place, and have constructed a reference column showing the succession of reversals through time (▶Fig. 12.21). This reference column resembles the bar code on a supermarket product. Knowledge of this bar code allows us to date sediments: by comparing the pattern of the reversals in a sequence of strata with the bar code of the reference column, a study known as **magnetostratigraphy**, geologists can determine the age of the sequence.

Fission tracks. In certain minerals, the ejection of an atomic particle during the decay of a radioactive isotope damages the nearby crystal lattice, creating a line called a **fission track.** This track resembles the line of crushed grass left behind when you roll a tire across a lawn. As time passes, more atoms undergo fission, so the number of fission tracks in the crystal increases (▶Fig. 12.22a, b). Therefore, the number of fission tracks in a given volume of a crystal represents the age of the crystal. Geologists have been able to measure the rate at which fission tracks are produced, and thus can determine the age of a mineral grain by counting the fission tracks within it.

12.10 ADDING NUMERICAL AGES TO THE GEOLOGIC COLUMN: DATING PERIODS

We have seen that radiometric dating can be used to date the time when igneous rocks formed and when metamorphic rocks metamorphosed, but not when sedimentary rocks were deposited. So how *do* we determine the numerical age of a sedimentary rock? We must answer this question if we want to add numerical ages to the geologic column—the column was originally constructed by studying only the *relative* ages of fossil-bearing sedimentary rocks.

Geologists obtain dates for sedimentary rocks by studying cross-cutting relationships between sedimentary rocks and datable igneous or metamorphic rocks. For example, if we find a sedimentary rock layer deposited unconformably on a datable igneous or metamorphic rock, we know that the sedimentary rock must be younger. If we find a datable igneous dike or pluton that cuts across beds of sedimentary rock, then the datable rock must be younger. And if a datable lava flow or ash layer spread out over a layer of sediment and then was buried by another layer of sediment, then the datable rock or ash must be younger than the underlying sediment and older than the overlying sediment.

Let's look at an example of dating sedimentary rocks (▶Fig. 12.23). Imagine that a layer of sandstone on Wingtip

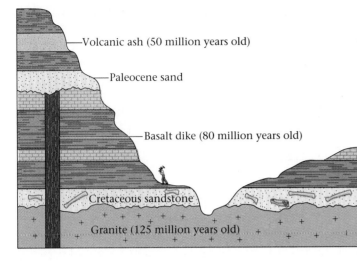

Volcanic ash (50 million years old)

Paleocene sand

Basalt dike (80 million years old)

Cretaceous sandstone

Granite (125 million years old)

FIGURE 12.23 The Cretaceous sandstone strata were deposited unconformably on a 125-million-year-old granite pluton, so they must be younger than the granite. The 80-million-year-old dike cuts across the sandstone, so the sandstone must be older than the dike. Similarly, the Paleocene bed was unconformably deposited over the dike and lies beneath a 50-million-year-old ash. Thus, the Paleocene bed must be between 80 and 50 million years old.

Mountain contains fossilized dinosaur bones. Geologists assign the fossils to the Cretaceous period by correlating them with fossils in known Cretaceous strata from elsewhere. At Wingtip Mountain, the sandstone was deposited unconformably over an eroded granite pluton whose numerical age determined by uranium-lead dating is 125 million years. This same layer of sandstone has been cut by a basalt dike whose numerical age based on potassium-argon dating is 80 million years. Our measurements mean that this Cretaceous sandstone bed was deposited sometime between 125 million and 80 million years ago. Note that the data provide an age range, not an exact age. Thus we know that the Cretaceous period includes the time interval of *at least* 125–80 million years ago.

Geologists have searched the world for localities where we can recognize cross-cutting relations between datable igneous rocks and sedimentary rocks; by radiometrically dating the igneous rocks, they have been able to provide numerical ages for the boundaries between all the geological periods. For example, work from around the world shows that the Cretaceous period actually began about 145 million years ago and ended 65 million years ago. The constant discovery of new data requires the numbers to change, which is why the term "numerical age" is preferred to "absolute age." In fact, around 1995, new research showed that the Cambrian-Precambrian boundary occurred at about 545 million years ago, in contrast to previous, less definitive studies that had placed the boundary at 570 million years

ago. ▶Figure 12.24 shows the currently favored numerical ages of periods and eras in the geologic column. This dated column is commonly called the **geologic time scale.** Because of the numerical constraints provided by the geologic time scale, when geologists say that the first dinosaurs appeared during the Triassic period, they mean that dinosaurs appeared about 245 million years ago.

12.11 THE AGE OF THE EARTH

In the eighteenth and nineteenth centuries, before the discovery of radiometric dating, scientists came up with a great variety of clever solutions to the question "How old is the Earth?" all of which have since been proven wrong. One approach assumed that sediments have been deposited at a constant rate of about 1 mm per year. Dividing the thickest known pile of sediments (about 20 km) by this number yields an age of about 20 million years. But this approach doesn't take unconformities into account, and neglects the fact that some sediments eventually transform into metamorphic rocks. Another approach assumed that the world's oceans started out as freshwater that then slowly accumulated salt. By measuring the rate of salt (whose elements come from the weathering of rock) flowing into the oceans (salt is carried to the ocean by rivers), you arrive at an age of about 90 million years. It turns out, however, that the salt content of the sea has remained fairly constant over time, because excess salt precipitates out of seawater to form evaporite minerals; so salt concentration is not a measure of the Earth's age. Lord William Kelvin, a nineteenth-century physicist renowned for his discoveries in thermodynamics, made the most influential scientific estimate of the Earth's age. Kelvin calculated how long it would take for the Earth to cool down from a temperature as hot as the Sun's, and concluded that the Earth is, again, about 20 million years old.

In sum, estimates of the Earth's age proposed before the twentieth century implied that the planet was not much older than 100 million years. This view contrasted with the idea being promoted by followers of Hutton and Darwin, who argued that if the concepts of uniformitarianism and evolution were correct, the Earth must be much older, because physical processes that shape the Earth and form its rocks and the process of natural selection that yields the diversity of species take a long time. Geologists and physicists debated the age issue for many years. The solution came in 1896, when physicist Henri Becquerel announced the discovery of radioactivity. Geologists immediately realized that the Earth's interior was producing heat from the decay of radioactive material. This realization uncovered the flaw in Kelvin's argument: Kelvin had assumed that no heat was added to the planet after it had cooled down from the origi-

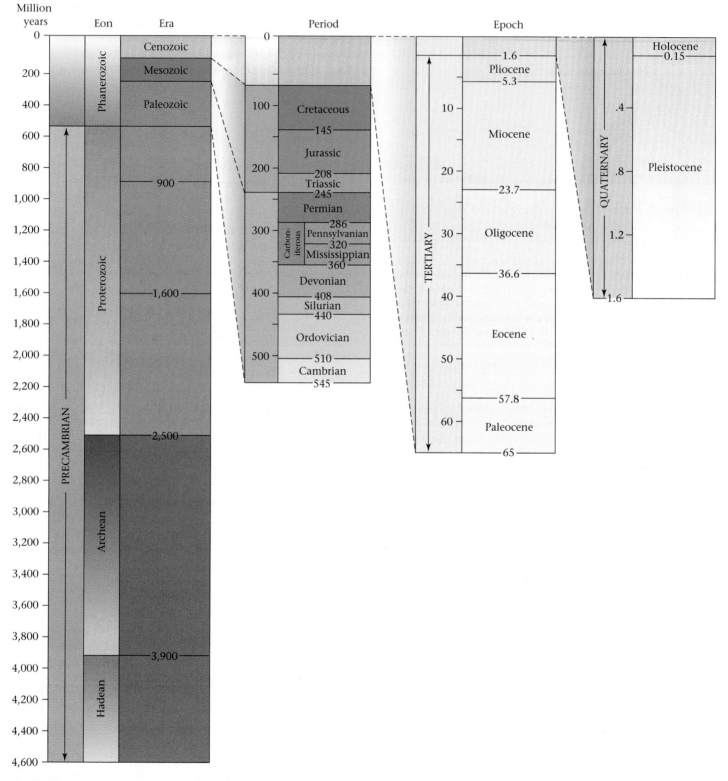

FIGURE 12.24 The geologic time scale assigns numerical ages to the intervals on the geologic column. Note that we have to change to a larger scale to portray the ages of intervals higher in the column.

nal solar nebula. Because radioactivity constantly generates new heat in the Earth, the planet has cooled down much more slowly than Kelvin had calculated. In 1904, British physicist Ernest Rutherford presented this discovery to an audience that included Kelvin, as he later recounted:

> I came into the room, which was half dark, and presently spotted Lord Kelvin in the audience and realized that I was in trouble at the last part of the speech dealing with the age of the Earth, where my views conflicted with his. To my relief, Kelvin fell fast asleep, but as I came to the important point, I saw the old bird sit up, open an eye and cock a baleful glance at me!

The discovery of radioactivity not only invalidated Kelvin's estimate of the Earth's age, it also led to the development of radiometric dating.

Since the 1950s, geologists have scoured the planet to identify its oldest rocks. Samples from several localities (Wyoming, Canada, Greenland, and China) have yielded dates as old as 3.96 billion years. And sandstones found in Australia contain clastic grains of zircon that reveal dates of 4.1–4.2 billion years, indicating that rock as old as 4.2 billion years did once exist (▶Fig. 12.25a). Models of the Earth's formation assume that all objects in the solar system developed at roughly the same time from the same nebula. Radiometric dating of meteors and Moon rocks have yielded ages as old as 4.6 billion years (▶Fig. 12.25b); geologists take this to be the approximate age of the Earth, leaving more than enough time for the rocks and life forms of the Earth to have formed and evolved.

We don't find 4.6 billion-year-old rocks in the crust because during the first half-billion years of Earth history, rocks in the crust remained too hot for the radiometric clock to start (their temperature stayed above the blocking temperature). Geologists have named the time interval between the birth of the Earth and the formation of the oldest dated rock the Hadean eon (see Fig. 12.24).

12.12 PICTURING GEOLOGIC TIME

The mind grows giddy gazing so far back into the abyss of time.
—JOHN PLAYFAIR (1747–1819, British geologist who popularized the works of Hutton)

The number 4.6 billion is so staggeringly large that we can't begin to comprehend it. If you lined up 4.6 billion pennies in a row, they would make an 87,400-km-long line that would wrap around the Earth's equator more than twice (▶Fig. 12.26). Notably, at the scale of our penny chain, human history is only about 100 city blocks long.

Another way to grasp the immensity of geologic time is to make a scale model, which we do by equating the entire 4.6 billion years to a single calendar year. On this scale, the oldest rocks on Earth date from early February, and the first bacteria appear in the ocean on February 21. The first shelly invertebrates appear on October 25, and the first amphibians crawl out onto land on November 20. On December 7, the continents coalesce into the supercontinent of Pangaea.

FIGURE 12.25 (a) One of the oldest dated rocks on Earth, this 3.96-billion-year-old gneiss comes from the Northwest Territories of Canada. (b) A Moon rock.

(a)

(b)

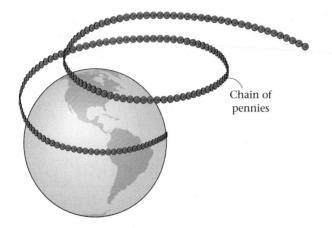

Chain of pennies

FIGURE 12.26 We can use the analogy of distance to represent the duration of geologic time.

The first mammals and birds appear about December 15, along with the dinosaurs, and the Age of Dinosaurs ends on Christmas Day. The last week of December represents the last 65 million years of Earth history, covering the entire Age of Mammals. The first human-like ancestor appears on December 31 at 3 P.M., and our species, *Homo sapiens,* shows up an hour before New Year's Eve. The last ice age ends a minute before midnight, and all of recorded human history takes place in the last 30 seconds. Put another way, human history occupies the last 0.000001% of Earth history.

CHAPTER SUMMARY

• The concept of geologic time, the span of time since the Earth's formation, developed during the eighteenth century, when early geologists (most notably James Hutton) suggested that the Earth must be very old if geologic features formed by same natural processes we see today. This idea—that the present is the key to the past—came to be known as the principle of uniformitarianism.

• Relative age specifies whether one geologic feature is older or younger than another; numerical age provides the age of a geologic feature in years.

• Using such principles as uniformitarianism, superposition, original horizontality, original continuity, cross-cutting relations, inclusions, and baked contacts, we can construct the geologic history of a region.

• A particular fossil species only occurs in a limited interval of strata. The relative age of this interval, with respect to the interval in which another species is found, is always the same. Because of this principle of fossil succession, geologists can define a succession of fossils—older species lie at the base, younger ones at the top—and can determine the age of a bed. Once a fossil species goes extinct, it never reappears higher in the succession.

• Strata are not deposited continuously at a location, because uplift of the land or a drop in global sea level exposes the region to erosion. An interval of nondeposition and/or erosion is called an unconformity. Geologists recognize three kinds: angular unconformity, nonconformity, and disconformity.

• A stratigraphic column shows the succession of formations in a region. A given succession of strata that can be traced over a fairly broad region is called a stratigraphic formation. The boundary surface between two formations is a contact. Several related formations in succession comprise a group. The process of determining the relationship between strata at one location and strata at another is called correlation. A geologic map shows the distribution of formations and the contacts between them.

• A composite stratigraphic column that represents the entirety of geologic time is called the geologic column. The geologic column's largest subdivisions, each of which represents a specific interval of time, are eons (Hadean, Archean, Proterozoic, Phanerozoic). Eons are further subdivided into eras, eras into periods, and periods into epochs.

• The numerical age of rocks can be determined by radiometric dating. This is because radioactive elements, found in small concentrations in many minerals, decay at a constant rate. During radioactive decay, parent isotopes transform into daughter isotopes. The decay rate for a given element is known as its half-life, the time it takes for half of the parent isotopes to decay. The ratio of parent to daughter isotopes in a mineral grain defines the age of the grain.

• The radiometric date of a mineral specifies the time at which the mineral cooled below a certain temperature, known as the blocking temperature. We can thus use radiometric dating to determine when an igneous rock solidified from a melt and when a metamorphic rock cooled from high metamorphic temperatures.

• Other methods for dating materials include counting rings in trees, growth rates of shells, ice layers in glaciers, and fission tracks in mineral grains. We can also compare the sequence of magnetic reversals in strata with the known reversal chronology.

• From the radiometric dating of meteors and Moon rocks, geologists have concluded that the Earth formed about 4.6 billion years ago. Our species, *Homo sapiens,* has been around for only 0.000001% of geologic time.

KEY TERMS

REVIEW QUESTIONS

1. How was the duration of a second originally defined? How is it defined now?

2. How do sidereal and tropical years differ?

3. Why do calendars get out of synch with the seasons?

4. What was Steno's insight into the nature of fossils?

5. Compare numerical age and relative age.

6. Describe seven principles that allow us to determine the relative ages of geologic events.

7. How does the principle of fossil succession allow us to determine the relative ages of geologic events?

8. How does an unconformity develop? What does it tell us about the pace of geologic events?

9. Describe the differences between the three kinds of unconformities.

10. Describe two different methods of correlating rock units. How was correlation used to develop the geologic column? What is a stratigraphic formation?

11. What does the process of radioactive decay entail?

12. How do geologists obtain a radiometric date? What are some of the pitfalls in obtaining a reliable one?

13. Why can't we date sedimentary rocks directly?

14. Why is carbon-14 dating used in archaeology, but useless for dating dinosaur fossils?

15. How are growth rings and ice cores useful in determining the ages of geologic events?

16. How are the reversals of the Earth's magnetic field useful in dating geologic events?

17. Why is it inappropriate to use the term "absolute age" for numerical dating?

18. Why did early scientists think the Earth was less than 100 million years old?

19. How did the discovery of radioactivity invalidate Kelvin's assumptions about the Earth's age and also provide a method for obtaining its true age?

20. How do you obtain a radiometric date? Why is knowledge of an element's half-life essential to the process?

21. What are you actually dating when you get a radiometric date? How can you determine numerical ages for sedimentary strata?

22. What is the age of the oldest rocks on Earth? What is the age of the oldest rocks known? Why is there a difference?

SUGGESTED READING

Berggren, W. A., D. V. Kent, M.-P. Aubry, and J. Hardenbol, eds. 1995. *Geochronology, Time Scales, and Global Stratigraphic Correlation.* SEPM Special Publication 54.

Berry, W. B. N. 1987. *Growth of a Prehistoric Time Scale.* 2nd ed. Palo Alto, Calif.: Blackwell Scientific Publications.

Burchfield, J. D. 1975. *Lord Kelvin and the Age of the Earth.* New York: Science History Publications.

Dalrymple, G. B. 1991. *The Age of the Earth.* Palo Alto, Calif.: Stanford University Press.

Dott, R. H., Jr., and D. R. Prothero. 1994. *Evolution of the Earth.* 5th ed. New York: McGraw-Hill.

Eicher, D. L. 1976. *Geologic Time.* 2nd ed. Englewood Cliffs, N.J.: Prentice-Hall.

Faul, H., and C. Faul. 1983. *It Began with a Stone: A History of Geology from the Stone Age to Plate Tectonics.* New York: Wiley.

Faure, G. 1986. *Principles of Isotope Geology.* 2nd ed. New York: Wiley.

Gould, S. J. 1987. *Time's Arrow, Time's Cycle.* Cambridge, Mass.: Harvard University Press.

Harland, W. B., et al. 1990. *A Geologic Time Scale 1989.* New York: Cambridge University Press.

Prothero, D. R. 1990. *Interpreting the Stratigraphic Record.* New York: Freeman.

Schoch, R. M. 1989. *Stratigraphy: Principles and Methods.* New York: Van Nostrand Reinhold.

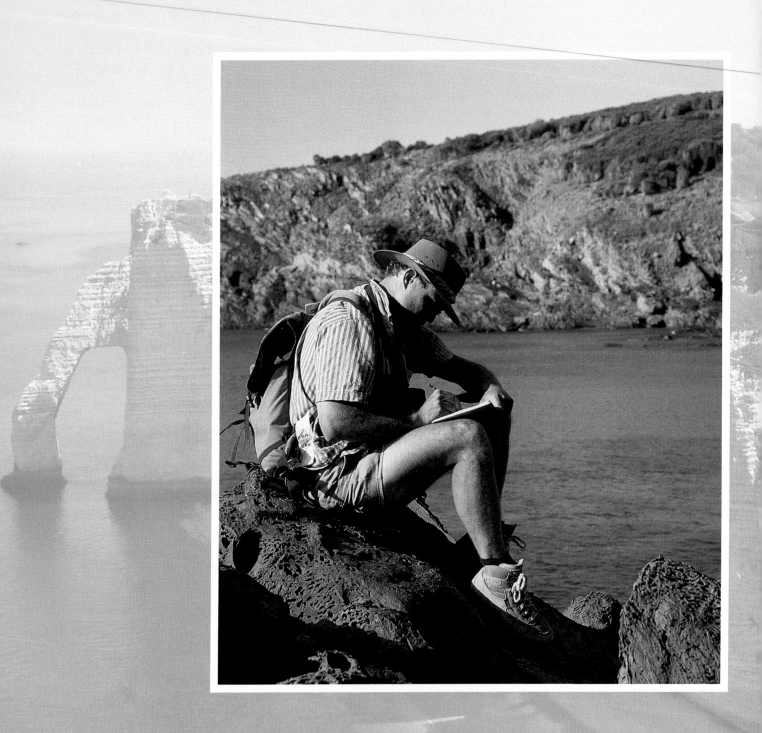

CHAPTER 13

A Biography of Earth

I weigh my words well when I assert that the man who should know the true history of the bit of chalk which every carpenter carries about in his breeches pocket, though ignorant of all other history, is likely, if he will think his knowledge out to its ultimate results, to have a truer and therefore a better conception of this wonderful universe and of man's relation to it than the most learned student who [has] deep-read the records of humanity [but is] ignorant of those of nature.

—THOMAS HENRY HUXLEY, from *ON A PIECE OF CHALK* (1868)

13.1 INTRODUCTION

In 1868, a well-known British scientist, Thomas Henry Huxley, presented a public lecture on geology to an audience in Norwich, England. Seeking a way to convey his fascination with Earth history to people with no previous geological knowledge, he focused his audience's attention on the piece of chalk he had been writing with (see epigraph above). And what a tale the chalk has to tell! Chalk, a type of limestone, consists of microscopic marine algae shells and shrimp feces. The specific chalk that Huxley held came from beds deposited in Cretaceous time (the name "Cretaceous," in fact, derives from the Latin word for "chalk") and now exposed along the White Cliffs of Dover (▶Fig. 13.1). Geologists in Huxley's day knew of similar chalk beds in outcrops throughout much of Europe, and had discovered that the chalk contains not only plankton shells but also fossils of bizarre swimming reptiles, fish, and invertebrates—species absent in the seas of today. Clearly, when the chalk was deposited, warm seas holding unfamiliar creatures covered much of what is dry land today.

Clues in his humble piece of chalk allowed Huxley to demonstrate to his audience that the geography and inhabitants of the Earth in the past differed markedly from those today, and thus that *the Earth has a history*. In the many decades since Huxley's lecture, studies of stratigraphy, radiometric dating, and rock chemistry, along with the results of mapping expeditions worldwide, have allowed geologists to develop an overall image of Earth's history. This chapter offers a concise geolog-

A geologist at work in the field, here on the shore of Kangaroo Island, Australia. Such field observations have allowed geologists to work out the history of Earth.

395

FIGURE 13.1 Chalk beds, created from layers of deep-sea sediment, exposed along the coast of England.

ical biography of the planet, from its birth 4.6 billion years ago to the present. After learning the methods geologists use to interpret the past, we see how continents came into existence and then waltzed across the globe. We also learn when mountain-building events took place and how Earth's climate and sea level have changed through time—in fact, the climate alternates from being relatively warm (greenhouse conditions) to being relatively cool (icehouse conditions). And while these physical changes have taken place, life has evolved. To simplify the discussion, the following abbreviations are used: Ga (for "billion years ago"), Ma ("million years ago"), and Ka ("thousand years ago").

13.2 METHODS FOR STUDYING THE PAST

When historians outline human history, they describe daily life, wars, economics, governments, leaders, inventions, and explorations. When geologists outline Earth history, we describe the distribution of depositional environments, mountain-building events (orogenies), past climates, life evolution, the changing positions of continents, the past configuration of plate boundaries, and changes in the composition of the atmosphere and oceans. Historians collect data by reading written accounts, examining relics and monuments, and, for more recent events, listening to recordings or watching videos. Geologists collect data by examining rocks, geologic structures, and fossils and, for more recent events, by studying sediments, ice cores, and tree rings.

Figuring out Earth's past hasn't been an easy task for geologists, because the available record isn't complete—Earth materials (the record of the past) don't form continuously through time, and erosion may destroy some materials. Not surprisingly, the record of early Earth history is vaguer than the record for more recent history, for older rocks are more likely to have been eroded away or metamorphosed, and because the uncertainty attached to a radiometric age measurement becomes larger as the rocks get older. Nevertheless, we can find enough of the record to at least outline major geologic events of the past. Following are a few examples of how we use observational data to study Earth history.

- *Identifying orogenies (mountain-building events):* We identify mountain belts (also called orogenic belts, or orogens) today by looking for regions of high, rugged topography. However, since it takes only about 50 million years to erode a mountain range entirely away, we cannot identify mountain-building events, or orogenies, of the past simply by studying topography; rather, we must look for the rock record they leave behind. Orogeny causes igneous activity, deformation (such as folds and faults), and metamorphism. Thus, we can recognize an ancient orogen by looking for a belt of crust containing these features (►Fig. 13.2), and we can date the orogen using radiometric dating.

FIGURE 13.2 Even after the topography of a mountain range has been eroded away, the deformed and metamorphosed rocks and plutons that formed within it can still be recognized. These complex rocks contrast with surrounding regions and define the position of the orogen.

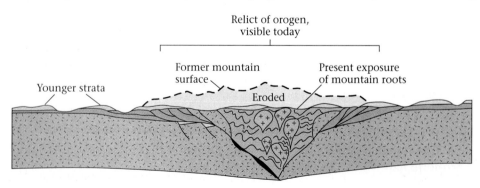

Orogeny also leads to the development of unconformities, for uplift exposes rocks to erosion, and of **foreland sedimentary basins** (located on the plains adjacent to the mountain front), for the weight of mountains pushes crust down, creating a depression that traps sediment. This sediment has been eroded from the mountains, so sequences of strata may provide a record of nearby uplift and erosion.

- *Recognizing the growth of continents:* Not all continental crust formed at the same time. In order to determine how a continent grew, geologists find the ages of different regions of the crust using modern radiometric dating techniques. They can figure not only when rocks originally formed from magmas rising out of the mantle, but also when the rocks were metamorphosed during a subsequent orogeny. The identities of the rock types making up the crust indicate the tectonic environment in which the crust formed.

- *Recognizing past environments:* As we learned in Chapter 12's discussion of the Grand Canyon, the environment at a particular location changes through time. To learn about these changes, we study successions of sedimentary rocks, for the environment controls both the type of sediment deposited at a location and the type of organisms that live there. For example, conglomerates containing land plants probably formed in an alluvial fan or a stream channel, while limestone containing coral fossils probably developed in a shallow sea.

- *Recognizing past changes in the relative sea level:* We can determine whether the sea level has gone up or down at a certain spot by looking for changes in the depositional environment. For example, where a marine limestone overlies an alluvial-fan conglomerate, the relative sea level rose at that site. By correlating sedimentary successions at many places around the world, we can identify a global sea-level change.

- *Recognizing past positions of continents:* To help us find out where a continent lay in the past, we have three sources of information. First, the study of apparent polar-wander paths can tell us the latitude of a continent in the past (see the discussion of paleomagnetism in Chapter 3; ▶Fig. 13.3a, b). Second, for more recent Earth history (post-Jurassic), we can reconstruct the change over time in the width of an ocean basin by studying marine magnetic anomalies. (All ocean floor older than Jurassic has been subducted, so we have no marine magnetic anomaly data for pre-Jurassic time; ▶Fig. 13.3c, d.) Third, we can compare fossils from different areas to see if the same species occur in these areas. This may be the case if the regions were adjacent.

- *Recognizing past climates:* We can gain insight into past climates by looking at fossils and rock types that formed at given latitudes. For example, if organisms requiring semitropical conditions lived near the poles during a given time period, then the atmosphere overall must have been warmer. In the past two decades, geologists have also learned how to use the ratios of isotopes for certain elements in fossil shells (for example, $^{18}O/^{16}O$) as a measure of past temperatures, for the ratio depends on the temperature.

- *Recognizing life evolution:* If we accept evidence supporting the principle of fossil succession, then progressive changes in the assemblage of fossils in a sequence of strata represent changes in the assemblage of organisms inhabiting Earth through time.

13.3 THE HADEAN EON: HELL ON EARTH

A viscid pitch boiled in the fosse below
and coated all the bank with gluey mire.
I saw the pitch; but I saw nothing in it
except the enormous bubbles of its boiling
which swelled and sank, like breathing, through all the pit.
—DANTE, *THE INFERNO*, CANTO XXI

The oldest mineral grains yet found on Earth come from sandstones in western Australia, and yield dates of 4.1–4.2 Ga (billion years ago). The oldest complete rocks are found in gneiss outcrops in Canada, Wyoming, Greenland, and China, with dates of 4.03 Ga. Yet meteorites have yielded dates of up to 4.6 Ga. If, as implied by the nebula hypothesis, the Earth formed at the same time as the meteorites, then apparently we have no direct record of the first 600 million years of Earth history. This lack of a record gives us an important clue to the nature of the newborn Earth: all materials at the surface must have been so hot during the first 600 million years that radiometric clocks could not start "ticking." Geologists call this interval of time the Hadean (from the Greek "Hades"), in acknowledgment of the hellish conditions that prevailed.

The Hadean eon began with Earth's formation out of an accumulation of planetesimals (see Chapter 1). As the planet grew, the collection and compression of matter into a ball generated substantial heat. In fact, each time the Earth collided with another planetesimal, it became warmer, as the kinetic energy of the planetesimal (meteor) transformed into heat. The newborn planet also warmed because it contained a relatively high concentration of radioactive atoms, which produce heat upon decay.

Eventually, the Earth became hot enough to begin melting, and when this happened, molten iron sank to the center to form the core, leaving behind the shell of rock that became the mantle. This event may have happened very fast—geologists refer to it as the **iron catastrophe.** The

(a)   (b)

(c)

(d)

FIGURE 13.3 Geologists use apparent polar-wander paths as clues to the past positions of continents. (a) This hypothetical example shows the movement of two continents (1 and 2). At 400 Ma (million years ago), the continents are separate. They collide at 300 Ma and move together as a supercontinent until 200 Ma. Then they rift apart, and by 100 Ma they have been separated by a substantial distance. Note that now the two continents each contain part of the former collisional orogen. (b) The apparent polar-wander paths, which geologists measure today, for the two continents. Note that the paths coincide when the continents move together as a supercontinent. (c) Geologists can use marine magnetic anomalies to define relative motions of continents for the past 200 million years. This map shows the age of sea floor (in Ma) corresponding with specific marine magnetic anomalies. If you remove the strip of sea floor between the coast and the 81-Ma anomaly (or between the coast and the 63-Ma anomaly, etc.), you will see the positions of the continents at 81 Ma (or 63 Ma, etc.). (d) This reconstruction of the position of Europe and Africa, relative to a fixed North America, is based on marine magnetic anomalies. Note how Europe and Africa progressively move away from North America with time.

FIGURE 13.4 A painting of the Hadean Earth. Note the magma ocean, with small crusts of solid basalt floating about. If the sky had been clear, streaks of meteors would have filled it, and the Moon would have appeared much larger than it does today, because it was closer. Probably, however, an observer on Earth could not have seen the sky, because the atmosphere contained so much water vapor and volcanic ash.

Hadean Earth grew so hot that a **magma ocean,** consisting of ultramafic melts (very rich in magnesium and iron) rising from the mantle, flooded its surface (▶Fig. 13.4). Volcanic activity provided gases that accumulated to form an unbreathable atmosphere of nitrogen (N_2), ammonia (NH_3), methane (CH_4), water (H_2O), carbon monoxide (CO), carbon dioxide (CO_2), and sulfur dioxide (SO_2). Perhaps a thin solid crust of ultramafic rock formed on the magma ocean, cooled by contact with the atmosphere; but if so, vigorous convection of the magma ocean soon swallowed and completely melted most of this crust, and frequent meteor impacts liquefied or pulverized any that remained. Thus, during the Hadean eon, Earth had no permanent crust, and because its surface temperature far exceeded the boiling point of water, it had no ocean.

13.4 THE ARCHEAN EON: THE BIRTH OF THE CRUST, THE OCEANS, AND LIFE

About 4 billion years ago, our planet had cooled sufficiently for the magma ocean on its surface to freeze for good, and the surface's outermost layer segmented into rigid plates, similar to modern oceanic lithosphere plates. When these plates subducted, partial melting yielded mafic-to-intermediate-composition magmas that rose and erupted at volcanic arcs. In addition, large piles of volcanic rocks probably formed at immense hot-spot volcanoes over mantle plumes. Both the volcanic arcs and the hot-spot volcanoes were too buoyant to be subducted, so they collided and stuck together to create the first continental crust. The oldest rocks on Earth come from these blocks, so with their formation, the Archean eon (from the Greek word for "ancient"), the first interval of Earth history for which we have a rock record (4.03–2.5 Ga), began.

Because the solid Earth remained quite hot during Archean time, convective movement in the mantle happened rapidly. Geologists speculate that this condition led to the formation of many mantle plumes and small, rapidly moving plates and thus to the creation of numerous volcanic island arcs and hot-spot volcanoes. Frequent collisions sutured volcanic arcs and hot-spot volcanoes together, creating progressively larger blocks called **protocontinents** (▶Fig. 13.5a, b). Some of these protocontinents developed rifts that filled with basalt. As the Earth gradually cooled, protocontinents became cooler and stronger, and by 2.7 Ga the first **cratons,** long-lived blocks of durable continental crust, had developed. By the end of the Archean eon, about 80% of continental crust had formed (▶Fig. 13.6).

Archean cratons contain five principal rock types: **gneiss** (relicts of Archean metamorphism in collisional zones), **greenstone** (metamorphosed relicts of ocean crust trapped between colliding blocks, or of basalts that had filled early continental rifts), **granite** (formed from magmas generated by the partial melting of the crust in continental volcanic arcs or above hot spots), **graywacke** (a mixture of sand and clay eroded from the volcanic areas and dumped into the nearby deep ocean), and **chert** (formed by the precipitation of silica in the deep sea). Archean shallow-water sediments are rare, either because

FIGURE 13.5 Construction of the first continental crust. (a) In the early Archean eon, plate tectonic processes had begun, and rocks that would eventually comprise continental crust began to form. At early subduction zones, volcanic island arcs, composed of relatively buoyant rocks, developed, and large shield volcanoes formed over hot spots. Larger crustal blocks underwent rifting, creating rift basins that filled with volcanic rocks, and the erosion of crustal blocks deposited graywacke (a mixture of sand and clay) on the sea floor. (b) Successive collisions brought all the buoyant fragments together to form a protocontinent. Melting at the base of the protocontinent may have caused the formation of plutons. Igneous rocks that had once comprised the ocean crust, or rift fill, became greenstone belts.

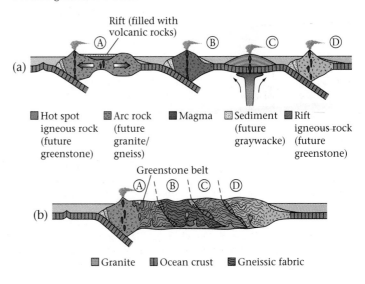

continents were so small that depositional environments in which such sediments could accumulate didn't exist, or because any that were present have since eroded away.

Volcanic activity during the Archean eon continued to supply gases, including water, that accumulated in the atmosphere. Some geologists speculate that the tails of comets that crossed Earth's orbit may also have contributed gases to the atmosphere. Eventually, as the Earth cooled, its surface dropped well below the boiling point of water, so the water vapor in the atmosphere condensed and rained onto the planet; and by about 3.8 Ga, a global ocean had accumulated and rivers flowed over the stark, unvegetated surface of the continents. Geologists reached this conclusion because sedimentary beds from this time contain clastic grains that were clearly rounded by transport in liquid water. Salts weathered out of rock and transported to the sea in rivers made the oceans salty.

Clearly, the Archean world saw many firsts in Earth history—the first plates, the first continents, and the first oceans. Archean rocks also contain the first record of life. Fossil evidence indicates that by about 3.8 Ga, **prokaryotic cells,** simple cells comprising bacteria and **cyanobacteria** (blue-green algae), had appeared on our planet, for chemical analyses of rocks of that age contain traces of **organic carbon,** carbon that has been incorporated in an organism.

FIGURE 13.6 As time progressed, the area of the Earth covered by continental crust increased, though not all geologists agree about the rate of increase. This model shows growth beginning about 3.9 Ga and continuing rapidly until the end of the Archean eon. The rate slowed substantially in the Proterozoic eon.

More direct evidence for life's appearance comes from 3.4-Ga chert beds in western Australia, which contain fossils that resemble cyanobacteria, and from 3.2-Ga rocks of South Africa, which include layered mounds of bacteria called **stromatolites.** Stromatolites form because the membranes of cyanobacteria secrete a sticky, mucous-like substance to which sediment sticks; as each layer of cyanobacteria gets buried by sediment, it colonizes the surface of the new sediment, building a mound upward (▶Fig. 13.7a–c).

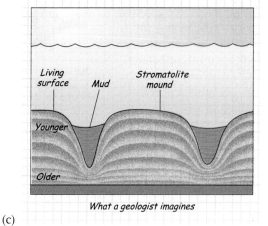

FIGURE 13.7 (a) The first fossils, found in 3.5-Ga rocks, look like strands of filamentous bacteria. (b, c) This outcrop shows a cross section through a stromatolite deposit. Stromatolites are mounds of bacteria that grow upward with time, with living bacteria making up the mounds' surface.

13.5 THE PROTEROZOIC EON: TRANSITION TO THE MODERN WORLD

The Proterozoic eon (from the Greek words meaning "first life") spans roughly 2 billion years, from about 2.5 billion years ago to the beginning of the Cambrian period, 545 million years ago—thus, it encompasses almost half of Earth's known history. The era received its now misleading name before fossil bacteria were discovered in Archean rock. During Proterozoic time, Earth's surface environment changed from the unfamiliar world of tiny, fast-moving plates, relatively small continents, and an oxygen-poor atmosphere to the more familiar world of relatively few large plates, large continents, and an oxygen-rich atmosphere.

First, let's look at changes to the lithosphere. New continental crust continued to form during the Proterozoic eon, but at progressively slower rates—in fact, by the middle of the eon, over 90% of the Earth's continental crust had formed. As the Earth continued to cool, plate tectonic processes slowed down, and oceanic plates grew larger. Also, collisions between the Archean cratons, as well as the accretion of new volcanic island arcs and hot-spot volcanoes, re-

sulted in the assembly of large cratons whose interiors were far from orogenic activity. Most of the large cratons that exist today had formed by about 1.8 Ga and remain intact (▶Fig. 13.8).

Let's look at the geology of a large craton a little more closely, by examining the interior of North America (▶Fig. 13.9). The Precambrian rocks comprising this craton crop out extensively in Canada. Geologists refer to this region as the **Canadian Shield.** A **shield** consists of a broad, low-lying region of exposed Precambrian rocks. In the United States, Phanerozoic strata bury most of the Precambrian rocks, forming a province called a **cratonic platform,** or **continental platform.** Here we see Precambrian rocks only where they have been exposed by erosion of the sedimentary cover. The Canadian Shield includes several Archean blocks that are sutured together along huge collisional belts. Of these belts, the largest is the Trans-Hudson orogen (▶Fig. 13.10). In the United States, Archean rocks underlie part of the Midwest and much of Wyoming, but most of the U.S. craton consists of crust that formed when a series of volcanic island arcs and continental slivers accreted (attached) to the southern margin of the Canadian Shield between 1.8 and 1.6 Ga. Such orogens, formed by the attachment of numerous buoyant slivers to an older, larger block, are called **accretionary orogens.**

FIGURE 13.8 The distribution of Precambrian crust. The cratons formed in Proterozoic time, either from magmas rising from the mantle or by the metamorphism and deformation of existing Archean crust.

FIGURE 13.9 On this map of North America, we see four different geological provinces: shield areas, where Precambrian rocks of the craton crop out extensively at the ground surface; platform areas, where Phanerozoic sedimentary rocks have buried Precambrian rocks of the craton; Phanerozoic orogenic belts, composed of rocks that formed or were deformed in mountain belts during the past half-billion years; and the coastal plain, low land buried by Cretaceous and Tertiary sediment.

In the Midwest, huge granite plutons intruded much of this region between 1.5 and 1.3 Ga.

Successive collisions ultimately brought together most continental crust on Earth into a single supercontinent, named **Rodinia,** by around 1 Ga. The last major collision during the formation of Rodinia produced a large collisional orogen called the **Grenville orogen.** Rocks that metamorphosed and deformed during this event crop out in eastern Canada and along the crest of the Appalachian Mountains in the eastern United States. If you look at a popular (though not totally proven) reconstruction of Rodinia, you can identify the crustal provinces that would eventually become the familiar continents of today (▶Fig. 13.11a). In this reconstruction, the future Antarctica and Australia lay somewhere along the west coast of North America, while fragments of the future South America lay to the east of Greenland and Canada.

Recent studies suggest that sometime between 800 and 600 Ma (million years ago), Rodinia "turned inside out," in that Antarctica, India, and Australia broke away from western North America and swung around and collided with the future South America, forming a new supercontinent that geologists refer to as **Pannotia** (▶Fig. 13.11b). The breakup of Rodinia resulted in the formation of rift basins and, eventually, a **passive-margin basin** (a thick accumulation of

sediment along a tectonically inactive coast) along the western edge of North America.

Not only did **paleogeography,** the past geography of the planet, undergo major changes during the Proterozoic eon, rock types also changed. In the sedimentary realm, shallow-marine sediments like limestone and quartz sandstone became common. Limestone became more abundant because calcite-secreting microorganisms (bacteria and algae) were more abundant.

FIGURE 13.10 The North American craton consists of a collage of different fragments, stitched together during orogenies of Precambrian time. (The explanation names these fragments and orogens.) We distinguish between collisional orogens, where large blocks joined, and accretionary orogens, composed of volcanic island arcs, hot-spot volcanoes, and continental slivers plastered onto the margin of a larger continent at a convergent margin.

☐ Phanerozoic orogen

☐ 1.1 Ga collisional orogen (G = Grenville)

⊞ 1.6–1.7 Ga accreted crust covered by 1.3–1.5 Ga granite and rhyolite, where patterned (GR = granite-rhyolite province)

☐ 1.6–1.7 Ga accreted crust (YM = Yavapai and Mazatzal)

☐ 1.8 Ga accreted crust (P = Penokean)

☐ 1.8 Ga collisional orogen (TH = Trans-Hudson; WP = Wopmay)

☐ 1.9 Ga collisional orogen (T = Thelon)

☐ Archean rocks, later deformed and metamorphosed in the Proterozoic (H = Hearn; R = Rae)

☐ Relics of Archean crust (WY = Wyoming; M = Mojave; S = Superior; N = Nain; SL = Slave)

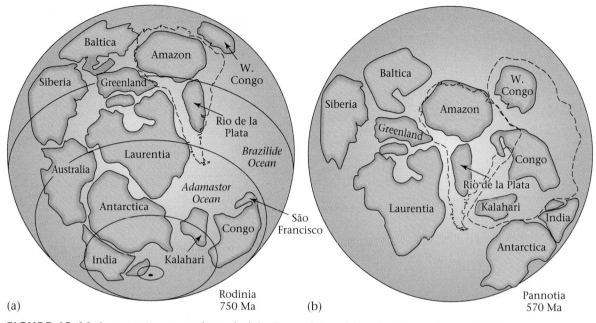

(a)   (b)
Rodinia   Pannotia
750 Ma   570 Ma

FIGURE 13.11 Supercontinents at the end of the Precambrian. (a) Rodinia formed around 1 Ga and lasted until about 700 Ma. (b) According to one model, by about 570 Ma, Antarctica, India, and Australia broke off the western margin of Rodinia and swung around and collided with the eastern margin of the future South America, to create a new, short-lived, supercontinent called Pannotia. Pannotia then broke apart at about 550 Ma.

Much of the transformation of the Earth's atmosphere from the oxygen-poor volcanic gas mix of the Archean eon to the oxygen-rich mix we breathe today occurred during the Proterozoic (▶Fig. 13.12; note that the present concentration of oxygen, 21%, was not attained until the Phanerozoic eon). This change probably reflects a large increase in the abundance of **photosynthetic** organisms, for such organisms remove gases like carbon dioxide, methane, and ammonia from the atmosphere, incorporate these gases into their tissues, and expel oxygen back to the atmosphere.

This new oxygen-rich atmosphere had a profound effect on the Earth, for it permitted a great diversification of life and the eventual conquest of the land by living organisms. Life could become more complex because oxygen-dependent metabolism produces energy much more efficiently than does oxygen-free metabolism—sulfide molecules may sustain bacteria, but they can't keep a clam or an elephant on the move! Life could eventually move onto land (in the Devonian period) because atmospheric oxygen provides the raw material for the production of ozone (O_3), a gas that screens out ultraviolet radiation from sunlight. Without the ozone in the atmosphere, skin and retinas would burn, and genetic mutations would occur at an accelerated rate.

The reason geologists have decided that the transformation to an oxygen-rich atmosphere happened during the Proterozoic (specifically, early Proterozoic) eon is that sedimentary sequences of this age contain thick **banded-iron formation** (**BIF**), a rock composed of alternating layers of hematite or magnetite (iron oxide) and chert (see Chapter 15). In the presence of abundant oxygen, iron becomes less soluble in seawater, so as oxygen grew progressively more concentrated in the atmosphere during the early Proterozoic eon, iron precipitated out of the sea and formed BIF. By the

FIGURE 13.12 The graph shows the change in the oxygen content of Earth's atmosphere through time. The vertical axis represents the percentage of the atmosphere that consists of oxygen. Today, oxygen accounts for about 21% of the atmosphere, but in Archean time it comprised only about 0.000000001%.

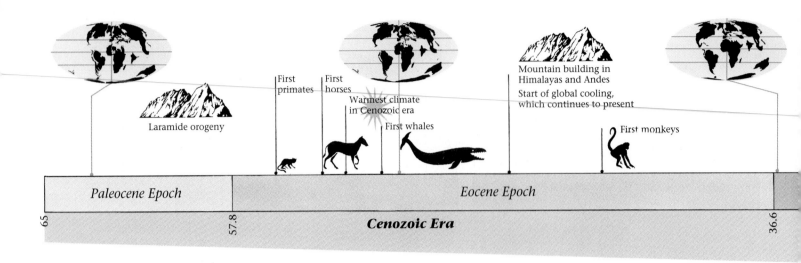

First primates

First horses

First whales

Warmest climate in Cenozoic era

Laramide orogeny

Mountain building in Himalayas and Andes

Start of global cooling, which continues to present

First monkeys

Paleocene Epoch	Eocene Epoch

65   57.8   **Cenozoic Era**   36.6

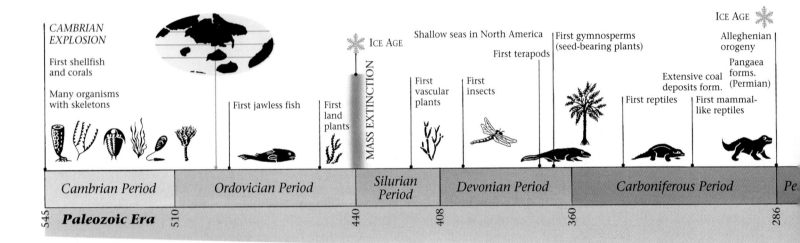

CAMBRIAN EXPLOSION

First shellfish and corals

Many organisms with skeletons

First jawless fish

First land plants

ICE AGE

MASS EXTINCTION

Shallow seas in North America

First vascular plants

First insects

First terapods

First gymnosperms (seed-bearing plants)

ICE AGE

Alleghenian orogeny

Extensive coal deposits form.

Pangaea forms. (Permian)

First reptiles

First mammal-like reptiles

Cambrian Period	Ordovician Period	Silurian Period	Devonian Period	Carboniferous Period	Pe

545   **Paleozoic Era**   510   440   408   360   286

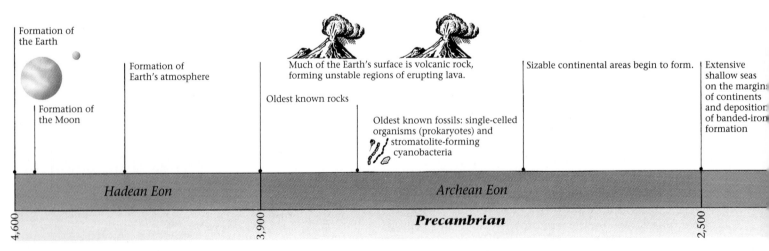

Formation of the Earth

Formation of the Moon

Formation of Earth's atmosphere

Oldest known rocks

Much of the Earth's surface is volcanic rock, forming unstable regions of erupting lava.

Oldest known fossils: single-celled organisms (prokaryotes) and stromatolite-forming cyanobacteria

Sizable continental areas begin to form.

Extensive shallow seas on the margins of continents and deposition of banded-iron formation

Hadean Eon	Archean Eon

4,600   3,900   **Precambrian**   2,500

The Evolution of Life

Earth has not had a static history; because of plate tectonics and its consequences (continental drift, sea-floor spreading, volcanism, etc.), the map of the Earth constantly changes. Distinct mountain-building events, or orogenies, have taken place during this process. The fossil record suggests that life first appeared within the first few hundred million years of our planet's existence, soon after a liquid-water ocean had accumulated, and like the planet itself has

constantly changed ever since. The progressive change of the assemblage of species of life on Earth is called evolution.

The earliest life forms were microscopic. By the end of the Precambrian, complex multicellular organisms had formed, and a burst of evolution at the Precambrian-Cambrian boundary yielded a diversity of invertebrates with shells. During the past half-billion years, several new major groups of organisms have appeared, and

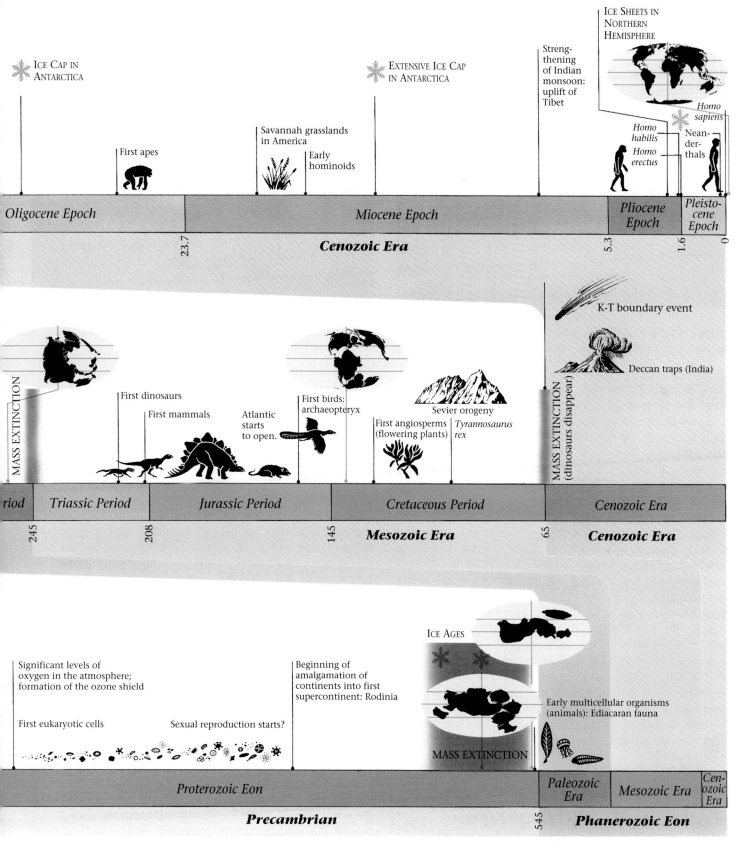

ICE CAP IN ANTARCTICA

First apes

Savannah grasslands in America

Early hominoids

EXTENSIVE ICE CAP IN ANTARCTICA

Strengthening of Indian monsoon: uplift of Tibet

ICE SHEETS IN NORTHERN HEMISPHERE

Homo habilis

Homo erectus

Homo sapiens

Neanderthals

Oligocene Epoch	Miocene Epoch	Pliocene Epoch	Pleistocene Epoch

23.7 — Cenozoic Era — 5.3 — 1.6 — 0

MASS EXTINCTION

First dinosaurs

First mammals

Atlantic starts to open.

First birds: archaeopteryx

Sevier orogeny

First angiosperms (flowering plants)

Tyrannosaurus rex

MASS EXTINCTION (dinosaurs disappear)

K-T boundary event

Deccan traps (India)

riod	Triassic Period	Jurassic Period	Cretaceous Period	Cenozoic Era

245 — 208 — 145 — **Mesozoic Era** — 65 — **Cenozoic Era**

Significant levels of oxygen in the atmosphere; formation of the ozone shield

First eukaryotic cells

Sexual reproduction starts?

Beginning of amalgamation of continents into first supercontinent: Rodinia

ICE AGES

Early multicellular organisms (animals): Ediacaran fauna

MASS EXTINCTION

Proterozoic Eon	Paleozoic Era	Mesozoic Era	Cenozoic Era

Precambrian — 545 — **Phanerozoic Eon**

Geological time (million years ago)

countless species have gone extinct. Evidence suggests that evolution is not a continuous, gradual process, but occurs in pulses, separated by intervals of time during which the assemblage of species is fairly stable.

The last few hundred million years of Earth history have seen life leave the ocean and spread across the land. During the Mesozoic era, dinosaurs roamed the Earth—then vanished abruptly 65

million years ago, perhaps as a result of a bolide colliding with the Earth. Since then, mammals have diversified into a great variety of species. And the last 100,000 years or so have witnessed the evolution of our own species, *Homo sapiens*. Considering the changes that people have brought about to the Earth system, the appearance of our species clearly represents a major event in the history of the planet.

(a)   (b)

FIGURE 13.13 Fossils of the Ediacaran fauna, an assemblage of soft-bodied organisms that appeared in late Proterozoic seas, had complex structures. (a) Dickinsonia; (b) Mawsonite.

end of the early Proterozoic, the seas contained too little dissolved iron for new BIF to develop.

Through most of the Proterozoic eon, only simple single-celled life forms existed. But around 1.5 Ga, **eukaryotic cells** appeared. These cells have a more complex internal structure and are capable of building multicellular organisms: protists, fungi, plants, and animals all consist of eukaryotic cells. The first of these were single-celled, but gradually multicellular organisms appeared. Eukaryotes can photosynthesize more efficiently, so the oxygen content of the atmosphere continued to increase, making it possible for more complex creatures to exist. Because of the incompleteness of the fossil record, we cannot track stages in the early evolution of eukaryotic organisms, but thanks to some remarkable outcrops in southern Australia and England, geologists can demonstrate that by 670 Ma, complex but shell-less invertebrates were inhabiting the seas. These organisms are called the **Ediacaran fauna** (▶Fig. 13.13a, b).

The diversification of Ediacaran fauna marks the end of the Precambrian interval, which saw the progression of the Earth from a hot ball covered by magma and encircled by an unbreathable atmosphere to a globe of land and sea encircled by an oxygen-rich atmosphere and inhabited by multicellular life.

13.6 THE PHANEROZOIC EON: LIFE DIVERSIFIES, AND TODAY'S CONTINENTS FORM

The Phanerozoic eon (Greek for "visible life") encompasses the last 545 million years of Earth history. Its name reflects the appearance of diverse organisms with hard,

mineralized skeletons that evolved into well-preserved fossils you can easily find in rock outcrops. Geologists divide the Phanerozoic into three eras—the Paleozoic (Greek for "ancient life"), the Mesozoic ("middle life"), and the Cenozoic ("recent life")—to emphasize the changes in Earth's living population throughout the eon. The solid Earth also changed during this time. Continental blocks rearranged, and new supercontinents formed, only to break apart again. Numerous orogenies built mountain ranges, the most recent of which persist today. Shell-secreting organisms, burrowing organisms, and plants invaded the sedimentary environment and changed the style of deposition. Below, we look at the changes in the planet's map (its paleogeography) and in its life forms during the three eras of the Phanerozoic eon.

13.7 THE PALEOZOIC ERA: FROM RODINIA TO PANGAEA

The Early Paleozoic Era (Cambrian-Ordovician Periods, 545–440 Ma)

Paleogeography. As the Proterozoic eon drew to a close, most continental crust had accumulated in a supercontinent called Pannotia, formed when Rodinia turned inside out. Recall that during this reorganization, Antarctica, India, and Australia rifted away from the future North America, and a passive-margin basin formed along North America's western margin. At the beginning of the Paleozoic era (Cambrian period), Pannotia broke up, yielding smaller continents including **Laurentia** (composed of North America and Greenland), **Gondwana** (South America, Africa, Antarctica, India, and

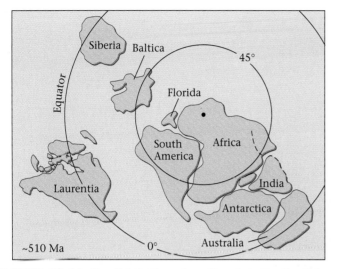

FIGURE 13.14 The distribution of continents in the Cambrian period (about 510 Ma). Note that Gondwana lay near the equator, and Laurentia straddled the equator. The map shows how the Earth might have looked as viewed from the South Pole. Note that Florida had not yet connected to North America (Laurentia).

FIGURE 13.15 A paleogeographic map of North America, showing the regions of dry land and shallow sea in the early Cambrian period. Note the edges of the passive margins, which define the borders of the continent. At this time, the regions between the edge of the passive margin and the coast did not yet exist. The transcontinental arch was a ridge of dry land.

Australia), **Baltica** (Europe), and **Siberia** (▶Fig. 13.14). As these continents drifted apart, Laurentia, Baltica, and Siberia stayed at low latitudes, but Gondwana drifted toward the South Pole, and for a brief interval in the Late Ordovician period, much of it became ice-covered.

Following the breakup of Pannotia, several more passive-margin basins formed around the globe, including one on the east coast of Laurentia (what is now eastern North America). In addition, the sea level rose, so that vast areas of continents were flooded with shallow seas called **epicontinental seas.** By the end of the Cambrian period, the only dry land in Laurentia was an oval region centered on Hudson's Bay, so most of what is now the United States lay beneath water (▶Fig. 13.15). In many places, water depths in epicontinental seas reached only a few meters, creating a well-lighted environment in which life abounded. The Moon was closer to the Earth in the Cambrian time, so tides in these shallow seas might have been enormous. Deposition in the seas yielded a layer of fossiliferous sediment that you can see today near the floor of the Grand Canyon. While a relatively thin layer of sediment formed in Laurentia's interior (the region that would become the cratonic platform), a much thicker layer accumulated in the passive-margin basins that fringed the continent.

The sea level, however, did not stay high for the entire early Paleozoic era. At the end of the Early Ordovician period, the sea level fell briefly, leading to a regression and the formation of an unconformity on top of preexisting strata. But the Middle Ordovician saw a second great flooding event, and a new sedimentary sequence blanketed the unconformity. This high sea level persisted for another 50 million years.

The geologically peaceful world of the early Paleozoic era in Laurentia abruptly came to a close in the Middle Ordovician period, for at this time its eastern margin rammed into a volcanic island arc. The resulting collision, called the **Taconic orogeny,** deformed and metamorphosed strata of the former passive-margin basin and created a mountain range (▶Fig. 13.16a, b). This event represents the first stage in the development of the Appalachian orogen.

Life evolution. Beginning with the Cambrian period, life on Earth left a clear record of evolution, because so many organisms developed shells of durable minerals. The fossil record indicates that life underwent remarkable diversification at this time, an event that paleontologists refer to as the **Cambrian explosion of life.** What caused this explosion? No one can say for sure, but considering that it occurred roughly at the time a supercontinent broke up, it may have something to do with the new ecological niches and the isolation of populations that resulted when small continents formed and drifted apart.

The first animals to appear in the Cambrian period had simple tube- or cone-shaped shells, but soon after, the shells became more complex. Small fossils called *conodonts*, which resemble teeth, are found in Cambrian strata. Their presence suggests that creatures with jaws had appeared at this time, hinting that shells evolved as a means of protection against

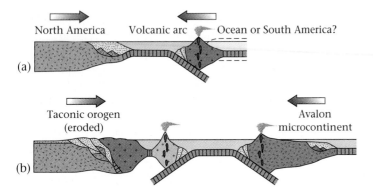

(a)

North America   Volcanic arc   Ocean or South America?

Taconic orogen
(eroded)

Avalon
microcontinent

(b)

FIGURE 13.16 (a) A volcanic island arc collided with the eastern margin of North America to cause the Taconic orogeny. Some geologists have suggested that South America lay behind the volcanic arc, so that at the end of the Taconic orogeny eastern North America bordered western South America. (b) Sometime after the collision, the high land of the Taconic orogen eroded away, and another block, the Avalon microcontinent, approached from the east. Its eventual collision with North America caused the Acadian orogeny (and the Caledonian orogeny of Europe).

FIGURE 13.17 This exposure of limestone in a quarry represents the deposits of a Silurian reef. Large reefs like this grew in the shallow seas in the interior regions of North America.

predators. By the end of the Early Cambrian period, *trilobites,* the first large animals, were grazing the sea floor. Trilobites shared the environment with mollusks, brachiopods, and echinoderms (see Interlude D). Thus, a complex food chain arose, which included plankton, bottom feeders, and at the top, giant predators like the 2-m-long *anomalocaris.* A type of sponge called an *archeocyathid* developed a mineral skeleton and grew in mounds that formed the first reefs. In the Ordovician period, new animal groups appeared, most notably the *graptolites* (floating organisms that contained a serrated mineral band), coral, crinoids, gastropods, and the first vertebrate animals, **jawless fish.** At the end of the Ordovician, mass extinction took place, perhaps because of the brief glaciation and associated sea-level lowering of the time.

But while the sea teemed with organisms during the early Paleozoic era, there were no land organisms for most of this time, except perhaps for simple fungi and algae, so the land surface was a stark landscape of rock and dust, subjected to extraordinarily rapid weathering and erosion rates. Our earliest record of primitive land plants comes from the Late Ordovician period, but these plants were very small and found only along water.

The Middle Paleozoic Era (Silurian-Devonian Periods, 440–360 Ma)

Paleogeography. As the world entered the Silurian period, the global climate warmed (leading to so-called **greenhouse conditions**), sea level rose, and the continents flooded once again. In some places, where water in the epicontinental seas was clear and could exchange with water

from the oceans, huge reef complexes grew, forming a layer of fossiliferous limestone on the land (▶Fig. 13.17). In other places, where the seas were stagnant and cut off from the oceans, evaporation rates were so high that thick deposits of gypsum and halite accumulated. For example, over 750 m (2,300 feet) of Silurian evaporites underlie Michigan and western New York. Mountains formed by the Taconic orogeny in eastern North America eroded entirely away, and their roots were buried beneath a thin layer of shallow-water limestone.

More orogeny took place during the middle Paleozoic era. For example, in eastern North America, the Appalachian region that had been affected by the Taconic orogeny underwent a collisional orogeny once again in the Devonian period, this time with a small continental mass called the Avalon block that drifted in from the east (Fig. 13.16b; ▶Fig. 13.18). Geologists refer to this event as the **Acadian orogeny.** Note that because of the Taconic and Acadian orogenies, the easternmost United States consists of crust that was attached to North America during the Paleozoic era. Large mountains rose during the Acadian orogeny, providing a source for the Catskill Delta, an enormous fan of conglomerates, sandstones, and shales that spread across western New York, Pennsylvania, Ohio, Kentucky, and Tennessee (▶Fig. 13.19). Similar rocks, called the Old Red Sandstone, crop out in the United Kingdom.

Throughout the middle Paleozoic era, the western margin of North America continued to be a passive-margin basin, filling with thick accumulations of sediment that can be seen today in Nevada or eastern California. Finally, in the Late Devonian period, the quiet environment of the west-coast passive margin ceased, possibly because of a collision between a large island arc and what is now Nevada. This event, known as the **Antler orogeny,** pushed slices of

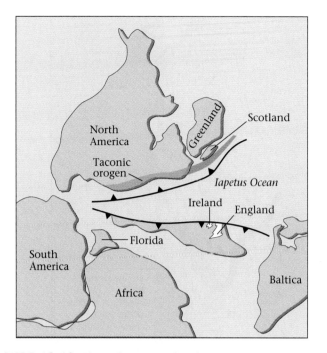

FIGURE 13.18 This paleogeographical map illustrates the movement of the Avalon microcontinent (including England and southern Ireland) toward its ultimate collision with the eastern margin of North America. During the process, the intervening Iapetus Ocean was subducted. The remnant of the Taconic orogen fringed the eastern margin of North America.

deep-marine strata eastward, and up and over the shallow-water strata. It was the first of many orogenies to affect the western margin of the continent.

Life evolution. Life on Earth underwent radical changes in the middle Paleozoic era. In the sea, new species of trilobites, eurypterids, gastropods, crinoids, and bivalves replaced species that had disappeared during the mass extinction at the end of the Ordovician period. On land, **vascular plants,** with woody tissues and seeds and veins for transporting water and food, rooted in the soil for the first time. With the evolution of veins and wood, plants could grow much larger, and by the Late Devonian period the land surface hosted swampy forests with tree-sized relatives of club mosses and ferns. Also at this time, spiders, scorpions, insects, and crustaceans came to exploit both dry-land and freshwater habitats, and jawed fish like sharks began to cruise the oceans. Finally, at the very end of the Devonian period, the first amphibians crawled out onto land and breathed with lungs.

The Late Paleozoic Era (Carboniferous–Permian Periods, 360–245 Ma)

Paleogeography. After the peak of the greenhouse conditions and high sea levels in the middle Paleozoic era, the climate cooled significantly and Earth entered a stage called the late Paleozoic **icehouse.** Seas gradually retreated from the continents, so that in the Carboniferous period regions that had hosted the limestone-forming reefs of epicontinental seas now became coastal swamps and river deltas, in which sand, shale, and organic debris accumulated. During the Carboniferous period, Laurentia lay near the equator (▶Fig. 13.20a, b), so it retained tropical and semitropical conditions that favored lush growth in **coal swamps;** this growth left thick piles of woody debris that transformed into coal upon burial. Much of Gondwana and Siberia, in contrast, lay at high latitudes. By the Permian period, when Earth's climate was cooler, ice sheets covered much of Gondwana and Siberia.

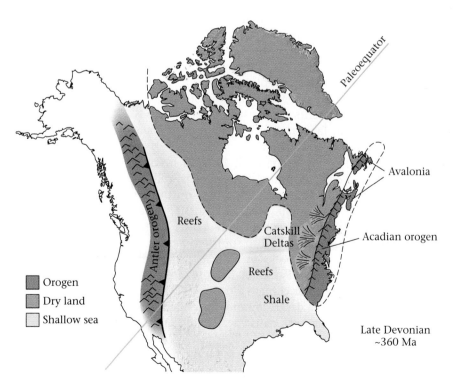

Late Devonian ~360 Ma

- Orogen
- Dry land
- Shallow sea

FIGURE 13.19 This paleogeographical map of North America depicts the distribution of land and sea and the position of mountain belts in the Late Devonian period. Note that the Acadian orogeny occurred in the east and, somewhat later, the Antler orogeny in the west. The Acadian orogen shed a vast delta, called the Catskill Delta, of sand and conglomerate into the eastern interior of North America.

FIGURE 13.20 (a) This map shows the distribution of the continents of Pangaea as viewed from the South Pole. (b) This paleogeographical map shows the distribution of land and sea and the location of the Alleghenian and Hercynian orogens. Extensive coal swamps lay along the interior coast. The Ancestral Rockies (late Paleozoic uplifts of the Rocky Mountain region) also developed at this time, as did smaller uplifts in the Midwest.

The late Paleozoic era also saw a succession of continental collisions, culminating in the formation of a single supercontinent, Pangaea, by the Mid-Permian period. The largest collision occurred in Carboniferous and Permian time, when Laurentia joined with Gondwana, an event called the **Alleghenian orogeny.** During this event, the final stage in the development of the Appalachians, eastern North America squashed against northwestern Africa, and what is now the Gulf Coast region of North America squashed against the northern margin of South America. A vast mountain belt grew, in which deformation generated huge faults and folds. We now see the eroded remnants of rocks deformed during this event in the Appalachian and Ouachita Mountains.

Along the continental side of the Alleghenian orogen, a wide band of deformation called the **Appalachian fold-thrust belt** formed. In this province, you can see a distinctive style of deformation, called **thin-skinned deformation,** characterized by displacement on thrust faults in sedimentary strata above the Precambrian basement (▶Fig. 13.21). At depth, the thrust system merges with a near-horizontal sliding surface, or detachment. Movement on the faults generates folds in overlying beds.

Forces generated during the Alleghenian orogeny were so strong that faults in the continental crust clear across North America moved, creating uplifts (local high areas) and sediment-filled basins in the Midwest and even in the present-day Rocky Mountains (Colorado, New Mexico, and Wyoming). Geologists refer to the late Paleozoic uplifts of the Rocky Mountain region as the **Ancestral Rockies.** Sediments eroded from these uplifts were deposited in alluvial fans and river systems and today form the red sandstones of the region.

The assembly of Pangaea involved a number of other collisions around the world as well. Notably, Africa collided with southern Europe to form the **Hercynian orogen,** a rift or small ocean in Russia closed to create the Ural Mountains, and parts of China along with other fragments of Asia attached to southern Siberia. Note that the arrangement of blocks in Pangaea differs markedly from that of Rodinia or Pannotia. Because of the immense size of Pangaea, its interior climate did not feel the moderating effect of the sea, so a desert of sand dunes, red mud, debris-choked channels, and salt pans covered large areas of the land surface.

Life evolution. The fossil record indicates that during the late Paleozoic era, plants and animals continued to evolve

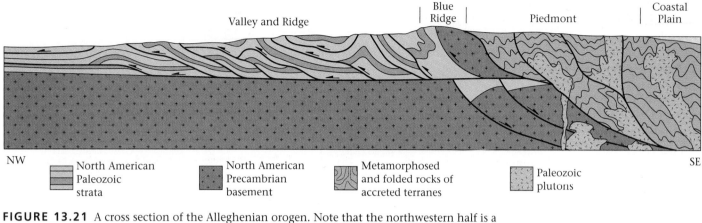

FIGURE 13.21 A cross section of the Alleghenian orogen. Note that the northwestern half is a thin-skinned fold-thrust belt, involving deformation of strata by thrust faults and folds above a sliding surface at the top of the Precambrian basement. The Blue Ridge exposes a slice of Precambrian North American basement. But in the metamorphic regions to the east (the piedmont and beneath the coastal plain), the crust consists of slices added on to North America during the Paleozoic era. In the coastal plain, these rocks have been buried by Cretaceous and Tertiary strata.

toward more familiar forms. In coal swamps, fixed-wing insects such as huge dragonflies flew through a tangle of ferns, club mosses, and scouring rushes, and by the end of the Carboniferous period insects such as the cockroach, with foldable wings, appeared. Forests containing **gymnosperms** ("naked seed" plants like conifers—pine trees and spruce) and **cycads** (trees with a palm-like stalk peaked by a fan of fern-like fronds) became widespread in the Permian period. Amphibians and, later, reptiles populated the land. The appearance of reptiles marked the evolution of a radically new component in animal reproduction: eggs with shells. Such eggs permitted reptiles to reproduce without returning to the water, and thus allowed the group to populate previously uninhabitable environments. Early members of the reptile group included the carnivorous, fin-backed *pelycosaurs* and *therapsids,* the forerunners of mammals. The late Paleozoic era came to a close with two major mass-extinction events, during which over 90% of marine species disappeared. Why these particular events occurred remains a controversy.

13.8 THE MESOZOIC ERA: WHEN DINOSAURS RULED

The Early and Middle Mesozoic Era (Triassic-Jurassic Periods, 245–145 Ma)

Paleogeography. Pangaea, the supercontinent formed in the late Paleozoic era, lasted for only about 100 million years, for during the Late Triassic and Early Jurassic periods, rifts formed and the supercontinent began to break up. Rifting started along the boundary between North America and Africa, and deep rift basins, which developed along the Appalachian margin and in North Africa and Europe, filled with river-channel sand and lake-bed mud—these sediments turned into red sandstone and shale. By the end of the Jurassic period, rifting had succeeded in creating the North Atlantic Ocean (▶Fig. 13.22). At first, the Atlantic was narrow and shallow, and evaporation made its water so salty that thick evaporite deposits, which now underlie much of the Gulf Coast region, accumulated.

According to the record of sedimentary rocks, Earth in the early Mesozoic era had icehouse conditions and low sea levels, so the interior of Pangaea remained a nonmarine environment. As a result, Triassic strata of the southwestern United States consist of red fluvial (river-lain) shales and sandstones that now crop out in the Petrified Forest National Monument, and Early Jurassic strata include relics of huge sand dunes that now form the spectacular cliffs of Zion National Park. However, the stratigraphic record indicates that by the Middle Jurassic period, icehouse climates began to disappear, and the sea level began to rise. During the resulting transgression, a shallow sea submerged much of the Rocky Mountain region.

On the western margin of North America, convergent-margin tectonics became the order of the day, and a convergent-margin orogen formed. Beginning with Late Permian and continuing through Mesozoic time, subduction created volcanic island arcs and caused them, along with microcontinents and hot-spot volcanoes, to collide with North America. The resulting orogenies included the **Sonoma orogeny** (in the Late Permian/Early Triassic periods) and the **Nevadan orogeny** (in the Late Jurassic), both of which created deformed and metamorphosed rocks. Thus, North America grew in land area by the addition (accretion) of crustal fragments and island arcs on its western margin (►Fig. 13.23). From the end of the Jurassic through the Cretaceous period, a major continental volcanic arc, the Sierran arc, formed along the western margin of North America; we'll learn more about this arc below.

Life evolution. During the early Mesozoic era, a variety of new species of already established plant and animal groups appeared, filling the ecological niches left vacant by the Late Permian mass extinction. New creatures such as swimming reptiles (*plesiosaurs,* for example) appeared in the sea. **Hexacorals** became the predominant reef builders of the day, and have remained so ever since. On land, gymnosperms (most notably the cycads) and reptiles diversified, and the Earth saw its first turtles and flying reptiles (*pterosaurs*). More dramatically, at the end of the Triassic period, the first true dinosaurs appeared. Dinosaurs differed from other reptiles in that their legs were positioned under their bodies rather than off to the sides, and they were probably warm-blooded. (The warm-blooded hypothesis has been supported by the recent discovery of a dinosaur heart; according to paleontologists, this heart closely resembles that of a bird.) By the end of the Jurassic period, gigantic *sauropod* dinosaurs (such as *seismosaurus,* which weighed up to 100 tons), along with other familiar ones like *stegasaurus* and the carnivorous *allosaurus,* thundered across the land-

Added in Cretaceous/Tertiary
Added in Cretaceous
Added in Jurassic
Added in late Paleozoic and early Mesozoic
North American basement

800 km

FIGURE 13.23 During the Mesozoic era, convergent-boundary tectonics took place along the western margin of North America. Volcanic island arcs, hot-spot volcanoes, and continental slivers accreted to the margin; the map shows the location of these blocks today.

FIGURE 13.22 The breakup of Pangaea began in the Late Triassic period, and by the Late Jurassic the North Atlantic had formed. Note that at this time, the South Atlantic remained closed, and southern Asia assembled out of a number of smaller continental fragments and volcanic island arcs.

scape, and the first feathered birds (*archaeopteryx*) took to the skies. True mammals appeared at the end of the Triassic period, in the form of small, rat-like creatures.

The Late Mesozoic Era (Cretaceous Period, 145–65 Ma)

Paleogeography. In the Cretaceous period, the Earth's climate continued to shift to warmer, greenhouse conditions, and the sea level rose significantly, reaching levels that had not existed for the previous 200 million years. Great seaways flooded most of the continents, producing the thick deposits of chalk in Europe that Thomas Henry Huxley discussed as well as layers of limestone and sandstone in the western interior of North America (►Fig. 13.24). In the latter part of the Cretaceous period, a shark could have swum from the Gulf of Mexico to the Arctic Ocean in the **Western Interior Seaway** of North America. The eastern and southern margins of North America were also submerged, creating an environment in which the sediments

FIGURE 13.24 Paleogeographical map of North America in the Late Cretaceous period, showing the distribution of land and sea and the convergent orogen of the cordillera. Note the position of the Sierran arc and the Sevier fold-thrust belt (inset). An interior seaway stretched from the Gulf of Mexico to the Arctic Ocean.

that now underlie the **coastal plain** collected. During this warm period, glaciers that once covered high-latitude regions of Gondwana and Siberia vanished.

The breakup of Pangaea continued through the Cretaceous period, with the opening of the South Atlantic Ocean and the separation of South America and Africa from Antarctica and Australia. India broke away from Gondwana and headed rapidly northward toward Asia (►Fig. 13.25). Along the continental margins of the newly formed Mesozoic oceans, new passive-margin basins developed that, like their predecessors of the early Paleozoic era, filled with great thicknesses of sediments. For example, along the Gulf Coast, a wedge of sediment over 15 km thick has accumulated.

Along western North America, the **Sierran arc,** a large continental volcanic arc that had initiated at the end of the Jurassic period, continued to be active. This arc resembled the one that currently lies along the Andes on the western edge of South America. Though the volcanoes of the arc have long since eroded away, we can see their roots in the form of the plutons that now comprise the granitic

batholith of the Sierra Nevada Mountains. A thick accretionary prism, formed from sediments and debris scraped off the subducting oceanic plate, was present to the west of the Sierran arc and now crops out in the Coast Ranges of California. Compressional forces along the western North American convergent boundary activated large thrust faults east of the arc, an event geologists refer to as the **Sevier orogeny.** Much of the Sevier orogen displays thin-skinned deformation, creating a fold-thrust belt whose remnants you can see today in the Canadian Rockies and in western Wyoming. The weight of the orogen pushed the surface of the continent down, forming a wide foreland basin that filled with sediment. This foreland basin comprises the western part of the Western Interior Seaway.

At the end of the Cretaceous period, continued compression along the convergent boundary of western North America caused large faults in the region of Wyoming, Colorado, eastern Utah, and northern Arizona to slip. In contrast to the faults of fold-thrust belts, these faults penetrated deep into the Precambrian rocks of the continent, and thus movement on them generated **basement uplifts:** overlying layers of Paleozoic strata warped into large monoclines, folds whose shape resembles the drape of a carpet over a step (►Fig. 13.26a, b). This event, which geologists call the **Laramide orogeny,** formed the present Rocky Mountains (►Fig. 13.26c). Some geologists have suggested that the contrast in the location of faulting between the Sevier and Laramide orogenies may reflect contrasts in the dip of the subducting plate—during the Laramide orogeny, the subducting plate entered the mantle at a shallower

FIGURE 13.25 By the Late Cretaceous period, the Atlantic Ocean had formed, and India moved rapidly northward, ultimately colliding with Asia.

(a)

(b)

(c)

FIGURE 13.26 (a) The geometry of a basement uplift before erosion. Note that the fault penetrates the basement, and that the overlying sediment warps into a fold that resembles a stair step. (b) After erosion, the Precambrian rocks in the core of the uplift crop out at the ground surface. (c) Here we see the shift of deformation eastward, from the Sevier fold-thrust belt to the belt of Laramide uplifts.

angle, and therefore scraped along the base of the continent farther inland. This change in dip angle did not occur in Canada, so during the Laramide orogeny the Canadian Rockies continued to grow eastward as a fold-thrust belt.

Fascinating new research suggests possible relationships between tectonic events, global climate, and sea-level changes in the Cretaceous period. For example, the breakup of Pangaea led to the formation of many mid-ocean ridges, and geologists have determined (by studying marine magnetic anomalies) that sea-floor spreading took place about three times faster on these ridges during the Cretaceous period than it does today. Thus, more of the oceanic crust was young and warm then than it is today. And since young sea floor lies at a shallower depth than older crust, mid-ocean ridges occupied more space than they do today. The ridges therefore displaced seawater, causing the sea level to rise and continents to flood (▶Fig. 13.27).

By mapping the sea floor, geologists have also discovered that huge submarine plateaus composed of hot-spot volcanic rocks formed during the Cretaceous period, and these plateaus would also have contributed to the displacement of seawater. The presence of hot spots implies that the Cretaceous was a time when particularly large mantle plumes, called **super plumes,** reached the base of the lithosphere. Hot-spot and mid-ocean-ridge activity may have influenced the climate, because the volcanoes they produced released huge quantities of carbon dioxide (CO_2) into the atmosphere, perhaps as much as eight times the

amount found today. An increase in the concentration of CO_2 in the atmosphere leads to an increase in atmospheric temperature, for CO_2 in the air traps infrared radiation rising from the Earth and causes the air to warm up. (In this way, CO_2 acts as a **greenhouse gas,** in that it plays the same role as glass panes in a greenhouse; see Chapter 23.) This increase would melt ice sheets, thus contributing to the rise of sea level.

FIGURE 13.27 A change in the rate of sea-floor spreading changes the volume of a mid-ocean ridge and therefore displaces seawater. At a slow ridge, a point on the sea floor has moved only to a distance D from the ridge axis at the time it reaches a depth of H, while at a fast ridge, the sea floor has reached a distance of 2D at the same depth (remember, the sea floor sinks as it moves away from the ridge because it cools and becomes denser).

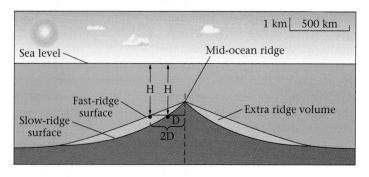

Life evolution. In the seas of the late Mesozoic world, modern fish, called *teleost fish,* appeared and became dominant. What set them apart from earlier fish were their short jaws, rounded scales, symmetrical tails, and specialized fins. Huge swimming reptiles and gigantic turtles (with shells up to 4 m across) preyed on the fish. On land, cycads largely vanished, and **angiosperms** (flowering plants), including hardwood trees, began to compete successfully with conifers for dominance of the forest. Angiosperms can produce seeds much more rapidly than conifers and can attract insects to help with pollination. Dinosaurs reached their peak of success at this time, inhabiting almost all environments on Earth. Social herds of grazing dinosaurs roamed the plains, preyed upon by the fearsome *Tyrannosaurus rex* (a Cretaceous, not Jurassic, dinosaur, despite what Hollywood says!). *Pterosaurs,* with wingspans of up to 11 m, soared overhead, and birds began to diversify. Mammals also diversified and developed larger brains and more specialized teeth, but they remained small and rat-like.

The "K-T boundary event." Geologists first recognized the K-T boundary (K stands for "Cretaceous" and T for "Tertiary") from nineteenth-century studies that identified an abrupt global change in fossil assemblages. Modern dating techniques indicate that this change happened almost instantaneously and that it represented the sudden extinction of most species on Earth. The dinosaurs, which had ruled the planet for over 150 million years, simply vanished, along with 90% of some plankton species in the ocean and up to 75% of plant species. What kind of catastrophe could cause such a sudden and extensive mass extinction? From data collected in the 1970s and 1980s, most geologists have concluded that the Cretaceous period came to a close, at least in part, as a result of the impact of a 10-km-wide **bolide** (a solid extraterrestrial object such as a meteorite, comet, or asteroid) at the site of the present-day Yucatán peninsula in Mexico (▶Fig. 13.28).

The discoveries leading up to this conclusion provide fascinating insight into how science works. The story began when Walter Alvarez, a geologist studying strata in Italy, noted that a thin layer of clay interrupted the deposition of deep-sea limestone precisely at the K-T boundary. Cretaceous plankton shells comprised the limestone below the clay layer, while Tertiary plankton shells made up the limestone just above the clay. Apparently, for a short interval of time at the K-T boundary, all the plankton died, so that only clay settled out of the sea. When Alvarez and his father, Luis (a physicist), and other colleagues analyzed the clay, they learned that it contained **iridium,** a very heavy element found only in extraterrestrial objects. Soon, geologists were finding similar iridium-bearing clay layers at the K-T boundary all over the world. Further study showed that the clay layer contained other unusual materials, such as

FIGURE 13.28 The painting illustrates the collision of a huge bolide with the Earth at the end of the Cretaceous period.

tiny glass spheres (or spherules), formed from the flash freezing of melt, wood ash, and **shocked quartz** (grains of quartz that had been subjected to intense pressure). Only an immense bolide impact could explain all these features. The glass spherules formed when melt sprayed in the air from the impact site, the ash resulted when forests were set ablaze by the impact, the iridium came from fragments of the asteroid, and the unusual quartz grains were created by the shock of the impact. The iridium-rich clay layer thus represents the debris of a huge impact, settling down through a sea robbed of its plankton.

Recent studies suggest that the impact not only formed a crater but probably also generated 2-km-high tsunamis that inundated the shores of continents. A blast of hot air from the impact would have set forests on fire, and the blast and the blaze together would have ejected so much debris into the atmosphere that for months there would have been perpetual night and winter-like cold. These conditions would cause photosynthesis to all but cease, and thus would break the food chain and probably trigger extinctions.

Geologists suggest that the bolide landed on the northwest coast of the Yucatán peninsula. Buried beneath younger sediment is a 100-km-wide by 16-km-deep scar called the **Chicxulub crater.** A layer of glass spherules up to 1 m thick occurs at the K-T boundary in strata near the site. And radiometric dating indicates that igneous melts in the crater formed at 65 ± 0.4 Ma, exactly the time of the K-T boundary event. The discovery of this event has led geologists to speculate that other such asteroid collisions may have helped punctuate the path of life evolution throughout Earth history.

Stratigraphic Sequences and Sea-Level Change

As we have seen in this chapter, there have been intervals in geologic history when the interior of North America was submerged beneath a shallow sea, and times when it was high and dry. This observation implies that sea level, relative to the surface of the continent, rises and falls through geologic time. When the continental surface was dry and exposed to the atmosphere, weathering and erosion ground away at previously deposited rock and, as a result, created a continent-wide unconformity. In the early 1960s, an American stratigrapher named Larry Sloss introduced the name **stratigraphic sequence** to refer to the strata deposited on the continent during periods of relatively high sea level. Such sequences are bounded above and below by regional unconformities.

We can picture the deposition of an idealized sequence as follows. As the starting condition, the entire surface of a continent lies above sea level, much like the continents of today, and an unconformity develops. Then, as the sea level rises and a transgression takes place, the margin of the continent floods first, then the interior. Thus, the preexisting unconformity gets progressively buried toward the interior, and the bottom layers of the newly deposited sequence tend to be older near the margin of the continent than toward the interior. The base of a sequence consists of terrestrial strata (river alluvium). As the sea level continues to rise, this sediment is buried by nearshore sediment, then deeper-water sediment (▶Fig. 13.29a). Eventually, the entire width of the continent, with the exception

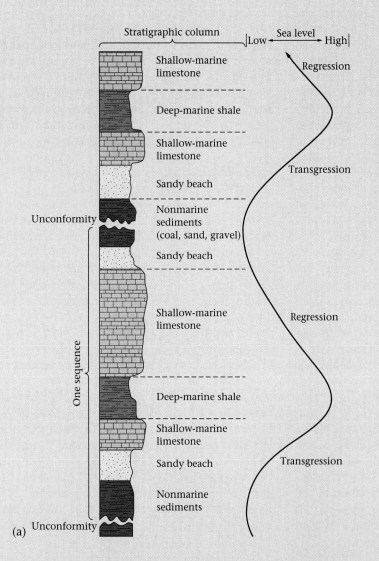

(a)

FIGURE 13.29 (a) Different types of strata are deposited as the sea level rises and falls. Unconformities develop when the sea level is low. A sequence is the interval between two major unconformities. (b) The major stratigraphic sequences of North America, as defined by Larry Sloss (1962). Stippled areas represent regions where there is deposition. The height of a blank interval indicates the duration of time during which there was no deposition. Note that in a given sequence, deposition occurs near the coasts of the continent first and in the interior later. Regional unconformities make up the contacts between sequences. (c) This graph, based on the study of stratigraphy in passive-margin basins, estimates the rise and fall of the sea level through time. Though there are probably some inaccuracies owing to flaws in the interpretation of the stratigraphic record, the overall patterns are likely correct.

of highlands and mountain belts, lies below sea level. When the sea level begins to fall, the shoreline regresses, first exposing the interior of the continent and then the margins. A new unconformity forms in the same order, first in the interior and only later along the margins. Many shorter-duration changes in sea level may happen during a single long-duration rise or fall.

Sloss recognized six major stratigraphic sequences in North America, and he named each after a nation of Native Americans (▶Fig. 13.29b). Each sequence represents deposition during an interval of time lasting tens of millions of years. In more recent decades, studies of strata along continental passive margins have provided an even more detailed record of sequences (because there is less erosion in these regions, and thus the record is more complete).

Taken together, this information provides insight into the history of the sea level's global rise and fall (▶Fig. 13.29c). Some sea-level changes may reflect changes in sea-floor-spreading rates and thus in the volume of mid-ocean ridges; some may reflect hot-spot activity; some may reflect variations in Earth's climate that caused the melting or formation of ice sheets; others may represent changes in areas of continents accompanying continental collisions; and still others may reflect the warping of continental surfaces as a result of mountain building. Rises in sea level generally indicate greenhouse conditions, and falls represent icehouse conditions. Thus, the sea level may reflect changes in the atmosphere's CO_2 concentration in response to volcanic activity, life activity, or rock weathering (weathering reactions absorb CO_2).

(b)

(c)

13.9 THE CENOZOIC ERA: THE FINAL STRETCH TO THE PRESENT

Paleogeography. During the last 65 million years, the map of the Earth has continued to change, gradually producing the configuration of continents we see today. The final stages of the Pangaea breakup separated Australia from Antarctica and Greenland from North America, and formed the North Sea between Britain and continental Europe. The Atlantic Ocean continued to grow, because of sea-floor spreading on the Mid-Atlantic Ridge, and thus the Americas moved west, away from Europe and Africa. Meanwhile, the continents that once comprised Gondwana drifted north as the intervening Tethys Ocean was consumed by subduction. Collisions of the former Gondwana continents with the southern margins of Europe and Asia resulted in the formation of the largest orogenic belt on the continents today, the **Alpine-Himalayan chain** (▶Fig. 13.30). India and a series of intervening volcanic island arcs and microcontinents collided with Asia to form the Himalayas and the Tibetan Plateau to the north, while Africa along with some volcanic island arcs and microcontinents collided with Europe to produce the Alps in the west and the Zagros Mountains of Iran in the east. Finally, collision between Australia and New Guinea led to orogeny in Papua New Guinea.

As the Americas moved westward, large convergent plate boundaries evolved along their western margins. In South America, convergent-boundary activity has built the Andes, which remains an active orogen to the present day. In North America, convergent-boundary activity continued without interruption until the Eocene epoch, yielding, as we have seen, the Laramide orogen, the eastern part of the North American Cordillera.

Then, because of the rearrangement of plates off the western shore of North America, beginning about 40 Ma, a transform boundary replaced the convergent boundary in the western part of the continent (▶Fig. 13.31a–c). When this happened, volcanism and compression ceased in western North America, the San Andreas fault system formed along the coast of the United States, and the Queen Charlotte fault system developed off the coast of Canada. Along the San Andreas and Queen Charlotte Faults today, the Pacific Plate moves north with respect to North America at a rate of about 6 cm per year. The Queen Charlotte Fault links to the Aleutian subduction zone along southern Alaska, where the Pacific Ocean floor undergoes subduction. In the western United States, convergent-boundary tectonics continues only in Washington, Oregon, and northern California where subduction of the Juan de Fuca Plate has yielded the Cascade volcanic chain.

FIGURE 13.30 The two main continental orogenic systems on the Earth today. The Alpine-Himalayan system formed when pieces of Gondwana (Africa, India, and Australia) migrated north and collided with Asia (inset). The Cordilleran and Andean systems are the consequence of convergent-boundary tectonism along the eastern Pacific.

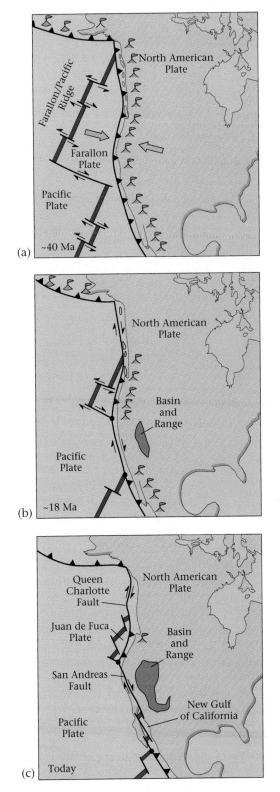

FIGURE 13.31 The western margin of North America changed from a convergent-plate boundary into a transform-plate boundary when the Farallon-Pacific Ridge was subducted. (a, b) The Farallon Plate was moving toward North America, while the Pacific Plate was moving parallel to the western margin of North America. (c) The Basin and Range Province opened as the San Andreas Fault developed. Subduction along the West Coast today only occurs where the Juan de Fuca Plate, a remnant of the Farallon Plate, continues to subduct.

When convergent tectonics ceased in the western United States south of the Cascades, the region began to undergo rifting (stretching) in roughly an east-west direction. The result was the formation of the **Basin and Range Province,** a broad continental rift that has caused the region to stretch to twice its original width (►Fig. 13.32a, b). The Basin and Range gained its name from its topography—the province contains long, narrow mountain ranges separated from each other by flat, sediment-filled basins. This geometry reflects the normal faulting resulting from stretching: crust of the region was broken up by faults, and movement on the faults created elongate depressions. The Basin and Range Province terminates at its north end just north of the Snake River Plain, the track of the hot spot that now lies beneath Yellowstone National Park.

Recall that in the Cretaceous period, the world experienced greenhouse conditions and the sea level rose so that extensive areas of continents were submerged. During the Cenozoic era, however, the global climate rapidly shifted to icehouse conditions, and by the early Oligocene epoch, Antarctic glaciers reappeared for the first time since the Triassic period. The climate continued to grow colder through the Late Miocene epoch, leading to the formation of **grasslands** in temperate climates. About 2.5 Ma, the Isthmus of Panama was created, separating the Atlantic completely from the Pacific, changing the configuration of oceanic currents, and allowing the Arctic Ocean to freeze over.

In the overall cold climate of the last 2 million years, the Quaternary period, continental glaciers have expanded and retreated across northern continents at least twenty times, resulting in the **Pleistocene ice age** (►Fig. 13.33). Each time the edge of the glacier advanced, surrounding regions shifted from dry to wet and from warm to cold; vegetation belts migrated southward; low-latitude year-round streams flowed through desert regions like the Sahara; and the sea level fell so much that the continental shelf became exposed to air. Also, during the Pleistocene epoch, a land bridge formed across the Bering Strait, west of Alaska, providing migration routes for animals and people from Asia into North America; a partial land bridge also formed from southeast Asia to Australia, allowing people to migrate into Australia. Erosion and deposition by the glaciers created much of the landscape we see today in northern temperate regions. About 11,000 years ago, the climate warmed, and we entered the interglacial time interval we are still experiencing today (see Chapter 22).

Life evolution. When the skies finally cleared in the wake of the K-T boundary catastrophe, plant life recovered, and soon forests of both angiosperms and gymnosperms reappeared. A new group of plants, the **grasses,** sprang up and began to dominate the plains in temperate and subtropical climates by the middle of the Cenozoic era. The dinosaurs, however, were gone for good, for no species of these great beasts survived. Mammals rapidly diversified into a variety of forms to

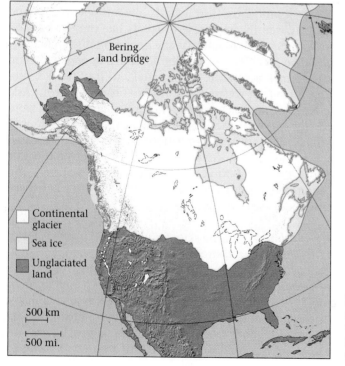

FIGURE 13.32 (a) The Basin and Range Province is a rift. The northern part has opened more than the southern part, causing the Sierran arc to swing westward and rotate (inset). The Rio Grande Rift is a small rift that links to the Basin and Range. The Colorado Plateau is a block of craton bounded by the Rio Grande Rift to the east and the Basin and Range to the west. (b) Before rifting, the region that was to become the Basin and Range Province was occupied by the Sevier fold-thrust belt.

take their place. In fact, most of the modern groups of mammals that exist today originated at the beginning of the Cenozoic era, giving this time the nickname "Age of Mammals." During the latter part of the era, remarkably huge mammals appeared (such as mammoths, giant beavers, giant bears, and giant sloths), but many went extinct during the past 10,000 years, perhaps because of human hunting.

It was during the Cenozoic era that our own ancestors first appeared. Ape-like primates diversified in the Miocene epoch (about 20 Ma), but *Australopithecus*, the first human-like primate, did not appear until about 4 Ma, followed by the first members of the human genus, *Homo*, at about 2.4 Ma. (Note that people and dinosaurs did *not* inhabit Earth at the same time!) Perhaps *Homo*, with its significantly larger brain, appeared when climate changes led to the spread of grasslands, allowing primates to leave the trees—life on the ground provides a longer time for infant development and the growth of a large brain. Fossil evidence, primarily from Africa, indicates that *Homo erectus*, capable of making stone axes, appeared 1.6 Ma, and the line leading to **Homo sapiens** (our species) diverged from *Homo neanderthalensis* (Neanderthal man) about 500,000 years ago. According to the fossil record, modern people appeared about 150,000 years ago. Thus, much of human evolution

FIGURE 13.33 North America during the maximum advance of the Pleistocene ice sheet. Because the sea level was so low (water had been transformed into the ice sheet), a land bridge formed across the Bering Strait, allowing people and animals to migrate from Asia to America.

Tectonic Events	Geologic Time	Evolution
	Quaternary	Human civiliztion Human-like ancestors
Pleistocene ice age		
San Andreas Fault / Basin and Range Province form.		Huge mammals
Himalayan orogeny	Tertiary	
Alpine orogeny		
Laramide orogeny ends.		
K/T boundary event		Dinosuars extinct/ mammals dominate
Laramide orogeny begins.	Cretaceous	
South Atlantic ocean forms.		
Sevier orogeny		
Convergence/accretion in western N. America	Jurassic	Angiosperms
North Atlantic Ocean forms.		
Pangaea starts to break apart.	Triassic	First mammals and birds First dinosaurs
Pangaea, a supercontinent, forms.	Carboniferous	Gymnosperms
Alleghenian orogeny (eastern N. America)		Coal swamps / first reptiles First amphibians
Antler orogeny (western N. America)	Devonian	First insects; first trees
Acadian orogeny (eastern N. America)	Silurian	Early land plants
Taconic orogeny (eastern N. America)	Ordovician	Jawless fish
Sea level rises and falls; continents submerged.	Cambrian	Many shelled invertebrates appear (trilobites/brachiopods)
Passive-margin basins on coast of N. America		
Pannotia breaks up.		
Pannotia, a supercontinent, forms.		Ediacaran fauna
Rodinia breaks up.	Proterozoic	
Rodinia, a supercontinent, forms.		Eukaryotes
Concentration of oxygen in atmosphere increases.		
Larger continents accumulate / cratons form.		First life
Ocean water accumulates.	Archean	
Continental crust begins to form.		
Earth forms from planetesimals.	Hadean	

FIGURE 13.34 A summary chart of the history of the Earth, as currently interpreted by geologists.

took place during the radically shifting climatic conditions of the Pleistocene epoch. (See ▶Fig. 13.34 for an overview of paleogeography and life evolution through Earth history.)

13.10 INTO THE FUTURE

In the near future, Earth's history will be dominated by changes reflecting glacial advances and retreats (see Chapter 22). We are scheduled to head into the next glacial period perhaps within the next 5,000 years, leading to ice sheets that could advance over populated regions of North America, Europe, and Asia. However, our contribution of carbon dioxide to the atmosphere may cause greenhouse warming. If so, then we might enter a "super-interglacial" period, in which global warming melts polar ice caps and causes a 50-m rise in the sea level. Such a rise would flood most of the coastal regions of the world—New York, London, Paris, Rome, and Tokyo would vanish beneath the waves.

In the longer term, the Earth will likely exist for another 5 billion years, barring a catastrophic collision with a planet-sized bolide. Astronomers have reached this conclusion by looking at the pattern of evolution for stars like our Sun elsewhere in the galaxy. The Sun is a mid-sized star, with an average lifetime of about 10 billion years. Some 5 billion years from now, it will run out of nuclear fuel. When this happens, it will begin to collapse inward, because the gravitational pull of its immense mass will no longer be countered by the outward push caused by intense heat. But as the Sun's gas squeezes into a smaller space, it will temporarily heat up again. This gravitationally induced heating will be short-lived, but when it occurs, the Sun will expand greatly, forming a **red giant** whose diameter will be close to that of Earth's orbit. As the expanding Sun's surface approaches Earth, our planet's surface will first heat up, so that the atmosphere and oceans will boil away. Then the planet itself will evaporate, and its atoms will mix with the gases of the Sun. This will probably be the end of the Earth. Gradually, the red giant Sun will cool and eject some gas into space as a nebula. The remainder will collapse inward to form a white dwarf star. Most likely, the nebula ejected by the dying Sun, along with the atoms of Earth that it contains (perhaps including the atoms that now reside in your body), will eventually form part of a new solar system.

Between now and the end of the Earth, we can expect a continuously changing globe as continents continue to drift, perhaps to combine in the future to form yet another supercontinent. Life, most likely, will continue to evolve in fits and starts, perhaps punctuated by future mass extinctions in the wake of another bolide collision, or by the destructive forces of human-made weapons and pollutants.

CHAPTER SUMMARY

• Earth formed about 4.6 billion years ago. During the first 600 million years of its history, the Hadean eon, the planet was so hot that its surface was a magma ocean. We have no rock record of this time interval because rocks were too hot then for the radiometric clock to work, but we can gain insight into it by studying the Moon and meteorites.

• The Archean eon began about 3.96 Ga, when the Earth had cooled sufficiently for permanent continental crust to form. The crust assembled out of volcanic arcs and hot-spot volcanoes that were too buoyant to subduct. Water oceans and stable continental blocks called cratons also formed. The atmosphere contained little oxygen, but the first life forms, bacteria, appeared, perhaps along mid-ocean ridges.

• In the Proterozoic eon, which began at 2.5 Ga, Archean cratons collided and sutured together along large orogenic belts. Volcanic island arcs and hot-spot volcanoes plastered onto the margins of continents, creating large Proterozoic cratons whose interiors were far from the orogenic belts. Photosynthesis by organisms added oxygen to the atmosphere, and iron precipitated out of the ocean to form banded-iron formation. Eukaryotic cells appeared at 1.5 Ga, greatly increasing the amount of oxygen in the atmosphere. With such an atmosphere, more complex creatures could develop, and by the end of the Proterozoic, complex but shell-less marine invertebrates populated the planet. Most continental crust accumulated to form a supercontinent called Rodinia at about 1 Ga. Rodinia broke apart and reorganized to form another supercontinent called Pannotia at the end of the Proterozoic eon.

• At the beginning of the Paleozoic era, about 545 Ma, Pannotia broke apart, yielding several smaller continents. The sea level rose and fell a number of times, creating sequences of strata in continental interiors. Continents began to collide and coalesce again, leading to orogenies (such as the Taconic, Acadian, and Alleghenian, of the Appalachian Mountains) and, by the end of the era, another supercontinent, Pangaea. Invertebrate organisms evolved shells, and life underwent diversification. Milestones in evolution include the appearance of trilobites, brachiopods, and large predators in the Cambrian period; jawless fish in the Ordovician period; land plants in the Silurian period; the first land animals (insects and amphibians), jawed fish, and seed plants in the Devonian period; and reptiles and gymnosperm trees in the Permian period.

• In the Mesozoic era, Pangaea broke apart and the Atlantic Ocean formed. Convergent-boundary tectonics dominated along the western margin of North America, causing orogenic events like the Sevier and Laramide orogenies. Dinosaurs appeared in Late Triassic time and became the dominant land animal through the Mesozoic era. During the Cretaceous period, the sea level was very high, and the continents flooded. Angiosperms (flowering plants) appeared at this time, along with modern fish. A huge mass extinction event (the K-T boundary event), which wiped out the dinosaurs, occurred at the end of the Cretaceous period, probably because of the impact of a large bolide with the Earth.

• In the Cenozoic era, continental fragments of Pangaea began to collide again. The collision of Africa and India with Asia and Europe formed the Alpine-Himalayan orogen. Convergent tectonics has persisted along the margin of South America, creating the Andes, but ceased in North America when the San Andreas Fault formed. Rifting in the western United States during the Cenozoic era produced the Basin and Range Province. Various kinds of mammals filled niches left vacant by the dinosaurs, and the human genus, *Homo,* appeared and evolved throughout the radically shifting climate of the Pleistocene epoch.

• The end of the Earth will probably happen 5 billion years in the future, when the Sun runs out of nuclear fuel, transforms briefly into a red giant, and vaporizes our planet.

KEY TERMS

REVIEW QUESTIONS

1. List some methods by which geologists study the past.

2. Why are there no rocks on Earth that yield radiometric dates older than 4 billion years?

3. Describe the condition of the crust, atmosphere, and oceans during the Hadean eon.

4. Describe the five principal rock types found in Archean protocontinents. Under what kind of environmental conditions did each type form?

5. What are stromatolites? How do they form?

6. How did the atmosphere and tectonic conditions change during the Proterozoic eon?

7. How do banded-iron formations tell us about atmospheric oxygen levels?

8. What evidence do we have that the Earth nearly froze over twice during the Proterozoic eon?

9. How did the Cambrian explosion of life change the nature of the living world?

10. How did the Taconic and Acadian orogenies affect the east coast of North America?

11. How did the Alleghenian and Ancestral Rocky orogenies affect North America?

12. Compare the typical sedimentary deposits of the early Paleozoic greenhouse Earth with those of the late Paleozoic/early Mesozoic icehouse Earth.

13. Describe the plate tectonic conditions that led to the formation of the Sierran arc and the Sevier fold-thrust belt.

14. How did the plate tectonic conditions of the Laramide orogeny differ from more typical subduction zones?

15. What caused the flooding of the continents during the Cretaceous period?

16. What could have caused the K-T extinctions?

17. What continents formed as a result of the breakup of Pangaea?

18. What are the causes of the uplift of the Himalayas and the Alps?

19. What events led to the end of the Mesozoic greenhouse and the development of glaciers on the Arctic and Antarctic during the Cenozoic era?

20. Describe some possible scenarios for the future evolution of the Earth.

SUGGESTED READING

Cloud, P. 1988. *Oasis in Space: Earth History from the Beginning.* New York: Norton.

Condie, K. C. 1989. *Plate Tectonics and Crustal Evolution.* New York: Pergamon.

———, ed. 1994. *Archean Crustal Evolution.* New York: Elsevier.

———, ed. 1994. *Proterozoic Crustal Evolution.* New York: Elsevier.

Dott, R. H., Jr., and D. R. Prothero. 1994. *Evolution of the Earth.* 5th ed. New York: McGraw-Hill.

Harland, W. B., and M. J. S. Rudwick. 1964. The great infra-Cambrian ice age. *Scientific American* 225: 28–36.

Hartmann, J., and R. Miller. 1991. *The History of the Earth: An Illustrated Chronicle of an Evolving Planet.* New York: Workman.

Hoffman, P. 1988. United Plates of America: The birth of a craton. *Annual Review of Earth and Planetary Sciences* 16: 543–604.

Holland, H. D., B. Lazar, and J. McCaffrey. 1986. Evolution of the atmosphere and oceans. *Nature* 320: 27–33.

Kirschvink, J. L. 1992. Late Proterozoic low-latitude global glaciation: The snowball earth. In *The Proterozoic Biosphere: A Multidisciplinary Study.* J. W. Schopf and C. Klein, eds. Cambridge: Cambridge University Press, pp. 51–52.

Nisbet, E. G. 1987. *The Young Earth: An Introduction to Archean Geology.* Boston: Allen and Unwin.

Rodgers, J. J. W. 1994. *A History of the Earth.* Cambridge: Cambridge University Press.

Stanley, S. M. 1999. *Earth System History.* New York: Freeman.

Windley, B. F. 1995. *The Evolving Continents.* 3rd ed. New York: Wiley.

Earth Resources

Many of the materials we use in our daily lives come from geologic materials—the Earth itself is, essentially, a natural resource. In Chapter 14, we look at the energy resources that come from the Earth. These include fossil fuels, such as oil and coal, as well as nuclear fuel and moving water. Chapter 15 focuses on nonenergy resources, particularly the mineral deposits from which we obtain metals. With the information we learn here, we can better evaluate political issues concerning these resources.

By the end of Part V, we'll realize that many natural resources are not renewable and thus must be conserved if we are to avoid shortages. Also, we'll see how our use of resources has made and continues to make an impact on the environment.

CHAPTER 14

Squeezing Power from a Stone: Energy Resources

14.1 INTRODUCTION

In the extreme chill of a midwinter night in northwestern Canada, a pan of water freezes almost instantly. But the low temperature doesn't stop a wolf from stalking its prey—the wolf's legs move through the snow, its heart pumps rapidly, and its body radiates heat. These life processes require energy. **Energy,** simply defined, means the capacity to do work, to cause something to happen, or to cause change in a system (see Appendix A). The wolf's energy comes from the metabolism of special chemicals like sugar, protein, and carbohydrates in its body, and these chemicals came from the food the animal ate. In order to survive, a wolf must catch and eat mice and rabbits, so to a wolf, these animals are "energy resources." In a general sense, we use the term **resource** for any item that can be employed for a useful purpose, and, more specifically, **energy resource** for something that can be used to produce heat, power muscles, produce electricity, or move automobiles. If the energy resource consists of matter that stores energy in a usable form, such as wood or oil, we can also call it a **fuel.**

The earliest humans needed about the same quantity of fuel per capita as a wolf, and thus could maintain themselves by hunting meat and gathering fruit. But when people discovered how fire could be used for cooking and heating, their need for energy resources began to exceed that of other animals, for they now needed fuel to feed the fires. Before the dawn of civilization, wood and dried dung provided adequate fuel. But as people began to congregate in towns, they also required energy for agriculture and transportation, and new resources such as animal power, wind, and flowing water came into use. By the end of the seventeenth century, energy-resource needs began to outpace the supplies available at the Earth's surface, for not only did the population continue to grow, but new industries like iron smelting became commonplace. In fact, to feed the smelting industry, woodcutters mowed down most European forests, so that

This inferno resulted when Kuwaiti oil wells were set ablaze at the end of the Gulf War. The flames speak of the energy the oil holds.

FIGURE 14.1 Historical drawing showing a giant nineteenth-century steam engine.

when the Industrial Revolution began in the eighteenth century, workers had to mine supplies of underground coal to feed the new steam engines that were driving factories (▶Fig. 14.1).

Since the Industrial Revolution, society's hunger for energy has increased almost unabated (▶Fig. 14.2). An American

FIGURE 14.2 The graph demonstrates how energy needs have increased in the past 150 years, and how different energy resources have been used to fill those needs. Hydrocarbons (oil and natural gas) together now account for more than half the world's energy usage.

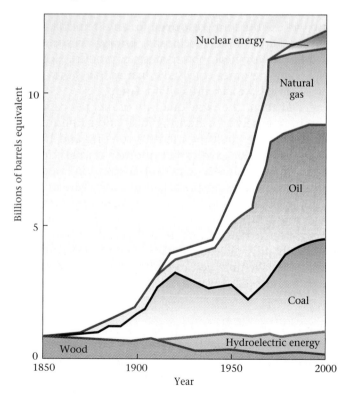

today, for example, uses more than 110 times the amount of energy needed by a prehistoric hunter-gatherer. Most energy for human consumption in the industrial world now comes from oil and gas. To a lesser extent, we also continue to use coal, wind, and flowing water, and in the last half century we've added nuclear energy, geothermal energy, and solar energy to the list.

Why is a chapter in a geology book devoted to such energy resources? Simply because these resources originate in geologic materials or processes: oil, gas, coal, and the fuel for atomic power plants come from rocks, geothermal energy is created by Earth's internal heat, and the movement of wind and water involves cycles in the Earth system. To understand the source and limitations of energy resources and to find new resources, we must understand their geology. That's why the multibillion-dollar-a-year energy industry employs tens of thousands of geologists—these are the people who find new energy-resource supplies.

In this chapter, we begin by looking at various types of energy resources on Earth. Then we focus on **fossil fuels** (oil, gas, and coal), fuels that have stored energy for geological time, in detail and other types of energy more briefly. The chapter concludes by outlining the dilemmas we will face as energy resources begin to run out, and as the products of energy consumption enter our environment.

14.2 SOURCES OF ENERGY IN THE EARTH SYSTEM

When you get down to basics, there are only five fundamental sources of energy on the Earth: (1) energy created by nuclear fusion in the Sun and transported to Earth via electromagnetic radiation; (2) energy created by the pull of gravity; (3) energy generated by nuclear fission reactions; (4) energy that has been stored in the interior of the Earth since the planet's beginning; and (5) energy stored in the chemical bonds of compounds. Let's look at the different ways these forms of energy become resources we can use (▶Fig. 14.3).

- *Energy directly from the Sun:* Solar energy, created by nuclear fusion reactions in the Sun, bathes the Earth's surface. It may be converted directly into electricity, using solar-energy panels, or it may be used to heat water or warm a house. No one has yet figured out how to produce controlled nuclear fusion on Earth, but we *can* produce uncontrolled fusion—this is the energy of a hydrogen bomb.

- *Energy directly from gravity:* The gravitational attraction of the Moon, and to a lesser extent the Sun, causes ocean **tides,** the daily up-and-down movement of the sea surface. The flow of water in and out of channels during tidal changes can drive turbines.

- *Energy involving both solar energy and gravity:* Solar radiation heats the air, which becomes buoyant and rises. As this happens, gravity causes cooler air to sink. The resulting air movement, **wind,** powers sails and windmills. Solar energy also evaporates water, which enters the atmosphere. When the water condenses, it rains, and water flows downhill in response to gravity. This moving water powers waterwheels and turbines.

- *Energy via photosynthesis:* Green plants absorb some of the solar energy that reaches the Earth's surface. Their green color comes from a pigment called chlorophyll. With the aid of chlorophyll, plants produce sugar through a chemical reaction called **photosynthesis.** In chemist's shorthand, we can write this reaction as

$$6CO_2 + 12H_2O + light \rightarrow 6O_2 + C_6H_{12}O_6 + 6H_2O.$$

carbon   water   oxygen   sugar   water
dioxide

Plants use the sugar produced by photosynthesis to manufacture more complex chemicals, or they metabolize it to provide themselves with energy. Animals that eat the plants convert the plant material into flesh.

Burning plant matter in a fire releases potential energy stored in the chemical bonds of organic chemicals. During burning, the molecules react with oxygen and break apart to produce carbon dioxide, water, and carbon (soot):

$$plant + O_2 \rightarrow burning \rightarrow CO_2 + H_2O + C \text{ (soot)}$$
$$+ \text{ other gases} + \text{heat energy}.$$

The flames you see in fire consist of glowing gases released and heated by this reaction.

FIGURE 14.3 The diverse sources of energy on Earth. Solar energy can be used directly, to drive wind or to cause water to evaporate and make rain and ultimately running water. It also provides energy for photosynthesis, which produces wood and plankton; these ultimately become fossil fuels (coal and oil, respectively). Radioactive material from the Earth powers nuclear reactors, and the Earth's internal heat creates geothermal resources. Gravity plays a role in producing wind and water power.

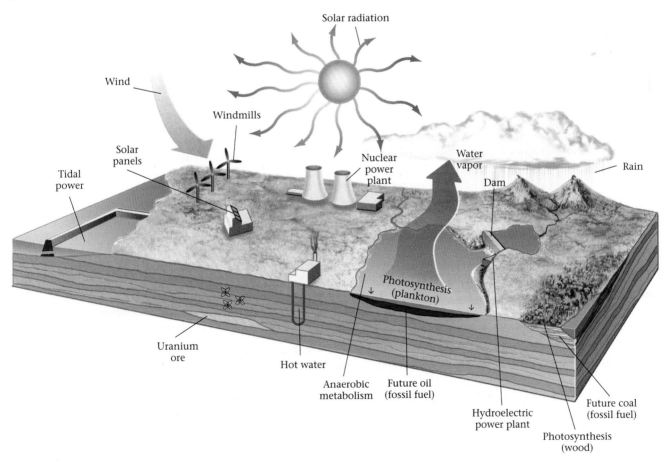

- *Energy from chemical reactions:* A number of inorganic chemical compounds can burn to produce light and energy. The energy results from exothermic (heat-producing) chemical reactions. A dynamite explosion is an extreme example of such energy production.

- *Energy from fossil fuels:* Oil and coal come from organisms that lived long ago, and thus store solar energy that reached the Earth long ago. We refer to these substances as **fossil fuels,** to emphasize that they have been preserved in rocks for ages. Burning fossil fuels produces energy in the same way that burning plants does.

- *Energy from nuclear fission:* Atoms of radioactive elements spontaneously split into smaller pieces, a process called nuclear fission (see Appendix A). During fission, a tiny amount of mass transforms into a large amount of energy, called nuclear energy. This type of energy runs nuclear power plants and nuclear submarines. Remember that fission is essentially the opposite of fusion.

- *Energy from Earth's internal heat:* Some of Earth's internal energy dates from the birth of the planet, while some is produced by radioactive decay. Some of this internal energy remains and heats water underground. The resulting hot water, when transformed to steam, serves as **geothermal energy** that can drive turbines.

14.3 OIL AND GAS

What Are Oil and Gas?

For reasons of economics and convenience, industrialized societies today rely primarily on oil (petroleum) and natural gas for their energy needs. Oil and natural gas consist of **hydrocarbons,** chain-like or ring-like molecules made of carbon and hydrogen atoms. For example, propane, the principal hydrocarbon of bottled gas, has the chemical formula C_3H_9. Chemists consider hydrocarbons to be a type of **organic chemical,** so named because similar chemicals make up living organisms.

Some hydrocarbons are gaseous and invisible, some resemble a watery liquid, some appear syrupy, and some are solid (▶Fig. 14.4). The **viscosity** (ability to flow) and the **volatility** (ability to evaporate) of a hydrocarbon product depend on the size of its molecules. Hydrocarbon products composed of short chains of molecules tend to be less viscous (they can flow more easily) and more volatile (they evaporate more easily) than products composed of long chains, because the long chains tend to tangle up with one another. Thus, short-chain molecules occur in gaseous form at room temperature, moderate-length-chain molecules occur in liquid form, and long-chain molecules occur in solid form (**tar**) (see Table 14.1).

FIGURE 14.4 The diversity of hydrocarbon products we use: natural gas piped to houses for heating and cooking, bottled gas (propane), gasoline for cars, kerosene for heating or illumination, diesel fuel for trucks, lubricating oil for motors, and solid tar, which melts when heated and can be used to make asphalt. Note that the sources are listed in order of increasing viscosity.

Why can we use hydrocarbons as fuel? Simply because hydrocarbons, like wood, burn—they react with oxygen to form carbon dioxide, water, and heat. For example, we can describe the burning of gasoline by the reaction

$$2C_8H_{18} + 25O_2 \rightarrow 16CO_2 + 18H_2O + \text{heat energy.}$$

TABLE 14.1 Sizes of Hydrocarbon Molecules

Natural gas	C_1— to C_4—
Gasoline	C_5— to C_{10}—
Kerosene	C_{11}— to C_{13}—
Heating oil	C_{14}— to C_{25}—
Lubricating oil	C_{26}— to C_{40}—
Tar	Larger than C_{40}—

FIGURE 14.5 Plankton, the living material out of which much of the world's oil forms.

During such reactions, the potential energy stored in the chemical bonds of the hydrocarbon molecules converts into usable heat energy.

Where Do Oil and Gas Form?

Many people entertain the false notion that hydrocarbons come from buried trees or the carcasses of dinosaurs. In fact, the primary sources of the organic chemicals in oil and gas are dead algae and plankton bodies. **Plankton,** as we have seen, are the tiny plants and animals (typically around 0.5 mm in diameter) that float in sea or lake water (▶Fig. 14.5).

When algae and plankton die, they settle to the bottom of a lake or sea. Because their cells are so tiny, they can only be deposited in quiet-water environments in which clay also settles, so typically the plankton cells mix with clay to create an organic-rich, muddy ooze. For this ooze to be preserved, it must be deposited in oxygen-poor water; otherwise, the organic chemicals in the ooze would react with oxygen or be eaten by aerobic (oxygen-using) bacteria, and thus would decompose quickly and disappear. But in some quiet-water environments, like lagoons and deep lakes, the dead algae and plankton get buried by still more sediment before being destroyed. Eventually, the muddy organic ooze lithifies and becomes **black organic shale** (in contrast to regular shale, which consists only of clay), which contains the raw materials from which hydrocarbons eventually form and is thus referred to as a **source rock.**

If black organic shale is buried deeply enough (2 to 4 km), it becomes warmer, since temperature increases with depth in the Earth. When the shale reaches about 100°C, chemical reactions slowly (taking up to millions of years) transform the organic material in the shale first into waxy molecules called **kerogen,** and then into a mixture of tar, oil, and gas (▶Fig. 14.6). Shale containing kerogen is called

FIGURE 14.6 Plankton and algae settle out of water and become progressively buried and compacted, gradually transforming into black organic shale. When heated for a long time, the organic matter in black shale transforms into oil shale, which contains kerogen. Eventually, the kerogen transforms into oil and gas. The oil may then start to seep upward out of the shale.

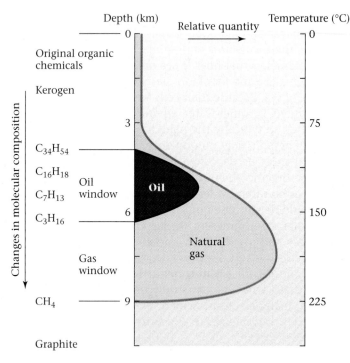

FIGURE 14.7 The oil window is the range of temperature conditions and depth at which hydrocarbons form. In regions with a geothermal gradient of 25°C per km, oil occurs only at depths of less than about 6.5 km. If the gradient is 50°C, oil is found at depths of less than about 4 km. Note that gas can be found at greater depths. The length of hydrocarbon chains decreases with increasing depth, because at higher temperatures longer chains break to form smaller ones.

oil shale. If the oil shale warms to temperatures of greater than about 80°C, the kerogen molecules break apart to form oil and natural gas molecules. At temperatures over about 160°C, any remaining oil breaks down to form natural gas, and at temperatures over 250°C, the remaining organic matter transforms into graphite. Thus, oil itself forms only in a relatively narrow range of temperatures, called the **oil window** (▶Fig. 14.7). For regions with low geothermal gradients, the oil window is located at depths of about 4 to 6.5 km (gas can exist down to 9 km), but for regions with high geothermal gradients, the oil window lies at shallower depths (2.5 to 4.5 km). In other words, hydrocarbon reserves can only exist in the topmost 15–20% of the crust.

14.4 MAKING AN OIL RESERVE

The oil-field fires of Kuwait blackening the sky remain one of the most vivid images of the 1991 Gulf War. As Iraqi troops retreated and left Kuwait, they set over 600 oil wells on fire, creating a conflagration that consumed over 6 million barrels of oil a day, an amount equal to Japan's total daily consumption.

Why was the Gulf War fought? As is the case for many wars, there were several complex reasons. The desire to control Kuwait's oil fields was probably one. An **oil field** is a region that contains a significant amount of accessible oil underground; the known supply of oil held underground in

FIGURE 14.8 The distribution of oil fields around the world. The largest fields are found in the region surrounding the Persian Gulf.

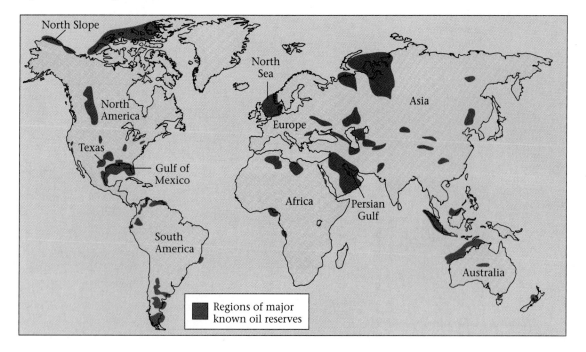

an oil field is an **oil reserve.** Countries bordering the Persian Gulf contain the world's largest oil fields and thus the most oil reserves. Oil reserves are not randomly distributed around the Earth (▶Fig. 14.8). In this section, we learn that the development of a reserve requires the existence of three geologic features: a source rock, a reservoir rock, and a trap.

Source Rocks

We saw above that the chemicals that become oil and gas start out in algae and plankton bodies. Their cells are deposited along with clay to form an organic ooze, which when lithified becomes black organic shale. Geologists refer to organic-rich shale as a **source rock** because it serves as the source for the organic chemicals that ultimately become oil. If black shale resides in the oil window, the organic material (kerogen) within transforms into oil.

Reservoir Rocks

Though in some places people dig up oil shale and heat it in a furnace to create oil, energy companies do not currently use much oil shale because the cost of producing oil from it remains prohibitively expensive. You can extract only about 40 gallons of oil per ton of shale, and the extraction itself takes energy. It is far cheaper for a company to extract oil from underground by drilling a well and pumping the oil out of the well. Wells drilled into source rocks, however, do not yield much oil because the kerogen can't flow easily from the rock into the well; any organic matter in an oil shale is trapped between the grains and can't move. So oil companies instead drill into **reservoir rocks,** rocks that contain (or could contain) an abundant amount of *easily accessible* oil, meaning oil that can be sucked out of the ground by a pump.

   To be a reservoir rock, a rock must have high porosity and permeability. **Porosity** refers to the amount of open space (pores) in a rock (▶Fig. 14.9a). Pore space exists either because the grains did not pack together tightly when the rock formed, or because cracking or dissolution of the rock later in its history made new openings. The space between the grains can hold liquid oil, much as the holes in a sponge hold water. It may come as a surprise that rocks contain a

(a)

Low porosity, low permeability
(b)

High porosity, low permeability
(c)

Fluid flow

(d)   High porosity, high permeability

(e)

Storage

Pump

Well

Impermeable   Permeable   Oil

FIGURE 14.9 The concepts of porosity and permeability. (a) Oil gathers in pore spaces (not as an underground lake). (b, c, d) The contrast between rocks with low porosity and low permeability, high porosity and low permeability, and high porosity and high permeability. In the latter, oil can flow through relatively easily and the rock can store a lot of oil, which makes it an ideal reservoir rock. (e) Oil must be able to flow into a well fast enough to make pumping economical. Thus, the rock must have high permeability.

FIGURE 14.10 (a) Initially, oil resides in the source rock. (b) Oil gradually migrates out of the source rock and rises into the overlying water-saturated reservoir rock. (c) Oil rises because it is buoyant relative to water—it tries to float on water. (d) The oil is trapped beneath a seal rock. If there is gas, it floats to the top of the oil.

certain amount of open space, but they do. For example, shale typically has a porosity of 10%, while sandstone has a porosity of 35%—that means that about a third of a block of sandstone actually consists of open space.

Permeability refers to the degree to which pore spaces connect to one another. Even if a rock has high porosity, it is not necessarily permeable (▶Fig. 14.9b–d). In a permeable rock, the holes and cracks (pores) are linked, so a fluid is able to flow slowly through the rock, following a tortuous pathway. Keeping the concepts of porosity and permeability in mind, we can see that a poorly cemented sandstone makes a good reservoir rock, because it is both porous and permeable. The greater the porosity, the greater the capacity of a reservoir rock to hold oil; and the greater the permeability, the easier it is for the oil to be extracted.

In an **oil well,** which is simply a deep hole drilled into the ground, oil flows from the permeable reservoir rock into the well and then up to the ground surface (▶Fig. 14.9e). If the oil in the rock is under natural pressure, it may move by itself, but usually producers set up a pump that literally sucks the oil up and out of the hole.

To fill the pores of a reservoir rock, oil must first *migrate* from the source rock into a reservoir rock, which it will do over millions of years of geologic time (▶Fig. 14.10). Why does oil migrate? Oil and gas are less dense than water, so they try to rise toward the Earth's surface to get above

groundwater, just as salad oil rises above the vinegar in a bottle of salad dressing. Natural gas, being less dense, ends up floating above oil. In other words, buoyancy drives oil and gas upward.

Traps and Seals

The existence of a reservoir rock alone does not create an oil reserve, because if oil can flow easily into a reservoir rock, it can also flow out. If oil or gas rises, escapes from the reservoir rock, and ultimately reaches the Earth's surface where it leaks away, there will be none left underground to pump. Thus, for an oil reserve to exist, oil and gas must be *trapped* underground in the reservoir rock, by means of a geologic configuration called an **oil trap.** An oil field contains one or more oil-bearing traps.

There are two components to an oil trap. First, a **seal rock,** a relatively impermeable rock such as shale, salt, or unfractured limestone, must lie above the reservoir rock and stop the oil from rising further. Second, the seal and reservoir rock bodies must be arranged in a geometry that collects the oil in a restricted area. Geologists recognize several types of oil-trap geometries, four of which are described in Box 14.1.

Note that when we talk about trapping oil underground, we are talking about a temporary process in the

BOX 14.1

THE REST OF THE STORY

Types of Oil Traps

Geologists who work for oil companies spend much of their time trying to identify underground oil traps. No two traps are exactly alike, but we can classify most into the following four categories.

- *Anticline trap:* In some places, sedimentary beds are not horizontal, as they are when originally deposited, but have been bent by the forces involved in mountain building. These bends, as we have seen, are called **folds.** An anticline is a type of fold with an arch-like shape (▶Fig. 14.11a; see Chapter 11). If the layers in the anticline include a source rock overlain by a reservoir rock, overlain by a seal rock, then we have the recipe for an oil reserve. The oil rises from the source rock, enters the reservoir rock, and rises to the crest of the anticline, where it is trapped by a seal.

- *Fault trap:* A fault is a fracture on which there has been sliding. If the slip on the fault crushes and grinds the adjacent rock to make an impermeable layer along the fault, then oil may migrate upward along bedding in the reservoir rock until it stops at the fault surface (▶Fig. 14.11b).

Alternatively, a fault trap develops if slip on the fault juxtaposes an impermeable rock layer against a reservoir rock.

- *Salt-dome trap:* In some sedimentary basins, the sequence of strata contains a thick layer of salt, deposited when the basin was first formed and seawater covering the basin was shallow and very salty. Sandstone, shale, and limestone overlie the salt. The salt layer is not as dense as sandstone or shale, so it is buoyant and tends to rise up through the overlying strata. Once the salt starts to rise, the weight of surrounding strata squeezes the salt out of the layer and up into a growing bulbous **salt dome.** As the dome rises, it bends up the adjacent layers of sedimentary rock. Oil in reservoir rock layers migrates upward until it is trapped against the boundary of the salt dome, for salt is not permeable (▶Fig 14.11c).

- *Stratigraphic trap:* In a stratigraphic trap, a tilted reservoir rock bed "pinches out" (thins and disappears along its length) between two impermeable layers. Oil migrating upward along the bed accumulates at the pinch-out (▶Fig. 14.11d).

FIGURE 14.11 (a) Anticline trap. The oil and gas rise to the crest of the fold. (b) Salt-dome trap. The oil and gas collect in the tilted strata on the flanks of the dome. (c) Fault trap. The oil and gas collect in tilted strata adjacent to the fault. (d) Stratigraphic trap. The oil and gas collect where the reservoir layer pinches out.

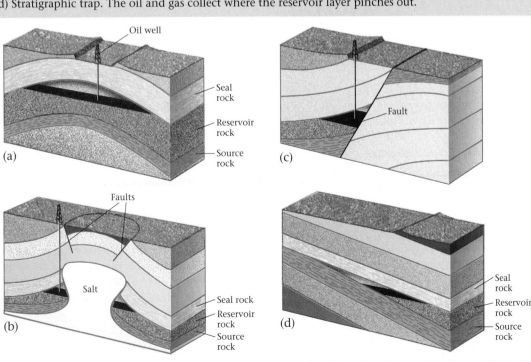

context of geologic time. Oil may be trapped for millions of years, but eventually it will manage to pass through a seal rock, because no rock is absolutely impermeable. Also, in some cases, microbes eat oil in the subsurface. Thus, innumerable oil fields that existed in the past have vanished, and the oil fields we find today, if left alone, will disappear millions of years in the future.

14.5 OIL EXPLORATION AND PRODUCTION

Birth of the Oil Industry

People have used oil since the dawn of civilization—as a cement, as a waterproof sealant, and even as a preservative to embalm mummies. But early on, the only available oil came from natural **seeps,** places where oil-filled reservoir rock intersects the ground surface or where fractures connect a reservoir rock to the ground surface, so that oil flows out on the ground on its own. Before the twentieth century, the discovery of an oil reserve was purely a fortuitous event, for geologists then did not know about oil traps.

In the United States during the first half of the nineteenth century, people collected "rock oil" (later called **petroleum,** from the Latin words *petra,* meaning "rock," and *oleum,* meaning "oil") at seeps and used it to grease wagon axles and to make patent medicine. But such oil was rare and expensive. In 1854, George Bissel, a New York lawyer, came to the realization that oil might have broader uses, particularly as fuel for lamps. Bissel and a group of investors contracted Edwin Drake, a colorful character who had drifted among many professions, to find a way to drill for oil in rocks beneath a hill near Titusville, Pennsylvania, where an oily film floated on the water of springs. Using the phony title "Colonel" to add respectability, Drake hired drillers, built a wooden drilling rig, and obtained a steam-powered drill. Work was slow and the investors became discouraged, but the very day that a letter arrived ordering Drake to stop drilling, his drillers found that the hole, which had reached a depth of 21.2 m, was filled with oil. They set up a pump, and on August 27, 1859, for the first time in history, pumped oil out of the ground. No one had given much thought to the issue of storing the oil, so workers dumped it into empty whisky barrels. This first oil well yielded 10–35 barrels a day, which sold for about $20 a barrel (1 barrel equals 42 gallons).

Within a few years, thousands of oil wells had been drilled in many states, and by the turn of the twentieth century civilization had begun its addiction to oil. Initially, most went into the production of kerosene for lamps. Later, as electricity took over from kerosene as the primary source for illumination, oil became the fuel of choice for the newly invented automobile, and was used to fuel electric power plants. In its early years, the oil industry was in perpetual chaos. When **wildcatters,** people who search for new oil

Spindletop

As demand for oil increased in the last decade of the nineteenth century, oil became "black gold," worth a fortune to anyone lucky enough to find it. One Texas mechanic, Patillo Higgins, not only dreamed of oil riches, but thought he knew a place to find them. Higgins had noticed a strange gas bubbling slowly up through little water springs on Spindletop Hill, where he took his Sunday school class for walks. When he lit the bubbles with a match, they burned. Did oil lie beneath the hill? Higgins struggled to convince investors to help him find out. Finally, in 1899, his arguments convinced Anthony Lucas, who brought a drilling crew to Spindletop.

Drilling began in late 1900, and soon after, Lucas and his crew reached a small oil pocket, which to their delight could yield 50 barrels a day, more than the average wells of Pennsylvania, where most oil wells were. But as long as drilling equipment was in place, Lucas decided to keep drilling. The operation proceeded routinely until January 10, 1901, when suddenly, without warning, the drilling platform began to vibrate violently. Workers fled for cover as the drilling pipe shot upward and out of the ground. As tons of equipment rained from the sky, a black fountain of oil erupted from the ground and blasted skyward to a height of almost 50 m. Oil from the Spindletop "gusher" rained over the pasture at the astounding rate of 75,000 barrels per day! No one had ever seen so much oil before.

News of the gusher chattered over telegraph lines, and pandemonium ensued. Tens of thousands of people flocked to Spindletop Hill, creating a boomtown of mud streets, plank shacks, and saloons. Land prices exploded from $10 per acre to almost $1 million per acre, and soon the hill looked like a pin cushion, punctured by over 200 wells. America's oil hysteria had moved to Texas.

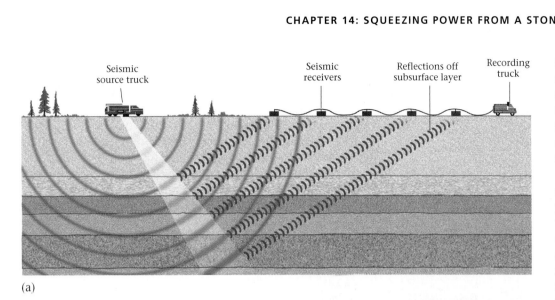

Seismic source truck

Seismic receivers

Reflections off subsurface layer

Recording truck

(a)

(b)

FIGURE 14.12 (a) The concept of seismic-reflection profiling. (b) A seismic-reflection profile. The colored bands are subsurface reflectors. The black line is an oil well. The vertical dimension is about 6 km, while the horizontal dimension is about 12 km.

tion, it may be possible to construct a preliminary **cross section,** depicting the geometry of the sedimentary layers underground as they would appear on an imaginary vertical slice through the Earth.

To add detail to the cross section, an exploration company makes a **seismic-reflection profile** of the region. To construct a seismic profile, a special vibrating truck or a dynamite explosion sends **seismic waves** (shock waves that move through the Earth) into the ground (▶Fig. 14.12a). The seismic waves reflect off contacts between rock layers, just as sonar waves sent out by a submarine reflect off the bottom of the sea. Reflected seismic waves then return to the ground surface, where their arrival is recorded by sensitive instruments. A computer measures the time between the generation of a seismic wave and its return, and from this information defines the depth to the contacts at which the wave reflected. With such information, the computer constructs an image of the configuration of underground rock layers (▶Fig. 14.12b). Unfortunately, seismic profiling is very expensive—it may cost millions of dollars to create one. Technological advances in the 1990s now allow geologists to create 3-D seismic-reflection profiles of the subsurface.

If geological studies identify an oil trap, and if the geologic history of the region indicates that there are good source rocks and reservoir rocks around, geologists make a recommendation to drill. (They do not make such recommendations lightly, as drilling a deep well may cost over $10 million.) Once the decision has been made, drillers go to work. These days, they use rotary drills to grind a hole down through rock. A rotary drill consists of a rotating pipe tipped by a **drill bit,** a bulb of metal studded with industrial diamonds or hard metal prongs (▶Fig. 14.13a, b). As the bit rotates, it scratches and gouges the rock, turning it into powder and chips. Drillers pump **drilling mud,** a slurry of water mixed with clay, down the center of the pipe. The mud squirts out of holes in the drill bit, cooling the bit and flushing rock cuttings up and out of the hole. The weight of the mud also keeps oil down the hole and prevents **gushers,** fountains of oil formed when underground pressure causes the oil to rise out of the hole on its own.

fields, discovered one, there would be a short-lived boom during which the price of oil could drop to pennies a barrel. In the midst of this chaos, John D. Rockefeller established the Standard Oil Company, which monopolized the production, transport, and marketing of oil; in the process, he became the richest man in America. The Supreme Court eventually broke Standard Oil down into several companies (including Exxon, Chevron, Mobil, Sohio, Amoco, Arco, and Conoco), and oil became a global industry governed by the complex interplay of politics, profits, supply, and demand.

The Modern Search for Oil

Wildcatters discovered the earliest oil fields either by blind luck or by searching for surface seeps. But when all known seeps had been drilled and blind luck became too risky, oil companies realized that finding new oil fields would require systematic exploration. The modern-day search for oil is a complex, sometimes dangerous, and often exciting procedure with many steps.

Source rocks are always sedimentary, as are most reservoir and seal rocks, so geologists begin their exploration by looking for a region containing appropriate sedimentary rocks. Then they compile a **geologic map** of the area, showing the distribution of rock units. From this informa-

(c)

FIGURE 14.13 (a) The derrick in this drilling platform is used to hoist the drill pipe, which comes in segments. (b) At the bottom of the pipe is a drill bit, typically consisting of three diamond-studded parts. The diameter of the bit is greater than the diameter of the pipe. Drilling mud passes into the hole inside the pipe, comes out through holes in the bit, and then rises between the pipe and the walls of the hole, thereby flushing cuttings out of the hole. (c) An offshore-drilling platform.

Drillers use derricks (tall spindly towers) to hoist the heavy drill pipe. To drill in an offshore oil reserve, one that occurs in strata of the continental shelf, the derrick must be constructed on an **offshore-drilling platform.** Offshore-drilling platforms can be immense (▶Fig. 14.13c); some tower more than 100 stories high from base (below sea level) to top. They serve as artificial towns for housing the drilling crew. In shallower water, platforms anchor on the sea floor, but in deeper water, platforms float on huge submerged pontoons (water-filled tubes). Drill holes can be aimed in any di-

rection (not just vertical), so drillers can reach many pockets of oil from the same platform.

Upon completion of a hole, workers remove the drilling rig and set up a pump. Some pumps resemble a bird pecking for grain; their heads move up and down to pull up oil that has seeped into the drill hole. Surprisingly, simple pumping gets only about 30% of the oil in a reservoir rock out of the ground. Thus, oil companies use **secondary recovery techniques** to coax up to 50% more oil out. For example, a company may drive oil toward a drill hole by forcing steam into the ground nearby—the steam heats the oil in the ground, making it less viscous, and pushes it along. In some cases, drillers create artificial fractures in rock around the hole so that there are easy routes for oil to follow from rock to well.

Presently, the vast majority of known oil reserves are distributed among only twenty-five fields, known as super-giant fields (Fig. 14.8). The largest occur around the Persian Gulf (in Saudi Arabia, Kuwait, Iraq, and neighboring countries). The United States is the largest consumer of oil (at a rate of 7 million barrels per day, about 25% of world consumption), but lost its position as the largest producer in the 1970s. Oil reserves in the United States now account for only about 4% of the world total. Thus, the United States must import more than half of the oil it uses.

Once extracted directly from the ground, **crude oil** flows first into storage tanks and then into a pipeline or tanker, which transports it to a refinery (▶Fig. 14.14a). At a **refinery,** workers distill crude oil into several separate components by heating it gently in a vertical pipe called a **distillation column** (▶Fig. 14.14b). Lighter molecules rise to the top of the column, while heavier molecules stay at the bottom. The heat may also "crack" larger molecules to make smaller ones. Chemical factories buy the largest molecules left at the bottom and transform them into plastics.

14.6 NATURAL GAS

Natural gas consists of volatile short-chain hydrocarbons, including methane, ethane, propane, and butane. It occurs in the pores of reservoir rock above oil, because it "floats" over the oil. Where temperatures in the subsurface are so high that oil molecules break apart to form gas, gas-only fields develop.

Gas burns more cleanly than oil (burning gas produces primarily carbon dioxide and water, while burning oil also produces complex organic pollutants), and is thus the preferred fuel for cooking and heating. But gas transportation requires expensive high-pressure pipelines or container ships, so even though gas is much more abundant than oil, the world's population still consumes more oil than gas.

(a)

FIGURE 14.15 Chunks of coal.

(b)

FIGURE 14.14 (a) The Alaska pipeline; (b) a distilling tower in an oil refinery.

14.7 COAL: ENERGY FROM THE SWAMPS OF THE PAST

Coal, a black, lumpy-looking sedimentary rock that burns, consists of elemental carbon mixed with minor amounts of organic chemicals, quartz, and clay (▶Fig. 14.15). Note that coal and oil do not have the same composition. In contrast to oil, coal forms from plant material (wood, stems, leaves) that once grew in swamps and adjacent forests. These **coal**

swamps resembled the wetlands and rain forests of modern tropical to semitropical coastal areas (▶Fig. 14.16a). Like oil and gas, coal is a fossil fuel because it stores solar energy that reached Earth long ago. Americans burn about 1 billion tons of coal per year, mostly at electrical-generating stations, yielding about 23% of the country's energy supply.

Significant coal supplies could not form until vascular land plants appeared in the late Silurian period, about 420 million years ago. The most extensive deposits of coal in the world occur in Carboniferous-age strata (deposited between 290 and 360 million years ago; ▶Fig 14.16b)—in fact, geologists coined the name "Carboniferous" because strata representing this interval of the geologic column contain so much coal. The abundance of Carboniferous coal reflects (1) the past position of the continents (during the Carboniferous period, North America, Europe, and northern Asia straddled the equator, and thus the region was warmer and vegetation flourished) and (2) the elevation of sea level (at this time, shallow seas bordered by coal swamps covered vast parts of continental interiors). Because of their antiquity, Carboniferous coal deposits contain fossils of long-extinct species like giant tree ferns, primitive conifers, and giant horsetails (▶Fig. 14.16c). Extensive coal deposits can also be found in strata of Cretaceous age (about 65 to 145 million years ago).

The Formation of Coal

How do the remains of plants transform into coal? The vegetation of an ancient swamp must fall and be buried in an *oxygen-poor* environment, such as stagnant water, so that it can be incorporated in a sedimentary sequence without first reacting with oxygen or being eaten. For example, in the oxygen-poor water of a swamp, dead vegetation decomposes so slowly that much of it gets buried. Compaction and partial decay of the vegetation transforms it into **peat.** Peat, which contains about 50% carbon, itself serves as a fuel in many parts of the world, where deposits formed

(a)

(b)

(c)

FIGURE 14.16 (a) This museum diorama depicts a Carboniferous coal swamp. (b) The graph shows the distribution of coal reserves in rocks of different ages. Notice that most reserves occur in late Paleozoic (Carboniferous) strata, a time when the continents were locked together to form Pangaea, a supercontinent that straddled the equator. (c) A fossil tree fern from a Carboniferous coal bed.

FIGURE 14.17 A peat bog.

from moss and grasses in bogs during the last several thousand years; it can easily be cut out of the ground, and once dried, it will burn (▶Fig. 14.17).

To transform peat into coal, the peat must be buried deeply (4–10 km) by sediment. Such deep burial can happen where the surface of the continent gradually sinks, creating a depression, or **sedimentary basin,** that can collect sediment. At any given time, the type of sediment deposited depends on the sea level, which rises and falls over geologic time. As the sea level rises, the shoreline migrates inland (**trangression** takes place), and the location of a coal swamp becomes submerged beneath deeper water and later buried by silt and mud. When the sea level later falls relative to the land, the shoreline migrates seaward (**regression** occurs) (▶Fig. 14.18).

FIGURE 14.18 Sea level transgresses and regresses over time, with the result that a coal swamp along the coast migrates inland and the swamp's deposits eventually get buried by other strata. Notice that the floor of the basin gradually sinks, so there is room for all the strata. (See Fig. 7.33 for the consequences of progressive transgression and regression.)

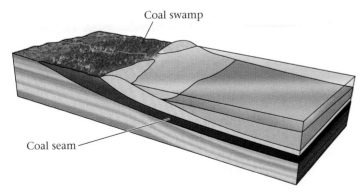

During successive transgressions and regressions in a sinking sedimentary basin (see Chapter 7), many kilometers of sediment containing numerous peat layers eventually accumulate. At depth in the pile, the weight of overlying sediment compacts the peat and squeezes out any remaining water. Then, because temperature increases with depth in the Earth, deeply buried peat gradually heats up. Heat accelerates chemical reactions that gradually destroy plant fiber and release elements like hydrogen, nitrogen, and sulfur in the form of gas. These gases seep out of the reacting peat layer, leaving behind a residue concen-

FIGURE 14.19 The evolution of coal from peat. (a) A coal swamp is underlain by a layer of peat. (b) Later, after the peat has been substantially buried, it is compacted, loses water, releases hydrogen, nitrogen, and sulfur, and becomes lignite. (c) Still later, after even more burial, the lignite compacts and alters still further to form bituminous coal. (d) In this mountain range, the coal bed has been folded (bent into curves), and faults (fractures on which sliding occurs) have transported warmer rock from greater depth over the coal. As a result, the coal loses still more hydrogen, nitrogen, and sulfur and turns into anthracite coal.

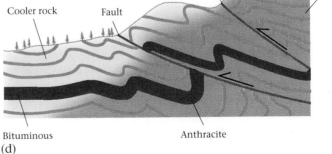

TABLE 14.2 Types of Coal

Material	% Carbon	Energy Content[a]	Rank
Peat	50	1,500 kcal/kg	
Lignite	70	3,500 kcal/kg	Low-rank coal
Bituminous coal	85	6,500 kcal/kg	Mid-rank coal
Anthracite coal	95	7,500 kcal/kg	High-rank coal

[a]1 kcal (kilocalorie) = 1,000 calories (a calorie is the heat needed to raise the temperature of 1 gram of water by 1°C). 1 kg (kilogram) = 2.2 pounds.

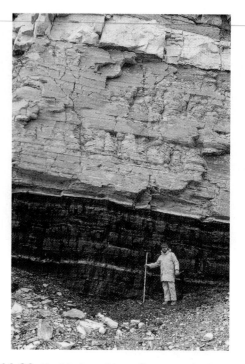

FIGURE 14.20 Coal is found in sedimentary beds interlayered with other strata (sandstone, shale, and limestone).

trated with carbon. Once the proportion of carbon in the residue exceeds about 70%, we have coal. With further burial and higher temperatures, chemical reactions remove additional hydrogen, nitrogen, and sulfur, yielding progressively higher concentrations of carbon.

The Classification of Coal

Geologists classify coal according to the concentration of carbon. Peat, with 50% carbon, transforms into a soft, dark-brown coal called **lignite.** At higher temperatures (about 100°–200°C), lignite becomes dull, black **bituminous coal,** and at still higher temperatures (about 200°–300°C), bituminous coal transforms into shiny, black **anthracite coal** (also called hard coal). The progressive transformation of peat to anthracite coal, which occurs as the coal layer is buried more deeply and becomes warmer, reflects the completeness of chemical reactions that remove water, hydrogen, nitrogen, and sulfur from the organic chemicals of the peat and leave behind carbon (▶Fig. 14.19a–d). As the carbon content of coal increases, we say the **coal rank** increases.

Of note, the formation of anthracite coal requires high temperatures that develop only on the borders of mountain belts, where mountain-building processes can push thick sheets of rock along thrust faults and over the coal-bearing sediment, so the sediment ends up at depths of 8–10 km, where temperatures reach 300°C. In the interiors of mountain belts, temperatures become even higher, and rock begins to undergo metamorphism, at which point almost all elements except carbon leave the coal. The remaining carbon atoms rearrange to form graphite, the gray mineral used to make pencils; thus, coal cannot exist in metamorphic rocks.

The burning of coal is a chemical reaction: C (carbon) + O_2 (oxygen) → CO_2 (carbon dioxide). Therefore, the different ranks of coal produce different amounts of energy when burned (Table 14.2). Anthracite contains more carbon per kilogram than lignite does, so it produces more energy.

14.8 FINDING AND MINING COAL

Finding Coal

Because the vegetation that eventually becomes coal was initially deposited in a sequence of sediment, coal occurs as sedimentary beds (**seams,** in mining parlance) interlayered with other sedimentary rocks (▶Fig. 14.20). To find coal, geologists search for sequences of strata that were deposited in tropical to semitropical, shallow-marine to terrestrial (fluvial or deltaic) environments—the environment in which a swamp could exist. The sedimentary strata of continents contain huge quantities of discovered coal, or **coal reserves.** For example, **economic seams** (beds of coal about 1–3 m thick, thick enough to be worth mining) of Cretaceous age occur in the U.S. and Canadian Rocky Mountain region, while economic seams of Carboniferous age are found throughout the midwestern United States (▶Fig. 14.21a, b).

Mining Coal

The way in which companies mine coal depends on the depth of the coal seam. If the coal seam lies within 100 m of the ground surface, **strip mining** proves to be most economic. In strip mines, miners use a giant shovel called a **drag line** to scrape off soil and layers of sedimentary rock above the coal seam (▶Fig. 14.22a, b). Drag lines are so big that they could swallow a two-car garage without a trace.

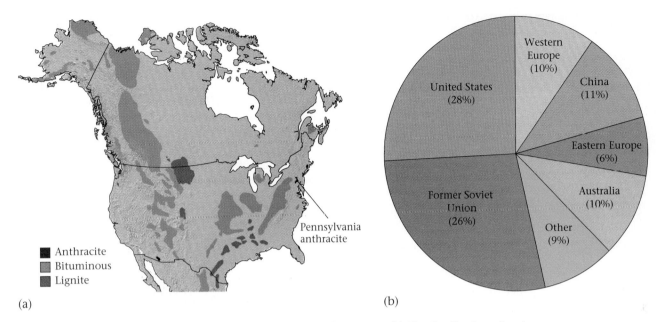

(a)

(b)

FIGURE 14.21 (a) The distribution of coal reserves in North America. (b) The distribution of coal reserves by region.

Once the drag line has exposed the seam, it then scrapes out the coal and dumps it into trucks or onto a conveyor belt. Before modern environmental awareness took hold, strip mining left huge scars on the landscape. Without topsoil, the rubble and exposed rock of the mining operation remained barren of vegetation. In contemporary mines, however, the drag-line operator separates out and preserves topsoil. Then, when the coal has been scraped out, the operator fills the hole with rock that had been stripped to expose the coal and covers the rock with topsoil. Within years, the former mine site can become a pasture or forest.

Deep coal can be obtained only by **underground mining.** Miners dig a shaft down to the depth of the coal seam and then create a maze of tunnels, using huge grinding machines that chew their way into the coal (▶Fig. 14.23). Underground coal mining can be very dangerous, not only because the sedimentary rocks forming the roof of the mine are weak and can collapse, but also because methane gas released by chemical reactions in coal can accumulate in the mine, leading to the danger of a small spark triggering a deadly mine explosion. As a result of mine collapses, explosions, and accidents, thousands of coal miners

FIGURE 14.22 (a) The configuration of a coal strip mine. (b) A drag line.

(a)

(b)

FIGURE 14.23 Machinery used in underground coal mining. The rotating head grinds away at the coal.

have died. Unless breathing through filters, underground miners also risk contracting **black-lung disease** from the inhalation of coal dust. The dust particles wedge into tiny cavities of the lungs and gradually cut off the oxygen supply or cause pneumonia.

14.9 NUCLEAR POWER

How Does a Nuclear Power Plant Work?

So far, we have looked at fuels (oil, gas, and coal) that release energy when burned—during burning, a chemical reaction between the fuel and oxygen releases the potential energy stored in the *chemical bonds* of the materials. Nuclear power comes from a different process: the fission, or breaking, of the *nuclear bonds* that hold protons and neutrons together in the nucleus provides the energy. Fission, the opposite of fusion, splits an atom into smaller pieces. Natural radioactive elements like uranium spontaneously undergo fission, or **radioactive decay.**

Nuclear power plants were first built to produce electricity during the 1950s. A **nuclear reactor,** the heart of the plant, commonly housed in a dome-shaped shell (containment building) made of reinforced concrete (▶Fig. 14.24a), contains **nuclear fuel,** pellets of concentrated uranium oxide or a comparable radioactive material, packed into metal tubes called **fuel rods.** Fission occurs when a neutron strikes a radioactive atom, causing it to split; for example, uranium-235 splits into barium-141 plus krypton-92 plus three neutrons plus energy (▶Fig. 14.24b). During fission, a tiny fraction of the matter composing the original atom transforms into a lot of thermal and electromagnetic

energy. Just one gram of nuclear-reactor fuel yields as much energy as 2.7 barrels of oil. The neutrons released during the fission of one atom strike other atoms, thereby triggering more fission in a self-perpetuating process called a **chain reaction.** Notably, if enough radioactive atoms cram into a small enough space, a **critical mass** forms: the chain reaction happens so fast that the mass explodes, making an atomic bomb. Such an explosion cannot occur in a reactor, because the fuel pile does not contain a critical mass.

In a nuclear power plant, pipes carry water close to the heat-generating fuel. The heat transforms the water into

FIGURE 14.24 (a) This nuclear power plant has two reactors, each in its own containment building. (b) A chain reaction involving the radioactive decay of uranium.

(a)

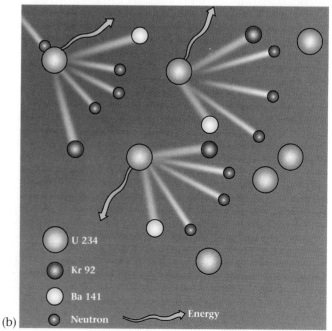

U 234

Kr 92

Ba 141

Neutron   Energy

(b)

(a)

(b)

FIGURE 14.25 (a) Nuclear fuel rods in a reactor. (b) A reactor can be used to generate electricity.

high-pressure steam. The pipes then carry this steam to a turbine, where it rotates fan blades; the rotation drives a dynamo that generates electricity. Eventually the steam goes into cooling towers, where it condenses back into water that can be reused in the plant or returned to the environment (▶Fig. 14.25a, b).

The Geology of Uranium

^{235}U, an isotope of uranium containing 143 neutrons, serves as the most common fuel for conventional nuclear power plants. ^{235}U accounts for only about 0.7% of naturally occurring uranium; most uranium consists of ^{238}U, an isotope with 145 neutrons. Thus, to make a fuel suitable for use in a power plant, the ^{235}U concentration in a mass of natural uranium must be increased by a factor of 2 or 3, an expensive process called **enrichment.**

Where does uranium fuel come from? The Earth's radioactive elements, including uranium, probably developed during the explosion of a supernova before the existence of the solar system. Uranium atoms from this explosion became part of the nebula out of which the Earth formed and thus were incorporated into the planet. They gradually rose into the upper crust in granitic magma.

Even though granite contains uranium, it does not contain very much. But nature has a way of concentrating uranium. Hot water circulating through the plutons after intrusion dissolves the uranium and precipitates it, along with other elements, in cracks. Uranium from veins is typically found in the mineral **pitchblende** (UO_2). Uranium may be further concentrated once plutons, and the associated uranium-rich veins, weather and erode at the ground surface. Sand derived from a weathered pluton washes down a stream, and as it does so, uranium-rich grains stay behind because they are so heavy, relative to quartz and feldspar grains. The world's richest uranium deposits, in fact, occur in ancient stream-bed deposits. Uranium deposits may also form when groundwater percolates through uranium-rich sedimentary rocks: the uranium dissolves in the water and moves with the water to another location, where it precipitates out of solution and fills the pores of the host sedimentary rock.

To find uranium deposits, prospectors use a Geiger counter, an instrument that detects radioactivity. These days, mining companies can quickly explore large regions using an instrument like a Geiger counter towed behind an airplane. If the instrument detects an unusually high concentration of radioactivity, land-based geologists explore the site further.

Nuclear Problems

Textbooks written in the 1950s touted nuclear power as a magical source that could meet our limitless energy needs cleanly for centuries to come. But in recent decades, many people have become concerned about plant safety and the disposal of radioactive waste. These concerns, along with vast cost overruns involved in the construction of nuclear plants, have slowed the growth of the nuclear power industry substantially.

To date, nuclear plants as a whole have been remarkably safe. But safety requires work. Operators must constantly cool the nuclear fuel with circulating water, and the rate of nuclear fission must be regulated by the insertion of boron-steel **control rods,** which absorb neutrons and thus decrease the number of collisions between neutrons and radioactive atoms. If left uncontrolled, so many neutrons begin to dash about in the fuel rods that the rate of fission becomes too fast and the fuel rods become too hot, in which case they melt, causing a **meltdown.** Were a meltdown to occur, the fuel could melt through pipes in the containment building and cause a steam explosion.

Only one near-meltdown has occurred since the beginning of the nuclear age—at the power plant in Chernobyl, Russia, in April 1986. Operators of the Chernobyl plant, while testing a procedure for cooling the reactor, shut down the safety devices. During the test, the fuel pile became too hot, triggering a steam explosion that raised the roof of the reactor and spread fragments of the reactor and its fuel around the plant grounds. Within six weeks, twenty people had died from radiation sickness. Further, about 10% of the radioactive material in the reactor entered the atmosphere and dispersed over Eastern Europe and Scandinavia; no one yet knows the effect of this fallout on the health of exposed populations. In the United States, the worst nuclear accident happened in the Three Mile Island plant near Harrisburg, Pennsylvania, in 1979. Because of faulty pumps, the reactor was not properly cooled for several hours and suffered damage. However, hardly any radioactivity entered the environment, and no one appears to have been directly injured.

Nuclear waste refers to the radioactive material produced in a nuclear plant. It includes spent fuel, which contains radioactive daughter products, as well as water and equipment that have come in contact with radioactive materials. Radioactive elements emit gamma rays and X-rays that can damage living organisms and cause cancer. Some radioactive material decays quickly (in decades to centuries), but some remains dangerous for thousands of years or more. **High-level waste** contains greater than 1 million times the safe level of radioactivity, **intermediate-level waste** contains between a thousand and 1 million times the safe level, and **low-level waste** contains less than a thousand times the safe level.

Nuclear waste cannot just be stashed in a warehouse or buried in the town landfill. If the waste were simply buried, groundwater passing through the dump site might transport radioactive elements into municipal water supplies or nearby lakes or streams. Ideally, waste should be sealed in containers that will last for thousands of years (the time needed for the short-lived radioactive atoms to undergo decay) and stored in a place where it will not come in contact with the environment. Finding an appropriate place is not easy. Geoscientists have suggested the following possibilities.

- Underground tunnels drilled into a mountain composed of solid, dry rock in a region safe from damage by earthquakes or volcanoes.
- The interiors of impermeable salt domes.
- Landfills surrounded by clay, for clay can absorb and trap radioactive atoms.
- Landfills in regions where the groundwater composition can react with radioactive atoms to form nonmovable minerals.
- The deep-ocean floor at a location where it will soon be subducted.

FIGURE 14.26 (a) This region can be used for geothermal energy production. Surface water sinks down into the ground as groundwater until it becomes heated by the hot ground (possibly by magma below). The hot water rises and, when it reaches shallow depths, turns to steam (as a result of decompression). The hot water may be pumped out. It turns to steam under ground-surface pressure, and runs turbines. (b) A geothermal energy plant.

(a)

(b)

So far, experts disagree about which is the best way to dispose of nuclear waste. The U.S. government favors storing its waste at Yucca Mountain, in the Nevada desert (see Fig. 19.26), which consists of fairly dry rhyolite. But because of continuing disagreement, much waste has been stored in temporary facilities.

14.10 GEOTHERMAL ENERGY

As the name suggests, **geothermal energy** refers to heat and electricity produced by using the internal heat of the Earth. Geothermal energy exists because the Earth grows progressively hotter with depth; recall that the **geothermal gradient,** the rate of temperature change, varies between 15° and 50°C per km in the upper part of the crust (Chapters 2, 6). Where the geothermal gradient is high, we find high temperatures at relatively shallow depths. Groundwater in such areas absorbs heat from the rock and becomes very hot.

We use geothermal energy to produce heat and electricity in two ways. In some places, hot groundwater is simply pumped out of the ground and run through pipes to heat houses and spas. Elsewhere, the groundwater is so hot that when it rises to the Earth's surface and decompresses, it turns to steam. This phenomenon, when it occurs naturally, produces geysers. Otherwise, pumps lift the hot water up from depth to create steam, which is then used to drive turbines and generate electricity (▶Fig. 14.26a, b).

In volcanic areas like Iceland or New Zealand, geothermal energy provides a major portion of energy needs. But on a global basis, it meets only a small proportion of energy needs because few cities lie near geothermal resources. Furthermore, even in geothermal regions the energy supplies can be destroyed, for if people pump groundwater out of the ground faster than it can be replenished, the hot-water supply diminishes.

14.11 HYDROELECTRIC POWER

As water flows downslope, its potential energy converts into kinetic energy. In a modern **hydroelectric power plant,** the water flow drives turbines, which in turn drive generators that produce electricity. In order to increase the rate and volume of water flow, engineers build dams to create a reservoir that retains water and raises it to a higher elevation—the water flows through pipes down to turbines at the foot of the dam (▶Fig. 14.27a).

At first glance, hydroelectric power seems ideal, because it produces no smoke or radioactive waste, and because reservoirs can also be used for flood control, irrigation, and recre-

(a)

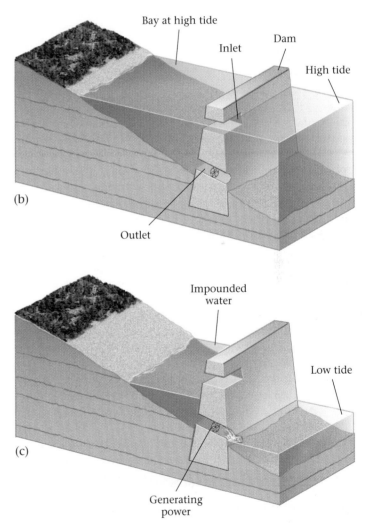

(b)

(c)

FIGURE 14.27 (a) A hydroelectric dam. (b, c) Oceanic tides can be used to generate electricity. At high tide, water fills the volume behind the dam. At low tide, water flows out via a turbine.

ation. But unfortunately, reservoirs may also bring unwanted changes to a region's landscape and ecology. Damming a fast-moving river may flood a spectacular canyon, eliminate exciting rapids, and destroy a river's ecosystem. Further, the reservoir traps sediment, so floodplains downstream lose their sediment and nutrient supply.

Not all hydroelectric power plants utilize river water. In a few places, engineers have employed the potential energy stored in ocean water at high tide (▶Fig. 14.27b, c). To do this, they build a floodgate dam across an inlet. Water flows into the inlet when the tide rises, only to be trapped when the gate is closed. After the tide has dropped outside the floodgate, the water retained by the floodgate flows back to sea via a pipe that carries it through an electrical-generating turbine.

14.12 ENERGY CHOICES, ENERGY PROBLEMS

The Crisis of the 1970s

Energy usage in industrialized countries grew with dizzying speed through the mid–twentieth century, and during this rise people came to rely increasingly on oil, which burns more efficiently than coal and, being a liquid, can be transported easily. Eventually, oil supplies within the borders of industrialized countries could no longer match the demand, and these countries began to import more oil than they produced themselves. Through the 1960s, oil prices remained low (about $1.80 a barrel), so this was not a problem.

In 1973, however, a complex tangle of politics and war led the Organization of Petroleum-Exporting Countries (OPEC) to limit their oil exports. In the United States, fear of an oil shortage turned to panic, and motorists began lining up at gas stations, in many cases waiting for hours to fill their tanks. The price of oil rose to $18 a barrel, and newspaper headlines proclaimed an "Energy Crisis!" Governments in industrialized countries instituted new rules to encourage oil conservation—speed limits dropped, manufacturers produced smaller cars, power companies switched back to coal, home insulation became tax deductible, and conservation slogans appeared next to light switches.

During the last two decades of the twentieth century, the oil market stabilized, though political events occasionally led to price jumps and short-term shortages—a shortage in 1979, for example, pushed the price of oil temporarily to over $30 a barrel. But was the oil crisis of the 1970s a harbinger of things to come? Will there come a day when shortages arise not because of an embargo but because there is no more oil to produce? To answer these questions, we need to examine the world oil supply. What we find there may be unnerving.

The Oil Crunch

To understand the issues involved in predicting the future of energy supplies, we must first classify energy resources. We call a particular energy resource **renewable** if nature can replace it within a short time span relative to a human life span (in months or, at most, decades). We call a resource **nonrenewable** if nature takes a very long time (hundreds to perhaps millions of years) to replenish it. Oil, gas, and coal are nonrenewable resources, in that the rate at which humans consume these materials far exceeds the rate at which nature replenishes them, so we will inevitably run out of oil. The question is, when?

Historians in the future will certainly refer to our time as the **Oil Age,** because so much of our economy depends on oil. How long will this Oil Age last? Geologists estimate that as of 2000, there were about 850–1,000 billion barrels of proven reserves (oil that had been found). There may be an additional 100–2,000 billion barrels not yet found. Thus, the world probably holds between 1,100 and 3,000 billion barrels of obtainable oil. Presently, we guzzle oil at a rate of about 25 billion barrels per year. At this rate of consumption, oil supplies will last until sometime between 2050 and 2150.

Unfortunately, problems may begin long before the supply runs out completely, because as oil becomes scarcer, it becomes harder to produce. Only the first third of the oil in a field flows easily into wells; another third may be urged into wells using secondary recovery techniques, but the remaining third stays put, trapped in isolated pores. Already, geologists see the beginning of the end of the Oil Age, for consumption now exceeds the rate of discovery of new oil by a factor of 3. In the end, the Oil Age will probably have lasted less than three centuries—on a time line representing the four thousand years since the construction of the Egyptian pyramids, this looks like a very short blip (▶Fig. 14.28a, b). We may indeed be living during a unique interval of human history.

Can Other Energy Sources Meet the Need?

Are there alternatives to oil? Perhaps. Certainly the world contains vast fossil-fuel supplies in other forms, enough to last for centuries if they can be exploited in an economical and environmentally sound way. These supplies include tar sand, oil shale, natural gas, and coal. **Tar sand** consists of sandstone reservoir rock in which less viscous oil and gas molecules have either escaped or been eaten by microbes, so that only tar remains. The tar is too viscous to pump, so it must be mined. Huge deposits of tar sands, known as the Athabaska Tar Sands, lie buried in western Canada; similarly large deposits occur in Venezuela.

All told, there are perhaps 1.5 trillion barrels of hydrocarbons trapped in tar sands and 3 trillion barrels trapped in oil shale, and there are coal deposits that could meet our en-

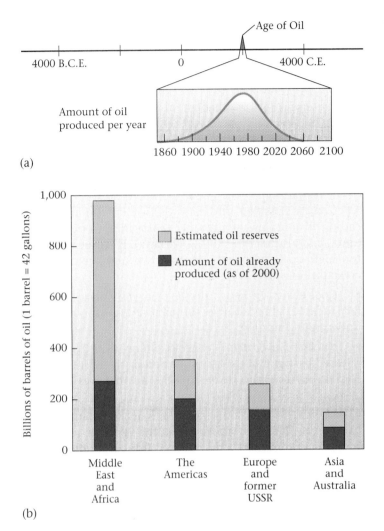

(a)

(b)

FIGURE 14.28 (a) The predicted history of the Oil Age, which may be a blip in human history like the Bronze Age. (b) The graph shows the amount of oil and gas reserves used and the estimated amounts still remaining.

ergy needs for the next few centuries. But extracting tar sand, oil shale, and coal requires the construction of huge mines, which might scar the landscape. Natural gas is probably our best alternative in the near future, but it remains very expensive to transport.

Can nuclear power or hydroelectric power fill the need? Vast supplies of uranium, the fuel of traditional nuclear plants, remain untapped, and in addition nuclear engineers have designed alternative plants, powered by **breeder reactors,** that essentially produce their own fuel. But many people view nuclear plants with suspicion, because of concerns about radiation, accidents, and waste storage, and these concerns have stalled the industry. A substantial increase in hydroelectric power production is not likely, as most appropriate rivers have already been dammed, and industrialized countries have little appetite for taming any

more. Similarly, the growth of geothermal-energy output seems limited.

Because of the problems that would result from relying more on coal, hydroelectric, and nuclear energy, researchers have been increasingly exploring **clean energy** options. One possibility is solar power. After all, as much solar energy reaches the Earth as could be produced by 115 million nuclear power plants. Light from the Sun can be converted to electricity by solar cells (special silicon-coated panels) or can be used to heat homes and water. Unfortunately, affordable technologies for large-scale solar energy do not yet exist, and the idea of covering the landscape with solar cells is an unattractive one. Similarly, we can turn to wind power in areas where strong gusts blow for small-scale energy production, but covering the landscape with windmills is no more appealing. Fusion power may be possible, but physicists and engineers have not yet figured out a way to harness it. Despite their inherent problems, natural gas, coal, and nuclear power will probably step in for oil when the Age of Oil comes to a close, but clearly society will be facing difficult choices in the not-so-distant future about where to obtain energy, and we will need to invest in the research needed to discover new alternatives.

Environmental Issues

Environmental concerns about energy resources begin right at the source. Oil drilling requires substantial equipment, whose use can damage the land. **Oil spills** from pipelines or trucks sink into the subsurface and contaminate groundwater, and oil spills from ships create a slick that spreads over the sea surface and fouls the shoreline (▶Fig. 14.29). Coal and uranium mining also scar the land and can lead to the production of **acid mine runoff,** a dilute solution of sulfuric acid produced when sulfur-bearing minerals such as

FIGURE 14.29 A marine oil spill.

pyrite (FeS_2) in mines react with rainwater. The runoff enters streams and kills fish and plants. Extensive coal mining may also cause the ground surface to subside. This happens when an underground mine collapses, so that the ground sinks. As a result, building foundations may crack. Coal-seam fires may also pose a problem. If ignited, a seam may smolder underground for years, releasing noxious gases that seep through cracks and pores to the ground surface, making the region above uninhabitable.

Numerous air-pollution issues also arise from the burning of fossil fuels, which sends soot, carbon monoxide, sulfur dioxide, nitrous oxide, and unburned hydrocarbons into the air, all pollutants that cause smog. Coal, for example, commonly contains sulfur, primarily in the form of pyrite, which enters the air as sulfur dioxide (SO_2) when coal is burned. This gas combines with rainwater to form sulfuric acid (H_2SO_4), or **acid rain.** For this reason, many countries now regulate the amount of sulfur that coal can contain when it is burned. Because of these regulations, **low-sulfur coal,** such as occurs in Cretaceous seams of the western United States, commands a higher price per ton than higher-sulfur coals of the Midwest.

But even if pollutants can be decreased, the burning of fossil fuels still releases carbon dioxide (CO_2) into the atmosphere. CO_2 is important because it traps heat in the Earth's atmosphere much like the glass traps heat in a greenhouse—this is the **greenhouse effect.** Too much CO_2 may lead to a global increase in atmospheric temperature (**global warming**), which in turn may radically alter the distribution of climatic belts. We'll learn more about this issue in Chapter 23.

CHAPTER SUMMARY

• Energy resources come in a variety of forms: energy directly from the Sun; energy from tides, flowing water, or wind; energy stored by photosynthesis (either in contemporary plants or in fossil fuels); energy from inorganic chemical reactions; energy from nuclear fission; and geothermal energy (from Earth's internal heat).

• Oil and gas are hydrocarbons, a type of organic chemical. The viscosity and volatility of a hydrocarbon depend on the length of its molecules. When hydrocarbons burn, they react with oxygen to release carbon dioxide, water, and heat.

• Oil and gas originally develop from the bodies of plankton, which settle out in a quiet-water, oxygen-poor depositional environment and form black organic shale. Later, chemical reactions at elevated temperatures convert the dead plankton into kerogen, then oil. Shale containing kerogen is oil shale.

• In order to create a usable oil reserve, oil must migrate from a source rock into a porous and permeable rock called a reservoir rock. Oil rises because it is more buoyant than water, so it tries to float above groundwater. Unless the reservoir rock is overlain by an impermeable seal rock, the oil will escape to the ground surface. The subsurface configuration of strata that leads to the entrapment of oil in a good reservoir rock is called an oil trap.

• For coal to form, the plant material must be deposited in an oxygen-poor environment, so that it does not completely decompose. Compaction near the ground surface creates peat, which, when buried deeply and heated, transforms into coal. Coal has a high concentration of carbon.

• Coal is classified into ranks, based on the amount of carbon it contains: lignite (low rank), bituminous (higher rank), and anthracite (still higher rank). If temperatures are too high, coal completely decomposes, and the carbon recrystallizes to form graphite.

• Coal occurs in beds, interlayered with other sedimentary rocks. Coal beds can be mined by either strip mining or underground mining.

• Nuclear power plants generate energy by using the heat released by the nuclear fission of radioactive elements. The heat turns water into steam, and the steam drives turbines.

• Some economic uranium deposits occur as veins in igneous rock bodies, some are found in sedimentary beds composed of grains eroded from the igneous rocks, and some occur in minerals precipitated from groundwater that passed through uranium-bearing rocks.

• Nuclear reactors must be carefully controlled to avoid overheating or meltdown. The disposal of radioactive nuclear waste can create environmental problems.

• Geothermal energy uses Earth's internal heat to transform groundwater into steam that drives turbines; hydroelectric power uses the potential energy of water; and solar energy uses solar cells to convert sunlight to electricity.

• We now live in the Oil Age, but oil supplies may last only for another century. Natural gas may become our major energy supply in the near future. Tar sand, oil shale, gas, and coal may also become sources of energy.

• Most energy resources have environmental consequences. Oil spills pollute the landscape, and the sulfur associated with some fuel deposits causes acid mine runoff. The burning of coal can produce acid rain, and the burning of coal and hydrocarbons produces smog and may cause global warming.

KEY TERMS

acid mine runoff (p. 449)
acid rain (p. 450)
anthracite coal (p. 442)

bituminous coal (p. 442)
chain reaction (p. 444)
coal swamps (p. 439)

critical mass (p. 444)
crude oil (p. 438)
energy resource (p. 427)
fossil fuels (p. 428)
fuel rods (p. 444)
geothermal energy (p. 447)
geothermal gradient (p. 447)
greenhouse effect (p. 450)
hydrocarbons (p. 430)
lignite (p. 442)
meltdown (p. 446)
nonrenewable resources (p. 448)
nuclear waste (p. 446)
oil reserve (p. 433)
oil shale (p. 432)
oil trap (p. 434)

organic shale (p. 431)
peat (p. 439)
permeability (p. 434)
photosynthesis (p. 429)
plankton (p. 431)
porosity (p. 433)
refinery (p. 438)
renewable resources (p. 448)
reservoir rock (p. 433)
seal rock (p. 434)
seismic-reflection profile
  (p. 437)
source rock (p. 431)
strip mining (p. 442)
tar sand (p. 448)
wildcatters (p. 436)

REVIEW QUESTIONS

1. Before the eighteenth century, where did people get most of their energy and fuel? How did this change during the Industrial Revolution?

2. What are the five fundamental sources of energy? How do they become eight sources of usable energy?

3. How does the length of a hydrocarbon chain affect its viscosity and volatility?

4. What is the source of the organic material in oil?

5. What is the oil window, and why does oil form only there?

6. How is organic matter trapped and transformed to create an oil reserve?

7. Why isn't oil shale widely used as an energy source at the present time?

8. How do porosity and permeability affect the oil-bearing potential of a rock?

9. Where is most of the world's oil found? At present rates of consumption, how long will it last?

10. How is coal formed?

11. Explain how coal is transformed in coal rank from peat to anthracite coal.

12. Describe the two main methods by which coal is mined.

13. What are some of the environmental drawbacks of mining and burning coal?

14. Describe how a nuclear reaction is initiated and controlled in a nuclear reactor.

15. Where does uranium form in the Earth's crust? Where does it usually accumulate in minable quantities?

16. What are some of the drawbacks of nuclear energy?

17. What is geothermal energy? Why is it not more widely used?

18. What is the difference between renewable and nonrenewable resources?

19. What is the likely future of oil production and use in the next century?

SUGGESTED READING

Abelson, P. H. 1987. Energy futures. *Scientific American* 75:584–93.

Brookins, D. G. 1981. *Earth Resources, Energy, and the Environment.* Columbus, Ohio: Merrill.

Craig, J. R., D. J. Vaughan, and B. J. Skinner. 1989. *Resources of the Earth.* Englewood Cliffs, N.J.: Prentice-Hall.

Department of Energy. 1991. *National Energy Strategy.* Washington, D.C.: Government Printing Office.

Energy for Planet Earth. 1990. Special issue of *Scientific American* (September 1990).

Energy Information Administration. 1994. *Annual Energy Review.* Washington, D.C.: Department of Energy.

Hunt, J. M. 1979. *Petroleum Geochemistry and Geology.* San Francisco: Freeman.

Hunt, V. D. 1982 *Handbook of Energy Technology.* New York: Van Nostrand Reinhold.

Jensen, M. L., and A. M. Bateman. 1981. *Economic Mineral Deposits.* New York: Wiley.

National Academy of Sciences. 1979. *Energy in Transition, 1985–2010.* San Francisco: Freeman.

North, F. L. 1990. *Petroleum Geology.* London: Unwin Hyman.

Petrick, A. 1986. *Energy Resource Assessment.* Boulder, Colo.: Westview Press.

Sawkins, F. J. 1984. *Mineral Deposits in Relation to Plate Tectonics.* New York: Springer Verlag.

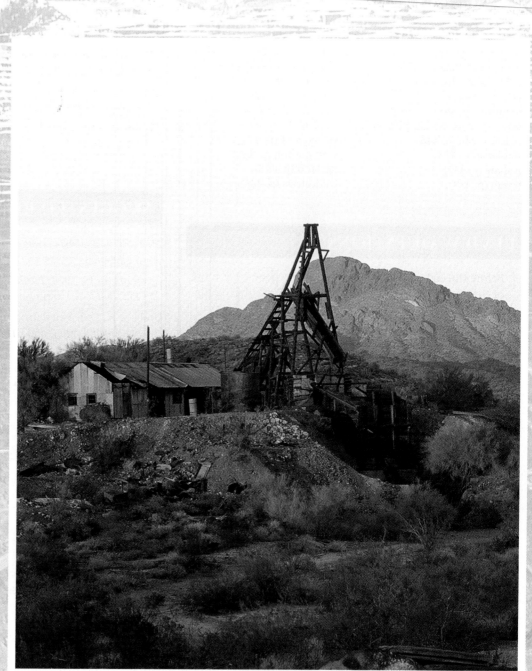

CHAPTER 15

Riches in Rock: Mineral Resources

15.1 INTRODUCTION

In June 1845, over a year after leaving Missouri, James Marshall arrived by horse at Sutter's Fort, in central California, to make a new life. After a stint as a rancher and a few months as a soldier, Marshall convinced Captain John Sutter, a former army officer, to finance the construction of a sawmill in the foothills of the Sierra Nevada Mountains. Marshall's scruffy crew finished the mill by the beginning of 1848. As Marshall stood admiring the new building, he noticed a glimmer of metal in the gravel that littered the bed of the adjacent stream. He picked it up, banged it between two rocks to test its hardness, and shouted, "Boys, by God, I believe I have found a gold mine!" For a short while, Marshall and Sutter managed to keep the discovery secret. But word of the gold soon spread, and within weeks the workers at Marshall's mill had disappeared into the mountains to seek their own fortunes.

All through that year, gold "fever" spread through California and eventually reached the East Coast. As a result, 1849 brought 40,000 prospectors to California. These **forty-niners,** as they were called, had abandoned their friends and relatives on the gamble that they could strike it rich. Many traveled for months by sea, sailing through stormy waters around the southern tip of South America, to reach San Francisco, a town of mud streets and plank buildings. In many cases, crews abandoned their ships in the harbor to join the scramble to the gold fields (▶Fig. 15.1). Although $20 million worth of gold was mined in 1849, few of the forty-niners became rich; most lost whatever wealth they found in the saloons and gambling halls that sprouted around the gold fields.

Gold is but one of the many **mineral resources**—minerals extracted from the Earth's upper crust—that civilization uses. Without these resources, industrialized societies could not function. Geologists divide mineral resources into two categories: **metallic mineral resources** (rocks containing gold, copper, aluminum, iron, etc.) and **nonmetallic mineral resources** (building stone, gravel, sand, gypsum, phosphate, salt, etc.). In this chapter, we look at the nature of mineral resources, the geologic phenomena responsible for their formation, and the process and problems of mining them. We conclude by considering limits to mineral reserves.

This abandoned mine in Arizona is all that's left of the sweat and toil of prospectors from decades past as they tried to pry copper and gold from the ground.

FIGURE 15.1 Forty-niners mining in the Sierra Nevada Mountains.

15.2 METALS AND THEIR DISCOVERY

What Is a Metal?

Metals are opaque, shiny, smooth solids that can conduct electricity and be bent, drawn into wire, or hammered into thin sheets—they look and behave quite differently from wood, plastic, meat, or rock. This is because, unlike other substances, the atoms that comprise metals are held together by **metallic bonds,** meaning that the outer electrons flow from atom to atom fairly easily. Despite the mobility of electrons, metals are solids, so their atoms lie fixed in a regular lattice defining a crystal structure. For example, the atoms in pure copper are arranged in wafer-like layers (▶Fig. 15.2a, b).

Not all metals behave the same way. Some are noted for their strength, others for their hardness, and others for their **malleability** (how easily they can be bent or molded). The behavior of a metal depends on the strength of bonds between atoms and on its crystalline structure. When you

FIGURE 15.2 (a) A copper crystal has a wafer-like appearance (the circles represent copper atoms). (b) When the crystal is deformed, one sheet of atoms slides past its neighbor.

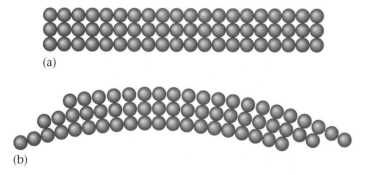

(a)

(b)

bend or stretch a strip of copper, for example, the wafer-like layers slip past one another quite readily. The shape and dimensions of crystals in a piece of metal are determined by how fast the molten metal cools and solidifies when removed from a furnace, by **tempering** (alternate heating and cooling), and by **cold working** (manipulating the shape of the metal after it is cooled). Before the development of the modern science of metallurgy, metalworkers learned how to obtain the qualities they wanted in a metal object simply by trial and error. Trade secrets of a skilled sword maker would be passed from master to apprentice for generations. But now engineers with a detailed knowledge of chemistry can control the properties of a metal.

The Discovery of Metals

Certain metals, namely copper, silver, and gold, can occur in rock as **native metals.** Native metals consist only of metal atoms, and thus look and behave like metal. Fifteen thousand years ago, ice-age hunters collected nuggets of native metal from stream beds and pounded them together with stone hammers to make arrowheads or scrapers (▶Fig. 15.3a–c). And because native metals were (and still are) rare and durable, people began to use them as money. Gold, in particular, attained popularity as a currency because of its unique warm yellow glow and its resistance to tarnish or rust. Though iron is not found as a native metal in rocks, chunks of iron fall to Earth from space in the form of meteorites; ancient peoples used this meteorite iron for tool making when they could find it, which was not often. Archaeologists have found implements fashioned from meteorite iron in the tombs of Egyptian pharaohs.

If we had to rely solely on native metals as our source of metal, we would have access to only a tiny fraction of our current supply. Most of the metal atoms we use today originated as ions bonded to nonmetallic elements in the lattice of a great variety of minerals that look nothing like the metals within. Only because of the chance discovery by some prehistoric genius that certain rocks, when heated to high temperatures in fire (in other words, **smelted**), decompose to yield metal plus a nonmetallic residue called **slag** do we now have the ability to produce sufficient metal for the needs of industrialized society.

Of the three principal metals in use today—copper, iron, and aluminum—copper came first, because copper smelting from sulfide minerals is relatively easy. Copper implements appeared as early as 4000 B.C.E., although pure copper has limited value because it is too soft to retain a sharp edge. Then around 2800 B.C.E., Sumerian craftsmen discovered that copper could be mixed with tin to produce bronze, an **alloy** whose strength exceeds that of either metal alone, and warriors came to rely on bronze for their swords (giving a name, the **Bronze Age,** to the next few hundred years).

(a)   (b)

heating. The widespread use of iron became possible only after someone discovered that metallic iron can be produced by heating iron-oxide minerals in the presence of carbon monoxide gas, which comes from burning charcoal (▶Fig. 15.4a, b). Chemists describe this reaction by the formula

$$Fe_2O_3 + 3CO \rightarrow 2\,Fe + 3CO_2.$$

FIGURE 15.4 (a) During smelting, iron ore and coke (carbon) placed in a furnace are heated by very hot air to temperatures of 1,250°–1,400°C. The iron ore separates into molten iron, which stays at the base of the furnace, and slag, which floats to the top. Slag consists of various silicates and metal oxides, and can be used for making roads and fertilizer. (Flux is a chemical added to make the ore melt at a lower temperature.) (b) Molten iron being poured.

(a)

(c)

FIGURE 15.3 (a) Gold nuggets come from quartz veins. At an early stage, much of the quartz remains. (b) With time, most of the quartz breaks away, leaving only gold. (c) The largest nugget on record.

(b)

Iron proves superior to copper or bronze for many purposes, because of its strength, hardness, and abundance. Still, people didn't start using iron widely until 1,500 years after they had begun to use bronze. The delay was due in part to the fact that iron has a very high melting temperature that can be difficult to reach. But in addition, the metal generally appears in the form of iron-oxide minerals (such as hematite, Fe_2O_3), and the liberation of iron metal from oxide minerals requires a chemical reaction, not just simple

More recently, people learned to make **steel,** an alloy of iron and carbon, and stainless steel, an alloy of iron, carbon, and chrome that resists corrosion.

Aluminum is abundant in rocks of the crust and in many ways is preferable to iron because it weighs less. But it does not occur in native form, and the extraction of aluminum from minerals requires complex methods that have only been economically practical relatively recently, so it has been in use only since the 1880s.

These days, in addition to iron, copper, tin, and aluminum, we use a vast array of different metals. Some are known as **precious metals** (gold, silver, and platinum) and others as **base metals** (copper, lead, zinc, and tin) because of the difference in their price. Curiously, of the sixty-three or so metals in use today, only nine (gold, copper, silver, mercury, lead, tin, antimony, iron, and arsenic) were known before 1700.

15.3 ORES, ORE MINERALS, AND ORE DEPOSITS

What Is an Ore?

As we have seen, metals occur in two forms—as native elements (copper, silver, or gold) or as ions mixed with nonmetallic elements in a great variety of minerals. If you pick up a chunk of common granite and analyze its mineral content, you will find that it is composed mostly of quartz and feldspar, with minor proportions of biotite (mica), and/or hornblende and/or garnet (see Chapter 5). Some of these minerals contain metals: potassium feldspar ($KAlSi_3O_8$), for example, contains about 8% aluminum. But even though common granite includes metal ions, we don't mine granite to produce these metals. Why? Simply because the minerals in granite contain relatively little metal, so it would cost more to separate the metals from granite than you would get if you were to sell them. To obtain the metals needed for industrialized society, we mine **ore,** rocks containing native metals or a concentrated accumulation of ore minerals.

Ore minerals (or **economic minerals**) are minerals that have metal in high concentrations and in a form that can be easily extracted. Galena (Pbs), for example, contains about 50% lead and so can be considered lead ore (▶Fig. 15.5a); the procedure for separating lead atoms from galena is relatively simple and inexpensive. We obtain most of our iron from the oxide minerals hematite and magnetite (▶Fig. 15.5b). Iron is also found in the common mineral pyrite (FeS_2), but we don't mine pyrite ("fool's gold") for iron because it's difficult to separate iron from sulfur.

Geologists have identified many different kinds of ore minerals (Table 15.1). As you can see from the chemical for-

(a)

(b)

FIGURE 15.5 (a) This block of limestone, containing galena, can be considered lead ore. (b) Less than 2% of granite consists of iron-oxide minerals, while iron ore may include over 80%.

mulas, many ore minerals are sulfides, in which the metal occurs in combination with sulfur (S), or oxides, in which the metal occurs in combination with oxygen (O). Of note, numerous ore minerals are colorful and come in interesting shapes, and some have a metallic luster (▶Fig. 15.6).

To be an ore, a rock must not only contain ore minerals, it must have a sufficient amount to make it worth mining. Iron constitutes about 6.2% of the continental crust's weight, while it makes up 30–60% of iron ore. The concentration of a useful metal in an ore determines the **grade** of an ore—the higher the concentration, the higher the grade. Whether or not an ore is worth mining depends on the price of metal in the market. For example, in 1880, copper-bearing rocks required at least 3% copper to considered economic ore, but in 1970, miners considered rock containing only about 0.3% to be economic. This change reflects new technology for mining and processing the ore.

TABLE 15.1 Some Common Ore Minerals

Metal	Mineral Name	Chemical Formula
Copper	Chalcocite	Cu_2S
	Chalcopyrite	$CuFeS_2$
	Bornite	Cu_5FeS_4
	Azurite	$Cu_3(CO_3)_2(OH)_2$
	Malachite	$Cu_2(CO_3)(OH)_2$
Iron	Hematite	Fe_2O_3
	Magnetite	Fe_3O_4
Tin	Cassiterite	SnO_2
Lead	Galena	PbS
Mercury	Cinnabar	HgS
Zinc	Sphalerite	ZnS
Aluminum	Kaolinite	$Al_2Si_2O_5(OH)_4$
	Corundum	Al_2O_3
Chrome	Chromite	$(Fe,Mg)(Cr,Al,Fe)2O_4$
Nickel	Pentlandite	$(Ni,Fe)_9S_8$
Titanium	Rutile	TiO_2
	Ilmenite	$FeTiO_3$
Tungsten	Wolframite	$FeWO_4$
Molybdenum	Molybdenite	MoS_2
Magnesium	Magnesite	$MgCO_3$
	Dolomite	$CaMg(CO_3)_2$
Manganese	Pyrolusite	MnO_2
	Rhodochrosite	$MnCO_3$

How Do Ore Deposits Form?

Ore minerals are not uniformly distributed through rocks of the crust—if they were, we would not be able to extract them economically. Fortunately for humanity, geologic processes concentrate these minerals in ore deposits. Simply put, an **ore deposit** is an economically significant accumulation of ore. The various kinds of ore deposits differ from one another in terms of which ore minerals they contain and which kind of rock body they occur in. Following are the principal types of ore deposits.

Magmatic deposits. When a magma cools, sulfide ore minerals crystallize before other minerals do and then, because sulfides tend to be dense, sink to the bottom of the magma chamber, where they accumulate; this accumulation is a **magmatic deposit.** When the magma freezes solid, the resulting igneous body may contain a solid mass of sulfide minerals at its base. Because of their composition, these masses are also a type of **massive-sulfide deposit** (▶Fig. 15.7).

Hydrothermal deposits. Hydrothermal activity simply refers to the circulation of hot-water solutions through a magma or through the rocks surrounding an igneous intrusion. These fluids dissolve metal ions. When a solution enters a region of lower pressure and/or lower temperature, the metals come out of solution and form ore minerals that precipitate in fractures and pores, creating a **hydrothermal deposit** (▶Fig. 15.8). Such deposits may take place within an igneous intrusion or in surrounding country rock. If the resulting ore minerals disperse through the intrusion, we call the deposit a **disseminated deposit,** but if they precipitate to fill cracks in preexisting rock, we call the deposit a **vein de-**

FIGURE 15.6 Colorful ore minerals containing copper. The blue mineral is azurite and the green is malachite.

FIGURE 15.7 Heavy, metal crystals can sink to the bottom of a magma chamber to form a massive-sulfide deposit. The sulfide concentrate may become an ore body in the future.

Heavy-ore crystals settling

Heavy-ore concentrate

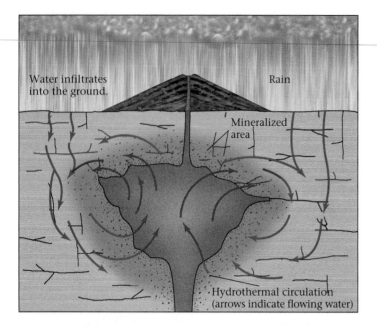

FIGURE 15.8 Water circulating through a granite pluton dissolves and redistributes metals.

(a)

(b)

(c)

FIGURE 15.9 (a) The difference between a vein deposit and a disseminated deposit; (b) ore containing veins; (c) a porphyry containing disseminated ore.

posit (veins are simply mineral-filled cracks; ▶Fig. 15.9a–c). Hydrothermal copper deposits commonly occur in porphyritic igneous intrusions; these are known as **porphyry copper deposits.** Typically, vein deposits include quartz in addition to the ore minerals. For example, native gold commonly appears as flakes in milky-white quartz veins.

In recent years, geologists have discovered that the hot water expelled at the submarine volcanoes along mid-ocean ridges contains high concentrations of dissolved metal and sulfur. When this hot water comes in contact with cold seawater, the dissolved components instantly precipitate as tiny crystals of metal sulfide minerals, creating black clouds known as **black smokers** (▶Fig. 15.10). The minerals in the cloud eventually sink and form a pile of nearly pure ore minerals around the vent. Since the ore minerals typically are sulfides, the resulting deposits are another type of massive-sulfide as well as hydrothermal deposit.

Secondary-enrichment deposits. Sometimes groundwater passes through ore-bearing rock long after the rock first formed. This groundwater dissolves some of the ore minerals and carries them away. When the water eventually flows into a different chemical environment (for example, one with a different amount of oxygen or acid), it precipitates new ore minerals, commonly in concentrations that exceed that of the original deposit. A new ore deposit formed from metals that were dissolved and carried away from a preexisting ore deposit is called a **secondary-enrichment de-**

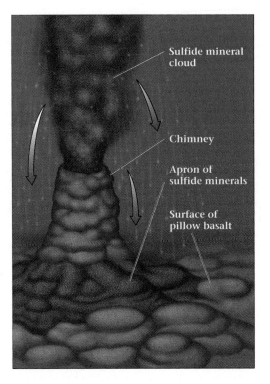

FIGURE 15.10 Aprons of sulfide ore precipitate around black smokers.

posit (▶Fig. 15.11a, b). Some of these deposits contain spectacularly beautiful copper-bearing carbonate minerals, like azurite and malachite (Fig. 15.6).

In recent years, geologists have recognized that some groundwater enters the Earth as rain in a mountain range, then flows several kilometers down into the crust before returning to the surface perhaps hundreds of kilometers away from the mountain range. At the depths reached by the groundwater, temperatures become high enough that the water dissolves metals. When the water returns to the surface and enters cooler rock, these metals precipitate in ore minerals. Ore deposits formed in this way, containing lead- and zinc-bearing minerals, appear in dolomite beds of the Mississippi Valley region, and thus have come to be known as **Mississippi Valley–type (MVT) ores.**

Sedimentary deposits. Some ore minerals accumulate in sedimentary environments under special circumstances. For example, between 2.0 and 2.5 billion years ago, the atmosphere, which previously had contained little oxygen, evolved into the oxygen-rich atmosphere we breathe today. In an oxygen-poor environment, very little iron can dissolve in seawater. Thus, when the Earth's atmosphere first became oxygen-rich, quantities of iron precipitated out of seawater and settled as sediment on the sea floor. As we learned in Chapter 13, the resulting iron-rich sedimentary layers are known as **banded-iron formation (BIF)** (▶Fig. 15.12a), because after lithification they consist of alternating beds of gray iron oxide (such as magnetite or hematite) and red beds of jasper (iron-rich chert). Microbial metabolism may have participated in the precipitation process.

The chemistry of seawater in some parts of the ocean leads to the deposition of manganese-oxide minerals on the sea floor. These minerals grow into lumpy accumulations known as **manganese nodules** (▶Fig. 15.12b). Mining companies have begun to explore technologies for vacuuming up these nodules, which geoscientists estimate contain

FIGURE 15.11 (a) In the process of secondary enrichment, water passing down through the ore body dissolves and carries the ore with it. (b) The ore then reprecipitates just below the water table (the level at which pores in the rock are filled with water).

(a)

(b)

(a)   (b)

FIGURE 15.12 (a) Banded-iron formation; (b) manganese nodules on the sea floor.

720 years' worth of copper and 60,000 years' worth of manganese, at current rates of consumption.

Residual mineral deposits. Recall from Chapter 7 that as rainwater sinks into the Earth, it leaches (dissolves) certain elements and leaves behind others, as part of the chemical weathering that forms soils. In rainy tropical environments, the residuum left behind in soils after leaching includes concentrations of iron or aluminum. Locally, these metals become so concentrated that the soil itself becomes an ore deposit (▶Fig. 15.13a, b). We refer to such deposits as **residual mineral deposits.** Most of the aluminum ore mined today comes from **bauxite,** a residual mineral deposit created by the extreme leaching of granite during soil formation.

Placer deposits. Ore deposits may develop when rocks containing native metals erode and create a mixture of sand grains and metal flakes or nuggets (pebble-sized fragments). The heavy metal grains (gold, for example) accumulate in sand or gravel bars along the course of rivers, for the moving water carries away lighter mineral grains but can't move the metal grains. Concentrations of metal grains in stream sediments are a type of **placer deposit** (▶Fig. 15.14). (The term is also used for concentrations of diamonds.) Panning further concentrates gold flakes or nuggets—by swirling water in a pan, the lighter sand grains wash away, leaving the gold behind. Placer deposits may eventually be buried and lithify to become part of a new sedimentary rock.

Where Are Ore Deposits Found?

The Inca Empire of fifteenth-century Peru boasted elaborate cities and temples, decorated with fantastic masks, jewelry, and sculpture made of gold. Then, about 1532, Spanish

FIGURE 15.13 (a) When rainwater sinks through the soil, it dissolves and removes many elements. (b) A thick soil forms, containing a residuum of iron or aluminum.

(a)   (b)

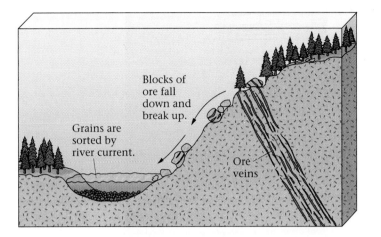

FIGURE 15.14 The process of forming a placer deposit. Ore-bearing rock is eroded, and clasts containing native metals fall into a stream. Sorting by the stream concentrates the metals.

conquistadors arrived in ships, led by commanders who quipped, "We Spaniards suffer from a disease that only gold can cure." The Incas, already weakened by civil war, were no match for the armor-clad Spaniards and their guns, and within six years the Inca Empire had vanished and Spanish ships were transporting Inca treasure back to Spain. Why did the Incas possess so much gold? Or to ask the broader question, what geologic factors control the distribution of ore? We can find the answer once again by considering the consequences of plate tectonics.

Several of the types of ore deposits mentioned above occur in association with igneous rocks. As we learned in Chapter 6, igneous activity does not happen randomly around the Earth, but rather concentrates at convergent plate boundaries (in the overriding plate of a subduction zone), at divergent plate boundaries (along mid-ocean ridges), at continental rifts, or at hot spots. Thus, magmatic and hydrothermal deposits (and secondary enrichment deposits derived from these) occur along plate boundaries, along rifts, or at hot spots. Placer deposits are typically found in the sediments eroded from magmatic or hydrothermal deposits.

Take the Inca gold, for example. The Inca Empire was situated in the Andes Mountains, which had formed as a result of compression and volcanic activity where the Pacific Ocean floor subducts beneath the South American Plate. As the mountains rose, erosion stripped away surface rocks to expose the large granite plutons that had intruded into the continental crust beneath. The magma that froze to make the granite brought gold, copper, and silver atoms with it. Some of the gold precipitated along with quartz to form veins in the plutons. Inca miners quarried these veins and separated the gold, or panned for gold in the streams choked with sediment eroded from the plutons. Plutons

that contain similar ore deposits developed in the western United States during the late Mesozoic and Cenozoic eras.

As noted earlier, some massive sulfide deposits accumulate at the base of a magma chamber, and some precipitate from black smokers along a mid-ocean ridge system. Thus, some massive sulfides occur in plutons at convergent margins or at rifts, while some are interlayered with sea-floor basalt. Miners can gain access to sea-floor deposits only in places where the collision of continents traps a sliver of sea floor and slides it up and onto continental crust. During the rifting of continents, large magma dikes form by the partial melting of the mantle along the axis of the rift. These bring many valuable metals with them, which may accumulate in magmatic or hydrothermal deposits.

Some ore deposits, notably sedimentary ores like banded-iron formation or residual mineral deposits, are not a direct result of plate tectonics activity, but are a consequence of unique environments at the surface of the Earth, and their existence emphasizes the interrelations in the Earth system between air, ocean, life, and rock. Residual ores like bauxite, the main source of aluminum, have developed in fairly recent times; bauxite can only form when plate motions move continents into equatorial realms where warm temperatures and monsoonal rains rapidly weather rocks.

15.4 ORE-MINERAL EXPLORATION AND PRODUCTION

We can all conjure up an image from the early days of exploring for ore minerals. An old prospector, for example, clanks through the desert with a broken-down donkey by his side, eyeing the hillsides for **shows** of ore (exposures of ore minerals at the ground surface). If he finds a show of minerals, he pries out chunks of the rock with a pick, and the poor donkey hauls the rock back to town for an **assay,** a test to determine how much extractable metal the rock contains. Mining laws permitted a prospector to, literally, "stake a claim" by marking off an area of ground with stakes. The prospector would then have the exclusive right to dig up ore at that spot and sell it. Old claims still litter the desert in Arizona and Nevada (▶Fig. 15.15a).

What does a show look like? Typically, prospectors looked for milky-white quartz veins and/or exposures in which rocks were stained green or red by the oxidation of metal-containing minerals (▶Fig. 15.15b). Some prospectors panned streams in search of gold flakes in the stream gravel. When a prospector did find ore, word usually spread fast and others rushed to stake neighboring claims.

These days, large mining companies employ geologists to systematically survey ore-bearing regions. The geologists focus their studies on rocks that developed in settings appro-

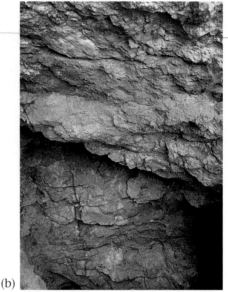

FIGURE 15.15 (a) An old mining claim in the Arizona desert. (b) Stained rock is an indicator of ore. The stain comes when ore rusts in the presence of air and water.

priate for ore formation. Once such a region has been identified, they measure the local strength of Earth's magnetic field and the local pull of gravity. These measurements lead them to ore bodies, because ore minerals tend to be denser and more magnetic than average rocks (▶Fig. 15.16a, b). Geologists also sample rocks and soils to test their metal content, and may even analyze plants in the area to detect traces of metals, for plants absorb metals through their roots. Once geologists have identified a possible ore body, they drill holes to sample subsurface rock and to determine the ore body's shape and extent. Ore-mineral exploration takes geologists into jungles, deserts, and tundras worldwide.

FIGURE 15.16 (a) On this magnetic-anomaly map, yellowish areas are regions with normal intensity, reddish areas are positive anomalies (greater-than-magnetic intensity), and purplish areas are negative anomalies (less-than-expected magnetic intensity). In some cases, ore-bearing rocks occur where there are positive anomalies. (b) The three-dimensional shape of an ore body underground, and the workings that miners dig to access the ore body. Note that shafts are vertical and adits (tunnels) are horizontal.

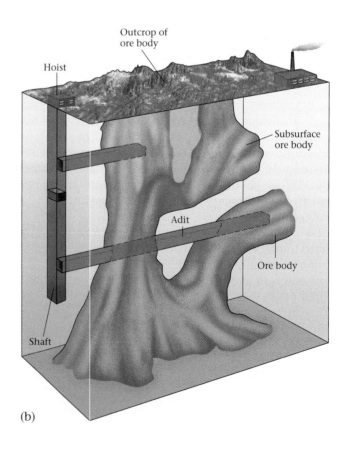

If calculations show that the mining of an ore deposit will yield a profit, and if environmental concerns can be accommodated, a company builds a mine. Mines can be below or above ground, depending on how close the ore deposit is to the surface. To make an **open-pit mine,** workers first drill a series of holes into the solid bedrock and then fill the holes with high explosives. They must space the holes carefully and must set off the charges in a precise sequence, so that the bedrock shatters into appropriate-sized blocks for handling. When the dust settles, large front-end loaders dump the ore into giant ore trucks, which can carry as much as 200 tons of ore in a single load (for comparison, a loaded cement mixer weighs about 70 tons). The tires on these trucks are so huge that a tall person only comes up to the base of the hub. The trucks transport **waste rock** (rock that doesn't contain ore) to a tailings pile and the ore to a crusher, a giant set of moving steel jaws that smash rock into small fragments. Workers then separate ore minerals from other minerals and send the ore-mineral concentrate to a processing plant, where workers smelt the ore or treat it with acidic solutions to separate metal atoms from other atoms. Eventually, the metal is melted and then poured into molds to make ingots (brick-shaped blocks) for transport to a manufacturing facility.

If the ore deposit lies more than about a hundred meters below the Earth's surface, miners must make an **underground mine.** Typically, the miners sink a vertical shaft in which they install an elevator. At the level in the crust where the ore body appears, they build a maze of tunnels, called **adits,** into the ore by drilling holes into the rock and then blasting. The rock removed must be carried back to the surface. Rock columns between the tunnels hold up the ceiling of the mine. The deepest mine on the planet currently reaches a depth of 3.5 km, where temperatures exceed 55°C, making mining there a very uncomfortable occupation. Although miners face the greatest danger from mine collapse, many have been killed or injured by **rock bursts,** sudden explosions of rock off the ceiling or walls of a mine adit. These explosions happen because the rock surrounding the adit is under such great pressure that it sometimes spontaneously fractures.

15.5 NONMETALLIC MINERAL RESOURCES

So far in this chapter, we've focused on resources that contain metal. But we use many other geological materials as well. From the ground we get the stone used to make roadbeds and buildings, the chemicals in fertilizers, the gypsum in drywall, the salt in salt shakers, and the sand used to make glass—the list is endless. Here, we look at a few of these geological materials and learn where they come from.

Dimension Stone

The Parthenon, a colossal stone temple rimmed by forty-six carved columns, has stood atop a hill overlooking the great city of Athens for almost 2,500 years. No wonder—**stone,** an architect's word for rock, outlasts nearly all other construction materials. Stone faces building walls, covers roofs, makes curbs and steps, and surfaces countertops and floors. It is used for its visual appeal as well as its durability. Usually the names architects give to various types of stone differ from the formal names used by geologists. For example, architects refer to any polished carbonate rock as marble, whether or not it has been metamorphosed. Likewise, they refer to any rocks containing feldspar and quartz to be granite, regardless of whether the rock has an igneous or metamorphic texture.

To obtain intact slabs and blocks of rock (granite or marble)—known as **dimension stone** in the trade—for architectural purposes, workers must carefully cut rock out of the walls of quarries (▶Fig. 15.17a, b). Note that a **quarry**

FIGURE 15.17 (a) An active quarrying operation, showing large blocks of cut stone; (b) polished building stone; (c) a crushed limestone quarry.

(a)

(b)

(c)

(a)

(b)

FIGURE 15.18 (a) Much of the land surface in a modern city is covered with concrete. (b) This thin-section photo of concrete shows the crystals comprising the cement.

provides stone, while a **mine** supplies ore. To cut stone slabs, quarry operators split rock blocks from bedrock by hammering a series of wedges into the rock, or by using either a wireline saw or a cutting jet. A **wireline saw** consists of a loop of braided wire moving between two pulleys. As the wire moves along the rock surface, the quarry operator spills abrasive (sand or garnet grains) and water onto the wire. The movement of the wire drags the abrasive along the rock and grinds a slice into it. A **cutting jet** looks like a long blowtorch, except that a ring of cold-water nozzles surrounds the flame at the end of the pole. The sudden heating and cooling of the rock by the simultaneous application of flame and water causes the rock to pulverize and allows the flame to cut through rock almost like a hot knife through butter.

Crushed Stone and Concrete

Crushed stone forms the substrate of highways and railroads and serves as the raw material for manufacturing cement, concrete, and asphalt. In crushed-stone quarries (▶Fig. 15.17c), operators use high explosives to break up bedrock into rubble that they then transport by truck to a jaw crusher, which reduces the rubble into usable-size fractions.

A great variety of rock types may be used for roadbeds, but cement making requires a precise mixture of sedimentary rocks. **Cement** forms by the precipitation of minerals out of a slurry containing water, lime (CaO), and other chemicals like silica (SiO_2), aluminum oxide (Al_2O_3), and iron oxide (Fe_2O_3). The mix of these chemicals must be correct to make good cement (▶Fig. 15.18a). Typically, lime accounts for 66% of cement, silica for about 25%, and the remaining chemicals for about 9%. A few special sedimentary rock units contain precisely the right mix of chemicals, but in general, manufacturers must mix minerals from several different rock units to create cement, much as a baker combines ingredients to create the right bread dough.

The lime used to make cement comes from the calcite ($CaCO_3$) in limestone; limestone is roasted in a furnace at high temperatures, where it decomposes into lime and car-

TABLE 15.2 Common Nonmetallic Resources

Limestone	Sedimentary rock made of calcite; used for gravel or cement.
Crushed stone	Any variety of coherent rock (limestone, quartzite, granite, gneiss).
Siltstone	Beds of sedimentary rock, used to make flagstone.
Granite	Coarse igneous rock, used for dimension stone.
Marble	Metamorphosed limestone; used for dimension stone.
Slate	Metamorphosed shale; used for roofing shingles.
Gypsum	A sulfate salt precipitated from saltwater; used for wallboard.
Phosphate	From the mineral apatite; used for fertilizer.
Pumice	Frothy volcanic rock; used to decorate gardens and paths.
Clay	Very fine mica-like mineral in sediment; used to make bricks or pottery.
Sand	From sandstone, beaches, or riverbeds; quartz sand is used for construction and for making glass.
Salt	From the mineral halite, formed by evaporating saltwater; used for food, melting ice on roads.
Sulfur	Occurs either as native sulfur, typically above salt domes, or in sulfide minerals; used for fertilizer and chemicals.

bon dioxide gas. The other elements generally come from shale and sandstone. Cement that's produced by mixing different rocks according to a recipe is known as **Portland cement,** because Isaac Johnson, the Englishman who first developed it in 1844, thought that it resembled rock near the town of Portland, England. Builders mix cement with sand or gravel to make **concrete,** an artificial clastic rock composed of fragments suspended in cement (▶Fig. 15.18b).

Geologic Materials in Your Home

We use an astounding variety of nonmetallic geologic resources (see Table 15.2) without ever realizing where they come from. Consider the materials in a house or apartment. The concrete foundation consists of cement, made from limestone mixed with sand or gravel. The bricks in the exterior walls originated as clay, formed from the chemical weathering of silicate rocks and perhaps dug from the floodplain of a stream. To make **bricks,** brick makers mold wet clay into blocks, which they then bake. Baking drives out

water and causes metamorphic reactions that recrystallize the clay.

The glass used to glaze windows consists largely of silica, formed by first melting and then freezing pure quartz sand from a beach or a sandstone formation. Gypsum board (drywall), used to construct interior walls, is made from a slurry of water and the mineral gypsum sandwiched between paper. Gypsum ($CaSO_4 \cdot 2H_2O$) is found in evaporite strata precipitated from seawater or saline lake water. Evaporites provide other useful minerals as well, such as halite.

The asbestos that was once used to make roof shingles comes from serpentine, a rock created by the reaction of olivine with water. The olivine may come from oceanic lithosphere, thrust onto continental crust during continental collisions. The copper in the house's electrical wiring most likely comes from the ores in a porphyry copper deposit; the iron in the nails probably comes from banded-iron formation; and the plastic used in everything from countertops to light fixtures comes from oil formed from the bodies of plankton that died millions of years ago. As you can see,

The Sidewalks of New York

Untold tons of concrete have gone into the construction of New York City. In fact, with the exception of a few city parks, all of the walking space in the city consists of concrete. And concrete skyscrapers tower above the concrete plain. Where does all this concrete come from?

Much of the sand used in New York concrete was deposited during the last ice age. As vast glaciers moved south over 14,000 years ago, they ground away the igneous and metamorphic rocks that comprised central and eastern Canada. These ancient rocks contained abundant quartz, and since quartz lasts a long time (it does not undergo chemical weathering easily), the sediment transported by the glaciers retained a large amount of quartz. Glaciers deposited this sediment in huge piles called moraines (see Chapter 22). As the glaciers melted, fast-moving rivers of meltwater washed the sediment, sorting sand from mud and pebbles. The sand was deposited in bars in the meltwater rivers, and these relict bars now provide thick lenses of sand that can be economically excavated.

What about the cement? Cement contains a mixture of lime, derived from limestone, and other elements (such as silica) derived from shale and sandstone. The bedrock of New York, though, consists largely of schist and gneiss, not sedimentary rocks. Fortunately, a source of rocks appropriate for making

cement lies up the Hudson River. A rock unit called the Rosendale Formation, which naturally contains exactly the right mixture of lime and silica needed to make a durable cement, crops out in low ridges just to the west of the river. Beginning in the late 1820s, workers began quarrying the Rosendale Formation for cement, creating a network of underground caverns. Quarry operators followed the Rosendale beds closely, making horizontal mine tunnels where the beds were horizontal, tilted mine tunnels where the beds tilted, and vertical mine tunnels where the beds were vertical. They then dumped the excavated rock into nearby kilns and roasted it to produce lime mixed with clay and quartz. The resulting powder was packed into barrels, loaded onto barges, and shipped downriver to New York. As demand for cement increased, operators eventually dug open-pit quarries in which they excavated other limestone and shale units, mixing them together in the correct proportion to make Portland cement.

The rocks comprising the Rosendale Formation consist of cemented-together shell fragments and small, reef-like colonies of organisms. In other words, the lime in the concrete of New York sidewalks was originally extracted from seawater by living organisms—brachiopods, crinoids, and bryozoans—over 350 million years ago.

geologic processes acting over millions of years have been at work to produce the materials making your home.

Chemicals employed for agricultural purposes also come from the ground. For example, phosphate (PO_4^{-3}) and potash (K_2CO_3) come from the minerals in evaporite deposits. Truly, without the geologic resources of the Earth, modern society would grind to a halt.

15.6 GLOBAL MINERAL NEEDS, RESERVES, AND POLITICS

How Long Will Reserves Last?

The average citizen of an industrialized country uses 25 kilograms (kg) of aluminum, 10 kg of copper, and 550 kg of iron and steel in a year's time (▶Fig. 15.19; Table 15.3). If you combine these figures with the quantities of energy resources and nonmetallic geologic resources a person uses, you get a total of about 15,000 kg (15 metric tons) of resources used per capita each year. Thus, the population of the United States consumes about 4 billion metric tons of geologic material per year, and to create this supply, workers must mine, quarry, or pump 18 billion metric tons. By comparison, the Mississippi River transports 190 million metric tons of sediment per year into the Gulf of Mexico.

Mineral resources, like oil and coal, are nonrenewable resources. Once mined, an ore deposit or a limestone hill disappears forever. Natural geologic processes do not happen fast enough to replace the deposits as fast as we use them. Geologists have calculated reserves for various mineral deposits just as they have for oil. Based on current definitions of reserves (which depend on today's prices) and rates of consumption, supplies of some metals may run out in only decades to centuries (see Table 15.4). But these esti-

mates may change as supplies become depleted and prices rise (making previously uneconomical deposits worth mining). And supplies could increase if geologists discover new reserves or if new ways of mining become available (providing access to nodules on the sea floor or to deeper parts of the crust). Further, increased efforts at conservation and recycling can cause a dramatic decrease in rates of consumption, and thereby stretch the lifetime of existing reserves.

Ore deposits do not occur everywhere, because their formation requires special geological conditions. As a result, some countries possess vast supplies, while others have none. In fact, no single country owns all the mineral resources it needs, so nations must trade with one another to

TABLE 15.3 Yearly Per Capita Usage of Geologic Materials in the USA

4,100 kg	Stone
3,860 kg	Sand and gravel
3,050 kg	Petroleum
2,650 kg	Coal
1,900 kg	Natural gas
550 kg	Iron and steel
360 kg	Cement
220 kg	Clay
200 kg	Salt
140 kg	Phosphate
25 kg	Aluminum
10 kg	Copper
6 kg	Lead
5 kg	Zinc

1 kg = 2.205 pounds.

FIGURE 15.19 We consume vast quantities of mineral resources in a year, as the diagram indicates.

Nonmetallic resources

4,100 kg Stone
3,860 kg Sand and gravel

360 kg Cement
220 kg Clay
200 kg Salt
140 kg Phosphate rock
480 kg Other nonmetals

Metallic resources

550 kg Iron and steel
25 kg Aluminum
10 kg Copper
6 kg Lead
5 kg Zinc
6 kg Manganese
9 kg Other metals

TABLE 15.4 Lifetimes (in years) of Currently Known Ore Supplies

Metal	World Resources	U.S. Resources
Iron	120	40
Aluminum	330	2
Copper	65	40
Lead	20	40
Zinc	30	25
Gold	30	20
Platinum	45	1
Nickel	75	less than 1
Cobalt	50	less than 1
Manganese	70	0
Chromium	75	0

maintain supplies, and global politics inevitably affects prices. Many wars have their roots in competition for mineral reserves, and not surprisingly, the outcome of some has hinged on who controls these reserves.

The United States worries in particular about supplies of so-called **strategic metals,** which include manganese, platinum, chromium, and cobalt—metals alloyed with iron to make the special-purpose steels needed in the aerospace industry. At present, the country must import 100% of the manganese, 95% of the cobalt, 73% of the chromium, and 92% of the platinum it consumes. Principal reserves of these metals lie in the crust of countries that have not always practiced open trade with the United States. As a defense precaution, the United States stockpiles these metals in case supplies are cut off.

Mining and the Environment

Mining leaves big footprints in the environment. Some of the gaping holes that open-pit mining creates in the landscape have become so big that astronauts can see them from space. Both open-pit and underground mining yield immense quantities of waste rock, which miners dump in **tailings piles,** some of which grow into artificial hills 200 m high and many kilometers long. Lacking soil, tailings piles tend to remain unvegetated for a long time. Mining also exposes ore-bearing rock to the atmosphere, and since many ore minerals are sulfides, they react with rainwater to produce **acid mine runoff,** which can severely damage vegetation downstream (▶Fig. 15.20a). Ore processing tends to release noxious chemicals that can mix with rain and spread over the countryside, damaging life. Before the installation of modern environmental controls, smoke from ore-process-

(a)

(b)

FIGURE 15.20 (a) The orange color in this acid mine runoff is from dissolved iron in the water. (b) This vegetation-free zone developed in response to acidic smelter smoke.

ing plants caused severe air pollution; plumes of smoke from the old smelters in Sudbury, Ontario, for example, created a wasteland for many kilometers downwind (▶Fig. 15.20b).

Clearly, a mine has the potential to become a scar on the landscape, the size of which depends on the efforts of miners to minimize damage. For this reason, many people object to mineral exploration in wilderness or scenic areas, and the prospect of mining in a region leads to battles between environmentalists and developers.

CHAPTER SUMMARY

• Industrial societies use many types of minerals, all of which must be extracted from the upper crust. We distinguish two general categories: metallic resources and nonmetallic resources.

• Metals are materials in which atoms are held together by metallic bonds. They are malleable and make good conductors. The use of metals began in prehistoric times but did not become widespread until the Bronze Age, around 4,500 years ago.

• Metals come from ore. An ore is a rock containing native metals or ore minerals (sulfide, oxide, or carbonate minerals with a high proportion of metal) in sufficient quantities to be worth mining. An ore deposit is an accumulation of ore.

• Magmatic deposits form when sulfide ore minerals settle to the floor of a magma chamber. In hydrothermal deposits, ore minerals precipitate from hot-water solutions. Secondary-enrichment deposits form when groundwater carries metals away from a preexisting deposit. (Mississippi Valley–type deposits precipitate from groundwater that has passed long distances through the crust.) Sedimentary deposits, like banded-iron formation, precipitate out of the ocean. Residual mineral deposits in soil are the result of severe leaching in tropical climates. Placer deposits develop when heavy metal grains accumulate in course sediment along a stream.

• Many ore deposits are associated with igneous activity in subduction zones, along mid-ocean ridges, along continental rifts, or at hot spots.

• The discovery of ore minerals requires the help of geologists. Economic and environmental concerns govern whether a particular ore deposit can be mined or not. Most mining companies today use open-pit techniques. Mining and processing ore can be a strain on the environment.

• Nonmetallic resources include dimension stone for decorative purposes, crushed stone for cement and asphalt production, clay for brick making, sand for glass production, and many others. A large proportion of materials in your home have a geological ancestry.

• Mineral resources are nonrenewable. Many are now or may soon become in short supply.

KEY TERMS

acid mine runoff (p. 467)
alloy (p. 454)
assay (p. 461)
banded-iron formation (p. 459)
base metals (p. 456)
bauxite (p. 460)
Bronze Age (p. 454)
cement (p. 464)
cold working (p. 454)
concrete (p. 465)
dimension stone (p. 463)
disseminated deposit (p. 457)
hydrothermal deposit (p. 457)
magmatic deposit (p. 457)
malleability (p. 454)
manganese nodules (p. 459)
massive-sulfide deposit (p. 457)
metallic bonds (p. 454)
metallic mineral resources (p. 453)
native metals (p. 454)
nonmetallic mineral resources (p. 453)
ore (p. 456)
ore deposit (p. 457)
ore grade (p. 456)
ore minerals (economic minerals) (p. 456)
placer deposit (p. 460)
porphyry copper deposit (p. 458)
Portland cement (p. 465)
precious metals (p. 456)
quarry (p. 463)
residual mineral deposit (p. 460)
secondary-enrichment deposit (p. 458)
slag (p. 454)
smelting (p. 454)
steel (p. 456)
strategic metals (p. 467)
tailings piles (p. 467)
tempering (p. 454)
vein deposit (p. 457)

REVIEW QUESTIONS

1. Describe how people have used copper, bronze, and iron throughout history.

2. Why do most rocks yield little or no useful metals?

3. What kinds of concentrations of metal are required for it to be economically minable?

4. Describe six different kinds of economic mineral deposits.

5. What procedures are used to locate and mine mineral resources today?

6. How is stone cut from a quarry?

7. Explain the differences between cement, Portland cement, and concrete.

8. How many kilograms of mineral resources does the average person in an industrial country use in a year?

9. Compare the estimated lifetimes of ore supplies (worldwide and in the United States) of iron, aluminum, copper, gold, and chromium.

10. How do political considerations affect the availability of strategic metals?

11. What are some environmental hazards of large-scale mining?

SUGGESTED READING

Brookins, D. G. 1981. *Earth Resources, Energy, and the Environment.* Columbus, Ohio: Merrill.

Carr, D. D., and N. Herz, eds. 1988. *Concise Encyclopedia of Mineral Resources.* Cambridge: MIT Press.

Clark, J. P., and F. R. Field III. 1985. How critical are critical minerals? *Technology Review* (August/September 1985).

Craig, J. R., D. J. Vaughan, and B. J. Skinner. 1989. *Resources of the Earth*. Englewood Cliffs, N.J.: Prentice-Hall.

Dorr, A. 1987. *Minerals: Foundations of Society*. Alexandria, Va.: American Geological Institute.

Guilbert, J. M., and C. F. Park, Jr. 1986. *The Geology of Ore Deposits*. New York: Freeman.

Jensen, M. L., and A. M. Bateman. 1981. *Economic Mineral Deposits*. New York: Wiley.

Kesler, S. E. 1994. *Mineral Resources, Economics, and the Environment*. New York: Macmillan.

National Research Council. 1990. *Competitiveness of the U.S. Minerals and Metals Industry*. Washington, D.C.: National Academy Press.

Sawkins, F. J. 1984. *Metal Deposits in Relation to Plate Tectonics*. New York: Springer Verlag.

Stone, I. 1956. *Men to Match My Mountains*. New York: Doubleday; reprinted 1982 by Berkley Books. See pp. 128–51.

*I*n the last part of this book, we focus on Earth's surface and near-surface, the outer 1.0% of our planet. This portion of the Earth system, which encompasses the boundaries between the lithosphere, hydrosphere, and atmosphere, displays great variability, for the dynamic interplay between internal processes (driven by Earth's internal heat) and external processes (driven by the warmth of the Sun) in Earth's gravitational field has resulted in a diverse array of landscapes. We examine five of these landscapes, plus groundwater and the atmosphere, and finally, in Chapter 23, see how forces at work in the Earth system cause the planet to constantly change.

Ever-Changing Landscapes and the Hydrologic Cycle

Talk of mysteries! Think of our life in nature—daily to be shown matter, to come in contact with it—rocks, trees, wind on our cheeks! the solid earth! the actual world! the common sense! Contact! Contact! Who are we? where are we?
—HENRY DAVID THOREAU (1817–1862)

E.1 INTRODUCTION

The Earth's surface is at once a place of endless variety and intricate detail. Observe the height of its mountains, the expanse of its seas, the desolation of its deserts, and you may be inspired, frightened, or calmed. It's no wonder that artists and writers across the ages have sought inspiration from the **landscape**—the character and shape of the land surface in a region—for landscapes have the diversity of human emotion (▶Fig. E.1a–d). Geologists, like artists and writers, savor the impression of a dramatic landscape, but on seeing one, they can't help but ask, "How did it come to be, and how will it change in the future?"

The subject of landscape development and evolution dominates many of the chapters in Part VI. Geologists who study landscape development are known as **geomorphologists.** This interlude, a general introduction to the topic, explains the driving forces behind landscape development, identifies factors that control which landscape develops in a given locality, and describes the hydrologic cycle, the pathway water molecules follow as they move from ocean to air to land and back to sea. We focus on the hydrologic cycle here because so many processes on or near Earth's surface involve water in its various forms.

E.2 THE BATTLE BETWEEN UPLIFT AND EROSION

Landscapes result from a never-ending battle between processes that uplift Earth's surface and processes that lower it. **Internal processes** such as subduction, hot-spot volcanism, collision between continents, and rifting—all consequences of plate motion or mantle flow—cause **uplift** (the vertical elevation of the land surface relative to a reference plane) and generate highlands. **External processes** such as **downslope movement** (the tumbling or sliding of rock and sediment from higher elevations to lower ones), **erosion** (the grinding away and removal of Earth's surface materials by moving water, air, or ice), and **deposition** (the settling of sediment out of a fluid) redistribute rocks and sediment and generate distinct shapes, or **landforms,** such as mesas, valleys, cliffs, and sand dunes (▶Fig. E.2a, b). All of these processes are ultimately the consequences of gravity and the interaction between the solid Earth and its fluid envelope (air and water).

The elevation of a particular region depends on the relative magnitude of the uplift rate in relation to the erosion rate. If the rate of uplift is greater, the land surface rises; if the rate of erosion is greater, the land surface sinks. Without uplift to rejuvenate elevation, Earth's surface would long ago have been beveled down to a low plain, and without erosion, highlands would have stayed high for all of Earth history. This dynamic, ever-changing landscape on Earth contrasts with the static, pockmarked landscape of our Moon. Because no plate tectonics occurs on the Moon, no new mountains form, and because no atmosphere or ocean

(a)

(b)

(d)

(c)

FIGURE E.1 (a) A tropical waterfall in Hawaii, where lush vegetation blunts the contours of the landscape, except where it's stripped clean by rushing water. (b) The desert ranges of the Mojave Desert, in southeastern California. Barren remnants of rocky cliffs protrude from fans of recently eroded and sparsely vegetated detritus. (c) The peaks of the Grand Tetons, in Wyoming. Fresh rock cliffs, carved by glaciers, penetrate the clouds. (d) A rock and sand seascape along the coast of Brazil, where the relentless pounding of Atlantic waves eats back into the land, grinds away at the massive "sugar loafs," and generates beach sand.

exists, there is no erosion from rivers, glaciers, or winds; thus the lunar surface has remained largely unchanged for over 2 billion years (▶Fig. E.3a, b).

How fast do uplift and erosion take place? The Earth's surface can rise by as much as 3 m during a single major earthquake, but uplift typically takes place very slowly; averaged over time, rates range between 0.01 and 10 mm per year (▶Fig. E.4a). Similarly, erosion can carve out several me-

ters of **substrate,** the material just below the ground surface, during a single flood, storm, or landslide (▶Fig. E.4b). But averaged over time, erosion rates also vary between 0.01 and 10 mm per year. While these rates seem small, a change in surface elevation of just 0.5 mm (the thickness of a fingernail) per year can yield a net change of 5 km in 10 million years. Uplift can build a mountain range and erosion can whittle one down to near sea level—it just takes time!

(a)

(b)

FIGURE E.2 Examples of landforms. (a) These buttes at Monument Valley, Arizona, are remnants of a once-continuous layer of red sandstone. Long-term erosion has removed the rest of the layer and has redistributed the sand elsewhere. (b) Uplift of the eastern margin of Australia created the highlands that could be eroded to form these steep cliffs, in Australia's Blue Mountains.

Notably, because the lithosphere floats on the asthenosphere (see the discussion of isostasy in Chapter 11), the removal of 10 cm off the surface of a mountain range causes the crust to rise by about 3 cm, just as the removal of heavy containers from the deck of a cargo ship makes the ship rise. Thus, since the crust is rising at the same time erosion is taking place, much more than 5 km of rock must erode from a 5-km-high mountain range to return the land surface to sea level.

Uplift is not the only kind of vertical movement to happen at the surface of the Earth. Regions of the crust can also move down; such motion is called **subsidence.** Subsidence happens, for example, in continental rifts, where the crust undergoes stretching and thinning. It also happens where a heavy load like a volcano, glacier, or pile of sediment builds up on the surface; the weight of such loads depresses the crust by about ⅓ km for every 1 km of load thickness. Subsidence also occurs where the lithosphere cools and as a result gets thicker and denser and thus sinks into the underlying asthenosphere. As it sinks, the asthenosphere flows away from underneath to make room. The depressions created as a consequence of subsidence are called **sedimentary basins;** here, thick sequences of sediment accumulate.

The battle between uplift and erosion, as well as the process of sediment deposition, produces a complex set of landforms, whose characteristic shapes depend on the re-

FIGURE E.3 (a) This lunar landscape is virtually the same landscape that would have been visible 2 billion years ago, because the Moon's surface is static. (b) A close-up view of the lunar landscape, with the lunar rover for scale.

(a)

(b)

(a)

(b)

FIGURE E.4 (a) Uplifted beach terraces along a coast appear where the coast is rising relative to sea level. Wave erosion eats into the land, creating a terrace. Eventually, the terrace rises out of the reach of the waves, which then cut a new one. (b) So much erosion can take place during a single hurricane that the foundations of houses built along the beach become undermined.

lief—the difference in elevation between high and low points (▶Fig. E.5a)—and on whether the eroding agent is wind, water, or ice. Geologists use the term **topography** to refer to the variations in elevation of a region; the topography defines the landform shapes.

We can portray topography by means of a **topographical map**, which uses contour lines to represent the variations in elevation (▶Fig. E.5b–d). A **contour line** is an imaginary line along which all points have the same elevation. For example, if you walk along the 200-m contour line on a hill slope, you stay at exactly the same elevation. The shoreline on a flat, calm body of water is a contour line. In other words, you can picture the contour line as the intersection between the land surface and an imaginary horizon-

FIGURE E.5 (a) Relief is simply the vertical elevation difference between two points (A and B) at the surface of the Earth. (b) A topographical map depicts the 3-D shape of the land surface on a 2-D map through the use of contour lines. The difference in elevation between two adjacent lines is the contour interval. (c) A topographical profile (along section line X-X′) shows the shape of the land surface as seen in a vertical slice. (d) You can picture a contour line as the intersection of a horizontal plane with the land surface. This block diagram shows the area of (a) in 3-D.

Wind transportation of moisture

The atmospheric reservoir

Cloud condensation

Evapotranspiration
(from vegetation,
trees, etc.)

Evaporation
of surface
ocean water

The organic reservoir

Precipitation
over oceans

Surface runoff
(returns to sea)

The ocean reservoir

tmospheric water vapor

Precipitation over land

The snow and ice reservoir

Moving glacier

Melting ice

Evaporation of surface waters

Rain

Infiltration

Percolation

The land reservoir

Soil waterflow

Groundwater flow

The subsurface reservoir

Emergence at a spring

Deep groundwater flow

The Hydrologic Cycle

Water circulates through a number of reservoirs in the Earth system. The largest reservoir by far is the ocean, which covers 71 percent of the Earth's surface. Water evaporates from the ocean and enters the atmosphere, where it may be stored for a while—the atmosphere is thus another reservoir. Atmospheric water gradually condenses and forms clouds that drop rain or snow onto the oceans or land.

The water that falls on land may be held in glacial ice or in surface water (lakes, rivers, streams, swamps, etc.); these bodies of ice and water comprise Earth's surface-water reservoir. Some water flows back to the ocean, some evaporates into the air, and some sinks into the ground. Water that sinks into the ground may remain temporarily on the surface of soil grains, while some water sinks deeper into the ground and may be trapped there for a while as groundwater.

Groundwater fills the holes and cracks between grains of rock or sediment. Thus, the subsurface of the land is also a water reservoir. Groundwater flows slowly. Some of it eventually bubbles back to the ground surface or into the bed of lakes, rivers, or streams; these outlets are called springs. But some groundwater flows all the way back to the coast and reaches the sea.

Not all of the reservoirs in the Earth system are inanimate—some water becomes part of living organisms, and returns to the atmosphere by transpiration from plants or respiration by animals. The overall circulation of water from reservoir to reservoir in the Earth system is called the hydrologic cycle.

tal plane. Contour lines form a closed loop around a hill, and they form a V shape that points upstream where they cross a river valley. Note that someone standing on the 300-m contour line is 100 m higher than someone standing on the 200-m contour line. The elevation difference between two adjacent contour lines on a topographical map is called the **contour interval.**

E.3 FACTORS CONTROLLING LANDSCAPE DEVELOPMENT

Imagine traveling across a continent. On your journey, you pass plains, swamps, hills, valleys, mesas, and mountains. Some of these features are **erosional landforms,** in that they result from the breakdown and removal of rock or sediment and develop where **agents of erosion** such as water, wind, or ice carve into the substrate. (Of these three agents of erosion, water has the greatest effect on a global basis.) Other features are **depositional landforms**, in that they result from the deposition of sediment where the medium carrying the sediment (again, water, wind, or ice) evaporates, slows down, or melts. The specific landforms that develop at a given locality, which together make up the landscape, reflect six factors.

- *Eroding or transporting agent:* Water, ice, and wind all can cause erosion and transport sediment. Landforms created by glacial erosion or deposition differ from landforms created by river erosion or deposition, and both differ from landforms created by the wind. The differences reflect the varying abilities of the water, ice, and wind to carve into the substrate and to carry debris.

- *Elevation:* In regions where the land surface has risen to a high elevation, rugged mountain ranges with deep valleys and steep cliffs can form, but in places where the land is a wide plain near sea level, valleys and cliffs do not exist. The elevation difference, or relief, between adjacent places in a landscape determines the height and steepness (angle) of the slopes. Steepness, in turn, controls the velocity of water or ice flow, and thus whether a locality is the site of erosion or the site of deposition. Steepness also determines whether rock or soil will stay in place or fall downslope—rock and soil stay put on a plain, but tend to tumble on a steep slope.

- *Climate:* The average mean temperature and the volume of precipitation in a region determines whether running water, flowing ice, or wind serves as the main agent of erosion or deposition. In cold climates, glaciers cause the most erosion and deposition, whereas in temperate climates, rivers play this role. In deserts, wind contributes substantially to landscape formation. If the climate of a region changes, the nature of the landscape changes.

(a)

(b)

(c)

FIGURE E.6 Human workings on a geological scale. (a) The pyramids of Egypt are essentially human-made hills that rise above the desert sands. They have lasted for thousands of years. (b) People routinely cut deep valleys through high ridges, like this highway road cut in Colorado, to make a level grade for a road. Think how long it would take a river to cut such a gorge. (c) This stone dam holds back a reservoir in Colorado. Think how long it would take a glacier to pile up so much sediment.

Also, if the change in climate makes the sea level rise, then the positions of landforms along the coast change.

- *Life activity:* Organisms ranging from bacteria to worms, trees, and elephants strongly affect landscape development. While some life activity weakens the substrate, some protects it. For example, bacteria contribute to the weathering of rock, while trees and grass cover and stabilize the land surface. The ecology of an area depends on the climate—a landscape that would consist of gently rolling, forested hills in a temperate climate might consist of barren escarpments in a desert climate.

- *Substrate composition:* A substrate's composition determines how it responds to erosion. Regions underlain by hard rock erode less easily than regions underlain by unconsolidated sediment. Thus, strong rocks can hold up steep cliffs, while soft sediment collapses to create gentle slopes.

- *Time:* Landscapes change in response to continued erosion and/or deposition acting over time. In erosional landscapes, more erosion happens as time passes, and high mountains gradually become low hills, while in depositional landscapes, more deposition takes place as time passes, and depositional landforms grow.

While water, wind, and ice create most landscapes, human activities have had an increasingly important impact on the Earth's surface. We have dug pits (mines) where once there were mountains, built hills (tailings piles and landfills) where once there were valleys, and made steep slopes gentle and gentle slopes steep (▶Fig. E.6a–c). By building concrete walls, we modify the shapes of coastlines, change the courses of rivers, and fill new lakes (reservoirs). In cities, buildings and pavements completely seal the ground and cause water that might once have seeped down into the ground to spill into streams instead, increasing their flow. And in the country, agriculture, grazing, water usage, and deforestation substantially alter the rates at which natural erosion and deposition take place. For example, agriculture greatly increases the rate of erosion, because for much of the year farm fields have no vegetation cover.

E.4 THE HYDROLOGIC CYCLE

As evident from the discussion above, water in its various forms (liquid, gas, and solid) plays a major role in erosion and deposition on Earth's surface. Our planet's water is found in certain distinct **reservoirs** (containers), namely: the oceans, glacier ice (today, mostly in Antarctica and Greenland), groundwater, lakes, soil moisture, living organisms, the atmosphere, and rivers (see Table E.1). Together, these comprise the **hydrosphere.** Water constantly flows from reservoir to reservoir, and this never-ending passage is called the **hydrologic cycle.** Without the hydrologic cy-

TABLE E.1 Major Water Reservoirs of the Earth

Reservoir	Volume of Water (km³) on the Earth	% of Total Water
Oceans	1.40×10^9	95.0
Glaciers and ice sheets	4.34×10^7	2.97
Groundwater	1.53×10^3	1.05
Lakes and rivers	1.27×10^3	0.009
Atmosphere	1.50×10^3	0.001
Living organisms and soil	2.00×10^9	0.0001

cle, the erosive forces of running water (rivers and streams) and flowing ice (glaciers) would not exist.

The *average* length of time that water stays in a particular reservoir during the hydrologic cycle is called the **residence time.** Different reservoirs have different residence times. For example, a typical molecule of water remains in the oceans for 4,000 years or less, in lakes and ponds for 10 years or less, in rivers for 2 weeks or less, and in the atmosphere for 10 days or less. Groundwater residence times are highly variable and depend on how deep the groundwater flows. Water can stay underground for anywhere from 2 weeks to 10,000 years before it inevitably moves on to another reservoir.

To get a clearer sense of how the hydrologic cycle operates, let's follow the fate of seawater that has just reached the surface of the ocean in an upwelling current. Solar radiation heats the water, and the increased thermal energy of the vibrating water molecules allows them to **evaporate** (break free from the liquid) and drift upward in a gaseous state to become part of the atmosphere. About 417,000 cubic km (102,000 cubic miles), or about 30% of the total ocean volume, evaporates every year. Atmospheric water vapor moves with the wind to higher elevations, where it cools, undergoes **condensation** (the molecules link together to form a liquid), and rains or snows. About 76% of this water **precipitates** (falls out of the air) directly back into the ocean. The remainder precipitates onto land; most of this becomes trapped temporarily in the soil or in plants or animals and soon returns directly to the atmosphere by what is called **evapotranspiration:** the sum of evaporation from bodies of water, evaporation from the ground surface, and **transpiration** (release as a metabolic byproduct) from plants and animals. Rainwater that did not become trapped in the soil or in living organisms either enters lakes or rivers and ultimately flows back to the sea as surface water, becomes trapped in glaciers, or sinks deeper into the ground to become groundwater. Groundwater also flows and ultimately returns to the Earth's surface reservoirs. In sum, during the hydrologic cycle, water moves between the ocean, the atmosphere, reservoirs on or below the land, and living organisms.

CHAPTER 16

Unsafe Ground: Landslides and Other Mass Movements

16.1 INTRODUCTION

It was Sunday, May 31, 1970, a market day, and thousands of people had crammed into the Andean town of Yungay, Peru, to shop. Suddenly they felt the jolt of an earthquake, strong enough to topple some masonry houses. But worse was to come. This earthquake also broke an 800-m-wide ice slab off the end of a glacier at the top of Nevado Huascarán, a nearby 6.6-km-high mountain peak. Gravity instantly pulled the ice slab down the mountain's steep slopes. As it tumbled down over 3.7 km, the ice disintegrated into a chaotic avalanche of chunks traveling at speeds of over 300 km per hour. Near the base of the mountain, most of the avalanche channeled into a valley and thickened into a moving sheet as high as a ten-story building that ripped up rocks and soil along the way. Friction transformed the ice into water, which when mixed with rock and dust created 50 million cubic meters of mud, a slurry viscous enough to carry boulders larger than houses. This mass, sometimes floating on a compressed air cushion that allowed it to pass without disturbing the grass below, traveled over 14.5 km in less than four minutes.

At the mouth of the valley, most of the mass overran the village of Ranrahica before coming to rest and creating a dam that blocked the Santa River. But part of it shot up the sides of the valley and became airborne for several seconds, flying over the ridge bordering Yungay. As the town's inhabitants and visitors stumbled

Tectonic forces uplift the Earth surface, and in Earth's gravity field what goes up must come down. As a consequence, rock and regolith forming the substrate of hill slopes occasionally give way and slide downslope. The results can be disastrous, as happened when this La Conchita, California, landslide buried nearby homes. Such mass movements contribute to shaping the Earth's surface.

481

(a)

(b)

FIGURE 16.1 (a) Before the May 1970 earthquake, the town of Yungay, Peru, perched on a hill within view of the ice-covered mountain Nevado Huascarán. (b) Three months after the earthquake, the town lay buried beneath debris. A landslide scar remains visible on the moutain.

out of earthquake-damaged buildings, they heard a deafening roar and looked up to see the churning mud cloud bursting above the nearby ridge. The town was completely buried under several meters of mud and rock. When the dust had settled, only the top of the church and a few palm trees remained visible to show where Yungay once lay (►Fig. 16.1); 18,000 people are forever entombed beneath the mass. Today, the site is a grassy meadow with a **hummocky surface** (irregular and lumpy), spotted by crosses left by mourning relatives.

Could the Yungay tragedy have been prevented? Perhaps. A few years earlier, climbers had recognized the instability of the glacial ice on Nevado Huascarán, and Peruvian newspapers published a warning, but alas, no one took no-

tice. In the aftermath of the event, geologists discovered that Yungay had been built on an ancient layer of debris, from past avalanches. The government has since prevented new towns from rising in the danger zone.

People often assume that the earth beneath them is *terra firma,* a solid foundation on which they can build their lives. But the catastrophe at Yungay says otherwise— much of the Earth's surface is **unstable ground,** land capable of moving downslope in a matter of seconds to weeks. Geologists refer to the gravitationally caused transport of rock, **regolith** (soil, sediment, and debris), snow, and ice downslope as **mass movement,** or **mass wasting.** Like earthquakes, volcanic eruptions, storms, and floods, mass movements are a type of **natural hazard,** meaning a natural feature of the environment that can cause damage to living organisms and to buildings. Unfortunately, mass movement becomes more of a threat every year, because as the world's population grows, cities expand into areas of unsafe ground. But mass movement also plays a critical role in the rock cycle, for it's the first step in the transportation of sediment. And it plays a critical role in the evolution of landscapes: it's the most rapid means of modifying the shapes of slopes.

In this chapter, we look at the types, causes, and consequences of mass movement, and precautions society can take to protect people and property from its dangers. You'll want to consider this information when selecting a site for a home or voting on land-use propositions for your community.

16.2 TYPES OF MASS MOVEMENT

Though in everyday language people commonly refer to all mass-movement events as landslides, geologists and civil engineers tend to distinguish different types of mass movement based on four factors: the type of material involved (rock, regolith, or snow and ice), the velocity of the movement (fast, intermediate, or slow), the character of the moving mass (chaotic cloud, slurry, or coherent body), and the environment in which the movement takes place (subaerial or submarine). Below, we look at mass movements roughly in order from slow to very fast.

Creep, Solifluction, and Rock Glaciers

In temperate climates, the upper few centimeters of ground freeze during the winter, only to thaw again the following spring. Because water increases in volume by about 9.2% when it freezes, the water-saturated soil and underlying fractured rock expand outward, and particles in the regolith move out perpendicular to the slope. During the spring thaw, water becomes liquid again, and gravity makes the particles sink vertically and thus migrate downslope slightly.

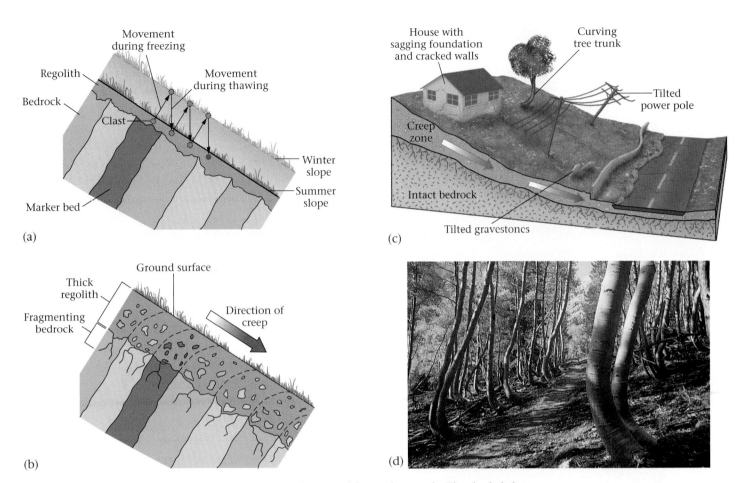

FIGURE 16.2 (a) Creep on hill slopes accompanies the annual freeze-thaw cycle. The shaded clast originally came from the shaded marker bed. It rises perpendicular to the slope when the slope freezes and drops down when the slope thaws. After three years, it has migrated downslope to the position shown. (b) As rock layers weather and break up, the resulting debris creeps downslope. (c) Soil creep causes walls to bend and crack, building foundations to sink, trees to bend, and power poles and gravestones to tilt. (d) Trees that grow in creeping soil gradually develop pronounced curves.

This gradual downslope movement of regolith is called **creep.** You can't see creep by staring at a hill slope because it occurs too slowly, but over a period of years creep causes trees, fences, gravestones, walls, and foundations built on a hillside to tilt downslope. Notably, trees that continue to grow after they have been tilted display a pronounced curvature at their base (▶Fig. 16.2a–d).

In Arctic or high-elevation regions, regolith freezes solid to great depth during the winter. In the brief summer thaw, only the uppermost 1–3 m of the ground thaws. Since meltwater cannot sink into permanently frozen ground, or **permafrost,** the melted layer becomes soggy and weak and flows slowly downslope in overlapping sheets. Geologists refer to this kind of creep, characteristic of **tundra** regions (cold, treeless plains), as **solifluction** (▶Fig. 16.3a). We can see another type of slow mass movement in cold regions in **rock glaciers,** which consist of a mixture of rock fragments and ice, with the rock fragments

comprising the major proportion (▶Fig. 16.3b). Rock glaciers develop where the volume of debris falling into a valley equals or exceeds the volume of glacial ice forming from snow. They move downslope as the ice component slowly recrystallizes.

Slumping

Near Pacific Palisades, along the coast of southern California, Highway 1 runs between the beach and a 120-m-high cliff. Between March 31 and April 3, 1958, a 1-km-long section of the highway disappeared beneath a mass of regolith that had moved down the adjacent cliff. When the movement stopped, the face of the cliff lay 200 m farther inland than it had before. It took weeks for bulldozers to uncover the road. During such **slumping,** a mass of regolith detaches from its substrate along a spoon-shaped sliding surface and slips semi-coherently downslope (▶Fig. 16.4a). We call the moving

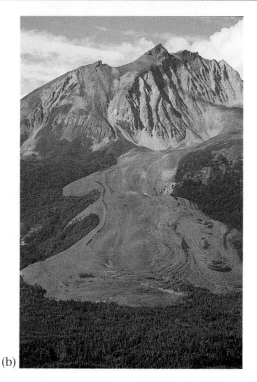

FIGURE 16.3 (a) Solifluction on a hill slope in the tundra. (b) A rock glacier in Alaska.

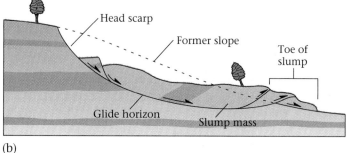

FIGURE 16.4 (a) A slump covering a highway in California. (b) Note the curving glide horizon in this slump. The dashed line indicates the slope's shape before slumping.

mass a **slump,** and the surface on which it slips a **glide horizon** (▶Fig. 16.4b). On average, the sea cliffs of southern California retreat (move inland) by up to a few meters a year because of slumping.

The distinct curve of the upslope edge of a slump, where the regolith detached, is called a **head scarp**. Immediately below the head scarp, the land surface sinks below its previous elevation. Farther downslope, at the toe or end of the slump, the ground elevation rises as the slump rides up and over the preexisting land surface. The toe may break into a series of slices that form curving ridges at the ground sur-

face. Slumps come in all sizes, from only a few meters across to tens of kilometers across. Slumps move at speeds of millimeters per day to tens of meters per minute. They typically break up as they move, and structures (such as houses, patios, and swimming pools) built on them crack and fall apart.

Mudflows and Debris Flows

Rio de Janeiro, Brazil, originally occupied only the flatlands bordering beautiful crescent beaches between towering sugarloaf mountains. But in recent decades, the population has grown so much that the city has expanded up the steep sides of the mountains, and in many places densely populated communities of makeshift shacks cover the slopes. These communities, which have no storm drains, were built on the thick regolith that resulted from rapid weathering in Brazil's tropical climate. In 1988, particularly heavy rains saturated the soil, which turned into a viscous slurry of mud that flowed downslope. Whole communities disappeared overnight, replaced by a hummocky muddle of mud and debris. And at the base of the cliffs, the flowing mud collapsed modern high-rise buildings (▶Fig. 16.5a).

In areas like the hill-slope communities of Rio, where neither vegetation nor drainage systems protect the ground from rainfall, water mixes with regolith to create a slurry that moves downslope. If the slurry consists of just mud, it's a **mudflow,** but if the mud is mixed with larger rock fragments, it's a **debris flow.** The speed at which mud or debris moves depends on the slope angle and on the water content. Flows move faster if they are wetter (that is, less viscous), and if they move on steeper slopes. On a gentle slope, viscous mud flows like molasses, but on a steep slope, low-viscosity mud may move at over 100 km per hour. Because mud and debris flows have greater viscosity than clear water, they can carry large rock chunks, as well as houses and cars. They typically follow channels downslope, and at the base of the slope spread out into a broad lobe.

FIGURE 16.5 (a) The aftermath of a mudflow in Rio de Janeiro. (b) A rapidly moving lahar at Mt. Saint Helens.

(a)

(b)

Particularly devastating mudflows spill down the river valleys bordering volcanoes. These mudflows, known as **lahars,** consist of a mixture of volcanic ash (from a currently erupting or previously erupted pyroclastic cloud) and water (from the snow and ice that melts in a volcano's heat or from heavy rains) (▶Fig. 16.5b; see Chapter 9). One of the most destructive recent lahars occurred on November 13, 1985, in the Andes Mountains in Colombia. That night, a major eruption melted a volcano's thick snowcap, creating hot water that mixed with ash. A scalding lahar rushed down river valleys and swept over the nearby town of Armero while most inhabitants were asleep (see Fig. 9.22e). Of the 25,000 residents, 20,000 perished.

Landslides (Rock and Debris Slides)

In the early 1960s, engineers built a huge new dam across a river on the north side of Monte Toc, in the Italian Alps, to create a reservoir for generating electricity. This dam, the Vaiont Dam, was an engineering marvel, rising 260 m above the valley floor—a concrete wall as high as an 85-story skyscraper (▶Fig. 16.6a). But the dam-side face of Monte Toc was underlain by dipping limestone beds interlayered with weak shale beds. These beds dipped parallel to the surface of the mountain, and curved under the reservoir (▶Fig. 16.6b). As the reservoir filled, the flank of the mountain cracked, shook, and rumbled, and local residents began to call Monte Toc *la montagna che cammina* (the mountain that walks).

After several days of rain, Monte Toc began to rumble so much that on October 9, 1963, engineers lowered the reservoir level. They thought the wet ground might slump a little into the reservoir, but no more than that, so no one ordered the evacuation of the town of Longarone, a few kilometers down the valley. Unfortunately, the engineers underestimated the problem. At 10:30 that evening, a huge chunk of Monte Toc—600 million tons of rock—detached from the mountain and slid downslope into the reservoir. Some debris rocketed up the opposite wall of the valley to a height of 260 m above the original reservoir level. The displaced water of the reservoir spilled over the top of the dam and rushed down into the valley below. When the flood had passed, nothing of Longarone and its 1,500 inhabitants remained. Though the dam itself still stands, it holds back only debris, and it has never provided any electricity.

Geologists refer to a sudden movement of rock and debris down a nonvertical slope as a **landslide.** If the mass consists only of rock, it may also be called a **rock slide** (the case in the Vaiont Dam disaster), and if it consists only of regolith, it may also be called a **debris slide.** Once a landslide has taken place, it leaves a **landslide scar** on the slope and forms a **debris pile** at the base of the slope.

Slides happen when bedrock and/or regolith detaches from a slope and shoots downhill on a glide horizon roughly parallel to the slope surface. Thus, landslides generally occur

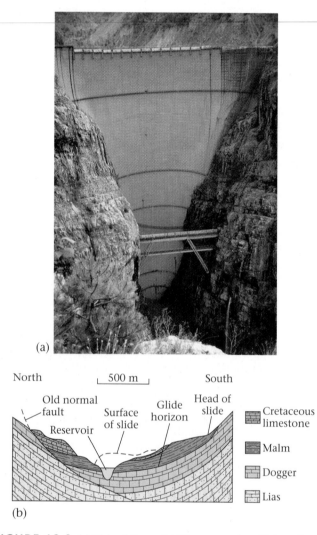

(a)

North   ⊢ 500 m ⊣   South

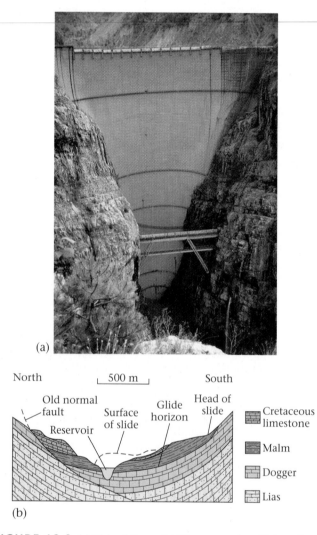

(b)

FIGURE 16.6 (a) Vaiont Dam. (b) The reservoir at Vaiont Dam before the landslide. Note the glide horizon at the base of the weak Malm Shale. The landslide completely filled the reservoir.

where a weak layer of rock or sediment at depth below the ground parallels the land surface. (At the Vaiont Dam, the glide horizon followed a weak shale bed.) Slides may move at speeds of up to 300 km per hour; they are particularly fast when a cushion of air gets trapped beneath the moving mass, so there is virtually no friction between the slide and its substrate, and the mass moves like a hovercraft. Rock and debris slides sometimes slam down into a valley and still keep enough momentum to climb the opposite side of the valley.

Landslides, like slumps, come at a variety of scales. Most are small, involving blocks up to a few meters across. Some, like the Vaiont slide, are large enough to cause a catastrophe. But even the Vaiont slide pales by comparison with some landslides for which evidence appears in the geologic record.

Avalanches

In the winter of 1999, an unusual weather system passed over the Austrian Alps. First it snowed, then the temperature warmed and the snow began to melt. But then the weather turned cold again, and the melted snow froze into a hard, icy crust. This cold snap ushered in a blizzard that blanketed the ice crust with tens of centimeters of snow, and at the mountaintops the wind built the snow into huge

FIGURE 16.7 (a) Aftermath of the 1999 avalanches in the Austrian Alps. (b) Trees that were flattened by an avalanche, now exposed after the snow melted.

(a)

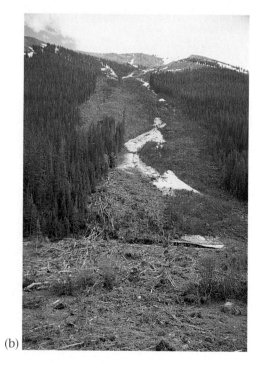

(b)

overhanging drifts, called **cornices.** Skiers delighted in the bounty of white, but not for long. Who knows how it started—perhaps a gust of wind or a sudden noise was enough—but the world witnessed the aftermath. With the frozen snow layer underneath acting as a glide horizon, the heavy layer of new snow began to slide down the mountain, accelerating as it moved and then disintegrating and mixing with air. It became a roaring cloud—an avalanche—traveling at hurricane speeds, and it flattened everything in its path. Trees and ski lodges toppled like toothpicks before the mass finally reached the valley floor and came to a halt (▶Fig. 16.7a, b). Unfortunately, many of those who survived the impact succumbed to suffocation in the minutes that followed, and the avalanche blocked roads into the region, tragically slowing rescue efforts.

Avalanches are turbulent clouds of debris mixed with air that rush down steep hill slopes at high velocity. If the debris consists of snow, like the Austrian avalanche, it's a **snow avalanche.** If it consists of fragments of rock and dust, it's a **debris avalanche.** The moving air-debris mass is denser than clear air and thus hugs the ground and acts like an extremely strong and viscous wind that can knock down and blow away anything in its path. As illustrated by the Austrian example, snow avalanches pose a particular threat when frozen snow layers get buried and thus can act as a glide horizon for the overlying snow. Typically, avalanches happen again and again in the same area, creating pathways, called **avalanche chutes,** in which no mature trees grow. Not all snow avalanches are the same—some are clouds of light powder traveling at up to 250 km per hour, while some are much slower flows of wet snow. The snow in an avalanche can break off a cornice, or it can start as a huge slab that detaches from its substrate in response to little more than the weight of a skier. Whatever their cause, avalanches are deadly. They buried 18,000 of Hannibal's troops and close to 80,000 soldiers in World War I.

Rock Falls and Debris Falls

Rock falls and **debris falls,** as their names suggest, occur when a mass free-falls from a steep (vertical) cliff (▶Fig. 16.8). Friction and collision with other rocks brings some blocks to a halt before they reach the bottom of the slope; these blocks pile up to form a **talus,** a sloping apron of rocks along the base of the cliff (▶Fig. 16.9). Rock or debris that has fallen a long way can reach speeds of 300 km per hour, and may have so much momentum that it keeps going when it touches bottom and triggers a debris avalanche that can cross a valley floor and rise up the other side. As with avalanches, large, fast rock falls push the air in front of them, creating a short blast of hurricane-like wind. For example, the wind alone from a 1996 rock fall in Yosemite National Park flattened over 2,000 trees. Commonly, rock falls happen when a rock separates from a cliff

FIGURE 16.8 Successive rock falls have littered the base of this sandstone cliff with boulders.

face along a joint. In the Yosemite case, rock separated from the mountain along an exfoliation joint (see Chapter 11).

Most rock falls involve only a few blocks detaching from a cliff face and dropping into the talus. But some falls dislodge immense quantities of rock. In September 1881, a 600-m-high crag of slate, undermined by quarrying, sud-

FIGURE 16.9 A talus pile at the foot of a cliff. Rocks fall from the cliff, but friction slows them to a halt before they reach the bottom.

denly collapsed onto the Swiss town of Elm in a valley of the Swiss Alps. Over 10 million cubic meters of rock fell to the valley floor, then rushed up the opposite slope to a height of 100 m. Elm and its 115 inhabitants lay buried to a depth of 10 to 20 m.

Rock falls typically take place along steep highway road cuts, leading to the posting of "Falling-rock zone" signs. Such rock falls occur with increasing frequency as the road cut ages, because frost wedging and/or root wedging pries fragments loose, and water infiltrates the outcrop and weakens clay-rich layers. In temperate regions, road-cut rock falls are more frequent in the spring, after the ice in cracks has melted and while rains are heavy.

Submarine Mass Movements

So far, we've focused on mass movements that occur subaerially, for these are the ones we can see and are affected by most. But mass wasting also happens underwater. The sedimentary record contains abundant evidence of submarine mass movements, because after they take place, they tend to be buried by younger sediments and are preserved.

Geologists distinguish three types of submarine mass movements, according to whether the mass remains coherent or disintegrates as it moves. The degree to which the mass comes apart during movement typically depends on the amount of water mixed with the sediment in the moving mass. In **submarine slumps,** semicoherent blocks slip downslope on weak mud detachments. The slump blocks, when buried and preserved, are known as **olistostromes** (►Fig. 16.10a). In some cases, the layers comprising the olistostrome become contorted as they move, like a tablecloth that has slid off a table. In **submarine debris flows,** the moving mass breaks apart to form a slurry containing larger clasts (pebbles to boulders) suspended in a mud matrix (►Fig. 16.10b). And in **turbidity currents,** sediment disperses in water to create a turbulent cloud of suspended sediment that avalanches downslope (►Fig. 16.10c). As the turbidity current slows, sediment settles out in sequence from coarse to fine, creating graded beds (see Chapter 7).

In recent years, marine geologists have used a sensitive type of sonar, called GLORIA (geologic long-range inclined asdic [sonar]), to map the sea floor. The instrument sees sideways and can map a 60-km-wide swath of the ocean floor all at once. GLORIA maps have documented huge near-surface submarine slumps. For example, huge slumps along the margins of the big island of Hawaii have substantially modified the shape of the island and have created a broad apron of hummocky sea floor around it (►Fig. 16.11). The slumping events probably happen at times when the volcano that forms Hawaii fills with magma and expands like a balloon, causing the surface of the island to crack and weak sliding surfaces to develop. Major slumping events, which seem to recur (on average) every 100,000 years, may

(a)

(b)

(c)

FIGURE 16.10 Submarine mass movements. (a) A semicoherent olistostrome. (b) A debris flow, in which the mass becomes a viscous slurry of chunks floating in a mud matrix. (c) A turbidity current, in which sediment is suspended like a cloud in water.

generate catastrophic tsunamis in the Pacific basin. Marine geologists have mapped similar near-surface slumps along the edge of North America's continental shelf. And archaeologists have recently proposed that a slump block the size of Scotland suddenly slid into the North Sea some 7,000 years ago and generated a tsunami that washed away most coastal Stone Age tribes in Europe.

Summary: A Variety of Mass Movements

We've seen that mass movements travel at a range of different velocities, from the slowest (creep) to the faster (slumps, mud and debris flows, and rock and debris slides), to the fastest (snow and debris avalanches and rock and debris falls). The velocity in turn depends on the steepness of the slope and the water or air content of the mass (►Fig. 16.12). In slides, the rock or debris remains coherent as it moves, and remains in contact with the underlying surface. In flows, the mud or debris becomes chaotically jumbled as it moves. In avalanches, debris mixed with air rushes down steep slopes like a strong wind. And in falls, rock or debris actually flies through the air in a free fall. Some mass movements start as one kind of event and change into another.

FIGURE 16.11 The irregular sea-floor surface surrounding Hawaii consists of a succession of huge slumps, here outlined; each slump is a mass of rock and sediment that slipped seaward.

16.3 SETTING THE STAGE FOR MASS MOVEMENTS

In order for the movements described above to take place, the stage must be set by the following phenomena: the development of relief, which creates slopes down which masses move, and fracturing and weathering, which weaken materials at Earth's surface so that they cannot hold up against the pull of gravity.

Relief and the Pull of Gravity

Mass movements could not take place without **relief,** the difference in elevation between one place and another. Where relief exists, gravity tries to pull materials at higher elevations down to lower elevations. If the force from the pull of gravity exceeds the force that holds the mass in place, the mass starts to move. Relief forms where one part of the crust moves up (undergoes uplift) or down (undergoes subsidence) relative to its surroundings, where material accumulates on the crust in a mound or pile, or where a pit or trough has been dug into the Earth. Here are some of the ways relief can be created (▶Fig. 16.13a–i):

FIGURE 16.12 Each type of mass movement has a range of velocities, depending on the steepness of the slope and the water content of the mass. In general, steeper slopes make for faster movements. Types of mass movements also differ in the degree to which they remain coherent (here, coherent movements are underlined).

Volcano

Head scarp

Sliding
surface

Solifluction

Tilted tree

Soil
creep

Slumping

Damaged
road

Mass Movement

In Earth's gravity field, what goes up must come down—
sometimes with disastrous consequences. Rock and regolith are
not infinitely strong, so every now and then slopes or cliffs give
way in response to gravity, and materials slide, tumble, or career

downslope. This downslope movement, called mass movement,
or mass wasting, is the first step in the process of erosion and
sediment formation. The resulting debris may eventually be
carried away by water, ice, or wind.

Deforested land

Rock slide

Rock avalanche

Rock fall

Debris flow

Lahar/mudflow

The kind of mass wasting that takes place at a given location reflects the composition of the slope (is it composed of weak soil, loose rock, or hard rock containing joints?), the steepness of the slope, and the climate (is the slope wet or dry, frozen or unfrozen?). Stronger rocks can hold up steep cliffs, but with time, rock breaks free along joints and tumbles or slides down weak surfaces. Coherent regolith may slowly slide down slopes, while water-saturated regolith may flow rapidly. Episodes of mass movement may be triggered by an oversteepened slope (when a river has cut away at the base of a cliff), a heavy rainfall that saturates the slope, an earthquake that shakes debris free, or a volcanic eruption, which not only shakes the ground but melts snow and ice to saturate regolith.

Geologists classify mass-wasting events by the rate and character of the movement. Soil creep accompanies seasonal freezing and thawing, which causes soil to gradually migrate downslope; if it creeps over a frozen substrate, it's called solifluction. Slumping involves semicoherent slices of earth that move slowly down spoon-shaped sliding surfaces, leaving behind a head scarp. Mudflows and debris flows happen where regolith has become saturated with water and moves downslope as a slurry. When volcanoes erupt and melt ice and snow at their summit, or if heavy rains fall during an eruption, water mixes with ash, creating a fast-moving lahar. Steep, rocky cliffs may suddenly give way in rock falls. If the rock breaks up into a cloud of debris that rushes downslope at high velocity, it is a rock avalanche. Snow avalanches are similar, but the debris consists only of snow.

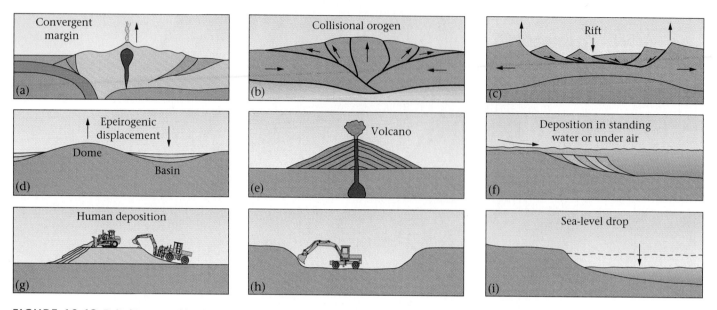

FIGURE 16.13 Relief is created by (a) uplift resulting from subduction at a convergent plate boundary; (b) uplift resulting from collisional orogeny; (c) the subsidence of a rift and the uplift of its margins; (d) epeirogenic movements of the crust, forming domes and basins; (e) the formation of a volcano; (f) the deposition of sediment; (g) human construction of a debris pile; (h) excavation; (i) a lowering of the sea level.

- Plate interaction at a convergent boundary or at a continent-continent collision zone thickens the crust and causes the surface of the crust to rise and form mountains.

- The faulting that accompanies continental rifting drops blocks of crust down relative to their surroundings.

- Changes in the temperature of the mantle or the tilting of large crustal blocks, among other reasons, lead to **epeirogenic movements** (the gradual uplift or subsidence of broad regions of the continent).

- The accumulation of extrusive rock at a volcanic vent builds a volcano that rises above the surroundings.

- The deposition of sediment in lakes creates a pile that slopes down into deeper water. Deposition on land (in sand dunes, for example) creates subaerial slopes.

- People build mounds of debris to make foundations for roads or buildings, mounds of garbage and sediment in landfills, and mounds of waste rock around a mine.

- Either flowing water or people excavate valleys or pits with steep walls.

- The sea level drops when water becomes trapped in glaciers during an ice age (see Chapter 22). When this happens, slopes form along the coast.

A car parked on horizontal ground (with no brakes) will not start rolling even if the ground is at a high elevation.

Rolling begins only if the car is parked on a **slope,** an inclined or tilted surface. The steeper the slope, the faster the car will move. The same relation holds for geological materials. Mass movements can occur only on slopes, and the steeper the slope, the faster the movement once it has begun.

Fragmentation and Weathering: Weakening the Surface

If the Earth's surface were covered by intact (unbroken) rock, mass movements would be of little concern, for intact rock has great strength and could form stalwart mountain faces that would never tumble, even if they were vertical. But Earth's surface instead consists of rock containing abundant joints, rock that has been broken up by faults during mountain building, or regolith created by the weathering of rock in Earth's corrosive, moisture-rich atmosphere. Fragmented rock and regolith are much weaker than intact rock and can collapse in response to Earth's gravitational pull (▶Fig. 16.14). Thus, jointing, faulting, and weathering make mass movements possible.

Why are fragmented rock and regolith so much weaker than intact rock? Intact rock is held together by the strong chemical bonds within mineral crystals, by mineral cement, or by the interlocking of grains. In contrast, a joint-bounded or fault-bounded block is held in place only by the weaker friction between the block and its surroundings. Regolith is **unconsolidated;** that is, it consists of unat-

FIGURE 16.14 Perfectly intact rock is rare at the surface of the Earth. Most outcrops, like this one, are highly jointed.

tached grains. Dry regolith holds together because of friction between adjacent grains or because weak electrical charges cause grains to attract each other. Slightly wet regolith holds together because of water's surface tension. (**Surface tension,** the phenomenon that makes water form drops, exists because water molecules have a positively charged side and a negatively charged side, so the molecules bond to mineral surfaces and attract each other. Because of surface tension, damp sand holds together to form a sand castle, while dry sand collapses into a shapeless pile.)

Slope Stability: The Battle Between Downslope Force and Resistance Force

Mass movements do not take place on all slopes, and even on slopes where such movements are possible, they occur only occasionally. Geologists distinguish between **stable slopes,** those on which sliding is unlikely, and **unstable slopes,** on which sliding will likely happen. When material starts moving on an unstable slope, we say that **slope failure** has occurred. Whether a slope fails or not depends on the balance between two forces—the **downslope force,** caused by gravity, and the **resistance force,** which inhibits sliding. If the downslope force exceeds the resistance force, the slope fails and mass movement is the consequence.

Imagine a block sitting on a slope. We can represent the gravitational attraction between this block and the Earth by an arrow (a vector) that points straight down, toward the Earth's center of gravity. This arrow can be separated into two components, one parallel to the slope (the downslope force) and one perpendicular to the slope (called the **normal force**). The resistance force can be represented by an arrow pointing uphill. If the downslope force is larger than the resistance force, then the block moves; otherwise, it stays in

place (►Fig. 16.15a, b). Note that for a given mass, the magnitude of the downslope force increases as the slope angle increases, so downslope forces are larger on steeper slopes.

What causes the resistance force? As we saw above, chemical bonds in mineral crystals, cement, and the jigsaw-puzzle-like interlocking of crystals hold intact rock in place, friction holds an unattached block in place, electrical charges and friction hold dry regolith in place, and surface tension holds slightly wet regolith in place.

Granular debris tends to pile up and create the steepest slope it can without collapsing. The angle of this slope is called the **angle of repose,** and for most dry unconsolidated materials (such as dry sand) it typically lies between 30° and 37°. The angle depends partly on the shape and size of grains, which determine the amount of friction across boundaries. For example, larger angles of repose (up to 45°) tend to form on slopes composed of large, irregularly shaped grains, for these grains interlock with one another (►Fig. 16.16a–c).

In many locations, rock or debris lies on top of a particularly weak surface. This surface can act as a glide horizon, separating a substrate below from unstable rock or debris above, especially if the surface is parallel or nearly parallel to the slope. Different materials have different kinds of glide horizons:

• In regolith, the glide horizon may be slippery wet clay. Clay consists of tiny flakes that easily slide past one another, especially if they are held apart by water.

FIGURE 16.15 (a) Gravity, represented by the black arrow, pulls a block toward the center of the Earth. The gravitational force has two components, the downslope force parallel to the slope and the normal force perpendicular to the slope. On gentle slopes, the normal force is larger than the downslope force. The resistance force, caused by friction and represented by an arrow pointing upslope, is larger in this example than the downslope force. (b) If the slope angle increases, the normal force becomes smaller than the downslope force. If the downslope force then becomes greater than the resistance force, then the block starts to move.

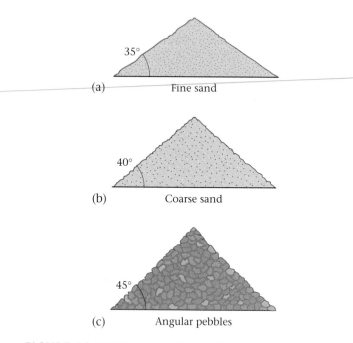

FIGURE 16.16 The angle of repose is the steepest slope that a pile of unconsolidated sediment can have and remain stable. Angles of repose depend on the size and shape of grains. (a) Fine, well-rounded sand has a small angle of repose. (b) Coarse, angular sand has a larger angle. (c) Large, irregularly shaped pebbles have a large angle of repose.

• In massive rocks, joints can act as glide horizons (▶Fig. 16.17a). Where rocks contain vertical joints, a block detaches along a joint surface to form a vertical cliff.

• The surfaces between beds in a sedimentary rock sequence tend to be significantly weaker than the rock within a bed, and thus can act as a glide horizon (▶Fig.

16.17b). Layers of shale or evaporite are especially weak, and may preferentially slip.

• In metamorphic rocks, glide horizons may develop along foliation planes (▶Fig. 16.17c), especially where foliation is defined by the parallel alignment of mica flakes (mica is a platy mineral with slippery faces).

Glide horizons that dip parallel to the slope are much more likely to fail than layers that dip into the slope, because in the first case the downslope force is parallel to the horizon, while in the second case it is at a high angle to the horizon (▶Fig. 16.18a, b). As an example, consider the 1959 landslide that occurred in Madison Canyon, in southwestern Montana. On August 17 of that year, shock waves from a strong earthquake jarred the region. The south wall of the canyon is underlain by metamorphic rock with a strong foliation dipping somewhat more steeply than the canyon walls. Nevertheless, the foliation provided a glide horizon, and when the ground vibrated, rock detached along a foliation plane and tumbled downslope, reaching speeds of 150 km per hour. Unfortunately, twenty-eight campers lay sleeping in the valley floor. They were awakened by the hurricane-like winds blasting in front of the moving mass, but seconds later were buried below 45 m of rubble.

16.4 FINGERS ON THE TRIGGER: FACTORS CAUSING SLOPE FAILURE

What triggers an individual mass-wasting event? In other words, what causes the balance of forces to change so that the downslope force exceeds the resistance force, and a

FIGURE 16.17 Different kinds of surfaces become glide horizons in different geologic settings. (a) In exfoliated massive granite, exfoliation joints become glide horizons. (b) In sedimentary rock, bedding planes become glide horizons. (c) In metamorphic rock, foliation planes, especially schistosity (the parallel alignment of mica flakes), become glide horizons.

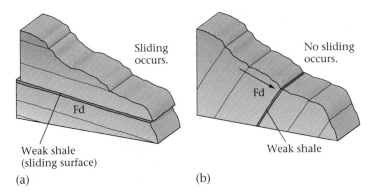

(a)   (b)

FIGURE 16.18 The orientation of a potential glide horizon affects its stability. (a) Glide horizons that dip in the same direction as the slope are unstable because the downslope force (Fd) parallels the horizon. (b) Glide horizons that dip in the opposite direction to the slope are stable because the downslope force is at a high angle to the horizon.

slope suddenly fails? Here, we look at various phenomena—natural and human-made—that trigger slope failure.

Shocks and Vibrations

Earthquake tremors, storms, the passing of large trucks, or blasting in construction sites may cause a mass that was on the verge of moving to actually start. For example, an earthquake-triggered slide dumped debris into Lituya Bay, in southeastern Alaska, in 1958. The debris displaced the water in the bay and sent it hurtling seaward as a 300-m-high wave, which washed the slopes on either side of the bay clean of their forest and carried fishing boats kilometers out to sea. The vibrations of such an earthquake break any remaining bonds that hold the mass in place and/or cause the mass and the slope to separate slightly, thereby decreasing friction. As a consequence, the resistance force decreases, and the downslope force sets the mass in motion.

FIGURE 16.19 (a) In a quick clay, before shaking, the grains stick together. (b) During shaking, the grains become suspended in water, and the once-solid mass becomes a movable slurry.

(a)   (b)

Shaking produces a unique effect in certain types of clay, called **quick clay.** Quick clay, which consists of damp clay flakes, behaves like a solid when still, for surface tension holds water-coated flakes together. But shaking separates the flakes from one another and suspends them in the water, thereby transforming the clay into a slurry that flows like a fluid (▶Fig. 16.19a, b). When this happens, a once-sticky clay layer becomes so weak that the mass above it can move. The disastrous 1964 slide at Turnagain Heights, in Alaska (see Chapter 10), occurred when a layer of quick clay liquefied, causing a large mass of sediment, along with seventy-five homes built on it, to slump into the sea.

Changing Slope Angles, Slope Loads, and Slope Support

The factors that make a slope steeper or heavier may ultimately cause it to fail, for these factors effectively increase the downslope force acting on the slope (▶Fig. 16.20a–c). For example, consider what happens when a truck dumps a load of waste rock at the tailings pile next to a mine: before the truckload is added, the slope of the pile is at the angle of repose, but afterward the angle of repose is exceeded and the slope fails. Similarly, when a soil-covered slope endures a steady rain, the water sinks into the soil, making it so heavy that the slope collapses under its own weight. Finally, consider a situation where a contractor cuts terraces into a hill slope to make a foundation platform for a house. By cutting into the hill slope, the contractor removes the support that holds up the higher part of the hill; later, a slump may take place, burying the new house.

Excavation at the foot of a hill or mountain, because of human activity or natural river erosion, is particularly dangerous when the excavation removes the support from rock or regolith that rests on a weak glide horizon (a potential sliding surface), as the excavation may trigger slip on the glide horizon. The largest observed landslide in U.S. history, the Gros Ventre slide, which took place in 1925 on the flank of Sheep Mountain, near Jackson Hole, Wyoming, illustrates this phenomenon (▶Fig. 16.21a–c). Almost 40 million cubic meters of rock, soil, and forest detached from the side of the mountain and slid 600 m down a slope, filling the valley and creating a 75-m-high natural dam across the Gros Ventre River.

In retrospect, the geology of the slide area made this landslide almost inevitable. The flank of Sheep Mountain is a **dip slope,** meaning that bedding parallels the face of the mountain. The Tensleep Formation, the stratigraphic unit exposed at the surface, consists of interbedded sandstone and shale. In the past, a thick sandstone layer spanned the valley and propped up sandstone farther up the side of the mountain. But river erosion cut down through the sand layer to a weak shale (the Amsden Shale) beneath. Rainfall in the weeks before the landslide made the ground heavier

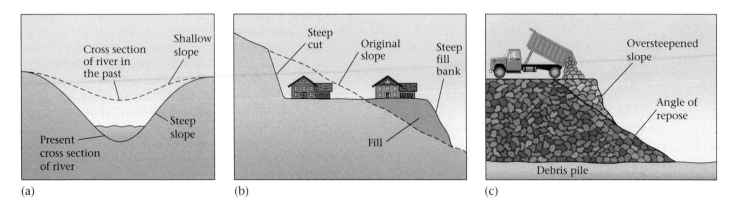

(a)   (b)   (c)

FIGURE 16.20 Slope angles may become steeper, making the slopes unstable. (a) A river can cut into the base of a slope, steepening the sides of the valley. (b) Cutting terraces in a hill slope creates a steeper slope. (c) Adding debris to the top of an unconsolidated sediment pile may cause the angle of repose to be exceeded.

than usual and increased the amount of water seeping into the shale layer, weakening it further. By June, the downslope force exceeded the restraining force, and a huge slab of rock upslope broke off and raced downhill at over 80 km per hour, using the wet shale layer as a glide horizon.

In some cases, excavation results in the formation of an overhang. When such **undercutting** has occurred, rock comprising the overhang breaks away from the slope and falls. Overhangs commonly develop above a weak horizontal layer that erodes back preferentially, or along seacoasts

FIGURE 16.21 (a) The huge Gros Ventre slide took place after heavy rains had seeped into the ground, weakening the Amsden Shale and making the overlying Tensleep Sandstone heavier. The slope was already unstable because the Gros Ventre River had cut down to the shale, and the bedding planes dipped parallel to the slope. (b) After the slide moved, it filled the river valley and dammed the river, creating a lake. A huge landslide scar formed on the hill slope. (c) The Gros Ventre slide.

(a)

(b)

FIGURE 16.22 (a) Undercutting by waves removes the support beneath an overhang. (b) Eventually, the overhang breaks off along joints, and a rock fall takes place.

and rivers where the water cuts into a fairly strong slope (▶Fig. 16.22a, b).

Changing the Slope Strength: The Effects of Weathering, Vegetation, and Water

The stability of a slope depends on the strength of the material comprising it. If the material weakens with time, the slope becomes weaker and eventually collapses. Three factors influence the strength of slopes: weathering, vegetation cover, and water.

With time, chemical weathering produces weaker minerals, and physical weathering breaks rocks apart. Thus, a once-intact rock composed of strong minerals transforms into a weaker rock or into regolith.

In the case of slopes underlain by regolith, vegetation tends to strengthen the slope, because the roots hold otherwise unconsolidated grains together. Also, plants absorb water from the ground, thus keeping it from turning into slippery mud. The removal of vegetation therefore has the net result of making slopes more susceptible to downslope movement. Deforestation in tropical rain forests, for example, leads to catastrophic mass wasting of the forest's substrate (▶Fig. 16.23).

We've seen that thin films of water create cohesion between grains. Water in larger quantities, though, decreases cohesion, because it fills pore spaces entirely and keeps grains apart (▶Fig. 16.24a, b). Though slightly damp sand makes a better sand castle than dry sand, a slurry of sand and water can't make a castle at all. Thus, the saturation of a regolith with water during a torrential rainstorm weakens the regolith so much that it may begin to move downslope

as a slurry. Rising floodwaters may saturate sediment at the base of a riverbank, causing the land to slump into the river. If the water weakens a specific subsurface layer, then the layer becomes a glide horizon and the material above slumps. Similarly, if the water table (the top surface of the

FIGURE 16.23 Gouges in the landscape are left by mud and debris flows in a deforested area.

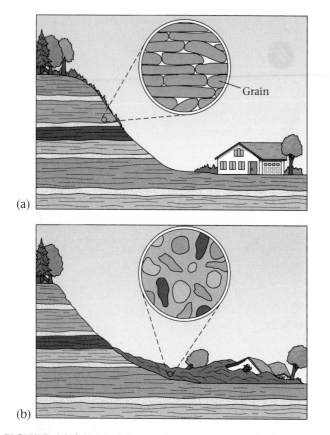

FIGURE 16.24 (a) In drier sediment, either the grains are interlocking, or surface tension in a thin fluid film holds the grains together. Tree roots may help keep the sediment dry. (b) When sediment is saturated, the grains are held apart by the water, and the mass moves to create a mud or debris flow.

groundwater layer) rises above a weak glide horizon after water has sunk into the ground, overlying rock or regolith may start to slide over the further weakened glide horizon.

Water infiltration has a particularly notable effect in regions underlain by **swelling clays.** These clays possess a mineral structure that allows them to absorb water—water molecules form sheets between layers of silica tetrahedra, which cause clay flakes to swell to several times their original size. Such swelling pushes up the ground surface, making it crack, and weakens the upper layer of the ground, making it susceptible to creep or slip. When the clay dries, it shrinks, and the ground surface subsides. This up-and-down movement wrinkles road surfaces and cracks foundations.

The saturation of the ground with rainwater may lead to the liquefaction of quick clays even in the absence of vibrations. Some quick clays are held together by salt that precipitated out of seawater when the clay was originally deposited. Fresh rainwater dissolves the salt, destroying this "cement" and transforming a solid mass of clay into a slurry of clay suspended in water. For example, beneath sand and gravel deposits in regions of Quebec, there is a layer of Pleistocene marine clay, deposited at the end of the last ice age. Heavy rains can cause this layer to fail, as happened when a mudflow of quick clay inundated the village of Saint-Jean Vianney in Quebec, killing thirty-one people.

16.5 PLATE TECTONICS AND MASS MOVEMENTS

The Importance of the Tectonic Setting

Most natural unstable ground on Earth ultimately owes its existence to the activity of plate tectonics. As we've seen, plate tectonics causes most major uplift, thus generating relief. Volcanic eruptions related to plate tectonics provide the atmosphere and water that weather and weaken the crust. And plate tectonics causes faulting, which further fragments the crust, and, of course, earthquakes on plate boundaries have triggered some of the world's most devastating landslides. Spend a day along the steep slopes of the Alpine Fault, a plate boundary that transects New Zealand, and you can hear mass movement in progress: during heavy rains, rockfalls and landslides clatter with astounding frequency, as if the mountains were falling down around you. Now, to see the interplay of plate tectonics and other factors, let's consider the example of mass movements in southern California.

A Case Study: Slumping in Southern California

Some Californians pay immense prices for the privilege of building homes on cliffs overlooking the Pacific. The sunset views from their backyard poolside patios are great. But the landscape is not ideal from the standpoint of stability. Slumps and mudflows on coastal cliffs in southern California have consumed many homes over the years, with a cost to their owners (or insurance companies) of untold millions of dollars (see Fig. 10.29a). Downslope movement is inevitable here—earthquakes, steep cliffs, weak substrates, lack of vegetation, heavy rains, and coastal wave erosion all contribute to the problem. What is special about the region that makes it so prone to mass wasting?

The main reason is that southern California is an active plate boundary. The coast borders the San Andreas Fault, a transform plate boundary accommodating the northward movement of the Pacific Plate with respect to North America. Faulting during the past 15 million years has shattered the rock of California's crust. The fractures act as planes of weakness, and they have allowed water to seep into the rock and cause chemical weathering; the resulting clay and other slippery minerals further weaken the rock. Also, the rocks in many areas are weak to begin with, because they are part of an accretionary prism, a chaotic mass of clay-rich sediment that was scraped off subducting oceanic lithosphere during the Mesozoic era. Though most of the movement between the North American and Pacific plates involves strike-slip displacement on the San Andreas Fault, compression across the fault also leads to uplift, and this uplift creates slopes. Since the uplifted region borders the coast, wave erosion steepens and in some places under-

cuts cliffs. Finally, because of the plate motion, numerous earthquakes rock the region, thus shaking regolith loose.

Another reason for southern California's susceptibility to mass movements is its climate. In general, the region is hot and dry and thus supports only semidesert flora. Development and brush fires remove much of this cover, leaving large areas with no dense vegetation. But since the region lies on the West Coast, it endures occasional heavy

---

BOX 16.1

THE HUMAN ANGLE

Los Angeles, a Mobile Society

During a year of abundant slumping in southern California, Art Buchwald wrote the following newspaper column.

I came to Los Angeles last week for rest and recreation, only to discover that it had become a rain forest.

I didn't realize how bad it was until I went to dinner at a friend's house. I had the right address, but when I arrived, there was nothing there. I went to a neighboring house where I found a man bailing out his swimming pool.

I beg your pardon, I said. Could you tell me where the Cables live?

"They used to live above us on the hill. Then, about two years ago, their house slid down in the mud, and they lived next door to us. I think it was last Monday, during the storm, that their house slid again, and now they live two streets below us, down there. We were sorry to see them go—they were really nice neighbors."

I thanked him and slid straight down the hill to the new location of the Cables' house. Cable was clearing out the mud from his car. He apologized for not giving me the new address and explained, "Frankly, I didn't know until this morning whether the house would stay here or continue sliding down a few more blocks."

Cable, I said, you and your wife are intelligent people, why do you build your house on the top of a canyon, when you know that during a rainstorm it has a good chance of sliding away?

"We did it for the view. It really was fantastic on a clear night up there. We could sit in our Jacuzzi and see all of Los Angeles, except of course when there were brush fires. Even when our house slid down two years ago, we still had a great sight of the airport. Now I'm not too sure what kind of view we'll have because of the house in front of us, which slid down with ours at the same time."

But why don't you move to safe ground so that you don't have to worry about rainstorms?

"We've thought about it. But once you live high in a canyon, it's hard to move to the plains. Besides, this house is built solid and has about three more good mudslides in it."

Still, it must be kind of hairy to sit in your home during a deluge and wonder where you'll wind up next. Don't you ever have the desire to just settle down in one place?

"It's hard for people who don't live in California to understand how we people out here think. Sure we have floods, and fire and drought, but that's the price you have to pay for living the good life. When Esther and I saw this house, we knew it was a dream come true. It was located right on the tippy top of the hill, way up there. We would wake up in the morning and listen to the birds, and eat breakfast out on the patio and look down on all the smog.

"Then, after the first mudslide, we found ourselves living next to people. It was an entirely different experience. But by that time we were ready for a change. Now we've slid again and we're in a whole new neighborhood. You can't do that if you live on solid ground. Once you move into a house below Sunset Boulevard, you're stuck there for the rest of your life.

"When you live on the side of a hill in Los Angeles, you at least know it's not going to last forever."

Then, in spite of what's happened, you don't plan to move out?

"Are you crazy? You couldn't replace a house like this in L.A. for $500,000."

What happens if it keeps raining and you slide down the hill again?

"It's no problem. Esther and I figure if we slide down too far, we'll just pick up and go back to the top of the hill, and start all over again; that is, if the hill is still there after the earthquake."

FIGURE 16.25 The Portuguese Bend slump viewed from the air.

winter rains. The water sinks quickly into the sparsely vegetated ground, adds weight to the mass on the slope, and weakens glide horizons.

Development is the final factor that triggers mass movements in southern California. Developers in some places have oversteepened slopes, overloaded slopes, and changed levels of groundwater. The consequences can be seen in an event like the Portuguese Bend slump of the Palos Verdes area, near Los Angeles. Between 1956 and 1985, this slump moved between 0.3 and 2.5 cm per day, and in the process destroyed over 150 homes (▶Fig. 16.25). The Portuguese Bend region borders steep cliffs along the Pacific coast, and is underlain by a thick, seaward-dipping layer of weak volcanic tuff, resting on shale. To provide a smooth foundation for homes and roads, developers deposited a 23-m-thick layer of fill over the ground surface. But elevating the ground surface made the water table rise above a shale glide horizon; water substantially weakened the glide horizon and decreased the resistance force, while the weight of the fill and the buildings increased the downslope force. The combination of these changes set the slump in motion once again in 1956. Almost thirty years later, the land had moved more than 200 m, but it was finally stopped thanks to an intense effort to stabilize the slump by draining water from the subsurface.

16.6 HOW CAN WE PROTECT AGAINST MASS-MOVEMENT DISASTERS?

Identifying Regions at Risk

Clearly, landslides, mudflows, and slumps are natural hazards we cannot ignore. Too many of us live in areas where mass wasting has the potential to kill people and destroy property. In many cases, the best solution is avoidance: don't build, live, or work in an area where mass movement is likely. But avoidance is only possible if we know where the hazards are.

To pinpoint dangerous regions, geologists look for landforms known to result from mass movements, for where these movements have happened in the past, they might happen in the future. Features such as slump head scarps, swaths of forest in which trees have been flattened and point downslope, piles of loose debris at the base of hills, and hummocky land surfaces all indicate recent mass wasting.

Geologists may also be able to detect regions that are beginning to move (▶Fig. 16.26). For example, roads, buildings, and pipes begin to crack over unstable ground. Power lines may be too tight or too loose because the poles to which they are attached move together or apart. Visible cracks form on the ground at the potential head of a slump, while the ground may bulge up at the toe of the slump, forming a **pressure ridge.** Subsurface cracks may drain the water from an area and kill off vegetation, while another area may sink and form a swamp. Slow movements cause trees to develop pronounced curves at their base. In some cases, the activity of land masses moving too slowly to be perceptible to people can be documented with sensitive surveying techniques that can detect a subtle tilt of the ground or changes in distance between nearby points.

If various clues indicate that a land mass is beginning to move, and conditions make accelerating movement likely (for example, persistent rain, rising floodwaters, or continuing earthquake aftershocks), then officials may order an evacuation. Evacuations have saved lives, and ignored warnings have cost lives. But unfortunately, some mass movements happen without any warning, and some evacuations prove costly but unnecessary.

Even if there is no evidence of recent movement, a danger may still exist, for just because a steep slope hasn't collapsed in the recent past doesn't mean it won't in the future. In recent years, geologists have begun to identify such potential hazards (by using computer programs that evaluate factors that trigger mass wasting) and create maps that portray the degree of risk for a certain location. These factors include the following.

- *Slope steepness:* Are slopes approaching or exceeding the angle of repose?

- *Strength of the substrate:* Is the land underlain by intact bedrock, highly fractured bedrock, or weak regolith? Are there potential glide horizons beneath the ground?

- *Degree of water saturation:* How wet is the rock or regolith forming the landscape? Can it transform into a slurry?

- *The dip of bedding, joints, or foliation relative to the slope:* Do potential weak surfaces lie parallel to the ground surface,

FIGURE 16.26 The features shown here indicate that a large slump is beginning to develop. Note the cracks at the site of the growing head scarp, which drain water and kill trees. Power-line poles crossing the unstable ground bend, and the lines become overtight. Fences and roads that straddle the scarp begin to break up. Houses that straddle the scarp begin to crack, and their foundations sag.

so they can become glide horizons, or at an angle to the ground surface?

- *Vegetation cover:* Is the land barren, so that water can quickly infiltrate the ground, or is there a buffer of vegetation to absorb water and bind regolith together?

- *Climate:* Is there a potential for heavy, long-lasting rain?

- *Undercutting:* Are waves or rivers undercutting slopes?

- *Seismicity:* Is the region in a seismically active locality, where earthquakes are likely to shake the ground?

From such hazard-assessment studies, geologists compile **landslide-potential maps,** which rank regions according to the likelihood that a mass movement will occur. In any case, common sense suggests that you should avoid building on or below particularly dangerous slide-prone slopes. In Japan, regulations on where to build in regions prone to mass wasting, careful monitoring of ground movements, and well-

designed evacuation plans have drastically reduced property damage and the number of fatalities.

Preventing Mass Movements

In areas where a hazard exists, people can take certain steps to remediate the problem and stabilize the slope (▶Fig. 16.27a–i).

- *Revegetation:* Since bare ground is much more prone to downslope movement than vegetated ground, stability in deforested areas will be greatly enhanced if owners replant the region with vegetation that sends down deep roots. The roots hold regolith together and prevent it from becoming oversaturated with water.

- *Regrading:* An oversteepened slope can be regraded so that it does not exceed the angle of repose. Some mass upslope could be moved to a lower region, or the ground

(a)

(b) Removed / Added

(c) High water table / High / Wet and weak potential glide horizon / Dry and somewhat stronger glide horizon / Lower reservoir level / Lower water table

(d) Unstable slope / Terrace steps (remove load, catch debris)

FIGURE 16.27 A variety of remedial steps can stabilize unstable ground. (a) Revegetation removes water, and tree roots bind regolith. (b) Redistributing the mass on a slope eases the load where necessary, adds support where necessary, and decreases slope angles. (c) Lowering the level of groundwater (the water table) may allow a glide horizon to dry out. (d) Terracing a steep slope may decrease the load and provide benches to catch debris. (e) Relocating a river channel stops undercutting, and filling the channel adds support. (f) Riprap absorbs wave energy along the coast. (g) A retaining wall traps falling rock. (h) Bolting rock to a steep cliff face holds loose blocks in place. (i) An avalanche shed diverts avalanche debris over a roadway.

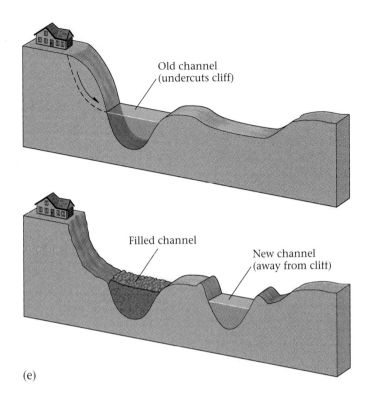

Old channel
(undercuts cliff)

Filled channel

New channel
(away from cliff)

(e)

Riprap absorbs
wave energy

(f)

Retaining
wall

(g)

Rock
bolts

(h)

Avalanche
shed

(i)

could be terraced so that the mass distribution overall becomes stable.

- *Reducing subsurface water:* Because water weakens material beneath a slope and adds weight to the slope, an unstable situation may be remedied either by improving drainage so that water does not enter the subsurface in the first place, or by removing water from the ground. In some cases, a potential hazard can be averted simply by stopping irrigation or by repairing leaky pools.

- *Preventing undercutting:* In places where a river undercuts a cliff face, engineers can divert the river. Similarly, along coastal regions they may build an offshore breakwater or pile **riprap** (loose boulders or concrete) along the beach to absorb wave energy before it strikes the cliff face.

- *Constructing safety structures:* In some cases, the best way to prevent mass wasting is to build a structure that stabilizes a potentially unstable slope or protects a region downslope from debris if a mass movement does occur. For example, civil engineers commonly build retaining walls to stabilize highway embankments. In some places, engineers bolt loose slabs of rock to more coherent masses in the substrate. The danger from rock falls can be decreased by covering a road cut with chainlink fencing or by excavating wide shoulders along the highway that can catch falling debris. Engineers also sometimes spray road cuts with **shotcrete,** a cement that coats the wall and prevents water infiltration and consequent freezing and thawing. Highways at the base of an avalanche chute should be covered by an **avalanche shed,** whose roof acts as a surrogate ground surface, thereby keeping debris off the underlying road.

- *Controlled blasting of unstable slopes:* When it is clear that unstable ground threatens a particular region, the best solution may be to blast the unstable ground loose at a time when its movement can do no harm. For example, officials may trigger a controlled avalanche after people have been evacuated from its path, by setting off explosives near a snow mass. The shock wave of the charge sets the mass loose before it has time to grow into a major hazard.

Clearly, the cost of preventing mass-wasting calamities is high, and people might not always be willing to pay the price. In such cases, they have a choice of avoiding the risky area, taking the chance that a calamity will not happen while they are around, buying appropriate insurance, or counting on relief agencies (or the general public) to come to the rescue if disaster does strike. Because mass movements can sometimes (not always) be predicted and/or prevented, some have generated immense lawsuits. Once again, geology and society cross paths.

CHAPTER SUMMARY

- Rock or regolith on unstable slopes has the potential to move downslope under the influence of gravity. This process, called mass movement, or mass wasting, plays an important role in the erosion of hills and mountains.

- Slow mass movement, caused by the freezing and thawing of regolith, is called creep. In places where slopes are underlain by permafrost, a thin, melted layer of regolith slowly flows down slopes. During slumping, a semicoherent mass of material moves down a spoon-shaped glide horizon. Mudflows and debris flows occur where regolith has become saturated with water and moves downslope as a slurry.

- Landslides (rock and debris slides) move very rapidly down a slope; the rock or debris breaks apart and tumbles. During avalanches, debris mixes with air and moves downslope as a turbulent cloud. And in a debris fall or rock fall, the material free-falls down a vertical cliff.

- Mass movements happen only where there is relief. And relief in most cases ultimately results from plate interactions. Unstable slopes may also develop when debris, sediment, or ash accumulates on the ground surface, or as a consequence of excavation by rivers or people.

- Intact, fresh rock is too strong to undergo mass movement. Thus, for mass movement to be possible, rock must be weakened by fracturing (joint formation) or weathering.

- Unstable slopes start to move when the downslope force exceeds the resistance force that holds material in place. The steepest that a slope of unconsolidated material can be without collapsing is the angle of repose.

- Downslope movement can be triggered by shocks and vibrations, a change in the steepness of a slope, or a change in the strength of a slope. Material comprising a slope becomes more susceptible to moving if it undergoes chemical weathering, if it is stripped of vegetation, or if it becomes saturated with water.

- The California coast is particularly susceptible to mass-movement events because it is tectonically active (so the region has slopes and feels earthquake shocks), because it is composed of weak material, because coastal erosion creates steep cliffs, because of the climate, and because of development.

- Geologists produce landslide-potential maps to identify areas susceptible to mass movement. Engineers can help prevent mass movements by revegetating, regrading, reducing subsurface water, preventing undercutting, bolting loose rocks to their substrate, covering outcrops with shotcrete, and blasting unstable slopes. Avalanche sheds may help protect roads that cross avalanche chutes.

KEY TERMS

angle of repose (p. 493)
avalanche (p. 487)
cornice (p. 487)
creep (p. 483)
debris fall (p. 487)
debris flow (p. 485)
debris slide (p. 485)
dip slope (p. 495)
downslope force (p. 493)
glide horizon (p. 484)
head scarp (p. 484)
hummocky surface (p. 482)
lahar (p. 485)
landslide (p. 485)
mass movement (wasting)
    (p. 482)
mudflow (p. 485)

normal force (p. 493)
olistostrome (p. 488)
pressure ridge (p. 500)
quick clay (p. 495)
regolith (p. 482)
resistance force (p. 493)
rock fall (p. 487)
rock glacier (p. 483)
rock slide (p. 485)
slope failure (p. 493)
slumping (p. 483)
solifluction (p. 483)
submarine slump (p. 488)
talus (p. 487)
turbidity current (p. 488)
undercutting (p. 496)

REVIEW QUESTIONS

1. What four factors distinguish the various types of mass movement?

2. How does a slump differ from creep? How does it differ from a mudflow or debris flow?

3. How does a rock or debris slide differ from a slump? How does it differ from a mudflow or debris flow?

4. How are submarine slumps similar to those above water? How are they different?

5. What factors cause the relief on the Earth's surface? How does relief affect mass movement?

6. How does a small amount of water between grains hold material together? How does this change when the sediment is oversaturated?

7. What force is responsible for downslope movement? What force helps resist that movement?

8. How does the angle of repose change with grain size? How does it change with water content?

9. What factors trigger downslope movement?

10. What factors contribute to the high frequency of landslides in coastal southern California? Why don't people just avoid building in these dangerous areas?

11. How do geologists predict whether an area is prone to mass wasting?

12. What steps can people take to reduce the risk of mass wasting?

SUGGESTED READING

Bloom, A. L. 1991. *Geomorphology.* 2nd ed. Englewood Cliffs, N.J.: Prentice-Hall.

Brabb, E. E., and B. L. Harrod, eds. 1989. *Landslides: Extent and Economic Significance.* Brookfield, Va.: Balkema.

Crozier, M. J. 1986. *Landslides: Causes, Consequences, and Environment.* Dover, N.H.: Croom Helm.

Costa, J. E., and V. R. Baker. 1981. *Surficial Geology.* New York: Wiley.

Costa, J. E., and G. F. Wieczorek. 1987. *Reviews in Engineering Geology.* Vol. 7, *Debris Flows, Avalanches: Process, Recognition, and Mitigation.* Boulder, Colo.: Geological Society of America.

Eckel, E. B., ed. 1958. *Landslides and Engineering Practice.* Highway Research Board Special Report 29. National Academy of Sciences, National Research Council Publication 544.

Fleming, R. W., and F. A. Taylor. 1980. *Estimating the Cost of Landslide Damage in the United States.* U.S. Geological Survey Circular 832.

Slosson, J. E., A. G. Keene, and J. A. Johnson, eds. 1993. *Reviews in Engineering Geology.* Vol. 9, *Landslides/Landslide Mitigation.* Boulder, Colo.: Geological Society of America.

Voight, B., ed. 1978. *Rockslides and Avalanches.* Vol. 1, *Natural Processes.* New York: Elsevier.

Zaruba, Q., and V. Mencl. 1969. *Landslides and Their Control.* New York: Elsevier.

CHAPTER 17

Streams and Floods: The Geology of Running Water

17.1 INTRODUCTION

By the 1880s, Johnstown, built along the Conemaugh River in scenic western Pennsylvania, had become a significant industrial town with numerous steel-making factories. Recognizing the attraction of the region as a summer retreat from the heat and pollution of nearby Pittsburgh, speculators built a mud and gravel dam across the river, upstream of Johnstown, to trap a pleasant reservoir of cool water. A group of industrialists and bankers bought the reservoir and established the exclusive South Fork Hunting and Fishing Club, a cluster of lavish fifteen-room "cottages" on the shore. Unfortunately, the dam had been poorly designed, and debris blocked its spillway (a passageway for surplus water), setting the stage for a monumental tragedy. On May 31, 1889, torrential rain drenched Pennsylvania, and the reservoir surface rose until water flowed over the dam and down its face. Despite frantic attempts to strengthen the dam, the soggy structure abruptly collapsed, and the reservoir emptied into the Conemaugh River Valley. A 20-m-high wall of water roared downstream and slammed into Johnstown, transforming bridges and buildings into twisted wreckage (▶Fig. 17.1). When the water subsided, 2,300 people lay dead, and Johnstown became the focus of national sympathy. Clara Barton mobilized the recently founded Red Cross, which set to work building dormitories, and citizens nationwide donated everything from clothes to beds. Nevertheless, it took years for the town to recover, and many residents simply picked up and left. Despite many lawsuits, no one paid a penny of restitution, but the South Fork Hunting and Fishing Club abandoned its property.

The unlucky inhabitants of Johnstown experienced one of the more destructive consequences of **running water,** or **runoff,** water that flows on the land surface. The Conemaugh River overflowed when the dam broke, causing a **flood,** an event during which the volume of water in a stream becomes so great that it covers areas outside the stream's normal limits. But even when not in flood, run-

---

Water that falls on land drains back to the sea in networks of rivers. The character of an individual river depends on a variety of factors, such as slope, water volume, and sediment load. The entwined channels of this braided stream in the Yukon Territories are reflecting the late-day sun.

FIGURE 17.1 During the disastrous 1889 flood in Johnstown, Pennsylvania, the force of the water was able to move and tumble sturdy buildings.

ning water relentlessly digs into and scrapes away at the surface of the Earth, carrying away debris and depositing it in other locations. Without running water, our planet would look very different.

Much of the work of running water takes place in **streams,** ribbons of water that flow down **channels,** or troughs, cut into the land. Streams remove, or drain, excess water from the landscape and carry it eventually to the sea (much like culverts, or ditches, drain water from parking lots), and in the process they have modified much of Earth's landscape. (Note that geologists call any channelized body of

flowing water a stream, regardless of its size.) In the process of draining the land, streams also transport nutrients and provide a home for living organisms. And for human society, streams supply avenues for commerce, water for agriculture, and sources for power.

In this chapter, we see how streams operate in the Earth system. We first learn about the origin of running water and the architecture of streams and how running water ultimately drops sediment. We then look at the landforms that erosion and/or deposition generate in response to stream flow, and conclude with a consideration of flooding and its effects on human society.

17.2 DRAINING THE LAND

Runoff, in the Context of the Hydrologic Cycle

Nothing that is can pause or stay—
 The moon will wax, the moon will wane,
 The mist and cloud will turn to rain,
 The rain to mist and cloud again,
 Tomorrow be today.

—Henry Wadsworth Longfellow

Perhaps without realizing it, Longfellow (1807–1882), an American poet fascinated with reincarnation, provided an ac-

FIGURE 17.2 Excess surface water comes from rain, melting ice or snow, and groundwater springs. Where the ground is flat, the water accumulates in puddles or swamps, but on sloping ground, it flows downslope, collecting in natural troughs called streams.

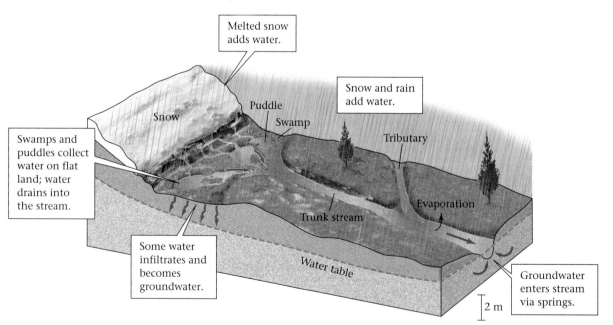

curate if somewhat romantic image of the hydrologic cycle (see Interlude E). Solar energy evaporates water from the Earth's surface (mainly from the sea); when the vapor rises, it cools and condenses to form rain or snow that falls back to Earth (**meteoric water**). More meteoric water falls on the land than can be trapped by soil or organisms, or can **infiltrate** (seep into) the ground to become groundwater. In freezing climates, this excess becomes ice, while elsewhere it becomes liquid **surface water.**

On horizontal ground, surface water collects in puddles or swamps, but on a sloping ground, it flows down the slope in response to gravity, following the steepest route possible. Melting ice and snow, as well as springs (places where groundwater seeps back out of the ground), contribute to the supply of running water (▶Fig. 17.2). Along its downslope journey, runoff may pause temporarily in lakes or reservoirs, but all *fresh* standing bodies of water have an outlet through which water escapes and continues to the sea. On a global basis, 36,000 cubic km of water becomes runoff every year, about 10% of the total volume that passes through the hydrologic cycle.

Forming Streams and Drainage Networks

Runoff begins its downslope journey as **sheetwash,** a film of water less than a few mm thick that covers the surface of the ground. You've seen sheetwash if you've watched water flowing down a sloping street during a rainstorm. Like any flowing fluid, sheetwash erodes its substrate (the material it flows over). The efficiency of such erosion depends on the velocity of the flow—faster flows erode more effectively. Since the ground is not perfectly planar, not all substrate has the same resistance to erosion, and the amount of vegetation that covers and protects the ground varies with location. Where the flow happens to be a bit faster, or the substrate a little weaker, erosion scours (digs) a channel into the substrate (▶Fig. 17.3a, b). As this channel is lower than the surrounding ground, sheetwash in adjacent areas starts to head toward it. With time, the extra flow deepens the channel relative to its surroundings, a process called **downcutting,** and creates a stream.

As its flow increases, a stream channel begins to lengthen up its slope, a process called **headward erosion** (▶Fig. 17.3c, d). Headward erosion occurs because the flow is more intense at the entry to the channel (upslope) than in the surrounding sheetwashed areas. At the same time, new channels form nearby; these merge with the main channel, because once a channel forms, the surrounding land slopes into it. Gradually, an array of linked streams evolves, with the smaller streams, or **tributaries,** flowing into a single larger stream, or **trunk stream.** The array of interconnecting streams together comprise the **drainage network.**

FIGURE 17.3 (a) Drainage on a slope first occurs when sheetwash, overlapping films or sheets of water, moves downslope. (b) Where the sheetwash happens to move a little faster, it scours a channel. (c) The channel grows upslope, a process called headward erosion, and new tributary channels form. The interconnecting streams comprise a drainage network. (d) Headward erosion in Canyonlands National Park, Utah, where the main canyon and its tributaries are cutting upstream slowly.

Picture the array of streets and highways in a town. In some regions, they define a grid, while in others they are arranged like the spokes of a wheel. But regardless of arrangement, they interconnect to form a network that allows you to start on the smallest lane and end up on a superhighway. Like transportation networks, drainage networks reach into all corners of a region to provide conduits for the removal of runoff. The configuration of tributaries and trunk streams defines the "map pattern" of a drainage network. This pattern depends on the shape of the landscape and the composition of the substrate. Geologists recognize several types of networks, based on their map pattern, including the following.

- *Dendritic:* When rivers flow over a fairly uniform substrate with a fairly uniform initial slope, they develop a **dendritic network,** which looks like the pattern of branches connecting to the trunk of a deciduous tree (▶Fig. 17.4a). In fact, the word "dendritic" comes from the Greek *dendros,* meaning "tree." The Mississippi drainage network displays a dendritic pattern. Streams in these networks join in a downstream-pointing V.

- *Radial:* Drainage networks forming on the surface of a cone-shaped mountain flow outward from the mountain peak, like spokes on a wheel. Such a pattern defines a **radial network** (▶Fig. 17.4b).

FIGURE 17.4 (a) Dendritic patterns resemble the branches of a tree and form on land with a uniform substrate. (b) A radial network drains a conical mountain; in this case, a volcano. (c) Rectangular patterns develop where a gridlike array of vertical joints controls drainage. (d) Trellis patterns (resembling a garden trellis) form where drainage networks cross a landscape in which ridges of hard rock separate valleys of soft rock. In this example, the alternation is due to folding of the rock layers.

(a) Dendritic

(b) Volcano — Radial

(c) Rectangular — Joint

(d) Resistant ridge — Trellis

- *Rectangular:* In places where a rectangular grid of fractures (vertical joints) breaks up the ground, channels form along the preexisting fractures, and streams join each other at right angles, creating a **rectangular network** (▶Fig. 17.4c).

- *Trellis:* In places where a drainage network develops across a landscape of parallel valleys and ridges, major tributaries flow down a valley and join a trunk stream that cuts across the ridges; the place where a trunk stream cuts across a resistant ridge is called a **water gap.** The resulting map pattern resembles a garden trellis, so the arrangement of streams comprise a **trellis network** (▶Fig. 17.4d).

Rectangular and trellis networks form where there is **structural control** on the development of a drainage network, where geologic structures such as folds, fractures, and faults affect the distribution of weaker and harder rocks—streams preferentially erode the softer rocks.

The streams that comprise a drainage network have distinctive parts—let's review the terms for these. The outlet, or termination, of a stream, where it discharges its contents into another stream or a lake or the sea, is called its **mouth,** while the upstream end, where it begins, is called its **headwaters,** or **source.** The floor of a stream is its **bed,** and the flat land on either side of the stream that becomes covered with water during a flood is its **floodplain.** The slope of the channel in the downstream direction is the **stream gradient.** The path a stream follows is called its **course,** and a specific segment of a stream is called a **reach**—upper reaches of the stream lie near its headwaters, while lower reaches lie near its mouth. In some places, a stream's channel can be straight, in others it may swing back and forth in snake-like curves called **meanders,** and in still others it may subdivide into numerous subchannels that merge and separate like hair in a braid. A reach that contains many meanders is a **meandering stream,** while a reach that consists of entwined subchannels is a **braided stream.**

Drainage Basins and Divides

A drainage network collects water from a broad region, variously called a **drainage basin, catchment,** or **watershed,** and feeds it into the trunk stream, which carries the water away (▶Fig. 17.5a, b). Note that a drainage basin can contain many small streams that are only visible on detailed maps. The highland, or ridge, that separates one watershed from another is a **drainage divide** (▶Fig. 17.6). A **continental divide** separates drainage that flows into one ocean from drainage that flows into another. For example, if you straddle the North American continental divide facing north and pour a cup of water out of each hand, the water in your right hand flows to the Atlantic, and the water in your left hand flows to the Pacific. This continental divide is not, however, the only important divide in North America. A divide runs along the crest of the Appalachians, separating Atlantic Ocean drainage from Gulf of Mexico drainage, and another one runs just south of the Canada-U.S. border, separating Gulf of Mexico drainage from Hudson Bay (Arctic Ocean) drainage. These three divides bound the Mississippi drainage network, which drains the interior of the United States (▶Fig. 17.7).

Many continental divides lie close to active continental margins (margins that coincide with plate boundaries), where major mountain ranges form; as a result, the Earth's major watersheds drain out to passive margins. For example, South America's continental divide follows the crest of the Andes—relatively small drainage networks flow to the Pacific west of the divide, while the immense Amazon River watershed drains into the Atlantic east of the divide. In fact, some of the source waters of tributaries into the Amazon lie within 100 km of the Pacific coast.

Ignoring Structure: Superposed and Antecedent Streams

In some locations, streams appear to ignore the structural controls of a landscape—they cut through resistant rock ledges instead of curving around them, or they carve deep canyons clear across high mountain ranges. Geologists have given two names to this kind of stream.

- *Superposed streams:* Imagine a region in which drainage forms on a layer of soft, flat strata that unconformably overlies folded strata (see Chapter 12 for a discussion of unconformities). Streams initially carve channels into the flat strata; when they eventually erode down through the unconformity and start to downcut into the folded strata, they maintain their earlier course, ignoring the structure of the folded strata. Geologists call such streams **superposed streams,** because their geometry has been superposed on the rock structure (▶Fig. 17.8a, b).

- *Antecedent streams:* In some cases, tectonic activity (such as subduction or collision) causes a mountain range to rise up beneath an already established stream. If the stream downcuts as fast as the range rises, it can maintain its course and will cut right across the range. Geologists call such streams **antecedent streams** (from the Greek *ante,* meaning "before"), to emphasize that they existed before the range uplifted. Note that if the range rises faster than the stream downcuts, the new highlands

FIGURE 17.5 (a) A drainage network collects water from a broad drainage basin, or watershed, via numerous tributaries. These carry water to a trunk stream and eventually to a standing body of water. In this example, the stream has carved a deep valley in the hills, whereas near its mouth it meanders across a flat floodplain. Points 1–5 refer to locations along the longitudinal profile. The cross-sectional profiles show how river valley shapes change along the length of a river. (b) A drainage network displays the same shape characteristics at all scales; an enlargement of a small part of the network resembles the whole network.

FIGURE 17.6 A drainage divide is a relatively high ridge that separates one drainage basin from another.

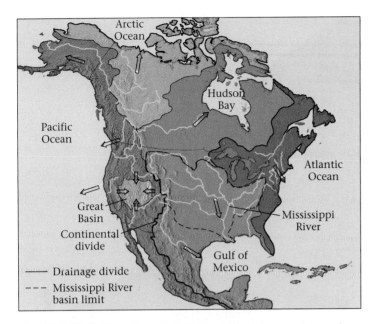

FIGURE 17.7 The Mississippi drainage basin is one of several drainage basins in North America. The continental divide separates basins that drain into the Atlantic (and waters connected to the Atlantic) from basins that drain into the Pacific.

divert (change) the stream's course so that it flows along the range face (▶Fig. 17.9a–d).

In some locations, antecedent streams display **incised meanders,** meanders that lie at the bottom of a steep-walled canyon; the canyon was carved out by the stream when uplift occurred in the region. The "Goosenecks" of the San Juan River, in southern Utah, illustrate this geometry (▶Fig. 17.10a–c).

Streams That Last, Streams That Don't: Permanent and Ephemeral Streams

Some streams flow all year long, while others flow for only part of the year; in fact, some flow only for a brief time after a heavy rain. The character of a stream depends on the depth of the **water table:** the boundary, approximately parallel to the Earth's surface, that separates substrate in which ground-water fills pores (the spaces between grains) from substrate in which air fills pores. Water-saturated substrate lies below the water table. If the bed, or floor, of a stream lies below the water table, then the stream flows year-round (▶Fig. 17.11a). In such **permanent streams,** found in humid or temperate climates, water comes not only from upstream or nearby surface runoff, but also from springs through which groundwater seeps. But if the bed of a stream lies above the water table, then water flows only after rainstorms or during spring thaws, when the rate at which water enters the stream channel exceeds the rate at which water infiltrates the ground below (▶Fig. 17.11b). In such **ephemeral streams,** found

FIGURE 17.8 (a) A superposed stream first establishes its geometry while flowing over uniform, flat layers above an unconformity. (b) The stream gradually erodes away the layers and exposes underlying rock with a different structure (in this example, the older strata are folded). The drainage is "superposed" (let down) on the folded rocks, and appears to ignore structural control.

in dry climates with intermittent rainfall and high evaporation rates, water gradually sinks into the ground, and the stream dries up when the supply of water stops. A dry ephemeral stream bed is called a **dry wash.**

17.3 DISCHARGE AND TURBULENCE

Geologists and engineers describe the amount of water a stream carries by its **discharge,** the volume of water passing a point on the stream bank in a second. We specify stream discharge either in cubic feet per second (ft^3/s) or in cubic meters per second (m^3/s). Stream discharge depends on two factors: the cross-sectional area of the stream (A_c; the area measured in a vertical plane perpendicular to the flow direction) and the average velocity at which water moves in the downstream direction (v_a). Thus, we can calcu-

Drainage before uplift

(a)

or

(b)

Antecedent drainage cuts through uplift.

New course

(c)

Older drainage is diverted by uplift.

(d)

FIGURE 17.9 (a) An antecedent stream flows across the land to the sea. (b) A mountain range develops across the path of the stream, in this example, as a consequence of subduction along the coast. If the stream erosion keeps pace with the rate of uplift, the stream cuts across the mountain range, and is an antecedent stream. (c) If uplift happens faster than erosion, the stream is diverted and flows along the front of the range. This stream is not antecedent. (d) An antecedent stream cutting through a rapidly uplifting ridge in Pakistan.

FIGURE 17.10 (a) A stream forms meanders while it flows across a plain. (b) Uplift of the land over which the stream flows causes the meanders to cut down and carve out canyons that meander like the stream. (c) The "Goosenecks" of the San Juan River, in Utah, are a frequently visited example of incised meanders.

FIGURE 17.11 (a) If the bed of a stream channel lies below the water table, then springs add water to the stream, and the stream contains water even during periods when there is no rainfall. Such streams are permanent. (b) If the stream bed lies above the water table, then the stream flows only during rainfall or spring melt, when water enters the stream faster than it can infiltrate into the ground. Such streams are ephemeral. In desert regions, a dry stream bed is called a dry wash.

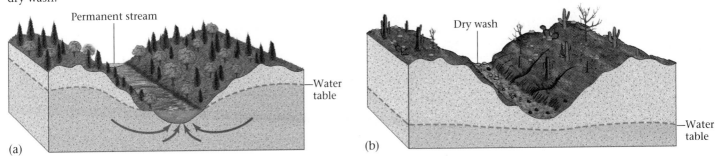

Badlands National Park

Badlands—the name conjures an image of barrenness, dryness, and waste. And rightly so, for technically, a **badlands** is an unvegetated region gullied by a dendritic network of channels separated by sharp-crested ridges. Badlands form where the runoff from sudden rainstorms cuts deeply into unprotected soft soil or weak sediment that then dries out between rainstorms. Perhaps the most famous example in North America is Badlands National Park, in South Dakota (►Fig. 17.12).

The geologic story of the park begins between 23 and 37 million years ago (Eocene to Oligocene time), when a network of streams and intervening marshland covered much of what is now the northern great plains. Occasionally, fine ash settling from huge volcanic eruptions mixed in with the mud and silt of this fluvial environment. Diverse mammals, including the first grass eaters, lived here. Sediments deposited in this environment, along with the carcasses of dead mammals, were eventually buried to form a succession of soft, fossiliferous strata in which alternations of color—tan, gray, yellow, and maroon—define the different beds.

Now let's jump forward to the last 14,000 years, approximately the time since the last ice age. Erosion has exposed the soft Eocene and Oligocene strata at the surface. During torrential rains, water does not have time to infiltrate into the ground before it starts flowing down the slopes, washing away vegetation in the process. The flow quickly cuts down into the soft strata and the land to form a myriad of small ephemeral streams. When the sun comes out, the flow stops, leaving the dendritic pattern of dry channels. Stripped of vegetation, this dramatically shaped surface exposes pastel strata, a visual effect so dramatic that in 1978 the region became a national park.

FIGURE 17.12 In Badlands National Park, a dendritic erosion pattern has carved into soft, colorful layers of strata.

FIGURE 17.13 To measure a stream's discharge, we need to know both the cross-sectional area of the stream and the average downstream velocity of the flow. Geologists obtain this information at a stream-gauging station. First, they make a survey of the channel so they know its shape. Then they measure its depth, using a well, and its velocity, using a current meter (a propeller that spins in response to moving water). The current meter can be moved back and forth to take measurements at various points in the stream, for velocity changes with location.

late stream discharge by using the simple formula $D = A_c \times v_a$. Stream discharge can be determined at a stream-gauging station, where instruments measure the velocity and depth of the water (▶Fig. 17.13).

Different streams have different discharges. For example, the Amazon River has the largest discharge in the world—about 200,000 m³/s, or 15% of the total amount of runoff on Earth. The next-largest stream, the Congo River, has an average discharge of 40,000 m³/s, while the "mighty" Mississippi's is only 17,000 m³/s. Also, the discharge of a given stream varies along its length. For example, the discharge in a temperate region increases in the downstream direction, because each tributary that enters the stream adds more water, while the discharge in an arid region may decrease downstream, as progressively more water seeps into the ground or evaporates. Finally, the discharge at a given location can vary with time: in a temperate climate, a stream's discharge during the spring may be double or triple the amount during a hot summer, and a flood may increase the discharge to more than a hundred times normal.

The average velocity of stream water (v_a) can be difficult to calculate, because the water doesn't all travel at the same velocity. Why not? First, friction at the channel surface slows the flow. Thus, water near the channel walls or the stream bed moves more slowly than water in the middle of the flow, and the fastest-moving part of the stream flow lies near the surface in the center of the channel. In a curving channel, centrifugal force causes the fastest flow to shift toward the outside curve, just as centrifugal force makes your car swerve to the outer edge of a curve on a highway. Therefore, the deepest part of a channel, its **thalweg,** lies near the outside

curve. In fact, as the water swings toward the outside wall of a curving channel, it begins a spiral motion. The amount by which friction slows the flow depends both on the roughness of the walls and bed and on the channel shape. A wide, shallow stream channel has a larger **wetted perimeter** (the area in which water touches the channel walls) than a semicircular channel, so water flows more slowly in the former than in the latter (▶Fig. 17.14a–c).

FIGURE 17.14 (a) In a straight semicircular channel, the maximum velocity occurs near the surface in the center of the stream. The deepest part of the channel is called the thalweg. (b) The maximum velocity also occurs in the center of a wide, shallow channel, but the maximum velocity of a shallow channel is less than that of a semicircular channel for a given cross-sectional area, and its discharge is less. That's because its wetted perimeter (where water touches the channel walls) is greater than that of the semicircular channel (even though its cross-sectional area is the same), so there is more friction between the water and the channel walls, which slows down the flow. (c) In a curving channel, the thalweg swings to the outer edge of the stream, because of centrifugal force. Also, the water follows a spiral path, set into motion by friction against the stream bed as the water moves diagonally across it.

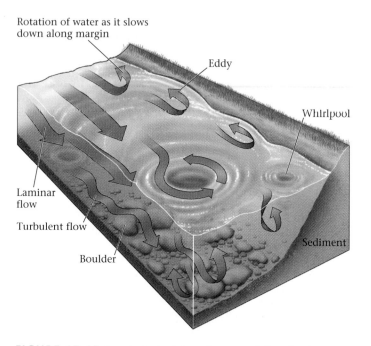

FIGURE 17.15 In a turbulent flow, because of shearing between parcels and movement over boulders, the water swirls in curving paths and becomes caught in eddies.

The second reason that parcels of water don't all flow at the same speed is turbulence. **Turbulence, or turbulent flow,** refers to the twisting, swirling motion that can create eddies (whirlpools) in which water curves and actually flows upstream, or circles in place (▶Fig. 17.15). Turbulence develops in part because the shearing motion of one parcel against its neighbor causes the neighbor to spin, and in part because obstacles like boulders deflect parcels, forcing them to move in a different direction. Other factors being equal, faster water is more turbulent than slower water, a shallow stream is more turbulent than a deep stream, and a stream flowing in a rough, gravelly channel is more turbulent than a stream flowing in a smooth, clay-lined channel.

17.4 THE WORK OF RUNNING WATER

Slicing Through the Earth: Erosional Processes

The energy that makes running water move comes from gravity. As water heads downslope from a higher to a lower elevation, the gravitational potential energy stored in water transforms into kinetic energy. About 3% of this energy goes into the work of eroding the walls and beds of stream channels.

Running water causes erosion in four ways.

- *Scouring:* Running water can remove loose fragments of sediment, a process called **scouring.**

- *Breaking and lifting:* In some cases, the push of flowing water can break chunks of solid rock off the channel floor or walls by pushing cracks open and making them grow until the rocks separate from the substrate. In addition, the flow of a current over a clast can cause the clast to rise, or "lift" off the substrate, in the same way that an airplane rises when a current of air flows over its wing.

- *Abrasion:* Clean water has little erosive effect, but sand-laden water acts like sandpaper and grinds or rasps away at the channel floor and walls, a process called **abrasion.** In places where turbulence creates long-lived whirlpools, abrasion by sand or gravel carves a bowl-shaped depression, called a **pothole,** into the floor of the stream (▶Fig. 17.16a, b).

- *Dissolution:* Running water dissolves soluble minerals as it passes, and carries the minerals away in solution.

The efficiency of erosion depends on the velocity and volume of water and on its sediment content. A large volume of fast-moving, turbulent sandy water causes more erosion than a trickle of quiet clear water. Thus, most erosion takes place during floods, which supply streams with large volumes of fast-moving, sediment-laden water.

Transport in Streams: Sediment Loads

The Mississippi River has received the nickname "Big Muddy" for a reason—its water can become chocolate brown, because of all the clay and silt it carries (▶Fig. 17.16c). All streams carry sediment, though not the same amount. Geologists refer to the total volume of sediment carried by a stream as its **sediment load.** The sediment load consists of three components (▶Fig. 17.16d):

- *Dissolved load:* Running water dissolves soluble minerals in the sediment or rock of its substrate, and groundwater seeping into a stream through the channel walls brings dissolved minerals with it. These ions comprise a stream's **dissolved load.** Typically, streams carry only about 120 ppm (parts per million—120 ions per million water molecules) of dissolved ions. But streams in arid climates, where water evaporates quickly, or streams flowing over particularly soluble minerals like halite or gypsum may contain over 1,000 ppm of dissolved ions.

- *Suspended load:* The **suspended load** of a stream consists of tiny solid grains (silt or clay size) that swirl along with the water without settling to the floor of the channel. The suspended load accounts for the bulk of the sediment carried by a river.

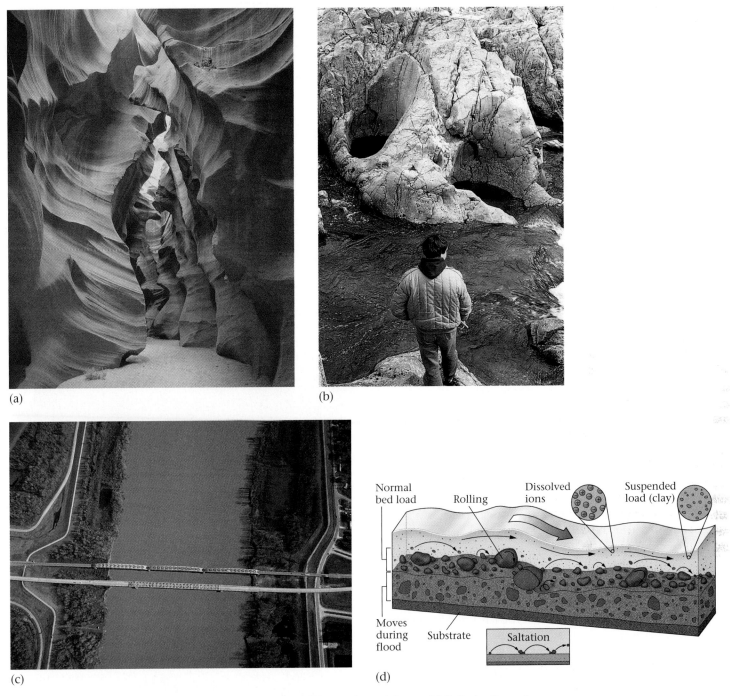

(a)

(b)

(c)

(d)

FIGURE 17.16 (a) A polished, red sandstone canyon wall in northern Arizona. (b) Potholes formed in a stream bed by the grinding power of swirling gravel. (c) The color of this river water comes from the sediment it carries. Here, we see the tan-brown waters of a mud-laden river. (d) Streams transport sediment in many forms. The dissolved sediment load consists of ions in solution. A suspended load consists of tiny grains distributed through the water. A bed load consists of grains that undergo saltation (they bounce along the bed) or grains that roll along the bed; part of the bed load stays in place during times of normal flow, but begins to move during a flood.

• *Bed load:* The **bed load** of a stream consists of large particles (such as sand, pebbles, or cobbles) that bounce or roll along the stream floor. Typically, bed-load movement involves **saltation,** a process during which grains on the channel floor get knocked into the water column momentarily, follow a curved trajectory downstream, and gradually sink to the bed again, where they knock other grains into the water column.

When describing a stream's ability to carry sediment, geologists measure its competence and capacity. The **competence** of a stream refers to the maximum particle size it carries; a stream with high competence can carry large particles, while one with low competence can carry only small particles. A fast-moving, turbulent stream has greater competence (it can carry bigger particles) than a slow-moving stream. If the density of stream water increases, because of an increase in a dissolved or suspended load, its competence also increases, for solids placed within the water feel a stronger buoyancy force. Thus, muddy water can carry larger fragments than can clear water, and saltwater can carry larger fragments than can freshwater. Considering that during floods water moves faster and is denser because it contains more suspended particles, floodwater carries much bigger clasts than does a normal stream flow—in fact, the huge boulders that litter the bed of a mountain creek move *only* during floods. The **capacity** of a stream refers to the total quantity of sediment it *can carry.* A stream's capacity depends on its competence and discharge.

When Streams Lose Their Loads: Depositional Processes

If you hike into the Catskill Mountains of New York, you cross layer after layer of sandstone, shale, and conglomerate, all of which were deposited either in a stream channel or along a stream bank (▶Fig. 17.17a–c). There are many such places on continents where the sedimentary record contains **fluvial deposits** (from the Latin *fluvius,* meaning "river"). Because a stream's competence increases the faster it flows, *any place where running water slows down,* the stream's competence decreases and *sediment is deposited.* As a stream slows, coarser components settle out, while finer components are carried farther downstream until they reach a spot where the water slows even more; this segregating of sediment by size is called **sediment sorting.** Geologists refer to sorted, stream-washed sediment as **alluvium.** Alluvium may collect in a sheet or in an elongate lens or mound called a **bar.**

What would cause water in a stream to slow down? For one thing, if water passes from a deeper part of the channel to a shallower part, or if it spreads out over a floodplain, friction between the bed and the water slows the water down. The average velocity of stream water also decreases where it

(a)

(b)

(c)

FIGURE 17.17 (a) These Devonian "redbeds" in the Catskill Mountains of New York consist of fluvial (river) deposits. (b) Recently deposited gravel in a stream bed. (c) Mudflats along a stream.

enters a standing body of water, or where the **gradient** of a stream (the downstream slope) decreases. Because maximum stream velocity varies with location, different-sized sediment accumulates in different locations in and along a stream.

17.5 HOW STREAMS CHANGE ALONG THEIR LENGTH

Longitudinal Profiles

In 1803, the United States under President Thomas Jefferson's leadership bought the Louisiana Territory, a vast tract of land encompassing the western half of the Mississippi drainage basin, for the bargain price of $27 million, from Napoleon, who was short of cash because of France's military campaigns. Suddenly, the western border of the United States jumped toward the Pacific by over 1,000 km. But the geography of the territory was a mystery. To fill the blank on the map, Jefferson asked Meriwether Lewis and William Clark to lead a voyage of exploration across the Louisiana Territory to the Pacific.

Lewis and Clark, along with about forty men, began their expedition at the mouth of the Missouri River where it joins the Mississippi. At this juncture, the Missouri is a wide, languid stream of muddy water. The group found the lower reaches of the Missouri, where the channel is deep and the water smooth, to be easy going. But the farther upstream they went, the more difficult their voyage became, for the stream gradient became progressively steeper, and its discharge became less. When Lewis and Clark reached the site of what is now Bismarck, North Dakota, they had to abandon their original boats and haul smaller vessels up **rapids,** particularly turbulent water that plunges over a steep, bouldery bed, and around **waterfalls,** where water drops over an escarpment. When they reached what is now southwestern Montana, they deserted these boats as well and followed stream valleys on foot or by horseback, struggling up steep gradients until they reached the continental divide.

If Lewis and Clark had been able to plot a graph showing their elevation above sea level relative to their distance along the Missouri, they would have found that the **longitudinal profile** of the Missouri, a cross-sectional image showing the variation in the river's elevation along its length, was roughly a concave-up curve (Fig. 17.5a). This curve illustrates that a stream's gradient is steeper near its headwaters (source) than near its mouth. Real longitudinal profiles are not perfectly smooth curves, but rather display little plateaus and steps, representing interruptions by lakes or waterfalls. Near its headwaters, an idealized stream flows down deep valleys or canyons, while near its mouth, it flows over nearly horizontal plains.

The Base Level

Streams progressively deepen their channels by downcutting, but there is a depth below which a stream cannot downcut. The lowest elevation a stream channel's floor can reach at a locality is called the **base level** of the stream. A **local base level** occurs upstream of a drainage network's mouth, while the **ultimate base level** (that is, the lowest elevation along the stream's longitudinal profile) is defined by sea level. The trunk stream cannot downcut deeper than sea level, for if it did, it would have to flow upslope to enter the sea.

Lakes or reservoirs can act as local base levels along a stream, for where the stream enters such standing bodies of water, it almost slows to a halt and cannot downcut further (▶Fig. 17.18a). A ledge of resistant rock can also act as a local base level, for the stream level cannot drop below the ledge until the ledge erodes away (▶Fig. 17.18b). Finally, where a tributary joins a larger stream, the channel of the larger stream acts as the base level for the tributary; thus, the mouth of a tributary lies at the same elevation as the stream that it joins at the point of intersection. Local base levels do not last forever, because running water eventually removes obstructions that create them.

FIGURE 17.18 (a) A lake acts as a local base level. The longitudinal profile of the stream upstream of the lake lies above the profile for a graded stream (one that deposits as much sediment as it removes). Eventually, headward erosion of the stream below the lake will cause the lake to drain. (b) A resistant rock ledge also acts as a local base level. With time, the ledge will erode and the waterfall will migrate upstream until the stream achieves grade. Sea level is the ultimate base level for a drainage network.

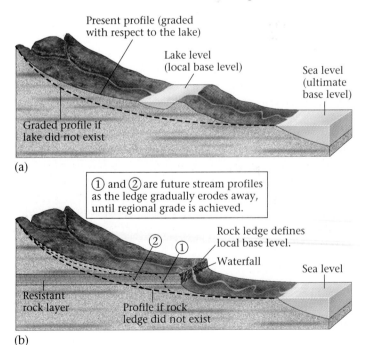

Streams attain their concave-up longitudinal profile gradually. During this process, steps or ledges defining local base levels along the stream erode away, and low areas fill in, until at any point along the stream there is no net erosion or deposition—the stream can carry all the sediment that has been supplied to it, and it deposits as much sediment as it removes. A stream that has reached this condition is called a **graded system.** When a stream's sediment supply, discharge, or base level changes, the stream erodes or deposits sediment until it once again becomes graded.

17.6 STREAMS AND THEIR DEPOSITS IN THE LANDSCAPE

Valleys and Canyons

About 10 million years ago, a large block of crust, the region now known as the Colorado Plateau (located in Arizona, Utah, Colorado, and New Mexico), began to rise. Before the rise, the Colorado River had been flowing over a plain not far above sea level, causing little erosion. But as the land uplifted, the river began to steadily downcut in its quest to achieve grade. Soon its channel lay as much as 1.6 km below the surface of the plateau, and the Grand Canyon was born (▶Fig. 17.19). The formation of the Grand Canyon illustrates a general phenomenon. In regions where the land surface lies well above the base level, a stream can create a deep trough, much deeper than the channel itself, in the land; if the walls of the trough slope gently, the landform is a **valley,** while if they slope steeply, the landform is a **canyon.**

FIGURE 17.19 Note the dendritic pattern of side canyons in this air view of the Grand Canyon region.

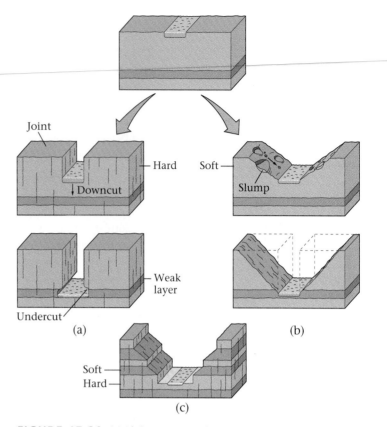

FIGURE 17.20 (a) If downcutting by a stream happens faster than mass wasting alongside the stream, as is typical when streams erode through strong rock, a slot canyon forms. The canyon widens with time as the stream undercuts the walls. (b) If mass wasting takes place as fast as downcutting occurs, a V-shaped valley develops. (c) In regions where the stream downcuts through alternating hard and soft layers, a stair-step canyon forms.

Whether stream erosion produces a valley or a canyon depends on the rate at which downcutting occurs relative to the rate at which mass wasting causes the walls on either side of the stream to collapse. In places where a stream downcuts through its substrate faster than the walls of the stream collapse, erosion creates a slot (steep-walled) canyon. Such canyons typically form in strong rock, which can hold up steep cliffs for a long time (▶Fig. 17.20a).

In places where the walls collapse as fast as the stream downcuts, landslides and slumps gradually cause the slope of the walls to approach the angle of repose. When this happens, the stream channel lies at the floor of a valley whose cross-sectional shape resembles the letter V (▶Fig. 17.20b); this is a **V-shaped valley.** Where the walls of the stream consist of alternating layers of hard and soft rock, the walls develop a stair-step shape such as that of the Grand Canyon (▶Fig. 17.20c).

FIGURE 17.21 The evolution of alluvium-filled valleys.
(a) Stream erosion creates a valley. (b) Later, a rise in the base level or a decrease in discharge allows the valley to fill with alluvium. (c) If the base level later falls or discharge increases, the stream downcuts through the alluvium and a new, lower floodplain develops. The remnants of the original alluvial plain remain as a pair of terraces, one on each side of the new floodplain. (d) Distinct episodes of renewed downcutting can create several pairs of terraces along a stream.

In places where active downcutting occurs, the valley floor remains relatively clear of sediment, for the stream— especially when it floods—carries away sediment that has fallen or slumped into the channel from the stream walls. But if the stream's base level rises or its discharge decreases, the stream fills the valley floor with sediment, creating an **alluvium-filled valley** (▶Fig. 17.21a–d). The surface of the alluvium becomes a broad floodplain.

In general, the size of a stream valley is proportional to the size of the stream—large streams produce large valleys, while small streams produce small valleys. Localities where a small stream flows in a large valley indicate that a larger stream produced the valley in the past, perhaps at a time when the local climate was wetter. Many **oversized stream valleys** formed at the end of the last ice age, when huge rivers carrying meltwater from the vast ice sheets cut large valleys.

Rapids and Waterfalls

When Lewis and Clark struggled up the Missouri River, they came to reaches that could not be navigated by boat because of **rapids,** particularly turbulent water with a rough surface (▶Fig. 17.22a–c). In rapids, the water depth is less than about twice the diameter of the coarsest sediment in the stream bed. When water flows over a submerged clast, it rises over the upstream side of the clast and plunges into a depression on the downstream side. If water collides with a protruding clast, it diverts to the side and may form a whirlpool on the downstream side. The turbulence resulting from flow over such a rough stream bed creates eddies and waves that roil and churn the water surface, in the process creating whitewater, a mixture of bubbles and water. **Standing waves,** whose crest and trough remain in place as the water moves through the waves, form above submerged boulders. In the channels between boulders, water deepens and accelerates. Modern-day whitewater boaters and rafters thrill to the unpredictable movement of a boat caught in rapids (▶Fig. 17.23).

A **waterfall** forms where the gradient of a stream becomes so steep that the water literally free-falls down the stream bed. The energy of falling water may scour a depression, called a **plunge pool,** at the base of the waterfall. Some waterfalls are found where a stream crosses a resistant ledge of rock, one with sufficient strength to sustain a steep cliff. Some form at the boundaries between a hard pluton and softer country rocks, while others form as a result of faulting, either because a steep escarpment develops along a fault scarp, or because a fault juxtaposes hard and soft rock. Finally, waterfalls occur where glacial erosion deepens the valley of a trunk stream to the extent that side valleys no longer intersect the trunk valley. Where this happens, a

FIGURE 17.22 (a) Rapids in a creek. (b) A waterfall in Brazil. (c) A waterfall emerging from a hanging valley.

FIGURE 17.23 Whitewater boaters distinguish among different classes of rapids (I to V) by the velocity of water, the steepness of the stream, the size of standing waves, the size of obstacles, and the difficulty of navigating past obstacles. Here we see classes II and V. Class V rapids should only be navigated by experts!

Class II

Class V

side valley becomes a **hanging valley,** and its stream must flow over a waterfall to reach the trunk stream (Fig. 17.22c).

Though a waterfall may appear to be a permanent entity of the landscape, all waterfalls eventually disappear as headward erosion (the lengthening of the stream channel up its slope) slowly eats back the resistant ledge until the stream reaches grade (it deposits as much sediment as it removes). We can see a classic example of headward erosion in Niagara Falls, at the juncture of Lakes Erie and Ontario. As water flows from Lake Erie to Lake Ontario, it drops over a 55-m-high ledge of hard Silurian dolostone (the Lockport Dolostone), which overlies a weak shale. Erosion of the shale leads to undercutting of the dolostone. Gradually, the overhang of dolostone becomes unstable and collapses, with the result that the waterfall migrates upstream. Before the industrial age, the edge of Niagara Falls cut upstream at a average rate of 1 m per year, but since then, the diversion of water from the Niagara River into a hydroelectric power station has decreased the rate of headward erosion to half that. Nevertheless, at this rate, Niagara Falls will cut all the way back to Lake Erie in about 60,000 years (▶Fig. 17.24a–d).

Alluvial Fans and Braided Streams

In arid or semi-arid regions, ephemeral streams abruptly drop their sedimentary load at a mountain front, where the gradient suddenly decreases, creating a gently sloping apron of sediment (sand, gravel, and cobbles) called an **alluvial fan** (▶Fig. 17.25a). The stream then divides into a series of

FIGURE 17.24 (a) Niagara Falls exists because Lake Erie lies at a higher elevation than Lake Ontario. The Lockport Dolostone, a resistant layer, serves as a local base level for Lake Erie. The Niagara Escarpment is located along the outcrop belt of the dolostone, and Niagara Falls first formed where the outlet of Lake Erie flowed over the escarpment. With time, the falls have cut upstream, at about 1 m per year or less, creating Niagara Gorge. When the falls reach Lake Erie, the lake will drain. (b) This cross section of the falls shows how undercutting of the soft shale layers eventually causes the resistant layers of dolostone to break off at joints. (c) The American Falls of Niagara Falls. (d) The American Falls with no water, showing the escarpment and the jumble of dolostone blocks that have broken off. The water was diverted from the falls so that geologists could investigate the erosion rate.

FIGURE 17.25 (a) An alluvial fan in Death Valley, California; (b) a braided stream.

small channels that spread out over the fan. During particularly strong floods, sheetwash smoothes out the fan's sediment. Water sinks into the porous sediment, so that the flow downstream of the fan is minimal.

Permanent streams flowing away from the foot of a mountain or draining a pile of sediment deposited by a glacier do not leave their sedimentary load in an alluvial fan, but rather carry large quantities of loose, coarse sediment out onto a plain, especially during floods. When the stream's gradient decreases, and/or when its discharge decreases, the sediment settles out. As a result, the channel becomes choked with sediment, and the stream splits into numerous strands weaving back and forth between elongate bars of gravel and sand. The result, as we saw earlier, is a **braided stream,** a network of channels that entwine like strands of hair (▶Fig. 17.25b). Because the gravelly sediment of a braided stream can't stick together, the stream cannot cut a deep channel with steep banks—the channel walls simply collapse, so the stream spreads out over a broad area.

Meandering Streams and Their Floodplains

A riverboat sailing along the lower reach of the Mississippi River cannot sail in a straight line, for the river channel winds back and forth in a series of snake-like curves called **meanders** (▶Fig. 17.26a). In fact, the boat has to cruise 500 km along the river channel to travel 100 km as the crow flies. **Meandering streams** like the lower Mississippi form where running water travels over a broad floodplain underlain by a soft substrate. The development of meanders increases the volume of the stream by increasing its length.

How do meanders form and evolve? Even if a stream starts out with a straight channel, random turbulence in the stream bed causes the fastest-moving current to swing back and forth from side to side. The water erodes the side of the stream more effectively where it flows faster, so it begins to cut away faster on the outer arc of the curve. Thus, each curve begins to migrate sideways and grow more pronounced until it becomes a meander. On the outside edge of a meander, erosion continues to eat away at the channel wall, creating a **cut bank,** while on the inside edge, water slows down so that its competence decreases and sediment accumulates, creating a wedge-shaped deposit called a **point bar.** (Mark Twain, who served as a riverboat pilot on the Mississippi River besides writing such books as *Huckleberry Finn,* took his pen name from the signals the mate of an old paddle-wheel steamer called out to the skipper to indicate water depth—"mark twain" means two fathoms, or about 4 m deep—so that the skipper could avoid running the boat aground on point bars.) With continued erosion, the meander may curve through more

than 180°, so that the cut bank at the meander's entrance approaches the cut bank at its end, leaving a **meander neck,** a narrow isthmus of land separating the portions of the meander.

People building communities along a riverbank may assume that the shape of a meander remains fixed for a long time—it doesn't. In a natural meandering river system, the river channel migrates back and forth across the floodplain. When erosion eats through a meander neck, a straight reach called a **cutoff** develops. The meander that has been cut off is called an **oxbow lake** if it remains filled with water, or an **abandoned meander** if it dries out (▶Fig. 17.26b).

Most meandering stream channels cover only a relatively small portion of a broad floodplain (▶Fig. 17.26c). Since many of the world's major cities border meandering streams in floodplains, it's worth understanding a few details of their landscape. In many cases, a floodplain terminates at its sides along a **bluff,** or escarpment. During a flood, water spills out from the stream channel onto the floodplain, and large floods may cover the entire region from bluff to bluff. As the water leaves the channel, friction between the ground and the thin sheet of water moving over the floodplain slows down the flow. This slowdown decreases the competence of the running water, so sediment settles out along the edge of the channel. Over time, the accumulation of this sediment creates a pair of low ridges, called **natural levees,** on either side of the stream. Natural levees may grow so large that the floor of the channel may become higher than the surface of the floodplain. In places where large natural levees exist, the region between the bluffs and the levees becomes a low, marshy **back swamp.** Also because of the levees, small tributaries may be blocked from joining the trunk stream; the tributaries, called **yazoo streams,** run in the floodplain parallel to the main river.

Deltas: Deposition at the Mouth of a Stream

At the mouth of the Nile in North Africa, the river divides into a fan of small streams called **distributaries,** and the stripe of fertile alluvium broadens into a triangular patch. The Greek historian Herodotus noted that this triangular patch resembles the shape of the Greek letter delta (Δ), and so the region became known as the Nile Delta.

Deltas develop where the running water of the stream enters standing water, the current slows, the stream loses competence, and sediment settles out (see Fig. 7.30). Geologists refer to any wedge of sediment formed at a river mouth as a delta, even though relatively few have the triangular shape of the Nile Delta (▶Fig. 17.27a–d). Some deltas define arc-like lobes, while others consist of many elongate lobes that protrude into the sea; the latter are

FIGURE 17.26 (a) A meandering stream. (b) Erosion occurs faster on the outer bank of a stream's curve, while deposition takes place on the inner curve. The meander becomes a progressively tighter curve until the stream cuts through the meander neck and forms a cutoff, thereby isolating an oxbow lake. (c) The landforms of a meandering stream. The detail of a stream channel shows a natural levee, the structure of a point bar, and floodplain deposits. Note that alluvium below the present-day channel includes ancient channels and point bars, surrounded by fine-grained floodplain deposits.

FIGURE 17.27 (a) The Nile is a Δ-shaped delta. (b) The Niger is an arc-like delta. (c) The Mississippi is a bird's-foot delta. (d) In this satellite photo of the end of the active tip of the Mississippi Delta, you can see the sediment beneath the surface of the water.

called **bird's-foot deltas,** because they resemble the scrawny toes of a bird. The existence of several toes indicates that the main course of the river in the delta has shifted on several occasions. These shifts occur when a toe builds so far out into the sea that the slope of the stream becomes too gentle to allow the river to flow. At this point, the river overflows a natural levee upstream and begins to flow in a new direction, an event called an **avulsion.** The distinct lobes of the Mississippi Delta, a bird's-foot delta, suggest that avulsion has happened several times during the past 9,000 years (▶Fig. 17.28a). New Orleans, built along one of the Mississippi's distributaries, may eventually

Delta deposits	Deposit age (years)
Ⓐ	4,400–5,300
Ⓑ	3,600–4,600
Ⓒ	2,700–3,900
Ⓓ	2,200–2,800
Ⓔ	700–900
Ⓕ	500–1,200
Ⓖ	0–500

(a)

(b)

FIGURE 17.28 (a) The map shows the different, dated lobes of the Mississippi Delta and the different channels that served as their source. Inset shows a current view of the delta, relative to Louisiana. A major flood could divert the main flow of the Mississippi into the channel of the Atchafalaya River, in which case a new delta would form to the west of the Mississippi's present mouth ("Future"), and New Orleans would no longer border a major river. Efforts by the U.S. Army Corps of Engineers have so far prevented this switch. (b) When a stream enters standing water, it deposits more sediment in the center of the channel than along the margins because the formerly fast-moving water at the center carried more sediment. The deposit builds a midstream bar, separating the stream into two distributaries. The same process happens at the mouth of each distributary, leading to further subdivisions into progressively smaller channels.

lose its riverfront, for a break in a levee upstream of the city could divert the Mississippi into the Atchafalaya River channel.

The shape of a delta depends on the quantity of sediment provided by the river, the current strength of the river, the degree to which ocean tides redistribute sediment, and the degree to which ocean currents flowing along the coast carry away sediment. Deltas that form where the strength of the river current exceeds that of ocean currents have a bird's-foot shape, since the sediment can be carried far offshore. In contrast, deltas that form where the ocean currents are strong have a Δ shape, for the ocean currents erode away lobes and redistribute them in bars running parallel to the shore. Deltas that develop where tides rework the sediment extensively evolve into arc-like lobes. And in places where waves and currents are strong enough to remove sediment as fast as it arrives, a river has no delta at all.

With time, the sediment of a delta compacts and the land beneath the delta sinks, so the delta becomes a low swampland called a **delta plain,** across which distributaries (known as bayous in Louisiana) sluggishly flow. In tropical climates, mangrove trees, which can grow in shallow tidal flats, cover the seaward edge of the delta plain.

Why do rivers divide into distributaries at their mouth? When a river reaches standing water, its velocity slows down and sediment settles out. If the river carries a sizable sediment load, and coastal erosion by waves does not carry all the sediment away, the sediment settles at the mouth to form a midstream bar. The presence of the bar causes the stream to split into two channels. Similar bars created at the mouths of these two subsidiary channels cause each to separate in turn, until eventually numerous distributary channels have formed (▶Fig. 17.28b).

River Systems

Rivers, or streams, drain the landscape of surface runoff. Typically, an array of connected streams called a drainage network develops, consisting of a trunk stream into which numerous tributaries flow. The land drained is the network's watershed. A stream starts from a source, or headwaters, in the mountains, perhaps collecting water from rainfall or from melting ice and snow. In the mountains, streams carve deep, V-shaped valleys, and tend to have steep gradients. For part of its course, a river may flow over a steep, bouldery bed, forming rapids. Locally, it may drop off an escarpment, creating a waterfall. Rivers gradually erode landscapes and carry away debris, so after a while, if there is no renewed uplift, mountains evolve into gentle hills. Through time, rivers can bevel once-rugged mountain ranges into nearly flat plains.

Transportation along the channel

Rapids

Braided channel

Cut bank

Meandering stream

Deposition

Terraced floodplain

Bank erosion

(present floodplain)

(oldest floodplain)

Deposition of point bar

Back swamps

Yazoo stream

Wide meanders

Neck

Oxbow lake

Natural levees

Cutoff

Wide floodplain

Collection of water in watershed

Headward erosion

Glaciers

Valleys with high relief

Melting ice

Lake

Dendritic drainage

Waterfall

Farther along its length, the river emerges from the mountains. If it is choked with sediment, it may split into numerous entwined channels separated from one another by gravel bars, creating a braided stream. Where a stream that is not choked by sediment flows over flat ground, it becomes a meandering stream, winding back and forth in snake-like curves called meanders. The current flows faster on the outer arc of a curve, so erosion takes place there, while the current flows more slowly on the inner arc, where it drops sediment. Because of erosion and deposition, a meandering stream changes shape over time. Occasionally a meander may be cut off, leaving a curving lake called an oxbow lake. A broad floodplain, covered with water only during floods, may develop on either side of the stream. Natural levees build up between the channel and the floodplain from sediment dropped as a flooding river starts to spill out of its channel. Eventually, a river reaches a standing body of water and slows down, and the sediment it carries gets deposited to form a delta. On a delta, the trunk stream divides into many smaller channels called distributaries.

Deposition at mouth

Delta

Distributaries

Natural levees

Swamps and marsh

Tidal flats

Bar

Banks

17.7 THE EVOLUTION OF DRAINAGE

Beveling Topography: Forming and Rejuvenating Peneplains

Imagine a place where continental collision uplifts a mountain range. In regions above the snow line, glaciers form and start to carve into the land (as we'll see in Chapter 22), but at lower elevations, rivers downcut rapidly, creating deep valleys. At first, rivers have steep gradients and drop over many rapids and waterfalls. But with time, mass-movement processes begin to wear down the mountains—rugged mountains become low, rounded hills, and once deep, narrow valleys broaden into wide floodplains, with less steep gradients. As more time passes, even the low hills are beveled down, becoming small mounds or even disappearing altogether. Ultimately, the streams meander across a broad surface that has hardly any relief at all. This nearly flat surface, which geologists refer to as a **peneplain** (from the Latin *pene*, which means "almost"), lies at an elevation close to that of a stream's base level. The original fluvial landscape thus changes or evolves through time, and extensive **denudation** (the removal of rock and regolith from the Earth's surface) occurs (▶Fig. 17.29a–d).

This image of a changing landscape was promoted by an American geomorphologist, William Morris Davis (1850–1934), at the end of the nineteenth century. Davis had been greatly influenced by Darwin's theory of biological evolution, and thought that a similar concept could apply to landscapes. According to Davis, rock or regolith at a high elevation in a mountain range has potential energy. Erosion transfers this material down to lower elevations, where it has less potential energy. When a landscape has been beveled so that it is a peneplain whose elevation approaches that of a stream's base level, it has the least potential energy. Davis's model is an oversimplification—plate tectonics can uplift the land again, and/or global sea-level change can lower the base level, so in reality peneplains rarely form before downcutting begins again. In many places around the world, streams *have* downcut into floodplains or peneplains. Geologists refer to such renewed downcutting as **stream rejuvenation.** Rejuvenation happens when the base level of a stream drops, when land is uplifted beneath a stream, or when the discharge of a stream increases. What is the evidence that rejuvenation has taken place? In the case of a stream flowing in an alluvium-filled valley, renewed downcutting allows the stream to create a new floodplain at a lower elevation than the original one (Fig. 17.21c). The younger floodplain tends to be narrower than the older, and the surface of the older floodplain becomes a **terrace** on either side of the new floodplain. In the case of a stream flowing on a peneplain, a drop of the base level causes the stream to downcut into the

Time 1:
Swampy, low-relief land
(a)

— Base level

Time 2:
Well-drained land
(b)

Time 3:
Valleys become broader.
(c)

— Reference plane

Time 4:
A new, low-relief landscape
(d)

FIGURE 17.29 (a) A fluvial landscape is first uplifted, so that the base level lies at a lower elevation than does the stream channel. (b) At first, the stream cuts down into the plain, leaving remnants of the plain as flat-topped mesas between valleys. (c) Later, the landscape consists of rounded hills dissected by tributaries that feed a trunk stream flowing on a floodplain near the base level. Valleys are V-shaped. (d) Still later, only a few remnant hills are left, for most of the landscape has been denuded to form a new peneplain, nearly at sea level. The height of the reference plane indicates the thickness of land that has eroded away.

FIGURE 17.30 The incised canyons of Canyonlands National Monument, in Utah, were created by stream rejuvenation.

peneplain. Initially, the landscape develops incised meanders. As time passes, so much of the old land surface erodes away that plateaus, hilltops, and ridge crests are all that remain of the peneplain. But the tops of these features all lie at the same elevation, for they once were part of the peneplain (▶Fig. 17.30).

Stealing Drainage: Stream Piracy

"Stream piracy" sounds like pretty violent stuff. In reality, **stream piracy,** or **stream capture,** simply refers to a situation in which headward erosion causes one stream to intersect the course of another, previously independent, stream. When this happens, the pirate stream "captures" the water in the stream it intersects, meaning that the captured stream starts flowing into the pirate stream (▶Fig. 17.31a, b). The trunk of the captured stream may dry up entirely downstream of the capture point, or it may become an undersized stream with respect to the valley in which it is flowing. The piracy of a stream that had been flowing through a water gap transforms the water gap into a **wind gap,** a dry pathway through a high ridge. In 1775, pioneer Daniel Boone blazed the "wilderness road" through the Cumberland Gap, a wind gap in the Appalachian Mountains at the border of Kentucky and Virginia, to provide other settlers with access to the Kentucky wilderness.

The Effects of Plate Tectonics and Climate Changes on Rivers

Earth is unique among the planets in having currently active drainage networks. Surface drainage couldn't exist at all were it not for plate tectonics, which generates volcanism and thus the gases from which water forms, and which

generates the gradients down which streams flow. In some cases, plate tectonics can even change the course of mighty rivers. For example, in the early Mesozoic era, when South America linked to Africa on its east coast, a "proto-Amazon" River flowed westward and drained the interior of Gondwana into the Pacific Ocean. Later, when South America separated from Africa, the Andes rose on South America's west coast. This event caused a **drainage reversal**—flow in the Amazon changed direction and began carrying water eastward to the newly formed Atlantic (▶Fig. 17.32a, b).

Movement on faults has a profound effect on drainage. For example, faulting can either raise or lower a stretch of river and thus change local base levels, causing streams to be filled with alluvium or to become rejuvenated, and in many cases actually determining the location of drainage. Major rivers like the Amazon, Mississippi, and Congo flow down the axis of ancient rifts (belts of stretched crust), for

FIGURE 17.31 (a) A drainage divide separates the Hades River drainage from the Persephone River drainage. Headward erosion by the Hades River gradually breaches the drainage divide, creating a water gap. (b) When the source of the Hades River reaches the channel of the Persephone River, Hades (the pirate stream) captures Persephone and carries off its water to the Styx Sea. As a result, the former channel of the Persephone River becomes a dry canyon.

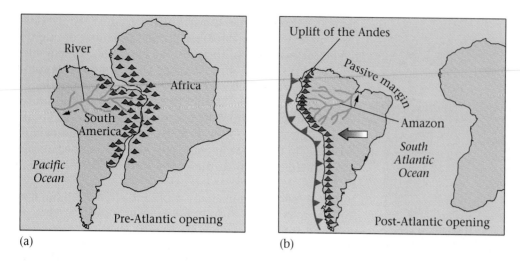

(a)   (b)

FIGURE 17.32 (a) In the early Mesozoic era, highlands existed along the boundary between Africa and South America, two continents that had collided during the formation of Gondwana. A river drained westward across South America to the Pacific. (b) In the late Mesozoic era, when the South Atlantic Ocean formed, South America began to drift westward. The Pacific Ocean floor was subducted beneath the continent, causing uplift of the Andes. When that happened, the Amazon River and its tributaries reversed course and drained water to the Atlantic. The river follows the axis of a rift.

with time, rifts cool and sink below the surrounding un-stretched crust, developing into an elongate depression.

Climate variations affect drainage in two ways. First, a stream's discharge increases in wetter climates and decreases in drier climates; an increase in discharge allows a stream to clear its channel of sediment and start downcutting, while a decrease allows the channel to fill with sediment. An extreme change to a drier climate can make a drainage network dry up altogether. Similarly, a change to a wetter climate may cause new streams to form. Second, climate changes also affect the base level of streams. For example, during ice ages, water becomes trapped as glacial ice on land, and the sea level drops. This drop rejuvenates streams on continents and leads them to grow in the direction of their mouth so that they downcut into the continental shelf. A rise in sea level causes stream valleys to fill with alluvium.

17.8 RAGING WATERS

[And Enhil, the ruler of the gods, said,] "The earth bellows like a herd of wild oxen. The clamor of human beings disturbs my sleep. Therefore, I want Adad [god of the skies] to cause heavy rains to pour down upon the Earth, both day and night. I want a great flood to come like a thief upon the Earth, steal the food of these people and destroy their lives."

—From the *EPIC OF GILGAMESH* (written in Sumeria, c. 2100 B.C.E.)

The Inevitable Catastrophe

As we saw at the beginning of this chapter, floods can be catastrophic—they can strip land of inhabitants and buildings, they can bury land in mud and silt, and they can submerge cities. A flood occurs when the volume of water flowing down a stream exceeds the volume of the stream channel, so water rises out of the normal channel and spreads out over the floodplain or delta plain (by breaking through levees), or it fills a canyon to a greater depth than normal. The news media may report that a river "crested at 9 feet [3 m] above **flood stage** at 10 P.M."—this means that at 10, the water surface in the stream was 3 m higher than the top of the normal channel, and after that time became lower. Because of its increased discharge, a stream in flood flows faster than it normally does, so it's more turbulent, has greater competence, and exerts more pressure on structures in its path. Muddy, viscous, fast-moving floodwaters can buoy and transport large quantities of sediment, as well as cars, buildings, and people.

Floods happen (1) during abrupt, heavy rains, when water falls on the ground faster than it can infiltrate; (2) after a long period of continuous rain, when the ground has become saturated with water and can absorb no more; (3) when heavy snows from the previous winter melt rapidly in response to a sudden hot spell; or (4) when an artificial or natural dam holding back a lake suddenly collapses and the contents of the lake suddenly flow downstream. All these conditions create excess runoff, which flows into the stream. Floods that occur regularly when rainfall is particularly heavy or when winter snows start to melt are called **seasonal floods.** Severe floods of this type take place in tropical regions that are drenched by monsoons. During the 1990 monsoon season of Bangladesh, for example, rain fell almost continuously for weeks and the delta plain became inundated; the resulting flood killed 100,000 people.

The floods of Bangladesh can also be called **delta-plain floods** because the water submerges the delta plain. **Floodplain floods** submerge a floodplain. Typically, such floods take time—hours or days—to develop. Thus, in many cases, authorities can evacuate potential victims and organize efforts to protect property. Nevertheless, so many people live on floodplains that these floods can cause a staggering loss of life and property. A 1931 flood of the Yangtze River in China led to a famine that killed 3.7 million people, and an 1887 flood of China's Hwang (Yellow) River (so named because of the yellow silt it carries) killed as many as 2.5 million. More recently, the flooding of Italy's Arno River submerged the treasures of Florence, and the flooding of North Dakota's Red River submerged the town of Fargo and threatened Winnipeg, Manitoba.

Events during which the floodwaters rise so fast that it may be impossible to escape from the path of the water are called **flash floods.** These happen during unusually intense rainfall or as a result of a dam collapse (as in the 1889 Johnstown flood). During a flash flood, a wall of water may slam downstream with great force, leaving devastation in its wake, but the floodwaters subside after a short time. Flash floods can be particularly unexpected in arid or semi-arid climates, where isolated thundershowers may suddenly fill the channel of an otherwise dry wash (a dry ephemeral stream bed), whose unvegetated ground can absorb little water; such a flood may even affect areas downstream that had not received a drop of rain.

A Floodplain Flood (Midwestern United States)

In the spring of 1993, the **jet stream,** the high-altitude (10–15 km high) wind current that controls weather systems, drifted southward (see Chapter 20). For weeks, the jet stream's cool, dry air formed an invisible wall that trapped warm, moist air from the Gulf of Mexico over the central United States. When the gulf air rose to higher elevations, it cooled, and the water it held condensed and fell as rain, rain, and more rain. In fact, almost a whole year's supply of rain fell in just that spring—some regions received 400% more than usual. Because the rain fell over such a short period, the ground became saturated and could no longer absorb additional water, so the excess entered the region's streams, which carried it into the Missouri and Mississippi Rivers. Eventually, the water in these rivers rose above the height of levees and spread out over the floodplain. By July, parts of nine states were underwater (▶Fig. 17.33a, b).

The roiling, muddy flood uprooted trees, cars, and even coffins (which floated up from inundated graveyards). All barge traffic along the Mississippi came to a halt, bridges and roads were undermined and washed away, and towns along the river were submerged in muddy water (▶Fig. 17.33c). For example, in Davenport, Iowa, the riverfront district and baseball stadium were covered with 4 m (14 feet) of water, and in Des Moines, Iowa, 250,000 residents lost their supply of drinking water when floodwaters contaminated the municipal water supply with raw sewage and chemical fertilizers. Rowboats replaced cars as the favored mode of transportation in towns where only the rooftops remained visible. In St. Louis, Missouri, the river crested 14 m (47 feet) above flood stage. When the water finally subsided, it left behind a thick layer of silt and mud, filling living rooms and kitchens in floodplain towns and burying crops of floodplain fields (▶Fig. 17.33d).

For seventy-nine days, the flooding continued. In the end, more than 40,000 square km of the floodplain had been submerged, fifty people died, at least 55,000 homes were destroyed, and countless acres of crops were buried. Officials estimated that the flood caused over $12 billion in damage. Once the waters had receded, the inhabitants of some ruined towns chose to rebuild in a new location on higher ground, and some regions passed legislation preventing further development on the floodplain. Yet many who had lost everything insisted on returning to the floodplain. Ultimately, another flood will come to remind us yet again that floodplains are for floods, not for houses.

A Flash Flood (Big Thompson Canyon)

On a typical sunny day in the Front Ranges of the Rocky Mountains, north of Denver, the Big Thompson River seems quite harmless. Clear water, dripping from melting ice and snow higher in the mountains, flows down its course through a narrow canyon, frothing over and around boulders. In places, vacation cabins, campgrounds, and motels line the river, for the pleasure of tourists escaping the heat of Denver. The landscape seems immutable, but, as is the case for so many geologic features, permanence is an illusion.

On July 31, 1976, easterly winds blew warm, moist air from the Great Plains toward the Rocky Mountain front. As this air rose over the mountains, towering thunderheads built up, and at 7:00 P.M. rain began to fall. It poured, in quantities that even old-timers couldn't recall. In a little over an hour, 7.5 inches (19 cm) drenched the watershed of the Big Thompson River. Water, unable to sink into the ground fast enough, flowed over the ground down to the river. The river's discharge grew to more than four times the maximum recorded at any time during the previous

FIGURE 17.33 (a) The Mississippi and Missouri Rivers during a time of drought as seen from high elevation. (b) The same area during the 1993 flood. Notice how the floodplains of the rivers are totally submerged. (c) Great Falls, Montana, submerged by floodwaters. (d) After floodwaters recede (here, from an orchard in Arizona), they leave a layer of mud and silt.

century. The river rose quickly, in places reaching depths several meters above normal. Turbulent water swirled down the canyon at up to 8 m per second and churned up so much sand and mud that it became a viscous slurry. Slides of rock and soil tumbled down the steep slopes bordering the river and fed the torrent with even more sediment. The water undercut house foundations and washed the houses away, along with their inhabitants. Roads and bridges dis-appeared (▶Fig. 17.34). Boulders that had stood like land-marks for generations bounced along in the torrent like beachballs, striking and shattering other rocks along the way; the largest rock known to be moved by the flood weighed 275 tons. Cars drifted downstream until they fi-nally wrapped like foil around obstacles. When the flood subsided, the canyon had changed forever, and 144 people had lost their lives.

FIGURE 17.34 During the 1976 Big Thompson River flood, this house was carried off its foundation and dropped on a bridge.

FIGURE 17.35 (a) The map illustrates Glacial Lake Missoula, blocked by an ice dam. When the dam melted away and finally broke, the lake drained westward, eroding away all sediment cover and leaving behind scoured basalt—a region now known as the channeled scablands. (b) The channeled scablands, in Washington State.

(a)

(b)

Ice-Age Floods (the Great Missoula Flood)

Perhaps the greatest floods recorded in the geologic record happen when natural dams burst. The Great Missoula Flood of about 11,000 years ago illustrates this phenomenon (see Box 22.2). This flood occurred at the end of the last ice age, when a glacier holding back a large lake, Glacial Lake Missoula, finally broke. Over the next few days, the lake's water roared over eastern Washington, eventually entering the Columbia River Valley and flowing on out to the Pacific Ocean. This flood created the **channeled scablands** of eastern Washington, a region where the soil and regolith have been stripped off the land surface, leaving barren, craggy rock (▶Fig. 17.35a, b).

The hypothesis that the channeled scablands formed as a consequence of catastrophic flooding was first proposed by J Harlan Bretz, who studied the landscape of the region in the 1920s. Initially, other geologists ridiculed Bretz because his idea seemed to violate the well-accepted principle of uniformitarianism (see Chapter 12). But Bretz steadily fought back, demonstrating that the scablands, an unglaciated region, are littered with boulders too large to have been carried by normal rivers, that hills in the region were giant ripples a thousand times larger than the ripples typically found in a stream bed, that the now-dry Grand Coulee was once a giant waterfall hundreds of times larger than Niagara Falls, and that deep pits in the region are actually huge potholes scoured by whirlpools. Ultimately the geologic community accepted the reality of the Great Missoula Flood and other such catastrophic events during Earth history.

Flood Prediction (and Prevention?)

Mark Twain once wrote that we "cannot tame that lawless stream, cannot curb it or confine it, cannot say to it, 'go here or go there,' and make it obey." Was Twain right? For over six decades, from the passage of the 1927 Mississippi River Flood Control Act (drafted after a disastrous flood took place that year), the U.S. Army Corps of Engineers

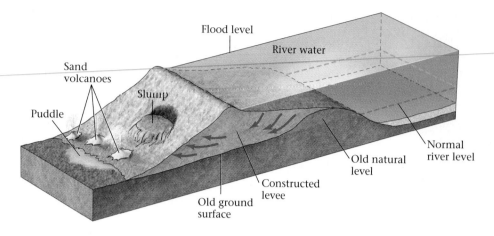

FIGURE 17.36 When the water level on the river side of the levee is much higher than on the dry floodplain, pressure causes water to infiltrate the ground and flow through this artificial levee. The water spurts out of the ground on the dry side of the levee, generating sand volcanoes. Water saturates the sediment making up the levee, so the face of the levee slumps. The passage of water so weakens the levee's foundation that the levee eventually collapses.

worked to control the Mississippi. First, engineers built about 300 dams along the river's tributaries so that excess runoff could be stored in the reservoirs and later be released slowly. And second, they built thousands of kilometers of sand and mud levees and concrete flood walls to increase the channel's volume. A typical **artificial levee,** which resembles the letter C, isolates a discrete area of the floodplain (▶Fig. 17.36).

But although the Corps' strategy worked for floods up to a certain size, it was insufficient to handle the 1993 flood. Because of the volume of water drenching the Midwest during the spring and early summer of that year, the reservoirs filled to capacity, and all additional water headed downstream. When this happened, the river rose until it spilled over the top of some levees and undermined others. Undermining occurs when rising floodwaters increase the water pressure on the river side of the levee, forcing water through sand under the levee. In susceptible areas, water begins to spurt out of the ground on the dry side of the levee, thereby washing away the levee's support. The levee finally becomes so weak that it collapses, and water fills in the area behind it.

Sooner or later a flood comes along that can breach a river's levees, allowing water to spread out over the floodplain. Defensive efforts merely delay the inevitable, for it is unfeasible and too expensive to build levees high enough to handle all conceivable floods. And in some cases, building levees may be counterproductive, since they constrain water to a smaller area and thus make floodwaters rise to a higher level than they would if they were free to spread over a wide floodplain. Those who build on floodplains must

face this reality and consider alternative ways to use the region that can accommodate occasional flooding. The cost of flood damage has quadrupled in recent years, despite the billions of dollars that have been spent on flood "control," because more people have settled in floodplains. Perhaps policies that encourage rebuilding in floodplains should be reconsidered.

Meanwhile, there are other ways to prevent floods besides building levees and reservoirs. For example, transforming portions of floodplains back into natural wetlands helps prevent floods, for wetlands absorb water like a sponge. A solution to some flooding may thus lie in the removal, rather than the construction, of levees. Property may also be kept safe by first mapping **floodways,** regions likely to be flooded, and then by moving or abandoning buildings located there. Even the simple act of moving levees farther away from the river and creating natural habitats in the resulting floodways would decrease flooding damage immensely (▶Fig. 17.37a, b).

Geologists and the news media rank floods in terms of their **recurrence interval,** the average time between successive floods of a given size. This interval—we speak of a 30-year flood or a 500-year flood, for example—represents the likelihood that a flood of a given size will happen in a given year. On average, a 500-year flood happens once during a 500-year period, so a river has a 1-in-500 chance of flooding in a given year. Note that the recur-

FIGURE 17.37 Concept of a floodway. (a) Building artificial levees directly on natural levees creates a larger channel for a river. (b) Building artificial levees at a distance from the river creates a floodway on either side of the river, an even larger channel than in (a), and a surface of wetland that can absorb floodwaters.

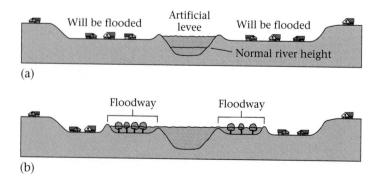

rence interval refers to the *average* time between comparable floods—a region may suffer a 500-year flood in two consecutive years. Some floodplain residents have been lured into thinking that their homes were safe because the 100-year flood happened last year, so it won't happen again "until long after I'm gone." But even if a 100-year flood did occur last year, that does *not* mean an equivalent or larger flood won't happen within the next 100 years—or even tomorrow.

The recurrence interval for a flood at a particular river depends on the size of the flood, as defined by its discharge.

To illustrate this relationship, geologists construct graphs that plot the recurrence interval on the horizontal axis and the flood discharge on the vertical axis (▶Fig. 17.38a–c). The larger the flood, the longer the recurrence interval and therefore the less likely the flood will occur in a given year. Government agencies and insurance companies use data on recurrence intervals to design flood-control facilities and to calculate insurance rates. If a levee has been constructed to withstand a 100-year flood, that means it can hold back a discharge equal to the discharge of a 100-year flood on a flood-frequency graph.

FIGURE 17.38 (a) A flood-frequency graph shows the relationship between the recurrence interval (the average time between floods of a given size) and discharge for an idealized river. The larger the flood (the greater the discharge), the longer the recurrence interval. (b) The peak discharge of the Mississippi River as measured at St. Louis, Missouri. Each bar represents the largest discharge of a given year. The horizontal line represents the discharge of a 100-year flood. Note that 100-year floods or larger occurred in 1844, 1903, and 1993. (c) A 100-year flood covers a larger area than a 2-year flood.

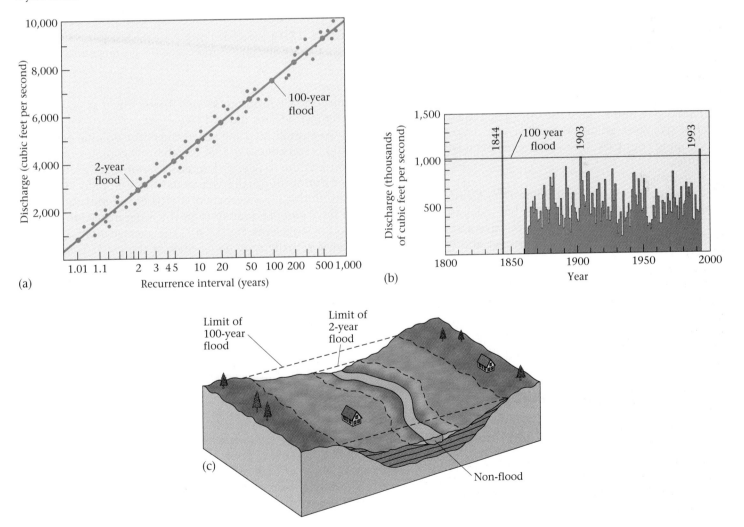

17.9 RIVERS: A VANISHING RESOURCE?

As *Homo sapiens* evolved from hunter-gatherers into farmers, rivers became attractive places to settle. Rivers serve as avenues for transportation and as sources for food, irrigation water, drinking water, power, recreation, and (unfortunately) waste disposal. Further, their floodplains provide particularly fertile soil for fields, replenished annually by seasonal floods. Considering the multitudinous resources that rivers provide, it's no coincidence that early civilizations gathered in river valleys and on floodplains: Mesopotamia sprung up around the Tigris and Euphrates Rivers, Egypt around the Nile, India in the Indus Valley, and China along the Hwang (Yellow) River. Over the millennia, rivers have killed millions of people in floods, but they have been the lifeblood for hundreds of millions more. Nevertheless, over time, humans have increasingly tended to abuse or overuse the Earth's rivers. Here we note three pressing environmental issues.

Pollution. The capacity of some rivers to carry pollutants has long been exceeded, transforming them into deadly cesspools (▶Fig. 17.39). Pollutants include raw sewage and storm drainage from urban areas, spilled oil, toxic chemicals from industrial sites, and excess fertilizer and animal waste from agricultural fields. Some pollutants directly poison aquatic life, some create algae blooms that strip water of its oxygen, and some settle out to be buried along with sediments. River pollution has become overwhelming in developing countries, where there are few waste-treatment facilities.

Dam construction. In 1950, there were about 5,000 large (over 15 m high) dams worldwide, but today there are over 38,000. Damming rivers has both positive and negative results. Reservoirs provide irrigation water and hydroelectric power, and they trap some floodwaters and create popular recreation areas. But sometimes their construction destroys "wild rivers" (the whitewater streams of hilly and mountainous areas) and alters the ecosystem of a drainage network by providing barriers to migrating fish, decreasing the nutrient supply to organisms downstream, removing the source of sediment for the delta, and eliminating seasonal floods that replenish nutrients in the landscape.

Overuse of water. Because of growing populations, our thirst for river water continues to increase, but the supply of water does not. The use of water has grown especially in response to the "Green Revolution" of the 1960s, during which huge new tracts of land came under irrigation. Today, 65% of the water taken out of rivers is used for agriculture, 25% for industry, and 9% for drinking and sewage

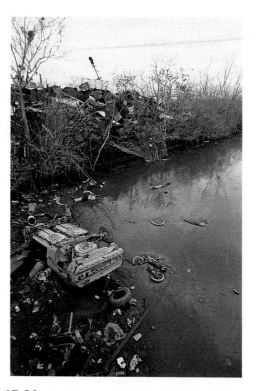

FIGURE 17.39 A polluted river. Note the bizarre color of the water, the oily film on its surface, and the floating garbage.

transport. Civilization needed three times as much river water in 1995 as it did in 1950.

As a result, in some places the entire volume of a river's water is consumed, so that the channel contains little more than a saline trickle, if that, at its mouth. For example, except during unusually wet years, the Colorado River contains almost no water where it crosses the Mexican border, for huge pipes and canals carry the water instead to Phoenix and Los Angeles (▶Fig. 17.40). The average discharge of the river is actually only about 90% of the water volume that cities and farmers have purchased. Perhaps the most dramatic consequence of river overconsumption can be seen in central Asia, where almost all the water in the Amu Darya and Syr Darya Rivers has been diverted into irrigation. These rivers once fed the Aral Sea, once the fourth-largest freshwater lake in the world. Now the sea has shrunk so much that "coastal" towns lie many kilometers from the coast, fishing trawlers rot in the desert (see Fig. 21.28c), and the catch of fish has dropped from 44,000 tons per year to zero tons per year.

The effects of urbanization and agriculture on discharge. While the consumption of water for agricultural and industrial purposes decreases the overall supply of river water, urbanization may actually increase the short-term supply. Cities cover the ground with impermeable concrete or

FIGURE 17.40 The Central Arizona Project canal, shunting water from the Colorado River to Phoenix.

blacktop, so rainfall does not soak into the ground but rather runs into storm sewers and then into streams, causing local flooding. Stream discharge during a rainfall thus increases much more rapidly than it would without urbanization (▶Fig. 17.41a, b). Similarly, while damming rivers decreases the amount of silt a river carries downstream, agriculture may increase the sediment supply: agriculture decreases the vegetative cover on the land, so that when it rains, soil washes into streams.

CHAPTER SUMMARY

- Streams are bodies of water that flow down channels and drain the land surface. Channels develop when sheetwash cuts into the substrate and concentrates the water flow; they grow by headward erosion. Some streams have straight reaches, others contain snake-like curves called meanders. Braided streams consist of many entwined channels. Streams carry water out of a drainage basin, or catchment; a drainage divide separates two adjacent catchments.

- Drainage networks consist of many tributaries that flow into a trunk stream. The networks are generally controlled by the structure of the substrate, but superposed and antecedent streams appear to ignore this structure.

- Permanent streams exist where the water table lies above the bed of the channel. Where the water table lies below the channel bed, streams are ephemeral and dry up between rainfalls to form dry washes.

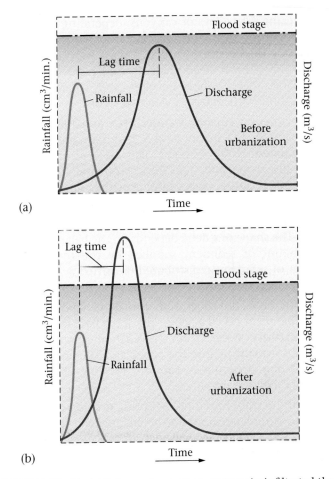

FIGURE 17.41 (a) Before urbanization, some rain infiltrated the ground, and vegetation slowed down sheetwash. As a consequence, the peak discharge resulting from a heavy rain was less, and the peak runoff occurred after some time (the lag time) had passed. (b) With urbanization, the peak discharge happens after a shorter lag time, and is great enough to reach flood stage (the level of the stream exceeds the top of the channel).

- The discharge of a stream is the volume of water passing a point in a second. Most streams are turbulent, meaning that their water swirls in complex patterns.

- Streams erode the landscape by scouring, lifting, abrading, and dissolving. The resulting sediment is divided among dissolved, suspended, and bed loads. The total quantity of sediment carried by a stream is its capacity. Capacity differs from competence, the maximum particle size a stream can carry. When stream water slows, it deposits fluvial sediment such as alluvium.

- Longitudinal profiles (the shape of a stream bed in cross section from its source to its mouth) of streams are concave up. Typically, a stream has a steeper gradients at its headwaters than near its mouth. Streams cannot cut below the base level. Sea level is the ultimate base level for a drainage network.

- Streams cut valleys or canyons, depending on the rate of downcutting relative to the rate at which the slopes on either side of the stream undergo mass wasting. Where a stream flows down steep gradients and has a bed littered with large rocks, rapids develop, and where a stream plunges off a vertical face, a waterfall forms.

- Meandering streams wander back and forth across a floodplain. Such a stream erodes its outer bank and builds out sediment into a point bar on the inner bank. Eventually, a meander may be cut off and turn into an oxbow lake. Natural levees form on either side of the river channel.

- Where streams flow into standing water, they deposit deltas. The shape of a delta depends on the balance between the amount of sediment supplied by the river and the amount of sediment redistributed or carried away by wave activity along the coast.

- With time, fluvial erosion can bevel landscapes to a nearly flat plain called a peneplain. If the base level drops or the land surface rises, stream rejuvenation causes the stream to start downcutting into the peneplain. The headward erosion of one stream may capture the flow of another, a process called stream piracy.

- If an increase in rainfall or spring melting causes more water to enter a stream than the channel can hold, a flood results. Some floods are seasonal, in that they accompany monsoonal rains. Some floods submerge broad floodplains or delta plains. Flash floods happen very rapidly. Officials try to prevent floods by building reservoirs and levees.

- Rivers are becoming a vanishing resource because of pollution, damming, and overuse. In some locations, the diversion of river water for agricultural uses and urban water supplies completely drains a river's channel.

KEY TERMS

abandoned meander (p. 526)
abrasion (p. 518)
alluvial fan (p. 524)
alluvium (p. 520)
antecedent stream (p. 511)
avulsion (p. 528)
base level (p. 521)
bed load (p. 520)
bird's-foot delta (p. 528)
braided stream (p. 511)
capacity (p. 520)
channeled scablands (p. 537)
competence (p. 520)
continental divide (p. 511)

cutoff (p. 526)
delta (p. 526)
delta-plain flood (p. 535)
dendritic network (p. 510)
discharge (p. 513)
dissolved load (p. 518)
downcutting (p. 509)
drainage basin (p. 511)
drainage divide (p. 511)
drainage network (p. 509)
drainage reversal (p. 533)
ephemeral stream (p. 513)
flash flood (p. 535)
floodplain flood (p. 535)
fluvial deposits (p. 520)

graded system (p. 522)
headward erosion (p. 509)
headwaters (p. 511)
longitudinal profile (p. 521)
meandering stream (p. 511)
natural levees (p. 526)
oxbow lake (p. 526)
peneplain (p. 532)
permanent stream (p. 513)
point bar (p. 526)
pothole (p. 518)
radial network (p. 510)
rectangular network (p. 511)
recurrence interval (p. 538)
runoff (p. 507)
saltation (p. 520)
scouring (p. 518)
seasonal flood (p. 535)

sediment load (p. 518)
sediment sorting (p. 520)
sheetwash (p. 509)
stream gradient (p. 511)
stream piracy (p. 533)
stream rejuvenation (p. 532)
superposed stream (p. 511)
suspended load (p. 518)
terrace (p. 532)
thalweg (p. 517)
trellis network (p. 511)
tributaries (p. 509)
trunk stream (p. 509)
turbulent flow (p. 518)
water gap (p. 511)
water table (p. 513)
wind gap (p. 533)
yazoo stream (p. 526)

REVIEW QUESTIONS

1. Why are continental divides often near active continental margins?

2. What is the continental divide in North America? Where does it run?

3. Describe the four different types of drainage networks. What factors are responsible for the formation of each?

4. How are superposed and antecedent drainages similar? How are they different?

5. What factors determine whether a stream is permanent or ephemeral?

6. How does discharge vary according to the stream's length, climate, and position along the stream course?

7. Why is average downstream velocity always less than maximum downstream velocity?

8. How does a turbulent flow differ from a laminar flow?

9. Describe how streams and running water erode the Earth.

10. What are three components of sediment load in a stream?

11. Distinguish between a stream's competence and its capacity.

12. Why is the average stream velocity actually slower in a turbulent mountain stream than in a broad flowing river farther downstream?

13. What factors determine the position of the base level?

14. What do lakes, rapids, and waterfalls indicate about the stream gradient and base level?

15. How does a braided stream differ from a meandering stream?

16. Describe how meanders form, develop, are cut off, and then abandoned.

17. Describe how deltas grow and develop.

18. What is stream piracy?

19. What human activities tend to increase flood risk and damage?

20. What is the recurrence interval of a flood? Why can't someone say that "the hundred-year flood happened last year, so I'm safe for another hundred years"?

21. How have humans abused and overused the resource of running water?

Restless Realm: Oceans and Coasts

18.1 INTRODUCTION

A thousand kilometers from the nearest shore, two scientists and a pilot wriggle through the entry hatch of the research submersible *Alvin,* ready for a cruise to the floor of the ocean and, hopefully, back. *Alvin* consists of a superstrong metal sphere embedded in a cigar-shaped tube (▶Fig. 18.1a). The sphere protects its crew from the immense water pressures in the deep ocean, and the tube holds motors and oxygen tanks. When the hatch seals, *Alvin* sinks at a rate of 1.8 km per hour mostly through utter darkness, for light penetrates only the top few hundred meters of ocean water. Upon reaching the bottom, at a depth of 4.5 km, the cramped explorers turn on outside lights to reveal a stark vista of loose sediment, black rock, and the occasional sea creature, and for the next five hours they photograph the vista and use a robot arm to collect samples. When finished, they release ballast and rise like a bubble, reaching the surface about two hours later.

Alvin dives began in the 1970s, but humans have explored the ocean for thousands of years, beginning with the Phoenician traders who had circumnavigated Africa by 590 B.C.E. and continuing with the incredible voyages of Polynesian sailors who used outrigger canoes to sail among South Pacific islands around 700 C.E. Mapmakers have known that the ocean spanned the entire globe since Ferdinand Magellan's round-the-world voyage of 1519–22, but could not systematically map the ocean until the late eighteenth century, when it became possible to determine longitude accurately by carrying a chronometer on board. Subsequently, naval officers gathered data on water depths in the ocean (using a **plumb line,** a lead weight on the end of a cable), and by 1839 the world knew that the greatest ocean depths could swallow the highest mountains without a trace.

A converted British navy ship, the HMS *Challenger,* made the first true ocean research cruise (▶Fig. 18.1b). On-board scientists spent four years, beginning in 1872, dredging rocks from the sea floor, analyzing water composition, collecting specimens of marine organisms, and measuring water depths and currents. But still our knowledge of the ocean remained spotty—in fact, we knew less about the ocean floor than we did about the surface of the Moon, for at least we could see the Moon with a telescope.

Over 70% of Earth's surface is covered by the oceans, which thus play a key role in the Earth system. Here, in the northwestern United States, we see the interface between land and sea, as the wind drives waves against the coast.

(a)

(b)

FIGURE 18.1 (a) The *Alvin,* a submersible used to explore the ocean floor. (b) The HMS *Challenger,* the first ship to undertake a cruise dedicated to ocean research.

FIGURE 18.2 (a) Researchers lowering sampling containers off the side of a modern oceanographic research vessel. (b) Sidescan sonar image of the sea floor (an area about 0.5 km wide) along a fracture zone in the Pacific, near the East Pacific Rise. You can see a flat, plateau-like surface along the top and a series of channels at the bottom. (c) In this traditional seismic-reflection profile of the sea floor, the darker lines represent the boundaries between sedimentary layers in the oceanic crust.

(a)

(b)

(c)

The field of **oceanography** (the study of ocean water and its movements), **marine geology** (the study of the ocean floor), and **marine biology** (the study of life forms in the sea) expanded rapidly in the latter half of the twentieth century, as new technology became available and a fleet of oceanographic research ships invaded the seas. These ships can tow instrument-laden "sleds" just above the sea floor; the sleds use sonar to generate detailed **bathymetric maps** illustrating the shape of the floor (▶Fig. 18.2a, b). Some ships send pulses of sound into the sea floor that reflect off layers in the subsurface and return to the ship to provide an image, called a **seismic-reflection profile,** of the layering in the oceanic crust (▶Fig. 18.2c). Other ships, like the *Joides Resolution,* drill holes as deep as 4 km into the sea floor and bring up samples of the oceanic crust. In addition, satellites circling the globe produce high-resolution maps of the ocean. And while ships and satellites research the open ocean, land-based geologists study its margins.

When seen from space, Earth glows blue, for the oceans cover 70.8% of its surface. The sea provides the basis for life, tempers Earth's climate, and spawns its storms. It serves as a vast reservoir for water and chemicals that cycle into the atmosphere and crust,

FIGURE 18.3 The bathymetric provinces of the sea floor. At a passive continental margin, a thick wedge of sediment accumulates in an ocean basin over continental lithosphere that had been stretched and thinned during the rifting that formed the basin. The flat surface of this sedimentary wedge creates a wide continental shelf. At an active continental margin, here a convergent plate boundary, a narrow continental shelf forms over an accretionary prism. Insets: These vertical slices through continental and oceanic lithosphere illustrate that oceanic lithosphere is thinner, and that the two kinds differ in composition.

and for sediment washed off the continents. In this chapter, we first learn the fundamental characteristics of ocean basins and seawater, and the role they play in the Earth system. Then we focus on the landforms that develop along the **coast,** the region where the land meets the sea (and where over 60% of the global population live today). Finally, we consider how to cope with the hazards of living on the coast.

18.2 LANDSCAPES BENEATH THE SEA

The oceans exist because oceanic lithosphere and continental lithosphere differ markedly from one another in terms of composition and thickness (▶Fig. 18.3; see Chapter 2). The surface of the denser and thinner oceanic lithosphere lies deeper than the surface of the relatively buoyant, thicker continental lithosphere, creating oceanic basins that collect water.

On the present-day map of the world, the ocean encircles the globe. For purposes of reference, however, cartographers divide the ocean into several major parts, with somewhat arbitrary boundaries and significantly dif-

ferent volumes (▶Fig. 18.4a). Curiously, most continental crust (81%) lies in the Northern Hemisphere today (▶Fig. 18.4b). Many smaller bays and seas link to the global ocean. Some of these, like the Mediterranean and Black Seas, consist of oceanic crust trapped between colliding continents, but others are underlain by different types of crust; for example, the North Sea and the northern Red Sea overlie rifted continental crust. The Sea of Japan and the Philippine Sea overlie back-arc basins (basins lying on the opposite side of a volcanic island arc from the trench), formed when sea-floor spreading occurs in the upper plate behind a convergent boundary.

Because of plate tectonics, the map of Earth's surface constantly changes. In fact, the oldest oceanic crust is only about 200 million years old, for subduction has consumed all older oceanic crust. Thus, though the percentage of Earth's surface area underlain by oceanic crust has probably remained much the same for the past 2 billion years, the present-day ocean basins have not always existed. For example, 200 million years ago, the Atlantic Ocean did not exist, and the Tethys Ocean lay to the east of Africa (see Chapter 13).

Have you ever wondered what the ocean floor would look like if all the water evaporated? Marine geologists can now provide a clear image of the ocean's **bathymetry,** or variation in depth, based on sonar measurements and more

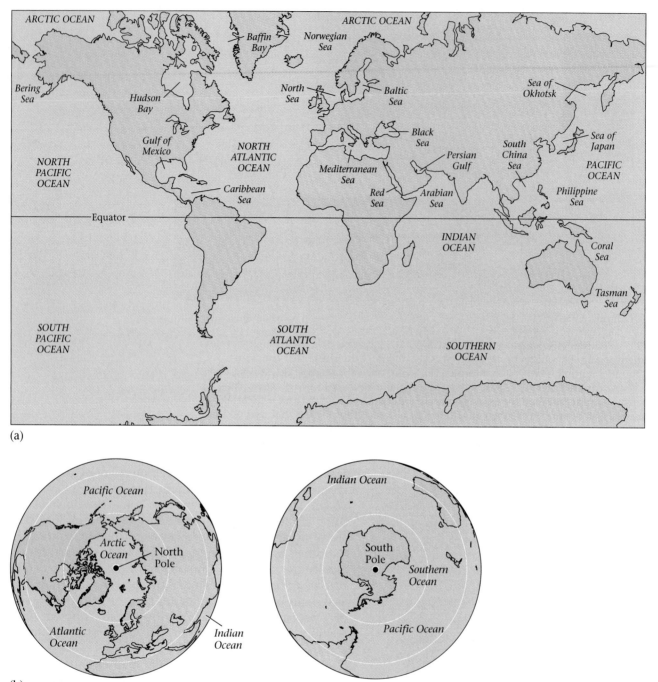

(a)

(b)

FIGURE 18.4 (a) The world's ocean basins and smaller seas. (b) The contrast between the "land hemisphere" and the "sea hemisphere." Most continental crust lies in the Northern Hemisphere.

recently on measurements made by satellites. Such studies indicate that the ocean contains broad **bathymetric provinces,** distinguished from one another by their water depth. Isolated submarine hills and mountains rise above these provinces. To learn how these bathymetric features formed, we look again to plate tectonics theory.

Continental Shelves, Slopes, and Rises

Imagine you're in a submersible cruising just above the floor of the western half of the North Atlantic. If you start at the shoreline of North America and head east, you will find that extending from the shoreline for about 200–500 km, a

continental shelf, a relatively shallow portion of the ocean in which water depth does not exceed 500 m, fringes the continent. Across the width of the shelf, the ocean floor slopes seaward at only about 0.3°, an almost imperceptible amount. At its eastern edge, the continental shelf stops at the **continental slope,** which descends down to depths of nearly 4 km at an angle of about 2°. From about 4 km down to about 4.5 km, a province called the **continental rise,** the angle decreases until at 4.5 km deep, you find yourself above a vast, nearly horizontal plain; the **abyssal plain.**

Broad continental shelves, like that of eastern North America, form along **passive continental margins,** margins that are not plate boundaries and thus host few earthquakes (▶Fig. 18.5a; see Chapter 4). Passive margins originate after rifting breaks a continent in two; when rifting stops and sea-floor spreading begins, the stretched crust at the boundary between the ocean and continent gradually cools and

sinks. Sediment washed off the continent, as well as the shells of marine creatures, buries the sinking crust, slowly creating a pile of sediment up to 20 km thick. The flat surface of this sedimentary pile comprises the continental shelf.

If you were to take your submersible to the west coast of South America and cruise out into the Pacific, you would find a very different continental margin. After crossing a narrow continental shelf, the sea floor falls off at the relatively steep angle of 3.5° down to a depth of over 8 km. South America does not have a broad continental shelf because it is an **active continental margin,** a margin that coincides with a plate boundary and thus hosts many earthquakes (▶Fig. 18.5b). In the case of South America, the edge of the Pacific Ocean is a convergent plate boundary. The narrow shelf along a convergent plate boundary forms where an apron of sediment spreads out over the top of an accretionary prism, the pile of material scraped off the

FIGURE 18.5 (a) This digital bathymetric map (constructed by computer from data on water depth) shows the surface of a passive continental margin, here a portion of the east coast of North America, with a broad continental shelf. (b) A digital bathymetric map of an active continental margin, here the subduction zone on the west coast of South America. (c) The sea floor slopes away from a mid-ocean ridge and gradually flattens out to become an abyssal plain. Sediment increases in thickness away from the ridge axis, because the sea floor gets older as it moves away from the ridge axis. (d) A digital bathymetric map of a segment of a mid-ocean ridge, showing transform faults that link segments of the ridge.

(a)

(b)

(c)

(d)

downgoing subducting plate. Here, the continental slope corresponds to the face of the accretionary prism.

At many locations, relatively narrow and deep valleys called **submarine canyons** dissect continental shelves and slopes (►Fig. 18.6a). The largest submarine canyons start offshore of major rivers, and for good reason: rivers cut into the continental shelf at times when the sea level was low and the shelf was exposed. But river erosion cannot explain the total depth of these canyons—some slice almost 1,000 m down into the continental margin, far greater than the maximum sea-level change. Submarine exploration demonstrates that much of the erosion of submarine canyons results from the flow of **turbidity currents,** avalanches of sediment mixed with water (see Chapter 9). Turbidity currents speed down the floor of a canyon at up to 60 km an hour, eroding as they go. When they finally reach the base of the continental slope, **turbidites** (composed of graded beds) accumulate and build up into a **submarine fan** (►Fig. 18.6b).

The Bathymetry of Oceanic Plate Boundaries

All three types of plate boundaries—divergent, convergent, and transform—affect the bathymetry of the ocean floor. Sea-floor spreading at a divergent boundary yields a **mid-ocean ridge,** a 2-km-high submarine mountain belt (►Fig. 18.5c). Because crust stretches and breaks as sea-floor spreading continues, the axis of a ridge includes escarpments, created by normal faulting. Oceanic transform faults, created by the strike-slip shear of one plate past another, typically link segments of mid-ocean ridges (►Fig. 18.5d). Transforms consist of narrow belts of steep escarpments that connect, along their length, with **fracture zones** that can be traced into the oceanic plate away from the ridge axis. Subduction

at convergent boundaries yields a **trench,** a deep, elongate trough bordering a volcanic arc. Many trenches are over 8 km deep—the deepest point in the ocean, –11,035 m, lies in the Mariana Trench of the western Pacific.

Abyssal Plains and Seamounts

Oceanic lithosphere older than about 80 million years cools so slowly that its rate of sinking becomes undetectable, and it effectively forms a flat surface. The **abyssal plains** are the regions of the ocean underlain by this old, cool, flat-surfaced lithosphere. A thick blanket of **pelagic sediment,** composed of microscopic plankton shells and fine flakes of clay, blankets the abyssal plain and further smoothes its surface. The thickness of pelagic sediment depends on the age of the sea floor—thicker deposits are found over older crust, because sediment has more time to accumulate.

Numerous islands rise from ocean basins away from plate boundaries. The islands, which formed as a result of hot-spot volcanic activity above a mantle plume (see Chapters 4 and 6), consist of mounds of basalt built on top of oceanic lithosphere. When plate movements carry such hot-spot volcanoes off the mantle plume, volcanic activity ceases, and erosion begins to grind down the volcano. Also, the plate carrying the now-extinct volcano sinks as it ages. Thus, the peak of the island gradually descends below sea level, and the island becomes a **seamount,** a submerged peak. Typically, several seamounts that developed above the same hot spot line up in a chain, with the oldest seamount at one end and the youngest at the other (see Fig. 4.22). Coral reefs grow on top of many seamounts, creating a flat-crested cap of limestone; these seamounts are called **guyots** (see Fig. 3.20).

FIGURE 18.6 (a) Submarine canyons along the East Coast of the United States. A particularly large one starts at the mouth of the Hudson River. (b) Submarine fans accumulate at the base of submarine canyons.

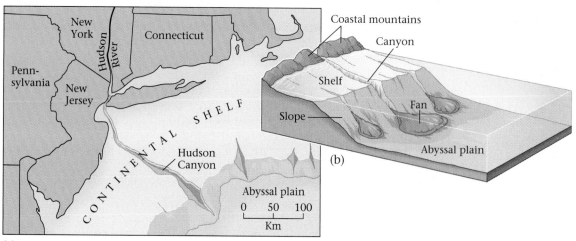

18.3 OCEAN WATER

Composition

If you've ever had a chance to swim in the ocean, you may have noticed that you float much more easily in ocean water than you do in freshwater. That's because ocean water contains an average of 3.5% dissolved salt ions (▶Fig. 18.7a); in contrast, typical freshwater contains only 0.02% salt. The dissolved ions fit between water molecules without changing the volume of the water, so adding salt to water increases the water's density, and you float higher in a denser liquid.

Leonardo da Vinci, the Renaissance artist and scientist, speculated that sea salt came from rivers passing through salt mines, but modern studies demonstrate that most cations in sea salt—sodium (NA^+), potassium (K^+), calcium (Ca^{2+}), and magnesium (Mg^{2+})—come from the chemical weathering of rocks, and the anions, chloride (Cl^-) and sulfate (SO_4^{-2}), from volcanic gases. Still, Leonardo was right in believing that dissolved ions get carried to the sea by flowing groundwater and river water—rivers deliver over 2.5 billion tons of salt every year.

There's so much salt in the ocean that if all the water suddenly evaporated, a 60-m-thick layer of salt would coat the ocean floor, consisting of about 75% halite (NaCl), with lesser amounts of gypsum ($CaSO_4 \cdot H_2O$), anhydrite ($CaSO_4$), and other salts (▶Fig. 18.7b). Oceanographers refer to the concentration of salt in water as **salinity.** Although ocean salinity averages 3.5%, measurements from around the world demonstrate that salinity varies with location, ranging from about 1.0% to about 4.1% (▶Fig. 18.8a). Away from polar regions, salinity reflects the balance between the addition of water from rivers or rain and the removal of freshwater by evaporation, for when seawater evaporates, salt stays behind. Thus, in regions with high evaporation rates and minimal rainfall, the oceans become saltier, reaching maximum saltiness in restricted seas that do not mix freely with the main ocean. Salinity decreases near the equator, because of high rainfall, and in rainy, high-latitude regions and near the mouth of large rivers.

The salinity of the ocean also changes with depth. Surface water is saltier in warm latitudes because of evaporation, and is less salty at high latitudes because of melting ice. A graph showing the variation in salinity with depth (▶Fig. 18.8b) indicates that such differences in salinity are only found in seawater down to a depth of about 1 km. Deeper water tends to be more homogenous around the world. Oceanographers refer to the boundary between surface-water salinities and deep-water salinities as the **halocline.**

Temperature

When the *Titanic* sank after striking an iceberg in the North Atlantic, most of the unlucky passengers and crew who

Magnesium	(Mg^{2+})	1.3g
Calcium	(Ca^{2+})	0.42g
Potassium	(K^+)	0.38g
All others		0.2g

(a)

(b)

FIGURE 18.7 (a) The composition of average seawater. The expanded part of the graph shows the proportions of ions in the salt of seawater. (b) A marine evaporite deposit in Newfoundland, formed from salt that precipitated out of seawater over a long period. Because of tectonic activity, the salt is now exposed on land.

jumped or fell into the sea died within minutes, because the seawater temperature at the site of the tragedy hovered at freezing, and cold water removes heat from a body very rapidly. Yet swimmers can play for hours in the Caribbean, where sea-surface temperatures reach 28°C (83°F). Though the *average* global sea-surface temperature hovers around 17°C, it ranges between freezing near the poles to almost 35°C in restricted tropical seas (▶Fig. 18.8c).

Because the intensity of solar radiation varies with the season, surface seawater temperature also varies with the season. But the difference is only around 2° in the tropics, 8° in the temperate latitudes, and 4° near the poles. (By contrast, the seasonal temperature change on land can be much greater—in central Illinois, for example, temperatures may reach 40°C [104°F] in the summer and drop to –32°C [–25°F]

FIGURE 18.8 (a) The variations in salinity in the world ocean. The contour lines represent regions of different salinity (numbers are percentage). (b) The variation of salinity with depth in the ocean depends on the latitude (the tropics are the Tropic of Cancer and the Tropic of Capricorn, about 23° north and south of the equator, respectively). The depth at which salinity changes rapidly is the halocline. (c) The variation in temperature with latitude. Contours are given in degrees Celsius. (d) The variation in temperature with depth. The depth interval where temperatures change rapidly is the thermocline. Note that the temperature variation is greatest at the equator, and that all ocean water approaches freezing at a depth of about 2.5 km. (e) The variation of density with depth. The depth interval at which density changes rapidly is the pycnocline.

in the winter.) The seasonal seawater temperature remains in a narrow range because of the high **heat capacity** of water: water can absorb or release large amounts of heat without changing temperature very much. Because of its high heat capacity, the ocean regulates the temperatures of coastal regions. For example, the air temperature in Vancouver, on the Pacific coast of Canada, rarely drops below freezing even though it lies farther north than Illinois.

Water temperature in the ocean varies markedly with depth (▶Fig. 18.8d). Warm, less dense water tends to float on top of cold water, so water heated by the Sun near the surface tends to remain near the surface. An abrupt **thermocline,** below which water temperatures decrease abruptly, reaching near freezing at the sea floor, appears at about 300 m deep in the tropics. There is no pronounced thermocline in polar seas, since surface waters there are so cold to start with.

Density

Both salinity and temperature affect the density of seawater. Specifically, density increases with salinity, so average seawater has a density of 1.025 g/cm^3 at 4°C, while freshwater has a density of 1.000 g/cm^3 at the same temperature. The effect of temperature on seawater density is a little more complicated. If you start to cool warm water, its molecules begin to squeeze more closely together, and thus its density increases, reaching a maximum density at 4°C. But if you cool water still further, ice crystals begin to form and the density decreases, for the spacing of molecules in the ice crystals is greater than in 4°C water. Thus, the density of ice is actually less than that of warm water, so ice floats. Density also varies with depth (▶Fig. 18.8e). The depth at which density changes rapidly is called the **pycnocline.** Later, we will see that these density contrasts play a major role in controlling the circulation of ocean water.

18.4 CURRENTS: RIVERS IN THE SEA

Since first setting sail on the open ocean, people have known that the water of the ocean does not stand still, but rather flows or circulates at velocities of up to several kilometers per hour in fairly well-defined streams called **currents.** Oceanographic studies made since the *Challenger* expedition demonstrate that circulation in the sea occurs at two levels: **surface currents** affect the upper hundred meters of water, and **deep currents** keep even water at the bottom of the sea in motion.

Surface Currents: A Consequence of the Wind

When the skippers of sailing ships planned their routes from Europe to North America, they paid close attention to

FIGURE 18.9 A satellite image of the Gulf Stream, a current of warm surface water flowing north along the East Coast. The colors represent different temperatures (red is warmer). Note the large eddies that form along the margin of the Gulf Stream.

the directions of surface currents, for sailing against a current slowed down the voyage substantially. If they headed due west at a high latitude, they would find themselves battling an east-flowing surface current, the Gulf Stream, that made the voyage nearly impossible (▶Fig. 18.9). Further, they found that the water moving in a surface current, like the water flowing in a river, does not flow smoothly but displays some turbulence. Isolated swirls or ring-shape currents of water called **eddies** form along the margins of currents and may keep a ship from making any progress at all.

Oceanographers now recognize major surface currents in all the world's oceans (▶Fig. 18.10). They result from interaction between the sea surface and the wind—as moving air molecules shear across the surface of the water, the friction between air and water drags the water along. If we look at a map that shows global wind patterns along with oceanic currents, we can see this relationship (see Fig. 18.10 inset). But the movement of water resulting from wind shear does not exactly parallel the movement of the wind. This is a consequence of Earth's rotation, which creates the Coriolis effect (see Box 18.1). This phenomenon causes surface currents in the Northern Hemisphere to veer toward the right and surface currents in the Southern Hemisphere to veer toward the left.

Because of the geometry of ocean basins and the pattern of wind directions, surface currents in the oceans today trace out large circular flow patterns known as **gyres,** clockwise in the northern seas and counterclockwise in the southern seas.

FIGURE 18.10 The major currents of the world's oceans. Inset: The relationship between the prevailing wind direction and the North Atlantic Current.

Within these gyres, individual currents have been given names (see Fig. 18.10). Notably, Northern and Southern Hemisphere gyres merge at the equator, creating an equatorial westward flow. Water in the center of a gyre is somewhat isolated from surface currents. Sailors refer to the center of the North Atlantic Gyre as the **Sargasso Sea,** for its relatively noncirculating waters accumulate sargassum, a tropical seaweed.

In the past, when continents were in different positions, the geometry of ocean currents was quite different—for example, the Circum-Antarctic Current that exists today did not appear until the Scotia Sea, between South America and Antarctica, opened 10 million years ago. Currents that move from pole to equator bring cool water down from the north, while currents that move from the equator toward the pole carry warm water northward—this transport of heat moderates the global climate. Thus, changes in the pattern of currents through geologic time affect the climate of the Earth system.

Upwelling, Downwelling, and Deep Currents

Surface currents are not the only means by which water flows in the ocean; it also circulates in the vertical direction. Oceanographers have now identified **downwelling zones,** places where near-surface water sinks, and **upwelling zones,** places where deep water rises.

What causes upwelling and downwelling? First, along coastal regions, these two phenomena exist because as the wind blows parallel to coastlines, it drags surface water along, but because of the Coriolis effect, water deflects at an angle to the wind (Box 18.1). If this deflection moves water toward the coast, then there is an oversupply of water along the shore and excess water must sink—that is, downwelling occurs (▶Fig. 18.13a). Alternatively, if deflection moves surface water away from the coast, then there is a deficit of water near the coast and water rises to fill in the gap—upwelling takes place (▶Fig. 18.13b).

The Coriolis Effect

BOX 18.1

SCIENCE TOOLBOX

Imagine you are spinning a playground merry-go-round counterclockwise around a vertical axis at a rate of 10 revolutions per minute. The circumference of the outer edge of the merry-go-round is 5 m. Thus, Emma, a child sitting at the outer edge, moves at a velocity of 50 m per minute, whereas David, a child sitting at the center, spins around an axis but moves at zero velocity. If Emma were to try throwing a ball at David by aiming directly along a radius, the ball would veer to the right of the radius and miss David, because the ball is not only moving in the direction parallel to a radius line, but also moving a little in the direction parallel to the edge of the circle. If David were to throw a ball along a radius to Emma, this ball would miss Emma because the revolution of the merry-go-round moves her relative to the ball's trajectory (▶Fig. 18.11a, b).

The rotation of the Earth creates the same phenomenon. Earth spins counterclockwise around its axis, so a cannon shell fired parallel to a line of longitude from the equator to the North Pole veers to the right (east), because as it moves north, it is traveling east faster than the land beneath it (▶Fig. 18.11c). Similarly, a cannon

shell fired from the equator to the South Pole veers to the left (east). A cannon shell fired along a line of longitude from the North Pole toward the equator veers to the right (west) because the Earth is moving faster to the east at the equator (▶Fig. 18.11d). German artillerymen learned this lesson during World War I, when shells they aimed at Paris from a distance of 100 km landed about 1 km to the right of their target.

In 1835, a French engineer named Gaspard Gustave de Coriolis (1792–1843) proposed that a similar effect would cause the deflection of winds and currents on the surface of the Earth. Because of this **Coriolis effect**, north-flowing currents in the Northern Hemisphere deflect to the east, while south-flowing currents deflect to the west. The opposite is true in the Southern Hemisphere (▶Fig. 18.11e).

FIGURE 18.11 The Coriolis effect. (a, b) The velocity of a point on the rim of this spinning merry-go-round is greater than the velocity at the center. A ball thrown from the center (D) to an observer at point E would follow a straight line, but while the ball is in the air, the catcher on the rim moves from point E to point E'. Relative to the surface of the merry-go-round, the ball looks like it follows a curved path—but remember, the ball goes straight, it's the surface that moves underneath the ball. A ball aimed from the rim to the center won't go straight to the center, because it moves parallel to the rim at the point of departure. Again, since the merry-go-round is moving under the ball, the ball appears to follow a curved path with respect to the surface of the merry-go-round. (c, d) The same phenomenon happens on Earth. A projectile shot from the equator to the pole in the Northern Hemisphere is deflected to the east, while a projectile shot from the pole to the equator is deflected to the west. (e) Thus, on the Earth, north-flowing currents in the Northern Hemisphere are deflected eastward, and south-flowing currents are deflected westward, while the opposite is true in the Southern Hemisphere.

In 1902, V. W. Ekman heard that the floating ice that had trapped the Norwegian explorer F. Nansen and his ship the *Fram* in the Arctic Ocean had drifted at an angle to the wind. To explain this phenomenon, Ekman proposed that wind shear in the Northern Hemisphere causes the thin layer of water immediately at the ocean's surface to move at an angle of about 45° to the right, because of the Coriolis effect. This layer of water, in turn, shears the next layer just beneath and makes it move also, though not as much (at an angle of 45° to the top layer), because some of the driving energy has been lost in friction. Progressing downward through the water column, the same effect takes place, until at a depth of about 100 m water actually moves at about 180° to the surface water, but the flow rate is less than 5% that of surface flow. Oceanographers refer to this spiral-like pattern of movement as the **Ekman spiral.** A mass of water in an Ekman spiral moves at an average of about 90° to the wind direction, an effect called the **Ekman transport** (▶Fig. 18.12). Because of the Ekman transport, winds parallel to a coast cause surface water to move away from or toward the coast, depending on the wind direction.

FIGURE 18.12 Because of the Coriolis effect, surface currents flow at an angle to the wind, each layer of water shears the layer below, and the underlying layer moves at an angle to the layer above. Thus, the angle of deflection increases with depth, a geometry called the Ekman spiral. When all the deflections are averaged together, the net flow of water is at 90° to the wind direction. This movement is called the Ekman transport.

As an example of upwelling and downwelling caused by surface currents, consider the west coast of South America. When strong winds blow north along this coast, the Humboldt Current transports water north and west. Surface water moves away from the shore and toward the center of the South Pacific Gyre, causing the upwelling of deep water. This deep water brings nutrients with it, feeding a large population of fish (▶Fig. 18.14).

FIGURE 18.13 Because currents are deflected relative to the wind direction, a wind blowing parallel to the coast can drive surface water offshore or onshore, depending on its direction. (a) Where surface water moves toward shore, it downwells to make room for more water. (b) Where surface water moves offshore, deep water upwells to replace the water that flowed away.

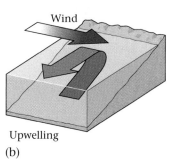

Downwelling

(a)

Upwelling

(b)

FIGURE 18.14 The major zone of upwelling in the Pacific along Peru's coast. The red color indicates a region of greater amounts of algae, fed by nutrients brought up with the water.

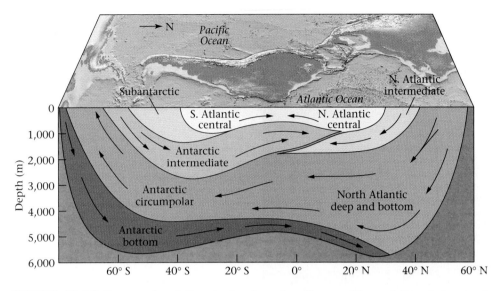

FIGURE 18.15 Because of variations in density, primarily caused by variations in temperature, the oceans are vertically stratified into moving water masses. Each mass has a name. Note that Antarctic bottom water, which sinks down from the surface along the chilly shores of Antarctica, flows north along the floor of the Atlantic at least as far as the equator.

Upwelling and downwelling can also be driven by contrasts in water density, caused by differences in temperature and salinity; the rising and sinking of water in this case is known as **thermohaline circulation.** During thermohaline circulation, cold salty water tends to sink, and warm, less salty water rises. As a result, the water in polar regions sinks and flows back along the bottom of the ocean toward the equator. This process divides the ocean vertically into a number of distinct **water masses,** which mix only slowly with one another. In the Atlantic Ocean, for example, the **Antarctic bottom water mass** sinks along the coast of Antarctica, and the **North Atlantic deep water mass** sinks in the north polar region (▶Fig. 18.15). The combination of surface currents and thermohaline circulation, like a conveyor belt, moves water and heat among the various ocean basins (▶Fig. 18.16).

FIGURE 18.16 The exchange between upwelling deep water and downwelling surface water creates a global conveyor belt that circulates water throughout the entire ocean; this takes hundreds of years to millennia.

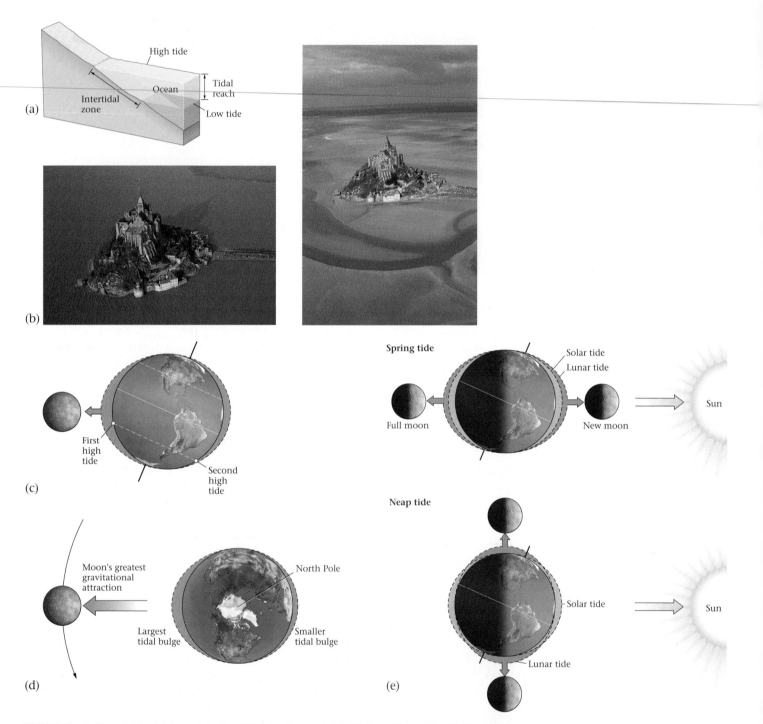

FIGURE 18.17 (a) The tidal reach is the difference between the high and low tide. (b) Mont-Saint-Michel, along the west coast of France, is an island during high tide, but at low tide it's surrounded by mudflats. (c) A larger tidal bulge appears on the side of the Earth closest to the Moon, and a smaller tidal bulge on the opposite side. Because the Earth spins beneath the bulges, each point on the planet generally experiences two high tides per day. Since the Earth's axis is tilted with respect to the Moon's orbit, the two high tides in a given day are not the same magnitude. (d) The tides as viewed looking down on the North Pole. The Moon exerts the greatest gravitational attraction, so the tidal bulge stays with the Moon while the Earth spins beneath it. A second bulge occurs on the opposite side of the Earth, where the Moon's attraction is weakest. (e) When the gravitational attraction of the Sun adds to that of the Moon (that is, at new moon or full moon), extra high tides, called spring tides, form. When the gravitational attraction of the Sun is at right angles to that of the Moon (at quarter moon), extra low tides, called neap tides, occur.

18.5 THE TIDES GO OUT . . . THE TIDES COME IN . . .

The sea level at a given point on the Earth rises and falls once or twice daily, a vertical movement called the **tide.** The difference in sea level between high tide and low tide is called the **tidal reach.** (When we talk about "sea level," we are actually referring to the **mean sea level**—the average between the high and low tide over the year.) The rising tide, or **flood tide,** causes the **shoreline,** the boundary between water and land, to migrate inland; while the falling tide, or **ebb tide,** causes the shoreline to migrate seaward. Where the coastal area consists of a broad, nearly horizontal plain, or **tidal flat,** the boundary may move a few kilometers, but along steep coasts the change in position can be minimal. In quiet water, the flood tide arrives as a visible wall of water, the **tidal bore,** moving inland at up to 35 km per hour, faster that a person can run. So beware! A beachcomber who wanders into the **intertidal zone** (across which the tide rises and falls) to investigate its fascinating ecosystem may become trapped and drown (▶Fig. 18.17a, b).

Tides are caused by a **tide-generating force,** which is due in part to the gravitational attraction of the Sun and Moon, and in part to centrifugal force caused by the revolution of the Earth-Moon system around its center of mass (▶Fig. 18.17c). The Moon contributes most to the tide-generating force because it lies so close to the Earth—the Sun's contribution is 46% that of the Moon's. The tide-generating force creates two bulges in the global ocean, and thus makes this envelope of water more elliptical (oval-shaped) than the solid Earth. One bulge, the sublunar bulge, lies on the side of the Earth closer to the Moon, and forms because the Moon's gravitational attraction is greatest at this point. The other, the secondary bulge, lies on the opposite (far) side of the Earth (12,000 km [the diameter of the Earth] farther from the Moon), and forms because the Moon's gravitational attraction is weakest at this point; here, the centrifugal force can push water outward (▶Fig. 18.17d). A depression in the global ocean surface separates the two bulges.

When a location lies under a bulge, it experiences a high tide, and when it passes under a depression, it feels a low tide. High tides arrive 50 minutes later each day, because of the difference between the time it takes for Earth to spin on its axis and the time it takes for the Moon to orbit the Earth. In addition, a number of factors influence the timing and reach of tides.

- *The Earth's tilt:* Because the Earth's axis tilts relative to the plane of the Moon's orbit, the two high tides in a given day do not reach the same level (Fig. 18.17c).

- *The Moon's orbit:* The track of the Moon over the Earth, relative to a location on Earth, varies with the seasons, so the distance between the crest of a tidal bulge and a location on Earth varies with the seasons. Also, because of the elliptical shape of the Moon's orbit, the gravitational attraction between the Earth and Moon changes seasonally.

- *The Sun's gravity:* When the angle between the direction to the Moon and the direction to the Sun is 90° (quarter moon), we have extra low tides (**neap tide**), because the Sun's attraction counteracts the Moon's. When the Sun is on the same side as the Moon (new moon or full moon), we have extra high tides (**spring tide**), because now the Sun's attraction *adds* to the Moon's (▶Fig. 18.17e).

- *Basin shape:* In the open ocean, the tidal reach is only about 0.5m, but in the Bay of Fundy, along the east coast of Canada, the tidal reach can be 20 m. In a bay that narrows to a point, like the Bay of Fundy, the flood tide brings a large volume of water into a small area, so the point experiences an especially large high tide.

- *Air pressure:* The effects of air pressure on tides can contribute to disaster. For example, during a hurricane, the air pressure drops radically and the sea surface rises; if the hurricane coincides with a high tide, the **storm surge** (water driven landward by the wind) can inundate the coast.

Interestingly, friction causes the movement of the tidal bulge to lag behind the movement of the Moon across the Earth. The Moon thus exerts a slight pull on the side of the bulge. This pull acts like a brake to slow the Earth's spin, so that days grow longer at a rate of about 0.002 seconds per century. (A day was only 21.9 hours long in the Middle Devonian period, 375 million years ago.) Also, as the spinning Earth slows, the Moon moves farther away. In the Archean eon (3.9 Ga), the Moon was 15,000 km closer, so the tidal reach was larger.

18.6 WAVE ACTION

Waves make the ocean surface restless, an ever-changing vista. They develop because of the shear between the molecules of air in the wind and the molecules of water at the surface of the sea. It may seem surprising that so much friction can arise between two fluids, but it can, as Benjamin Franklin demonstrated back during the American Revolution. Franklin noted that oily waste spilled from ships on a windy day made the water surface smoother. He proposed that the oily coating on the water decreased frictional shear between the air and the water, and therefore prevented waves from forming.

When you watch a wave travel across the open ocean, you may get the impression that the whole mass of water comprising the wave moves with the wave. But drop a cork overboard and watch it bob up and down and back and forth; it does not move along with a wave. Within a wave, away from shore, a particle of water moves in a circular mo-

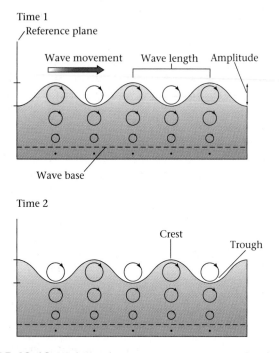

FIGURE 18.18 Within a deep-ocean wave, water molecules follow a circular path. The diameter of the circle decreases with depth to the wave base, below which the wave has no effect. When a wave passes, the shape of the water surface changes, but water does not move as a mass.

tion, as viewed in cross section. The diameter of the circle is greatest at the ocean's surface, where it equals the amplitude of the wave. With increasing depth, though, the diameter of the circle decreases until, at a depth equal to about half the **wavelength** (the horizontal distance between two wave

troughs), there is no wave movement at all (▶Fig. 18.18). Submarines traveling below this **wave base** cruise through smooth water, while ships toss about above.

The character of waves in the open ocean depends on the strength of the wind (how fast the air moves) and on the **fetch** of the wind (over how long a distance it blows). When the wind first begins to blow, it creates **ripples** in the water surface, pointed waves whose **amplitude** (the height from crest to trough) and wavelength are small. With continued blowing over a long fetch, **swells,** larger waves with amplitudes of 2–10 m and wavelengths of 40–500 m, begin to build. Hurricane wave amplitudes may grow to over 25 m. The largest documented swell in the open ocean was 35 m high, taller than an ocean liner.

Swells may travel for thousands of kilometers across the ocean, well out of the region where they were created, and thus may interact with waves formed elsewhere. The overlap, or **interference,** between comparable-sized swells leads to the formation of larger waves, and some of these can be truly mountainous. One of the most disastrous instances of wave interference took place in 1979 off the coast of Ireland, where over 300 yachts were competing in the Fastnet Race between the Isle of Wight, off the south coast of England, and Fastnet Rock, off the south coast of Ireland. Waves generated by an east-blowing gale collided with waves generated by a west-blowing gale and built up to an amplitude of over 15 m, capsizing twenty-three yachts and killing fifteen sailors.

Waves have no effect on the ocean floor, as long as the floor lies below the wave base. However, near the shore, where the wave base just touches the floor, it causes a slight back-and-forth motion of sediment. Closer to shore, as the water gets shallower, friction between the wave and the sea floor slows the deeper part of the wave, and the back-and-

FIGURE 18.19 This profile shows the various landforms of a beach, as well as a cross section of a barrier island. As a wave approaches the shore, it touches the bottom of the sea, at a depth of about half the wavelength. Because of friction, the wave slows down and the wavelength decreases, so the wave height must increase. As the bottom of the wave moves slower than the top, the wave builds up into a breaker that carries water up onto the beach, with the top of the wave falling over the bottom. The water washing up on the beach is swash, and the water rushing back is backwash (indicated by arrows).

FIGURE 18.20 (a) Wave refraction occurs when waves approach the shore at an angle. The part of the wave that touches bottom first slows down, then the rest of the wave catches up. As a result, the wave bends so that it's nearly parallel with the shore. However, because the wave hits the shore at an angle, water moving parallel to the shore creates a longshore current. (b) Wave refraction on a beach. (c) Like a lens, wave refraction focuses wave energy on a headland, so erosion occurs; and it disperses wave energy in embayments, so deposition occurs.

forth motion in the wave becomes more elliptical. Eventually, water at the top of the wave curves over the base, and the wave becomes a **breaker,** ready for surfers to ride. Breakers crash onto the shore in the **surf zone,** sending a surge of water up the beach. This upward surge, or **swash,** continues until friction brings motion to a halt. Then gravity draws the water back down the beach as **backwash** (▶Fig. 18.19).

Waves may make a large angle with the shoreline as they're coming in, but they bend as they approach the shore, a phenomenon called **wave refraction**; right at the shore, their crests make no more than about a 5° angle with the shoreline. To see why this happens, imagine a wave ap-

proaching the shore so that its crest makes an angle of 45° with the shoreline. The end of the wave closest to the shore touches bottom first and slows down because of friction, while the end farther offshore still continues to move at its original velocity, swinging the whole wave around so that it's parallel with the shoreline.

Though refraction decreases the angle at which a wave rolls onto shore, the wave may still arrive at an angle. When the water returns seaward in the backwash, however, it must flow straight down the slope of the beach in response to gravity. Overall, this sawtooth-like flow results in a **longshore current,** which flows parallel to the beach (▶Fig. 18.20a, b). Also because of wave refraction, wave energy is

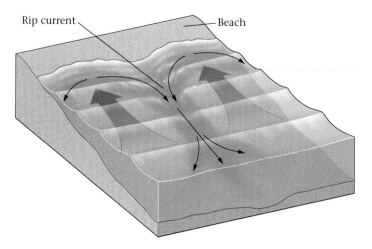

FIGURE 18.21 Waves bring water up on shore. The water may return to sea in a narrow rip current perpendicular to the shore.

focused on **headlands** (places where higher land protrudes into the sea), and is weaker in **embayments** (set back from the sea). Thus, erosion happens at headlands, forming a cliff, while deposition takes place in an embayment, forming a beach (▶Fig. 18.20c).

Waves pile water up on the shore incessantly. Where the excess water finds a way back to sea, it creates a strong, localized seaward flow perpendicular to the beach called a **rip current** (▶Fig. 18.21). Rip currents are the cause of many drownings every year along beaches, because they suddenly carry unsuspecting swimmers away from the beach.

18.7 WHERE LAND MEETS SEA: COASTAL LANDFORMS

Tourists along the Amalfi coast of Italy thrill to the sound of waves crashing on rocky shores. But on the Gulf Coast of Florida, sunbathers lie on endless white sand beaches next to calm seas. Large dome-like mountains rise directly from the sea in Rio de Janeiro, Brazil, but a 100-m-high vertical cliff marks the boundary between the Nullarbor Plain of South Australia and the Great Southern Ocean (▶Fig. 18.22a–d). And New Orleans sprawls over a vast plain of former swamp. As these examples illustrate, **coasts,** the belt of land bordering the sea, vary dramatically in terms of topography and associated landforms (▶Fig. 18.23a–g).

Sandy Shores

When World War II generals designed the D-Day invasion at Normandy, on the coast of France, they searched for a place where the shore was fringed by a gently sloping band of

(a)

(b)

(c)

(d)

FIGURE 18.22 (a) A rocky shore in eastern Italy. (b) A flat, sandy beach along the Gulf Coast of Florida. (c) The sugar loafs (rounded mountains) rising out of the sea at Rio de Janeiro, Brazil. (d) The abrupt edge of the Nullarbor Plain in South Australia.

FIGURE 18.23 A wide variety of coastal landforms have developed on Earth. (a) Drowned river valleys, formed where sea level rises and floods valleys, create complex, irregular coastlines. (b) Uplifted terraces develop where the coastline rises relative to sea level, and creates escarpments. (c) Swampy deltas form where a sediment-laden stream deposits sediment along the coast. (d) Along sandy coastal plains, large beaches and offshore bars appear. (e) Glacial fjords develop where the sea level rises and floods a glacially carved valley. (f) Coastal dunes form where there is a large sand supply and strong wind. (g) In tropical environments, mangrove swamps grow along the shore, protected from wave action by offshore coral reefs.

(a)   (b)

FIGURE 18.24 (a) A cobble beach. (b) The Cape Cod national seashore, a sandy beach.

sediment—in other words, a **beach.** The beach they found provided a platform for landing craft and a firm surface for soldiers to walk on when they left the water. Fortunately, today most people seek out beaches for more pleasant pastimes, as they provide a platform for sunbathing and a pathway into the surf.

Though some beaches consist of pebbles or boulders, most consist of sand grains (▶Fig. 18.24a, b). This is no accident, for waves winnow out finer sediment like silt and mud and carry it to quieter water, where it settles. And storm waves, which can smash coarse grains against one another with enough force to shatter them, have little effect on sand, for sand grains can't collide with enough energy to crack. (Thus, boulder or pebble beaches exist only where nearby cliffs continuously supply large rock fragments.)

The composition of sand itself varies from beach to beach, because different sands come from different sources. Sands derived from the weathering and erosion of silicic-to-intermediate rocks consist mainly of quartz; other minerals in these rocks tend to chemically weather to form clay, which washes away in waves. Beaches made from the erosion of limestone or of recent corals and shell beds consist of carbonate sand, including lots of sand-sized chips of shells. And beaches derived by the recent erosion of basalt may have black sand, made of tiny basalt grains.

A **beach profile,** a cross section drawn perpendicular to the shore, illustrates the shape of a beach (Fig. 18.20). Starting from the sea and moving landward, a beach consists of a **foreshore zone,** or **intertidal zone,** across which the

tide rises and falls. The **beach face,** a steeper, concave part of the foreshore zone, forms where the swash of the waves actively scours the sand. The **backshore zone** extends from a small step, or escarpment, cut by high-tide swash to the front of the dunes or cliffs that lie farther inshore. The backshore zone includes one or more **berms,** horizontal to landward-sloping terraces that received sediment during a storm.

Geologists commonly refer to beaches as "rivers of sand," to emphasize that beach sand moves along the coast over time—it is not a permanent substrate. Wave action at the shore moves an **active sand** layer on the sea floor on a daily basis. **Inactive sand,** buried below this layer, moves only during severe storms or not at all. Where waves hit the beach at an angle, the swash of each successive wave moves active sand up the beach at an angle to the shoreline, but the backwash moves this sand down the beach parallel to the slope of the shore. This sawtooth motion causes sand to gradually migrate along the beaches, a process called **beach drift.** Beach drift, which happens in association with the longshore drift of water, can transport sand hundreds of kilometers in a matter of centuries. Where the coastline indents landward, beach drift stretches beaches out into open water to create a **sand spit.** Some sand spits grow across the opening of a bay, to form a **baymouth bar** (▶Fig. 18.25).

The scouring action of waves piles sand up in a narrow ridge away from the shore called an **offshore bar,** which parallels the shoreline. In regions with an abundant sand

FIGURE 18.25 Beach drift can generate sand spits and baymouth bars. Sedimentation fills in the region behind a baymouth bar. As a result, the shoreline gets smoother with time.

supply, offshore bars rise above the mean high-water level and become **barrier islands,** common along the coast of North Carolina and Texas (▶Fig. 18.26a). The water between a barrier island and the mainland becomes a quiet-water **lagoon,** a body of shallow seawater separated from the open ocean.

Though developers have covered some barrier islands with expensive resorts, in the time frame of centuries to millennia barrier islands are temporary features. For example, wind and waves pick up sand from the ocean side of the barrier island and drop it on the lagoon side, causing the island to migrate landward—tough luck for any real estate

owners on the island. Storms may breach barrier islands and create an inlet (a narrow passage of water). Finally, beach drift gradually transports the sand of barrier islands and modifies their shape.

Tidal flats, regions of mud and silt exposed or nearly exposed at low tide but totally submerged at high tide, develop in regions protected from strong wave action (Fig. 18.17b; ▶Fig. 18.26b). They are typically found along the margins of lagoons or on shores protected by barrier islands. In tidal flats that provide a home for burrowing organisms like clams and worms, **bioturbation** ("stirred by life") mixes sediments together.

Because of the movement of sediment, the **sediment budget** (the proportion of sand supplied to sand removed) plays an important role in determining the long-term evolution of a beach. Let's look at how the budget works for a small segment of beach (▶Fig. 18.27). Sand may be supplied to the segment from local rivers or by wind from nearby dune fields; it may also be brought from just offshore by waves or from far away by beach drift. (In fact, the large quantity of sand along beaches of the southeastern United States may have originated in Pleistocene glacial outwash far to the north.) Some of the sand from a stretch of beach may be removed by beach drift, while some gets carried offshore by waves, where it either settles locally or tumbles down a submarine canyon into the deep sea. If the lost sand cannot be replaced, the beach segment grows narrower,

FIGURE 18.26 (a) The barrier islands off the coast of North Carolina. (b) Tidal flats are broad muddy areas submerged only at high tide. This tidal flat, along the coast of Wales, is at low tide.

(a)

(b)

FIGURE 18.27 The sediment budget along a coast. Sediment is brought into the system by rivers, by the erosion of cliffs and moraines, and by wind. Sediment moves along the coast as a result of beach drift. And sediment leaves the system by being blown off the beach, by sinking into deeper water, or by being carried out by the longshore current.

together called **wave erosion,** gradually undercut a cliff face and make a **wave-cut notch** (▶Fig. 18.29b, c). Undercutting continues until the overhang becomes unstable and breaks away at a joint, creating a pile of rubble at the base of the cliff that waves immediately attack and destroy. By this process, wave erosion cuts away at a rocky coast, so that the cliff gradually migrates inland. Such **cliff retreat** leaves behind a **wave-cut bench,** or **platform,** at low tide (▶Fig. 18.29d).

Other processes besides wave erosion break up the rocks along coasts. For example, salt spray coats the cliff face above the waves and infiltrates into pores. When the water evaporates, salt crystals grow and push apart the grains, thereby weakening the rock. Biological processes also contribute to erosion, for plants and animals in the intertidal zone bore into the rocks and gradually break them up.

Many rocky coasts start out with an irregular coastline, with headlands protruding into the sea and embayments set back from the sea. Such ir-

while if the supply of sand exceeds the amount that washes away, the beach becomes wider.

In temperate climates, winter storms tend to be stronger and more frequent than summer ones. The larger, shorter-wavelength waves of winter storms wash beach sand into deeper water and thus make the beach narrower, while the smaller, longer-wavelength summer waves bring sand in from offshore and deposit it on the beach (▶Fig. 18.28a, b).

Rocky Coasts

More than one ship has met its end smashed and splintered in the spray and thunderous surf of a **rocky coast,** where bedrock cliffs rise directly from the sea (Fig. 18.22a; ▶Fig. 18.29a). Lacking the protection of a beach, rocky coasts feel the full impact of ocean breakers. The water pressure generated during the impact of a breaker can pick up boulders and smash them together until they shatter, and it can squeeze air into cracks, creating enough force to widen them. Further, because of its turbulence, the water hitting a cliff face carries suspended sand, and thus can abrade the cliff. The combined effects of shattering, wedging, and abrading,

FIGURE 18.28 (a) In the winter, when waters are stormier, sand moves offshore, and the beach narrows and may become stonier. (b) During the summer, waves bring sand back to replenish the beach.

(a)

(b)

(a)

(c)

(d)

FIGURE 18.29 (a) The major landforms of a rocky shore include cliffs, sea caves, wave-cut notches, sea stacks, sea arches, wave-cut benches, and tombolos. Beaches tend to collect in embayments, while erosion happens at headlands. (b) Erosion by waves creates a wave-cut notch. Eventually, the overhanging rock collapses into the sea to form gravel on the wave-cut bench. (c) A wave-cut notch exposed along a rocky shore. (d) A wave-cut bench at the foot of the cliffs at Etrétat, France.

FIGURE 18.30 The erosion of a headland. (a) At first, wave refraction causes wave energy to attack the sides of a promontory, making a sea cave on either side. (b) Gradually erosion breaks through the promontory to create a sea arch. (c) The arch finally collapses, leaving a sea stack.

regular coastlines tend to be temporary features in the context of geologic time: wave energy focuses on headlands and disperses in embayments (a result of wave refraction), so that debris is removed by erosion at headlands and accumulates in embayments (Fig. 18.20c); thus, over time the shoreline becomes less irregular.

A headland erodes in stages (▶Fig. 18.30a–c). Because of refraction, waves curve and attack the sides of a headland, slowly eating through it to create a **sea arch** connected to the mainland by a narrow bridge (▶Fig. 18.31a). Eventually the arch collapses, leaving isolated **sea stacks** just offshore (▶Fig. 18.31b). The unusual shapes of sea stacks may seem lifelike, so people have invented colorful names for them. For example, the Twelve Apostles are a popular tourist attraction on the south coast of Australia. Once formed, a sea stack protects the adjacent shore from waves. Therefore, sand collects in the lee of the stack, slowly building a **tombolo,** a narrow ridge of sand that links the sea stack to the mainland.

Coastal Wetlands

Let's move now from the crashing waves of rocky coasts to the gentlest type of shore, the **coastal wetland:** a flat-lying stretch of the coast that floods with shallow water but does not feel the impact of strong waves. Coastal wetlands, which can build out into lagoons, estuaries (where seawater and river water mix), or coastal plains, include **swamps** (wetlands dominated by trees), **marshes** (wetlands domi-

FIGURE 18.31 (a) A sea arch exposed along a rocky coast. (b) These sea stacks along the south coast of Australia are a tourist attraction called the Twelve Apostles.

(a)

(b)

nated by grasses; ▶Fig. 18.32a), and **bogs** (wetlands dominated by moss and shrubs). So many marine species use wetlands to spawn that despite their relatively small area when compared with the oceans as a whole, wetlands account for 10–30% of marine organic productivity.

In tropical or semitropical climates (between 30° north and 30° south of the equator), **mangrove swamps** thrive (▶Fig. 18.32b). Mangrove tree roots can filter salt out of water, so the trees have evolved the ability to survive in freshwater, saltwater, or brine (in between). Some mangrove species form a broad network of roots above the water surface, making the plant look like an octopus standing on its tentacles, and some send up small protrusions (pneumatophores) from roots that rise above the water and allow the plant to breathe. Dense stands of mangroves dampen the effects of stormy weather and thus prevent coastal erosion.

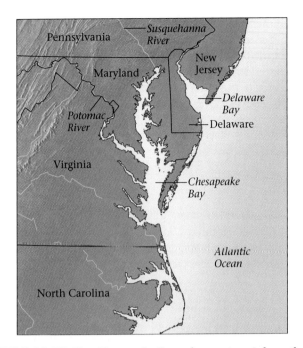

FIGURE 18.33 The Chesapeake Bay, a large estuary along the East Coast of the United States, formed when the sea level rose and flooded the Potomac and Susquehanna river valleys.

FIGURE 18.32 Examples of coastal wetlands: (a) a salt marsh, (b) a mangrove swamp.

In temperate climates, most wetlands consist of salt marshes. Healthy ones contain cordgrass (*Spartina*), while polluted ones contain plume grass (*Phragmites*), plants that can withstand winter freezes. Salt-marsh plants can trap sediments, keeping nearby seas clear.

Estuaries

Along some coastlines, the rising sea level or subsidence of the coast has allowed the sea to flood the valley and create an **estuary,** where seawater and river water mix. You can recognize an estuary on a map by its jagged coastline, mimicking the complexity of a river-carved landscape. The Chesapeake Bay, for example, illustrates this pattern (▶Fig. 18.33). Oceanic and fluvial water interact in two ways within an estuary. In quiet estuaries, protected from wave action or river turbulence, the water becomes stratified, with denser oceanic saltwater flowing upstream as a wedge beneath less dense fluvial freshwater. Such saltwater wedges migrate about 100 km up the Hudson River in New York, and about 40 km up the Columbia River in Oregon. In turbulent estuaries like the Chesapeake Bay, oceanic and fluvial water combine to create nutrient-rich brackish water (**brine**) with a salinity between that of oceans and rivers. Estuaries are complex ecosystems inhabited by unique species of shrimp, clams, oysters, worms, and fish that can tolerate large changes in salinity.

Sediment in an estuary may be supplied both from the river and from the ocean. Clay suspended in river water tends to **flocculate** (clump together into bunches that are large enough to settle out) in the brine of an estuary, forming mudflats and salt marshes. Therefore, estuaries used as harbors must be dredged, a process that unfortunately stirs up pollutants like PCBs (polychlorinated biphenyls, a dangerous organic chemical once used in electrical transformers), which settle with the clay.

Fjords

During the last ice age, glaciers carved deep valleys in coastal mountain ranges. When the ice age came to a close, the glaciers melted away, leaving deep, U-shaped valleys (see Chapter 22). The water stored in the glaciers, along with the water within the vast ice sheets that covered continents during the ice age, flowed back into the sea and caused the sea level to rise. The rising sea filled the deep valleys, creating **fjords,** or flooded glacial valleys. Coastal fjords are fingers of the sea surrounded by mountains; because of their deep-blue water and steep walls of polished rock, they command distinctively beautiful coastal landscapes (▶Fig. 18.34a–c). Some of the world's most spectacular fjords decorate the west coasts of Norway, British Columbia, and New Zealand. Smaller examples appear along the coast of Maine and southeastern Canada.

Coral Reefs

In the Undersea National Park of the Virgin Islands, visitors swim through colorful growths of living coral (▶Fig. 18.35a).

FIGURE 18.34 (a) The subsurface shape of a fjord, a drowned U-shaped glacial valley. (b) A satellite image shows that the west coast of Norway is very irregular, owing to the abundance of fjords. (c) Fjords are associated with spectacular scenery.

(a)

(b)

(c)

FIGURE 18.35 (a) Corals and other organisms make up a reef. (b) A coral reef bordering an island in the western Pacific. (c) The distribution of coral reefs on Earth today.

Some corals look like brains, others like elk antlers, still others like delicate fans. Sea anemones, sponges, and clams grow on and around the coral. Though at first glance coral looks like a plant, it is actually a colony of **cnidarians,** tiny invertebrates related to jellyfish. An individual coral animal, or **polyp,** has a tube-like body with a head of tentacles. Corals obtain part of their livelihood by filtering nutrients out of seawater; the remainder comes from algae (*zooxanthellae*) that live on the corals' tissue. Corals have a **symbiotic** (mutually beneficial) relationship with the algae, in that the algae photosynthesize and provide nutrients and oxygen to the corals, while the corals provide carbon dioxide and nutrients for the algae.

Coral polyps secrete calcite shells, which gradually build into a mound of solid limestone, whose top surface lies just below the low-tide level. At any given time, only the surface of the mound lives—the mound's interior consists of shells from previous generations of coral. The realm of shallow water underlain by coral mounds, associated organisms, and debris is called a **coral reef** (▶Fig. 18.35b). Reefs absorb wave energy and thus serve as a living buffer zone that protects coasts from erosion. Corals need clear, well-lit, warm (18°–30°C) water with normal oceanic salinity, so coral reefs only grow along clean coasts at latitudes of less than about 30° (▶Fig. 18.35c).

Marine geologists distinguish three different kinds of coral reef, based on their geometry (▶Fig. 18.36a–c). A **fringing reef** forms directly along the coast, a **barrier reef** develops offshore (separated from the coast by a lagoon), and an **atoll** makes a circular ring surrounding a lagoon. As Charles Darwin first recognized back in 1859, coral reefs associated with islands in the Pacific start out as fringing reefs and then later become barrier reefs and finally atolls. Darwin suggested, correctly, that this progression reflects the continued growth of the reef as the island around which it formed gradually subsides. Finally, the reef itself sinks too far below sea level to remain alive and becomes the cap of a guyot.

FIGURE 18.36 The progressive change from a fringing reef around a young volcanic island to a ring-shaped atoll. (a) The reef begins to grow around the volcano. (b) The volcano subsides as the sea floor under it ages, so the reef is now a ring, separated from a small island (the peak of the volcano) by a lagoon. (c) The volcano has subsided completely, so that all that remains is an atoll surrounding a lagoon. When the lagoon and atoll finally sink, the result is a guyot.

18.8 CAUSES OF COASTAL VARIABILITY

Plate Tectonic Setting

The present or recent plate tectonic setting of a coast determines whether the coast has steep slopes or a broad plain. Along an active margin, compression squeezes the crust and pushes it up, creating steep slopes at the shore like those along the west coast of South America. Rugged topography may persist for tens of millions of years after the compression ceases. Along a passive margin, the cooling and sinking of the lithosphere may create a broad **coastal plain,** a flat land that merges with the continental shelf, as exists along the Gulf Coast and southeastern Atlantic coast of the United States. But not all passive margins have coastal plains. At some, the margin of the rift that gave birth to the passive margin remains at a high elevation, even tens of millions of years after rifting ceased. For example, highlands formed during recent rifting border the Red Sea, while highlands formed during Cretaceous rifting persist along portions of the Brazilian coast (Fig. 18.22c).

Relative Sea-Level Changes (Emergent and Submergent Coasts)

The sea level, relative to the land surface, changes during geologic time. If the change is due to the uplift or subsidence (sinking) of land only along a particular coastal region, then only that region displays the effects of the sea-level change. Such vertical movement of the land may reflect plate tectonic processes or the addition or removal of a load (such as a glacier) on the crust. Some local changes in sea level reflect human activity. When people remove groundwater, for example, the pores between grains in the sediment beneath the ground collapse, and the land surface sinks (see Chapter 19). If this happens in a coastal area, the sea floods the area.

Some relative sea-level changes, however, are due to a *global* rise or fall of the ocean surface. Such **eustatic sea-level changes** may reflect changes in the voume of mid-ocean ridges—an increase in the number or size of ridges, for example, displaces water and causes the sea level to rise. Alternatively, eustatic sea-level changes may reflect changes in the volume of glaciers (see Chapter 22).

FIGURE 18.37 The development of uplifted terraces on an emergent coast. (a) Wave erosion creates a wave-cut bench. (b) The land rises, and the bench becomes a terrace.

Geologists refer to coasts where the land is rising relative to the sea level, or the sea level is falling relative to the land, as **emergent coasts.** At emergent coasts, landscapes typically include steep slopes. Interestingly, a series of step-like terraces form along some emergent coasts (▶Fig. 18.37a, b). These terraces reflect episodic changes in the relative sea level. Along coasts where wave erosion occurs (such as southern California), each terrace originated as a wave-cut bench; along coasts where beaches form (such as the Hudson Bay), each terrace originated as a wedge of beach sediment; and along coasts where coral reefs grow (such as Papua New Guinea), the terraces represent uplifted reefs.

Those coasts at which the land is sinking relative to sea level are **submergent coasts** (▶Fig. 18.38a, b). At submergent coasts, landforms include estuaries and fjords that developed when the sea flooded coastal valleys. Many of the coastal landforms of eastern North America represent the consequences of submergence (▶Fig. 18.38c).

Sediment Supply and Climate

The quantity and character of sediment supplied to a shore affects its character. That is, coastlines where the sea washes sediment away faster than it can be supplied (**erosional coasts**) recede landward and may become rocky, while coastlines that receive more sediment than erodes away (**accretionary coasts**) grow seaward and develop broad beaches. On accretionary coasts, the nature of the sedimentary accumulation depends on the kind of sediment supplied. For example, coastlines that receive quartz sand develop sandy beaches, while coastlines that receive mud may be rimmed by broad mudflats.

Climate also affects the character of a coast. Shores that enjoy generally calm weather erode less rapidly than those constantly subjected to ravaging storms. A sediment supply large enough to generate an accretionary coast in a calm environment may be insufficient to prevent the development of an erosional coast in a stormy environment. The climate also affects biological activity along coasts. For example, in the warm water of tropical climates, mangrove swamps flourish along the shore, and coral reefs form offshore. The reefs may build into a broad carbonate platform such as appears in the Bahamas today. In cooler climates, salt marshes develop, while in arctic regions, the coast may be a stark environment of lichen-covered rock and barren sediment.

18.9 COASTAL PROBLEMS AND SOLUTIONS

Contemporary Sea-Level Changes

People tend to view a shoreline as a permanent entity. But in fact, shorelines are ephemeral geologic features. On a time scale of hundreds to thousands of years, a shoreline moves inland or seaward depending on whether the relative sea level rises or falls. In places where the sea level is rising today, shoreline towns will eventually be submerged. For example, the Persian Gulf now covers about twice the area that it did 4,000 years ago. And if present rates of sea-level rise along the East Coast of the United States continue, major coastal cities like Washington, New York, and Philadelphia may be inundated within the next millennium (▶Fig. 18.39). In the Gulf Coast region, the extraction of groundwater and a cutoff of sediment supply has led to the flooding of portions of the Texas and Louisiana coasts.

Beach Destruction—Beach Protection?

In a matter of hours, a hurricane can radically alter a landscape that took centuries or millennia to form. The back-

FIGURE 18.38 The development of a submerged coast. (a) A coast before the sea level rises. Rivers drain valleys onto a coastal plain. (b) The sea level rises and floods the valleys, and waves erode the headlands. (c) During the last ice age, North America's continental shelf lay exposed to the air, the United Kingdom and Ireland were not islands, and the Mediterranean was cut off from the Atlantic Ocean.

wash of storm waves sweeps vast quantities of sand seaward, leaving the beach a skeleton of its former self (▶Fig. 18.40a, b). The surf submerges barrier islands and shifts them toward the lagoon. Waves and wind together rip out mangrove swamps and salt marshes and fragment coral reefs, thereby destroying the organic buffer that normally protects the coast and leaving it vulnerable to erosion for years to come. Of course, major storms also destroy human constructions: erosion undermines shore-side buildings, causing them to collapse into the sea; wave impacts smash buildings to bits; and the **storm surge**—very high water levels created when storm winds push water toward the shore—floats buildings off their foundations and carries them inshore (▶Fig. 18.41a, b).

But even less dramatic events, such as the loss of river sediment, a gradual rise in sea level, the evolution of the shape of a shoreline, or the destruction of coastal vegetation, can alter the balance between sediment accumulation and sediment removal from a beach, leading to **beach erosion.** In some places, beaches retreat landward at rates of 1–2 m per year, forcing homeowners to pick up and move in the same direction. Even large lighthouses have been moved to keep them from washing away.

In many parts of the world, beachfront property has great value; but if a hotel loses its beach sand, it probably won't stay in business. Thus, property owners often construct artificial barriers to "protect" their stretch of coastline, or shelter the mouth of a harbor from waves. These

FIGURE 18.39 A possible sea-level rise in the future may flood major cities of the northeastern United States. The Washington–New York corridor would lie underwater.

FIGURE 18.40 A beach before (a) and after (b) a hurricane.

FIGURE 18.41 (a) Damaged beachfront homes after a hurricane in Florida. (b) Storms eroded away the foundation of this house, causing it to collapse.

Before   After

(a)   Groin

(b)   Jetty   Sandbar

(c)   Breakwater

FIGURE 18.42 (a) The construction of groins creates a sawtooth beach. (b) Jetties extend a river farther into the sea, but may result in the deposition of a sand bar at the jetties' ends. (c) A breakwater causes the beach to build out in the lee.

barriers alter the natural movement of sand in the beach system and thus change the shape of the beach, sometimes with undesirable results. For example, people may build **groins,** concrete or stone walls protruding perpendicular to the shore, to prevent beach drift from removing sand (▶Fig. 18.42a). Sand accumulates on the updrift side of the groin, forming a long triangular wedge, but sand erodes away on the downdrift side. Needless to say, the property owner on the downdrift side doesn't appreciate this process.

A pair of walls called **jetties** may protect the entrance to a harbor (▶Fig. 18.42b). But jetties erected at the mouth of a river channel effectively extend the river into deeper water, and thus may lead to the deposition of an offshore delta. Engineers may also build an offshore wall called a **breakwater,** parallel or at an angle to the beach, to prevent the full force of waves from reaching a harbor. With time, however, sand builds up in the lee of the breakwater and the beach grows seaward, clogging the harbor (▶Fig. 18.42c). And to protect expensive shoreside homes, people build **seawalls,** out of **riprap** (large stone or concrete blocks) or reinforced concrete, on the landward side of the backshore zone. Seawalls reflect wave energy that crosses the beach back to sea. Unfortunately, this process increases

the rate of erosion at the foot of the seawall, and thus during a large storm the seawall may be undermined so that it collapses (▶Fig. 18.43).

In some places, people have given up trying to decrease the rate of beach erosion, and instead have worked to increase the rate of sediment supply. To do this, they truck or ship in vast quantities of sand to replenish a beach. This procedure, called **beach nourishment,** can be hugely expensive and at best provides only a temporary fix, for the backwash and beach drift that removed the sand in the first

FIGURE 18.43 A seawall protects the sea cliff under most conditions, but during a severe storm the wave energy reflected by the seawall helps scour the beach. As a result, the wall may be undermined and collapse.

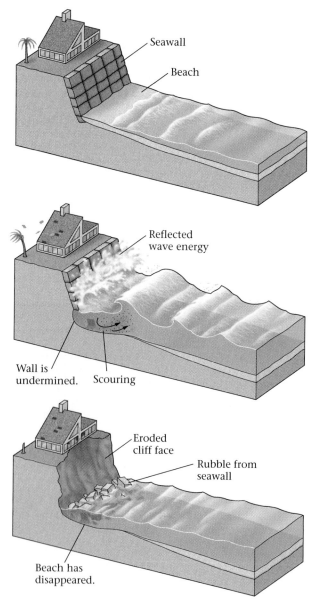

Seawall

Beach

Reflected wave energy

Wall is undermined.   Scouring

Eroded cliff face

Rubble from seawall

Beach has disappeared.

place continue unabated as long as the wind blows and the waves break. Clearly, beach management remains a controversial issue, for beachfront properties are expensive, but the shore is, geologically speaking, a temporary feature whose shape can change radically with the next storm.

Bad cases of **beach pollution** create headlines. Because of beach drift, garbage dumped in the sea in an urban area may drift along the shore and be deposited on a tourist beach far from its point of introduction. For example, in the last decade, hospital waste from New York City washed up on beaches tens of kilometers to the south. **Oil spills**—most commonly from ships that flush their bilges but also from tankers that have run aground or foundered in stormy seas—have contaminated shorelines at several places around the world. The oil coats beach sand and gravel and later, after the volatile component of the oil has evaporated, leaves behind tar, which, when mixed in with sand and pebbles, creates natural asphalt.

Pollution and the Destruction of Organic Coasts

Coasts in which living organisms control landforms along the shore are sometimes called **organic coasts.** These coasts are particularly susceptible to changes in the environment. Their loss can increase a coast's vulnerability to erosion and, considering that they provide spawning grounds for marine organisms, can upset the food chain of the global ocean.

In wetlands and estuaries, sewage, chemical pollutants, and agricultural runoff create havoc. Toxins settle along with flocculating clay and concentrate in the sediments, where they contaminate burrowing marine life (such as oysters) and then move up the food chain. Fertilizers and sewage that enter the sea with runoff increase the nutrient content of water, creating algae blooms that absorb oxygen and therefore kill animal and plant life. Coastal wetlands face destruction by development—they have been filled or drained to be converted into farmland or suburbia, and have been used as garbage dumps. In most parts of the world, between 20 and 70% of coastal wetlands have been destroyed in the last century.

Reefs, which depend on the health of delicate coral polyps, can be devastated by even slight changes in the environment. Pollutants and hydrocarbons, for example, will poison them. Organic sewage fosters algae blooms that rob water of dissolved oxygen and suffocate the coral. And agricultural runoff or suspended sediment introduced to coastal water during beach-nourishment projects reduces the light, killing the algae that live in the coral, and clogs the pores that coral polyps use to filter water. Changes in water temperature or salinity caused by dumping waste water from power plants into the sea or by global warming of the atmosphere also destroy reefs, for reef-building organisms

are very sensitive to temperature changes. People can destroy reefs directly by dragging anchors across reef surfaces, by touching reef organisms, or by quarrying reefs to obtain construction materials. In the last decade, marine biologists have noticed that reefs around the world have lost their color and died. This process, called **reef bleaching,** may be due to the removal or death of *zooxanthellae,* in response to the warming of seawater triggered by El Niño, or may be a result of the dust carried by winds from desert or agricultural areas.

CHAPTER SUMMARY

- The landscape of the sea floor depends on the character of the underlying crust. Wide continental shelves form over passive-margin basins, while narrow continental shelves form over accretionary prisms. Continental shelves are cut by submarine canyons. Abyssal plains develop on old, cool oceanic lithosphere. Seamounts and guyots form above hot spots.

- The salinity, temperature, and density of seawater vary with location and depth.

- Water in the oceans circulates in currents. Surface currents are driven by the wind and are deflected in their path by the Coriolis effect. The vertical upwelling and downwelling of water creates deep currents. Some of this movement is thermohaline circulation, a consequence of variations in temperature and salinity. Surface currents and thermohaline circulation transport heat in the oceans.

- Tides—the daily rise and fall of sea level—are caused by a tide-generating force, mostly driven by the gravitational pull of the Moon.

- Waves are caused by friction where the wind shears across the surface of the ocean. Water particles follow a circular motion, in a vertical plane, as a wave passes. Waves refract (bend) when they approach the shore because of frictional drag with the sea floor. Water moving parallel to the shore creates a longshore current.

- Sand on beaches moves the swash and backswash of waves. If there is a longshore current, the sand gradually moves along the beach and may extend outward from headlands to form sand spits. Beaches contain a number of distinctive landforms.

- At rocky coasts, waves grind away at rocks, yielding such features as wave-cut beaches and sea stacks. Some shores are wetlands, where marshes or mangrove swamps grow. Coral reefs grow along coasts in warm, clear water.

- The differences in coasts reflect their plate tectonic setting, whether the sea level is rising (to form a submergent coast) or falling (to form an emergent coast), the sediment supply, and the climate.

• To protect beach property, people build groins, jetties, breakwaters, and seawalls, but these structures may produce undesired consequences.

• Human activities have led to the pollution of coasts. And in the last decade, reef bleaching has become dangerously widespread.

KEY TERMS

abyssal plain (p. 550)
active continental margin (p. 549)
amplitude (p. 560)
atoll (p. 571)
backshore zone (p. 564)
barrier island (p. 565)
barrier reef (p. 571)
bathymetry (p. 547)
beach drift (p. 564)
berm (p. 564)
bioturbation (p. 565)
breakwater (p. 576)
cliff retreat (p. 566)
continental rise (p. 549)
continental shelf (p. 549)
continental slope (p. 549)
Coriolis effect (p. 555)
downwelling zone (p. 554)
ebb tide (p. 559)
eddies (p. 553)
Ekman spiral (p. 556)
Ekman transport (p. 556)
embayment (p. 562)
estuary (p. 569)
eustatic sea-level change (p. 572)
fjord (p. 570)
flood tide (p. 559)
foreshore (intertidal) zone (p. 564)
fracture zone (p. 550)
fringing reef (p. 571)
guyot (p. 550)
gyre (p. 553)
halocline (p. 551)
headlands (p. 562)

jetty (p. 576)
longshore current (p. 561)
mid-ocean ridge (p. 550)
neap tide (p. 559)
passive continental margin (p. 549)
pelagic sediment (p. 550)
pycnocline (p. 553)
rip current (p. 562)
salinity (p. 551)
sea arch (p. 568)
sea stack (p. 568)
seamount (p. 550)
sediment budget (p. 565)
spring tide (p. 559)
submarine fan (p. 550)
thermocline (p. 553)
thermohaline circulation (p. 557)
tidal bore (p. 559)
tidal flat (p. 565)
tidal reach (p. 559)
tombolo (p. 568)
trench (p. 550)
turbidite (p. 550)
turbidity current (p. 550)
upwelling zone (p. 554)
water mass (p. 557)
wave base (p. 560)
wave-cut bench (platform) (p. 566)
wave-cut notch (p. 566)
wavelength (p. 560)
wave refraction (p. 561)

REVIEW QUESTIONS

1. How much of the Earth's surface is covered by oceans? How much of the world's population lives near a coast?

2. Describe the typical topography of a passive continental margin, from the shoreline to the abyssal plain.

3. How do the shelf and slope of an active continental margin differ from those of a passive margin?

4. Where does the salt in the ocean come from?

5. How does the salinity in the ocean vary?

6. How does the temperature of the ocean vary?

7. How does the density of seawater vary?

8. What factors control the direction of surface currents in the ocean?

9. What is the Coriolis effect, and how does it affect oceanic circulation?

10. What is the Ekman spiral, and how does it affect oceanic circulation?

11. How do currents determine where there is oceanic upwelling or downwelling?

12. What causes the tides?

13. Describe the motion of water molecules in a wave.

14. How does wave refraction cause longshore currents?

15. What is a rip current?

16. How does beach sand migrate as a result of longshore currents?

17. Describe how waves affect a rocky coast.

18. What is an estuary? Why is it such a delicate ecosytem?

19. Describe how a reef system on a seamount evolves from a fringing reef to an atoll.

20. How do plate tectonics, sea-level changes, sediment supply, and climate change affect the shape of a coastline?

21. How does a sea-level change affect human activities in coastal regions?

22. How does human interference with beach drift cause problems?

SUGGESTED READING

Anderson, R. 1986. *Marine Geology.* New York: Wiley.

Barnes, R. S. K., ed. 1977. *The Coastline.* New York: Wiley.

Beer, T. 1983. *Environmental Oceanography.* New York: Pergamon Press.

Bird, E. C. F. 1969. *Coasts.* Cambridge: MIT Press.

Carter, W. 1988. *Coastal Environments.* New York: Academic Press.

Cone, J. 1991. *Fire Under the Sea.* New York: Morrow.

Davis, R. A. 1994. *The Evolving Coast.* New York: Freeman.

Hardisty, J. 1990. *Beaches: Form and Process.* New York: HarperCollins.

Heezen, B. C., and C. D. Hollister. 1971. *The Face of the Deep.* New York: Oxford University Press.

Heikoff, J. M. 1980. *Marine Shoreland Resources Management.* Ann Arbor, Mich.: Ann Arbor Science.

Kennett, J. P. 1982. *Marine Geology.* Englewood Cliffs, N.J.: Prentice-Hall.

Pethick, J. 1984. *An Introduction to Coastal Geomorphology.* London: Edward Arnold.

Scrutton, R. A., and M. Talwani, eds. 1982. *The Ocean Floor.* New York: Wiley.

Seibold, E., and W. H. Berger. 1982. *The Sea Floor.* New York: Springer Verlag.

Shepard, F. P. 1977. *Geological Oceanography.* New York: Crane, Russak.

CHAPTER 19

A Hidden Reserve: Groundwater

19.1 INTRODUCTION

Imagine Rosa May Owen's surprise when, on May 8, 1981, she looked out her window and discovered that a large sycamore tree in the backyard of her Winter Park, Florida, home had suddenly disappeared. It wasn't a particularly windy day, so the tree hadn't blown over—it had just vanished! When Owen went outside to investigate, she found that more than the tree had disappeared. Her whole backyard had become a deep, gaping hole. The hole continued to grow for a few days until finally it swallowed Owen's house and six other buildings, as well as the deep end of the municipal swimming pool (and all of the pool's water), part of a road, and the stock of expensive Porsches in a car dealer's lot (►Fig. 19.1a).

What had happened in Winter Park? The bedrock beneath the town consists of limestone, a fairly soluble rock, as does the bedrock beneath much of Florida. **Groundwater,** the water that resides under the surface of the Earth, gradually dissolved the limestone, carving open rooms, or caverns, underground. In fact, unbeknown to its inhabitants, Winter Park had been built atop a network of underground caverns. On May 8, the roof of the cavern underneath Owen's backyard began to collapse, forming a circular depression called a **sinkhole.** The sycamore tree and the rest of the neighborhood simply dropped down into the sinkhole. It would have taken too much effort to fill in the hole with soil, so the community allowed it to fill with water, and now it's a circular lake, the centerpiece of a pleasant municipal park. Similar lakes appear throughout central Florida (►Fig. 19.1b).

The Winter Park sinkhole serves as one of the more dramatic reminders that significant quantities of water reside underground. While we can easily see Earth's surface water (in lakes, rivers, streams, marshes, and oceans) and atmospheric water (in clouds and rain), groundwater lies hidden beneath the surface in the pores and cracks found within sediment or rock. We see groundwater only when it seeps out of a spring, when it has been pumped up in a well, or where it flows on the floor of a cavern large enough for us to enter. Nevertheless, groundwater has increasingly become a major supply of water for homes, agriculture, and industry (►Fig. 19.2).

This chapter provides a basic picture of where groundwater comes from, how it flows, and how it interacts with the rock and sediment it flows through. We also

Groundwater can turn a desert green, as shown by these irrigated fields sprouting in the sands of Jordan.

(a) (b)

FIGURE 19.1 (a) Winter Park, Florida, sinkhole; (b) sinkhole lakes in central Florida, as seen from a high altitude.

find out why geysers erupt and how caverns form. Much of the world's population depends on groundwater, but unfortunately we have put this hidden reserve at risk by extracting groundwater faster than it can be replenished or by contaminating it with toxic waste.

19.2 WHERE DOES GROUNDWATER RESIDE?

The Underground Reservoir of the Hydrologic Cycle

As we saw in Interlude E, water moves between various reservoirs (the ocean, the atmosphere, rivers and lakes,

FIGURE 19.2 Circular groundwater-irrigated fields in the relatively dry high plains of the United States.

groundwater, living organisms and soil, and glaciers) during the **hydrologic cycle.** When atmospheric moisture rains or snows, about 76% falls into the ocean, and the remainder falls onto land. Of the moisture that falls on land, some evaporates directly back into the atmosphere, some gets trapped in glaciers, and some becomes runoff that enters a network of streams and lakes that drains to the sea. The remainder sinks, or **infiltrates,** into the ground, because the upper part of the crust behaves like a giant sponge that can soak up water. How much water infiltrates into the crust at a certain location depends on the vegetation cover (vegetation absorbs water) and on the composition of materials making up the surface layer of the Earth—loose sand lets water sink in easily, while hard-packed clay, concrete, or dense, unfractured rock does not.

Of the water that does infiltrate, some descends only into the soil and wets the surfaces of grains and organic material making up the soil. This water, called **soil moisture,** later evaporates back into the atmosphere or gets sucked up by the roots of plants and then transpires back into the atmosphere. But some water sinks deeper and fills available spaces in cracks and between grains of sediment or rock; we call this water **groundwater.** Groundwater slowly flows underground for anywhere from a few months to tens of thousands of years before returning to the surface to pass once again into other reservoirs of the hydrologic cycle. (Note: The age of groundwater is determined by the radiocarbon dating of atmospheric carbon dioxide dissolved in the groundwater, if we can assume that the CO_2 dissolved before the water infiltrated into the ground.) At any given time, groundwater accounts for about two-thirds of the world's freshwater supply.

When discussing groundwater, geologists are usually referring to freshwater that started as rain and then infiltrated into the ground. But notably, some groundwater began as seawater that was trapped with sediment during deposition and did not escape when the sediment was buried. Keep in mind that groundwater exists as well in the sediment and rock comprising the crust beneath the ocean floor, for seawater can infiltrate downward too.

Porosity: The Home of Groundwater

Where does groundwater reside in the subsurface? Contrary to popular belief, only a small proportion flows freely in the underground lakes and streams of cavern networks. Most groundwater is found within the pore space of what might at first glance look like solid rock or sediment. Generally speaking, a **pore** is any open space (as opposed to solid material) within sediment or rock, and **porosity** refers to the total volume of empty space in a material, usually expressed as a percentage. For example, if we say that a piece of sandstone has 30% porosity, then 30% of a block of what looks like solid sandstone actually consists of open space. Table 19.1 shows the wide range of porosity in geologic materials.

Geologists further distinguish between primary and secondary porosity. **Primary porosity** is the space that remains between solid grains or crystals immediately after sediment accumulates or rocks form. To see why solid pieces do not fit together perfectly, fill a glass with marbles. The marbles contact one another at many points, but no matter how you shake the glass to encourage the marbles to settle together, open spaces still exist between the marbles except at contact points.

Let's look at different types of primary porosity (▶Fig. 19.3a–g). In unlithified sediment, like sand or gravel, primary porosity exists for the simple reason that rounded clasts can't fit together tightly. In such sediment, porosity depends on the size and sorting of clasts (see Chapter 7)—the overall porosity of a poorly sorted sediment (a mixture of different-sized grains) is less than that of a well-sorted sediment, because in a poorly sorted sediment smaller grains fill in spaces between larger grains. The primary porosity of a

FIGURE 19.3 Various kinds of primary porosity in rock. (a) Vesicles create variable porosity in basalt. (b) Uncemented spaces create moderate porosity in fossiliferous limestone. (c) Shale has packed together clay flakes, resulting in very low porosity. (d) Well-sorted sandstone has high porosity. (e) Poorly sorted sandstone has lower porosity. (f) Metamorphic rock (such as marble) has very low porosity. (g) Igneous rocks (such as granite) have very low porosity.

TABLE 19.1 Porosity of Some Geologic Materials

Uncompacted mud	40–70%
Highly fracture rock	up to 60%
Pumice	40–60%
Unconsolidated sand	25–50%
Poorly cemented sandstone	5–30%
Shale	less than 10%
Unfractured igneous rock	less than 5%
Unfractured marble	less than 3%
Unfractured obsidian	less than 1%

sediment tends to decrease with greater burial depth, because the weight of overburden pushes the sediment together. During the transition from sediment to sedimentary rock, primary porosity decreases further because some space between sediment grains fills with cement (minerals precipitated from groundwater). Nevertheless, a significant amount of primary porosity typically remains in sedimentary rock.

FIGURE 19.4 Fractures in a rock provide secondary porosity.

In most crystalline igneous and metamorphic rocks, grains interlock as they grow, so the amount of primary porosity is small. In fine-grained and glassy igneous rocks, vesicles (gas bubbles trapped when the rock freezes from lava), if they exist, provide primary porosity. Frothy rocks, like pumice, contain so much primary porosity that they float.

Secondary porosity refers to new pore space in rocks, created some time after the rock first forms. When rocks fracture, the opposing walls of the fracture do not fit together tightly, so a narrow space remains in between. Thus, joints and faults may provide openings for water (▶Fig. 19.4). Faulting may also fracture rock to form breccia, a jumble of angular fragments, with the space between the fragments providing another type of secondary pore space. Finally, as groundwater passes through rock, it may dissolve and remove some minerals, creating secondary porosity where once there was solid. Such solution cavities are common in limestone, a soluble rock.

Not all water within a pore can move freely. Some water chemically bonds to minerals, because water molecules feel an electrostatic attraction to mineral surfaces. This bonded water is called **immobile water.** Water molecules that do not feel this electrostatic attraction, and thus can move, make up **mobile water.** Pore spaces or cracks must be greater than 0.05 mm across to hold mobile water—if the space is narrower, all the water in it is immobile.

19.3 PERMEABILITY: THE EASE OF FLOW

If solid rock completely surrounds a pore, the water in the pore cannot flow to another location. For groundwater to flow, therefore, pores must be linked by conduits (openings). The ability of a material to allow fluids to pass through an interconnected network of pores is a characteristic known as **permeability** (▶Fig. 19.5a, b). A permeable material is one through which water flows easily, whereas an impermeable material is one through which water flows slowly or not at all. The permeability of a material depends on several factors.

- *Number of available conduits:* As the number of conduits increases, the permeability increases.

- *Size of the conduits:* More fluids can travel through wider conduits than through narrower ones.

- *Straightness of the conduits:* Water flows more rapidly through straight conduits than it does through crooked ones. In crooked channels, the distance a water molecule actually travels is many times the straight-line distance between the two end points. The process of meandering through tiny crooked channels is called **percolation.**

Note that the factors that control permeability in rock or sediment resemble those that control the ease with which traffic moves through a city. Traffic can flow quickly where there are many straight, multilane boulevards, while it can only flow slowly along a few narrow, crooked streets.

Porosity and permeability are not the same. A material whose pores are isolated from one another can have high porosity but low permeability. For example, if you plug the bottom of a water-filled tube with a cork, the water remains in the tube almost indefinitely, even though cork is very porous; this is because woody cell walls separate adjacent pores in the cork and prevent communication between them, making cork impermeable.

FIGURE 19.5 (a) Isolated, nonconnected pores in an impermeable material; (b) pores connected to one another by a network of conduits in permeable material.

(a)

|— 1 mm |

(b)

In some cases, defining the permeability of a region depends on the scale of observation. For example, water cannot easily pass through solid granite, but it *can* pass along joints. Thus, a region underlain by granite that contains many joints may have high permeability, though the granite between the joints is impermeable.

Aquifers and Aquitards

With the concept of permeability in mind, **hydrogeologists** (geologists who study groundwater) distinguish between **aquifers,** sediment or rocks that transmit water easily, and **aquitards,** sediment or rocks that do not transmit water easily and therefore retard the motion of water. **Aquicludes** do not transmit water at all. When hydrogeologists talk about the **principal aquifer** of a region, they are referring to the geologic unit that serves as the primary source of groundwater in that region. For example, a Cretaceous sandstone formation called the Dakota Sandstone serves as the principal aquifer for much of the high plains region of the United States. Gravel deposited during the last ice age, in some cases completely filling up old river valleys, provides the principal aquifers for the northern Midwest. Aquifers that intersect the surface of the Earth are called **unconfined aquifers,** because water can percolate directly from the surface down into the aquifer, and water in the aquifer can rise to the surface. Aquifers that are separated from the surface by an aquitard are called **confined aquifers,** as the water they contain is isolated from the surface (▶Fig. 19.6).

FIGURE 19.6 An aquifer is a high-porosity, high-permeability rock. If it has access to the ground surface, it's an unconfined aquifer. If it's trapped below an aquitard, a rock with low permeability, it's a confined aquifer.

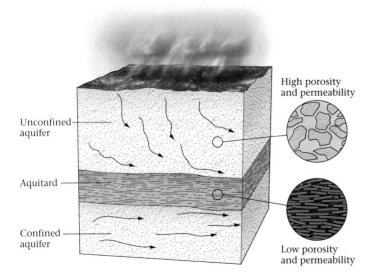

19.4 THE WATER TABLE

We've seen that groundwater resides in subsurface pore space. But are pores filled with water right up to the ground surface everywhere? The answer is no. Geologists define a boundary, called the **water table,** above which pore spaces contain mostly air and below which they contain only water (▶Fig. 19.7a). In technical jargon, the region of the subsurface above the water table is called the **unsaturated zone,** or the **zone of aeration,** and the region below the water table is the **saturated zone,** or **zone of saturation.** Material above the water table can be damp, but the dampness (soil moisture, if in soil) reflects the presence of immobile water. Typically, the electrostatic attraction (surface tension) of water molecules to mineral surfaces causes water to seep up from the water table (just as water rises in a thin straw), filling pores in the **capillary fringe,** a thin layer between the saturated and unsaturated zones.

How deep the water table lies in the subsurface varies greatly with location. The surface of a permanent stream, lake, or marsh, for example, defines the water table at that location, for water saturates the soil or rock below lakes, streams, and marshes (▶Fig. 19.7b, c). Elsewhere, the water table lies hidden below the ground surface: in humid regions, it typically lies within a few meters of the ground surface, whereas in arid regions, it may lie tens of meters to over 200 m below the surface.

Rainfall affects the water-table depth—the level sinks during a dry season or drought (▶Fig. 19.7d). If the water table drops below the floor of a river or lake, the river or lake dries up, because the water it contains infiltrates into the ground. In arid regions, there are no permanent streams, as the water table lies far below the stream bed. Streams in such regions flow only during storms, when rain falls at a faster rate than the water can infiltrate underlying rock or sediment.

Note that "water table" refers to the position of the *top* of the groundwater reservoirs beneath land. Defining the bottom of the groundwater reservoirs proves to be more controversial. As depth in the crust increases, the pressure of overlying rocks squeezes pores together so that the pore volume decreases, and the pores that remain tend to fill with minerals precipitated out of groundwater. Thus, hydrogeologists estimate that more than half of the groundwater on the planet is found in the upper kilometer of the crust. In regions where the geothermal gradient is low, deep-drilling tests have located water in rock at least down to a depth of 10 km. Below 10–15 km, rocks are too weak to keep pore spaces open (minerals plastically flow and close up any space), and/or minerals in rocks begin to undergo metamorphic reactions that involve water, so almost no mobile groundwater can exist there.

(a)

(b)

(c)

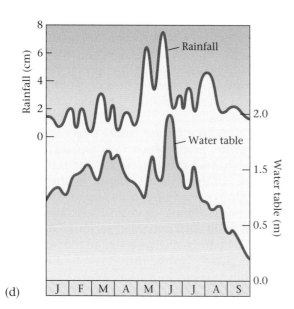

(d)

FIGURE 19.7 (a) The geometry of the water table, illustrating the saturated zone, the unsaturated zone (zone of aeration), and the capillary fringe. (b) The surface of a permanent pond is the water table. (c) During the dry season, the water table can drop substantially, causing the pond to dry up. (d) The graph shows the relation between the height of the water table and rainfall during one year (January–September).

FIGURE 19.8 The shape of a water table beneath hilly topography. The "head," or elevation of the water table at a given location, is higher beneath a hill. The head at h_1 is greater than at h_2.

Topography of the Water Table

In hilly regions, the water table is not a planar surface, but rather its shape mimics, in a subdued way, the shape of the overlying topography (▶Fig. 19.8). This means that the water table lies at a higher elevation beneath hills than it does beneath valleys. But the relief (the vertical distance between the highest and lowest elevations) of the water table is not as great as that of the overlying land, so the surface of the water table tends to be smoother than that of the landscape. The elevation of the surface of the water table above a reference horizon is called the **head** (h) of the water table.

At first thought, it may seem surprising that the elevation of the water table varies as a consequence of ground-surface topography; after all, when you pour a bucket of water into a pond, the surface of the pond immediately adjusts to remain horizontal. The elevation varies because groundwater moves so slowly through rock and sediment that it cannot quickly assume a horizontal surface. When it rains on a hill and water infiltrates down to the water table, the water table rises a little.

While it doesn't rain, the water table sinks slowly, but so slowly that rain will fall again and make the water table rise before it has had time to sink down to the level of the water table in an adjacent valley.

Perched Water Tables

In some locations, layers of strata are discontinuous, meaning that they pinch out at their sides. As a result, lens-shaped layers of impermeable rock (such as shale) may lie within a thick aquifer. A mound of groundwater accumulates above this aquitard. The result is a **perched water table,** a quantity of groundwater that lies above the regional water table because an underlying lens of impermeable rock or sediment prevents the water from sinking down to the regional water table (▶Fig. 19.9).

19.5 GROUNDWATER FLOW: FROM RECHARGE TO DISCHARGE

Groundwater Flow Paths

Most groundwater does not stand still, but rather flows, sometimes moving great distances underground. In the zone of aeration, water percolates straight down, as you might expect for a fluid feeling the force of gravity. But in the zone of saturation, the water flow is a bit more complex, for in addition to the downward pull of gravity, the water feels differences in pressure.

To better understand how pressure variations control groundwater flow, imagine that you have two water-filled

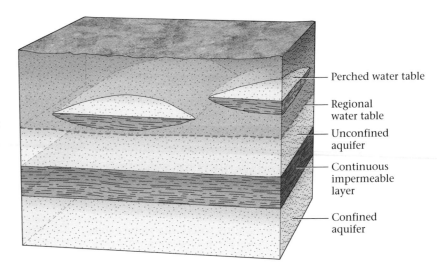

FIGURE 19.9 The configuration of a perched water table. A lens of groundwater lies above, and the regional water table lies at greater depth.

Perched water table

Regional water table

Unconfined aquifer

Continuous impermeable layer

Confined aquifer

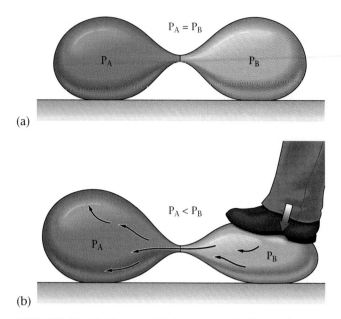

(a)

(b)

FIGURE 19.10 This model of connected balloons shows how differences in pressure drive the flow of groundwater. (a) The pressure (P) is the same in both balloons, so water doesn't flow. (b) If you increase pressure in balloon B, water flows into balloon A. Note that the pressure can even cause water to flow upward.

balloons connected at their necks (▶Fig. 19.10a, b). If the water pressure in the two balloons is the same, the water doesn't move. But if you squeeze one balloon so that it feels greater pressure, water in that balloon flows into the other until the pressure in both balloons is the same. If you squeezed one balloon hard enough, you could even get the water to flow up, against the pull of gravity.

Pressure in the groundwater of an unconfined aquifer is caused by the weight of overlying water. (The weight of overlying rock does not cause the pressure, for this weight is borne at the contact points between grains.) Thus, the magnitude of pressure in groundwater depends on the depth below the water table—the thicker the layer of water above, the greater the pressure. If the water table is horizontal, the pressure acting on an imaginary horizontal plane at some depth below the water table will be the same everywhere. But if the water table is not horizontal, the pressure on this imaginary horizontal plane below varies with location. For example, picture a landscape in which a valley lies between two hills. Groundwater at a specific elevation underground feels greater pressure under the hills, where the water table is higher, than under the valley, where the water table is lower. This difference in pressure, combined with the force of gravity, which always pulls straight down, causes groundwater to follow a curved path from regions where the water table is high (under a hill) to regions where the water table is low (below a valley). The shape of the curved path varies with location—some water flows nearly parallel to the water table down the slope of the water table, while some flows

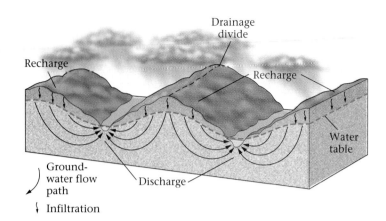

FIGURE 19.11 The flow lines from the recharge area to the discharge area curve through the substrate. In fact, some groundwater descends to great depth and then rises back to the surface.

nearly straight down, curves, and then rises nearly vertically up. Remember, pressure differences drive this upward flow.

The location where water enters the ground (that is, where the flow direction has a downward trajectory) is called the **recharge area,** and the place where groundwater flows back up to the surface is called the **discharge area** (▶Fig. 19.11). Locally, the recharge area occurs at the top of a hill, whereas the discharge area is located in the valley floor. But in large sedimentary basins (broad regions containing a thick sequence of sediment or sedimentary rock), flow lines from the recharge to the discharge area may carry the water all the way from one side of the basin to the other (▶Fig. 19.12). For example, the major recharge area of eastern Australia lies in the mountains along Australia's east coast; this water flows westward and resurfaces at a discharge area in central Australia, nearly 1,000 km away.

FIGURE 19.12 Cross section showing regional-scale groundwater flow in a sedimentary basin. The arrows indicate the flow paths of groundwater. Note that recharge occurs in the hills and discharge in the basins. Some of the water flows through fractures in the basement.

Curving flow lines mean that groundwater may flow to great depths before rising back to the surface. In fact, in a large sedimentary basin, water may reach a depth of 7–10 km before rising back. The deeper the water goes, the longer it takes for the water to cover the distance from the recharge to the discharge area.

Rates of Groundwater Flow

Flowing water in an ocean current moves at about 3 km per hour (over 26,000,000 m per year), and water in a steep river channel can reach speeds in excess of 30 km per hour (over 260,000,000 m per year). In contrast, groundwater moves at a snail's pace—typical rates range between 0.01 and 1.4 m per day (about 4 to 500 m per year). Groundwater, in general, moves much more slowly than open water because it percolates through a complex network of tiny conduits; it must follow a crooked path and travel a great distance. In addition, friction between water and conduit walls slows down the water flow.

Not all groundwater flows at the same rate. The rate at a given location depends on two factors. First, the permeability of the material containing the groundwater: groundwater flows faster in material with greater permeability than in material with less permeability. Second, the flow rate depends on the slope of the water table: the steeper the slope, the faster the water moves, because the greater the pressure difference between the recharge and discharge areas. Hydrogeologists refer to the slope of the water table as the **hydraulic gradient** (▶Fig. 19.13). Written as an equation, the hydraulic gradient between two points = $\Delta h/d$. In this equation, $\Delta h = (h_1 - h_2)$, where h_1 is the head (elevation of the water table) at one point and h_2 is the head at the other point, and d is the horizontal distance between the two points.

Hydrogeologists may measure the rate of groundwater flow in a region by "tagging" water with dye or a radioactive element. They then inject the water in one well and time how long it takes for the water to reach another well down the hydraulic gradient. They may also calculate the rate by using radiocarbon dating to measure the age of groundwater. If the distance that the groundwater has traveled to the discharge area can be determined, the rate of flow can be calculated by applying the equation: rate = distance / the age of the groundwater.

Darcy's Law

Now let's consider the practical problem of determining the volume of water that will pass through an area below the ground during a specified time. In 1856, this problem occupied a French engineer, Henri Darcy, who wanted to find out whether the city of Dijon in central France could supply its water needs from groundwater. Darcy carried out a series of experiments in which he measured the rate of water flow through sediment-filled tubes tilted at varying angles. Darcy found that the volume of water that passed through a specified area in a given time, a quantity he called the discharge (Q), can be calculated from the equation

$$Q = K(\Delta h/d)A,$$

where $\Delta h/d$ is the hydraulic gradient, A is the area through which the water is passing, and K is a coefficient called the **hydraulic conductivity.** The hydraulic conductivity takes into account the permeability as well as the fluid's viscosity ("stickiness"). Hydrogeologists refer to this equation as **Darcy's law.**

In words, Darcy's law simply states that groundwater flow is faster through very permeable rocks than it is through impermeable rocks, and is faster where the water table has a steep slope than where the water table has a shallow slope. Not surprisingly, therefore, groundwater moves very slowly (0.5 to 3.0 cm per day) through a well-cemented layer of the Dakota Sandstone in the Great Plains, but moves relatively quickly (over to 15 cm per day) through a steeply dipping layer of unconsolidated gravel on the side of a steep mountain.

FIGURE 19.13 A hydraulic gradient (HG) simply represents the slope of the water table (d is the horizontal distance between points h_1 and h_2). Groundwater near the water table flows roughly parallel to the water table, down the gradient.

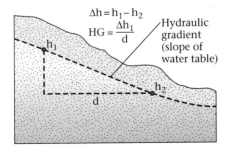

19.6 THE CHEMISTRY OF GROUNDWATER

Groundwater is not distilled water; rather, it is a dilute solution of ions derived by the dissolution of the rock through which the groundwater passes. The quantity of ions that groundwater can carry (that is, the *concentration* of ions in solution) depends on a variety of factors, including temperature and pressure. Warm water can contain more dissolved salt than cold water, and water under high pressure at depth can hold more dissolved carbon dioxide than water standing at the Earth's surface.

Groundwater that carries as many ions as possible under given environmental conditions is called a **saturated solution.** If the water has the capacity to dissolve more ions, it is an **undersaturated solution.** When environmental conditions change so that the capacity of saturated groundwater to carry ions decreases, it temporarily becomes **oversaturated.** When undersaturated groundwater passes through rock, it dissolves minerals until it becomes saturated. This reaction may create secondary porosity in the rock. If saturated groundwater then rises into cooler rock, it becomes oversaturated and precipitates minerals. This process may lead to the formation of new cement or veins (mineral-filled cracks). In some cases, groundwater becomes a complex solution out of which minerals like clay, feldspar, and a variety of ore minerals can precipitate.

The presence of dissolved ions in water affects the value of the water for human use. Groundwater that has passed through salt-containing strata becomes quite salty, and thus unsuitable for irrigation or drinking. Groundwater that has passed through limestone or dolomite contains dissolved calcium (Ca^+) and magnesium (Mg^+) ions; this is called **hard water,** and it can be a problem because carbonate minerals precipitate to form the "scale" that clogs pipes. Also, it can be difficult to generate soap lather when washing with hard water. Groundwater that has passed through iron-rich rocks may contain dissolved iron oxide, which precipitates to form an orange stain in water containers. Typically, water that has been underground longer has had more time to dissolve minerals and tends to be richer in dissolved chemicals.

19.7 TAPPING THE GROUNDWATER SUPPLY

Groundwater can be brought to the ground surface at wells or springs. **Wells** are holes that people dig or drill to obtain water. **Springs** are natural outlets from which groundwater flows. Wells and springs provide a welcome source of water but must be treated with care if they are to last.

Wells

In an **ordinary well,** the base of the well penetrates an aquifer below the water table. Water from the pore space in the aquifer seeps into the well and fills it; the water surface in the well is the water table (▶Fig. 19.14). Drilling a well into an aquitard, or into rock that lies above the water table, will not supply water, and thus yields a **dry well.** Some ordinary wells are **seasonal wells** in that they only function during the rainy season, when the water table rises; during the dry season, the water table lies below the base of the well.

FIGURE 19.14 In this ordinary well, the water has to be brought up by bucket.

Ideally, well diggers want an ordinary well to be as shallow as possible (to decrease its cost). So hydrogeologists use maps (produced by government agencies) to search for particularly porous and permeable aquifers in which the water table lies near the surface. Contrary to legend, a dowser with a forked stick cannot help find water—when the charlatans who practice dowsing do strike water, either they have enjoyed dumb luck or they have geological knowledge about the water table in the area.

To obtain water from an ordinary well, you either pull water up in a bucket or pump the water out. As long as the rate of supply in the well is greater than the rate at which water is removed, the level of the water table near the well remains about the same. However, if users pump water out of the well too fast, then the water table sinks down around the well, a process called **drawdown,** so that the water table becomes a downward-pointing, cone-shaped surface called a **cone of depression** (▶Fig. 19.15). If the drawdown causes the point of the cone to drop below the base of a well, the well runs dry.

An **artesian well,** named for a region in France where they occur, penetrates confined aquifers. In such a well, water rises on its own to a level above the surface of the aquifer. If this level lies below the ground surface, the well is a **nonflowing artesian well.** But if the level lies above the ground surface, the well is a **flowing artesian well,** and water actively fountains out of the ground (▶Fig. 19.16a). Artesian wells are found in special situations where a confined aquifer lies beneath a sloping aquitard. Because of this configuration, the water pressure in the confined aquifer is great enough to push the water up.

We can understand why artesian wells exist if we look first at the configuration of a city water supply (▶Fig. 19.16b). Water companies pump water into a high tank that has a significant head relative to the surrounding areas. If

FIGURE 19.15 The base of an ordinary well penetrates below the water table. If groundwater is extracted faster than it can be replaced, a cone of depression forms around the well. Pumping by the big well, in this example, may lower the water table sufficiently to cause the small well to become dry.

(a)

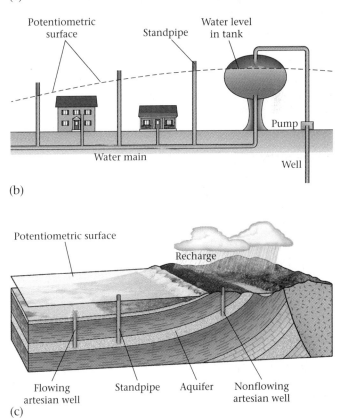

(c)

FIGURE 19.16 (a) A flowing artesian well. (b) The configuration of a city water supply. Water will rise in vertical pipes up to the level of the potentiometric surface. (c) The configuration of an artesian system. Artesian wells flow if the potentiometric surface lies above the ground surface. Nonflowing artesian wells form where the potentiometric surface lies below the ground.

the water were connected by a water main to a series of vertical pipes, pressure caused by the elevation of the water in the high tank would make the water rise in the pipes until it reached an imaginary surface, called a **potentiometric surface,** that lies above the ground. This pressure drives water through water mains to your household water system without requiring pumps. In an artesian system, water enters a tilted, confined aquifer that intersects the ground in the hills of a high-elevation recharge area (▶Fig. 19.16c). The confined groundwater flows down the hydraulic gradient to the adjacent plains, which lie at a lower elevation. The potentiometric surface to which the water would rise were it not confined lies above this aquifer; in fact, it may

lie above the ground surface of the plains. Pressure in the confined aquifer pushes water up a well. Where the potentiometric surface lies below ground, the well will be a nonflowing artesian well, but where the surface lies above the ground, the well will be a flowing artesian well.

Springs

More than one town has grown up around a spring, a place where groundwater naturally flows or seeps onto the Earth's surface, for in effect springs provide fresh, clear groundwater for drinking or irrigation, without the expense of drilling or digging. Springs form under a variety of conditions.

- Where the ground surface intersects the water table in a discharge area (▶Fig. 19.17a): such springs typically emit water upward into rivers, ponds, or valley floors.
- Where a perched water table intersects the surface of a hill (▶Fig. 19.17b).
- Where downward-percolating water runs into a continuous impermeable layer and migrates along the top surface of the layer to a hillslope (▶Fig. 19.17c).
- Where a particularly permeable layer or zone intersects the surface of a hill (▶Fig. 19.17d): water percolates down through the hill and then migrates along the permeable layer to the hill face.
- Where a network of interconnected fractures channels groundwater to the surface of a hill (▶Fig. 19.17e).
- Where flowing groundwater collides with a steep impermeable barrier, and pressure pushes it up to the ground along the barrier (▶Fig. 19.17f): faulting can create such barriers by juxtaposing impermeable rock against permeable rock.
- **Artesian springs** form if the ground surface intersects a natural fracture (joint) that taps a confined aquifer in which the pressure is sufficient to drive the water to the surface (▶Fig. 19.17g).

FIGURE 19.17 Springs form (a) where groundwater rises in a discharge area; (b) where a perched water table intersects the surface of a hill; (c) where groundwater has been forced to migrate along an impermeable barrier; (d) where a particular permeable layer transmits water to the surface of a hill; (e) where a network of interconnected fractures channels water to the hill face; (f) where groundwater collides with a steep impermeable barrier, and pressure pushes it up to the ground along the barrier. (g) An artesian spring forms where water from a confined aquifer migrates up a joint.

BOX 19.1

THE HUMAN ANGLE

Oases

The Sahara Desert of northern Africa is now one of the most barren and desolate places on Earth, for it lies in a climatic belt where rain seldom falls. But it wasn't always that way. During the last ice age, when glaciers covered parts of northern Europe on the other side of the Mediterranean, the Sahara enjoyed a more temperate climate, and the water table was high enough that permanent streams dissected the landscape. Using ground-penetrating radar, geologists can detect these streams today—they look like ghostly valleys beneath the sand. When the climate warmed after the ice age, rainfall diminished, the water table sank below the floor of most stream beds, and the streams dried up.

Today, the water of the Sahara region lies locked in a vast underground aquifer composed of porous sandstone. Recharge into this aquifer comes from highlands bordering the desert, from occasional downpours, and from the Nile River (particularly Lake Nassar, created by the Aswan High Dam). In general, the water of the aquifer can only be obtained by digging deep wells, but locally, water spills out at the surface—either because folding brings the aquifer particularly close to the ground so that valley floors intersect the water table, or because artesian pressure pushes groundwater up along joints of faults (▶Fig. 19.18a). In either case, the aquifer feeds springs that quench the thirst of desert and tropical plants and create an **oasis,** an island of green in the sand sea (▶Fig. 19.18b). Oases became important stopping points along caravan routes, allowing both people and camels to replenish water supplies.

In some oases, settlements developed, which used the groundwater to irrigate date palms and other crops. For example, the Bahariya oasis, about 400 km southwest of Cairo, Egypt, hosted a town of perhaps 30,000 between 300 B.C.E. and 300 C.E. During that time, the water table lay only 5 m below the ground and could be easily accessed by shallow wells. Today, as a result of changing climates and centuries of use, the water table lies 1,500 m below the ground, almost out of reach. Bahariya's glorious past came to light in 1996, quite by accident. A guard from a temple ruin in the oasis was riding his donkey in the nearby desert when the ground beneath the donkey suddenly caved in. The guard had inadvertently opened the roof into a tomb filled with over 150 mummies, along with thousands of well preserved artifacts. The site has since come to be known as the "Valley of the Mummies."

FIGURE 19.18 (a) This subsurface configuration of aquifers leads to the formation of an oasis, where groundwater reaches the surface. (b) An oasis in the Sahara Desert.

19.8 HOT SPRINGS AND GEYSERS

Hot springs, springs that emit water that ranges in temperature from about 30° to 104°C, are found in two geologic settings. First, they occur where very deep groundwater, heated in warm bedrock at depth, is able to flow up to the ground surface. This water brings heat with it as it rises. Such hot springs only form in places where faults or fractures provide a high-permeability conduit for deep water, or where the water emitted in a discharge region has followed a trajectory that first carried it deep into the crust. Second, hot springs develop in **geothermal regions,** places where volcanism currently takes place or has occurred recently, so that magma and/or very hot rock resides close to the Earth's surface. In hot springs, groundwater is a steaming tea of water and dissolved minerals. Hot groundwater contains more dissolved minerals because water becomes a more effective solvent when hot. People use the water emitted at hot springs to fill relaxing mineral baths (▶Fig. 19.19).

Numerous distinctive geologic features form in geothermal regions as a result of the eruption of hot water (▶Fig. 19.20a–d). In places where the hot water rises into soils rich in volcanic ash and clay, a viscous slurry forms and fills bubbling **mud pots.** Bubbles of steam rising through the slurry cause it to splatter about in goopy drops. Where geothermal waters spill out of natural springs and cool, dissolved minerals in the water precipitate, forming colorful mounds or terraces of travertine and other minerals. Geothermal waters may accumulate in brightly colored pools—the gaudy greens, blues, and oranges of these pools come from exotic bacteria that thrive in hot water and metabolize the sulfur-containing minerals dissolved in the groundwater.

FIGURE 19.19 In the city of Bath, England, the Romans built an elaborate spa around an artesian hot spring. The spring formed along a fault that tapped a supply of deep groundwater that still flows today. The baths of Bath have remained popular through the ages, and are currently undergoing renovation.

(a)

(b)

(c)

(d)

FIGURE 19.20 Features of geothermal regions. (a) Mud pots in Yellowstone National Park; (b) travertine terraces at Mammoth Hot Springs, Yellowstone; (c) colorful bacteria-laden pools, Yellowstone; (d) geysers at Yellowstone.

(a)

(b)

(c)

FIGURE 19.21 (a) Geysers and hot springs form where groundwater, heated at depth, rises to the surface. (b) Geysers in the geothermal region of Rotorua, New Zealand. (c) A warning sign in Rotorua, where steam can obscure visibility on highways.

The most spectacular consequence of geothermal waters is a **geyser** (from the Icelandic word for "gush"), a fountain of steam and hot water that erupts periodically from a vent in the ground (▶Fig. 19.21a). To understand why geysers erupt, we first need a picture of the underground plumbing. Beneath a geyser lies a network of irregular fractures in very hot rock; groundwater sinks and fills these fractures. Adjacent hot rock then "superheats" the water: it raises the temperature above the temperature at which water at a pressure of 1 atmosphere will boil. Eventually, this superhot water rises through a conduit to the surface. When some of this water transforms into steam, the resulting expansion causes water higher up to spill out of the conduit at the ground surface. When this spill happens, pressure in the conduit, from the weight of overlying water, suddenly decreases. A sudden drop in pressure causes the superhot water at depth to instantly turn to steam, and this steam quickly rises, ejecting all the water above it out of the conduit. Thus, the entire contents of the conduit quickly spew out at the surface as a mix of steam and boiling water, creating a geyser eruption. Once the conduit is

empty, the eruption ceases, and the conduit fills once again with water that gradually heats up, starting the eruptive cycle all over again. In a geyser system, the conduit must contain constrictions, so that water cannot easily reach the surface—straight conduits without constrictions yield hot springs instead of geysers.

Hot springs are found in many localities around the world: at Hot Springs, Arkansas, where deep groundwater rises to the surface; in Yellowstone National Park, above the magma chamber of a continental hot spot; around the Salton Sea in southern California, where the mid-ocean ridge of the Gulf of California merges with the San Andreas Fault; in Iceland, which has grown on top of an oceanic hot spot along the Mid-Atlantic Ridge; and in Rotorua, New Zealand, which lies in an active volcanic field above a subduction zone. Some geothermal regions are populated, though they present unique natural hazards. For example, in Rotorua, signs along the road warn of steam, which can obscure visibility, and steam indeed spills out of holes in backyards and parking lots (▶Fig. 19.21b, c). But all this hot water does offer a benefit: in Rotorua, waters circulate through pipes to provide home heating, and in geothermal areas worldwide, steam provides a relatively inexpensive means of generating electricity.

19.9 GROUNDWATER USAGE PROBLEMS

Since prehistoric times, groundwater has been an important resource that people have relied on for drinking, irrigation, and industry. Groundwater feeds the lushness of desert oases in the Sahara Desert, the amber grain in the North American high plains, and the growing cities of the sunbelt. Agricultural and industrial usage accounts for about 93% of all water usage, so as once-empty land comes under cultivation and countries become increasingly industrialized, demands on the groundwater supply soar. Globally, groundwater provides only about 20% of the water we use, but this number has increased as surface-water resources decrease. Though groundwater accounts for about 95% of the liquid freshwater on the planet, accessible groundwater cannot be replenished quickly in important locations, and this leads to shortages. In the last century, the problem has been exacerbated by the contamination of existing groundwater. Such pollution, caused when toxic wastes and other impurities infiltrate down to the water table, may be invisible to us but may ruin a water supply for generations to come. In this section, we'll take a look at the use and misuse of groundwater supplies.

Depletion of Groundwater Supplies

Is groundwater a renewable resource? In a time frame of 10,000 years, the answer is yes, for the hydrologic cycle will eventually resupply depleted reserves. But in a time frame of 100 to 1,000 years—the span of a human lifetime or a civilization—groundwater in many regions may be a nonrenewable resource. By pumping water out of the ground at a rate faster than nature replaces it, people are effectively "mining" the groundwater supply. In fact, in portions of the desert Southwest region of the United States, supplies of young groundwater have already been exhausted, and deep wells now extract 10,000-year-old groundwater. But such ancient water has been in rock so long that it has become too mineralized to be usable. A number of other problems accompany the depletion of groundwater.

• *Lowering the water table:* When we extract groundwater from wells at a rate faster than it can be resupplied, the water table drops. First, a cone of depression forms around the well; gradually the water table becomes lower in a broad region. As a consequence, existing wells, springs, and rivers dry up (▶Fig. 19.22a, b). To continue tapping into the water supply, we must drill progressively deeper. The lowering of water tables has become a serious problem in the desert Southwest of North America.

The water table can also drop when people divert surface water from the recharge area. Such a problem has developed in the Everglades of southern Florida, a huge swamp where, before the expansion of Miami and the development of agriculture, the water table lay at the ground surface (▶Fig. 19.22c, d). Diversion of water from the Everglades' recharge area into canals has significantly lowered the water table. As a result, parts of the Everglades have dried up.

• *Reversing the flow direction of groundwater:* The cone of depression that develops around a well creates a local slope to the water table. The resulting hydraulic gradient may be large enough to reverse the flow direction of nearby groundwater (▶Fig. 19.23a, b). Such reversals can lead to the contamination of a well, from pollutants seeping out of a septic tank.

• *Saline intrusion:* In coastal areas, fresh groundwater lies in a layer above saltwater that entered the aquifer from the adjacent ocean (▶Fig. 19.23c, d). (Saltwater is denser than freshwater, so the fresh groundwater floats above it.) If people pump water out of a well too quickly, the boundary between the saline water and the fresh groundwater rises. And if this boundary rises above the base of the well, then the well will start to yield useless saline water.

• *Pore collapse and land subsidence:* When groundwater fills the pore space of a rock, it holds the grains of the rock or regolith apart, for water cannot be compressed. The extraction of water from a pore eliminates the support holding the grains apart, because the air that replaces the water *can* be compressed. As a result, the grains pack more closely together. Such **pore collapse** permanently decreases the porosity and permeability of a rock, and thus lessens its value as an aquifer (▶Fig. 19.23e).

Pore collapse also decreases the volume of the aquifer, with the result that the ground above the aquifer sinks. Such **land subsidence** may cause fissures at the surface to develop and the ground to tilt (▶Fig. 19.23f). Buildings constructed over regions undergoing land subsidence may themselves tilt, or their foundations may crack. The Leaning Tower of Pisa, in Italy, tilts because the removal of groundwater caused its foundation to subside (▶Fig. 19.23g).

In the San Joaquin Valley of California, the land surface subsided by 9 m between 1925 and 1975, because water was removed to irrigate farm fields (▶Fig. 19.23h). In coastal areas, land subsidence may even

(a)

(b)

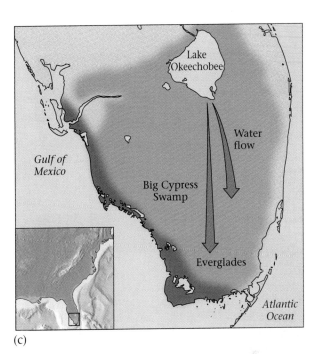

(c)

FIGURE 19.22 (a) Before a water table is lowered, a large swamp exists. (b) Pumping by a nearby city causes the water table to sink, so the swamp dries up. (c) The Everglades, in Florida, before the advent of urban growth and intensive agriculture. Water flowed south from Lake Okeechobee, creating a fast swamp in which the water table lay just above the ground surface. In effect, the Everglades is a "river of grass." (d) Channelization and urbanization have removed water from the recharge area, disrupting the groundwater flow path in the Everglades. Since less water enters the Everglades, the water table has dropped, so locally the swamp has dried up. The decrease in the supply of freshwater has also led to saltwater (saline) intrusion along the coast.

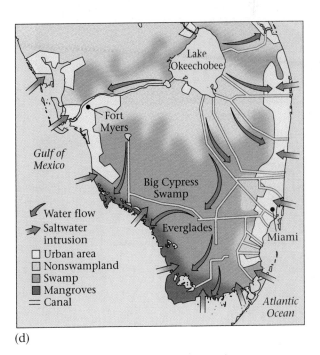

(d)

make the land surface sink below sea level. The flooding of Venice, Italy, is due to land subsidence accompanying the withdrawal of groundwater (▶Fig. 19.23i).

To avoid such problems, communities have sought to prevent groundwater depletion either by directing surface water into recharge areas, or by pumping surface water back into the ground. For example, some communities have excavated to lower the land surface of a park, then configured storm sewers so that they drain onto its grassy surface; the park then acts as a catchment for storm water (▶Fig. 19.24). Normally, the park's lawns are dry and available for recreation, but when it rains heavily, the park temporarily floods. Over a period of days after the flood, the water infiltrates down to the water table below the park.

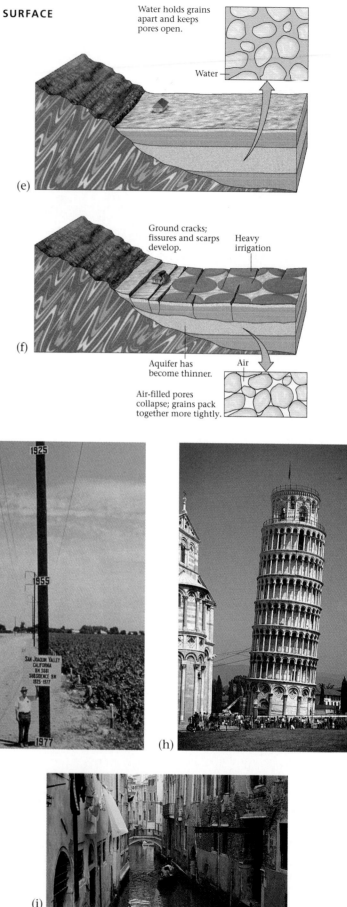

FIGURE 19.23 (a) Before pumping, effluent from a septic tank drifts west with the regional groundwater flow. (b) After pumping, it drifts east into the well, in response to the local slope of the water table. (c) Before pumping, fresh groundwater forms a large lens over salty groundwater. (d) Pumping too fast sucks saltwater from below into the well. (e, f) Pore space collapses when water is removed. The pore collapse makes the land subside, as indicated by fissures and cracked houses. (g) Evidence of land subsidence in the San Joaquin Valley, California. Former ground elevations are marked on the pole. (h) The Leaning Tower of Pisa's foundation was destabilized by groundwater removal. (i) The canals of Venice provide transportation routes in a city whose foundations have subsided below sea level.

Drain
Recharge basin

FIGURE 19.24 Sketch of an enhanced recharge catchment in a city.

Groundwater Contamination

Rocks and sediment are natural filters capable of efficiently removing suspended solids (mud and solid waste) from groundwater, for these solids get trapped in the tiny pathways between pores. Thus, groundwater tends to be clear when it emerges from the ground in a spring or well. Nevertheless, dissolved, invisible organic and inorganic chemicals may be carried along with flowing groundwater; some dissolved chemicals are toxic (such as arsenic, mercury, and lead), while others are not (salt, iron, lime, and sulfur). Even nontoxic chemicals can make groundwater unusable—saltwater harms plants and animals, sulfur makes water smell like rotten eggs, iron stains surfaces, and lime makes water hard. In addition, liquids like gasoline that do not dissolve or mix thoroughly with water can be pushed through the subsurface by flowing groundwater. The addition of such substances in quantities that make the groundwater dangerous to use represents **groundwater contamination.** And once groundwater has been contaminated, it may stay so as long as the water resides underground.

Only tiny quantities of some chemicals are sufficient to make groundwater undrinkable. In fact, the concentration of contaminants is measured in **parts per million** (ppm). If we say that the concentration of molecule A in a glass of water is 1 ppm, then we mean that 1 molecule of A coexists with every 999,999 molecules of water in the glass; in other words, 0.0001% of the water consists of molecule A. To be drinkable (potable), water must contain less than 500–1000 ppm of most nontoxic dissolved components. For some chemicals (such as salt), the concentration is even less—water with 100 ppm of sodium will taste salty—while for others (such as calcium carbonate) the permissible concentration is a little more. In the case of particularly dangerous elements, only 0.1 ppm can render groundwater undrinkable.

Some contaminants in groundwater are natural. For example, sulfur, iron, calcium carbonate, methane, and salt can all be introduced to groundwater directly from the rock through which it is flowing. But in recent decades, contaminants have increasingly been introduced into aquifers through human activity (▶Fig. 19.25a). These include agricultural waste (pesticides, fertilizers, and animal sewage), industrial waste (dangerous organic and inorganic chemicals), effluent from "sanitary" landfills and septic tanks (including bacteria and viruses), petroleum products, radioactive waste (from weapons manufacture, power plants, and hospitals), and acids leached from sulfide-rich minerals in coal and metal mines. Some of these contaminants seep into the ground from subsurface tanks, some infiltrate from the surface, and some are intentionally forced through **injection wells,** wells in which a liquid is pumped down into the ground under pressure so that it passes from the well back into the pore space of the rock or regolith. The cloud of contaminated groundwater that moves away from the source of contamination is called a **contaminant plume** (▶Fig. 19.25b, c). Staggering quantities of contaminating liquids (trillions of gallons in the United States alone) enter the groundwater system every year.

The best way to avoid contamination is to prevent contaminants from entering groundwater in the first place. This can be done by locating potential sources of contamination on impermeable bedrock so that they are isolated from the aquifer. If such a site is not available, the storage area should be lined with a thick layer of clay, for the clay not only acts as an aquitard, it can absorb contaminants into its mineral structure. For this reason, environmental engineers commonly place landfills on top of a liner of packed clay, or place waste in durable, sealed containers. Government agencies have studied various options to safely store the containers. One option involves stockpiling them in tunnels cut into salt domes, for salt is impermeable. Another option is hiding waste containers in tunnels above the water table. For example, American nuclear waste may be stored in a network of tunnels 300 m beneath Yucca Mountain, Nevada. This mountain consists of dry, fairly impermeable welded tuff, below which the water table lies at great depth (▶Fig. 19.26a). Some geologists, however, are concerned that water may trickle into Yucca Mountain's containment rooms from fractures in the overlying bedrock, or that the climate may change within the next 10,000 years (the length of time that the stored waste must be kept isolated), causing the water table to rise and flood the containment room.

Fortunately, in some cases, natural processes can clean up groundwater contamination. Chemicals may be absorbed by minerals like clay, oxygen in the water may oxidize the chemicals, and bacteria in water may metabolize the chemicals, thereby turning them into harmless substances. Where contaminants do make it into an aquifer, environmental engineers drill test wells to determine which way and how fast the contaminant plume is flowing; once

FIGURE 19.25 (a) The various sources of groundwater contamination. (b) A contaminant plume as seen in cross section. (c) A map of the contaminant plume emanating from a military waste dump on Cape Cod, Massachusetts. The darkness of the color represents the concentration of pollutant. The dashed lines indicate that the water table slopes to the south. Note that the plume is also moving south. Pollutants are no longer being added to the aquifer, so the greatest concentration of pollution is now south of the source.

they know the flow path, they can close wells in the path and thereby prevent consumption of contaminated water. Alternatively, engineers attempt to clean the groundwater by drilling a series of extraction wells to pump it out of the ground. If the contaminated water does not rise fast enough, engineers drill injection wells to force clean water or steam into the ground beneath the contaminant plume (►Fig. 19.26b, c). The injected fluids then push the contaminated water up into the extraction wells. More recently, environmental engineers have begun exploring techniques of **bioremediation**: injecting oxygen and nutrients into a contaminated aquifer to foster growth of bacteria that can ingest and break down molecules of contaminants. Need-

less to say, such cleaning techniques are extremely expensive and generally only partially effective.

Unwanted Effects of Rising Water Tables

We've seen that the negative consequences of sinking water tables, but what happens when the water table rises—is that necessarily good? Sometimes, but not always. If the water table rises above the floor of a house's basement, water seeps through the foundation and floods the basement floor. Catastrophic damage occurs when a rising water table weakens the base of a hill slope. The addition of water into pore space

(a)

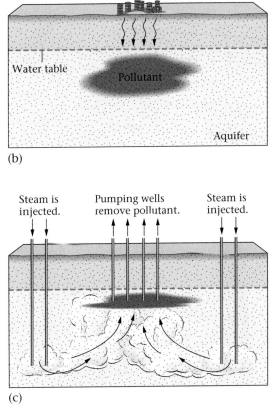

(b)

(c)

FIGURE 19.26 (a) Diagram of the Yucca Mountain radioactive waste storage facility. Note that the tunnels containing the waste lie above the water table. (b) Leaky drums of chemicals introduce pollutants into the groundwater. (c) Steam injected beneath the contamination drives the contaminated water upward in the aquifer, where pumping wells remove it.

FIGURE 19.27 When the water table rises, material above a weak sliding surface begins to slump, and a landslide may result.

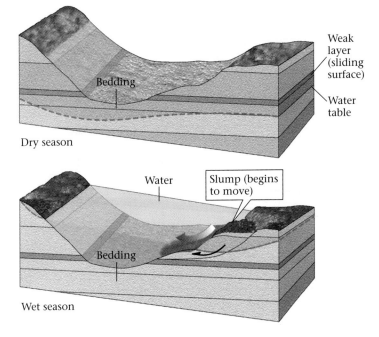

buoys up the overlying rock or regolith, and therefore makes the slope unstable and liable to slip. Thus, rising water tables can trigger landslides and slumps (▶Fig. 19.27).

19.10 CAVES AND KARST: A SPELUNKER'S PARADISE

Dissolution and the Development of Caves

In 1799, as legend has it, a hunter by the name of Hutchins was tracking a bear through the woods of central Kentucky when the bear suddenly disappeared on a hill slope. Baffled, Hutchins plunged through the brambles trying to sight his prey. Suddenly he felt a draft of surprisingly cool air flowing down the slope from uphill. Now curious, Hutchins climbed up the hill and found a dark portal into the hill slope beneath a ledge of rocks. Bear tracks were all around—was the creature inside? Hutchins returned later with a lantern and cautiously stepped into the passageway. It led into a large, open room. Hutchins had discovered Mammoth Cave, an immense network—over 540 km in total

Natural bridge

Disappearing
stream

Sinkhole

Collapsed
breccia

Stalagmite

Stalactite

Dissolved
joint

Soda straw

Flowstone

Cavern

Stalactite

Limestone
column

Underground
stream

Underground pool

Corridor

Emerging
spring

Caves and Karst Landscapes

Limestone, a sedimentary rock made of the mineral calcite, is soluble in acidic water. Much of the water that falls to the ground as rain, or seeps through the ground as groundwater, tends to be acidic, so in regions of the Earth where bedrock consists of limestone, there are signs of dissolution. Underground openings that develop by dissolution are called caves or caverns. Some of these may be large open rooms, while others are long, narrow passages. Underground lakes and streams may form on the floor. A cave's location depends on the orientation of bedding and joints, for these features localize the flow of groundwater.

Caves originally form at or near the water table (the subsurface boundary between rock or sediment in which pores contain air and rock or sediment in which pores contain water). As the water table drops, caves empty of most water and become filled with air. In many locations, groundwater drips from the ceiling of a cave or flows along its walls. As the water evaporates and thus loses its acidity (because of the evaporation of dissolved carbon dioxide), new calcite precipitates. Over time, this calcite builds into cave formations, or speleothems, such as stalactites, stalagmites, columns, and flowstone.

Distinctive landscapes, called karst landscapes, develop at the Earth's surface over limestone bedrock. In such regions, the ground may be rough where rock has dissolved along joints, and where the roofs of caves collapse, sinkholes develop. If a surface stream flows through an open joint into a cave network, we say that the stream is "disappearing." The water from such streams may flow underground for a ways and then reemerge elsewhere as a spring. In some places, the collapse of subsurface openings leaves behind natural bridges.

length—of tunnel-like openings connecting spacious underground rooms.

Large networks like Mammoth Cave developed primarily in limestone bedrock, as a consequence of the dissolution of limestone by groundwater. Slightly acidic groundwater reacts with calcite to produce HCO_3^{-1} and Ca^{+2} ions, which readily dissolve and move away from the site of dissolution in flowing groundwater. Of note, groundwater develops acidity because the water that ultimately becomes groundwater absorbed carbon dioxide (CO_2) from the atmosphere when it fell as rain, and from organic-rich soil as it percolated down to the water table. And when it absorbs CO_2, water transforms into carbonic acid. Limestone dissolves quickly in the warm water of temperate or tropical climates, but not in the cold water of Arctic climates.

Geologists have debated about the depth at which limestone caves form in the subsurface. Acidic rainwater clearly dissolves limestone near the ground surface, as indicated by the presence of extensive pitting on limestone pavements (exposed bedrock surfaces) and along joints in limestone bedrock. And some caves may also form at depth below the water table. It appears, however, that most dissolution takes place in limestone that lies just below the water table, for in this interval the acidity of the groundwater remains high, the mixture of groundwater and newly introduced rainwater is undersaturated, and groundwater flow is fastest. The association between cave formation and the water table helps explain why individual openings in a cave system align along the same horizontal plane.

The formation of major caves requires the following conditions: (1) a thick interval of limestone bedrock, (2) significant rainfall, (3) uplift of the land surface above sea level, and (4) temperate to tropical warmth. Without thick limestone, the caves will be too small to be noticed; without significant rainfall, not enough carbonic acid will flow through the subsurface; unless the land surface lies above sea level, the water table will not be underground; and unless the climate is warm enough, dissolution progresses too slowly to have much of an effect.

Cave Networks

Caves in limestone usually occur not in isolation, but rather as part of a network. Networks include **rooms,** or **chambers,** which are large, open spaces sometimes with cathedral-like ceilings, and tunnel-shaped or slot-shaped **passages.** (See art on pp. 602–603.) Some caves follow a straight course, while others include many right-angle bends and zigzag back and forth.

The shape of the cave network reflects variations in permeability and in the composition of the rock from which the caves formed. Larger open spaces develop where the limestone was most soluble and where groundwater flow was fastest. Thus, in a sequence of strata, caves develop preferentially in the more soluble limestone beds, so bedding

More soluble bed
Less soluble bed

FIGURE 19.28 Joints act as conduits for water in cave networks. Thus, caves and passageways lie along joints.

partially controls the shape of a cave network. In fact, in regions where beds of more soluble limestone alternate with beds of less soluble sandstone or shale, the cave network may contain separate levels. Passages in cave networks typically follow preexisting joints, for the joints provide secondary porosity along which groundwater can move quickly (▶Fig. 19.28). Because joints commonly occur in orthogonal systems (consisting of two sets of joints oriented at right angles to each other; see Chapter 11), passages are typically grid-like.

In some caves, very large rooms and passages develop—the largest known room is 700 m long, 300 m wide, and 70 m high. Such size does not simply reflect dissolution in groundwater. Large rooms develop where blocks fall down from the roof, land on the floor, and eventually dissolve.

Precipitation and the Formation of Speleothems

When the water table drops below the level of a cave, the cave becomes an open space filled with air. Acidic, calcite-containing water then emerges from the rock above the cave and drips from the ceiling or trickles down the walls. As soon as this water reenters the air, it evaporates a little and releases some of its dissolved carbon dioxide. As a result, calcium carbonate (limestone) precipitates out of the water. Rock formed by such precipitation is called **dripstone,** and the various intricately shaped formations that grow in caves by the accumulation of dripstone are called **speleothems.**

Cave explorers, or **spelunkers,** and geologists have developed a detailed nomenclature for different kinds of speleothems (▶Fig. 19.29). Where water drips from the ceiling of the cave, the precipitated limestone adds to the tip

FIGURE 19.29 Speleothems.

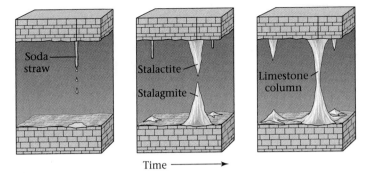

FIGURE 19.30 The evolution of a soda straw stalactite into a limestone column.

of an icicle-like cone called a **stalactite.** Initially, calcite precipitates around the outside of the drip, forming a delicate, hollow stalactite called a **soda straw.** But eventually, the soda straw fills up, and water migrates down the margin of the cone to form a more massive, solid stalactite. Where the drips hit the floor, the resulting precipitate builds an upward-pointing cone called a **stalagmite.** If the process of dripstone formation in a cave continues long enough, stalagmites merge with overlying stalactites to create **limestone columns** (▶Fig. 19.30). If the groundwater flows along the surface of a wall, it drapes the wall with cloth-like sheets of limestone called **flowstone.** Thin sheets of dripstone and flowstone tend to be translucent and, when lit from behind, glow with an eerie amber light.

The Formation of Karst Landscapes

Limestone bedrock underlies most of the Kras Plateau in Slovenia, along the east coast of the Adriatic Sea. The name "Kras," which means "bare, rocky ground," is apt because of the rough, unvegetated land surface in the region. Throughout the area, the landscape contains deep circular-to-elongate depressions, or **sinkholes,** which, as we have seen, form where the roof of a cave collapses. The sinkholes are separated from one another by hills or walls of bedrock (▶Fig. 19.31a–d). Locally, most of a cave collapses, leaving a **natural bridge** that spans the cave remnant, and sinkholes fill to become lakes. A vast network of caverns lies under the ground. Where surface streams intersect cracks or

FIGURE 19.31 (a) Numerous sinkholes of a karst landscape. (b) The Arecibo Radio telescope in Puerto Rico was built in a sinkhole. (c) Natural Bridge, Virginia. (d) A disappearing stream.

(a)

(b)

(c)

(d)

WT—

Caves form just below the water table.

(a)

Old caves empty; speleothems grow.

WT—

Water table sinks; new caves form.

(b)

Sinkhole

Caves collapse; karst landscape develops.

WT—

New caves get bigger.

(c)

FIGURE 19.32 The formation of caves and a karst landscape. (a) Dissolution takes place near the water table (WT) in an uplifted sequence of limestone. (b) Downcutting by an adjacent river lowers the water table, and the caves empty. Speleothems grow on the cave walls. (c) After roof collapse, the landscape becomes pockmarked by sinkholes.

holes that link to these caverns, the water disappears into the subsurface and becomes an underground stream. Such **disappearing streams** reemerge from a cavern entrance downstream.

Geologists refer to regions such as the Kras Plateau as **karst landscapes,** from the Germanized version of "kras." Karst landscapes form (in response to the dissolving action of water) in a series of stages (▶Fig. 19.32a–c).

• *The establishment of a water table in limestone:* The story of a karst landscape begins after the formation of a thick interval of limestone. Limestone forms in seawater, and thus initially lies below sea level. If the sea level drops and the limestone emerges above it, a water table can develop in the limestone below the surface. If, however, the limestone gets buried deeply, it must first be uplifted and exposed by erosion before it can contain the water table.

• *The formation of a cave network:* Once the water table has been established, dissolution begins and a cave network develops. The geometry of this network reflects both the distribution of soluble beds in the stratigraphic sequence and the distribution of joints.

• *A drop in the water table:* If the water table later becomes lower, either because of a decrease in rainfall or because nearby rivers cut down through the landscape and drain the region, newly formed caves dry out. Once downward-percolating groundwater emerges from the roofs of the caves, dripstone and flowstone precipitate.

• *Roof collapse:* If rocks fall off the roof of a cave for a long time, the roof eventually collapses entirely. Such collapse creates sinkholes and troughs, leaving behind hills, ridges, and natural bridges of limestone—that is, a karst landscape.

Some karst landscapes contain many round sinkholes separated by hills. The giant Arecibo Radio telescope in

(a)　　(b)

FIGURE 19.33 (a) Tower karst in China; (b) painting of tower karst by an unknown Chinese artist.

Puerto Rico, for example, consists of a dish formed by smoothing the surface of a 300-m-wide round sinkhole (Fig. 19.31b). In regions where vertical joints control roof collapse, steep-sided residual bedrock towers remain between sinkholes. A karst landscape with such spires is called **tower karst.** The surreal collection of pinnacles comprising the tower karst landscape in the Guilin region of China has inspired generations of artists to portray them on scroll paintings (▶Fig. 19.33a, b).

With the concept of karst landscape evolution in mind, let's reconsider the events that occurred in Winter Park, Florida, where a sinkhole formed in response to the collapse of an underground cave. Geologists have suggested that the region's cave network developed during the last ice age, when the sea level had dropped substantially, and Florida became a plateau about 100 m above sea level. During the past 10,000 years, the sea level has gone up, leading to a rise in the water table and flooding of the caves. The water helped support the roofs of the caves. Then in 1981, drought conditions developed and the water table temporarily dropped; as a result, the cave roofs lost their support and collapsed.

Life in Caves

Despite their lack of light, caves are not sterile, lifeless environments. Caves that are open to the air provide a refuge for bats as well as for various insects and spiders. Similarly, fish and crustaceans enter caves where streams flow in or out. Species living in caves have evolved some unusual characteristics. For example, cave fish lose their pigment and in some cases their eyes (▶Fig. 19.34a). Recently, explorers discovered caves in Mexico in which warm, mineral-rich groundwater currently flows. Colonies of bacteria metabolize sulfur-containing minerals in this water, and create thick mats of living ooze in the complete darkness of the cave. Long gobs of this bacteria slowly drip from the ceiling. Because of the mucus-like texture of these drips, they have come to be known as **snotites** (▶Fig. 19.34b).

Caves have played an important part in human history as well, for they provide shelter—originally for hunter-gatherers and more recently for monks, bandits, and fugitives. In the nineteenth century, Jesse James, the famed bank robber, hid in the caves of the Ozark region in Missouri. Caves have been

(a)   (b)

FIGURE 19.34 Life in caves: (a) blind fish and (b) snotites (gobs of bacteria).

particularly important in wartime, for they allow the local combatants to disappear from view. In Vietnam, Vietminh troops were able to elude enemies for decades by hiding in the cave network of a vast karst landscape.

CHAPTER SUMMARY

• During the hydrologic cycle, water infiltrates into the ground and fills the pores in rock and sediment. This subsurface water is called groundwater. The amount of open space in rock or sediment is its porosity, and the degree to which pores are interconnected, so that water can flow through, defines its permeability.

• Geologists classify rock and sediment according to their permeability. Aquifers are relatively permeable, and aquitards (and aquicludes) are relatively impermeable.

• The water table is the surface in the ground above which pores contain mostly air, and below which pores are filled with water. The shape of a water table is a subdued imitation of the shape of the overlying land surface; water tables are higher beneath hills and lower beneath valleys.

• Groundwater flows wherever the water table has a slope, or hydraulic gradient. Groundwater flows from recharge areas to discharge areas. The rate of flow is very slow; Darcy's law shows that this rate depends on permeability and on the hydraulic gradient.

• Groundwater contains dissolved ions. These ions may come out of solution to form the cement of sedimentary rocks or to fill veins.

• Groundwater can be extracted in wells. An ordinary well penetrates below the water table, but in an artesian well, water rises on its own. Pumping water out of a well too fast causes drawdown, yielding a cone of depression. At a spring, groundwater exits the ground on its own.

• Hot springs and geysers release hot water to the Earth's surface. This water may have been heated by residing very deep in the crust, or by the proximity of a magma chamber or recently formed volcanic rock. Geysers periodically release a fountain of steam and water.

• Groundwater is a precious resource, used for municipal water supplies, industry, and agriculture. In recent years, some regions have lost their groundwater supply because of overuse or contamination.

• When limestone dissolves just below the water table, underground caves are the result. Soluble beds and joints determine the location and orientation of caves. If the water table drops, caves empty out. Limestone precipitates out of water dripping from cave roofs, and creates speleothems (such as stalagmites and stalactites). Regions that contain abundant caves, some of which have collapsed to form sinkholes, are called karst landscapes.

KEY TERMS

aquiclude (p. 585)	bioremediation (p. 600)
aquifer (p. 585)	capillary fringe (p. 585)
aquitard (p. 585)	cone of depression (p. 590)
artesian well (p. 590)	confined aquifer (p. 585)

REVIEW QUESTIONS

1. What factors affect how much water infiltrates into the groundwater in a given region?

2. How does submarine groundwater help keep the Earth temperate?

3. What types of materials are most porous? Which are least porous?

4. Distinguish between primary and secondary porosity.

5. What three factors affect the permeability of a substance?

6. How do porosity and permeability differ? Give examples of substances with high porosity but low permeability.

7. What factors affect the level of the water table?

8. What factors affect the flow direction of the water below the water table?

9. How does the rate of groundwater flow compare with that of moving ocean water or river currents?

10. How do hydrogeologists measure the rate of water flow?

11. What does Darcy's law tell us about how the gradient and permeability affect discharge?

12. Why is "hard water" hard?

13. How does excessive pumping affect the local water table?

14. How is an artesian well different from an ordinary well?

15. Explain what makes a geyser erupt.

16. Is groundwater a renewable or nonrenewable resource? Explain how the difference in time frame changes this answer.

17. Describe some of the ways in which human interference can adversely affect the water table.

18. What are some sources of groundwater contamination? How can it be prevented?

19. What four factors are typical of regions where caves form?

SUGGESTED READING

Dolan, R., and H. G. Goodell. 1986. Sinking cities. *American Scientist* 74:38–47.

Dunne, T., and L. B. Leopold. 1978. *Water in Environmental Planning.* San Francisco: Freeman.

Fetter, C. W. 1988. *Applied Hydrogeology.* Columbus, Ohio: Merrill.

Francko, D. A., and R. G. Wetzel. 1983. *To Quench Our Thirst.* Ann Arbor: University of Michigan Press.

Frederick, K. D. 1986. *Scarce Water and Institutional Change.* Washington, D.C.: Resources for the Future.

Freeze, R. A., and J. A. Cherry. 1979. *Groundwater.* Englewood Cliffs, N.J.: Prentice-Hall.

Jennings, J. N. 1983. Karst landforms. *American Scientist* 71:578–86.

———. 1985. *Karst Geomorphology.* Oxford, Eng.: Basil Blackwell.

Jermar, M. K. 1986. *Water Resources Management.* New York: Elsevier.

Leopold, L. B. 1974. *Water: A Primer.* San Francisco: Freeman.

Price, M. 1985. *Introducing Groundwater.* London: Allen and Unwin.

Rinehart, J. S. 1980. *Geysers and Geothermal Energy.* New York: Springer Verlag.

CHAPTER 20

An Envelope of Gas: Earth's Atmosphere and Climate

20.1 INTRODUCTION

On March 21, 1999, Bertrand Piccard, a Swiss psychiatrist, and Brian Jones, a British balloon instructor, became the first people to circle the globe nonstop in a balloon (►Fig. 20.1). They began their flight on March 1, following more than twenty unsuccessful attempts by various balloonists during the previous eighteen years. Their airtight gondola, which could float in case they had to ditch in the sea, contained a heater, bottled air, food and water, and instruments for navigation and communication. The complex nature of their equipment hints at the challenges the balloonists faced, and why it took so many years before anyone finally succeeded in circling the globe.

Balloons can rise from the Earth only because an **atmosphere,** a layer consisting of a mixture of gases called **air,** surrounds our planet. Any object placed in the air feels a buoyancy force, and if the object is less dense than air, then the buoyancy force can lift it off the ground. The gas (either helium or hot air) in a balloon has such low density that the buoyancy force acting on it can lift not only the balloon but a payload as well.

Balloonists control their vertical movements by changing either the buoyancy of the balloon or the weight of the payload, but they cannot directly control their horizontal motions—balloons float with the **wind,** the flow of air from one place to another. In order to reach their destination before running out of supplies, long-distance balloonists must find a fast wind flowing in the correct direction. Thus, round-the-world balloonists study the **weather,** the physical conditions (the temperature, pressure, moisture content, and wind velocity and direction) of the atmosphere at a given time and location, in great detail to decide when and where to take off and how high to fly. Different winds may blow at different elevations, so balloonists vary their elevation to catch the

The spiral of clouds in this photograph, taken from the space shuttle Discovery, *cloaks the driving winds of hurricane Elena. Hurricanes are one of the most powerful phenomena to affect Earth's atmosphere.*

FIGURE 20.1 The balloon and gondola used by Piccard and Jones during their successful attempt to circle the globe in March 1999.

best wind. Typically, the fastest winds blow at high elevations (8–15 km), but these heights make survival difficult, as the density and temperature of air decrease with elevation. On leaving their launching point in the Swiss Alps, Piccard and Jones entered a strong wind flowing from west to east. Despite a few problems with communications, fuel supplies, and heaters, the balloonists circled the globe, past the longitude of the Alps, and finally touched down in Egypt.

In this chapter, we explore the envelope of air—the atmosphere—through which Piccard and Jones traveled. We begin by learning where the gases came from and how the atmosphere evolved in the context of Earth system. Then we look at the structure of the atmosphere and learn that it consists of discrete layers. Next, we observe the global-scale and local-scale circulation of the lowermost layer; this circulation ultimately controls the weather we experience daily, and can lead to the growth of storms. We conclude by exploring the variability of climate, the average weather conditions during the year, around the globe.

20.2 THE FORMATION OF THE ATMOSPHERE

By about 4.5 Ga (billion years ago), Earth had cooled sufficiently to retain an atmosphere consisting mainly of gases released by volcanic activity, including water vapor (H_2O), carbon dioxide (CO_2), and sulfur dioxide (SO_2) along with small amounts of nitrogen (N_2) and other gases. But the atmosphere we have today clearly does *not* have the same composition as volcanic gas, which would quickly kill modern multicellular life forms. This is because through time, the atmosphere has interacted with the solid Earth and with the hydrosphere (the liquid water of the surface and near surface), and eventually with life.

The first major change happened around 3.8 Ga, when the Earth finally became cool enough for liquid water to exist at its surface, and downpours from the early, moisture-rich atmosphere filled the oceans and left the atmosphere with far less H_2O gas. Once oceans formed, the concentration of CO_2 dropped as well, because this gas dissolved in the oceans and then precipitated to create limestone. Significant amounts of CO_2 were also absorbed by rocks during chemical weathering. The removal of CO_2 and H_2O from the atmosphere meant that the proportion of N_2 gradually increased, for nitrogen doesn't react with other chemicals, and it stayed behind and accumulated.

It took the appearance of life on Earth to add sizable amounts of oxygen (O_2) to the air, for O_2 is produced by photosynthesis. The first photosynthetic organisms, cyanobacteria (blue-green algae), appeared on Earth between 3.8 and 3.5 Ga and began to add O_2 to the atmosphere. By 2 Ga, Earth's atmosphere contained about 1% of its present oxygen level. Oxygen concentrations increased very slowly, so that it took another 1.5 billion years for oxygen to reach 10% of its present level. As more species of photosynthetic organisms appeared, still more oxygen entered the atmosphere, with oxygen achieving its present concentration by about 0.4 Ga.

Oxygen is important not only because it allows complex multicellular organisms to breathe, but also because it supplies the raw components for the production of **ozone** (O_3), a gas that absorbs harmful ultraviolet (short-wavelength) radiation from the sun. Ozone, which accumulates primarily at an elevation of about 30 km, forms by a two-step reaction:

(1) O_2 + energy (from the Sun) → 2O (2 oxygen atoms that are not bonded together);

(2) $O_2 + O \rightarrow O_3$.

Only when enough ozone had accumulated in the atmosphere could life leave the protective blanket of seawater, which also absorbs ultraviolet radiation, and propagate in the air itself. When this happened, terrestrial plants and animals evolved, organisms that themselves have interacted with and modified the atmosphere ever since.

In sum, Earth's atmosphere today consists of volcanic gas modified by interactions with the land and life. It must be maintained, though—if all life were to vanish and all volcanic activity to shut off, much of the gas would leak into space in only a few million years.

20.3 THE ATMOSPHERE IN PERSPECTIVE

The Components of Air

Gases can consist of only one kind of element, or can contain molecules of several different elements or compounds as long as the molecules don't chemically react with one another. Completely dry air is made up of 78% nitrogen and 21% oxygen. The remaining 1% includes several gases in trace amounts.

Trace gases are important in the Earth system. They include carbon dioxide (CO_2) and methane (CH_4), which are **greenhouse gases** that regulate Earth's atmospheric temperature; greenhouse gases allow solar radiation from the Sun to pass through, but trap infrared radiation rising from the Earth's surface. Trace gases also include ozone (O_3), which protects the surface from ultraviolet radiation. Not all trace gases are good, though. **Radon,** produced by the natural decay of uranium in rocks, is radioactive and can cause cancer; it can seep from the Earth and collect in dangerous concentrations in the basements of houses.

In addition to gases, the air contains trace amounts of **aerosols.** These tiny particles (less than 1 micrometer) of liquid or solid material are so small that they remain suspended in the air, just as fine mud remains suspended in river water. Aerosols include tiny droplets of water and acid and microscopic particles of sea salt, volcanic ash, clay, soot, and pollen (▶Fig. 20.2).

FIGURE 20.2 A recent forest fire produces smoke that adds aerosols, as well as CO_2 gas, to the atmosphere. Aerosols (like microscopic particles of soot) contributed by such smoke will form nuclei around which water droplets condense in clouds.

Atmospheric Pollutants

During the past two centuries, human activity has added substantial amounts of pollutants (both gases and aerosols) to the air, primarily through the burning of fossil fuels and through industrial operations. Pollutants include sulfate (SO_4^{-2}) and nitrate (NO_x), which react with water to make a weak acid that then falls from the sky as **acid rain.** Where acid rain falls, lakes, streams, and the ground become more acidic and thus toxic to fish and vegetation (particularly conifer trees). Burning fossil fuels has significantly increased the amount of CO_2 in the atmosphere. Because CO_2 is a greenhouse gas, this increase might lead to **global warming,** a rise in the average atmospheric temperature. Recently, atmospheric scientists have discovered that certain pollutants, notably chlorofluorocarbons (CFCs), react with ultraviolet light from the Sun to release chlorine atoms, which in turn react with ozone and break it down. These reactions appear to happen mainly in high clouds above polar regions during certain times of the year, thus preferentially removing ozone from these regions to create an **ozone hole.** We'll look further at the human impact on the atmosphere in Chapter 23.

Pressure and Density Variations

Air is not uniformly distributed in the atmosphere, because air is a mixture of gases, and gases can be compressed (here, meaning squeezed together to occupy a smaller volume). In the Earth's gravity field, the weight of air at higher elevations pushes down on and compresses air at lower elevations. **Air pressure,** the push that air can exert on its surroundings, caused by the weight of the overlying atmosphere, increases toward the surface of the Earth (▶Fig. 20.3). As air gets compressed, its density increases, so air at sea level has a greater density than air on a mountaintop. Because the density of a gas reflects the number of gas molecules in a given volume, a gulp of air on the top of Mt. Everest, where the air pressure is about one-third that at sea level, contains about a third as many O_2 molecules as air at sea level. Therefore, most climbers seeking to reach Mt. Everest's summit must breathe bottled oxygen. Similarly, the cabin of an airliner must be pressurized to provide adequate oxygen for normal breathing. We measure air pressure in units called atmospheres (atm), where one **atm** is approximately the pressure exerted by the atmosphere at sea level (about 14.7 pounds per square inch, or 1,035 grams per square centimeter). (Another commonly used measurement is the **bar;** 1 bar is about 0.986 atm.)

Because of the decrease in air density with elevation, 50% of the atmosphere's molecules lie below an elevation of 5.6 km, 90% lie below 16 km, and 99.99997% lie below 100 km. Thus, even though the outer edge of the atmo-

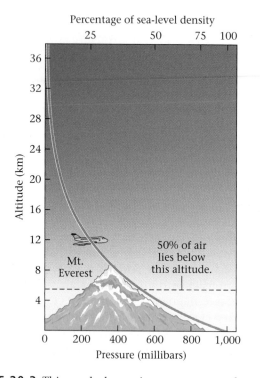

FIGURE 20.3 This graph shows air pressure versus elevation on the Earth. Note that climbers on top of Mt. Everest breathe an atmosphere that contains only about 33% of the atmospheric gases at sea level. A nonsupersonic commercial airliner flies through air that is only 25% as dense as the air at sea level.

sphere, a vague boundary where the gas density becomes the same as that of interplanetary space, lies as far as 10,000 km from the Earth's surface, most of the atmosphere's molecules lie within a shell only 0.5% as wide as the solid Earth.

Heat and Temperature

The molecules that comprise the atmosphere, or any gas, are not standing still but are constantly moving. We refer to the *total* kinetic energy (energy of motion; see Appendix A) resulting from the movement of molecules in a gas as its thermal energy, or **heat.** Note that heat and temperature are not the same—a gas's **temperature** is a measure of the *average* kinetic energy of its molecules. A volume of gas with a small number of rapidly moving molecules has a higher temperature but may contain less heat than a volume with a large number of slowly moving molecules. If we add heat to a gas, its molecules move faster and its temperature rises, and the gas will try to expand to occupy a larger volume.

Relations Between Pressure and Temperature: Adiabatic Change

When air moves from a region of higher pressure to a region of lower pressure, without adding or subtracting heat, it ex-

pands. When this happens, the gas molecules slow down, and the air temperature decreases. Such a process is called **adiabatic cooling** (from the Greek *adiabatos,* meaning "impassable"; air cools at 6°–10°C per kilometer that it rises). The reverse is also the case: if air moves from a region of lower pressure to a region of higher pressure, without adding or subtracting heat, it contracts; the molecules speed up, and the air temperature increases. Such a process is called **adiabatic heating.** Adiabatic cooling and heating are important processes in the atmosphere, because pressure changes with elevation—when air near the ground surface (where pressure is higher) flows up to higher elevations (where pressure is less), it undergoes adiabatic cooling, but when air from high elevations flows down and compresses, it undergoes adiabatic heating.

Water in the Air: Relative Humidity and Latent Heat

Earlier, we examined the percentages of gases in completely dry (water-free) air. In reality, however, air contains variable amounts of water—from 0.3% above a hot desert to 4% in a rainforest during a heavy downpour.

Meteorologists, scientists who study the weather, specify the water content of air by a number called the **relative humidity,** the ratio between the measured water content and the maximum possible amount of water the air could hold, expressed as a percentage. The maximum possible amount varies with temperature—warmer air can hold more water than colder air. When air contains as much water as possible, it is **saturated,** while air with less water is **undersaturated.** If we say that air at a given temperature has a relatively humidity of 20%, we mean the air contains only 20% of the water that it could hold at that temperature when saturated. Such air feels dry. Air with a relative humidity of 100% is saturated and feels very humid, or damp.

Because cold air can't hold as much water as warm air, air that is undersaturated when warm may become saturated when cooled, without the addition of any new water. The temperature at which the air becomes saturated is called the **dewpoint temperature;** dew forms when undersaturated air cools at night and becomes saturated, so that water condenses on surfaces. When the dewpoint temperature is below freezing, frost develops. And when moist air rises and adiabatically cools, its moisture condenses to form a **cloud,** a mist of tiny droplets (▶Fig. 20.4).

When water in the air changes from liquid to gas or vice versa (a process called a **change of state**), the temperature of the air also changes. That is, when water evaporates, it absorbs heat, so that molecules can break free from the liquid. The condensation of water reverses this process and therefore releases heat. The heat released during condensation was "hidden," in the sense that it comes only from the change of state and does not require an ex-

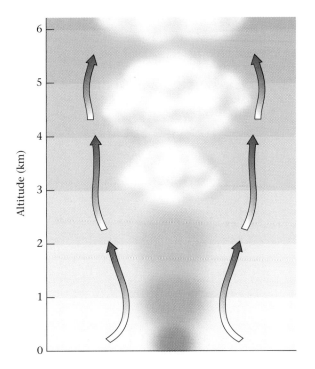

FIGURE 20.4 As air rises and enters regions of lower pressure, it expands and adiabatically cools. If the air contains moisture, the moisture condenses (it turns into water droplets or ice crystals) after the air has risen high enough. Here, we see moist air rising and becoming less dense; its moisture condenses at elevations above 3 km and produces a cloud.

ternal energy source. Thus, it is called the **latent heat of condensation.**

Atmospheric Layers

Temperature, in contrast to pressure, does not decrease continuously from the surface of the Earth to the outer edge of the atmosphere. In fact, starting from the surface and going up, temperature decreases, then increases, then decreases, then increases. Elevations where temperatures stop decreasing and start increasing or vice versa are called **pauses.** Atmospheric scientists divide the Earth's atmosphere into four layers, separated by pauses (▶Fig. 20.5).

• *Troposphere:* This layer starts at the surface of the Earth and rises to an elevation of 9 km at the poles and 12 km at the equator. Within this layer, the temperature decreases progressively from an average of 18°C at the surface to about –55°C at the top, a boundary called the **tropopause.** The name **troposphere** comes from the Greek *tropos,* which means "turning." It is an appropriate name, because air in the troposphere constantly flows in response to **convection** (the heat transfer that results when warmer, less dense material rises while cooler denser material sinks; see Appendix A). The heat that initiates

movement in the troposphere comes primarily from infrared radiation rising from the Earth's surface. The radiation heats air at the base of the troposphere—in effect, the air is baked from below. This air then rises, and cold air sinks to take its place. As we will see, this movement causes most weather phenomena, so the troposphere can also be thought of as the "weather layer."

• *Stratosphere:* Beginning at the tropopause and continuing up for about 10 km, the temperature stays about the same. Then it slowly rises, reaching a maximum of about 0°C at an elevation of about 47 km, a boundary called the **stratopause.** The layer between the tropopause and the stratopause is the **stratosphere,** so named because it doesn't convect and is thus a stable, stratified layer. The stratosphere doesn't mix with the underlying troposphere, because at the tropopause hotter (less dense) air already lies on top of cooler (denser) air. Most of the ozone in Earth's atmosphere resides in the stratosphere.

• *Mesosphere:* The temperature decreases in the interval from 47 to 82 km. This interval is called the **mesosphere,** and its top is the **mesopause.** At the mesopause, the temperature has dropped to –85 C. The

FIGURE 20.5 The principal layers of the atmosphere are separated from each other by pauses. At a pause, the temperature gradient in the atmosphere changes direction.

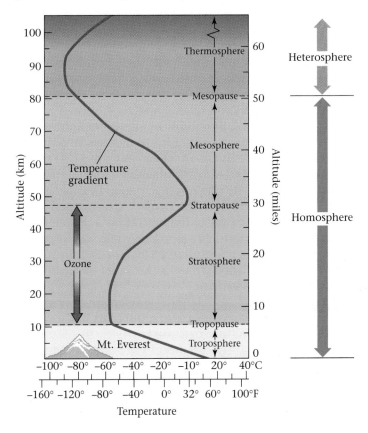

mesosphere does not absorb much solar energy and thus cools with increasing distance from the hotter stratosphere below. Most meteors (shooting stars) begin burning here and have vaporized by the time they reach an altitude of 25 km.

- *Thermosphere:* The outermost layer of the atmosphere, the **thermosphere,** contains very little of the atmosphere's gas (less than 1%). The temperature increases with elevation in this layer, because gases of the thermosphere absorb short-wavelength solar energy (the layer is broiled from above). Because the thermosphere has so little gas, it contains very little heat, even though it registers a high temperature. Thus, an astronaut walking in space at an elevation of 200 km doesn't feel hot.

The densities of gases in the lower three layers of the atmosphere are great enough that moving atoms and molecules frequently collide. Like billiard balls, they bounce off each other and shoot off in different directions. This constant, chaotic motion stirs the gases sufficiently to make a homogeneous mixture, so that the air in the lower three layers has essentially the same proportion of different gases regardless of location. For this reason, atmospheric scientists refer to the troposphere, stratosphere, and mesosphere together as the **homosphere.** In contrast, atoms and molecules in the low-density thermosphere collide so infrequently that this layer does not homogenize. Rather, gases separate into distinct layers based on composition, with the heaviest (nitrogen) on the bottom, followed in succession by oxygen, helium, and at the top, hydrogen, the lightest atom. To emphasize this composition, atmospheric scientists refer to the thermosphere as the **heterosphere.**

So far, we've distinguished atmospheric layers according to their thermal structure (troposphere, stratosphere, mesosphere, and thermosphere) and according to the degree their gases mix (homosphere and heterosphere). We need to add one more "sphere" to our discussion. The **ionosphere** is the interval between 50 and 400 km and thus includes most of the mesophere and the lower part of the thermosphere. It was given its name because in this layer, short-wavelength solar energy strips nitrogen molecules and oxygen atoms of their electrons and transforms them into positive ions. The ionosphere plays an important role in modern communication in that, like a mirror, it reflects radio transmissions from Earth so that they can be received over great distances.

The ionosphere also hosts a spectacular atmospheric phenomenon, the auroras (**aurora borealis** in the Northern Hemisphere and **aurora australis** in the Southern) which look like ghostly curtains of varicolored light in the night sky (▶Fig. 20.6). They appear when charged particles (protons and electrons) ejected from the Sun, especially during times of solar flares (solar storms), reach the Earth and interact with the ions in the ionosphere, making them

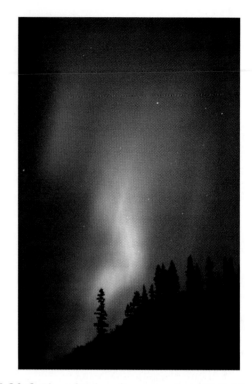

FIGURE 20.6 The splendor of an aurora borealis lights up the night sky in Arctic Canada. The colors result when particles emitted from the Sun interact with atoms in the thermosphere of the Earth.

release energy. Auroras occur primarily at high latitudes because Earth's magnetic field traps solar particles and carries them to the poles.

20.4 WIND AND GLOBAL CIRCULATION IN THE ATMOSPHERE

A gusty breeze on a summer day, the steady "trades" that blew clipper ships across the oceans, and a fierce hurricane all are examples of the **wind,** the movement of air from one place to another. We can feel the wind, because of the impacts of air molecules as they strike us. The existence of wind illustrates that the lower part of the atmosphere is in constant motion, swirling and overturning at rates between a fraction of a kilometer and a few hundred kilometers per hour. This circulation happens on two scales, local and global. **Local circulation** is the movement of air over a distance of tens to a thousand kilometers. **Global circulation** is the movement of volumes of air in paths that ultimately carry it around the planet. (We can picture local circulation as eddies in global-scale "rivers" of air.) To understand both kinds of circulation, we must first see what drives air from one place to another, and examine energy inputs into the atmosphere.

Lateral Pressure Changes and the Cause of Wind

The air pressure of the atmosphere not only changes vertically, it also changes horizontally at a given elevation, because of changes in the atmosphere's temperature with location. Warmer air is less dense and therefore expands upward. Thus, at a given elevation, there is more air above you where it's warm than where it's cold—as a consequence there is a progressive decrease in pressure as you fly from warmer to colder air without changing elevation. The rate of pressure change over a given horizontal distance, called a **pressure gradient,** can be represented by the slope of a line on a graph plotting pressure on the vertical axis (specified in bars) and distance on the horizontal axis (▶Fig. 20.7a).

Winds form wherever there is a pressure gradient. Air always flows from a high-pressure region to a low-pressure region; in other words, it flows "down" a pressure gradient. For example, if you step on one end of a long cylindrical balloon filled with air, you increase the pressure at that end, so the air flows toward the other end (▶Fig. 20.7b).

We can use a map to represent air pressure at a given elevation. A line on a map along which the air has a specified pressure is called an **isobar** (▶Fig. 20.7c). In other words, the pressure is the same all along an isobar. Isobars can never touch, because they represent different values of pressure. A difference in pressure exists between one isobar and the next, so the air starts flowing perpendicular to these lines. As we will see, however, wind direction is modified by the Coriolis effect.

Energy Input into the Atmosphere: Convection

Ultimately, air circulation, like the movement of water in a heated pot, results from convection: heated air expands and becomes less dense, so it rises to be replaced by sinking cooler, denser air. In the case of Earth's atmosphere, the energy comes from solar radiation. Solar energy constantly bathes the Earth. Of this energy, 30% reflects back to space (off of clouds, water, and land); air or clouds absorb 19%; and land and water absorb 51%. The energy absorbed by land and water later reradiates as infrared radiation and thus bakes the atmosphere from below. As noted earlier, part of this reradiated energy is absorbed by greenhouse gases before it can return to space.

Because the Earth is a sphere, not all areas receive the same amount of incoming solar energy, or **insolation:** portions of the Earth's surface hit by direct rays of the Sun receive more energy per square meter than portions hit by oblique rays. We can simulate this contrast with a flashlight; if you point a flashlight beam straight down, you get a small but bright spot of light on the ground, but if you aim the beam so that it hits the ground at an angle of 45°, the spot covers a broader area but does not appear as bright (▶Fig. 20.9a). Higher latitudes thus receive less energy than lower latitudes (▶Fig. 20.9b). Because of the tilt of Earth's axis, the amount of solar radiation that any point on the surface receives changes during the year, which is why we have seasons (see Box 20.2).

The contrast in the amount of solar radiation received by different latitudes means that polar regions are cooler at the surface than are equatorial regions. In 1735, George Hadley, a British mathematician, realized that this contrast could cause global air to circulate by convection. Specifically, he suggested that warm air at the equator would rise

FIGURE 20.7 (a) This graph shows a profile from location X to location Y. Notice that the pressure gradient, the slope of the line, is greater between X and C than between C and D. (b) Air flows from a high-pressure region to a low-pressure region to cause wind. If you step on one end of a balloon, for example, you increase the pressure there, so that the air flows toward the unsqueezed end. (c) Isobars on a map are lines of equal pressure. Every place along the 1,005 isobar is experiencing a pressure of 1,005 millibars (1 millibar = 0.001 bar). By moving in any direction not parallel to the isobars, you will feel a pressure change. For example, if you walk from X to Y, you will experience a decrease in pressure.

(a) Distance

(b) Lower pressure   Higher pressure

(c) Pressure (millibars)

BOX 20.1

THE HUMAN ANGLE

Why Is the Sky Blue?

When astronauts standing on the Moon looked up, even during the day, they saw a black sky filled with stars and a nearly white Sun. On Earth, when we look up during the day, we see a blue sky when there are no clouds and a white sky when there are. The color we seen in the sky is the result of the dispersal of energy that occurs when light interacts with particles in the atmosphere, a process called **scattering.** When light is scattered, a single ray divides into countless beams, each heading off in a different direction. It's similar to what happens when you shine a spotlight on a mirror ball over a dance floor.

Sunlight consists of a broad spectrum of electromagnetic radiation. Different components of the atmosphere scatter different wavelengths of this light, because the ability of a particle to scatter light depends on its size relative to the wavelength. Aerosols (soot, water droplets, dust, etc.) are larger and scatter all wavelengths of light, while gas molecules are smaller and scatter short-wavelength (blue) light but not long-wavelength (red) light. Since scattered light heads in all directions, some of it returns back to space; this light is called **backscattered light.** Because of backscattering, the intensity of light received on Earth's surface is less than it would be if Earth had no atmosphere. It's also because of scattering that shadows are not completely dark—light is able to enter regions (beneath trees, for example) that are blocked from the Sun.

When you look at an object, the color you see is the color of light either emitted from the object or reflected off it—a plum appears violet because it reflects violet light and absorbs other wavelengths, and a traffic light appears green because it emits green light. On a clear day, with the Sun high in the sky, the gas in the atmosphere scatters primarily blue light. When we look up, we are seeing the blue light scattered off the molecules (▶Fig. 20.8a). (Other wavelengths of light pass through the atmosphere without scattering, so we don't see that light until it reflects off objects on the ground.) Because blue light is scattered, not all of it reaches the Earth (some has been backscattered to space), so the Sun appears yellow rather than white (if you subtract a little blue light from white light, you get yellow light). On a cloudy or hazy day, there are more aerosols (such as water droplets) in the atmosphere, and these scatter all wavelengths of light. Therefore, the sky appears white unless the clouds are so dense that they do not let light through, in which case they appear gray.

At the end of the day, when the Sun is low on the horizon, light passes through a thicker amount of the atmosphere, and so much of the blue light scatters back to space that the sunlight that makes it through contains mostly red wavelengths (if you subtract a lot of blue light from white light, you are left with red light). Thus, the Sun appears red, as does the light that reflects off clouds (▶Fig. 20.8b).

FIGURE 20.8 (a) Scattering of light by gas molecules leads to the brilliant blue of a clear sky at noon. (b) So much scattering happens when sunlight hits the atmosphere at a low angle that all that's left is red light.

(a)

(b)

FIGURE 20.9 (a) A flashlight beam aimed straight down produces a narrower and less intense beam than a flashlight aimed obliquely. Thus, the area under the straight beam heats up more than the area under the oblique beam. (b) Sunlight hitting the Earth near the equator provides more heat per unit area of surface than sunlight hitting the Earth at a polar latitude. That is why the poles are colder.

and flow toward the pole, to be replaced by cool polar air, which would flow to the equator at lower elevations (▶Fig. 20.11a). Hadley's proposal, however, did not take into account an important factor, namely the Earth's rotations and the resulting Coriolis effect. The **Coriolis effect,** as we learned in Chapter 18, refers to the deflection that an object is subject to as it moves from the circumference to the center of a rotating disk or from the equator to the pole of a rotating sphere (and vice versa) (▶Fig. 20.11b).

The moving atmosphere is also subject to the Coriolis effect. Northward-moving high-altitude air in the Northern Hemisphere deflects to the east, so by the latitude of 30°N, it is basically moving due east and cannot make it the rest of the way to the pole. By the time it reaches this latitude, the air has also cooled significantly and thus starts to sink. This means that a global convection cell (a current that looks like a loop in cross section) develops that conveys warm air north from the equator to the subtropics (latitude 30°), where it cools and sinks. When the sinking air reaches low elevations, it divides, some moving back toward the equator near the surface and some moving north near the surface. A place where sinking air separates into two flows that move in opposite directions is called a **divergence zone.**

Meanwhile, cool air from the polar region moves south near the surface and deflects to the west. At about a latitude of 60°, the near-surface polar air collides, or converges, with the northward-moving, near-surface mid-latitude air. This air must rise, because there is nowhere for the extra air to go but up. A place where two surface air flows meet so that air has to rise is called a **convergence zone.** The convergence zone at latitude 60° is called the **polar front.** The air that rises along the polar front divides at the top of the troposphere, with some air heading toward the equator at high altitude to complete a mid-latitude convection cell, and some heading toward the pole to create a polar convection cell. Because of seasonal changes on Earth, the exact position of the polar front changes during the year.

In sum, because of the Coriolis effect, circulating air in the troposphere splits into three globe-encircling convection cells in each hemisphere. The low-latitude cells, extending from the equator to a latitude of about 30°, are called **Hadley cells,** in honor of George Hadley. The mid-latitude cells are called **Ferrel cells,** in honor of the American meteorologist, William Ferrel, who proposed them. The high-latitude cells are simply called **polar cells** (▶Fig. 20.11c).

In this three-cell-per-hemisphere model of atmospheric circulation, several belts of high and low pressure form around the Earth. The equatorial regions are marked by a belt of convergence called the **intertropical convergence zone.** In these regions, intense insolation makes air hot, so that it rises. The upward flow of hot, rising air leaves an area of low pressure in its wake; thus, the convergence zone at the

The Earth's Tilt: The Cause of Seasons

If the Earth's axis were not tilted, we would not experience seasons, for the amount of solar energy received at a given latitude would stay the same all year. But at present, the axis tilts at 23.5° (this number changes with time, because of the axis's wobble; see Chapter 22), so the amount of radiation received at a given latitude varies during the year. As a result, the average daily temperature changes during the year, and this change, along with associated variations in amounts of precipitation, brings the seasons.

At any given time, half the Earth has daylight, and half experiences night. The boundary between these two hemispheres is called the **terminator.** Because of the Earth's tilt, the terminator does *not* always pass through the North and South Poles (▶Fig. 20.10). On June 21, a special day called a **solstice,** the terminator lies 23.5° away from the poles, as measured along the Earth's surface. The line of latitude at this position is called the **Arctic Circle** in the Northern Hemisphere and the **Antarctic Circle** in the Southern Hemisphere. On June 21, a person standing on the Arctic Circle sees the **midnight sun** (the Sun is visible for an entire day), while anyone south of the Antarctic Circle has night for a full twenty-four hours. Regions north of the Arctic Circle see the midnight sun for more than a day, and the North Pole itself experiences the midnight sun for a full six months. Similarly, regions south of the Antarctic Circle experience twenty-four-hour nights for more than a day, and the South Pole itself sees nothing but night for a full six months. December 21 is another solstice, but on this day all regions south of the Antarctic Circle see

the midnight sun, while all regions north of the Arctic Circle have perpetual night.

On the June 21 solstice, the Sun's rays are exactly perpendicular to the Earth (the Sun is directly overhead) at latitude 23.5°N, the **Tropic of Cancer,** and on December 21 the Sun's rays are exactly perpendicular to the Earth at 23.5°S, the **Tropic of Capricorn.** On September 22 and March 21, each known as an **equinox,** the Sun is directly overhead at the equator. Note that north of the Tropic of Cancer and south of the Tropic of Capricorn, the Sun never lies directly overhead, even at noon.

The two solstices and two equinoxes divide the year into four astronomical seasons. In the Northern Hemisphere, for example, winter comes between the winter solstice on December 21 and the spring equinox on March 21. Spring lasts from the spring equinox to the summer solstice, summer from the summer solstice to the fall equinox, and fall from the fall equinox to the winter solstice. During the summer, the Northern Hemisphere tilts toward the Sun and enjoys longer days, and the Sun rises higher in the sky; thus, the atmosphere heats up. During the winter, the Northern Hemisphere tilts away from the Sun and experiences short days, and the Sun rises lower in the sky; the atmosphere then cools down. The reverse timing of seasons occurs in the Southern Hemisphere. In reality, of course, the seasons as we experience them begin and end a bit earlier than the astronomical calendar specifies. For example, the Northern Hemisphere usually begins to deal with cold weather up to a month before the winter solstice.

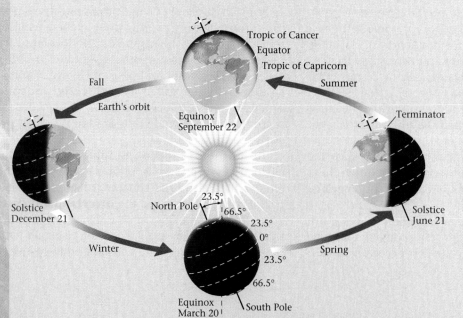

FIGURE 20.10 Because of the tilt of the Earth's axis, we have seasons. At the Northern Hemisphere's summer solstice, the noonday sun is directly overhead at the Tropic of Cancer and the region above the Arctic Circle sees the midnight sun, while at the winter solstice, the noonday sun is directly overhead at the Tropic of Capricorn and the region south of the Antarctic Circle sees the midnight sun. At the equinoxes, the noonday sun is directly overhead at the equator.

(a)

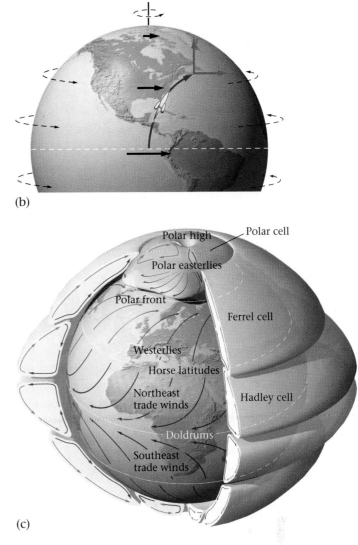

(b)

(c)

FIGURE 20.11 (a) If the Earth did not rotate, two simple convection cells would be established in the atmosphere, one stretching from the equator to the North Pole and one stretching from the equator to the South Pole. (b) Because of the Coriolis effect, a rocket sent north from the equator curves to the right (east). This is because the rocket not only has a northward velocity driven by its engines but also has an eastward velocity caused by the Earth's rotation at the launch site. (c) Because of the Coriolis effect, atmospheric circulation breaks into three convection cells within each hemisphere, named, from equator to pole, Hadley, Ferrel, and polar.

equator is also called the **equatorial low** (▶Fig. 20.12a–c). To help understand why there is low pressure at the equator, imagine a large truck driving down the road. The truck pushes on the air in front of it, making that air flow in the direction the truck is moving; in front of the truck is a region of high pressure, because the air has been compressed. Behind the truck, there is a space that has less air than before; this is a region of low pressure. Air rushes in from either side to fill the space. In the case of the Earth, the low-pressure area in equatorial regions fills with cooler air flowing in from higher latitudes at the base of the Hadley cells.

At the latitude of about 30° and at a high altitude, a Hadley cell converges with a Ferrel cell, causing cooler, dense air to sink. This downward movement forms a belt of high pressure called a **subtropical high.** (A subtropical high can also be called a **subtropical divergence zone,** because the downward-flowing air diverges when it reaches the Earth's surface.) Finally, the rise of air where the surface

flow of a polar cell converges with the surface of a Ferrel cell (at the latitude of 60°) creates a **subpolar low,** and the sinking of air at the pole itself creates a **polar high.**

Recall that where warm, moist air rises, it cools adiabatically. Cooler air can hold less moisture, so this air becomes oversaturated. The moisture condenses and forms clouds, which produce rain. Therefore, the equatorial lows are regions of heavy rainfall, which lead to the growth of tropical rainforests. In contrast, in high-pressure belts where cool, dry air sinks, it contracts and heats adiabatically. The resulting hot air can absorb moisture, so that it rains only rarely. Thus, as we see in Chapter 21, the subtropical regions at latitude 30° include some of the Earth's major deserts.

Prevailing Surface Winds

The global-scale flow in the six major convective cells on Earth creates belts in which surface winds generally flow in

(a)   Higher   Lower   Higher

(b)   Lower   Higher   Lower

(c)

FIGURE 20.12 (a) In places where hot air rises, the pressure beneath the rising column of air is less than in surrounding regions. (b) In places where cold air sinks, the pressure beneath the cold air is greater than in surrounding regions. (c) When a truck speeds down a highway, it pushes air in front of it, creating a high-pressure zone (as is felt beneath downward-flowing air), while it leaves a space of low pressure air behind it (as is felt beneath upward-flowing air). Note that air flows into the low-pressure zone.

a consistent direction—such winds are called **prevailing winds.** (Note that here we're examining *surface* winds, the base of the convective cell.) These simple patterns tend to be disrupted by local-scale winds. When describing winds, meteorologists label them according to the direction the air is coming from. Thus, a westerly wind blows from west to east.

Let's start our tour of prevailing winds at the base of the Hadley cell in the Northern Hemisphere. Near-surface winds start to flow from 30°N to the south and are deflected west. Thus between the equator and 30°N, surface winds comes out of the northeast, and are called the **northeast**

tradewinds, so named because they carried trading ships westward from Europe to the Americas. Tradewinds in the Southern Hemisphere, which start flowing northward and then deflect to the west, end up flowing from southeast to northwest, making these the **southeast tradewinds.** Where the southeast and northeast tradewinds merge at the equator, they are almost flowing due west (▶Fig. 20.13). But winds along the equator are very slow, because the air is mostly rising. Ships tended to be becalmed in this belt, which came to be called the **doldrums.**

At the base of a Ferrel cell, at mid-latitudes, surface air starts to move toward the north, but because of the Coriolis effect it curves to the east. Thus, throughout much of North America and Europe, the prevailing surface winds come out of the west or southwest and are known as the **surface westerlies.** In the subtropical high itself, where air flow is primarily down, winds are weak and tend to shift in different directions. In the past, these conditions inhibited the progress of sailing ships. Perhaps because so many horses being transported by ship died of heat exhaustion in the subtropical high, the region came to be known as the **horse latitudes.**

Finally, at the base of polar cells, surface air starts by flowing from the pole southward, but deflects to the west. The resulting prevailing winds, which are known as the **polar easterlies,** flow from the polar high to the subpolar low, and converge with the westerlies of mid-latitudes at the polar front.

FIGURE 20.13 If the Earth had a uniform surface, distinct high- and low-pressure zones would form on its surface. But surface winds flowing from high-pressure zones to low-pressure zones are deflected by the Coriolis effect, so that for much of their course they flow almost parallel to pressure zones. Compare this figure with Figure 20.11c.

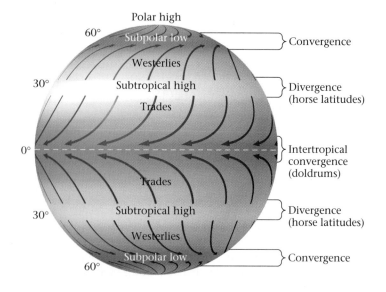

Note that prevailing winds, for much of their course, follow paths almost parallel to the high- and low-pressure belts that surround the Earth. In effect, they flow nearly parallel to the isobars defining the belts. This situation seems to contradict the statement made earlier that wind flows from regions of high pressure to regions of low pressure, and thus should flow at almost right angles to isobars. This apparent contradiction is a consequence of the Coriolis effect.

High-Altitude Winds in the Troposphere: The Jet Streams

In the upper atmosphere, a global-scale pressure gradient exists because of temperature differences between the equator and the pole. At the equator, air is warmer and thus expands. This causes the top of the troposphere there to rise relative to the top of the troposphere in polar regions, so, as noted earlier, the troposphere is thicker over the equator than over the poles. As a consequence, the air pressure at a given elevation above the equator is greater than at the same elevation above the poles. This pressure gradient causes overall high-altitude air to flow north; in fact, some air from the top of the Hadley cell spills over and moves north over the top of the Ferrel cell (▶Fig. 20.14a). Once again, the Coriolis effect comes into play and makes this air deflect to the east, and so in the Northern Hemisphere we have generally westerly winds at the top of the troposphere. These are called the **high-altitude westerlies.**

In two special places, over the polar front and over the horse latitudes, the pressure gradient at the top of the troposphere is particularly steep. Because of the steepness of the gradient, high-altitude westerlies flow particularly fast. These zones of fast movement, where winds typically flow at speeds of between 200 and 400 km per hour, are called **jet streams** (▶Fig. 20.14b, c). The polar jet stream is faster

and more important than the subtropical jet stream. Both can be viewed as fast-flow zones within an overall westerly high-altitude flow.

The position of the polar front tends to undulate over time in wave-like motions. Thus, as viewed on a map, the positions of the polar front and therefore of the jet stream follow large curving trajectories that change with time, sometimes bringing the jet stream down to southern lati-

(a)

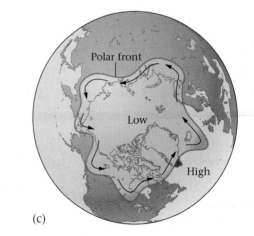

(b)

FIGURE 20.14 (a) In an equator-to-pole cross section of the atmosphere (simplified by removing the curvature of the Earth), we see that the troposphere varies in thickness. The atmosphere above the troposphere, however, is essentially uniform in thickness from equator to pole, so the tropopause is an isobaric surface (all points on the surface have the same pressure). Therefore, at a given elevation within the troposphere, there is a pressure gradient from equator to pole, causing a northward flow of air at the top of the troposphere. Because of the Coriolis effect, high-altitude westerly winds develop in the Northern Hemisphere. At places where the pressure gradient is particularly steep (at the boundary between convection cells), the air moves very fast, creating the jet streams. (b) The polar jet stream meanders with time, and changes its overall position with the seasons; it tends to be farther south in the winter than in the summer. (c) Looking down on the Earth from the North Pole, we see the irregular shape of the polar front, along which the polar jet stream flows.

(c)

TABLE 20.1 Weather Extremes

Lowest recorded air pressure at sea level	870 millibars (25.7 inches); during Typhoon Tip (1979), Pacific Ocean
Highest air pressure at sea level	1,084 millibars (32.02 inches); Agata, Siberia
Lowest air temperature (world)	–89°C (–129°F); Vostok, Antarctica
Lowest air temperature (North America)	–63°C (–81°F); Yukon, Canada
Highest air temperature (world)	58°C (136°F); Libya
Highest air temperature (North America)	57°C (134°F); Death Valley, California
Highest wind speed (world)	372 km per hour (231 mph); Mt. Washington, New Hampshire
Highest wind (hurricane)	325 km per hour (202 mph)

tudes of North America and sometimes running it across northern Canada. Further, the average position varies with the season. The jet stream, because of its high steady winds, affects airplane flights, for modern jet aircraft fly at the height of the jet stream and the high-altitude westerlies. Planes flying east have shorter flying times than those flying west, because the former are helped along by strong tailwinds, while the latter battle headwinds that slow them down.

20.5 WEATHER AND ITS CAUSES

If the surface of the Earth were a perfectly uniform sphere, air flow—driven by convection and modified by the Coriolis effect—might follow the global pattern just described. But in fact, the Earth's surface is heterogeneous—parts are covered with water, parts with land. In the oceans, water currents affect the temperature of the sea surface, and on land, there are variations in altitude and vegetation cover. All these factors ensure that the heat that rises into the base of the atmosphere from the heated surface of the Earth varies with time and location in a complex way, and that moving air must interact with obstacles and either be forced to rise or be shoved to the side. Thus, it's not surprising that on a *local scale,* atmospheric flow can be quite turbulent.

The term **weather** refers to local-scale conditions as defined by temperature, air pressure, relative humidity, and wind speed (see Table 20.1 for the range of weather conditions that can exist on Earth). A specific set of weather conditions, reflecting the configuration of air movement in the atmosphere, that affects a region for a period of time is called a **weather system.** Weather systems can move across the surface of the Earth, carried by prevailing winds. In this section, we'll briefly review some of the more general aspects of weather, and see how it results from the interaction of air masses along fronts.

Air Masses and Fronts

Air that remains or passes over a certain region for a length of time takes on characteristics that reflect both its interaction with the Earth's surface and the amount of insolation it receives. For example, air that has hovered over a warm sea tends to become warm and moist, while air stuck over cold land becomes cold and dry. A body of air, at least 1,500 km across, that has recognizable physical characteristics is called an **air mass.** Air masses move within the overall global circulation of the atmosphere, and their paths are controlled by prevailing winds. Meteorologists refer to air masses that originate over land in polar regions as **Arctic air masses,** and those that originate over tropical or subtropical oceanic regions as **maritime tropical air masses.** (See ▶Fig. 20.15 for names of other air masses.) The weather of a region

FIGURE 20.15 Air masses that form in different places have been assigned different names. The arrows indicate the average directions in which the air masses move.

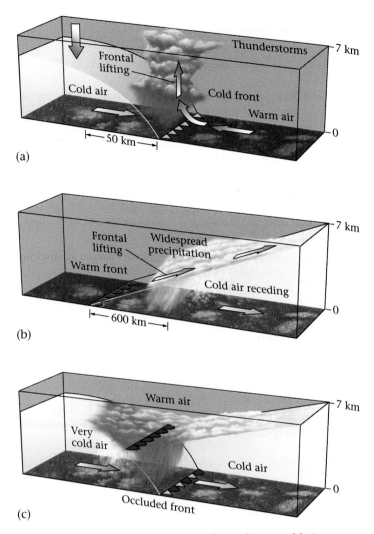

FIGURE 20.16 (a) A cold front develops where a cold air mass moves under a warm air mass. Clouds and thunderstorms form above the front. (b) A warm front develops where a warm air mass moves into a cold one. Note that the slope of a warm front is much gentler than that of a cold front. Clouds form over a broad area. (c) An occluded front develops where a fast-moving cold front overtakes a warm front and lifts the base of the warm front off the ground.

changes drastically when one air mass replaces another. For example, a summer day in a Midwestern state will be cool and dry under an Arctic air mass but hot and humid under a maritime tropical air mass.

The boundary between two air masses is called a **front.** Meteorologists recognize three kinds of fronts. At a **cold front,** a cold air mass pushes underneath a warm air mass (▶Fig. 20.16a). As a consequence, the warm, moist air flows up to higher elevations, where it expands and cools adiabatically. The moisture it contains then condenses to form large clouds from which heavy rains fall. As a **warm front** moves into a region, the warm air slowly rises over the cool

air (▶Fig. 20.16b). Again, the air expands and cools, and the moisture condenses, so clouds develop over the boundaries between the two air masses. It's important to note that even though a front can be portrayed on a map as a line, it is in fact a sloping surface. Cold fronts have much steeper surfaces than warm fronts.

Not all air masses move at the same velocity. Typically, cold fronts move faster than warm fronts and overtake them. Where this happens, the warm front is lifted up so that it no longer intersects the ground surface. Meteorologists refer to the geometry that results when a cold front pushes underneath a warm front as an **occluded front** (▶Fig. 20.16c).

Cyclonic and Anticyclonic Flow

An air mass may be a region of particularly high or low pressure. Thus, a pressure gradient exists along the margin of an air mass; and anywhere that a pressure gradient forms, winds begin to blow. Because air moves from high pressure to low pressure, winds tend to flow toward the center of a low-pressure mass, where they converge and create an excess that must flow up. Thus, air rises at the center of a low-pressure mass. In contrast, air spreads outward along the margins of a high-pressure mass, creating a deficit of air at the center that pulls air down from above. Because rising air expands and cools and its moisture condenses to form clouds, it may rain as a low-pressure mass moves by. But sinking air gets compressed, warms up, and can absorb more moisture, so high-pressure masses tend to mean fair (clear and dry) weather.

If the Earth did not spin, air flow into a low-pressure mass and out of a high-pressure mass would be radial: that is, the air would move in a direction perpendicular to the center of the air mass, like the spokes of a wheel. But because of the Coriolis effect, the wind is deflected and moves in a spiral around the air mass, eventually flowing nearly parallel to isobars. In the Northern Hemisphere, the flow becomes counterclockwise (creating a **cyclone**) around a low-pressure mass, and clockwise (creating an **anticyclone**) around a high-pressure mass (▶Fig. 20.17). (Note that "cyclone" here does not refer to a tornado or hurricane.)

In mid-latitudes, latitudes that encompass the United States and much of Canada as well as much of Europe, the weather commonly reflects the movement of a large low-pressure air mass that moves from west to east. Air circulates counterclockwise around the mass, creating a **mid-latitude cyclone,** or **wave cyclone** (▶Fig. 20.18a). Wave cyclones develop when air on one side of a cold front shears sideways past air on the other side (▶Fig. 20.18b). Just as the shear of air along the surface of the sea creates water waves, the shear of air along the cold front warps the face of the front into the shape of a wave. When this happens, warm air starts to

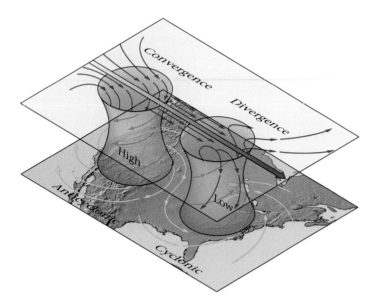

FIGURE 20.17 Air spirals downward and clockwise (creating an anticyclone) at a high-pressure mass and upward and counterclockwise (creating a cyclone) at a low-pressure mass.

move north, up and over the cold air mass, creating a warm front, while cold air circles around and starts to move south and downward, pushing the cold front forward. The two fronts meet at a V, the point of which lies near the center of the low-pressure mass. In a satellite image, a mid-latitude cyclone is a huge spiral mass of clouds, rotating counterclockwise; the clouds develop along both fronts and are centered on the low-pressure mass. The cold front of a wave cyclone

tends to move faster than the warm front, so eventually the warm front becomes occluded, and the cyclone dies out.

Clouds and Precipitation

Let's now look at clouds a little more closely (▶Fig. 20.19). As noted earlier, a **cloud** is a region of the atmosphere where moisture has condensed to form tiny water droplets (about 20 micrometers across, less than a third of the diameter of a human hair; 1 micrometer = 0.001 mm), or has crystallized to form tiny grains of ice. Clouds that form at ground level we call **fog.** Because clouds reflect and scatter incoming sunlight, they keep the ground cooler during the day, but at night, they prevent infrared radiation from escaping, and thus keep the ground warmer.

Droplets or ice grains in clouds form by condensation or precipitation, respectively, when the air becomes saturated with water vapor. During cloud formation, water condenses on **condensation nuclei,** preexisting solid or liquid particles, such as aerosols. Air can become saturated (so that clouds develop) when evaporation provides additional water or when the air cools so that its capacity to hold water decreases. Cooling may take place at night, simply because of the loss of sunlight, or any time that air rises and adiabatically cools. Meteorologists recognize several conditions that cause air to rise, and they refer to these as **lifting mechanisms.**

• *Convective lifting:* This process occurs where air on the ground warms, becomes buoyant, and rises. Clouds may form by convective lifting over an island that heats up

FIGURE 20.18 (a) A fully developed mid-latitude (wave) cyclone in the middle of its west-to-east trek across North America. Note that clouds form in association with the fronts and with the large low-pressure zone that develops where the two fronts merge. (b) The evolution of a mid-latitude cyclone. 1: Two air masses shear past each other. 2: An indentation, or "wave," develops along the front. 3: The indentation becomes more pronounced, and air starts circulating in a counterclockwise path (it becomes a cyclone). 4: When fully developed, the warm front is occluded. 5: When occlusion is complete, the cyclone dissipates.

FIGURE 20.19 Clouds can create a dramatic spectacle. Here, towering cumulonimbus clouds form.

(a) Convective lifting

(b) Convergence lifting

(c) Orographic lifting

FIGURE 20.20 (a) Convective lifting occurs where particularly warm air starts to rise. (b) Convergence lifting takes place where winds merge—the air has nowhere to go but up. (c) Orographic lifting occurs where winds off the sea run into a mountain range and are forced up.

during the day, relative to the surrounding sea (▶Fig. 20.20a).

- *Frontal lifting:* Frontal lifting takes place along the fronts between air masses (Fig. 20.16). At cold fronts, warm air pushes up and over a steep wall of cold air, rising rapidly to form large clouds. At warm fronts, the advancing warm air rides up the gentle slope of the front and condenses. At occluded fronts, clouds develop above the V at the intersection of the warm front and cold front.

- *Convergence lifting:* Where air converges or pushes together, as happens where air spirals up into a low-pressure zone or where two winds that have been deflected around an obstacle meet again, the air has nowhere to go but up, and thus rises and cools, forming clouds (▶Fig. 20.20b).

- *Orographic lifting:* This type of lifting happens where moisture-laden wind blows toward a mountain range, and on meeting the mountain range can go no farther and must rise. As a result, clouds form above the mountain range (▶Fig. 20.20c).

Rain, snow, and sleet precipitate from clouds in two ways, depending on the temperature of the cloud. In warm clouds, rain develops by **collision and coalescence,** during which the tiny droplets that compose the cloud collide and stick together to create a larger drop (▶Fig. 20.21a). Eventually, water drops that are too big to be held in suspension by circulating air start to fall, incorporating more droplets as they go. Depending on how far the drops have fallen and on how much moisture is available, drops reach different sizes before they hit ground. Typical raindrops have a diameter of 2 mm and fall at a velocity of about 20 km per hour. Any drops larger than 5 mm tend to break into

smaller ones because of air resistance. If rain falls through colder air near the ground, it freezes to become **sleet.**

In cold clouds, the mist contains a mixture of very cold water droplets and tiny ice crystals. The water droplets evaporate faster than the ice (because water molecules are less tightly bound to water than to ice) and provide moisture that condenses onto preexisting ice crystals, growing familiar hexagonal snowflakes. If the air below the cloud is very cold, the snow falls as powder-like flakes; if the air is

(a)

(b)

FIGURE 20.21 (a) Some rain forms when droplets collide and coalesce. Once a drop becomes large enough to fall, it incorporates more drops on the way down. Note that drops are flattened at their base because of air resistance (real raindrops are not teardrop-shaped). When a drop gets too big, it splits in two. (b) During the Bergeron process, water drops evaporate, releasing vapor that attaches to growing snowflakes, which then fall. If the air below a cloud is warm enough, the snowflakes melt and turn to rain before hitting the ground.

close to the melting temperature, large, wet clumps of snowflakes fall; and if the air is warmer than 0°C, the snow transforms to rain before it hits the ground. This kind of precipitation, involving the growth of ice crystals in a cloud at the expense of water droplets, is called the **Bergeron process,** after Tor Bergeron, the Swedish meteorologist who discovered it (▶Fig. 20.21b).

Many kinds of clouds form in the troposphere. It wasn't until 1803, however, that Luke Howard, a British naturalist, proposed a simple terminology for describing clouds (▶Fig. 20.22). First, we divide clouds into types based on their shape: puffy, cotton-ball- or cauliflower-shaped clouds are

cumulus (from the Latin word for "stacking"). Clouds that occur in relatively thin, stable layers and thus have a sheet-like or layered shape are called **stratus.** Clouds that have a wispy shape and taper into delicate, feather-like curls are called **cirrus.** We can then add a prefix to the name of a cloud to indicate its elevation: high-altitude clouds (above about 7 km) take the prefix "cirro," mid-altitude clouds take the prefix "alto," and low-altitude clouds (below 2 km) do not have a prefix. Finally, we add the suffix "nimbus" or the prefix "nimbo-" if the cloud produces rain.

Applying this cloud terminology, we see that a nimbostratus is a layered, sheet-like raincloud, and a **cumulonimbus** is a rain-producing puffy cloud. Cumulonimbus clouds can be truly immense, with their base lying at less than 1 km high and their top butting up against the tropopause at an elevation of over 14 km. These clouds are considered **vertically developed,** because they grow across altitude divisions. Large cumulonimbus clouds spread laterally at the tropopause to form a broad, flat top; flat-topped clouds are called **anvil clouds.**

The differences in cloud types depend on whether the clouds develop in stable or unstable air. **Stable air** does not have a tendency to rise, because it is colder than its surroundings. **Unstable air** has a tendency to rise, because it is warmer than its surroundings. Cumulus clouds, which in time-lapse photography look like they're boiling, form in unstable air. They billow because of **updrafts** (upward-moving air) and **downdrafts** (downward-moving air)—plane flights through these clouds will be rather bumpy. In contrast, stratus clouds indicate stable air.

20.6 STORMS: NATURE'S FURY

The rain from an overcast sky may be inconvenient for a picnic, but it won't threaten life or property and will be appreciated by farmers. In contrast, a storm poses a threat. A **storm** is an episode of severe weather, when winds, rainfall, snowfall, and in some cases lightning become strong enough to be bothersome and even dangerous (▶Fig.

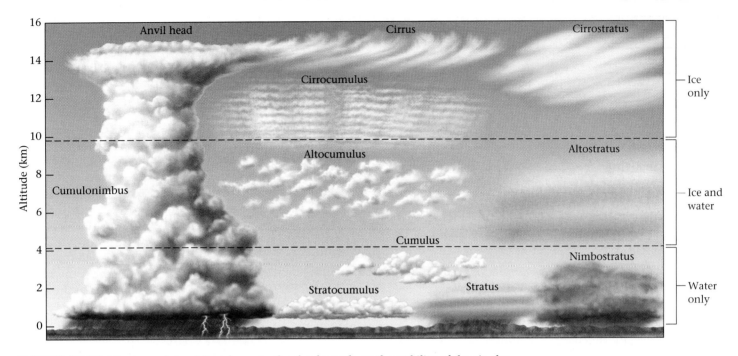

FIGURE 20.22 The type of cloud that forms in the sky depends on the stability of the air, the elevation at which moisture condenses, and the wind speed. Note that cumulonimbus clouds develop vertically, in that they grow across elevation boundaries.

20.23). Storms happen when the contrast between air masses across a front is particularly pronounced, with large differences in temperatures and pressures, or when prevailing conditions cause positive feedback, in that a storm, once established, can absorb hot, moist air and become stronger. As the rising moist air cools, water condenses and releases latent heat, which makes the air even warmer and more buoyant, so it can rise still higher.

Thunderstorms

A **thunderstorm** is a local episode of intense rain accompanied by strong, gusty winds, lightning, and sometimes hail. Over 2,000 are occurring at any given time somewhere on the Earth, and over 100,000 take place in the United States every year. Thunderstorms form where a cold front moves into a region of particularly warm, moist air (as happens, for example, in North America's mid-latitudes during the summer where cold polar air masses collide with warm gulf air masses); where convective lifting is driven by solar radiation in a region with an immense supply of moisture (as occurs over tropical rainforests); or where orographic lifting causes clouds to form over mountains.

A typical thunderstorm has a relatively short life, lasting from under an hour to a few hours (▶Fig. 20.24). The storm begins when a cumulonimbus cloud, fed by a steady supply of warm, moist air, grows large. The rising hot air,

kept warm by the addition of energy as the latent heat of condensation is released, creates updrafts that cause the cloud to stack or billow upward toward the tropopause. When this air adiabatically cools, precipitation begins.

If updrafts in the cloud are strong enough, ice crystallizes in the higher levels of the cloud, where temperatures are below freezing, building into ice balls known as **hail** (or

FIGURE 20.23 A thunderstorm can drench an area with rain and attack it with lightning.

FIGURE 20.24 A thunderstorm evolves in three stages. First, warm, unstable moist air rises and builds a cumulus cloud. Second, updrafts, fueled by more warm air at the surface and the heat released by condensation, cause the cloud to billow higher. Heavy rain falls, and lightning flashes; the downpour creates cold downdrafts. Finally, the cloud reaches the top of the troposphere, but downdrafts cut off the supply of warm air at the ground surface, and without this fuel, the thunderstorm dissipates.

Once precipitation begins, a thunderstorm has reached its "mature stage." Now, falling rain pulls air down with it, creating strong downdrafts. By this stage, the top of the cloud has reached the top of the troposphere and begins to spread laterally to form an anvil cloud. Because of the simultaneous occurrence of updrafts and downdrafts, a mature thunderstorm produces gusty winds and the greatest propensity for lightning. Eventually, downdrafts become the overwhelming wind; their cool air cuts off the supply of warm, moist air, so the thunderstorm dissipates.

Lightning accompanies thunderstorms because electrical charges separate in a cloud (▶Fig. 20.25a). Surprisingly, meteorologists still aren't sure why this happens, but they speculate that the rubbing of air and water molecules together in a cloud creates positively charged hydrogen ions and/or ice crystals that drift to the top of the cloud, and negatively charged OH⁻ ions and/or water droplets that sink to the base. The negative ions at the base of the cloud repel negative ions on the ground below,

hailstones). A discrete mass of hail may fall from a cloud over a few minutes to form a **hail streak** on the ground, typically 2 by 10 km and elongated in the direction that the storm moves. Though most hailstones are pea-sized, the largest recorded hailstone reached a diameter of 14 cm and weighed 0.7 kilograms.

FIGURE 20.25 (a) Lightning flashes when a charge separation develops in a cloud, with a negative charge at the base and a positive charge at the top. The negative charge repels negative charges on the ground, so positive charges develop on the ground. A leader begins to descend from the cloud. (b) As the leader grows downward, positive charges begin to flow upward from an object on the ground. (c) When the current is complete, the return stroke carries positive charges rapidly from the ground to the cloud, creating the main part of the flash.

creating a zone of positive charge on the ground. Air is a good insulator, so the charge separation can become very large until a giant spark or pulse of current, a **lightning flash,** jumps across the gap—essentially, lightning is like a giant short circuit across which a huge (30-million-volt) pulse of electricity flows.

Lightning flashes can jump from one part of a cloud to another, or from a cloud to the ground. In the case of cloud-to-ground lightning, the flash begins when electrons leak from the negatively charged base of the cloud incrementally downward across the insulating air gap, creating a conductive path, or **leader.** While this happens, positive charges flow upward to the cloud through conducting materials, like trees or buildings (▶Fig. 20.25b). When the connection is completed, a strong current carries positive charges up into the cloud; this upward-flowing current is the **return stroke** (▶Fig. 20.25c).

Thunder is the cracking or rumbling noise that accompanies lightning. We hear it because the immense energy of a flash almost instantaneously heats the surrounding air to a temperature of 8,000 to 33,000°C, and this abrupt expansion, like an explosion, creates sound waves that travel through the air to our ears. But because sound travels so much more slowly than light, we hear thunder after we see lightening. A five-second time delay between the two means that the lightning is about 1.6 km (1 mile) away.

Over eighty people a year die from lightning strikes in the United States alone, and many more are seriously burned or shocked. Lightning that strikes trees heats the sap so quickly that the tree can literally explode. Lightning can spark devastating forest fires and set buildings on fire. You can reduce the hazard to buildings by installing lightning rods, upward-pointing iron spikes that conduct electricity directly to the ground so that it doesn't pass through the building.

Tornadoes

Some thunderstorms grow to be quite violent, and spawn one or more tornadoes. A **tornado** is a near-vertical, funnel-shaped cloud in which air rotates extremely rapidly around the axis (center) of the funnel (▶Fig. 20.26a, b). In other words, a tornado is a vortex beneath a severe thunderstorm. The word probably comes from the Spanish *tonar,* meaning "to turn," or *tronar,* meaning "thunder" (perhaps referring to the loud noise generated by a tornado).

In the mid-latitudes of the Northern Hemisphere, where most tornadoes take place, air in the funnel rotates counterclockwise around the center and spirals upward. Air in the fiercest tornadoes probably moves at speeds of up to 500 km per hour (about 300 mph). The diameter of the base of the funnel in a small tornado may be only 5 m across, while in the largest tornadoes it may be as wide as 1,500 m across. Because of the upward movement of air, air pressure within

(a)

(b)

FIGURE 20.26 (a) A tornado touches down. (b) The intense winds of a tornado rip through houses and scatter debris.

a tornado can drop by as much as 0.1 bars (by comparison, average sea-level air pressure is 1.013 bars).

In North America, tornadoes drift with a thunderstorm from southwest to northeast, because of the prevailing wind direction, traveling at speeds of 0 to 100 km per hour. They tend to hopscotch across the landscape, touching down for a stretch, then rising up into the air for a while before touching down again. This characteristic leads to a bizarre incidence of damage—one house may be blasted off its

TABLE 20.2 Fujita Scale for Tornadoes

Scale	Category	Wind Speed km/h (mph)	Path Length; Path Width	Typical Damage
F0	Weak	64–116 (40–72)	0–1.6 km; 0–17 m	Branches and windows broken.
F1	Moderate	117–180 (73–112)	1.6–5.0 km; 18–55 m	Trees broken; shingles peeled off; mobile homes moved off their foundations.
F2	Strong	181–253 (113–157)	5–16 km; 56–175 m	Large trees broken; mobile homes destroyed; roofs torn off.
F3	Severe	254–332 (158–206)	16–50 km; 76–556 m	Trees uprooted; cars overturned; well-constructed roofs and walls removed.
F4	Devastating	333–418 (207–260)	50–160 km; .56–1.5 km	Strong houses destroyed; buildings torn off foundations; cars thrown; trees carried away.
F5	Incredible	419–512 (261–318)	160–500 km; 1.5–5.0 km	Cars and trucks carried more than 90 m; strong houses disintegrated; bark stripped off trees; asphalt peeled off roads.

foundation while its next-door neighbor remains virtually unscathed. Small tornadoes may cut a swath less than a kilometer long, but large tornadoes raze the ground for tens of kilometers, and the largest have left a path of destruction up to 500 km long. In 1925, for example, one of the most enormous tornadoes on record ripped across Missouri, Illinois, and Indiana, killing 689 people before it dissipated. In some cases, two or three tornadoes may erupt from a single thunderstorm. Massive thunderstorm fronts may produce a **tornado swarm,** dozens of tornadoes out of the same storm. In April 1974, a single frontal thunderstorm system generated a swarm of at least 148 individual tornadoes, which killed 307 people over eleven states.

Tornadoes cause damage both because of the force of their rapidly moving wind and because of their low air pressure. The wind lifts trucks and tumbles them for hundreds of meters, uproots trees, and flattens buildings—particularly large tornadoes can even rip asphalt off the highway. The low air pressure around a tornado can make windows pop out of buildings and may cause airtight buildings to explode, as the air inside suddenly expands relative to the air outside. In some cases, tornadoes cause strange kinds of damage: they have been known to drive straw through wood, lift cows and carry them unharmed for hundreds of meters, and raise railroad cars right off the ground.

Because of the range of damage a tornado can cause, T. T. Fujita, of the University of Chicago, proposed a scale that distinguishes among tornadoes based on wind speed, path dimensions, and possible damage. (The wind speeds are estimates, based on the damage they cause, as no one has yet succeeded in measuring winds directly in a tornado.) The **Fujita scale** is increasingly referred to in news reports about tornadoes (see Table 20.2).

Tornadoes in North America form where strong westerlies exist at high altitudes while strong southeast winds de-

velop near the ground surface (▶Fig. 20.27a–c). These opposing winds shear the air between them, so that it begins to rotate in a horizontal cylinder. Tornado watchers search for the resulting funnel cloud, as a precursor to a true tornado. As the associated thunderstorm matures, updrafts tilt one end of the cylinder up and downdrafts push the other end down, until the air in the cylinder starts spiraling inward and upward, gaining speed like spinning figure skaters who pull their arms inward. At this stage, the cloud consists only of rotating moisture, and may look grayish white. But if the process continues, the low end of the funnel touches ground, and at the instant of contact dirt and debris get sucked into the tornado, giving it a dark color.

The special weather conditions that lead to tornadoes exist where cold polar air from Canada collides with warm tropical air from the Gulf of Mexico. These conditions happen most frequently during the months of March–September in the Midwestern states and in Florida; the frequency of tornadoes in particular along the line from Texas to Indiana has given this region the unwelcome nickname "tornado alley" (▶Fig. 20.28). During a thirty-year span, the number of reported tornadoes per year ranged between about 420 and 1,100 in the United States (with an average of 770 per year; fewer than twenty strike Canada annually). About eighty people a year die in tornadoes—on occasion, a single F5 event may kill hundreds.

Because of the threat tornadoes pose to life and property, meteorologists have worked hard to be able to forecast them. First they search for appropriate weather conditions—a major cold front along with westerly winds aloft and southeasterly winds at the surface. If these conditions exist, meteorologists issue a "tornado watch." If observers spot an actual tornado forming, they issue a "tornado warning" for the region in its general path (the exact path can't be predicted). If you hear a warning, it's best to take cover immediately in a

FIGURE 20.27 (a) Tornadoes initiate at intense fronts where high-altitude westerlies flow over low-altitude southeasterlies. The resulting shear creates a horizontal cylinder of rotating air in a cloud. (b) Updrafts and downdrafts in the cloud eventually tilt the cylinder, creating a tornado. (c) When the tornado touches down, destruction follows. Note that the spiraling winds of the funnel also circulate inside the main cloud; the tornado we see protruding from the base of the cloud is only the tip of a much larger flow.

basement, or at least in an interior room away from windows. With the invention of Doppler radar, meteorologists can often detect large tornadoes without even going outside. They can detect rain in clouds by bouncing radar beams off the rain, and, by taking into account the Doppler effect (see Chapter 1), can determine if the rain is moving in strong

FIGURE 20.28 North American tornadoes are most common in "tornado alley," a band extending from Texas to Indiana, where the polar mass collides with the Gulf Coast maritime tropical air mass. The storm systems are urged eastward by the jet stream and related high-altitude westerlies.

Number of tornadoes per year (per 26,000 sq. km, for a 27-year period)

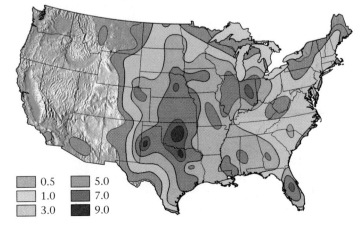

0.5	5.0	
1.0	7.0	
3.0	9.0	

winds. On a Doppler radar image, a tornado appears as a hook shape, typically developing along the western edge (the back side) of a storm that is drifting to the east.

Nor'easters

In some cases, large mid-latitude (wave) cyclones of North America affect the Atlantic coast. Because the cold, counterclockwise winds of these cyclones come out of the northeast, they are called **nor'easters.** Some nor'easters are truly phenomenal storms. One of the strongest occurred at the end of October 1991 (the "perfect storm" made famous by the book and movie of that name). During this storm, winds were not as strong as those of a hurricane, but they covered such a large area that waves in the open ocean built to a height of over 11 m, meaning sure disaster for ships. When the waves reached shore, they eroded huge stretches of beach. The rainfall from the storm caused flooding inland.

Hurricanes

During the summer and early fall, cyclonic wind systems called **tropical disturbances** develop off the west coast of Africa, near the Cape Verde Islands (latitude 20°N). In these low-pressure regions, air converges and rises and, because of the Coriolis effect, begins to circulate counterclockwise. But because these storms form over the warm tropical waters of the central Atlantic, the air that rises within them is particularly warm and moist. As it rises, it cools, and moisture condenses, releasing latent heat of condensation. This latent heat is a fuel that provides energy to the storm—it causes the air to rise still higher, creating even lower pressure at the Earth's surface, which in turn sucks up even more warm, moist air. Thus, the storm grows in strength until it becomes a **tropical depression,** in which winds may reach 61 km per hour.

As long as there continues to be a supply of warm, moist air, the storm continues to grow broader, and air within begins to rotate still faster. When the sustained wind speed exceeds 119 km per hour, a hurricane has been born. In other words, a **hurricane** (named for the Carib god of evil) is a huge rotating storm, resembling a giant spiral in map view, in which sustained winds are greater than 119 km (74 miles) per hour (▶Fig. 20.29a). Within a hurricane, air pressure becomes much lower, because of the upward flow of air. Before the days of satellite forecasting, mariners closely watched

FIGURE 20.29 (a) A hurricane approaches the Florida peninsula. Note the spiral shape of the storm and the relatively small eye. (b) The tracks of several important Atlantic hurricanes show how most begin at latitudes of 15°–20° off the west coast of Africa, then drift westward and northward. (c) Some Atlantic hurricanes make it across Central America and lash the eastern Pacific. Similar storms occur in the western Pacific, where they are known as typhoons, and in the Indian Ocean, where they are known as cyclones; these storms also originate at latitudes of about 20°.

Cool dry air

Eye

Warm water vapor

Spiraling bands
of storm clouds

Eye wall

Spiraling winds

Warm ocean water

FIGURE 20.30 This cutaway diagram of a hurricane shows the spirals of clouds, the eye, and the eye wall. Dry air descends in the eye, creating a small region of calm weather.

their barometers, knowing that an extreme drop signaled the approach of a hurricane.

Traditionally, the designation "hurricane" applies to storms born in the east-central Atlantic that drift first westward and then northward, with the prevailing winds. The path, or **hurricane track,** that they follow allows them to inflict damage to land regions in the Caribbean and the Gulf Coast and East Coast of North America (▶Fig. 20.29b). Occasionally, hurricanes make it across Central America and then thrash the Pacific coast of Mexico and the western United States; some may even drift northwestward to Hawaii. Because hurricanes require warm water (warmer than 27°C), they only develop at latitudes south of about 20°. They do not form close to the equator, because there is not enough atmospheric motion or Coriolis effect at that latitude. Similar storms that form at latitude 20°N and 20°S in the western Pacific are called **typhoons;** those at latitude 20°N in the Indian Ocean are called **cyclones** (a second use of the word; ▶Fig. 20.29c).

A typical hurricane consists of several spiral arms, called **rain bands,** extending inward to a central zone of relative calm known as the hurricane's **eye.** A rotating vertical cylinder of clouds, called the **eye wall,** surrounds the eye (▶Fig. 20.30). The entire width of a hurricane ranges from 100 to 1,500 km, with the average around 600 km. Winds spiral toward the eye, and thus their angular momentum, and therefore speed, increases toward the center; as a result, the greatest rotary wind velocity occurs in the eye wall.

Hurricanes as a whole move along their track because of the prevailing motion of the atmosphere. Typically, a hurricane's velocity along its track, the **storm-center velocity,** ranges from 0 to 60 km per hour (on rare occasions, as fast as 100 km per hour). Because of the rotary motion of wind around the eye, winds on one side of the eye move in the opposite direction to those on the other side. On the side of the hurricane where winds move in the same direction as the storm's center, the storm-center velocity adds to the rotary motion, making surface winds particularly fast. On the

FIGURE 20.31 Hurricane Andrew caused immense damage in southern Florida before crossing the Gulf of Mexico and slamming into Louisiana.

other side of the hurricane, the storm-center velocity subtracts from the rotary velocity, so the winds are slower.

Hurricanes cause damage in several ways (▶Fig. 20.31).

- *Wind:* Hurricane-force winds may reach such intensity that buildings cannot stand up to them. Hurricanes can tear off branches, uproot trees, rip off roofs, collapse walls, force vehicles off the road, knock trains from their tracks, and blast mobile homes off their foundations. And they can send debris hurtling through the air.

- *Waves:* The force of hurricane winds shearing across the sea surface generates huge waves, sometimes tens of meters high. Out in the ocean, these waves can swamp or capsize even large ships. Near shore, they batter beachside property, rip up anchored boats and carry them inland, and strip beaches of their sand.

- *Storm surge:* Coastal areas may be severely flooded by a **storm surge**, excess water that is carried landward by the hurricane. In the portion of a hurricane where winds blow onshore, water piles up and becomes deeper over a region of 60 to 80 km, allowing waves to break at a higher elevation than they otherwise would. Storm surges during hurricanes are exacerbated by the very low air pressure in a hurricane, which allows the sea surface to rise still higher—if a hurricane hits at high tide, the damage may be even worse. A storm surge during the 1900 hurricane that hit Galveston, Texas, flooded the city and killed 6,000 people. Storm surges in Bangladesh cause horrendous loss of life because the Indian Ocean submerges the low-lying delta of the Ganges. During one 1970 cyclone, the death toll reached 500,000.

- *Rainfall:* The intense rainfall during a hurricane (in some cases, half a meter of rain has fallen in a single day) causes streams far inland to flood. The flooding itself can submerge towns and bring death. Intense rains can also weaken the soil of steep slopes, especially in deforested areas. For example, the rains of a 1998 hurricane that struck Central America created innumerable mudslides.

On average, five hurricanes happen every year. But in 1995, there were eleven—in fact, three or four tropical depressions and hurricanes existed at a single time, forming a deadly procession that marched across the Atlantic. Because hurricanes are nourished by warm ocean water, some atmospheric scientists fear that a warming of the climate may

TABLE 20.3 Saffir-Simpson Scale for Hurricanes

Scale	Category	Wind Speed (km/h)	Air Pressure in Eye (millibars)	Damage
1	Minimal	119–153	989 or more	Branches broken; unanchored mobile homes damaged; some flooding of coastal areas; no damage to buildings; storm surge of 1.2–1.5 m.
2	Moderate	154–177	965–979	Some roofs, doors, and windows damaged; mobile homes seriously damaged; some trees blown down; small boat moorings broken; storm surge of 1.6–2.4 m.
3	Extensive	178–209	945–964	Some structural damage to small buildings; large trees blown down; mobile homes destroyed; structures along coastal areas destroyed by flooding and battering; storm surge of 2.5–3.6 m.
4	Extreme	210–250	920–944	Some roofs completely destroyed; extensive window and door damage; major damage and flooding along coast; storm surge of 3.7–5.4 m. Widespread evacuation of regions within up to 10 km of the coast may be necessary.
5	Catastrophic	over 250	less than 920	Many roofs and buildings completely destroyed; extensive flooding; storm surge greater than 5.4 m. Widespread evacuation of regions within up to 16 km of the coast may be necessary.

lead to more and fiercer hurricanes in the future. Many Atlantic hurricanes veer northward when they bump into high-pressure air masses over North America. This sends the storms up the Eastern Seaboard. Hurricanes die out when they lose their source of warm, moist air—when they cross onto land or move over the cold waters of high latitudes.

Once a storm reaches tropical-depression status, it is assigned a name. Names for less significant hurricanes may be reused, but the names of particularly notorious hurricanes are retired. Meteorologists classify hurricanes according to the Saffir-Simpson scale (Table 20.3). Intense hurricanes have sustained winds of over 230 km per hour.

20.7 GLOBAL CLIMATE

When we talk about the weather, we're referring to the atmospheric conditions at a certain location at a specified time. But when we speak of the average weather conditions, and the range of conditions, of that region over a year, we use the term **climate.** For example, on a given summer day in Winnipeg, central Canada, it may be sunny, hot, and humid, and on a given winter day in south Florida, it may drop below freezing. But averaged over a year, the weather in Winnipeg is more likely to be cooler and drier than the weather in south Florida. The variables that characterize the climate of a region include its temperature (both the yearly average and yearly range), its humidity, its precipitation (both the yearly amount and the distribution during the year), its wind conditions, and the character of its storms.

Climate Controls and Belts

Climatologists, scientists who study the Earth's climate, suggest that several distinct factors control the climate of a region.

- *Latitude:* This is perhaps the most significant factor, for the latitude determines the amount of solar energy a region receives, as well as the contrasts between seasons. Polar regions, which receive much less solar radiation over the year, have colder climates than equatorial regions. And the contrast between winter and summer is greater in mid-latitudes than at the poles or the equator. We can easily see the influence of latitude by examining the global distribution of temperature, represented on a map by **isotherms,** lines along which the temperature is exactly the same (▶Fig. 20.32). Because land and sea do not heat up at the same rate, because the distribution of clouds is not uniform, and because ocean currents transfer heat across latitudes, isotherms are not perfect circles.

- *Altitude:* Because temperature decreases with elevation, cold climates exist at high elevations even at the equator.

FIGURE 20.32 Isotherms of January roughly parallel lines of latitude, owing to contrasts in amounts of insolation, except where they are distorted by oceanic currents. The Gulf Stream in the Atlantic Ocean deflects isotherms northward, granting the United Kingdom, Ireland, and Scandinavia milder climates than they might otherwise have. The pattern of isotherms is different in July.

Hiking from the base of a high mountain at the Andes to its summit takes you through the same range of climate belts you would pass through on a hike from the equator to the pole.

- *Proximity of water:* Land and water have very different **heat capacities** (ability to absorb and hold heat). Land absorbs or loses heat quickly, while water absorbs or loses heat slowly. Also, water can absorb and hold on to more heat than land can because water is semi-transparent; sunlight heats water down to a depth of up to 100 m, while sunlight heats land down to a depth of only a few centimeters. Thus, the proximity of the sea tempers the climate of a region: as a rule, locations in the interior of a continent (such as Winnipeg) experience a much greater range of weather conditions than regions along the coast (like Vancouver). Also, because the ocean provides moisture for rain, coastal regions tend to be wetter than inland regions.

- *Proximity to ocean currents:* Where a warm current flows offshore, it may warm the overlying air, and where a cold current flows offshore, it may cool down the overlying

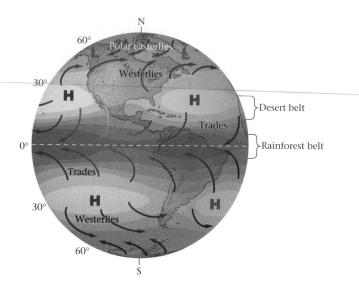

FIGURE 20.33 Because of land masses, atmospheric pressure belts vary in width to create lens-shaped, semipermanent high- and low-pressure cells. Compare this figure with Figure 20.13.

air. For example, the Gulf Stream brings warm water north from the Gulf of Mexico and keeps Ireland, the United Kingdom, and Scandinavia much warmer than they would be otherwise.

• *Proximity to orographic barriers:* An **orographic barrier** is a landform (such as a mountain range) that diverts air flow upward or laterally. This diversion affects the amount of precipitation and wind a region receives.

• *Proximity to high- or low-pressure zones:* Zones of high and low pressure, roughly parallel to the equator, encircle the planet (Fig. 20.13). Because land and sea have different heat capacities, they modify the zones, so that high-pressure zones tend to be narrower over land. Meteorologists refer to the resulting somewhat elliptical regions of high or low pressure as **semipermanent pressure cells** (▶Fig. 20.33). And because pressure zones and cells influence the prevailing wind direction of a region, as well as its long-term relative humidity, they strongly influence its climate.

Climatologists who have studied the distribution of climatic conditions around the globe have developed a classification scheme for climates, based on such factors as the average monthly and annual temperatures and the total monthly and yearly amounts of precipitation. The vegetation of a region proves to be an excellent indicator of climate, because plants are sensitive to temperature and to the amount and distribution of rainfall. Table 20.4 lists the principal types of climate belts (▶Fig. 20.34).

Climate Variability: Monsoons and El Niño

The climate at a certain location may change during the course of one or more years. Here, we look at two important

TABLE 20.4 Climate Types of the Earth

Climate Type	Regions and Characteristics
Tropical rainy	**Tropical rainforests** experience rain throughout the year, commonly in the form of afternoon thunderstorms (rainforests tend to straddle the equator). **Tropical savanna** (grasslands with brush and drought-resistant trees), on either side of a rainforest, have a rainy season and a dry season. These regions are subjected to **tropical monsoons.**
Dry	Dry regions include **deserts** (regions with very little moisture or vegetation cover; vegetation that does exist is adapted to long periods without moisture). **Steppe regions** (vast grassy plains with no forest) border the desert and have somewhat more precipitation; some steppe regions occur at high elevations, in latitudes where the climates would otherwise be more humid.
Humid mesothermal	This category includes **humid subtropical climates,** with moist air and warm temperatures for much of the year and mixed deciduous-coniferous forest; **Mediterranean climates,** coastal regions with most rainfall in the winter, very hot summers, and scrub forests; and **marine west-coast climates,** where the sea tempers the climate and may create a coastal temperate rainforest.
Humid microthermal	These higher-latitude temperate climates, which occur only in the Northern Hemisphere, include **humid continental regions with long summers** (as in the U.S. Midwest and mid-Atlantic states), with deciduous forest; **humid continental regions with short summers,** characterized by mixed deciduous-coniferous forest or coniferous-only forest; and **subarctic climates,** with very short, cool summers and coniferous forest that becomes lower and scrubbier at higher latitudes.
Polar	These cold climates include **tundra,** regions with no summer and an extremely cold winter, and inhabited only by low, cold-resistant plants (moss, lichen, and grass); much of the ground in tundra is permafrost (permanently frozen ground). In **ice-cap regions,** near the poles where the climate is sub-freezing year around, any land not covered by ice has essentially no vegetation cover. **Highlands** lie at lower (non-polar) latitudes but at such a high elevation that they have polar-like climates; when you enter a region above the treeline, you have entered a highland polar climate.

Ice cap
Tundra; subpolar
Highland
Continental
Subtropical
Tropical
Dry land (desert; steppe)

FIGURE 20.34 The basic climate belts on Earth. Note that highland areas (above snow line) are considered to have polar climates.

examples of climate variability that affect the human population in notoriously significant ways.

A **monsoon** is a major reversal in the wind direction that causes a shift from a very dry season to a very rainy season. In southern Asia, home to about half the world's population, people depend on monsoonal rains to bring moisture for their crops.

The Asian monsoon develops primarily because Asia is so large that it includes vast tracks of land far from the sea. Further, a substantial part of this land, the Tibet Plateau, lies at a high elevation. During the winter, central Asia becomes very cold, much colder than coastal regions to the south. This coldness creates a stable high-pressure cell over central Asia. Dry air sinks and spreads outward from this cell and flows southward over southern Asia, pushing the intertropical convergence zone (ITCZ) out over the Indian Ocean, south of Asia (▶Fig. 20.35a). Thus, during the winter, southern Asia experiences a dry season. During the summer, central Asia warms up dramatically. As warm air rises over central Asia, a pronounced low-pressure cell develops, and the intertropical convergence zone moves north. When this happens, warm air flows northward from the Indian Ocean, bringing with it substantial moisture, and the summer rains begin. Rainfalls are especially heavy on the southern slope of the Himalayas, because orographic lifting leads to the production of huge cumulonimbus clouds (▶Fig. 20. 35b).

Long before the modern science of meteorology became established, fishermen from Peru and Ecuador who ventured into the coastal waters west of South America knew that in late December the fish population that provided their livelihood diminished. Because of the timing of this event, it came to be known as **El Niño,** Spanish for "the Christ child." Why did the fish vanish? Fish are near the top of a food chain that begins with plankton, which live off nutrients in the water. These nutrients increase when cold water upwells from the deep along the coast of South America. During El Niño, warm water currents flow eastward from the central Pacific, and the cold, nutrient-rich water that supports the marine food chain remains at depth. With less nutrients, there are less plankton, and without the plankton, the fish migrate elsewhere.

To understand why El Niño occurs, we need to look at atmospheric flow and related surface ocean currents in the equatorial Pacific (▶Fig. 20.36a, b). Normally (when El Niño is not in progress), a major equatorial low-pressure cell exists in the western Pacific over Indonesia and Papua New Guinea, while a high-pressure cell forms over the eastern Pacific, along the coast of equatorial South America. This geometry means that a large east-west convection cell exists in the troposphere over the equatorial Pacific—air rises in the western Pacific, flows east, sinks in the eastern Pacific, and then flows west at the surface. The easterly surface winds blow warm surface water westward, so that it pools in

FIGURE 20.35 In the monsoonal climate of Asia, each year can be divided into a dry season and a wet season. (a) During winter, the dry season, a large high-pressure cell develops over central Asia, and the intertropical convergence zone lies well south of Asia. (b) During summer, the wet season, a low-pressure cell develops over Asia, so warm, moisture-laden water from the Indian Ocean flows landward. Orographic lifting along the Himalayas leads to cloud formation and intense rainfall.

FIGURE 20.36 El Niño exists because of a change in winds and currents in the central Pacific. (a) During La Niña ("normal" time), a low-pressure cell lies over the western Pacific, and surface trade winds blow to the west. These winds drive warm surface water westward, so cold water rises along the west coast of South America to replace it. (b) During El Niño, the low-pressure cell moves eastward, and the westward flow stops, so that cold water no longer upwells. The drift of the low-pressure cell back and forth across the Pacific is called the southern oscillation.

the western Pacific. Cold water from the deep ocean rises along South America, to replace the warm water that moved west. It is this rising cold water that brings nutrients to the surface. During El Niño, the low-pressure cell moves eastward over the central Pacific, and a high-pressure cell develops over Indonesia; so two convective cells develop. As a result, surface winds start to blow east in the western Pacific, driving warm surface water back to South America. This warm surface water prevents deep cold water from rising, sending the fish away. In effect, pressure cells oscillate back and forth across the Pacific, an event now called the **southern oscillation.**

El Niño gained world notoriety in late 1982 and early 1983, when a particularly large low-pressure cell developed in the eastern part of the Pacific. As a result, the jet streams stayed farther north than is typical. In effect, El Niño caused a temporary climate change worldwide. Drought conditions persisted in the normally rainy western Pacific, while unusually heavy rains drenched western South America. In North America, rains swamped the southern United States and storms battered California, winters were warmer than usual in Canada, and snowfalls were heavier in the Sierra Nevada Mountains, leading to spring floods.

Climatologists have been working intensely to understand the periodicity of El Niño. It is clear that strong El Niños take place around once every four years, with even stronger ones possibly happening at other intervals. Meteorologists have come to refer to only the strong events with global climatic effects as El Niño, and intervening "normal" years as **La Niña.**

CHAPTER SUMMARY

• The early atmosphere of the Earth contained high concentrations of water, carbon dioxide, and sulfur dioxide, gases erupted by volcanoes.

• After the oceans formed, much of the carbon dioxide dissolved in the ocean and then became trapped in limestone. When photosynthetic organisms evolved, they produced oxygen, and the concentration of this gas gradually increased.

• Air, the mixture of gases comprising the atmosphere, consists mostly of nitrogen (78%) and oxygen (21%). Several other gases, including carbon dioxide and methane, occur in trace amounts. The atmosphere also contains aerosols, tiny solid particles or liquid drops.

• Air pressure decreases with elevation. Thus, 90% of the air in the atmosphere occurs below an elevation of 16 km.

• When air rises, it expands and cools, a process called adiabatic cooling. If air is compressed, it heats up, a process called adiabatic heating.

• Air generally contains water. The ratio between the measured water content and the maximum possible amount of water that the air can hold is its relative humidity.

• The atmosphere is divided into layers, separated from each other by pauses. In the lowest layer, the troposphere, temperature decreases with elevation. The troposphere convects—its air movement causes weather. The other layers are the stratosphere, the mesosphere, and, at the top, the thermosphere.

• Air circulates on two scales, local and global. Winds blow because of pressure gradients: air moves from regions of higher pressure to regions of lower pressure.

• High latitudes receive less solar energy, or insolation, than low latitudes. This contrast in temperature initiates convection in the atmosphere: warm air rises at the equator, and cold air sinks at the pole. Because of the Coriolis effect, air moving north from the equator to the pole deflects to the east, and in each hemisphere, three convection cells develop (the Hadley, Ferrel, and polar cells). Divergence or convergence zones lie between the cells.

• Prevailing surface winds—the northeast tradewinds, the surface westerlies, and the polar easterlies—develop because of circulation in global convection cells.

• Air pressure at a given latitude decreases from the equator to the pole, causing a poleward flow of air. In the Northern Hemisphere, the Coriolis effect deflects this flow to generate high-altitude westerlies. Over the polar front and the horse latitudes, the pressure gradient is particularly steep, so the high-altitude winds are particularly strong; these strong winds are known as jet streams.

• Weather refers to the temperature, air pressure, wind speed, and relative humidity at a given location and time. Much of the weather that we enjoy (or endure) reflects the interaction of air masses, bodies of air with recognizable physical characteristics. The boundary between two air masses is a front.

• Air sinks in high-pressure air masses and rises in low-pressure air masses. Because of the Coriolis effect, the air begins to rotate around the center of the mass as a consequence. In a cyclone, air rises and circulates counterclockwise around a low-pressure mass. In an anticyclone, air sinks and circulates clockwise around a high-pressure mass.

• Clouds, which consist of tiny droplets of water or tiny crystals of ice, form when the air is saturated with water and contains condensation nuclei on which water condenses. Clouds commonly develop where air rises and adiabatically cools. Cumulus clouds are puffy, stratus clouds are sheet-like, and cirrus clouds are wispy and delicate.

• Thunderstorms begin when cumulonimbus clouds grow large. Friction between air and water molecules separates positive and negative charges. Lightning flashes when a giant spark jumps across the charge separation.

• Tornadoes, rapidly rotating funnel-shaped clouds, develop in violent thunderstorms. Nor'easters are large storms associated with wave cyclones. Hurricanes, huge rotating storms, originate over oceans where the water temperature is over 27°C. A hurricane is nourished by warm, moist air, because when this air rises, moisture condenses and releases latent heat of condensation.

• The climate refers to the average weather conditions in a region over a year. Climate is controlled by latitude, altitude, and the proximity of water and ocean currents, orographic barriers, and high- or low-pressure zones. Climate classes can be recognized by the vegetation they support.

• Monsoonal climates occur where there is a seasonal shift in the wind direction. El Niño, a temporary shift in climate triggered by shifts in the position of high- and low-pressure cells in the Pacific, makes an appearance every few years.

KEY TERMS

acid rain (p. 613)
adiabatic cooling, heating
 (p. 614)
aerosol (p. 613)
air mass (p. 624)
air pressure (p. 613)
anticyclone (p. 625)
atmosphere (p. 611)
aurora australis (p. 616)
aurora borealis (p. 616)
backscattered light (p. 618)
Bergeron process (p. 628)
cirrus cloud (p. 628)
cold front (p. 625)
convergence zone (p. 619)
Coriolis effect (p. 619)
cumulonimbus cloud
 (p. 628)
cumulus cloud (p. 628)
cyclone (p. 625)
dewpoint temperature
 (p. 614)
divergence zone (p. 619)
doldrums (p. 622)
downdraft (p. 628)
El Niño (p. 639)
equatorial low (p. 621)
equinox (p. 620)
eye wall (p. 635)
Ferrel cell (p. 619)
global warming (p. 613)
greenhouse gas (p. 613)
Hadley cell (p. 619)
heterosphere (p. 616)
high-altitude westerlies
 (p. 623)
homosphere (p. 616)
horse latitudes (p. 622)
insolation (p. 617)
ionosphere (p. 616)
isobar (p. 617)
isotherm (p. 637)

jet stream (p. 623)
La Niña (p. 641)
latent heat of condensation
 (p. 615)
lifting mechanism (p. 626)
mesopause (p. 615)
mesosphere (p. 615)
mid-latitude (wave) cyclone
 (p. 625)
monsoon (p. 639)
nor'easter (p. 633)
northeast tradewinds (p. 622)
occluded front (p. 625)
ozone hole (p. 613)
polar cell (p. 619)
polar easterlies (p. 622)
polar front (p. 619)
polar high (p. 621)
pressure gradient (p. 617)
prevailing winds (p. 622)
radon (p. 613)
relative humidity (p. 614)
solstice (p. 620)
southeast tradewinds (p. 622)
southern oscillation (p. 640)
storm surge (p. 636)
stratopause (p. 615)
stratosphere (p. 615)
stratus cloud (p. 628)
subpolar low (p. 621)
subtropical high (p. 621)
surface westerlies (p. 622)
thermosphere (p. 616)
tropical depression (p. 634)
tropical disturbance (p. 633)
tropopause (p. 615)
troposphere (p. 615)
typhoon (p. 635)
updraft (p. 628)
warm front (p. 625)
weather system (p. 624)
wind (p. 616)

REVIEW QUESTIONS

1. Where does the ozone in the atmosphere come from, where does it mostly occur, and why is it important?

2. Describe the composition of air (considering both its gases and its aerosols). Why are trace gases important?

3. How does air pressure change with elevation? Does the density of the atmosphere also change with elevation? Explain why or why not.

4. Describe the atmosphere's structure from base to top. What characteristics define the boundaries between layers?

5. What is the relative humidity of the atmosphere? What is the latent heat of condensation, and what is its relevance to a thunderstorm or hurricane?

6. Explain the relationship between the wind and variations in air pressure.

7. Why do changes in atmospheric temperature depend on latitude and the seasons? Why does global circulation break into three distinct convection belts in each hemisphere, separated by high-pressure or low-pressure belts?

8. Why do prevailing winds develop at the Earth's surface? Why are there prevailing westerly high-altitude winds, such as the jet stream?

9. Explain the origin of cyclones and anticyclones, and note their relationship to high-pressure and low-pressure air masses. What is a mid-latitude cyclone?

10. How does a cold front differ from a warm front and from an occluded front?

11. Why do clouds form? (Include a discussion of lifting mechanisms.) What are the basic categories of clouds?

12. Under what conditions do thunderstorms develop? What provides the energy that drives clouds to the top of the troposphere? Why does lightning occur in such storms?

13. What conditions lead to the formulation of a tornado? Where do most tornadoes appear?

14. Describe the stages in the development of a hurricane. Describe a hurricane's basic geometry.

15. What factors control the climate of a region? What special conditions cause monsoons? El Niño?

SUGGESTED READING

Ahrens, C. D. 1998. *Essentials of Meteorology: An Invitation to the Atmosphere*. New York: Wadsworth.

Bluestein, H. B. 1999. *Tornado Alley: Monster Storms of the Great Plains*. New York: Oxford University Press.

Burroughs, W. J. 1999. *The Climate Revealed*. Cambridge: Cambridge University Press.

Davies, P. 2000. *Inside the Hurricane*. New York: Holt.

Firor, J. 1990. *The Changing Atmosphere: A Global Challenge*. New Haven, Conn.: Yale University Press.

Graedel, T. E., and P. J. Crutzen. 1993. *Atmospheric Change: An Earth System Perspective*. New York: Freeman.

Longshore, D. 2000. *Encyclopedia of Hurricanes, Typhoons, and Cyclones*. New York: Checkmark.

Lutgens, F. K., and E. J. Tarbuck. *The Atmosphere*. 7th ed. Englewood Cliffs, N.J.: Prentice-Hall.

CHAPTER 21

Dry Regions: The Geology of Deserts

21.1 INTRODUCTION

For generations, nomadic traders have used camels to traverse the Sahara Desert in northern Africa (▶Fig. 21.1) The Sahara, the world's largest desert, receives so little rainfall that it has little if any surface water or vegetation. So camels must be able to walk for up to three weeks without drinking or eating. They can survive these journeys because they sweat relatively little, thereby conserving their internal water supply; they have the ability to metabolize their own body fat (up to 20 kg of fat comprise the animal's hump alone) to produce new water; and they can withstand severe **dehydration** (loss of water). Most mammals die after losing only 10–15% of their body fluid, for their blood plasma dries up, but camels can survive 30% dehydration with no ill effects. Camels do get thirsty, though—after a marathon trek across the desert, a camel may guzzle up to twenty-seven gallons of water in less than ten minutes.

The survival challenges faced by a camel emphasize that deserts are lands of extremes—extreme dryness, heat, cold, and, in some places, beauty. Desert vistas include everything from sand seas to sagebrush plains, cactus-covered hills to endless stony pavements. While less populated than other regions on Earth, deserts cover a significant percentage (about 25%) of the land surface, and thus comprise an important component of the Earth system. In this chapter, we take a look at the desert landscape. We learn why deserts occur where they do, and how erosion and deposition shape their surface. We conclude by exploring life in the desert and by examining the problem of desertification, the gradual transformation of temperate lands into desert.

21.2 WHAT IS A DESERT?

Formally defined, a **desert** is a region that is so **arid** (dry) that it contains no permanent streams, except for rivers that bring water in from temperate regions elsewhere, and supports vegetation on no more than 15% of its surface. In general,

The flowing shapes of sand dunes in the Sahara Desert of Libya form when strong, dry winds blow sand. In a desert, there is very little vegetation, because rain is so rare.

FIGURE 21.1 Camels, as in this Sahara Desert caravan in Mauritania, survive harsh desert conditions by storing water in fat reserves and by sweating little, if at all.

FIGURE 21.2 The shimmering in this desert mirage may look like water, but it's not. Mirages result from the interaction of light with a thin layer of hot air just above the ground surface.

desert conditions exist where less than 25 cm per year (10 inches per year) of rain falls, on average. But rainfall alone does not determine the aridity of a region. Aridity also depends on rates of evaporation and on whether rainfall occurs only sporadically or more continuously during the year. If all the rain in a region falls during isolated downpours once every few years, the region becomes a desert, because the intervals of drought last so long that plants and permanent streams cannot survive. Similarly, if high temperatures and dry air cause evaporation rates from the ground to exceed the rate at which rainfall wets the ground, then the region becomes a desert even if there is more than 25 cm per year of rain.

Note that the definition of a desert depends on a region's aridity, not on its temperature. Geologists distinguish between **cold deserts,** where temperatures generally stay below about 20°C, and **hot deserts,** where summer daytime temperatures exceed 35°C. Cold deserts exist at high latitudes where the Sun's rays strike the Earth obliquely and thus don't provide much energy, at high elevations where the air is too thin to hold much heat, or in lands adjacent to cold oceans, where the cold water absorbs heat from the air above. Hot deserts are found at low latitudes where the Sun's rays strike the desert at a high angle, at low elevations where dense air can hold a lot of heat, and in regions distant from the cooling effect of cold ocean currents. The hottest recorded temperatures on Earth occur in low-latitude, low-elevation deserts—58°C (136°F) in Libya and 56°C (133°F) in Death Valley, California.

Notably, the ground surface absorbs so much heat in hot deserts that a layer of very hot air (up to 77°C, or 170°F) forms just above the ground. This layer refracts sunlight, creating a **mirage,** a wavering pool of light, on the ground. Mirages make the dry sand of a desert wasteland look like a shimmering lake and distant mountains look like islands

(▶Fig. 21.2). Heat also contributes to aridity by increasing the rate of evaporation. In fact, evaporation rates in hot deserts may be so great that even when it rains, the ground stays dry because raindrops evaporate in mid-air. Significantly, even the hottest of hot deserts become cold at night: because of the dryness of the air, the lack of cloud cover, and the lack of foliage, deserts re-radiate their heat back into space at night. As a consequence, the desert air temperature at the ground surface may change by as much as 80°C in a single day.

The aridity of deserts causes different weathering, erosion, and depositional processes from those of temperate or tropical regions. Without plant cover, rain and wind batter and scour the ground, and during particularly heavy rains water accumulates into flash floods of immense power. Without abundant water, chemical reactions happen slowly, and humus (organic matter) does not collect on the ground. Thus, thick soils do not form in deserts, so the land surface consists of either rough bedrock, accumulations of clasts, caliche (see Chapter 7), or windblown sand. Overall, therefore, desert landscapes tend to be harsher and more rugged than temperate or tropical ones; if eastern North America were a desert, the Appalachians would not be gentle, forested hills but rather would consist of stark, rocky ridges.

21.3 TYPES OF DESERTS

Each desert on Earth has unique characteristics of landscape and vegetation that distinguish it from others. Geologists group deserts into five different classes, based on the environment in which the desert forms (▶Fig. 21.3).

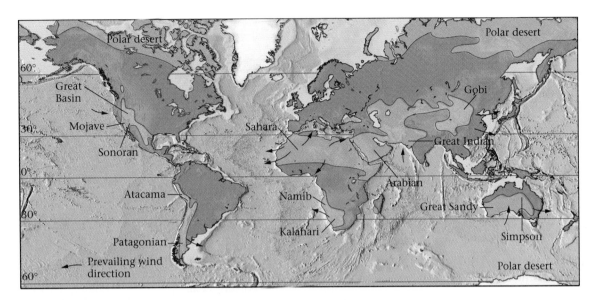

FIGURE 21.3 The global distribution of deserts. Note that the largest lie in the subtropical belts.

FIGURE 21.4 Rising air at the equator loses its moisture by raining over rainforests. When the air sinks over the subtropics, it warms and absorbs water. Thus, rainfall rarely occurs in the subtropics.

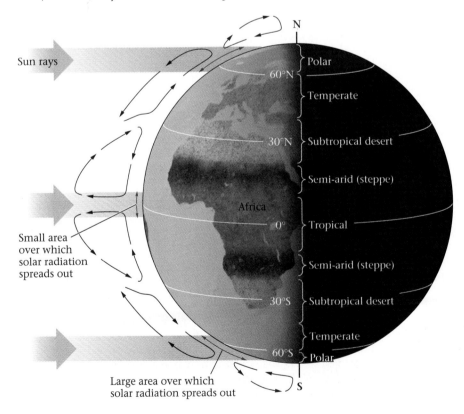

• *Subtropical deserts:* The world's largest deserts (for example, the Sahara, Arabian, Kalahari, and Australian) exist because they lie within climatic belts where cloud cover and rain rarely occur and, under the intense rays of the Sun, the daytime air gets very hot. Climatologists refer to these belts as the **subtropics,** because they lie on either side of the equatorial tropics (between the 20° and 30° lines of latitude north and south of the equator).

Subtropical deserts form because of the pattern of convection cells in the atmosphere (▶Fig. 21.4; see Chapter 20). At the equator, the air becomes warm and humid, for sunlight is intense and water rapidly evaporates from the ocean. The hot, moisture-laden air rises to great heights above the equator. As this air rises, it expands and cools, and can no longer hold as much moisture. Water condenses and falls in downpours that feed the lushness of the equatorial rainforest. The now-dry air high in the troposphere spreads laterally north or south. When this air reaches a latitude of 20° to 30°, it has become cold and dense enough to sink. Because the air is dry, no clouds form, and intense solar radiation strikes the Earth's surface. The sinking dry air condenses and heats up, soaking up any moisture present. In the regions swept by this air on its journey back to

FIGURE 21.5 Moist air, when forced to rise by mountains, cools. As this happens, the moisture condenses and rain falls, nourishing coastal rainforests, so by the time the air reaches the inland side of the mountains, it no longer holds enough moisture to rain. Deserts thus form in the rain shadow of mountains.

the equator, evaporation rates greatly exceed rainfall rates, and thus the regions become deserts.

In subtropical deserts that border the sea, episodes of high sea level flood coastal areas. The eventual evaporation of stranded seawater leaves broad regions called **sabkhas,** where a salt crust covers a mire of organic-rich mud.

• *Deserts formed in rain shadows:* As air flows over the sea toward a coastal mountain range, the air must rise (▶Fig. 21.5). As the air rises, it expands and cools. The water it contains condenses and falls as rain on the seaward flank of the mountains, nourishing a coastal rain forest. When the air finally reaches the inland (lee) side of the mountains, it has lost all its moisture and can no longer provide rain. As a consequence, the inland side of a mountain range becomes an arid region called a **rain shadow.** Such rain-shadow deserts can be found east of the Cascade Mountains in Washington and east of the Sierra Nevadas in Nevada.

• *Coastal deserts formed along cold ocean currents:* Cool ocean water cools the overlying air by absorbing heat, and decreases the capacity of the air to hold moisture. For example, the cold Humboldt Current, which carries water northward from the coast of Antarctica and runs along the west coast of South America, absorbs water from the breezes that blow east, over the coast. Thus, rain rarely falls on the coastal areas of Chile and Peru. As a result, this region hosts a desert landscape, including one of the driest deserts in the world, the Atacama (▶Fig. 21.6a–c). Portions of this narrow (less than 200 km wide) desert, which lies between the Pacific coast on the west and the Andes on the east, received no rain at all between 1570 and 1971.

• *Deserts formed in the interiors of continents:* As air masses move across a continent, they lose moisture by dropping rain, even in the absence of a coastal mountain range. Thus, when an air mass reaches the interior of a particularly large continent like Asia, it has grown quite

FIGURE 21.6 (a) Currents bringing cold water up from the Antarctic cool the air along the southwest coasts of South America and Africa. (b) The cool, dry air absorbs moisture from the adjacent coastal land, keeping it dry, so coastal deserts form. (c) The Atacama Desert of South America is the driest place in the world.

(a)

(b)

(c)

dry, so the land beneath becomes arid. The largest example of such a **continental-interior desert,** the Gobi, lies in central Asia, over 2,000 km away from the nearest ocean.

- *Deserts of the polar regions:* Precipitation in much of Earth's polar regions (north of the Arctic Circle, at 66°30'N, and south of the Antarctic Circle, at 66°30'S) is so low that these areas are, in fact, arid. Polar regions are dry, in part, for the same reason that the subtropics are dry (the global pattern of air circulation means that the air flowing over these regions is dry) and, in part, for the same reason that coastal areas along cold currents are dry (cold air holds little moisture).

Plate Tectonics and the Global Distribution of Deserts

The distribution of deserts around the world through geologic time reflects the process of plate tectonics, for plate movements determine the latitude of land masses, the position of land masses relative to the coast, and the proximity of land masses to a mountain range, and thus whether or not land lies in a region of dry air. Because of continental drift, regions that were deserts in the past are temperate regions now, and vice versa. With plate tectonics in mind, let's consider the origin of some major deserts today.

- The Sahara Desert exists because North Africa now lies in subtropical latitudes. The Gobi Desert exists because it lies in the interior of a large continent. But the Gobi's interior setting only developed when India collided with Asia 40–50 million years ago and thus added land to the southern margin of Asia.

FIGURE 21.7 Jointing controls the physical (mechanical) erosion of a cliff face in a desert.

- Regions that now lie in rain shadows develop because tectonic activity has uplifted a coastal mountain range. In western North America, the Cascade Mountains exist because Pacific Ocean floor currently subducts beneath the west coast of the continent. Such convergent-margin tectonics also led to the uplift of the Andes, which creates the rain shadow of the Pampean Desert in Chile and Argentina.

- The rain shadow east of the Sierra Nevadas exists because of rifting in the adjacent Basin and Range Province—the margins of rifts go up when the floor of the rift goes down, and the Sierra Nevadas make up the margin of the Basin and Range rift. Indeed, the elongate desert mountains and sediment-filled desert basins that define the Basin and Range Province of Nevada, Utah, and Arizona formed because of rifting (see Chapter 11).

21.4 WEATHERING AND EROSIONAL PROCESSES IN DESERTS

Without the protection of foliage to catch rainfall and slow the wind, and without roots to hold regolith in place, rain and wind attack and erode the land surface of deserts so efficiently that thick soil does not develop. The result, as we have noted, is that desert landscapes contrast markedly with those of temperate regions.

Weathering and Soil Formation

In the desert, as in temperate climates, physical weathering happens primarily when joints (natural fractures) split rock into pieces (▶Fig. 21.7). Joint-bounded blocks eventually break free of bedrock and tumble down slopes, fragmenting into smaller pieces as they fall. In temperate climates, thick soil forms over bedrock, so it lies buried beneath the surface. In deserts, however, the jointed bedrock commonly remains exposed at the ground surface, creating rugged, rocky escarpments.

Chemical weathering happens more slowly in deserts than in temperate or tropical climates, because there is less water available to react with rock. Still, rain or dew provides enough moisture for *some* weathering to occur. This water seeps into rock and **leaches** (dissolves and carries away) calcite and quartz, minerals that commonly form the cement, or matrix, of the rock. Leaching effectively rots the rock, by transforming it into a poorly cemented aggregate. Over time, the rock crumbles to form a pile of unconsolidated sediment, susceptible to transport by water or wind. If water passing through the rock contains dissolved salts, salt crystals may develop in the rock where the water dries out. The growth of such crystals pushes neighboring crystals apart and can also greatly weaken rock.

(a)
(b)

FIGURE 21.8 (a) Desert varnish, made from iron oxide and manganese oxide, is a dark coating on desert rock surfaces. Native American artists created petroglyphs, images on the rock surface, by chipping through the varnish to reveal the lighter rock beneath. (b) The Painted Desert of Arizona. The different colors of the rock layers come, in part, from the presence of oxidized or reduced iron in the rock.

Chemical weathering in desert climates commonly leads to the formation of shiny **desert varnish,** a dark, rusty brown coating of iron oxide and magnesium oxide that accumulates when water from rain or dew seeps into a rock, dissolves iron and magnesium ions, and carries the ions back to the surface of the rock by capillary action. Once at the surface, the water evaporates and leaves behind a residue of iron and magnesium. Such varnish won't form in humid climates, because rain washes the ions away too fast. Recent studies suggest that bacteria in the near-surface pore space of the rock may play a role precipitating the varnish.

Desert varnish takes a long time to form. In fact, the thickness of a desert varnish layer provides an approximate estimate of how long a rock has been exposed at the ground surface. In past centuries, Native Americans used desert-varnished rock as a medium for art: by chipping away the varnish to reveal the underlying lighter-colored rock, they were able to create to light-colored figures or symbols on a dark background. The resulting drawings are called **petroglyphs** (▶Fig. 21.8a).

Reactions between water (from dew or rain), oxygen (from the atmosphere), and iron-containing minerals in rocks yield oxidized iron, a rusty red material that, because of the lack of plant cover in deserts, colors rock. Slight variations in the concentration of iron, or in the degree of iron oxidation (iron that is not oxidized lends a greenish hue to rock), in adjacent beds result in spectacular color bands in rock layers and the thin soils derived from them. The Painted Desert of northern Arizona got its name from the brilliant and varied hues of oxidized iron (▶Fig. 21.8b).

We've noted that thick, organic-rich soils don't develop in deserts. But in flat-lying areas that aren't washed clean during rains, thin, clay-rich, stony soils do form. During heaving rains, downward-percolating water dissolves soluble minerals like calcite and carries them deeper into the soil or alluvium. The minerals then precipitate out to form cement that binds the unconsolidated grains comprising regolith together. As a result, a concrete-like material called **caliche** develops. In some regions, caliche forms so rapidly that it may incorporate abandoned tools and cans left behind by prospectors less than a century ago.

Water Erosion

Although rainfall is relatively rare in deserts, when it does come, it can radically alter a landscape in a matter of minutes. Since deserts lack plant cover, rainfall, sheetwash, and stream flow all are extremely effective agents of erosion. It may seem surprising, but water causes more erosion than does wind in most deserts.

Water erosion begins with the impact of raindrops, which eject sediment into the air. On a hill, the ejected sediment lands downslope, and thus during a rain it gradually migrates to lower elevations. The ground quickly becomes saturated with water during a heavy rain, so water starts flowing across the surface, carrying the loose sediment with it. Within minutes after a heavy downpour begins, dry stream channels fill with a turbulent mixture of water and sediment, which rushes downstream as a flash flood. When the rain stops, the water sinks into the stream bed's gravel and disappears—such

FIGURE 21.9 Gravel, sand, and boulders are left behind on the floor of a dry wash after a flash flood. The wash has steep walls because downcutting happens so fast.

FIGURE 21.10 Dust clouds form in deserts when turbulent air carries very fine sediment in suspension.

streams are called **intermittent, or ephemeral streams** (see Chapter 17). Because of the relatively high viscosity of the water (owing to its load of suspended sediment) and the velocity and turbulence of the flow, flash floods in deserts cause intense erosion—they undercut cliffs and transport huge boulders downstream. As rocks roll and tumble along, they strike one another and shatter, creating smaller pieces that can be carried still farther. Between floods, the stream floor consists of gravel, littered with boulders.

Flash floods carve steep-sided channels into the ground. Scouring of the canyon walls by sand-laden water may polish the walls and create stream-lined grooves. Dry stream channels in desert regions of the western United States are called **dry washes, or arroyos,** and in the Middle East and North Africa they are called **wadis** (▶Fig. 21.9).

Wind Erosion

In temperate and humid regions, plant cover protects the ground surface from the wind, but in deserts, the wind has direct access to the ground. In hot deserts, gusts of hot air feel like blasts from a furnace. Wind, just like flowing water, can carry sediment both as suspended load and as bed load. **Suspended load** (fine-grained sediment such as dust and silt held in suspension) floats in the air and moves with it (▶Fig. 21.10). The suspended sediment can be carried so high into the atmosphere (up to several kilometers above the Earth's surface) and so far downwind (tens to hundreds of kilometers) that it may move completely out of its source region. In fact, recent studies have demonstrated that dust lifted from the Sahara in northern Africa travels with the wind across the width of the Atlantic Ocean and finally settles out over the coral reefs

of the Caribbean. In some cases, tiny vortices can churn up dust. In general, these vortices are very small (centimeters to meters high). But in some cases, they become "dust devils" up to 100 m high, and look like miniature tornadoes. Cars driving down dirt roads in deserts have the same effect, generating plumes of suspended sediment.

Moderate to strong winds can roll and bounce sand grains along the ground, a process called **saltation** (▶Fig. 21.11). Saltating sand comprises the wind's **bed load.** Saltation begins when turbulence caused by wind shearing along the ground surface lifts sand grains. The grains move downwind, following an asymmetric, arch-like trajectory, but eventually return to the ground where they strike other sand grains, causing the new grains to bounce up and drift or roll downwind (like a billiard ball hitting other balls). The collisions between sand grains make the grains rounded and

FIGURE 21.11 During saltation, sand grains roll and bounce along the ground surface. As they bounce, they follow parabolic paths.

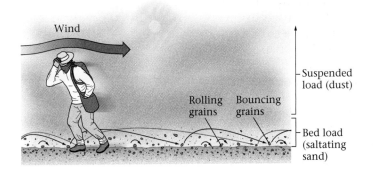

Wind

Rolling grains   Bouncing grains

Suspended load (dust)

Bed load (saltating sand)

frosted. Saltating grains generally rise no more than 0.5 m. But where sand bounces on bedrock during a desert sandstorm, they may rise 2 m, and can strip the paint off a car.

Wind cannot move grains larger than coarse sand. Thus, wind erosion does an effective job of sorting sediment, sending dust-sized particles skyward and sand-sized particles bouncing along the ground, and leaving pebbles and larger grains behind. The coarser sediment left behind is called a **lag deposit.** In such deposits, larger clasts may gradually settle down and fit together, creating a natural mosaic-like stone surface called **desert pavement** (▶Fig. 21.12a, b). Desert pavements armor the ground against further attack by wind or water, and so once they've formed, they may be stable for a long time. Pavements typically develop a coating of shiny desert varnish. (Notably, pavements may also form when a litter of stones on the desert surface gradually break apart, and settle adjacent to one another, creating a mosaic.

FIGURE 21.12 The progressive development of a lag deposit. (a) Pebbles in a desert are distributed through a matrix of finer sediment. With time, wind blows the finer sediment away, and the pebbles concentrate on the ground surface and may settle together to create a tight mosaic, a desert pavement. The pavement acts like armor, protecting the substrate from further erosion. (b) A well-developed desert pavement in Arizona.

(a)

(b)

(a) (b)

(c)

(d)

FIGURE 21.13 The progressive development of a ventifact. (a) Wind abrades the face of a rock. (b) Slowly, erosion carves a smooth surface called a facet. (c) Later, the wind shifts direction, and a new facet forms. The two facets join at a sharp edge. (d) A ventifact.

Just as sand blasting cleans the grime off the surface of a building, windblown sand and dust grind away at surfaces in the desert. Over long periods of time, such **wind abrasion** creates smooth faces, or facets, on pebbles, cobbles, and boulders. If a rock rolls or tips relative to the prevailing wind direction after it has been faceted on one side, or if the wind shifts direction, a new facet with a different orientation forms, and the two facets join at a sharp edge. Rocks whose surface has been faced by the wind are called **faceted rocks,** or **ventifacts** (▶Fig. 21.13a–d). Wind abrasion also gradually polishes and bevels down irregularities on a desert pavement and polishes the surfaces of desert-varnished outcrops, giving them a reflective sheen.

In places where a resistant layer of rock overlies a softer layer of rock, wind abrasion may create mushroom-like columns with a resistant block perched on an eroding column of softer rock. These unusual features are called

FIGURE 21.14 Yardangs are small landforms sculpted by the wind, where a resistant rock layer overlies a softer rock layer. They may be elongate, aligned with the wind direction. These yardangs formed in the Sahara Desert of Egypt.

FIGURE 21.16 This NASA photo shows the wind abrasion of clasts and small sand dunes on the surface of Mars. Clearly, the surface of the "red planet" now experiences desert conditions.

yardangs (▶Fig. 21.14). If a strong wind blows in only one direction, the yardangs become elongate, aligned with the wind direction.

Over time, wind in regions where the substrate consists of soft sediment picks up and removes so much sediment that the land surface sinks. The process of lowering the land surface by wind erosion is called **deflation.** Shrubs can stabilize a small patch of sediment with their roots, so after deflation a forlorn shrub with its residual pedestal of soil stands isolated above a lowered ground surface. In some places, the shape of the land surface channels the wind into a turbulent vortex that causes enough deflation to scour a deep, bowl-like depression called a **blowout** (▶Fig. 21.15).

FIGURE 21.15 A blowout, made where wind erosion has hollowed out a topographic depression.

Satellite exploration in the last two decades shows that wind affects the desert-like surface of Mars just as it does the deserts on Earth. In fact, during the Martian fall and spring, when differences in temperature between the ice-covered poles and the ice-free equator are large, strong winds blow from the poles to the equator, transporting so much dust that the planet's surface, as seen from Earth, visibly changes. At times, the entire planet becomes enveloped in a cloud of dust. Close-up images taken by spacecraft that have landed on Mars show that rocks have been abraded by saltating sand and that sand has accumulated on the lee side of the rocks (▶Fig. 21.16).

21.5 DEPOSITIONAL ENVIRONMENTS IN DESERTS

We've seen that erosion relentlessly eats away at bedrock and sediment in deserts. Where does the debris go? Below, we examine the various desert settings in which sediment accumulates.

Talus Aprons

With time, joint-bounded blocks of rock break off rock ledges and cliffs on the sides of hills. Under the influence of gravity, the resulting debris tumbles downslope and accumulates as a **talus apron** at the base of a hill. Talus aprons can survive for a long time in desert climates, so we typi-

FIGURE 21.17 This talus apron along the base of a desert cliff formed from rocks that broke off and tumbled down the cliff.

cally see them fringing the base of cliffs in deserts (►Fig. 21.17). The angular clasts comprising talus aprons gradually become coated in desert varnish.

Alluvial Fans

Flash floods can carry sediment downstream in an ephemeral stream channel. When the turbulent water flows out into a plain at the foot of mountains, it slows, and as a consequence, the sediment it has carried settles out. The resulting lenses of sediment cause the channel that has emerged from the mountains to subdivide into a number of subchannels (distributaries) that spread outward in a

FIGURE 21.18 The sediment comprising this alluvial fan in Death Valley was carried to the mountain front during flash floods. The water drops the sediment at the mountain front, because the change in slope slows the water down.

broad fan. The fan of distributaries spreads the sediment, or alluvium, out into a broad **alluvial fan,** a wedge- or apron-shaped pile of sediment (►Fig. 21.18). Alluvial fans emerging from adjacent valleys may merge and overlap along the front of a mountain range, creating an elongate wedge of sediment called a **bajada.** Over long time periods, the sediment of bajadas fills in adjacent valleys to depths of several kilometers.

Playas and Salt Lakes

Water from a flash flood may make it out to the center of an alluvium-filled basin, but if the supply of water is relatively small, it quickly sinks into the permeable alluvium without accumulating into a standing body of water. During a particularly large storm or an unusually wet spring, however, a temporary lake may develop over the low part of a basin. During drier times, such desert lakes evaporate entirely, leaving behind a dry flat lake bed known as a **playa** (►Fig. 21.19a). Over time, a smooth crust of clay and various salts (halite, gypsum, borax, and other minerals) accumulates on the surface of playas— some of these minerals have industrial uses and thus have been mined. (The salts had dissolved in low concentrations in the stream water that fed the lake, and were then concentrated by the process of evaporation.) Notably, when it rains slightly, clay-covered playa surfaces become very slippery. Racetrack Playa, in California, becomes so slippery, in fact, that the wind sends stones sliding out across the surface, leaving grooves behind them to mark their path (►Fig. 21.19b).

Where sufficient water flows into a desert basin, it creates a permanent lake. If the basin is an **interior basin,** with no outlet to the sea, the lake becomes very salty, because although its water escapes by evaporation in the desert sun, its salt cannot. The Great Salt Lake, in Utah, exemplifies this process. Even though the streams feeding the lake are fresh enough to drink, their water contains trace amounts of dissolved ions. Because the lake has no outlet, these ions have become concentrated in the lake over time, making it saltier than the ocean.

Deposition from the Wind

As mentioned earlier, wind carries two kinds of sediment loads—a suspended load of dust-sized particles and a bed load of sand. Much of the dust is carried out of the desert and accumulates elsewhere, forming layers of fine-grained sediment called **loess.** Sand, however, cannot travel far, and accumulates within the desert in piles called **dunes,** ranging in size from less than a meter to over 300m high. In favorable locations, dunes accumulate to form vast sand seas hundreds of meters thick. We'll look at dunes in more detail later in the chapter.

(a)

FIGURE 21.19 (a) A playa, a basin covered with clay and salt, develops where a shallow desert lake dries up. (b) Rocks slide along the slippery clay surface of California's Racetrack Playa when strong winds blow.

(b)

21.6 DESERT LANDSCAPES

The popular media commonly portray deserts as endless seas of sand, piled into dunes that hide the occasional palm-studded oasis. In reality, vast sand seas are merely one type of desert landscape. Some deserts are vast rocky plains, others sport a stubble of cacti and other hardy desert plants, and still others contain intricate rock formations that look like medieval castles. Explorers of the Sahara, for example, traditionally distinguished between **hamada** (barren, rocky highlands), **reg** (vast stony plains), and **erg** (sand seas in which large dunes form). In this section, we'll see how the erosional and depositional processes described above lead to the formation of such contrasting landscapes.

Rocky Cliffs and Mesas

In hilly desert regions, the lack of soil exposes rocky ridges and cliffs. As noted earlier, cliffs erode when rocks split away along vertical joints. When this happens, the cliff face retreats but retains roughly the same form. The process, commonly referred to as **cliff retreat,** or **scarp retreat,** occurs in fits and starts—a cliff may remain unchanged for decades or centuries, and then suddenly a block of rock falls off and crumbles into rubble at the foot of the cliff (▶Fig. 21.20a). Cliff height depends on bed thickness: in places where particularly thick resistant beds crop out, tall cliffs develop (for example, massive sandstone cliffs in Canyonlands National Park exceed 100 m in height). This is because large, widely spaced joints form in thick beds, so the collapse of a portion of the wall generates huge blocks. In thinly bedded shales, joints are small and closely spaced, so shale beds erode to make an overall gradual slope, consisting of many little stair steps. Thus, cliffs formed from stratified rocks (such as beds of sandstone and shale) develop a step-like shape; strong layers (sandstone or limestone) become vertical cliffs, and weak layers (shale) become rubble-covered slopes (▶Fig. 21.20b). (This landscape contrasts with landscapes in humid climates, where thick soils form; ▶Fig. 21.20c.) Typically, as cliff retreat takes place, the base of the cliff evolves into a broad, nearly horizontal bedrock surface known as a **pediment.**

With continued erosion and cliff retreat, a plateau of rock slowly evolves into a cluster of isolated hills, ridges, or columns (▶Fig. 21.20d). Flat-lying strata or flat-lying layers of volcanic rocks erode to make flat-topped hills. These go by different names, depending on their size. Large examples (with a top surface area of several square km) are **mesas,** from the Spanish word for "table." Medium-sized examples are **buttes** (▶Fig. 21.20e). Small examples, whose height greatly exceeds their top surface area, are **chimneys.** In North America, the best examples of such landforms can be seen on the Colorado Plateau, a broad region of generally flat-lying strata in northern Arizona, southern Utah, and western Colorado. Erosion of this strata has resulted in the skyscraper-like buttes of Monument Valley, Arizona, and the stark cliffs of Canyonlands National Park. Bryce Canyon National Park of Utah contains countless chimneys of brightly colored shale and sandstone—locally, these chimneys are called **hoodoos** (▶Fig. 21.20f). **Natural arches,** such as those of Arches National Monument, form when erosion along joints leaves narrow walls of rock. When the lower part of the wall erodes while the upper part remains, an arch results. (See art on pp. 660–61.)

FIGURE 21.20 (a) The process of cliff retreat in a desert. Rocks break off the cliff along joints that run parallel to the cliff face. The cliff face therefore maintains the same shape and orientation, even though erosion causes the cliff to retreat from point A to point B. (b) Stair-step cliffs appear where thick beds of strong rock are interlayered with thin beds of weak rock. Joints are more closely spaced in the thin layers. (c) If the sequence of rocks shown in (b) were to occur in a temperate or humid climate, the slope would instead be smooth and underlain by thick soil. (d) Because of cliff retreat, a once-continuous layer of rock evolves into a series of isolated remnants. (e) These buttes were carved from massive red sandstone layers in Monument Valley, Arizona. (f) These hoodoos, in Bryce Canyon, Utah, were made from the erosion of multicolored layers of sandstone, siltstone, and shale.

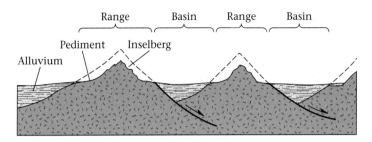

FIGURE 21.22 Inselbergs are small islands of rock surrounded by pediments. Alluvium gradually covers the pediments.

In places where bedding dips at an angle to horizontal, flat-topped mesas and buttes don't form; rather, asymmetric ridges called **cuestas,** or **hogbacks,** develop. A joint-controlled cliff forms the steep front side of a cuesta, while the tilted top surface of a resistant bed forms the gradual slope on the backside (▶Fig. 21.21). Because the angle of the gradual slope is the same as the dip angle of the bed (the angle the bed surface makes with respect to horizontal), it is called a **dip slope.** If desert hills consist of homogeneous rock like granite, rather than stratified rock, they typically erode to make a pile of rounded blocks (see Fig. 7.10c).

With progressive cliff retreat on all sides of a hill, finally all that remains of the hill is a relatively small island of rock, surrounded by a pediment or by alluvium. Geologists refer to such islands of rock by the German word **inselberg** ("is-land mountain"; ▶Fig. 21.22). Depending on the rock type or the orientation of stratification in the rock, and on rates of erosion, inselbergs may be sharp-crested, plateau-like, or loaf-shaped (steep sides and a rounded crest). Inselbergs with a loaf geometry, as exemplified by Uluru (Ayers Rock) in central Australia (Box 21.1), are also known as **bornhardts.**

Stony Plains

The coarse sediment eroded from desert mountains and ridges washes into the lowlands and builds out to form gently sloping alluvial fans. The surfaces of these gravely piles are strewn with pebbles, cobbles, and boulders, and are dissected by dry washes (wadis or arroyos). Portions of these stony plains evolve into desert pavements or playas.

Seas of Sand: The Geometry of Dunes

In places where abundant sand accumulates, sand seas (ergs) bury the landscape. The wind builds the sand in these ergs into dunes that come in a variety of shapes and sizes, depending on the character of the wind and the sand supply (▶Fig. 21.24). Where the sand supply is relatively scarce and the wind blows steadily in one direction, beautiful crescents called **barchan dunes** develop, with the tips of the crescents pointing downwind. If the wind shifts direction frequently, a group of crescents pointing in different directions overlap one another, creating a constantly changing **star dune.** Where enough sand accumulates for the ground surface to be completely buried but only moderate winds blow, sand piles into simple, wave-like shapes called **transverse dunes.** The crests of transverse dunes lie perpendicular to the wind direction. Strong winds may break through transverse dunes and change them into **parabolic dunes,** whose ends point in the upwind direction. Finally, if there is abundant sand and a strong, steady wind, the sand streams into **longitudinal dunes** (also called **seif dunes,** after the Arabic word for "sword"), whose axis lies parallel to the wind direction. In the southern third of the Arabian Peninsula, a region called the

FIGURE 21.21 Asymmetric ridges called cuestas appear where the strata in a region are not horizontal. A joint-controlled cliff forms the steep side, while a dip slope makes up the gentle side. The surface of a dip slope, by definition, is parallel to the bedding of strata beneath.

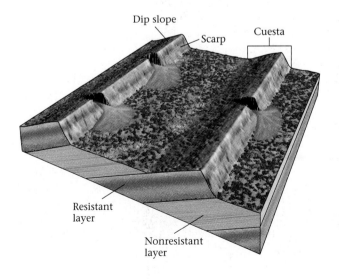

Empty Quarter because of its total lack of population, a vast erg called the Rub al Khali contains seif dunes that stretch for almost 200 km and reach heights of over 300 m.

In a sand dune, sand saltates up the windward side of the dune, blows over the crest of the dune, and then settles on the steeper, lee face of the dune. The slope of this face attains the angle of repose, the slope angle of a freestanding pile of sand. As sand collects on this surface, it may become unstable and slide down the slope—thus, geologists refer to the lee side of a dune as the **slip face.** As progressively more sand accumulates on the slip face, the crest of the dune migrates downwind, and former slip faces become

BOX 21.1

THE REST
OF THE
STORY

Uluru (Ayers Rock)

Much of central Australia comprises an immense desert. Parts are barren and sandy, but much of the land is covered with scrub brush, which provides meager grazing for cattle and kangaroo. Uluru, also known by its English name, Ayers Rock, is a huge bornhardt that towers 360 m above the desert plain (▶Fig. 21.23a, b). This rock mass, 3.6 km long, and 2 km wide, consists of nearly vertical dipping sandstone beds. It makes up one limb of a huge regional syncline. The other limb is also a bornhardt, known locally as the Olgas, and the entire area in between is buried in alluvium. Uluru formed because its sandstone resisted erosion, while adjacent rock formations did not. Thus, through geologic time, alluvium buried the surrounding landscape, but Uluru remained high. Because its strata dip vertically, it has not developed the stair-step shape of mesas and buttes on the Colorado Plateau.

Because of its grandeur, Uluru plays a sacred role in Australian Aboriginal traditions. In the dreamtime (time of creation) legends of the Aboriginal people, erosional features on the surface of the rock record a fierce battle between ancient clans. For example, the legends say that large holes in the side of the rock represent places hit by spears. In recent years, the rock has attracted tourists from around the world, who risk life and limb to climb to its top. Plaques at the base of the rock record the names of those who slipped and fell from its steep side.

FIGURE 21.23 (a) Originally Uluru (Ayers Rock), in Australia, was part of a large syncline beneath a mountain. Erosion has beveled the mountain away, until all that's left now above a sediment-covered plain are the resistant sandstone limbs of the syncline; softer units are covered by sediment. Uluru is the vertical limb, while the Olgas are the gently dipping limb. (b) Uluru is a dramatic landform.

(a)

(b)

Barchan

Star

Transverse

Parabolic

Longitudinal (seif)

FIGURE 21.24 The various kinds of sand dunes.

preserved inside the dune. In cross section, these slip faces appear as cross beds (▶Fig. 21.25a–c). The surfaces of dunes are not, in general, smooth surfaces, but rather are covered with delicate ripples.

With the exception of star and longitudinal dunes, sand dunes migrate downwind as the wind continuously picks up sand from the gently dipping windward slope and drops it onto the leeward side, or slip face. Rates of migration can exceed 25 m per year. Because of moving sand on an active dune, vegetation can't grow there. If a change in climate brings more rain, however, plant cover may grow and stabilize the dunes. At the end of the last ice age, for example, the "sand hills region" of western Nebraska was a vast dune field, but in the past 11,000 years it has been covered by grasslands, and the dunes have become stabilized.

21.7 LIFE IN THE DESERT

In the midst of a large erg, there seems to be nothing growing or moving at all. But most desert landscapes do include plants and animals. These organisms must possess special characteristics to enable them to survive in the desert: they must be able to withstand extremes of temperature—oppressive heat during the day and chilling cold at night—and to survive without abundant water.

Plants have evolved a number of different means to survive desert conditions. Some produce thick-skinned seeds that last until a heavy rainfall, then quickly germinate, grow, and create new seeds only while water is available. The new generation of seeds then waits until the next rainfall to start the cycle over again. Other plants have evolved the ability to send roots down to find deep groundwater; these plants have very long taproot systems. Still others have shallow root systems that spread over a broad area so that they can efficiently soak up water when it does rain.

Many desert plants have thick, fleshy stems and leaves. These plants, known as **succulents,** can store water for long periods of time (▶Fig. 21.26). Because succulents may be the only source of water during a time of drought, they have developed threatening thorns or needles to keep away thirsty animals. Plant life is much more diverse in desert **oases,** the verdant islands that crop up where natural springs spill groundwater onto the surface (see Chapter 19). The nearly year-round supply of water in an oasis nourishes a variety of palms and other nonsucculent plants.

Animal life in the desert includes scavengers like the coyote and dingo, plant eaters like the jack rabbit and kangaroo, and rodents, cactus-dwelling birds, vultures, insects, reptiles, and amphibians. Animals face the same challenge as plants: to retain water and survive extreme temperatures. To accomplish these goals, desert animals have also evolved numerous strategies. Frogs, for example, burrow beneath the ground and remain dormant for months, waiting until the next rain. Reptiles escape the midday heat by crawling into dark cracks between rocks. Rodents forage for food only during the cool night. And kit foxes, jack rabbits, and mule deer have disproportionately large ears through which they efficiently lose body heat. Many desert mammals, such as camels, retain body water by not sweating.

Humans are not meant to live in the desert. The loss of body moisture in extreme heat is so rapid that a person will die in less than 24 hours unless shaded from the Sun and supplied with at least 8 liters of water per day. Nevertheless, all but the most barren deserts are inhabited, though sparsely. Before civilization provided water wells, pipelines, and mechanized transportation, desert peoples lived in small nomadic groups, spaced far enough apart that they could live off the land; deserts have too little water to sustain agriculture or husbandry. Australian Aboriginal groups, for example, rarely included more than a dozen individuals

Sierra Nevada

Range (exposed rock)

Basin (alluvium-filled)

Colorado Plateau

The Desert Realm

Playa lake

Flat-lying strata

Alluvial fan

Normal fault

Granite

The desert of the Basin and Range Province in Utah, Nevada, and Arizona consists of alternating basins (grabens or half-grabens) separated by narrow ranges (tilted fault blocks). The Sierra Nevadas, underlain largely by granite, border the western edge of the province, while the Colorado Plateau, underlain by flat-lying sedimentary strata, borders the eastern edge. The overall climate of the region is dry. Because of the great variety of elevations and rock types, the region hosts different desert landscapes.

Except in the case of large rivers, like the Nile or Colorado, which bring water into a desert region from a more temperate region, streams in deserts fill with water only after heavy rains. At other times, the stream channels are dry. These channels are called dry washes, arroyos, or wadis. When there is a heavy rain, water cannot be absorbed into the ground fast enough, so runoff enters dry washes and fills them very quickly, creating a flash flood. The turbulent, muddy water of a flash flood can transport even large boulders. This flash flood is rushing down a stream in the Sonoran Desert of Arizona.

Flash flood

Barchan dune

Cross beds

Where there is a large supply of sand, a variety of sand dunes develop. The geometry of a particular sand dune (e.g., barchan, longitudinal, or star) depends on the sand supply and the wind. Inside sand dunes, we find cross beds.

Pediment

Alluvial apron with dry channels

Inselberg

Pediment

Alluvial apron

In places where ranges consist of granitic rock, they tend to be bordered by pediments. The isolated mountains that remain become inselbergs. Sediment derived by the erosion of the mountains fills the basins between the mountains. Dry washes (arroyos) channel both the pediments and the alluvium.

Pediment

In places where flat-lying strata crop out in deserts, beautiful cliffs, chimneys, buttes, and arches can form. Typically, strong rocks (like sandstone) underlie the steep cliffs, while weaker rocks underlie gentler slopes. Sediment is washed out of valleys during floods to create alluvial fans. Some of the debris carried out of the highlands breaks up and settles together to form desert pavements (stony plains).

Butte

Headward erosion

Desert plateau

Mesa

Chimney cap rock

Hard sandstone

Canyon

Alluvial fan

Formation of a pedestal

Rocky desert pavement

Natural arch

Shale

Playa lake

Yardangs and pedestal rocks

Water from flash floods flows into depressions in the adjacent valleys, to temporarily fill playa lakes. When these dry up, they leave behind salt pans. Wind, carrying sand and dust, can be an effective agent of erosion in the desert, sculpting features such as yardangs.

Dune formation

Barchans

Star dunes

Eolian sand deposit on top of sandstone

Transverse dunes

(a)

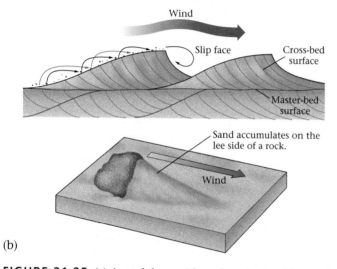

Wind

Slip face

Cross-bed surface

Master-bed surface

Sand accumulates on the lee side of a rock.

Wind

(b)

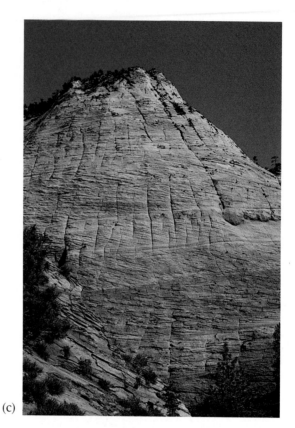

(c)

FIGURE 21.25 (a) A sand dune with surface ripples; (b) cross bedding inside a dune; (c) cross beds exposed in the Mesozoic rocks of Zion National Park, in Utah.

FIGURE 21.26 These plants in the Sonoran Desert, Arizona, are well adapted for dry conditions.

FIGURE 21.27 Aboriginal people of central Australia. Before the modern era, inhabitants of Australia's central desert (the "Red Center," named for its red soil and rocks) lived in small nomadic groups. There was not enough water to supply permanent communities until the advent of technology.

(►Fig. 21.27). In the past, nomadic desert dwellers either built temporary shelters out of local materials or traveled with tents. Locally, people carved underground dwellings in loess or soft rock, for rock is such a good insulator that a few meters below the ground surface it stays close to the region's mean temperature year-round.

21.8 DESERTS IN THE MODERN ERA: DESERTIFICATION

The historical era has seen a remarkable change in desert margins. Natural droughts, aggravated by overpopulation, overgrazing, careless agriculture, and diversion of water supplies, have transformed semi-arid grasslands into true deserts, leading to tragic famines that have killed millions of people. **Desertification,** the process of transforming nondesert areas to desert, has accelerated.

The consequences of desertification have devastated the Sahel, the belt of semi-arid land that fringes the southern margin of the Sahara Desert. In the past, the Sahel provided sufficient vegetation to support only a small population of nomadic people and animals. But during the second half of the twentieth century, large numbers of people migrated into the Sahel to escape overcrowding in central Africa. The immigrants began farming, and maintained large herds of cattle and goats. Plowing and overgrazing removed soil-preserving grass and allowed the soil to dry out. In addition, the trampling of animal hooves compacted the ground so it could no longer soak up water. In the 1960s and again in the 1980s, a series of natural droughts hit the region, bringing catastrophe (►Fig. 21.28a, b). Wind erosion stripped off the remaining topsoil. Without vegetation, the air grew drier, and the semiarid grassland of the Sahel has now become desert, with mass starvation as the result.

Other regions on Earth are developing the same problem. The Aral Sea in Kazakstan, for example, has started to dry up. Diversion of the rivers that once fed the sea have so diminished its water supply that the area of the sea has shrunk dramatically. Ships that once plied its waters now lie as rusting hulks in a sea of sand (►Fig. 21.28c).

Desertification does not only happen in less industrialized nations. People in the western Great Plains of the United States and Canada suffered from the problem beginning in 1933, the fourth year of the Great Depression. Banks had failed, workers had lost their jobs, the stock market had crashed, and hardship burdened all. No one needed yet another disaster—but that year, even nature turned hostile. All through the fall, so little rain fell in the plains of Texas and Oklahoma that the region's grasslands and croplands browned and withered, and the topsoil turned to powdery dust. Then, on November 12 and 13, strong storms blew eastward across the plains. Without vegetation to pro-

tect it, the wind lapped at the ground, stripped off the topsoil, and sent it skyward to form rolling black clouds that literally blotted out the sun (►Fig. 21.28d). Anyone caught in this **dust storm** found themselves choking and gasping for breath. When the dust finally settled, it had buried houses and roads under huge drifts, and dirtied every nook and cranny. The dust blew east as far as New England, where it turned the snow brown. What had once been a rich farmland in the southwestern plains turned into a wasteland that soon acquired the nickname "Dust Bowl."

For several more years, the drought persisted, leading to starvation and bankruptcy. Many of the region's residents were forced to move to wherever they could find work. Hundreds of "Okies," as the natives of Oklahoma were then called, piled into jalopies and drove old Route 66 out to California, looking for jobs in the state's still-green agricultural regions. Many were subjected to exploitation and persecution once they arrived in California. John Steinbeck dramatized this staggering human tragedy, which came to symbolize the Depression, in his novel *The Grapes of Wrath.*

Why did the fertile soils of the southern Great Plains suddenly dry up? The causes were complex; some were natural and some were human-induced. Drought, a period of unusually low rainfall, episodically visits certain regions, but drought alone doesn't inevitably lead to dust-bowl conditions. They may result from human meddling. Typically, the Great Plains region has a semi-arid climate in which only thin soil develops. But the plains were settled in the 1880s and 1890s, unusually wet years in the climate cycle. Not realizing its true character, far more people moved into the region than it could really sustain, and the land was farmed too intensively. When farmers used steel plows, they destroyed the fragile grassland root systems that held the thin soil in place. And when the inevitable 1930s drought came, it was catastrophic.

Desertification can be reversed, but at a price. Planting and irrigation may transform desert pavement into farm fields, orchards, forests, or lawns. But water to nourish the plants has to come from somewhere, and farmers obtain it by diverting rivers or by pumping groundwater, activities that create their own set of problems. River diversion robs regions downstream of their water supply, and pumping out too much groundwater lowers the water table so substantially that the pore space in aquifers collapses and the ground surface subsides. In short, people will need to rethink land-use policies in semi-arid lands to avoid catastrophe.

The Dust Bowl of the 1930s and the fate of the Sahel in Africa remind us of how fragile the Earth's green blanket of vegetation really is. Global climate changes can shift climatic belts sufficiently to transform agricultural regions into deserts. Some 5,000 years ago, the swath of land between the Nile Valley in Egypt and the Tigris-Euphrates Valley of Mesopotamia was known as the "fertile crescent"; here, people first abandoned their nomadic ways and settled in agri-

(a)

(b)

(c)

(d)

FIGURE 21.28 (a) The Sahel is the semi-arid land along the southern edge of the Sahara Desert. As a result of grazing and agriculture, the vegetation in this region has vanished, and large parts have undergone desertification. (b) Drought in the Sahel has brought deadly consequences. Here, residents seek water from a dwindling pond. (c) Desertification is a problem in many places around the world. Here, ships are stranded in the desert lands that once lay beneath the Aral Sea, in central Asia. (d) During the "Dust Bowl" days in central Oklahoma (ca. 1930s), dust storms stripped valuable topsoil off the land.

cultural communities. Yet the original land of "milk and honey" has become a barren desert, in need of intensive irrigation to maintain any agriculture. The change in landscape reflects a change in climate—the beginning of Western civilization occurred during the warmest and wettest period of global climate since before the last ice age. So much water drenched the Middle East and North Africa that rivers flowed where Saharan sands now blow. Unfortunately, if the current trend in climate change continues, our present agricultural belts could someday become new Saharas.

CHAPTER SUMMARY

• Deserts receive so little rain (less than 25 cm per year) that no more than 15% of their surface is covered by vegetation. Permanent streams cannot survive there.

• Subtropical deserts form between 20° and 30° latitude, rain-shadow deserts are found on the inland side of mountain ranges, coastal deserts are located on the land adjacent to cold ocean currents, continental-interior deserts exist in land-locked regions far from the ocean, and polar deserts form in polar regions (at high latitudes).

• In deserts, chemical weathering happens slowly, so rock bodies tend to erode primarily by the physical weathering of joints. Nevertheless, some chemical weathering does occur, leading to the formation of desert varnish. Some soils contain caliche.

• Water causes significant erosion in deserts, mostly during heavy downpours. Flash floods can carry large quantities of coarse sediments, including boulders, down ephemeral streams. When the rain stops, these streams dry up, leaving a steep-sided wash (also called an arroyo or wadi).

• Wind causes significant erosion in deserts, for it picks up dust and silt and carries them as suspended load, and causes sand to saltate (bounce along the ground). Where wind blows away finer sediment, a lag deposit remains, which may later settle into a mosaic called desert pavement. Wind-blown sediment abrades the ground, creating a variety of features such as ventifacts and yardangs.

• Talus aprons form when rock fragments accumulate in an apron at the base of a steep slope. Alluvial fans form at a mountain front where water in ephemeral streams deposits its sediment load. When temporary desert lakes dry up, they leave a playa.

• In some desert landscapes, erosion causes cliff (scarp) retreat of high areas, eventually resulting in the formation of inselbergs ("island mountains"), in some cases surrounded by a pediment. The erosion of stratified rock yields such landforms as buttes, mesas, and cuestas.

• Where there is abundant sand, the wind builds it into dunes. Common types include barchan, star, transverse, parabolic, and longitudinal (seif) dunes.

• Deserts contain a great variety of plant and animal life. All are adapted to survive extremes in temperature and without abundant water.

• Changing climates and land abuse may cause desertification, the transformation of semi-arid land into deserts.

KEY TERMS

bajada (p. 654)
barchan dune (p. 657)
blowout (p. 653)
bornhardt (p. 657)
butte (p. 655)
chimney (p. 655)
cliff (scarp) retreat (p. 655)
continental-interior desert (p. 649)
cuesta (p. 657)
deflation (p. 653)
desert (p. 645)
desert pavement (p. 652)
desert varnish (p. 650)
desertification (p. 663)
dry wash (p. 651)
erg (p. 655)
hoodoo (p. 655)
inselberg (p. 657)
interior basin (p. 654)
lag deposit (p. 652)
longitudinal (seif) dune (p. 657)
mesa (p. 655)
parabolic dune (p. 657)
pediment (p. 655)
petroglyph (p. 650)
playa (p. 654)
rain shadow (p. 648)
saltation (p. 651)
slip face (p. 658)
star dune (p. 657)
transverse dune (p. 657)
ventifact (p. 652)
wadi (p. 651)
yardang (p. 653)

REVIEW QUESTIONS

1. What factors determine whether a region can be classified as a desert or not?

2. Explain why deserts form.

3. Have today's deserts always been deserts? Keep in mind the consequences of plate tectonics.

4. How do weathering processes in deserts differ from those in temperate or humid climates?

5. Describe how water modifies the landscape of a desert. Be sure to discuss both erosional and depositional landforms.

6. Explain the ways in which desert winds transport sediment.

7. Explain how the following features form: (a) desert varnish; (b) desert pavement; (c) ventifacts; (d) yardangs.

8. Describe the process of formation of the different types of depositional landforms that develop in deserts.

9. Describe the process of cliff (scarp) retreat and the landforms that result from it.

10. What are the various types of sand dunes? What factors determine which type of dune develops?

11. Discuss various adaptations that life forms have evolved in order to be able to survive in desert climates.

12. What is the process of desertification, and what causes it?

SUGGESTED READING

Abrahams, A. D., and A. J. Parsons, eds. 1994. *Geomorphology of Desert Environments*. London: Champman and Hall.

Dregne, H. E. 1983. *Desertification of Arid Lands*. Chur, Switz.: Harwood Academic.

Tchakerian, V. P., ed. 1995. *Desert Aeolian Processes*. London: Chapman and Hall.

CHAPTER 22

Amazing Ice: Glaciers and Ice Ages

22.1 INTRODUCTION

There's nothing like a good mystery, and one of the most puzzling in the annals of geology came to light in northern Europe early in the nineteenth century. When farmers of the region prepared their land for spring planting, they occasionally broke their plows by running them into large boulders that appeared to be buried randomly through otherwise fine-grained soil. Many of these boulders did not consist of local bedrock, but rather came from outcrops hundreds of kilometers away. Because the boulders had apparently traveled so far, they came to be known as **erratics** (from the Latin *errare,* "to wander").

The mystery of the wandering boulders became a subject of great interest to early nineteenth-century geologists, who realized that deposits of *un*sorted sediment (containing boulders distributed in fine sediment) were not examples of typical stream alluvium, because running water sorts sediment by size, creating separate deposits of gravel, sand, and mud. Nevertheless, most attributed the deposits to a vast flood that they imagined had spread a slurry of boulders, sand, and mud across the continent. In 1837, however, a young Swiss paleontologist named Louis Agassiz suggested a radically different interpretation. Agassiz often hiked among **glaciers,** slowly flowing sheets or rivers of ice that survived the summer melt, in the Alps near his home. He observed that glacial ice could carry enormous boulders as well as sand and mud, because solid water has the strength to support the weight of rock. Agassiz realized that because ice does not sort sediment as it flows, glaciers leave behind a mixture of boulders, cobbles, sand, and mud when they melt. Based on these observations, he proposed that the mysterious unsorted sediment and erratics of Europe represented the deposits of **ice sheets,** vast glaciers that had once covered much of the continent. In Agassiz's mind, Europe had at one time been in the grip of an **ice age,** a time when the climate was significantly colder (▶Fig. 22.1a, b).

Glaciers are rivers or sheets of ice. They carve beautiful landscapes and deposit hills of sediment. This is the Surprise Glacier, near Prince William Sound, Alaska. The black stripe down its axis consists of sediment falling from adjacent mountains. During an ice age, glaciers like these—and larger—covered vast areas of continents.

(a)

(b)

FIGURE 22.1 (a) Louis Agassiz. (b) Agassiz envisioned that much of North America and Europe were covered with ice, as Antarctica now is. The blue ice of this glacier almost submerges the mountains in the distance.

Agassiz's radical story faced intense criticism for the next two decades. But he didn't back down, and even challenged his opponents to visit the Alps and examine the sedimentary deposits that glaciers had left behind. By the late 1850s, most doubters had changed their minds, and the geological community concluded that the notion that Arctic-like climates had once gripped much of Europe was correct. Later in life, Agassiz traveled to the United States and documented many glacier-related features in North America's landscape, proving that the ice age had affected the entire Northern Hemisphere.

Glaciers, which have many forms, cover only about 10% of the land on Earth today (▶Fig. 22.2a, b), but during the most recent ice age, which ended only about 11,000 years

ago, as much as 30% of the continents had a coating of ice. New York City, Montreal, and many of the great cities of Europe now occupy land that once lay beneath hundreds of meters to a few kilometers of ice. Further, when water transferred from the ocean into glaciers during the ice age, the sea level dropped substantially, and continental shelves became dry land.

The work of Louis Agassiz brought the subject of glaciers and the ice age into the realm of geologic study and led people to recognize that major climate changes happen in Earth history, even on a time scale measured in thousands of years. In this chapter, after looking at the nature of ice—it is a rock, after all—we see how glaciers form and why they move. Next, we consider how glaciers modify landscapes by erosion and deposition. A substantial portion of the chapter is devoted to the Pleistocene ice age, for its impact on the landscape can still be seen today. But ice ages happened earlier in Earth history too. We conclude by considering hypotheses to explain why ice ages occur.

FIGURE 22.2 (a) A piedmont glacier near the coast of Greenland. (b) A large valley glacier in Pakistan.

(a)

(b)

22.2 ICE: A ROCK MADE FROM WATER

Pure ice, solid water, has the transparency of glass, but if ice contains tiny air bubbles or cracks that disperse light, it becomes milky white. Like glass, ice has a high **albedo,** meaning that it reflects light well—so well, in fact, that if you walk on ice without eye protection, you risk losing your eyesight from the glare. Ice differs from most other familiar materials in that its solid form is less dense than its liquid form—the architecture of an ice crystal holds water molecules apart. Ice, therefore, floats on water. If ice didn't float, ice in oceans would sink, leaving room for new ice to form until the oceans froze solid.

We can consider a single ice crystal to be a mineral: it is a naturally occurring, inorganic solid, with a definite chemical composition (H_2O) and a regular crystal structure. Ice crystals have a hexagonal form, so snowflakes grow into six-pointed stars (▶Fig. 22.3a). We can think of a layer of fresh snow as a layer of sediment, and a layer of snow that has been compacted so that the grains stick together as a layer of sedimentary rock (▶Fig. 22.3b). We can also think of the ice that appears on the surface of a pond as an igneous rock, for it forms when "molten ice" (liquid water) solidifies; this ice consists of individual crystals that have grown together (▶Fig. 22.3c). Glacier ice, in contrast, is in essence a metamorphic rock. It develops when preexisting ice recrystallizes in the solid state, meaning that the molecules in solid water rearrange to form new crystals without first transforming into liquid (▶Fig. 22.3d).

22.3 THE NATURE OF GLACIERS

Categories

Glaciers are streams or sheets of recrystallized ice that last all year long. They highlight coastal and mountain scenery in Alaska, western North America, the Alps of Europe, the Southern Alps of New Zealand, the Himalayas of Asia, and the Andes of South America, and they cover most of Greenland and Antarctica. Geologists distinguish between two main categories, mountain glaciers and continental glaciers.

Mountain glaciers (also called **Alpine glaciers**) exist in or adjacent to mountainous regions (▶Fig. 22.4a). Topographical features of the mountains control their shape, and the slope of the substrate controls their flow direction. Mountain glaciers include **cirque glaciers,** which fill bowl-shaped depressions, or **cirques,** on the flank of a mountain; **valley glaciers,** rivers of ice that flow down valleys (Fig. 22.2b); **mountain ice caps,** mounds of ice that submerge peaks and ridges at the crest of a mountain

(a)

(b)

(c)

(d)

FIGURE 22.3 (a) The hexagonal shape of snowflakes. (b) Snow drifts on a winter day. (c) Frost on a glass window, showing the large crystals of water. (d) Thin section of glacial ice, showing the texture of ice crystals and air bubbles.

FIGURE 22.4 (a) The various kinds of mountain glaciers. (b) A valley glacier. (c) A mountain ice cap. Note the ice flowing out of the ice cap in a big valley glacier. (d) A piedmont glacier.

range; and **piedmont glaciers** (Fig. 22.2a), fans or lobes of ice that form where a valley glacier emerges from a valley and spreads out into the adjacent plain (▶Fig. 22.4b–d). Mountain glaciers range in size from a few hundred meters to a few hundred kilometers long.

Continental glaciers (also called **ice sheets**) are vast layers of ice that spread over thousands of square kilometers of continental crust. Presently, continental glaciers are found only on Antarctica and Greenland (▶Fig. 22.5a, b). (Remember that Antarctica is a continent, so the ice beneath the South Pole is part of a continental glacier; in contrast, the ice beneath the North Pole is part of a thin sheet of sea ice floating on the Arctic Ocean.) Continental glaciers flow outward from their thickest point (up to 3.5 km thick) and thin toward their margins, where they may be only a few hundred meters thick. The front edge of the glacier may contain several tongue-shaped **lobes,** because not all of the glacier flows at the same speed.

Forming a Glacier

In order for a glacier to form, three conditions must be met: first, the local climate must be sufficiently cold that winter snow does not melt entirely away during the summer; second, there must be sufficient snowfall for a large amount to accumulate; and third, the slope of the surface on which the snow accumulates must be gentle enough that the snow does not slide away in avalanches and protected enough that the snow doesn't blow away.

Glaciers exist in polar regions because even though relatively little snow falls, temperatures remain so low that snow stays all year. Glaciers develop in mountains, even at low latitudes, because temperature decreases with elevation; at high elevations, the mean temperature stays low enough for snow to remain all year. Since the temperature of a region depends on latitude, the specific elevation at which mountain glaciers form also depends on latitude. In Earth's

FIGURE 22.5 (a) A map and cross section of the Greenland ice sheet. (b) A map and cross section of the Antarctic ice sheet. Note that the Transantarctic Mountains divide the ice sheet in two. The Ross Ice Shelf lies between the East Antarctic and West Antarctic sheets. Valley glaciers carry ice from the ice sheets down to the shelf.

FIGURE 22.6 The snow line is a function of latitude.

present-day climate, glaciers occur at elevations above 5 km between 0° and 30° latitude, at elevations above 2.5–3.5 km between 30° and 60° latitude, and down to sea level at 60°–90° latitude (▶Fig. 22.6). Thus, you can see Alaskan glaciers from a cruise ship, but you have to climb the highest mountains of the Andes to find glaciers at the equator. Mountain glaciers tend to develop on the lee side of the mountain and on the side that receives less sunlight. Glaciers do not exist on slopes greater than about 30°, because avalanches clear such slopes.

The transformation of snow to glacier ice takes place slowly, as older snow gets progressively buried by younger. Freshly fallen snow consists of delicate hexagonal crystals with sharp points. The crystals do not fit together tightly, so this snow contains about 90% air. With time, the points of the snowflakes become blunt, because they either **sublimate** (evaporate directly into vapor) or melt, and the snow packs more tightly. As snow becomes buried, the weight of the overlying snow increases pressure, which causes remaining points of contact between snowflakes to melt, just as pressure causes ice to melt beneath the blade of an ice skate. This process of melting at points of contact where the pressure is greatest is called **pressure solution.** Gradually, the snow transforms into a packed granular material called **firn,** which contains only about 25% air (▶Fig. 22.7a, b). Pressure solution melts grains of firn at contact points, producing water that crystallizes in the spaces between grains until eventually the firn transforms into a solid mass of interlocking ice crystals. Such glacial ice, which may still contain up to 20% air trapped in bubbles, tends to absorb red light and thus has a bluish color. The transformation of fresh snow to glacier ice can take as little as tens of years in regions with abundant snowfall, or as long as thousands of years in regions with little snowfall.

The Movement of Glacial Ice

When Louis Agassiz became fascinated by glaciers, he decided to find out how fast they flowed, so he hammered stakes into an Alpine glacier and watched the stakes move during the year. More recently, researchers have observed glacial movement with the aid of time-lapse photography, which shows the movement of a glacier over several years in a movie that lasts a few minutes. In such movies, you can

(a)   (b)

FIGURE 22.7 (a) Snow compacts and melts to form firn, which recrystallized to make ice. (b) The size of ice crystals increases with depth in a glacier, where some crystals grow at the expense of others.

actually see a glacier trundle across the screen, making its nickname "river of ice" seem perfectly appropriate.

How do glaciers move? Some move when meltwater accumulates at the base of a glacier, so that the mass of the glacier slides on a layer of water or on a slurry of water and sediment. During this kind of movement, known as **basal sliding,** the water or wet slurry holds the glacier above the underlying rock and eliminates friction between the glacier and its substrate. Basal sliding is the dominant style of movement for **wet-bottom,** or **temperate glaciers,** where pressure causes water to form at their base (▶Fig. 22.8a).

Glaciers also move by means of **internal flow,** during which the mass of ice slowly changes shape internally without breaking apart or completely melting. By studying ice deformation with a microscope, geologists have determined that internal flow involves two processes. First, ice crystals

(c)

FIGURE 22.8 (a) Wet-bottom glaciers move by means of basal sliding. (b) Dry-bottom glaciers flow in the solid state (internally). In some cases, the ice crystals stretch and rotate, while in other cases, the ice recrystallizes and some crystals slide past each other, especially if there are thin films of water between grains. (c) Crevasses form in the brittle ice above the brittle-plastic transition in glaciers.

become plastically deformed, by the rearrangement of water molecules within the crystal lattice (during this process, ice crystals change shape, old crystals disappear, and new ones form); second, if conditions allow very thin films of water to form on the surfaces of ice crystals, the crystals can slide past one another. Internal flow is the dominant style of movement for **dry-bottom,** or **polar, glaciers,** those that are so cold that their base remains frozen to their substrate (▶Fig. 22.8b).

Note that the plastic deformation of ice in glaciers resembles the plastic deformation of metamorphic rock in collisional mountain belts. However, metamorphic rock deforms plastically only at depths greater than 10–15 km in the crust, while ice deforms plastically at depths below about 60 m (▶Fig. 22.8c). In other words, the **brittle-plastic transition** in ice lies at a depth of about 60 m—above this depth, ice can form large cracks, while below this depth, ice cannot, because it is too plastic. A large crack that develops by brittle deformation in the top 60 m of a glacier is called a **crevasse.** In large glaciers, crevasses can be hundreds of meters long and tens of meters deep. Tragically, several explorers have met their death by falling into crevasses (▶Fig. 22.9).

Glaciers generally flow between 10 and 300 meters per year—far slower than a river, but far faster than silicate rock. The velocity of a particular glacier depends on the steepness of the slope down which it flows (glaciers flow faster down steeper slopes), and whether or not water is present at the base of the glacier (ice sliding on a sheet of water or on a slurry of wet sediment moves faster than ice frozen to its substrate). Thus, temperate glaciers (which move as much as a few hundred meters per year) move faster than polar glaciers (which move only tens of meters per year).

Not all parts of a glacier move at the same rate. For example, friction between rock and ice slows a glacier, so the cen-

FIGURE 22.9 Crevasses.

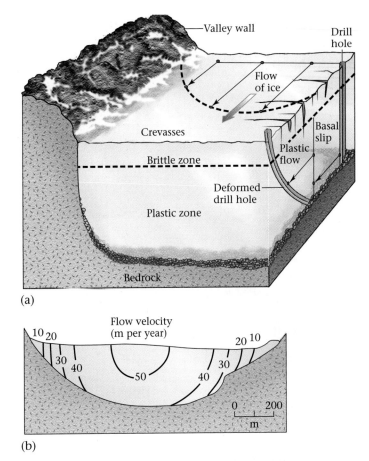

(a)

(b)

FIGURE 22.10 (a) Different parts of a glacier may flow at different velocities. The vector lengths indicate the velocity of flow. (b) This cross section of a glacier shows measured velocities of flow.

ter of a valley glacier moves faster than its margins, and the top of a glacier moves faster than its base (▶Fig. 22.10a, b). And because water at the base of a glacier allows it to travel more rapidly, portions of a continental glacier that flow over water or wet sediment travel ten to a hundred times faster than adjacent dry-bottom portions of the glacier. Such fast-moving parts are called **ice streams.** Similarly, if water builds up beneath a valley glacier to the point where it lifts the glacier off its substrate, the glacier accelerates and flows much faster for a limited time (rarely more than a few months), until the water escapes. Such a pulse of rapid flow is called a **surge.** During surges, glaciers have been clocked at speeds of 20–110 m per day.

Why do glaciers move? Ultimately, because of the pull of gravity (▶Fig. 22.11a–c). Valley glaciers flow down a slope much like water does, though more slowly—gravity simply pulls the ice downhill. Continental ice sheets, however, "spread" laterally across the landscape. When a thick pile of ice builds up, gravity causes the top of the pile to push down on the ice at the base until the basal ice can no longer

FIGURE 22.11 (a) The downslope motion of glaciers is driven by gravity, one component of which (g_n) is parallel to the slope. (b, c) The gravitational spreading of an ice sheet is similar to the spreading of honey.

dough spreads out laterally to form a pancake shape. Similarly, when you pour honey onto a plate, the honey can't build up into a narrow column because it's too weak; rather, it flows laterally away from the point where it lands to form a wide, thin layer. This process is called **gravitational spreading**—it occurs in continental glaciers because ice is so weak.

Glacial Advance and Retreat

Glaciers resemble bank accounts: snowfall adds to the account, while **ablation**—the removal of ice by sublimation (the evaporation of ice into water vapor), melting (the transformation of ice into liquid water, which flows away), and **calving** (the breaking off of chunks of ice at the edge of the glacier)—subtracts from the account (▶Fig. 22.12). Snowfall adds to the glacier in the **zone of accumulation,** while ablation subtracts in the **zone of ablation;** the boundary between these two zones is the **snow line.** The zone of accumulation occurs where the temperature remains cold enough year around so that winter snow does not melt or sublimate away entirely during the summer. Elevation determines the position of the snow line for a mountain glacier (the zone of accumulation lies at altitudes above the snow line), while latitude determines the position of the snow line for continental glaciers (the zone of accumulation lies at latitudes higher than that at which the snow line intersects the surface of the glacier).

The leading edge or margin of a glacier is called its **toe,** or **terminus** (▶Fig. 22.13a). If the rate at which ice accumulates in the zone of accumulation exceeds the rate at which ablation occurs below the snow line, then the toe moves forward into previously unglaciated regions. Such a change is called a **glacial advance** (▶Fig. 22.13b). In the case of mountain glaciers, the position of a toe moves downslope during an advance, while in the case of continental glaciers, the toe moves outward, away from the glacier's origin and into unglaciated lands. If the rate of ablation below the snow line equals the rate of accumulation, then the position of the toe remains fixed. But if the rate of ablation exceeds the rate of accumulation, then the posi-

support the weight of the overlying ice and begins to deform plastically. When this happens, the basal ice starts squeezing out to the side, carrying the overlying ice with it. The greater the volume of ice that builds up, the wider the sheet of ice becomes. You've seen a similar process if you've ever pressed down on a ball of dough—as you press, the

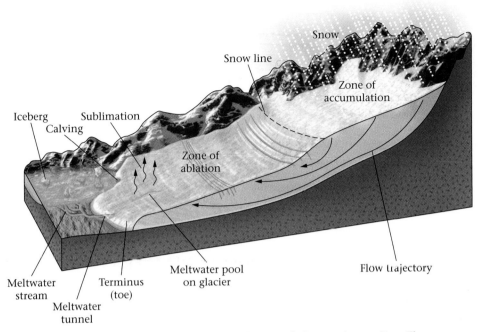

Snow

Snow line

Zone of
accumulation

Iceberg

Sublimation

Calving

Zone of
ablation

Meltwater pool
on glacier

Flow trajectory

Meltwater
stream

Terminus
(toe)

Meltwater
tunnel

FIGURE 22.12 The zone of ablation, zone of accumulation, and snow line. The arrows illustrate how a grain of ice flows down in the zone of accumulation and up in the zone of ablation, so that it follows a curved trajectory.

tion of the toe moves back toward the origin of the glacier; such a change is called a **glacial retreat** (▶Fig. 22.13c). During a mountain glacier's retreat, the position of the terminus moves upslope, while during a continental glacier's retreat, the position of the terminus moves to higher latitudes. It's important to realize that when a glacier retreats, it's only the *position* of the toe that moves back toward the origin, for ice continues to flow toward the toe. Glacial ice cannot flow uphill or back toward the origin.

One final point before we leave the subject of glacial flow. Beneath the zone of accumulation, a crystal of ice gradually moves down toward the base of the glacier as new ice accumulates above it, while beneath the zone of ablation, a crystal of ice gradually moves up toward the surface of the glacier, as overlying ice

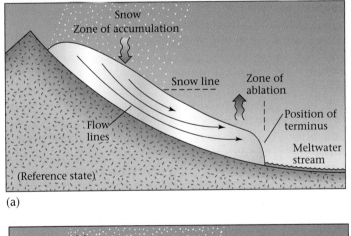

Snow

Zone of accumulation

Snow line

Zone of
ablation

Flow
lines

Position of
terminus

Meltwater
stream

(Reference state)

(a)

Less snow

Higher snow line

More ablation

Retreat

Glacial retreat

(c)

More snow

Lower snow line

Less ablation

Advance

Glacial advance

(b)

FIGURE 22.13 (a) The position of a glacier's toe represents the balance between the amount of ice that forms beneath the zone of accumulation and the amount of ice lost in the zone of ablation. (b) Glacial advance and (c) glacial retreat. Notice that ice always flows toward the toe of the glacier regardless of whether the toe advances or retreats.

(a)

(b)

FIGURE 22.14 (a) Beneath the zone of ablation, ice crystals move up to the surface. Long-buried meteorites also collect at the surface. (b) A meteorite sitting on the surface of a glacier.

ablates. Thus, as a glacier flows, ice crystals follow curved trajectories (Fig. 22.13). For this reason, rocks picked up by ice at the base of the glacier may slowly move to the surface. Interestingly, the upward flow of ice where the Antarctic ice sheet collides with the Transantarctic Mountains brings up meteorites long buried in the ice. Every year, researchers survey Antarctic ice fields to harvest these meteorites (▶Fig. 22.14).

Ice in the Sea

In coastal areas at high latitudes, some valley glaciers flow out into the sea and become protruding **ice tongues.** Because four-fifths of the ice lies beneath the sea surface (because of isostasy), the base of a glacier remains in contact with the sea floor if the water is shallow, but it floats where the water becomes deep enough. Large blocks calve (break) off the front of the glacier and drop into the sea to create **icebergs,** like the one that slashed the hull of the *Titanic*.

Icebergs may carry sediment, called **ice-rafted sediment,** out into the deep sea. Rocks that drop to the sea floor when the ice melts are called **drop stones** (▶Fig. 22.15a).

Not all ice floating in the sea originated as glaciers on land. In polar climates, the surface of the sea itself freezes, forming **sea ice.** Some sea ice, such as that found in the interior of the Arctic Ocean, is free floating. But some protrudes outward from the shore; such an **ice shelf** surrounds Antarctica (▶Fig. 22.15b, c). The Antarctic ice shelf has been diminishing rapidly in the past few decades, possibly because of global warming. During this process, immense, flat-topped icebergs, some over 100 km long, have separated from the shelf (▶Fig. 22.15d).

22.4 CARVING AND CARRYING BY ICE

Glacial Erosion and Its Products

The Sierra Nevada Mountains of California consist largely of granite that intruded during the Mesozoic era into the crust beneath a volcanic arc. About 10–20 million years ago, the region was uplifted, and erosion stripped away overlying rock to expose the granite in a terrane of rounded domes. Then, during the last ice age, valley glaciers cut deep, steep-sided valleys into the range. In the process, some of the domes were cut in half, leaving a rounded surface on one side and a steep cliff on the other. Half Dome, in Yosemite National Park, formed in this way (▶Fig. 22.16a); its steep cliff has challenged many rock climbers. Such glacial erosion also produces the knife-edge ridges and pointed spires of high mountains (▶Fig. 22.16b), and the broad expanses where rock outcrops have been stripped of overlying sediment and polished smooth (▶Fig. 22.16c). Glacial erosion in the mountains can lower a valley floor by over 100 m, and continental glaciation during the last ice age stripped up to 30 m of material off the land in northern Canada.

Glaciers erode their substrate in several ways (▶Fig. 22.17a, b). During **glacial plowing,** flowing ice bulldozes and moves loose sediment, and during **glacial incorporation,** ice surrounds and incorporates the debris. The combination of plowing and incorporation may remove all the regolith in an area and thereby expose the underlying bedrock. During **glacial plucking** (or **glacial quarrying**), a glacier breaks off and then carries away fragments of bedrock. Plucking may leave wedge-shaped indentations, called **chatter marks,** on rock surfaces (▶Fig. 22.17c).

As glaciers flow, clasts embedded in ice act like the teeth of a giant rasp and grind away the substrate. This process, called **glacial abrasion,** produces very fine sediment called **rock flour,** just as sanding wood produces sawdust. Rasping by embedded sand yields shiny, **glacially polished sur-**

(a)

(c)

(d)

March

September

Polynya

(b)

FIGURE 22.15 (a) Where a glacier reaches the sea, the ice stays grounded in shallow water and floats in deep water. Icebergs calve off the front. Sediment known as drop stones fall off the base of icebergs and collect on the sea floor. (b) The Antarctic ice shelf at two different times of year. Sometimes large holes called polynyas form in the shelf. (c) Icebreakers have shallowly sloped, reinforced hulls, which allow the ship to ride up on the ice. The weight of the ship then crushes the ice. (d) Huge, flat-topped icebergs floating in the sea off the coast of Antarctica.

faces. Long gouges, grooves, or scratches (1 cm to 1 m across) called **glacial striations** are commonly cut into glacially abraded surfaces. Striations form when the base of the ice contains scattered large clasts that dig deeply into the substrate (▶Fig. 22.17d). As you might expect, striations run parallel to the flow direction of the ice.

Let's now look more closely at the erosional features associated with a mountain glacier (▶Fig. 22.18a–f). Freezing and thawing during the fall and spring help fracture the rock bordering the **head** of the glacier (the ice edge high in the mountains). This rock falls on the ice or gets picked up at the base of the ice, and moves downslope with the glaciers. As a consequence, a bowl-shaped depression, or **cirque,** develops on the side of the mountain. If the ice later melts, a lake called a **tarn** may form at the base of the cirque. An **arête** (French for "ridge") is a residual knife-edge ridge of rock that separates two adjacent cirques. A pointed mountain peak surrounded by at least three cirques is called a **horn.** The Matterhorn, a peak in Switzerland, is a particularly beautiful example of a horn.

Glacial erosion severely modifies the shape of a valley. For example, look along the length of a river in unglaciated

(a)

(b)

(c)

FIGURE 22.16 Products of glacial erosion. (a) Half Dome, in Yosemite National Park, California. Before glaciation, the mountain was a complete dome; then glacial erosion truncated one side. (b) Glaciated mountains with sharp peaks. (c) A glacially polished surface along the shore of Lake Superior. The bedrock is so smooth that someone can slide down it.

(a)

(b)

(c)

(d)

FIGURE 22.17 The various kinds of glacial erosion. (a) Plowing and incorporation; (b) plucking and abrasion; (c) chatter marks on a glacially eroded surface in Switzerland; (d) a glacially striated and polished surface next to Crater Lake, Oregon.

(a)

(b)

(c)

(d)

(e)

(f)

FIGURE 22.18 (a) A mountain landscape before glaciation. The V-shaped valleys are the result of river erosion; the floor of the tributary valleys are the same elevation as the trunk valley where they intersect. (b) During glaciation, the valleys are filled with ice. (c) After glaciation, the region contains U-shaped valleys, hanging valleys, truncated spurs, and horns. (d) A U-shaped glacial valley. (e) The Matterhorn in Switzerland. (f) A waterfall spilling out of a hanging valley.

mountains, and you'll see that it flows down a **V-shaped valley,** with the river channel forming the point of the V. The V develops because river erosion occurs only in the channel, and mass wasting causes the valley slopes to approach the angle of repose. If the climate changes and a glacier fills the valley and flows down it, the combined processes of glacial abrasion and plucking not only lower the floor of the valley, they also bevel its sides. As a consequence, glacial erosion produces a **U-shaped valley,** with very steep walls.

Glacial erosion in mountains also modifies the intersections between tributaries and the trunk valley. In a river system, tributaries cut side valleys that merge with the trunk valley, so that the mouths of the tributary valleys lie at the same elevation as the trunk valley. The ridges ("spurs") between valleys taper to a point when they join the trunk valley floor. During glaciation, tributary glaciers flow down side valleys into a trunk glacier. But the trunk glacier cuts the floor of its valley down to a depth that far exceeds the depth cut by the tributary glaciers. Thus, when the glaciers melt away, the mouths of the tributary valleys perch at a higher elevation than the floor of the trunk valley. Such side valleys are called **hanging valleys.** The water in post-glacial streams that flow down a hanging valley must cascade over a spectacular waterfall to reach the post-glacial trunk stream. Trunk glaciers, as they erode, also chop off the ends of spurs, creating **truncated spurs.**

Now let's look at the erosional features produced by continental ice sheets. To a large extent, these depend on the nature of the preglacial landscape. Where an ice sheet spreads over a region of low relief, like the Canadian Shield, glacial erosion creates a vast region of polished, striated surfaces. Where an ice sheet spreads over a hilly area, it deepens valleys and smoothes hills. In central New York, for example, continental glaciers carved the deep valleys that now cradle the Finger Lakes, and in Maine glaciers smoothed and streamlined the granite and metamorphic rock hills of Acadia National Park. Notably, glacially eroded hills end up being elongate in the direction of flow and are asymmetric; glacial rasping smoothes and bevels the upstream part of the hill, creating a gentle slope, while glacial plucking eats away at the downstream part, making a steep slope. Ultimately, the hill's profile resembles that of a sheep lying in a meadow—such hills are called **roche moutonnée,** from the French for "sheep rock" (▶Fig. 22.19).

Fjords: Submerged Glacial Valleys

As noted earlier, where a valley glacier meets the sea, the glacier's base remains in contact with the ground until the water depth exceeds about four-fifths of the glacier's thickness, at which point the glacier floats. Thus, glaciers can continue carving a U-shaped valley even below sea level. In

FIGURE 22.19 A roche moutonnée.

addition, during an ice age, water extracted from the sea becomes locked in the ice sheets on land, so the sea level drops significantly. Therefore, the floors of valleys cut by coastal glaciers during the last ice age were cut much deeper than present sea level. Today, the sea has flooded these deep valleys, producing **fjords** (see Chapter 18). In the spectacular fjord-land regions along the coasts of Norway, New Zealand, Chile, and Alaska, the walls of submerged U-shaped valleys rise straight from the sea as vertical cliffs up to 1,000 m high (▶Fig. 22.20). Fjords also develop where a dammed glacial valley fills to become a lake.

The Glacial Conveyor: The Transport of Sediment by Ice

Glaciers can carry sediment of any size and, like a conveyor belt, transport it in the direction of flow (that is, toward the

FIGURE 22.20 One of the many spectacular fjords of Norway.

toe). The sediment load either falls onto the surface of the glacier from bordering cliffs or gets plucked and lifted from the substrate and incorporated in the moving ice. Because of the curving flow lines of glacial ice, rocks plucked off the floor may eventually reach the surface.

Sediment dropped on the glacier's surface moves with the ice and becomes a stripe of debris along the margin of the glacier. Such stripes are called **lateral moraines** (**moraine** was a local term used by Alpine farmers and shepherds for piles of rock and dirt). When the glacier finally melts, lateral moraines lie stranded along the side of the glacially curved valley, like a bathtub ring. If flowing water runs along the edge of the glacier and sorts the sediment of a lateral moraine, a stratified sequence of sediment called a **kame** forms. Where two valley glaciers merge, the debris comprising two lateral moraines merges to become a **medial moraine** in the interior of the composite glacier (▶Fig. 22.21a, b). Trunk glaciers created by the merging of tributary glaciers contain several medial moraines. Glaciers passing through ranges with excessively high erosion rates may be completely buried by rocky debris. In some cases, a glacier incorporates so much rock that geologists refer to it as a **rock glacier.** Sediment that is transported to a glacier's toe by the glacial "conveyor" accumulates in a pile as the toe builds up an **end moraine** (▶Fig. 22.21c).

22.5 DEPOSITION ASSOCIATED WITH GLACIATION

Types of Glacial Sedimentary Deposits

If you drill through the soil throughout much of the upper midwestern and northeastern United States and adjacent parts of Canada, the drill penetrates a layer of sediment de-

FIGURE 22.21 (a) Lateral and medial moraines on a glacier. (b) A glacier with lateral and medial moraines. (c) The surface load plus the internal load accumulates at the toe of the glacier to comprise an end moraine; in effect, glaciers act like conveyor belts, constantly transporting more sediment toward the toe, regardless of whether the glacier advances or retreats.

(a)

(c)

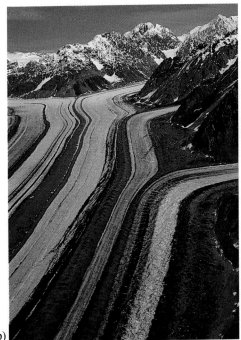

(b)

posited during the last ice age. A similar story holds for much of northern Europe. Thus, many of the world's richest agricultural regions rely on soil derived from sediment deposited by glaciers during the ice age. Notably, this sediment buries a pre-ice-age landscape, as frosting fills the irregularities on a cake. Pre-glacial valleys may be completely filled with sediment.

Several different kinds of sediment can be deposited in glacial environments, all of which together comprise **glacial drift.** (The term dates from pre-Agassiz studies of glacial deposits, when geologists thought that the sediment had "drifted" into place during the biblical Noah's flood.) Glacial drift includes the following.

• *Till:* Sediment transported by ice and deposited beneath or at the toe of a glacier is called **glacial till.** Since the

solid ice of glaciers can carry clasts of all sizes, glacial till is unsorted (▶Fig. 22.22a).

• *Erratics:* Glacial **erratics** are boulders that have been dropped by a glacier. Some protrude from till piles, others rest on glacially polished surfaces.

• *Glacial marine:* Where a sediment-laden glacier flows into the sea, icebergs calve off the toe and raft clasts out to sea. As the icebergs melt, they drop the clasts, which settle into the muddy sediment on the sea floor. Pebble- and larger-sized clasts that are deposited in this way, as we have seen, are called **drop stones.** Sediment consisting of ice-rafted clasts mixed with marine sediment is called **glacial marine.**

• *Glacial outwash:* Till deposited by a glacier at its toe may be picked up and transported by meltwater streams that

FIGURE 22.22 (a) The unsorted sediment comprising glacial till in Ireland. (b) Braided streams choked with glacial outwash near the toe of a glacier. (c) In this cross section of varved glacial lake bed sediment, note the alternating light and dark layers. (d) A small escarpment cut into glacial loess deposits of Illinois.

(a)

(b)

(c)

(d)

sort the sediment. The coarse clasts are deposited by a braided stream network in a broad area of gravel and sand bars called an **outwash plain.** This sediment is known as **glacial outwash** (▶Fig. 22.22b).

• *Glacial lake-bed sediment:* Fine clasts, including rock flour, are transported away from the glacial front and settle in **meltwater lakes,** forming a thick layer of **glacial lake-bed sediment.** This sediment commonly contains varves. A **varve** is a pair of thin layers, one consisting of silt brought in during spring floods, and the other of clay deposited in winter when the lake's surface freezes over and the water is still (▶Fig. 22.22c).

• *Loess:* When the warmer air above ice-free land rises, the cold, denser air from above glaciers rushes in to take its place; strong winds called **catabatic winds** form at the margin of a glacier. This wind picks up fine clay and silt (including rock flour) that collected along the margins of an outwash plain and transports it away from the glacier's toe. Where the winds die down, the sediment settles and forms a thick layer. This sediment, called **loess,** sticks together because of the electrical charges on clay flakes, so when loess is subsequently eroded by streams, steep escarpments develop (▶Fig. 22.22d).

FIGURE 22.23 (a) The depositional landforms resulting from glaciation. (b) The setting in which various types of moraines form.

Till, which contains no layering, is sometimes called **unstratified drift,** while glacial sediments that have been redistributed by flowing water are called **stratified drift.**

Depositional Landforms of Glacial Environments

Picture a hunter, dressed in deerskin, standing at the toe of a continental glacier in what is now southern Canada, waiting for an unwary woolly mammoth to wander by. It's a sunny summer day 12,000 years ago, and milky, sediment-laden streams gush from tunnels and channels at the base of the glacier and pour off the top as the ice melts. No mammoths today, so the bored hunter climbs to the top of the glacier for a view. The climb isn't easy, partly because of the incessant catabatic wind, and partly because deep crevasses inter-

rupt his path. Reaching the top of the ice sheet, the hunter looks north, and the glare almost blinds him. Squinting, he sees the white of snow, and where the snow has blown away, he sees the rippled, glassy surface of bluish ice (Fig. 22.1b). Here and there, a rock protrudes from the ice. Now looking south, he surveys a stark landscape of low, sinuous ridges separated by hummocky (bumpy) plains. Braided streams, which carry meltwater out across this landscape, flow through the hummocky plains and supply a number of lakes. The air is dusty because of the wind.

All the landscape features observed by the hunter as he looks south form by deposition in a glacial environment, both mountain and continental (▶Fig. 22.23a, b). The low, sinuous ridges, called **end moraines,** develop when the terminus of a glacier stalls in one position for a while; the ice keeps flowing to the terminus, and, like a giant conveyor

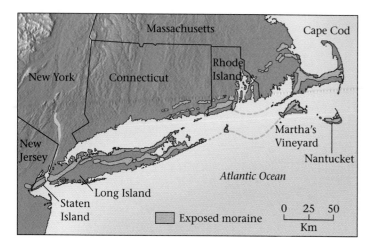

FIGURE 22.24 The moraines that comprise Long Island and Cape Cod.

(a)

(b)

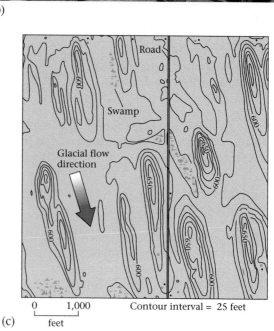

(c)

FIGURE 22.25 (a) The formation of a drumlin beneath a glacier. (b) Drumlins near Rochester, New York. (c) Topographic map emphasizing the shape of drumlins in New York State.

belt, transports sediment with it, which accumulates in a pile at the terminus. The end moraine at the farthest limit of glaciation is called the **terminal moraine.** (The ridge of sediment that makes up Long Island, New York, and continues east-northeast into Martha's Vineyard, Nantucket, and Cape Cod, Massachusetts, is part of the terminal moraine of the ice sheet that covered New England and eastern Canada during the last ice age; ▶Fig. 22.24.) The end moraines that form when a glacier stalls for a while as it recedes are called **recessional moraines.**

If the glacier overrides and bulldozes an end moraine as it advances, it smears out the wet till like cake frosting and creates a flat layer of **lodgment till** over the ground. Clasts in lodgment till may be aligned and scratched during their movement or as ice flows over them. Where the till is drier, the flow of the glacier molds it into streamlined, elongate hills called **drumlins** (from the Gaelic word for "hills"). Drumlins tend to be asymmetric along their length, with a gentle downstream slope, tapered in the direction of flow, and a steeper upstream slope (▶Fig. 22.25a–c). The till left behind during rapid recession forms a thin, hummocky layer on the land surface; this till, together with lodgment till, is called **ground moraine.**

The hummocky surface of moraines reflects partly the variations in the amount of sediment supplied by the glacier, and partly the occurrence of **kettle holes,** circular depressions made when blocks of ice calve off the toe of the glacier, become buried by till, and then melt (▶Fig. 22.26a). A land surface with many kettle holes separated by round hills of till displays **knob-and-kettle topography** (▶Fig. 22.26b, c).

Sediment-choked water pours out of tunnels in the ice. This water feeds large braided streams that sort till and redeposit it as glacial outwash, stratified layers and lenses of gravel and sand. Thus, outwash plains form between recessional moraines and beyond the terminal moraine. Some of

(a)

Freshly fallen ice block
Ice block buried by till
Ice block surrounded by outwash

Kettle lake
Kettle hole
Knob-and-kettle topography

(b)

(c)

FIGURE 22.26 (a) When this ice block melts away, a kettle hole will form. (b, c) Knob-and-kettle topography, the hummocky surface of a moraine.

the water pools in lowlands between recessional moraines or between the glacier's terminus and the nearest recessional moraine, creating meltwater lakes in which varved sediment accumulates. Lakes along the edge of the glacier are called **ice-margin lakes.** Even long after the glacier has melted away, the lowlands between recessional moraines persist as lakes or swamps. Glacial lake beds provide particularly fertile soil for agriculture. When a glacier eventually melts away, ridges of sorted sand and gravel, deposited in sub-glacial meltwater tunnels, snake across the ground moraine. These ridges are called **eskers** (▶Fig. 22.27).

By observing the landscape, we can identify the boundary between the region that was eroded by North American Pleistocene ice sheet and the region where the ice sheet deposited the material it had eroded. In a roughly circular region centered in northeastern Canada and ending just south of Hudson Bay, Pleistocene glaciers scoured deeply into the crust. In a ring around this central area, extending a bit south of the U.S.-Canadian border, the glacier eroded and polished the land surface. Farther south, the glacier deposited the sediment it had eroded in a network of moraines. In portions of the Midwest, the sediment was so wet

FIGURE 22.27 Eskers are snake-like ridges of sand that form when sediment fills meltwater tunnels at the base of a glacier.

Terminal moraine
Esker
Ground moraine

Cross section

Glaciers and Glacial Landforms

Continental ice sheet

Crevasses

Ice shelf

Lower sea level

Dropstones

Iceberg

Mountain (alpine) glaciation

Horn

Valley glacier

Lateral moraine

Mountain ice cap

Cirque glacier

Arête

Medial moraine

Meltwater lake

U-shaped valley

Outwash plain

Drumlin

Braided stream

Esker

Recessional moraine

End moraine

Kettle hole

Striations

Roche moutonnée

tic

Glaciers are rivers or sheets of ice that last all year and slowly flow. Continental glaciers, vast sheets of ice up to a few kilometers thick, covered extensive areas of land during times when Earth had a colder climate. Continental glaciers form when snow accumulates at high latitudes, then, when buried deeply enough, packs together and recrystallizes to make glacial ice. Ice, though solid, is weak, and thus ice sheets spread over the landscape, like syrup over a pancake. At the peak of the ice age, almost all of Canada, much of the United States, northern Europe, and parts of Russia were covered by ice sheets.

The upper part of a sheet is brittle and may crack to form crevasses. Because so much of the Earth's water is stored in ice sheets, the sea level becomes lower during an ice age. When a glacier reaches the sea, it becomes a floating ice shelf. Rock that has been plucked up by the glacier along the way is carried out to sea with the ice; when the ice melts, the rocks fall to the sea floor as dropstones. At the edge of the shelf, icebergs calve off and float away.

A second class of glaciers, called mountain, or alpine, glaciers, exist in mountainous areas because snow can last all year at high elevations. During the ice age, mountain glaciers grew and flowed out onto the land surface beyond the mountain front. The glacier at the right has started to recede after once advancing and covering more of the land. Glacial recession happens when the climate warms, so ice melts away faster at the toe (terminus) of the glacier than can be added at the source. In front of the glacier, you can find consequences of glacial erosion such as striations on bedrock and roche moutonnée.

When the glacier pauses, till (unsorted glacial sediment) accumulates to form an end moraine. Meltwater lakes gather at the toe. Streams carry sediment and deposit it as glacial outwash. Sediment that accumulates in ice tunnels, exposed when the glacier melts, make up sinuous ridges called eskers. Even when the toe is fixed in position for a while, the ice continues to flow, and thus molds underlying sediment into drumlins. Ice blocks buried in till melt to form kettle holes.

In the mountains, the glacier is confined to a valley. Sediment falling on it from the mountains creates lateral and medial moraines. Glaciers carve distinct landforms in the mountains, like cirques, arêtes, horns, and U-shaped valleys.

that it behaved like a slurry, which the glacier spread out like frosting across the countryside, resulting in the flat plains we see today.

22.6 OTHER CONSEQUENCES OF CONTINENTAL GLACIATION

Ice Loading and Glacial Rebound

When a large ice sheet (more than 50 km in diameter) grows on a continent, its weight causes the surface of the lithosphere to sink. In other words, **ice loading** causes **glacial subsidence.** Lithosphere, the relatively rigid outer shell of the Earth, can sink because the underlying asthenosphere is soft enough to flow slowly out of the way.

You can see how this works by conducting a simple experiment. First, fill a bowl with honey and then place a thin rubber sheet over the honey (▶Fig. 22.28a). The rubber represents the lithosphere, and the honey the asthenosphere. If you place an ice cube on the rubber sheet, the sheet sinks because the weight of the ice pushes it down; the honey flows out of the way to make room. Because of ice loading, much of Antarctica and Greenland now lie below sea level (Fig. 22.5), so if their ice were instantly to melt away, these continents would be flooded by a shallow sea.

What happens when continental ice sheets do melt away? Gradually, the surface of the underlying continent rises back up, a process called **glacial rebound,** and the asthenosphere flows back underneath to fill the space (▶Fig. 22.28b–e). In the honey and rubber example, when you remove the ice cube, the rubber sheet slowly returns to its original shape. This process doesn't take place instantly, be-

So You Want to See a Glacier?

Though the last of the Pleistocene continental ice sheet that once covered much of North America vanished about 6,000 years ago, you can find evidence of its power quite easily. The Great Lakes, along the U.S.-Canadian border, the Finger Lakes and drumlins of New York, the low-lying moraines and outwash plains of Illinois, and the polished outcrops of southern Canada all formed in response to movement of this glacier. But if you want to see continental glaciers in action today, you must trek to Greenland or Antarctica.

Mountain glaciers are easier to reach. A trip to the mountains of western North America (including Alaska), the Alps of France or Switzerland, the Andes of South America, or the mountains of southern New Zealand will bring you in contact with live glaciers. You can even spot glaciers from the comfort of a cruise ship.

Some of the most spectacular glacial landscapes in North America formed during the Pleistocene epoch, when mountain glaciers were more widespread. These are now on display in national parks.

- *Glacier National Park (Montana):* This park, which borders Waterton Lakes National Park in Canada, displays giant cirques, U-shaped valleys, hanging valleys, terminal moraines, and fifty small relicts of once-larger glaciers, all in a mountainous terrane that reaches elevations of over 3 km.

- *Yosemite National Park (California):* A huge U-shaped valley, carved into the Sierra Nevada granite batholith, makes up the centerpiece of this park. Waterfalls spill out of hanging valleys bordering the valley.

- *Voyageurs National Park (Minnesota):* This park lacks the high peaks of mountainous parks, but shows the dramatic consequences of glacial scouring and deposition on the Canadian Shield. The low-lying landscape, dotted with lakes, contains abundant polished surfaces, glacial striations, and erratics, along with moraines, glacial lake beds, and outwash plains.

- *Acadia National Park (Maine):* During the ice age, the continental ice sheet overrode low bedrock hills and flowed into the sea along the coast of Maine. This park provides some of the best examples of the consequences. Its hills were scoured and shaped into large roches moutonnées by glacial flow. Some of the deeper valleys have now become small fjords.

- *Glacier Bay National Park (Alaska):* In Glacier Bay, huge valley glaciers reach the sea, creating immense ice cliffs along the water that dramatically calve off icebergs. Cruise ships bring tourists up to the toes of these glaciers. More adventurous visitors can climb the coastal peaks and observe lateral and medial moraines, crevasses, and the erosional and depositional consequences of glaciers that have already retreated up the valley.

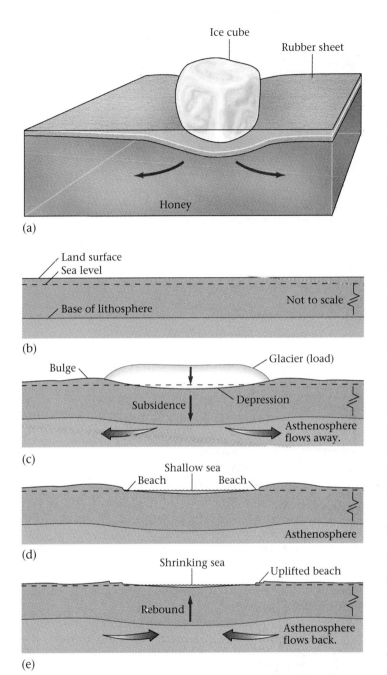

FIGURE 22.28 (a) An ice cube placed on the surface of a rubber sheet floating on honey illustrates the concept of glacial loading. (b) Cross sections illustrating the concept of glacial rebound. Before glaciation, the surface of the lithosphere is flat. (c) The weight of the glacier pushes the lithosphere down below sea level. The asthenosphere flows out of the way below, and lithosphere on either side bulges up. (d) When the glacier melts, the depression fills with water. (e) During glacial rebound, the floor of the shallow sea rises, and beaches along its shore are uplifted.

cause the honey can only flow slowly. Similarly because the asthenosphere flows *so* slowly (at rates of a few millimeters per year), it takes thousands of years for ice-depressed continents to rebound. Thus, glacial rebound is still taking place in some regions that were burdened by ice during the last ice age. Where rebound affects coastal areas, beaches along the shoreline rise several meters above sea level and become terraces.

Sea-Level Changes: The Glacial Reservoir in the Hydrologic Cycle

More of the Earth's surface and near-surface freshwater is stored in glacial ice than in any other reservoir. In fact, glacial ice accounts for 2.15% of Earth's total water supply, while lakes, rivers, soil, and the atmosphere together contain only 0.03%. During the ice age, when glaciers covered almost three times as much land area as they do today, they held significantly more water (70 million cubic km, as opposed to 25 million cubic km today). In effect, water from the ocean reservoir transferred to the glacial reservoir and remained trapped on land. As a consequence, the sea level dropped by as much as 100 m, and extensive areas of continental shelves were exposed as the coastline migrated seaward, in places by more than 100 km (▶Fig. 22.29a–c). People and animals migrated into the newly exposed coastal plains; in fact, fishermen dragging their nets along the Atlantic Ocean floor off New England today occasionally recover artifacts. The drop in sea level also created land bridges across the Bering Strait between North America and northeast Asia and between Australia and Indonesia, providing convenient migration routes.

If today's ice sheets in Antarctica and Greenland were to melt and the crust were then to undergo glacial rebound, global sea level would rise substantially, and low-lying areas of other continents would undergo flooding. In the United States, large areas of the coastal plain along the East Coast and Gulf Coast would flood, and cities like Miami, Houston, New York, and Philadelphia would disappear beneath the waves. In Canada, substantial parts of the shield would flood.

Ice Dams, Drainage Reversals, and Lakes

When ice freezes over a sewer opening in a street, neither meltwater nor rain can enter the drain, and the street floods. Ice sheets play a similar role in glaciated environments. The ice may block the course of a river, leading to the formation of a lake. In addition, the weight of a glacier changes the tilt of the land surface and therefore the gradient of streams, and glacial sediment may fill preexisting valleys. In sum, continental glaciation destroys preexisting drainage networks. While the glacier exists,

(a)

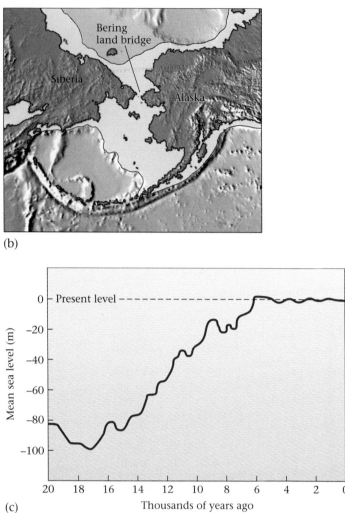

(b)

(c)

FIGURE 22.29 (a) This map shows the coastline of North America during the last ice age, and the coastline should present-day ice sheets melt. Note that much of the coastal plain was exposed during the ice age. (b) A land bridge existed across the Bering Strait between Asia and North America during the last ice age. (c) The graph shows the change in sea level during the past 20,000 years.

streams find different routes and carve out new valleys; when the glacier melts away, these new streams have become so well established that old river courses may remain abandoned.

Glaciation during the last ice age profoundly modified North America's drainage. Before the ice age, several major rivers drained much of the interior of the continent to the north, into the Arctic Ocean (▶Fig. 22.30a, b). The ice sheet buried this drainage network and diverted the flow into the Mississippi-Missouri network, which became larger. Then after the ice receded, regions covered by knob-and-kettle topography became spotted with thousands of small lakes, as now occur in northern Minnesota, while regions of the Midwest covered by a smooth frosting of till became swampland, with little or no drainage at all. In the Canadian Shield, scouring left innumerable depressions that have now become lakes. A huge ice-margin lake, Lake Agassiz, developed in south-central Canada, while in the United States, glacial

scouring created or enlarged the Great Lakes and carved New York's Finger Lakes.

Inevitably, the ice dams that held back large ice-margin lakes melted and broke. In a matter of hours to days, the contents of the lake drained, creating immense floodwaters that stripped the land of soil and left behind huge ripple marks. The flood unleashed by the draining of Glacial Lake Missoula, in Montana, scoured eastern Washington, creating a barren, soil-free landscape called the **channeled scablands** (see Box 22.2).

Pluvial Features

During ice ages, regions to the south of continental glaciers were wetter than they are today. Fed by enhanced rainfall, lakes accumulated in low-lying land even at a great distance from the ice front. The largest of these **pluvial lakes** (from the Latin *pluvia*, for "rain") in North America flooded inte-

FIGURE 22.30 (a) When a glacier advances on the course of a river, the glacier blocks the drainage and causes a new stream to form. After the glacier melts away, the river remains diverted. (b) In North America, the major river systems that flowed north before the last ice age have been destroyed.

rior basins of the Basin and Range Province in Utah and Nevada (▶Fig. 22.32a). These basins, formed as a result of normal faulting during continental rifting, received drainage from the adjacent ranges but had no outlet to the sea, so they filled with water. The largest pluvial lake, Lake Bonne-ville, covered almost a third of western Utah (▶Fig. 22.32b). When this lake suddenly drained after a natural dam holding it back broke, it left a bathtub ring of shorelines rimming the Wasatch Mountains near Salt Lake City. Today's Great Salt Lake itself is but a small remnant of Lake Bonneville.

The Great Missoula Flood

In the 1920's, geologist J Harlan Bretz became fascinated by the landscape of eastern Washington, a barren region known as the channeled scablands (▶Fig. 22.31). Bedrock in this region consists of 15-million-year-old basalt flows. Surprisingly, little or no soil covered these rocks, and even though the region is now semi-arid, huge vertical-sided valleys called **coulees** had been scoured into the basalt. These features could not have been created by glaciers because there were no glacially polished surfaces, striations, or till accumulations. Most geologists simply attributed them to normal stream erosion in the past. Bretz was not convinced. To him, only catastrophic floodwaters could have scoured basalt so deeply over such a broad area. But this notion seemed counter to the uniformitarian views held by most geologists, and for thirty years Bretz's ideas were scorned or ignored. Part of the skepticism stemmed from his inability to identify a source for such an enormous volume of water.

In the 1950s, however, the first aerial photos of eastern Washington became available, and they revealed 7-m-high sediment ripples (ripples in normal rivers are 3 cm high) and a huge dry waterfall, further evidence of a massive flood. The mystery was solved when geologists realized that valleys in western Montana have numerous stair-step lake terraces on their sides, indicating that the valleys once confined huge volumes of water. After geologists had mapped glacial deposits, it became clear that North America's continental glacier had blocked off the valley, creating a huge lake now called Glacial Lake Missoula. When the glacier receded, the ice dam gave way and the lake drained in a matter of days, unleashing an immense flood. A turbulent sheet of water about 250 m deep and 75 km wide churned at velocities of 50–75 km per hour (30–50 mph) over eastern Washington and into the Columbia River drainage system. Finally, the water reached the Columbia River Valley and disappeared into the Pacific, leaving behind a permanently changed landscape. Bretz had been right. The **Great Missoula Flood** now serves as a model for floods that appear to have washed over the surface of Mars in the distant past.

FIGURE 22.31 The path of the Great Missoula Flood across the channeled scablands of eastern Washington.

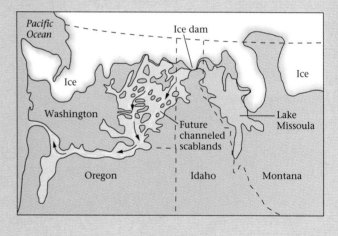

22.7 PERIGLACIAL ENVIRONMENTS

In polar latitudes today, and in regions adjacent to the front of continental glaciers during the ice age, the mean annual temperature stays low enough (below –5°C) that soil moisture and groundwater freeze and, except in the upper few meters, stay solid all year. Such permanently frozen ground, or **permafrost,** may extend to depths of 1,500 m below the ground surface. Regions with widespread permafrost but without a blanket of snow or ice are called **periglacial environments** (the Greek *peri* means "around," or "encircling"; periglacial environments appear around the edges of glacial environments) (▶Fig. 22.33a).

The upper few meters of permafrost may defrost during the summer months, only to refreeze again when winter comes. As it freezes, the ground contracts and splits into pentagon or hexagon shapes, creating a landscape called **patterned ground** (▶Fig. 22.33b). Water fills the gaps between the cracks and freezes to create wedge-shaped walls of ice. In some places, freeze and thaw cycles in permafrost gradually push cobbles and pebbles up from the subsurface. Because the expansion of the ground is not even, the stones gradually collect in ridges between adjacent bulges to form **stone rings.**

Permafrost presents a unique challenge to people who live in polar regions or who work to extract resources from

(a)

(b)

FIGURE 22.32 (a) The distribution of pluvial lakes in the Basin and Range Province during the last ice age. (b) The shoreline of Lake Bonneville along a mountain near Salt Lake City. The Great Salt Lake is a small remnant of this once-huge pluvial lake.

(b)

(c)

FIGURE 22.33 (a) The present-day distribution of periglacial environments in the Northern Hemisphere. (b) The patterned ground of permafrost. (c) The trans-Alaska pipeline must be suspended above the ground to prevent the warmth of the oil from melting the permafrost.

these regions. For example, heat from a building may warm and melt underlying permafrost, creating a mire into which the building settles and breaks apart. For this reason, buildings in permafrost regions must be placed on stilts, so that cold air can circulate beneath them to keep the ground frozen. When geologists discovered oil on the north coast of Alaska, oil companies faced the challenge of shipping the oil to markets outside of Alaska. After much debate over the environmental impact, the trans-Alaska pipeline was built, and now it carries oil for 1,000 km to a seaport in southern Alaska (▶Fig. 22.33c). The oil must be warm during transport, or it would be too viscous to flow, so to prevent the warm pipeline from melting underlying permafrost, it had to be built on a frame that holds it above the ground for its entire length.

22.8 THE PLEISTOCENE ICE AGE

The Pleistocene Glaciers

Today, most of the land surface in New York City lies hidden beneath concrete and steel, but in Central Park it's still possible to see land in a seminatural state. If you stroll through the park and study the rock outcrops, you'll find that their top surfaces are smooth and polished, and in places have been grooved and scratched. Here and there, a large boulder perches on the polished surface (▶Fig. 22.34). You are seeing evidence that an ice sheet once scraped along this ground, rasping and gouging the bedrock as it moved and leaving behind erratics when it melted away. Geologists estimate that the ice sheet that overrode the New York City area may have

FIGURE 22.34 A glacially polished surface and an erratic in Central Park, New York City.

been 250 m thick, enough to bury the Empire State Building up to the 75th floor.

Glacial features like those on display in Central Park first led Louis Agassiz to propose the idea that vast continental glaciers advanced over substantial portions of North America, Europe, and Asia during a great ice age. Since Agassiz's day, thousands of geologists, by mapping out the distribution of glacial deposits and landforms, have gradually defined the extent of ice-age glaciers and a history of their movement.

The fact that these glacial features are found at the surface of the Earth today means that the most recent ice age occurred fairly recently during Earth history. This ice age, responsible for the glacial landforms of North America and Eurasia, happened mostly during the Pleistocene epoch, which began 1.8 million years ago (see Chapter 13), so it is commonly known as the **Pleistocene ice age.** (Recent studies demonstrate that the glaciations of the ice age actually began between 3.0 and 2.5 million years ago, during the Pliocene epoch, and continued through the Pleistocene; but in general, the time interval of the Pleistocene is taken to refer to the ice age.) Geologists use the name **Holocene** to refer to the time since the last glaciation (that is, to the last 11,000 years).

Based on their mapping of glacial striations, geologists have determined where the great Pleistocene ice sheets originated and flowed. In North America, the **Laurentide ice sheet** started to grow over northeastern Canada, then merged with the **Keewatin ice sheet,** which originated in northwestern Canada. Together, these ice sheets eventually covered all of Canada east of the Rocky Mountains and extended southward across the border as far as southern Illinois (▶Fig. 22.35a, b). At their maximum, they attained a thickness of 2–3 km. The Laurentide ice sheet also eventually merged with the Greenland ice sheet to the northeast and the **Cordilleran ice sheet** to the west; the Cordilleran covered the mountains of western Canada, as well as the southern third of Alaska.

Mountain ice caps and valley glaciers also grew in the Rocky Mountains and the Sierra Nevada and Cascade Mountains. In Eurasia, a large ice sheet formed in northernmost Europe and adjacent Asia, and gradually covered all of Scandinavia and northern Russia. This ice sheet flowed south across France until it reached the Alps and merged with Alpine mountain glaciers; it also covered all of Ireland and almost all of the United Kingdom. A smaller ice sheet grew in eastern Siberia and expanded in the mountains of central Asia. In the Southern Hemisphere, Antarctica remained ice-covered, and mountain ice caps expanded in the Andes, but there were no continental glaciers in South America, Africa, or Australia.

In addition to continental ice sheets, sea ice in the Northern Hemisphere expanded to cover all of the Arctic Ocean and parts of the North Atlantic. Sea ice surrounded

(a)

(b)

(c)   Today

(d)   Last glaciation

FIGURE 22.35 (a) The distribution of major ice sheets during the Pleistocene epoch. The arrows indicate the flow trajectories of the ice sheets. (b) Top: A continental glacier scours and erodes the land surface beneath its center, while at the margins it deposits sediment. Map: The Laurentide ice sheet scoured and eroded the land in northern and eastern Canada, and deposited sediment in western Canada and the midwestern United States. (c) Sea temperatures (°C) and the limit of sea ice today. (Colors indicate belts of different temperatures, with purple the coldest and red the warmest.) (d) Sea temperatures and the limit of sea ice during a glaciation.

Iceland and approached Scotland, and also fringed most of western Canada and southeastern Alaska (▶Fig. 22.35c, d).

Life and Climate in the Pleistocene World

During the Pleistocene ice age, all climatic belts shifted southward (▶Fig. 22.36a, b). Geologists can document this shift by examining fossil pollen, which can survive for thousands of years, preserved in the sediment of bogs. Presently, the southern boundary of North America's **tundra,** a treeless region supporting only low shrubs, moss, and lichen ca-

FIGURE 22.37 Examples of now-extinct large mammals that roamed the countryside during the Pleistocene epoch.

FIGURE 22.36 (a) Pleistocene climate belts in North America; (b) Pleistocene climate belts in Europe.

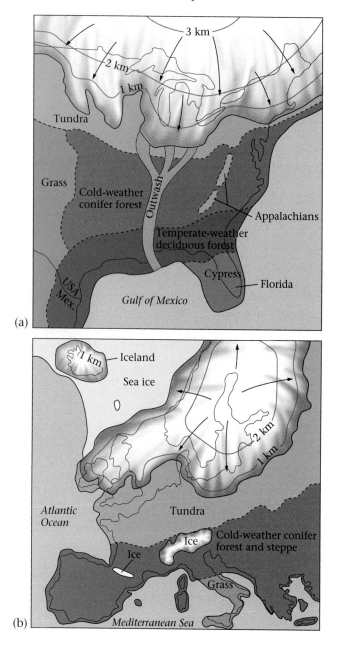

pable of living on permafrost, lies at a latitude of 68°N; during the ice age, it moved down to 48°N. Much of the interior of the United States, which now has temperate, deciduous forest, harbored cold-weather spruce and pine forest during the ice age. Ice-age climates also changed the distribution of rainfall on the planet: increased rainfall in North America led to the filling of pluvial lakes in Utah and Nevada, while decreased rainfall in equatorial regions led to shrinkage of the rain forest. Overall, the contrast between colder glaciated regions and warmer unglaciated regions created windier conditions worldwide. These winds sent glacial rock flour skyward, creating a dusty atmosphere (and, presumably, great sunsets). The dust settled to create extensive deposits of loess. And because so much ice was trapped in glaciers, as we have seen, the sea level lowered.

Numerous species of now-extinct large mammals inhabited the Pleistocene world (▶Fig. 22.37). Giant mammoths and mastodons, relatives of the elephant, along with woolly rhinos, musk oxen, reindeer, giant ground sloths, bison, lions, saber-toothed cats, giant cave bears, and hyenas wandered forests and tundra in North America. Early human-like species were already foraging in the woods by the beginning of the Pleistocene epoch, and by the end modern *Homo sapiens* lived on every continent except Antarctica, and had discovered fire and invented tools. Rapidly changing climates may have triggered a global migration of early humans, who gained access to the Americas, Indonesia, and Australia via land bridges that became exposed when the sea level dropped.

Chronology of the Pleistocene Ice Age

Louis Agassiz assumed that only one ice age had affected the planet. But close examination of the stratigraphy of glacial deposits on land revealed that **paleosol** (ancient soil preserved in the stratigraphic record), as well as fossils of warmer-weather animals and plants, separated distinct layers of glacial sediment. This observation suggested that between

FIGURE 22.38 Pleistocene deposits in the United States.

episodes of glacial deposition, glaciers had receded and temperate climates had prevailed. In the second half of the twentieth century, when modern methods for dating geological materials became available, the difference in ages between the different layers of glacial sediment could be confirmed. Clearly, ice-age glaciers had advanced and then retreated more than once. Times during which the glaciers grew and covered substantial areas of the continents are called glacial periods, or **glaciations,** and times between glacial periods are called interglacial periods, or **interglacials.**

Using the on-land sedimentary record, geologists recognized five Pleistocene glaciations in Europe (named, in order of increasing age: Würm, Riss, Mindel, Gunz, and Donau) and four in the midwestern United States (Wisconsinan, Illinoian, Kansan, and Nebraskan, named after the southernmost states in which their till was deposited; ►Fig. 22.38). With the advent of radiometric dating in the mid-twentieth century, the ages of the younger glaciations were found by dating wood trapped in glacial deposits. The ages of the older glaciations are determined by identifying fossils of known age found in the deposits. Because of their greater age, their deposits have been thoroughly weathered and dissected.

The four-stage chronology of North American glaciation was turned on its head in the 1960s, when geologists began to study submarine sediment containing the fossilized shells of microscopic marine plankton. Because the assemblage of plankton species living in warm water is not the same as the assemblage living in cold water, geologists can track changes in the temperature of the ocean by studying plankton fossils. Researchers found that in sediment of the last 2 million years, assuming that cold water indicates a glacial period and warm water an interglacial period, there

were twenty to thirty different glacial advances during the Pleistocene epoch. The four traditionally recognized glaciations probably represent only the largest of these—sediments deposited on land by other glaciations were eroded and redistributed during subsequent glaciations, or were eroded by streams and wind during interglacials.

Geologists refined their conclusions about the frequency of Pleistocene glaciations by examining the composition of fossil shells. Shells of many plankton species consist of calcite $(CaCO_3)$. The oxygen in the shells includes two isotopes, a heavier one (^{18}O) and a lighter one (^{16}O). The ratio of these isotopes tells us about the water temperature in which the plankton grew; this is because as water gets colder, plankton incorporate a higher proportion of ^{18}O into their shells (see Chapter 23). Thus, intervals in the stratigraphic record during which plankton shells have a large ratio of $^{18}O/^{16}O$ define times when Earth had a colder, glacial climate. The record indicates that twenty to thirty of these times occurred during the last 3 million years (►Fig. 22.39).

FIGURE 22.39 This time column shows the variations in oxygen-isotope ratios from marine sediment that define twenty to thirty glaciations and interglacials during the Pleistocene epoch. The right-hand column indicates the approximate boundaries of the traditional four glaciations that have been recognized on land in North America.

Older Ice Ages During Earth History

So far, we've focused on the Pleistocene ice age, because of its importance in developing Earth's present landscape. Was this the only ice age during Earth history, or do ice ages happen frequently? To answer such questions, geologists study the stratigraphic record and search for ancient glacial deposits that have hardened into rock. These deposits, called **tillites,** consist of larger clasts distributed through a matrix of sandstone and mudstone (▶Fig. 22.40a). The poor sorting of tillites indicates that they were glacially deposited, for glaciers can carry sediment of all sizes. In many cases, tillites are deposited on glacially polished surfaces.

By using the stratigraphic principles described in Chapter 12, geologists have determined that tillites were deposited about 280 million years ago (in Permian time; these are the deposits Alfred Wegener had in mind when he argued in favor of continental drift; ▶Fig. 22.40b), about 600 to 700 million years ago (at the end of the Proterozoic eon), about 2.2 billion years ago (near the beginning of the Proterozoic), and perhaps about 2.7 billion years ago (in the Archean eon). Strata deposited at other times in Earth history do not contain tillites. Thus, it appears that glacial advances and retreats have not occurred steadily throughout Earth history, but rather are restricted to specific time intervals, or ice ages, of which there are four or five: Pleistocene, Permian, late Proterozoic, early Proterozoic, and perhaps Archean. Of particular note, some tillites of the late Proterozoic event were deposited at equatorial latitudes, suggesting that for at least a short time the continents worldwide were largely glaciated, and the sea may have been covered worldwide by ice. This condition is known as "snowball Earth."

22.9 ICE AGES: THE CAUSES

The discovery of the complexity of the glacial-interglacial record from marine sediment not only gave us a better idea of the ice ages, it also led to the discovery of their causes. Ice ages happen when it snows heavily and the Earth becomes cool enough, at least in higher latitudes, for snow to remain year-round. They take place only during restricted intervals of Earth history, hundreds of millions of years apart, but within an ice age glaciers advance and retreat with a frequency measured in tens of thousands to hundreds of thousands of years. Thus, there must be both long-term and short-term controls on glaciation. What are they?

Long-Term Causes

Plate tectonics probably provides some long-term controls on glaciation. First, for an ice age to happen, substantial areas of continents must have drifted to high latitudes; if all continents sat along the equator, the land would be too warm for snow to accumulate. Second, glaciations can only take place when most continents lie well above sea level—in other words, when global sea level is low; sea-level changes may be controlled by changes in the volume of mid-ocean ridges and thus changes in rates of sea-floor spreading. Finally, ice ages can't develop when oceanic currents carry heat to high latitudes; and currents, in turn, are controlled by positions of continents and volcanic arcs, as determined by plate motions.

The concentration of carbon dioxide in the atmosphere also determines whether or not an ice age can occur. Carbon dioxide is a greenhouse gas—it traps infrared radiation rising from the Earth—so if the concentration of CO_2 increases, the atmosphere becomes warmer. Ice sheets sim-

FIGURE 22.40 (a) This time column shows pre-Pleistocene glaciations during Earth history. (b) The distribution of Permian glacial features on a reconstruction of Gondwana, the southern part of the supercontinent that existed at the time.

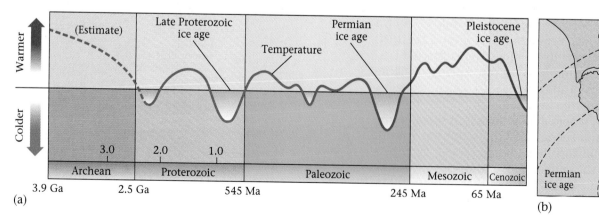

ply cannot form during periods when the atmosphere has a relatively high concentration of CO_2, even if other factors favor glaciation. But what might cause long-term changes in CO_2 concentration? Possibilities include changes in the number of marine organisms that extract CO_2 to make shells, changes in the amount of chemical weathering on land (determined by the abundance of mountain ranges; weathering absorbs CO_2), changes in the amount of volcanic activity, and the appearance of coal swamps (it takes CO_2 to make the plants from which coal forms). At present, there are no clear-cut answers.

Short-Term Causes

Now we've seen how the stage could be set for an ice age to happen, but why do glaciers advance and retreat periodically *during* an ice age? In 1920, Milutin Milankovitch, a Serbian astronomer and geophysicist, came up with an explanation. Milankovitch studied how the Earth's orbit changes shape and how its axis changes orientation through time, and calculated the frequency of these changes. In particular, he evaluated three aspects of Earth's movement around the Sun.

- *Orbital eccentricity:* Milankovitch showed that the Earth's orbit gradually changes from a more circular shape to a more elliptical shape. This **eccentricity cycle** takes around 100,000 years (▶Fig. 22.41a).

- *Tilt of Earth's axis:* We have seasons because the Earth's axis is not perpendicular to the plane of its orbit. Milankovitch calculated that, over time, the tilt angle varies between 22.5° and 24.5°, with a frequency of 41,000 years (▶Fig. 22.41b).

- *Precession of Earth's axis:* If you've ever set a top spinning, you've probably noticed that its axis gradually traces a

FIGURE 22.41 The Milankovitch cycles affect the amount of insolation (exposure to the Sun's rays) at high latitudes. (a) Variations in insolation caused by changes in orbital shape; (b) variations caused by changes in the tilt angle of Earth's axis; (c) variations caused by the precession of Earth's axis. (d) Eccentricity, tilt, and precession all affect the amount of insolation, but with different periodicities. When the effects are combined, we see that there are distinct warm and cold periods; cold periods occur when there is less insolation.

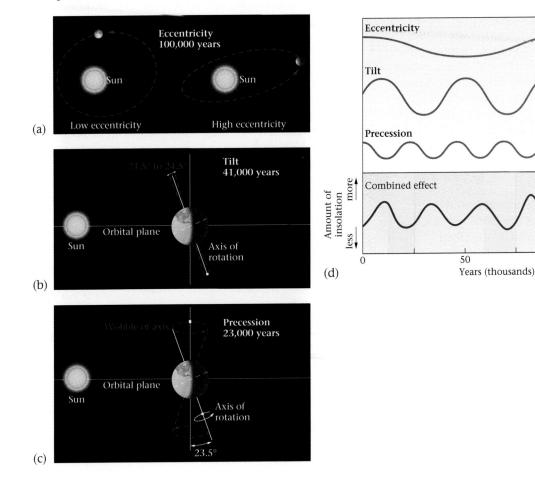

conical path. This motion, or "wobble," is called **precession** (▶Fig. 22.41c). Milankovitch determined that the Earth's axis wobbles like that of a top over the course of about 25,000 years; right now, the Earth's axis points toward Polaris, making it the "north star," but 12,000 years ago the axis pointed to Vega. Precession determines the relationship between the timing of the seasons and the position of Earth along its orbit around the Sun. (If summer happens when Earth is closer to the Sun, then we have a warm summer, but if it happens when Earth is farther away from the Sun, then we have a cool summer.)

Milankovitch showed that precession, along with variations in orbital eccentricity and tilt, combine to affect the total annual amount of **insolation** (exposure to the Sun's rays) and the seasonal distribution of insolation that the Earth receives at the high latitudes (such as 65°N). For example, high-latitude regions receive more insolation when the Earth's axis is almost perpendicular to its orbital plane than when its axis is greatly tilted. According to Milankovitch, glaciers tend to advance during times of cool summers at 65°N, which happen roughly 100,000, 40,000, and 20,000 years apart. When geologists began to study the climate record, they found climate cycles with the frequency predicted by Milankovitch. These climate cycles are now called **Milankovitch cycles** (▶Fig. 22.41d).

The discovery of Milankovitch cycles in the geologic record strongly supports the contention that changes in the Earth's orbit and tilt help trigger short-term advances and retreats during an ice age. But orbit and tilt changes cannot be the whole story, because they could cause only about a 4°C temperature decrease (relative to today's temperature), and during glaciations the temperature decreased 5°–7°C along coasts and 10°–13°C inland. Geologists suggest that several other factors come into play in order to trigger a glacial advance.

- *A changing albedo:* When snow remains on land through the year, the albedo (reflectivity) of the Earth increases; so Earth's surface reflects incoming sunlight and thus becomes even cooler.

- *Interrupting the global heat conveyor:* As the climate cools, evaporation rates from the sea decrease, so seawater does not become as salty. And decreasing salinity might stop the system of thermohaline currents that brings warm water to high latitudes (see Chapter 18). Thus, the high latitudes become even colder than they would otherwise.

- *Biological processes that change CO_2 concentration:* Several kinds of biological processes may have amplified climate changes by altering the concentration of carbon dioxide in the atmosphere. For example, a greater amount of plankton growing in the oceans could absorb more

carbon dioxide and thus remove it from the atmosphere. Other biological processes include the death of coral reefs in response to a glacially caused drop in sea level (coral produces CO_2 as it grows, so if it dies, the amount of CO_2 emitted to the atmosphere decreases) and the growth of large coal swamps (plants extract CO_2 to produce wood and leaves).

The three processes described above are called **positive-feedback mechanisms:** they enhance the process that causes them. Because of positive feedback, the Earth cools more than it would otherwise.

A Model for Pleistocene-Ice-Age History

Long-term cooling in the Cenozoic era. Taking all of the above "causes" into account, we can now propose a scenario for the events that led to the Pleistocene glacial advances. Our story begins in the Eocene epoch, about 55 million years ago (▶Fig. 22.42). At that time, climates were warm and balmy, not only in the tropics but even above the Arctic Circle. At the end of the Middle Eocene (37 million years ago), the climate began to cool, and by Early Oligocene time (33 million years ago), Antarctica became glaciated. The Antarctic ice sheet came and went until the middle of the Miocene epoch (15 million years ago), when an ice sheet formed that has never disappeared. Ice sheets did not appear in the Arctic, however, until 2–3 million years ago, when the Pleistocene ice age began.

The climate changes that occurred in Oligocene and Miocene time indicate that the Pleistocene ice age came at the culmination of a long-term cooling of the climate. Though several factors may have caused this long-term change, perhaps the most important was the architecture of oceanic currents, controlled by plate tectonics. Today, a huge oceanic current known as the Circum-Antarctic Current circulates around Antarctica, maintaining a band of

FIGURE 22.42 The graph shows the gradual cooling of Earth's atmosphere since the Cretaceous period.

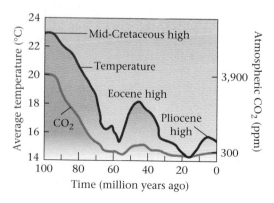

cold water along the coast. The current did not exist before Early Eocene time, because Australia and South America were still connected to Antarctica, so waters in the southern Pacific and Atlantic flowed to the equator and back. This flow of warm water kept the climate in polar regions relatively warm. By the Early Oligocene, Australia had separated from Antarctica, and the new current configuration may have triggered cooling. The Circum-Antarctic Current became complete at the end of the Oligocene, when South America pulled away from Antarctica, setting the stage for overall cooler climates.

Another possible event that led to a long-term change in the Cenozoic climate occurred in southern Asia. During the Mesozoic and early Cenozoic eras, a tropical sea known as the Tethys Sea ran all the way from the Straits of Gibraltar to Indonesia (see Chapter 13), permitting warm tropical waters to circulate. In the Eocene epoch, however, India collided with southern Asia, an event that chopped the Tethys Sea in half and ended its influence on oceanic circulation. As the collision proceeded in the Oligocene and Miocene, the mighty Himalayas and Tibet rose to become the world's highest range. Computer modeling of global climates has shown that the uplift of the Himalayas and Tibet changed atmospheric temperatures and circulation and could have contributed to long-term global cooling. The uplift also exposed more rock to chemical weathering, and thus could have removed CO_2 from the atmosphere, further contributing to cooling.

So far, we've examined hypotheses that explain long-term cooling since 55 million years ago, but what caused the sudden appearance of the Arctic ice cap about 2–3 million years ago? This event coincides with another well-known plate-tectonic event, the closing of the gap between North and South America by the growth of the Isthmus of Panama. When this Panama land bridge formed, it separated the waters of the Caribbean from those of the tropical Pacific for the first time. A number of oceanographic studies have shown that when this happened, warm currents that previously flowed out the Caribbean into the Pacific were blocked and diverted northward to merge with the Gulf Steam. This current transfers warm water from the Caribbean up the Atlantic Coast of North America and ultimately to the British Isles. As the warm water moves up the Atlantic Coast, it generates warm, moisture-laden air that provides a source for the snow that falls over New England, eastern Canada, and Greenland. In other words, the Arctic has long been cold enough for ice caps, but until the Gulf Stream was diverted north by the growth of Panama, there was no source of moisture to make abundant snow and ice.

Short-term advances and retreats in the Pleistocene epoch. Once the Earth's climate had cooled overall, short-term processes such as the Milankovitch cycles led to periodic advances and retreats of the glaciers. To understand how, let's look at the case history of a single advance and retreat of the Laurentide ice sheet.

- *Stage 1:* During the overall cooler climates of the late Cenozoic era, the Earth reaches a point in the Milankovitch cycles when the average mean temperature in temperate latitudes drops. Since the continents had been ice free, the surface of northern Canada has risen to an altitude of several hundred meters above sea level, and with the lower temperatures and higher elevations, not all of winter's snow melts away during the summer. Eventually, the snow covers the entire region of northern Canada even during the summer. Because of the snow's high albedo, it reflects sunlight, so the region grows still colder (a positive-feedback effect) and even more snow accumulates. Precipitation rates are high, because evaporation off the Gulf Stream provides moisture. Finally, the snow at the base of the pile turns to ice, and the ice begins to spread outward under its own weight. A new continental glacier has been born.

- *Stage 2:* The ice sheet continues to grow as more snow piles up in the zone of accumulation. And as the ice sheet grows, the atmosphere continues to cool because of the albedo effect. But now, the weight of the ice loads the continent and makes it sink, so the elevation of the glacier decreases, and its surface approaches the snow line. Also, the temperature becomes cold enough that the Atlantic Ocean in high latitudes begins to freeze. As the ocean is covered with sea ice, the amount of evaporation decreases, so the source of snow is cut off and the amount of snowfall diminishes. The glacial advance pretty much chokes on its own success. The decrease in the glacier's elevation (leading to warmer summer temperatures) on the ice surface as well as the decrease in snowfall causes ablation to occur faster than accumulation, and the glacier begins to retreat.

- *Stage 3:* As the glacier retreats, temperatures gradually increase, and the sea ice begins to melt. The supply of water to the atmosphere from evaporation increases once again, but with the warmer temperatures and lower elevations, this water precipitates as rain during the summer. The rain drastically accelerates the rate of ice melting, and the retreat progresses quite rapidly.

22.10 WILL THERE BE ANOTHER GLACIAL ADVANCE?

What does the future hold? Considering the periodicity of glacial advances and retreats during the Pleistocene epoch, we may be living in an interglacial period. Pleistocene interglacials lasted about 10,000 years, and since the present in-

FIGURE 22.43 Skaters (c. 1600) on the frozen canals of Holland during the little ice age.

terglacial began about 11,000 years ago, the time seems ripe for a new glaciation. If a true ice age on the scale of the Laurentide ice sheet were to develop, major cities and agricultural belts would be overrun by ice, and their populations would have to migrate south. Long before the ice front arrived, though, the climate would become so hostile that the cities would already be abandoned.

The Earth actually had a brush with ice-age conditions between the 1300s and the mid-1800s, when average annual temperatures in the Northern Hemisphere fell sufficiently for mountain glaciers to advance significantly. During this period, now known as the **little ice age,** sea ice surrounded Iceland, and canals froze in the Netherlands, leading to that country's tradition of skating (▶Fig. 22.43). But during the past 150 years, temperatures have warmed, and most mountain glaciers have retreated significantly. We no longer see the icebergs that once threatened Atlantic shipping lanes and sank the *Titanic,* and large slabs frequently calve off the Antarctic ice sheet. Some researchers suggest that this global-warming trend is due to the addition of carbon dioxide to the atmosphere from burning forests and fossil fuels (see Chapter 23). Global warming could conceivably cause a "super-interglacial."

If the climate were to become significantly warmer than it is today, the ice sheet of West Antarctica might begin to float and then break up rapidly. If all of today's ice caps melted, global sea level would rise by 70 m (230 feet), extensive areas of coastal plains would be flooded, and major coastal cities like New York, Miami, and London would be submerged (Fig. 22.31). Instead of protruding from ice, the tip of the Empire State Building would protrude from the sea. Ice house or greenhouse? We may never know which scenario will play out in the future until it happens.

CHAPTER SUMMARY

- Glaciers are streams or sheets of recrystallized ice that survive for the entire year. Mountain (Alpine) glaciers exist in very high regions, and continental glaciers (ice sheets) spread over substantial areas of the continents.

- Mountain glaciers fill cirques and valleys. Some may form mountain ice caps, and some may spread out into lobes on the land next to the mountains.

- Glaciers form when snow accumulates over a long period of time. With progressive burial, the snow first turns to firn, and then to ice.

- Wet-bottom glaciers move by basal sliding over water or wet sediment. Dry-bottom glaciers move by internal flow. In general, glaciers move tens of meters per year, but during a surge, they may flow up to 110 m per day.

- Valley glaciers move because gravity pulls them down slopes, while continental glaciers move because of gravitational spreading.

- Whether the terminus (toe) of a glacier stays fixed in position, advances farther from the glacier's origin, or retreats back toward the origin depends on the balance between the rate at which snow builds up in the zone of accumulation and the rate at which glaciers melt or sublimate in the zone of ablation.

- Icebergs break off glaciers that flow into the sea. Sea ice forms when the ocean surface freezes. Sea ice grounded along the coast is called an ice shelf.

- Glacial ice can plow over sediment, incorporate sediment, and pluck sediment. The clasts embedded in glacial ice act like a rasp that abrades the substrate.

- Mountain glaciers carve numerous landforms, including cirques, arêtes, horns, U-shaped valleys, hanging valleys, and truncated spurs. Glacially carved valleys that fill with water when the sea level rises after an ice age are called fjords.

- Glaciers can transport sediment of all sizes. Glacial drift includes till, glacial marine, glacial outwash, lake-bed mud, and loess. Lateral moraines accumulate along the sides of valley glaciers, and medial moraines form down the middle of a glacier. End moraines accumulate at a glacier's terminus, and lodgment till smears out beneath a glacier.

- Numerous depositional landforms develop in glacial environments or during glacial retreat. These include moraines, knob-and-kettle topography, drumlins, kames, eskers, meltwater lakes, and outwash plains.

- The sinking of continental crust as a result of ice loading is glacial subsidence. When the glacier melts away, the crust rises, a phenomenon called glacial rebound.

- When water is stored in continental glaciers, the sea level drops. When glaciers melt, the sea level rises.

- Continental glaciers, as they recede, can dam valleys and cause huge lakes to accumulate. When the ice dams break, a large flood ensues. Such a flood carved the channeled scablands of eastern Washington.

- During ice ages, the climate in regions south of the continental glaciers was wetter, and pluvial lakes formed in regions that are now desert. Permafrost (permanently frozen ground) exists in periglacial environments.

- During the Pleistocene ice age, large continental glaciers covered much of North America, Europe, and Asia. Modern humans first appeared at this time.

- The stratigraphy of Pleistocene glacial deposits preserved on land records five European and four U.S. glaciations, times during which ice sheets advanced, separated from each other by interglacials, times during which ice sheets retreated. The record preserved in marine sediments records twenty to thirty such events.

- Long-term causes of ice ages include plate tectonics and changes in the concentration of CO_2 in the atmosphere. Short-term causes include the Milankovitch cycles (caused by periodic changes in Earth's orbit and tilt), a changing albedo, changes in ocean currents, and changes in plankton productivity.

KEY TERMS

ablation (p. 674)
albedo (p. 669)
arête (p. 677)
basal sliding (p. 672)
calving (p. 674)
catabatic winds (p. 683)
channeled scablands (p. 690)
cirque (p. 669)
continental glacier (ice sheet) (p. 670)
crevasse (p. 673)
drop stones (p. 676)
drumlin (p. 684)
eccentricity cycle (p. 699)
end moraine (p. 681)
erratics (p. 682)
esker (p. 685)
firn (p. 671)
fjord (p. 680)
glacial drift (p. 682)
glacial rebound (p. 688)

glacial striations (p. 677)
glacial till (p. 682)
glaciations (p. 697)
glacier (p. 669)
Great Missoula Flood (p. 692)
ground moraine (p. 684)
hanging valleys (p. 680)
horn (p. 677)
ice age (p. 667)
ice loading (p. 688)
ice-margin lake (p. 685)
ice shelf (p. 676)
interglacials (p. 697)
internal flow (p. 672)
kame (p. 681)
kettle hole (p. 684)
knob-and-kettle topography (p. 684)
lateral moraine (p. 681)
little ice age (p. 702)
lodgment till (p. 684)

loess (p. 683)
medial moraine (p. 681)
meltwater lake (p. 683)
Milankovitch cycles (p. 700)
moraine (p. 681)
mountain (alpine) glacier (p. 669)
outwash (p. 683)
patterned ground (p. 692)
periglacial environment (p. 692)
permafrost (p. 692)
Pleistocene ice age (p. 694)
pluvial lakes (p. 690)
positive-feedback mechanism (p. 700)
precession (p. 700)

recessional moraine (p. 684)
roche moutonnée (p. 680)
rock flour (p. 676)
snow line (p. 674)
sublimation (p. 671)
terminal moraine (p. 684)
terminus (toe) (p. 674)
tillite (p. 698)
truncated spurs (p. 680)
tundra (p. 696)
U-shaped valley (p. 680)
V-shaped valley (p. 680)
varve (p. 683)
zone of ablation (p. 674)
zone of accumulation (p. 674)

REVIEW QUESTIONS

1. What evidence did Louis Agassiz offer to support the idea of an ice age?

2. How do mountain glaciers and continental glaciers differ in terms of dimensions, thickness, and patterns of movement?

3. Describe the transformation from snow to ice.

4. Explain how arêtes, cirques, and horns form.

5. Describe the mechanisms that allow glaciers to move, and explain why they move.

6. How fast do glaciers normally move? How fast can they move during a surge?

7. Explain how the balance between ablation and accumulation determines whether a glacier advances or retreats.

8. How can a glacier continue to flow downhill even though its toe is retreating?

9. How does a glacier transform a V-shaped river valley into a U-shaped valley?

10. Discuss how hanging valleys can develop and why they cannot be explained by the concept of a graded stream.

11. Describe the various kinds of glacial deposits. Be sure to note the materials from which the deposits are made and the landforms that result from deposition.

12. How do the crust and mantle respond to the weight of glacial ice?

13. How was the world different during the glacial advances of the Pleistocene ice age? Be sure to mention the relation between glaciations and sea level.

14. How was the standard four-stage chronology of U.S. glaciations developed? Why was it so incomplete? How was it modified with the study of marine sediment?

15. Were there ice ages before the Pleistocene? If so, when?

16. What are some of the long-term causes that lead to ice ages?

17. Describe the three kinds of Milankovitch cycles. How does each affect the amount of sunlight received on Earth?

SUGGESTED READING

Anderson, B. G., and H. W. Borns, Jr. 1994. *The Ice Age World.* Oslo-Copenhagen-Stockholm: Scandinavian University Press.

Barth, M. C., and J. J. Titus, eds. *The Greenhouse Effect and Sea-Level Rise.* New York: VNR.

Broecker, W. S., and G. H. Denton. 1990. What drives glacial cycles? *Scientific American* (January): 48–56.

Covey, C. 1984. The Earth's orbit and the ice ages. *Scientific American* (February): 58.

Dawson, A. G. 1992. *Ice Age Earth: Late Quaternary Geology and Climate.* New York: Routledge, Chapman, and Hall.

Denton, G. H., and T. J. Hughes. 1981. *The Last Great Ice Sheets.* New York: Wiley.

Drewry, D. J. 1986. *Glacial Geologic Processes.* London: Edward Arnold.

Embleton, C., and C. A. M. King. 1975. *Glacial and Periglacial Geomorphology.* London: Edward Arnold.

Eyles, N., ed. 1983. *Glacial Geology.* New York: Pergamon Press.

Flint, R. F. 1971. *Glacial and Quaternary Geology.* New York: Wiley.

Goldthwait, R. P., ed. 1975. *Glacial Deposits.* New York: Dowden Hutchinson and Ross.

Grove, J. M. 1988. *The Little Ice Age.* London: Methuen.

Hambrey, M. J. 1994. *Glacial Environments.* Vancouver, B.C.: UBC Press.

Hambrey, M. J., and J. Alean. 1992. *Glaciers.* Cambridge: Cambridge University Press.

Imbrie, J., and K. P. Imbrie. 1979. *Ice Ages: Solving the Mystery.* Short Hills, N.J.: Enlow Press.

John, B. S. 1977. *The Ice Ages: Past and Present.* London: Collum.

———. 1979. *The Winters of the World.* London: David and Charles.

LaChapelle, E. R., and A. S. Post. 1971. *Glacier Ice.* Seattle: University of Washington Press.

Price, R. J. 1973. *Glacial and Postglacial Landforms.* New York: Macmillan.

Schneider, S. 1990. *Global Warming: Are We Entering the Greenhouse Century?* San Francisco: Sierra Club Books.

Sharp, R. P. 1988. *Living Ice: Understanding Glaciers and Glaciation.* Cambridge: Cambridge University Press.

CHAPTER 23

Global Change in the Earth System

23.1 INTRODUCTION

Would the Earth's surface have looked the same to a Jurassic "Moonling" as it did to the *Apollo* astronauts? Definitely not—the Earth of 200 million years ago differs from that of today in many ways. In the Jurassic period, the North Atlantic Ocean was a narrow sea and the South Atlantic Ocean didn't exist at all, so most dry land connected to form a single vast continent (▶Fig. 23.1a). Today, both parts of the Atlantic are wide oceans, and the Earth has seven separate continents (▶Fig. 23.1b). In the Jurassic, southern Europe was a broad platform underlain by sediment. Today, the region encompasses the rugged Alps, and its Jurassic sediments are now complexly folded metamorphic rocks. And in the Jurassic, the call of the wild rumbled from the throats of dinosaurs, while today, the largest land animals are mammals. In essence, what we see of the Earth today is just a snapshot, an instant in the life of a constantly changing planet. This idea, which grew from James Hutton's proposal of the principle of uniformitarianism in the late eighteenth century (see Chapter 12), arguably stands as geology's greatest philosophical contribution to humanity's understanding of its surroundings.

Why does the Earth change so much through time? Ultimately, it's because the Earth's asthenosphere is warm and soft enough to flow, because the Sun is close enough to heat the Earth's surface, and because gravity causes heavier objects to fall and buoyant ones to rise (▶Fig. 23.2). If the asthenosphere could not flow and gravitational force did not exist, internal processes such as sea-floor spreading, subduction, and continental drift would not occur. Without the warmth of the Sun, there could be no liquid water, advanced life, wind, rivers, or glaciers. Without gravity, wind, rivers, and glaciers would not move. And without wind, rivers, or glaciers, there would be no external processes like weathering and erosion and thus no landscape modification or sediment formation. No other planet in our solar system has a mobile asthenosphere and a surface whose temperature straddles the freezing point of water, so no other planet undergoes the kinds of changes that Earth does. The Moon, for example, changes so little through time that it looked essentially the same to a Jurassic dinosaur as it does to you.

Recent studies suggest that Earth's climate is getting warmer. In the worst-case scenario, regions that are now temperate woodlands will transform into deserts, and forests will die.

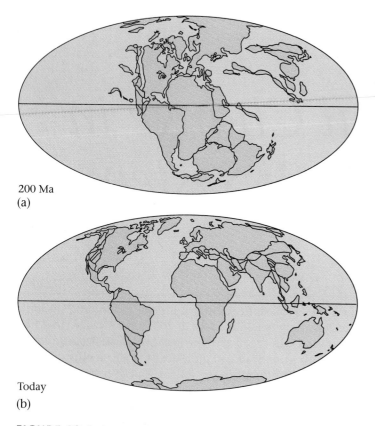

200 Ma
(a)

Today
(b)

FIGURE 23.1 A comparison of (a) a map of the Earth 200 million years ago with (b) a map of today's Earth emphasizes the change that has resulted from continental drift.

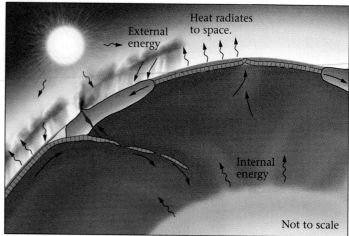

FIGURE 23.2 Internal processes, driven by internal energy (heat inside the Earth), and external processes, driven by external energy (heat from the Sun), combine to make the Earth a dynamic planet. Plate interactions lead to uplift, while atmospheric circulation and flowing water or ice lead to erosion.

The changes that take place on Earth also reflect complex interactions between geologic and biological phenomena. For example, photosynthetic organisms affect the composition of the atmosphere by providing oxygen, and atmospheric composition in turn determines the nature of chemical weathering in rocks. For purposes of discussion, we refer to the global interconnecting web of physical and biological phenomena on Earth as the **Earth system,** and we define **global change** as the transformations or modifications of both physical and biological components of the Earth system through time. (Of note, some researchers suggest that the Earth system, overall, is analogous to a complex living entity, and they use the name **Gaia,** the Greek word for the Earth goddess, for the Earth system.)

Geologists distinguish among different types of global change, based on the rate or way in which change progresses with time. **Gradual change** takes place over long periods of geologic time (millions to billions of years), while **catastrophic change** takes place instantly (seconds) or rapidly in geologic time (centuries). **Unidirectional change** refers to transformations that never repeat, while **cyclic change** repeats the same steps over and over, though not necessarily with the same results. Some types of

cyclic change are "periodic," in that the cycles happen with a definable frequency.

In this chapter, we begin by reviewing examples of global change, both unidirectional and cyclic, involving phenomena discussed earlier in the book. Then we look at an example of a **biogeochemical cycle,** the exchange of chemicals between living and nonliving reservoirs; some kinds of global change are due to changes in the proportions of chemicals held in different reservoirs through time. Finally, we focus on **global climate change** (transformations or modifications in Earth's climate through time) and on **anthropogenic** (human-caused) contributions to global change. (Note that we use "global change" here more broadly than do the popular media, who commonly employ the term as a synonym only for "recent global climate change.") We conclude this chapter, and the book, by recalling hypotheses that describe the ultimate global change— the end of the Earth.

23.2 UNIDIRECTIONAL CHANGES

The Evolution of the Solid Earth

Recall from Chapter 1 that Earth began as a fairly homogenous mass, the consequence of planetesimals coalescing. But the homogeneous proto-Earth did not last long— within about 0.5 billion years of its birth, the planet began to melt, yielding a liquid iron alloy that sank rapidly to the center to form the core (▶Fig. 23.4a, b). Thus, within a few hundred million years of its birth, Earth underwent a major unidirectional change: it differentiated into a layered, onion-like planet, with an iron alloy core surrounded by a rocky mantle.

At this point, Earth looked vastly different than it does today. The surface probably consisted of a thin "skin" of frozen mantle, pockmarked by meteorite impacts. Subduction cycled this skin back into the still-hot mantle, where it melted completely. The probable collision of a Mars-sized body with the Earth caused a catastrophic change—a significant portion of the mantle fragmented and vaporized, creating a ring that quickly coalesced to form the Moon (▶Fig. 23.4c–e). Gradually, however, Earth cooled sufficiently so that only minerals with a lower melting temperature (silicic minerals) melted. Magmas formed by partial melting of the mantle were more silicic than the mantle itself. The rocks that solidified from these magmas at or near the surface were buoyant and could not be subducted, so they accumulated to create the continental crust. The Earth's surface had changed again: it now had distinct continents and ocean basins, and plate tectonics began operating.

The Evolution of the Atmosphere and Oceans

Like the Earth's surface, its atmosphere has also changed through time. In the young Earth, melting released large quantities of carbon dioxide (CO_2) and water vapor (H_2O). These two gases, along with gases belched from volcanoes (and possibly left by comets), accumulated to make up Earth's early atmosphere. When the planet's surface cooled sufficiently, water condensed and collected to create oceans. Once formed, the oceans absorbed other gases from the atmosphere (notably CO_2). Nitrogen (N_2), which doesn't react with other

---

BOX 23.1

THE REST OF THE STORY

The Goldilocks Effect

Like Baby Bear's porridge in the tale of *Goldilocks and the Three Bears,* Earth is not too hot, and it's not too cold . . . it's just right, as far as complex life is concerned (▶Fig. 23.3). This condition is known as the **Goldilocks effect.**

Two factors play the key roles in determining Earth's surface temperature: the distance between the Earth and the Sun, and the concentration of carbon dioxide (CO_2) and other greenhouse gases in the atmosphere. If Earth were just 13% closer to the Sun, the heat would be so intense that liquid water could not exist. Without liquid water, CO_2 would not be stored as limestone and coal but rather would remain in the atmosphere. As a result, the atmosphere would contain so much CO_2 that it would become hot enough to melt lead, a condition that now exists on Venus. On the other hand, if Earth were significantly farther from the Sun so that sunlight was weaker, or if the atmosphere contained less CO_2 with which to trap heat, our planet's surface would be so cool that the oceans would freeze solid and complex life would die.

FIGURE 23.3 (a) Venus is too hot, (b) Mars too cold, and (c) Earth just right.

(a)   (b)   (c)

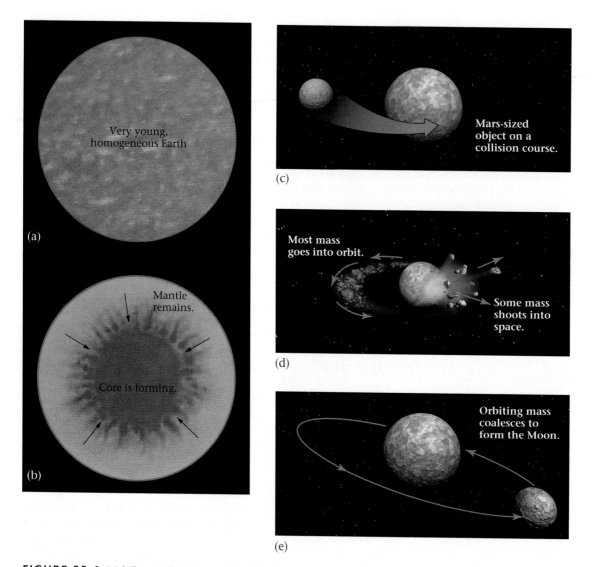

FIGURE 23.4 (a) When it first formed, the Earth was probably homogeneous. (b) Soon thereafter, the iron in the Earth melted and sank to the center. (c–e) Then, a Mars-sized body collided with Earth, sending off fragments that coalesced to form the Moon. All these phenomena radically changed the Earth.

chemicals, was left behind. Thus, the atmosphere's composition changed to become dominated by nitrogen. And when photosynthetic organisms appeared, the atmosphere changed once again, becoming rich in oxygen (O_2).

The Evolution of Life

During the Hadean eon, Earth's surface was probably lifeless, for carbon-based organisms could not survive the high temperatures of the time. The fossil record indicates that life, in the form of cyanobacteria, had appeared by 3.8 billion years ago and has undergone unidirectional change (evolution) in fits and starts ever since (see Interlude D). Though simple organisms like viruses and bacteria still exist, life evolution during the Phanerozoic eon yielded multicellular plants and animals (▶Fig. 23.5). Life now inhabits regions from a few kilometers below the surface to a few kilometers above—this is the **biosphere.**

23.3 PHYSICAL CYCLES

The Supercontinent Cycle

During Earth history, the map of the planet's surface has constantly changed. At times, almost all continental crust has merged to form a single supercontinent, but usually the crust

FIGURE 23.5 New species of life have evolved through geologic time. Though some of the simplest still exist, more complex organisms have appeared more recently.

is distributed among several smaller continents. The process of change during which supercontinents form and later break apart is called the **supercontinent cycle** (▶Fig. 23.6). Geologists have found evidence that at least three times during the past two billion years of Earth history, supercontinents existed—most recently at the end of the Paleozoic era, yielding the land mass Alfred Wegener called Pangaea.

The supercontinent cycle begins with the merging of numerous smaller continents when the oceanic lithosphere that separates them subducts. Once formed, a supercontinent gradually absorbs heat from the underlying mantle and weakens. Later, rifting begins, during which the supercontinent separates into several smaller continents. During breakup, new mid-ocean ridges develop, and sea-floor spreading produces new oceans. An ocean grows in width until subduction begins along one or both of its margins, at which time the ocean gets narrower until it has been entirely consumed. Then the continents collide again, producing a new supercontinent. And after a while, the new supercontinent rifts apart and the cycle begins all over again. Recall that plates move only 1–10 cm per year, so one passage through the supercontinent cycle takes at least a few hundred million years. Though this description of the cycle sounds as though oceans open and close like accordions, in reality plate motions are more complex, so the land is never arranged exactly the same way through two supercontinent cycles.

The Sea-Level Cycle

Global sea level has risen and fallen by as much as 300 m during the Phanerozoic eon, and likely did the same in the Precambrian. When the sea level rises, the shoreline migrates inland, and low-lying plains in the continents become submerged. During periods of particularly high sea level, more than half of Earth's continental area can be covered by shallow seas; at this time, continental regions receive a blanket of sedimentary rock that permanently changes their surface (▶Fig. 23.7a). When the sea level falls, the continents are exposed again, and regional unconformities develop. For example, the sedimentary strata of the midwestern United States record at least six continent-wide advances and retreats of the sea, each of which left behind a blanket of sediment called a **sedimentary sequence;** unconformities define the boundaries between the sequences (▶Fig. 23.7b). Of note, the sequence deposited during the Pennsylvanian period contains at least thirty shorter repeated intervals, called **cyclothems,** each of which contains a specific succession of sedimentary beds. At the base of each cyclothem, you'll find sandstone, and in the middle you'll find coal (▶Fig. 23.7c). Cyclothems represent short-term cycles of sea-level rise and fall.

After studying sedimentary sequences around the world, geologists at Exxon Corporation pieced together a

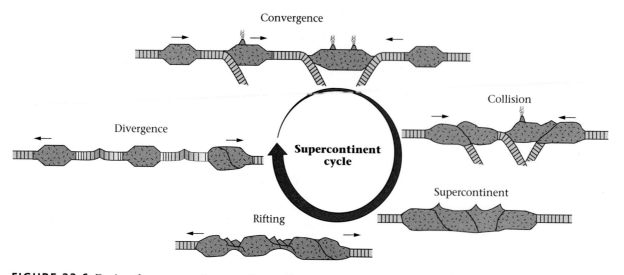

FIGURE 23.6 During the supercontinent cycle, smaller continents coalesce to form a supercontinent, which then later rifts and break apart, only to recombine later on. Since continents drift around the surface of the Earth, collisions do not necessarily bring previously adjacent continents back together again.

chart defining the succession of global transgressions and regressions during the Phanerozoic eon. The global **sedimentary cycle chart,** sometimes referred to as the "Vail curve" (named for Peter Vail, the leader of the team that developed it), may largely reflect the cycles of **eustatic** (worldwide) **sea-level change** during the Phanerozoic (▶Fig. 23.7d). However, the chart probably does not give us an exact image of sea-level change, because the sedimentary record reflects other factors as well, such the rise or fall of the land surface resulting from plate interactions, and changes in sediment supply.

Geologists recognize three "orders" of cycles on this chart. First-order cycles (lasting 200 million to 400 million years) are the longest and may be related to stages in the supercontinent cycle, which affects the area of continental crust. Second-order cycles (10 million to 100 million years) may be controlled by changes in the spreading rate at mid-ocean ridges, which in turn affect the volume of mid-ocean ridges. Third-order cycles (1 million to 10 million years) may reflect ice ages. And fourth-order cycles (less than 500,000 years) may be governed by the Milankovitch cycle of changes in Earth's orbit and tilt, which cause glacial advances and retreats within an ice age (▶Fig. 23.7e; see Chapter 22).

The Rock Cycle

We learned early in this book that the crust of the Earth consists of three rock types: igneous, sedimentary, and metamorphic. Atoms comprising the minerals of one rock type may later become part of another rock type. In effect, rocks are simply reservoirs of atoms, and the atoms move from reservoir to reservoir through time. As we learned in Interlude B, this process is called the **rock cycle.**

Each stage in the rock cycle changes the Earth by redistributing and modifying material. For example, when the asthenosphere partially melts to form a basaltic magma, which then rises, the magma extracts elements from the asthenosphere, thus changing its composition. And if the magma erupts at the surface and solidifies into a lava flow, the crust changes because it now includes a layer of new rock. If the lava flow weathers and turns to clay, which is then deposited in the sea, the crust changes again because new sediment has developed; and if this sediment later gets buried and turns to schist, the crust changes yet again because new metamorphic rock has formed.

23.4 BIOGEOCHEMICAL CYCLES

A **biogeochemical cycle** involves the passage of a chemical among nonliving and living reservoirs in the Earth system, mostly on or near the surface. Nonliving reservoirs include the atmosphere, the crust, and the ocean, while living reservoirs include plants, animals, and microbes. Although a great variety of chemicals (water, carbon, oxygen, sulfur, ammonia, phosphorous, and nitrogen) are involved in biogeochemical cycles, here we look at only two: water (H_2O) and carbon (C).

Some stages in a biogeochemical cycle may take only hours, some may take thousands of years, while others may take millions of years. Because chemicals can cycle

(a) The sea level has changed significantly over geologic time. For example, large parts of North America were once submerged by shallow seas. (The continental outline is shown for reference only—the continent did not have its present shape in the past.) Note that during Pennsylvanian time, large coal swamps developed along the shore, a region that is now the Midwest.

Cambrian Silurian Devonian Cretaceous Pennsylvanian

◼ Land ▪ Submerged continental shelf □ Shallow sea ▪ Swamp Eq. = Equator

(a)

FIGURE 23.7 (a) The sea level has changed significantly over geologic time. For example, large parts of North America were once submerged by shallow seas. (The continental outline is shown for reference only—the continent did not have its present shape in the past.) Note that during Pennsylvanian time, large coal swamps developed along the shore, a region that is now the Midwest. (b) The stratigraphic record shows that sedimentary sequences are deposited during long-term transgressions and regressions. Transgression starts near the edge of a continent and then moves inland, so land near the margin is submerged longer and accumulates more sediment. (c) During the Pennsylvania period, short-term transgressions and regressions created cyclothems. (d) To a large extent, the transgressions and regressions indicated by this sedimentary sequence chart, developed by researchers at Exxon, reflect global (eustatic) sea-level change, though other factors, like displacement of the land surface and a change in sediment supply, can influence the record. Here, two versions of the sequence chart are shown, each produced by a different author—the shape of the curve is still a subject for research. Note that during the Cretaceous period, the sea level was much higher than it is today. (e) At a finer time scale, there are many short-term ups and downs in sea level. This graph shows sea-level changes during the last 150,000 years.

rapidly, the transfer of a chemical from reservoir to reservoir during these cycles doesn't really seem like a "change" in the Earth in the way that the movement of continents or the metamorphism of rock seems like change. In fact, for intervals of time, biogeochemical cycles attain a **steady-state condition,** meaning that the proportions of a chemical in different reservoirs remains fairly constant, even though there is a constant **flux** (flow) of the chemical between reservoirs. When we speak of global change in a biogeochemical cycle, we mean a change in the relative proportions of a chemical held in different reservoirs at a given time; in other words, a change in the steady-state condition.

The Hydrologic Cycle

As we learned in Interlude E, the **hydrologic cycle** refers to the movement of water from reservoir to reservoir on or near the surface of the Earth. The hydrologic cycle is an example of a biogeochemical cycle, in that a chemical (H_2O) passes through both nonliving and living entities—the oceans, the atmosphere, surface water, groundwater, glaciers, soil, and living organisms. Global change in the hydrologic cycle can occur when there is a change in global climate that alters the ratio between the amount of water held in the ocean and the amount held in continental ice sheets. For example, during an ice age, water that had been stored in oceans moves into glacial reservoirs. Thus, the continents become covered with ice, and the sea level drops. When the climate warms, more water is again stored in oceans, and the sea level rises.

The Earth System

External energy

Sun

Thunderhead

Lightning

Mountain uplift

Rain and snow

Continental glacier

City

Ocean

Rocky coastline

Desert

Valley

Arid mountains

Mining

Lakes

Deciduous forest

Beach

Forested mountains

The Earth's surface represents the interface between the solid Earth (lithosphere); the ice and liquid water of oceans, lakes, streams, groundwater, and glaciers (the hydrosphere); and the planet's gaseous envelope (the atmosphere). Countless species of life, ranging from nearly invisible bacteria to giant whales and trees, make up the complex ecosystems of Earth's biosphere. All of these components—the lithosphere, hydrosphere, atmosphere, and biosphere—interact with one another and comprise the Earth system.

Various materials cycle among living and nonliving components of the Earth system. In the hydrologic cycle, for example, water evaporates from the sea, rains on the land, and eventually flows back to the sea. During this process, water may be temporarily trapped in living organisms, clouds, subsurface pores, or ice sheets. Carbon dioxide can be stored in the air, dissolved in water, or trapped in plants, coal, or limestone. Some limestone forms when coral extracts ions from water. Meanwhile, the atoms that make up minerals, over the vastness of geologic time, pass through the rock cycle. New elements from the mantle may enter the cycles of the Earth system at volcanoes or black smokers. Elements at the surface may be carried back into the mantle at subduction zones. Some atoms escape from the atmosphere into space.

Tropical rain forest

Coral reef

Shark

Internal energy

Jet stream

Cirrus clouds

Ice and snow

oniferous
orest

Industrial
pollution

Wind system

Evaporation

Volcanic islands

Cold surface current

Field pattern

Delta

Swamps

Surface waters

Warm surface current

Twilight zone

Abyssal zone

Whale

Sea floor

Bacteria and
plankton

Giant squid

Deep-sea
current

Black
smokers

Two key sources of energy fuel the dynamic Earth system. External energy comes from solar radiation, which drives the hydrologic cycle and the circulation of the atmosphere and oceans. These, in turn, cause erosion of the land surface. Internal energy, a relict of Earth formation and a product of radioactive decay, drives volcanism and the uplift of mountains.

Because of cycles, features of the Earth system undergo cyclic change. For example, in the time frame of centuries to millions of years, climate changes and sea-level changes have markedly affected the character of the planet's surface. Plate interactions constantly, though slowly, change the map of the planet. And over geologic history, there has been an evolution of life and the atmosphere. Despite its immensity, the Earth system is fragile; human activity, which helps create global change, has had major effects on the Earth system.

The Carbon Cycle (the Flux of a Greenhouse Gas)

Most carbon in the near-surface realm of Earth originally bubbled out of the mantle in the form of CO_2 gas released at volcanoes (▶Fig. 23.8). Once it enters the atmosphere, it can be removed in various ways. Some dissolves in seawater to form carbonate ($-CO_3^{-2}$) ions, while some is absorbed by photosynthetic organisms that convert it into sugar and other organic chemicals by the reaction

$$6CO_2 + 6H_2O \rightarrow C_6H_{12}O_6 + 6O_2.$$

This carbon enters the food chain and ultimately makes up the flesh, fat, and sinew of animals. In fact, about 63 billion tons of carbon move from the atmosphere into life forms every year. Carbon, in the form of CO_2, may also be removed from the atmosphere when it is absorbed by chemical weathering. For example,

weathering of limestone: $CaCO_3 + CO_2 + H_2O \rightarrow$ $Ca^{2+} + 2HCO^{3-}$;

weathering of calcium silicate: $CaSiO_3 + 2CO_2 + H_2O \rightarrow$ $Ca^{2+} + 2HCO^{3-} + SiO_2.$

Some carbon returns directly to the atmosphere by the respiration of animals (again as CO_2), by the flatulence of animals (as methane [CH_4]), or by the decay of dead organisms. But some can be stored for long periods of time in fossil fuels (oil and coal) or in limestone. Fossil fuel deposits and limestone contain most of the carbon in the near-surface realm of Earth and can hold on to it for long periods of time. But this carbon either returns to the atmosphere in the form of CO_2, as a result of the burning of fossil fuels and the metamorphism of rocks containing carbonate, or returns to the sea after being dissolved in river water or groundwater.

The concentration of carbon dioxide in the atmosphere plays an essential role in controlling Earth's climate because, as we saw in Chapter 20, CO_2, along with several other trace gases (such as water and methane), is a greenhouse gas—an increase in CO_2 concentration warms the atmosphere, while a decrease cools it down.

FIGURE 23.8 In the carbon cycle, carbon transfers between various reservoirs at or near the Earth's surface. Carbon occurs in the atmosphere as CO_2, in the ocean as dissolved $-CO_3^{-2}$ ions, in the crust as limestone, coal, or oil, and in living organisms as complex organic chemicals. Chemical weathering absorbs CO_2, while metamorphism releases it.

23.5 GLOBAL CLIMATE CHANGE

How often have you seen a newspaper headline proclaim "Record High Temperatures!" Does this mean that the climate, the average range of weather conditions for a given region, is changing? Is it something we should be worried about. Perhaps. **Global climate change,** the transformations or modifications in Earth's climate through time, is indeed important because it affects the sea level and the distribution and character of climatic belts and, therefore, habitats, agricultural lands, and landscapes. A rise in the average global temperature of only a few degrees might bring drought conditions that could melt enough of the polar ice sheets to cause a sea-level rise that would in turn flood coastal population centers.

For purposes of discussion, we distinguish between **long-term climate change,** which takes place over millions to tens of millions of years, and **short-term climate change,** which takes place over hundreds to hundreds of thousands of years. If the average atmospheric and sea-surface temperature rises, we have **global warming,** while if it falls, we have **global cooling.** Some changes are great enough to cause large regions of continents to be submerged by shallow seas or covered by ice, while others are subtle, creating only a slight latitudinal shift in vegetation belts and a sea-level change measured in meters or less.

Methods of Study

Geologists and climatologists are working hard to define the kinds of climate changes that can occur, the rates at which these changes take place, and the effects they may have on society. There are two basic approaches to studying global climate change: (1) researchers measure past climate change, as indicated by the stratigraphic record, to document the magnitude of change that is possible and the rate at which such change occurred; (2) researchers make computer programs to calculate how factors like at-

FIGURE 23.9 Changes in the assemblage of pollen in sediment indicate a shift in climate belts. (a) Spruce pollen from a cold-climate coniferous forest. (b) Hemlock pollen from a warm-climate deciduous forest. (c) This model shows how the proportion of tree pollen relative to grass pollen can change in a sedimentary sequence through time. Tree pollen indicates cooler and drier conditions, while grass pollen indicates warmer and wetter conditions.

mospheric composition, topography, ocean currents, and Earth's orbit affect the climate. The resulting **climate-change models** provide insight into when and why changes took place in the past and whether they will happen in the future.

Let's look first at how geologists study the **paleoclimate** (past climate), so as to document climate changes through Earth history. Any feature whose character depends on the climate and whose age can be determined serves as a clue to defining the paleoclimate.

• *The stratigraphic record:* The nature of sedimentary strata deposited at a certain location reflects the climate at that location. For example, an outcrop exposing cross-bedded sandstone, overlain successively by coal and glacial till, indicates that the site of the outcrop has endured different climates (desert, then tropical, then glacial) through time. The amount of organic carbon in lake sediment provides a clue to the quantity of vegetation that has grown in a region, and more vegetation indicates a wetter climate. And the proportion of rock fragments in sea-floor sediment hints at the number of icebergs that have floated above, for icebergs carry the fragments to sea and then drop them.

• *Paleontological evidence:* Different assemblages of species survive in different climatic belts. Thus, the succession of species in a sedimentary sequence provides clues to the changes in climate at that site. For example, a record of short-term climate change can be obtained by studying the succession of plankton fossils in sea-floor sediments, for cold-water species of plankton are different from warm-water species.

Fossil pollen also yields clues to the paleoclimate. **Pollen,** tiny grains involved in plant reproduction, looks like dust to the unaided eye. But under a microscope, each grain has a distinctive structure, and grains of one species look different from grains of another species (▶Fig. 23.9a, b). Further, pollen grains have a tough coating and can survive burial. By studying pollen in sediment, **palynologists** (scientists who study pollen) can determine whether the sediment accumulated in a cold-climate conifer forest or in a warm-climate deciduous forest. And by recording changes in the pollen assemblage found in successive layers of sediment, palynologists can track the movement of climate belts over the landscape (▶Fig. 23.9c). For example, the overall warming since the retreat of the last glacier can be tracked by the range of spruce pollen (▶Fig. 23.10a, b).

• *Oxygen-isotope ratios:* Recall that two isotopes of an element have the same atomic number but different atomic weights (see Appendix A). For instance, oxygen occurs as ^{16}O (8 protons and 8 neutrons) and ^{18}O (8 protons and 10 neutrons). Geologists have found that the ratio of $^{18}O/^{16}O$ in glacial ice indicates the atmospheric

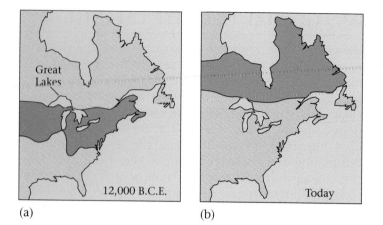

FIGURE 23.10 Warming after the last ice age led to the migration of spruce forests. (a) About 12,000 years ago, spruce forests lay south of the Great Lakes. (b) Today, they are found north of the Great Lakes.

temperature in which the snow that made up the ice formed: the ratio is larger in snow that forms in warmer air, while the ratio is smaller in snow that forms in colder air. Because of this relationship, the isotope ratio measured in a succession of ice layers in a glacier indicates temperature change through time. Researchers have now obtained ice cores down to a depth of almost 3 km in Antarctica and in Greenland; this record shows variations in temperature back to about 160,000 years B.C.E. (▶Fig. 23.11a).

Study of the $^{18}O/^{16}O$ ratio in the calcium carbonate ($CaCO_3$) making plankton shells gives us an indication of ocean temperature and of the amount of water stored in glaciers. The ratio indicates ocean temperature because when the water is colder, plankton incorporate more ^{18}O in their shells, so the ratio of $^{18}O/^{16}O$ is larger. And the ratio reflects the amount of water stored in glaciers for the following reason. Water containing the lighter isotope evaporates more easily than water containing the heavier isotope; therefore, atmospheric water vapor is richer in ^{16}O than it is in ^{18}O. When snow precipitates out of this vapor, it contains a proportionally larger amount of ^{16}O. When glaciers grow, the ^{16}O becomes trapped in glacial ice and cannot return to the sea, so the ratio of $^{18}O/^{16}O$ in seawater increases. But when glaciers shrink, the ^{16}O in glacial ice returns to the sea and the ratio of $^{18}O/^{16}O$ decreases. Because such changes are clearly recorded in plankton shells, by studying the isotope ratios in plankton fossils from a core of marine sediment, we can track changes in water temperature (and therefore atmospheric temperature) through time (▶Fig. 23.11b, c).

• *Bubbles in ice:* Bubbles in ice trap the air present at the time the ice forms. By analyzing these bubbles, geologists can measure the concentration of CO_2 in the atmosphere

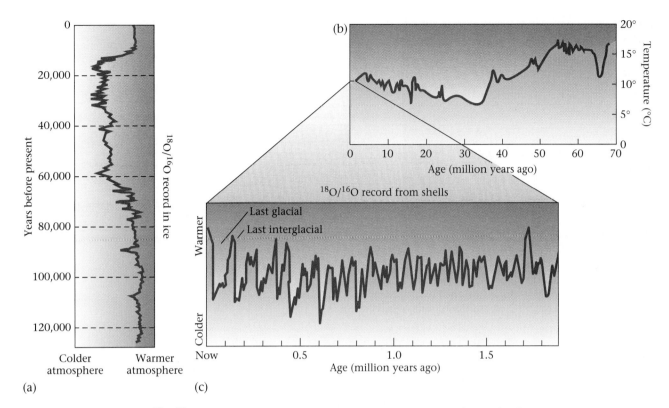

FIGURE 23.11 (a) The $^{18}O/^{16}O$ ratio in a 2-km-long ice core drilled in the ice cap of Greenland varies with depth in the core, indicating that atmospheric temperature varies over time. Smaller ratios mean a colder atmosphere. Note that the ratio is smaller between about 65,000 and 15,000 years ago, a time corresponding to the last glacial advance of the ice age. (b) The ratio of $^{18}O/^{16}O$ in the calcite of fossil plankton shells of deep-marine sediment shows variations in temperature for the past 70 million years. These changes show up partly because the $^{18}O/^{16}O$ ratio incorporated by plankton when making shells depends on the temperature of the sea, and partly because water enriched in ^{16}O will be stored in glacial ice sheets and thus is removed from the sea. (c) The detailed plankton record of temperature for the past 2 million years. The decreases in the $^{18}O/^{16}O$ ratio correspond with glacial advances. As the record becomes more detailed, geologists have discovered that changes in temperature may happen very quickly—the transition from a low to a high may take as little as 10 years.

back through time. This information can be used to correlate CO_2 concentration with past atmospheric temperature. Because geologists have found a strong correlation between temperature and CO_2 concentration, some researchers use CO_2 concentration alone as an indicator of temperature. Bubbles with high CO_2 concentration, for example, developed when atmospheric temperature was warm.

- *Growth rings:* If you've ever looked at a tree stump, you'll have noticed the concentric rings visible in the wood. Each ring represents one year of growth, and the thickness of the ring indicates the rate of growth in a given year. Trees grow faster during warmer, wetter years and more slowly during cold, dry years (▶Fig. 23.12a); thus, the succession of ring widths provides an easily calibrated record of climate during the lifetime of the tree.

Bristlecone pines may supply a record back through 4,000 years. To go further into the past, **dendrochronologists** (scientists who study tree rings) look at the record of rings in logs dated by the radiocarbon technique or in logs whose ages overlap with that of the oldest living tree. Growth rings in corals and shells can provide similar information.

- *Human history:* Researchers have made careful, direct measurements of climate changes only for the past few decades. This record is not long enough to display long-term climate change. But history, both written and archaeological, contains important clues to climates at times in the past. Periods of unusual cold or drought leave an impression on people, and they are recorded in paintings, stories, and records of crop success or failure (▶Fig. 23.12b).

(a) (b)

FIGURE 23.12 (a) Tree rings provide a record of climate, for more growth happens in wet years than in dry years. (b) The stories of great floods recorded by artists and historians provide clues to the timing of climate changes. This woodcut depicts a flood in medieval England.

Long-Term Climate Change

Using the variety of techniques described above, geologists have reconstructed an approximate record of global climate, represented by mean temperature and rainfall, for geologic time. The record shows that at some times in the past, the Earth's atmosphere was significantly warmer than it is today, whereas at other times it was significantly cooler. The warmer periods have come to be known as **greenhouse** (or **hot-house**) **periods** and the colder as **ice-house periods.** (The more familiar term, "ice age," refers to the times during an ice-house period when the Earth was cold enough for ice sheets to advance and cover substantial

areas of the continents.) As the chart in ▶Figure 23.13a shows, there have been at least five major ice-house periods during geologic history (early Proterozoic, late Proterozoic, Late Ordovician, Permo-Carboniferous, and late Cenozoic).

Let's look a little more closely at the climate record of the last 100 million years, for this time interval includes the transition between a greenhouse and ice-house period. Paleontological and other data suggest that the climate of the Mesozoic era, the Age of Dinosaurs, was much warmer than the climate of today. At the equator, average annual temperatures may have been 2°–6°C warmer, while at the poles, temperatures may have been 20°–60°C warmer. In fact, during the Cretaceous period, dinosaurs were able to live at

BOX 23.2

THE HUMAN ANGLE

Global Climate Change and the Birth of Legends

Some geologists argue that myths passed down from the early days of civilization may have their roots in global climate change. For example, recent evidence suggests that earlier than 7,600 years ago, the region that is now the Black Sea contained a much smaller freshwater lake surrounded by settlements. After the most recent ice-age glacial advance, the ice sheets melted and the sea level rose, and the Mediterranean eventually broke through a dam at the site of the present Bosporous Strait. Researchers suggest that seawater from the Mediterranean spilled into the Black Sea

basin, via a waterfall two hundred times larger than Niagara Falls. This influx of water caused the lake level to rise by perhaps 10 cm per day, and within a year 155,000 square km (60,000 square miles) of populated land had become submerged beneath hundreds of meters of water. This flooding presumably led to a huge human migration, and its timing has suggested to some researchers that it may have inspired the Babylonian *Epic of Gilgamesh* (ca. 2000 B.C.E.) and, later, the biblical epic of Noah's Ark.

FIGURE 23.13 (a) The chart shows the timing of ice-house and greenhouse (or hot-house) periods during Earth history. (b) Continental drift and the growth of volcanic arcs affect the pattern of currents in the ocean. Here, we see that currents changed when the Isthmus of Panama (a volcanic arc) developed. (c) The maps show that the proportion of the Earth's surface at high elevations, where it is exposed to chemical weathering, has changed through time as a consequence of mountain building. For example, the collision of India with Asia about 40 Ma created a broad high plateau.

high latitudes, and there were no polar ice caps on Earth. But starting about 80 million years ago, the Earth's atmosphere began to cool. We entered an ice-house period about 33 million years ago, and the climate reached its coldest condition about 2 million years ago, during the Pleistocene ice age (see Chapter 22).

What caused long-term global climate change? The answer may lie in the complex relationships between the vari-

ous geologic and biogeochemical cycles described earlier. Following are some particularly notable influences on global change:

- *Positions of continents:* Continental drift influences the climate by controlling the pattern of oceanic currents, which redistribute heat around the planet's surface (▶Fig. 23.13b). Drift also determines whether the land is at high or low latitudes (and thus how much solar radiation strikes it), and whether or not there are large continental interior regions where extremely cold winter temperatures can develop.

- *Volcanic activity:* A long-term global increase in volcanic activity may contribute to long-term global warming, because it increases the concentration of greenhouse gases in the atmosphere. For example, during the Cretaceous period when Pangaea broke up, numerous rifts formed, and sea-floor-spreading rates were particularly fast, so volcanoes were more abundant than they are today; volcanic activity may thus have triggered Cretaceous greenhouse conditions.

- *The uplift of land surfaces:* Tectonic events that lead to the long-term uplift of the land affect atmospheric CO_2 concentration, because such events expose land to weathering, and chemical-weathering reactions absorb CO_2. Thus, uplift decreases the greenhouse effect and causes global cooling. For example, uplift of the Himalayas and Tibet may have triggered Cenozoic ice-house conditions (▶Fig. 23.13c). (Of note, such uplift will also affect atmospheric circulation and rainfall rates; see Chapter 20.)

- *The formation of coal, oil, or limestone:* At various times during Earth history, environments suitable for coal swamps and limestone reefs have been particularly widespread. Because coal and limestone contain carbon, their formation removes CO_2 from the atmosphere and stores it underground. For example, global cooling in the Carboniferous period, which ultimately led to Permo-Carboniferous glaciations, may correlate with the development of vast coal swamps on Pangaea.

Many of the effects described above control the climate because they either add or subtract CO_2 from the atmosphere—as noted earlier, factors that increase CO_2 cause global warming, while those that decrease CO_2 cause global cooling. In fact, probably the major cause of long-term global climate change is a long-term change in the distribution of CO_2 between various reservoirs in the carbon cycle.

It is important to note that feedback among components of the Earth system helps regulate the amount of CO_2 in the atmosphere. **Negative feedback** slows a process down or even reverses it. For example, as global temperature rises because of an increase in CO_2, rates of evaporation and therefore amounts of precipitation increase. As a consequence, weathering rates (which absorb CO_2) increase, so the concentration of CO_2 then goes down. **Positive feedback,** on the other hand, makes a process continue or even accelerate. For example, positive feedback on Venus may have led to a **runaway greenhouse effect,** through the following steps:

1. Because Venus is closer to the Sun than Earth is, it receives more solar radiation. In the past, the high heat caused any surface water to evaporate until the planet became enveloped in clouds.

BOX 23.3

THE REST OF THE STORY

The Faint Young Sun Paradox

The intensity of sunlight striking the Earth has likely changed substantially during the history of the solar system, because the composition of the Sun changes with time. The Sun's energy comes primarily from the fusion reaction that bonds four hydrogen atoms together to form one helium atom. Through time, the proportion of helium in the Sun increases. Calculations suggest that because one helium atom takes up less space than four separate hydrogen atoms, the production of helium should allow the inside of the Sun to contract. This contraction raises the Sun's temperature, which in turn increases the rate of fusion reactions and therefore the amount of energy the Sun produces. Overall, the Sun may be 30% brighter today than it was when the Earth first formed.

If this is the case, the Earth's average temperature *should have been* over 20°C less in the Archean eon than it is today, and Archean oceans *should have been* frozen solid. But fossil evidence suggests that liquid water has existed on our planet's surface almost continuously since at least 3.8 billion years ago, so the Earth couldn't have been so cold. This apparent conflict between calculation and observation is called the **faint young Sun paradox.**

Most researchers agree that the paradox can be explained by remembering that earlier in Earth history, before the widespread appearance of photosynthetic life, the atmosphere contained more carbon dioxide than it does today. The greenhouse effect caused by this excess CO_2 could have increased the temperature of the Earth sufficiently to counter the effect of the faint young Sun, so that surface temperatures remained above freezing.

FIGURE 23.14 (a) The past 15,000 years (the Holocene epoch) experienced several periods of warming and cooling. (b) Coastal Greenland was settled by the Vikings during one of the warmer periods, when the region could support agriculture. (c) Glaciers formed during the little ice age persisted into the nineteenth century. Here we see a glacier in the French Alps as it appeared in 1850. (d) By the second half of the twentieth century (1966), the glacier had almost disappeared.

2. The water vapor, a greenhouse gas, did not allow infrared radiation to escape, so the atmosphere became still hotter, approaching 1,500°C. At these extremely high temperatures, water molecules break apart, forming hydrogen and oxygen gas. The hydrogen escaped to space, and the oxygen reacted with surface rocks to oxidize (rust) them.

3. Without water, CO_2-absorbing chemical weathering ceased, so CO_2 built up in the atmosphere, making the temperature rise even more. This dense hot atmosphere persists on Venus today.

Short-Term Climate Change

The record of the past million years gives a sense of the magnitude and duration of short-term climate change. During this period, there have been about five major and twenty to thirty minor episodes of glaciation, separated by interglacial periods. If we focus on the last 15,000 years, we see that overall the temperature has increased, but there are still notable ups and downs (▶Fig. 23.14a). In fact, by studying evidence for icebergs in the ocean, researchers have identified cycles in this period that lasted centuries to millennia.

As a result of the warming that began 15,000 years ago, the glaciers retreated for about 4,500 years. Then there was a return to colder conditions for a few more thousand years. This interval of cooler temperature is named the **Younger Dryas,** after an Arctic flower that became widespread during the time. The climate then warmed, reaching a peak at 5,000–6,000 years ago, a period called the **Holocene climatic optimum,** when average temperatures were about

2°C above temperatures of today. This warming peak led to increased evaporation and therefore precipitation, making the Middle East unusually wet and fertile—conditions that may partially account for the rise of civilization in this part of the world.

The temperature dipped to a low about 3,000 years ago, before returning to a high during the Middle Ages, a time called the **medieval warm period.** During this time, when temperatures were 0.5 to 0.8° above those of today, it was possible for the Vikings to establish self-supporting agricultural settlements along the coast of Greenland (▶Fig. 23.14b). The temperature dropped again from 1500 to about 1800, a period known as the **little ice age,** when Alpine glaciers advanced and the canals of Holland froze over in winter (▶Fig. 23.14c, d; see Fig. 22.43). The climate, overall, has warmed since the end of the little ice age, though there are shorter-duration variations superimposed on this overall trend.

Geologists have focused on four factors to explain short-term climate change.

- *Fluctuations in solar radiation:* The amount of energy produced by the Sun varies with the **sunspot cycle.** This cycle refers to the appearance of large numbers of sunspots (black spots thought to be magnetic storms on the Sun's surface) about every 9 to 11.5 years (▶Fig. 23.15a, b). There may be longer-term cycles that have not yet been identified. This variation in energy may affect the climate.

- *Changes in Earth's orbit and tilt:* As Milankovitch first recognized in 1920, the change in the tilt of Earth's axis over a period of 41,000 years, the Earth's 25,000-year precession cycle, and changes in the eccentricity of its orbit over a period of 100,000 years together cause the amount of summer heat in high latitudes to vary, and cause the overall amount of heat reaching Earth to vary (see Chapter 22). These changes correlate with observed ups and downs in atmospheric and oceanic temperature.

- *Changes in the reflectivity (albedo) of Earth:* Not all of the sunlight that reaches the Earth penetrates its atmosphere and warms the ground. Some is reflected by the atmosphere or by the Earth's surface. The degree of reflectivity, or **albedo,** of the atmosphere increases if cloud cover increases, or if the concentration of aerosols in the atmosphere increases. Earth's albedo may also be affected by changes in the proportion of the surface covered by snow or ice, which reflect solar energy efficiently.

The short-term effect of volcanism on global temperature is abundantly clear. For example, the year following the 1815 eruption of Mt. Tambora in the western Pacific became known as the "year without a summer," for the dust that erupted encircled the Earth

(a)

(b)

FIGURE 23.15 (a) The appearance of sunspots correlates with increased energy production on the Sun. (b) There may be cycles in sunspot activity that could influence the climate.

and blocked the Sun—snow fell in Europe throughout the spring, and the entire summer was cold.

- *Changes in ocean currents:* Recent studies suggest that the configuration of currents can change quite quickly, and that this configuration affects the climate. The Younger Dryas may have resulted when a layer of freshwater from melting glaciers spread out over the North Atlantic and prevented thermohaline circulation in the ocean, thereby shutting off the Gulf Stream (see Chapter 18).

Catastrophic Climate Change and Mass-Extinction Events

The changes that happen on Earth almost instantaneously are called **catastrophic changes.** For example, a volcanic eruption, an earthquake, a tsunami, or a landslide can change a local landscape in seconds or minutes. But such

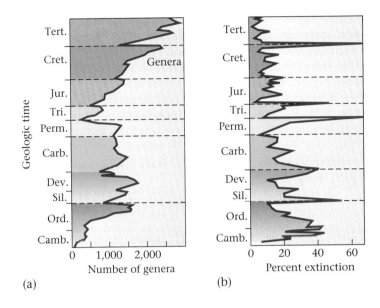

(a)

(b)

events affect only relatively small areas. Can such catastrophes happen on a global scale? In the past decades, geoscientists have come to the conclusion that the answer is yes. The stratigraphic record shows that Earth history includes several **mass-extinction events,** when large numbers of species abruptly vanished (▶Fig. 23.16a). Some of these define boundaries between geologic periods. A mass-extinction event decreases the **biodiversity** (the number of different species that exist at a given time) of life on Earth (▶Fig. 23.16b). It takes millions of years after a mass-extinction event for biodiversity to increase again—and significantly, the new species that appear are different from the ones that vanished, for evolution is unidirectional.

Geologists speculate that some mass-extinction events reflect a catastrophic change in the planet's climate, possibly brought about by unusually voluminous volcanic eruptions or the impact of a **bolide** (meteorite, comet, or asteroid) with the Earth (▶Fig. 23.16c). Either of these events could eject enough debris into the atmosphere to block sunlight. Without the warmth of the Sun, winter-like or night-like conditions would last for weeks to years, long enough to disrupt the food chain. Either event, in addition, could eject aerosols that would turn into global acid rain, scatter hot debris that would ignite forest fires, or give off chemicals that, when dissolved in the ocean, would make the ocean either toxic or so nutritious that an oxygen-consuming algae bloom could thrive.

As an example, consider the Permian-Triassic boundary. This boundary was defined back in the nineteenth century precisely because the assemblage of fossils below the boundary is radically different from that above; perhaps 90% of the species on the planet disappeared at the time of the boundary. Radiometric dating indicates that the boundary coincides with the eruption of vast amounts of flood basalt

(c)

FIGURE 23.16 (a) During a mass-extinction event, biodiversity on Earth, as indicated by the number of fossil species, suddenly decreases. Many of the important boundaries in the geologic column correlate with these events. (b) Paleontologists can calculate the percentage of species that went extinct during a given event. Notice that most existing species went extinct at the boundary between the Permian and Triassic periods. (c) The K-T (Cretaceous-Tertiary) boundary event may have been caused by the impact of a large bolide. The evidence is a spike in the iridium concentration of sediments. Here, we see an artist's rendition of the impact and its aftermath.

in Siberia. By the end of these eruptions, over 2.5 million square km of continental crust lay buried beneath basalt. The flood basalts are so extensive that geologists attribute their source to a **superplume,** a mantle plume many times larger than the plume that currently underlies Hawaii. Volcanic gases erupted at this time could have changed the climate. Another mass-extinction event, called the **K-T boundary event,** happened at the end of the Cretaceous period, 65 million years ago, when virtually all species of dinosaurs as well as innumerable species of marine life disappeared. As discussed in Chapter 13, this event may have been due to the impact of a 10-km-wide asteroid at the site of what is now Yucatán, in Mexico.

Both the volcanic eruption hypothesis and the impact hypothesis are controversial; some geologists have been convinced of their merit, but others remain skeptical. There may be additional ways of causing major, short-term climate change that could lead to mass extinction. Or perhaps mass extinction happens for reasons other than climate change. Some geologists suggest that mass-extinction events are periodic (about once every 100 million years, or even as frequently as once every 26 million years).

23.6 ANTHROPOGENIC CHANGES IN THE EARTH SYSTEM

When Stone Age artists decorated the caves of Lascaux, in France (►Fig. 23.17a), the human population worldwide was less than 10 million. By the dawn of civilization, 4000 B.C.E., it was still, at most, a few tens of millions. But by the

beginning in the nineteenth century, revolutions in industrial methods, agriculture, medicine, and hygiene had substantially lowered death rates and raised living standards, so that the population began to grow at accelerating rates—it took tens of thousands of years to reach 1 billion, in 1850, but it took only eighty years to double again, reaching 2 billion in 1930. The growth rate increased during the twentieth century, with the population reaching 4 billion in 1975. Now, the doubling time is only forty-four years, and the population passed the 6 billion mark just before the year 2000 (►Fig. 23.17b).

As the population grows and the standard of living improves, per capita usage of resources increases; we use land for agriculture and grazing, forests for wood, rock and dirt for construction, oil and coal for energy or plastics, and ores for metals. Without a doubt, our usage of resources has impacted the Earth system profoundly, and thus humanity has become a major agent of global change. Here, we examine some of these anthropogenic impacts.

The Modification of Landscapes

Every time we pick up a shovel and move a pile of dirt, we are redistributing a portion of the Earth's crust, an activity that before civilization was undertaken largely by rivers, the wind, rodents, and worms. In the last century, the pace of Earth movement has accelerated, for now we have drag lines in coal mines that can move 300 cubic m of coal in a single scoop, trucks that can carry 200 tons of ore in a single load, and tankers that can transport 50 million liters of oil during a single journey. The extraction of rock during mining, the building of levees and dams along rivers or of

FIGURE 23.17 (a) When these Stone Age paintings were drawn on cave walls of Lascaux in France, the world's human population was less than 10 million. (b) The population has increased dramatically during the past two centuries. Currently, it doubles every forty-four years.

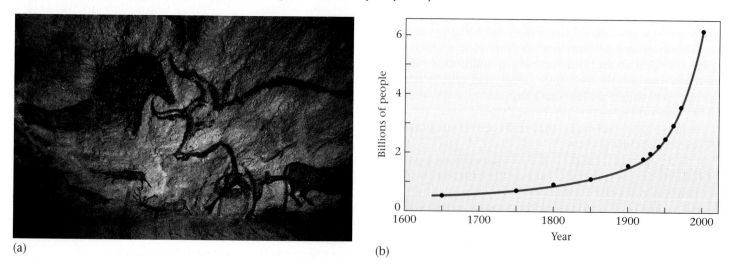

(a)

(b)

sea walls along the coast, and the construction of highways and cities all involve the redistribution of Earth materials (►Fig. 23.18a). In addition, people clear and plow fields, drain and fill wetlands, and pave over the land surface (►Fig. 23.18b). All these activities change the shape of the landscape.

FIGURE 23.18 (a) A giant, power-driven shovel can move more dirt and rock in a day than a stream can move in a decade. (b) Agriculture and urbanization radically change the landscape of a region.

(a)

(b)

Landscape modification has side effects, though. For example, it may make the ground unstable and susceptible to landslides. And it may expose the land to erosion, thereby changing the volume of sediment transported by natural agents (such as running water and wind). Locally, flood-control projects may diminish the sediment supply, also with unfortunate consequences; for example, the damming of the Nile by the Aswan High Dam has cut off the sediment supply to the Nile Delta, so ocean waves along the Mediterranean coast of the delta have begun to erode the coastline back by more than 1 m per year.

The Modification of Ecosystems

In undisturbed areas, the **ecosystem** of a region (its overall environment and inhabitants) is the product of evolution for an extended period of time. The ecosystem's flora (plant life) includes species that have adapted to living together in that particular climate and on the substrate available, while its fauna (animal life) is capable of surviving local climate conditions and utilizing local food supplies. Human-caused deforestation, overgrazing, agriculture, and urbanization disrupt ecosystems and lead to a decrease in biodiversity.

Archaeological studies have found that the earliest example of human modification of an ecosystem occurred in the Stone Age, when hunters played a major role in causing the mass extinction of many species of large mammals (mammoths, giant sloths, giant bears) that had populated the Earth since the last continental glaciers had receded. Today, less than 5% of Europe retains its original habitats (►Fig. 23.19a). The same numbers can be applied to the eastern United States, which lost its original forest and prairie (►Fig. 23.19b). Tropical rainforests cover less than about half the area worldwide that they covered before the dawn of civilization, and they are disappearing at a rate of about 1.8% per year (►Fig. 23.20a). Much of this loss comes from **slash-and-burn agriculture,** during which the forest is cut and the vegetation burned to make open land for farming (►Fig. 23.20b); but unfortunately, the heavy rainfall of tropical forests removes nutrients from the soil, making the soil useless in just a few years. Overgrazing by domesticated animals (the breeds of which only appeared on Earth during the past few millennia) can remove vegetation so completely that some grasslands have undergone desertification. And urbanization replaces the natural land surface with concrete or asphalt, a process that completely destroys an ecosystem.

Human-caused changes to the ecosystem affect the broader Earth system, because they modify biogeochemical cycles and Earth's albedo (surface reflectivity). For example, deforestation increases the CO_2 concentration in the atmosphere, as much of the carbon that was stored in trees is burned and rises. And the replacement of forest cover with concrete or fields increases Earth's albedo.

(a)

(b)

FIGURE 23.19 (a) The forests of England have now been replaced by tidy fields and pastures, or treeless grasslands. (b) The prairie of the midwestern United States has now been replaced by giant fields of corn and soybeans.

Pollution

The environment has always contained contaminants such as soot, dust, and the byproducts of organisms. But when human populations grew, urbanization, industrial and agricultural activity, the production of electricity, and modern modes of transportation greatly increased both the quantity and diversity of contaminants that entered the air, surface water, and groundwater. These contaminants, or **pollution,** including both natural and synthetic materials (in liquid, solid, or gaseous form), have become a problem because they are produced at a faster rate than can be absorbed or modified by the Earth system. For example, while small quantities of sewage can be absorbed by clay minerals in the soil or destroyed by bacterial metabolism, large quantities overwhelm natural controls and can accumulate into

(a)

(b)

FIGURE 23.20 (a) The area of the Earth covered by rainforest has shrunk steadily during the past century. On this map, the green areas refer to existing forest (including high-latitude scrub forest), while the brown areas refer to regions that were forested 8,000 years ago. Note how tropical rainforests are shrinking. (b) Much of the loss is due to slash-and-burn agriculture.

destructive concentrations. Further, because many of the contaminants are not produced in nature, they are not easily removed by natural processes. Pollution of the Earth system is a type of global change, because it represents a redistribution and reformulation of materials. Some key problems associated with this change include the following.

- *Smog:* The term was originally coined to refer to the dank, dark air that resulted when smoke from the burning of coal mixed with fog in London and other industrial cities. Another kind of smog, called **photochemical smog,** is the ozone-rich brown haze that blankets cities when exhaust from cars and trucks reacts in the presence of sunlight.

- *Water contamination:* We dump a great variety of chemicals into surface water and groundwater, including gasoline, other organic chemicals, radioactive waste, acids, fertilizers—the list could go on for pages. These chemicals change the biodiversity and concentration of life in a region.

- *Acid runoff:* Dissolving sulfide-containing ores or coal in groundwater or stream water makes the water acidic (it increases the concentration of hydrogen ions in the water) and toxic to life forms.

- *Acid rain:* When rain passes through air that contains sulfur-containing aerosols (emitted from power plants), the water dissolves the sulfur, creating sulfuric acid, or **acid rain.** Wind can carry aerosols far from a power plant, so acid rain can damage a broad region. Midwestern power plants, for example, cause acid rain to fall in New England (▶Fig. 23.21).

- *Radioactive materials:* Nuclear weapons, nuclear energy, and medical waste transfer radioactive materials from rock to Earth's surface environment. Also, human-caused nuclear reactions produce new, nonnatural radioactive isotopes, some of which have relatively short half-lives. Thus, society has changed the distribution and composition of radioactive material worldwide.

- *Ozone depletion:* Human-produced chemicals, most notably chlorofluorocarbons (CFCs), when emitted into the atmosphere react with ozone in the stratosphere. This reaction, which happens most rapidly on the surfaces of tiny ice crystals in polar stratopheric clouds, destroys ozone molecules, thus creating an **ozone hole** over high-latitude regions, particularly during the winter (▶Fig. 23.22a, b). (The ozone hole is much more prominent in the Antarctic than in the Arctic, because a current of air [the polar vortex] circulates around the land mass of Antarctica and traps the air above the continent, with its CFCs, from mixing with air from elsewhere). Ozone holes have dangerous consequences, for they affect the ability of the atmosphere to shield the Earth's surface from

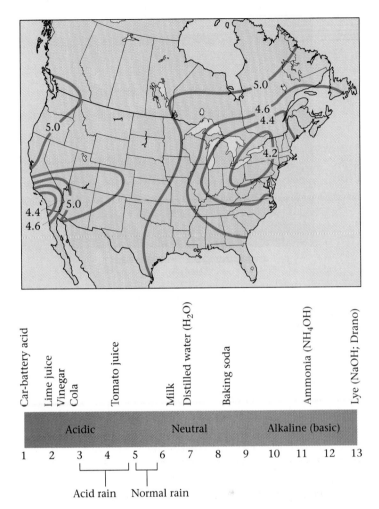

FIGURE 23.21 Acid rain has affected large portions of the United States. The map contours pH numbers, indicating the concentration of hydrogen ions in a solution (according to the formula $pH = -\log[H^+]$; the square brackets mean "concentration"). Note that very acidic rain falls in the Northeast and Southwest. The scale gives a sense of what the numbers mean. A solution with a pH of 7 is neutral; acidic solutions have a pH less than 7, while alkaline solutions have a pH greater than 7.

harmful ultraviolet radiation. This radiation can cause cancer. In 1987, a summit conference in Montreal proposed a global reduction of CFC emissions that destroy the ozone layer.

Effects on the Climate: The Global-Warming Issue

During the past two centuries, industry, energy production, and agriculture have significantly altered the rate at which greenhouse gases, such as carbon dioxide (CO_2) and methane (CH_4), have been added to the atmosphere. In fact, the rate of addition exceeds the rate at which these gases can be absorbed by dissolution in the ocean or by chemical-

and the flatulence of cows have released enough methane in recent years to measurably change the atmosphere (▶Fig. 23.23c).

We might expect this increase in greenhouse gases to cause global warming, and indeed global mean temperature rose by about 0.7° between 1880 and 1995 (▶Fig. 23.24a).

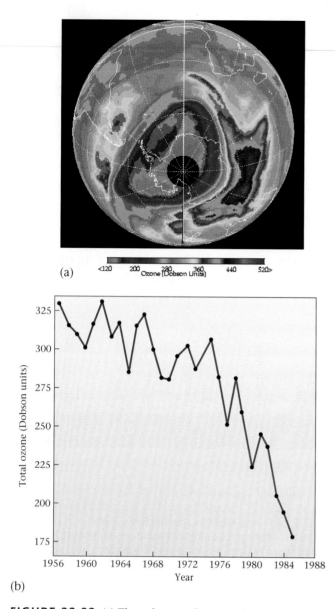

(a)

FIGURE 23.22 (a) The colors on this map show the quantity of ozone in the atmosphere. A significant ozone hole has formed over the Antarctic region, where the amount of ozone has decreased by as much as 50%. (b) The graph shows the decrease in ozone above Antarctica, specified in Dobson units (named for a scientist who helped identify the ozone hole). One Dobson unit is the amount of ozone needed to make a 0.01-mm-thick layer at the surface of the Earth.

FIGURE 23.23 (a) By studying bubbles in ice cores, the record of CO_2 in the atmosphere can be traced back in time. Note that there has been a significant increase since the start of the Industrial Revolution. (b) Direct measurements of the concentration of CO_2 in the atmosphere, taken since 1958, show that the concentration changes with the seasons and that the average concentration has increased steadily during the past several decades. The wavy lines indicate seasonal changes; the smooth curve is the average. (c) The concentration of methane (CH_4) has also increased steadily during the past few centuries. The dotted line is an extrapolation into the past.

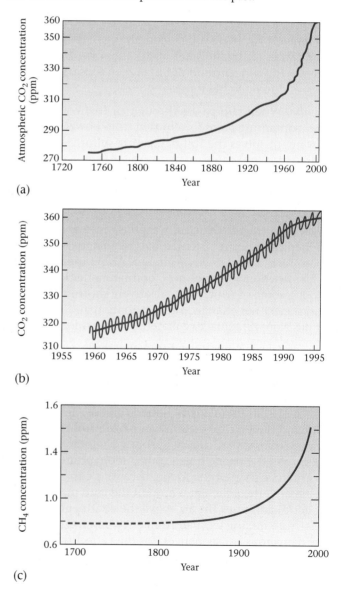

weathering reactions. For example, the carbon contained in oil and coal recombines with atmospheric oxygen during burning and becomes CO_2; by burning fossil fuels at the furious rate that we do, we transfer CO_2 from underground reservoirs (oil and coal deposits) back into the atmosphere. In 1800, the mean concentration of CO_2 in the atmosphere was 280 parts per million (ppm), in 1900 it was 295 ppm, and by 1995 it had reached 360 ppm (▶Fig. 23.23a, b). At the same time, the decay of organic material in rice paddies

(a)

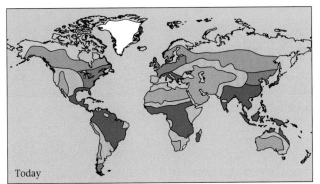

(b)

FIGURE 23.24 (a) The average global atmospheric temperature varies year by year, but overall, there has been a noticeable increase since about 1920. Here, the average global temperature is plotted as a function of time. (b) This map shows the results of a complicated computer model predicting the annual mean temperature change (°C) between 1795 and 2050 if the concentration of greenhouse gases continues to increase. Notice that some regions of the world will cool, but others will warm. The warming effect is most pronounced in the Northern Hemisphere.

This may not seem like much, but by comparison, the magnitude of change between the ice age and now was only a few degrees. In fact, the decade of the 1990s was the warmest decade of the century, and the winter of 1999–2000 was the warmest winter on record in the United States. Similarly, mean ocean temperature has been rising: in 2000, temperatures at the North Pole were the warmest in four centuries, and a 15-km-by-5-km patch of open water appeared at the pole. Not all researchers agree, however, that measurements of global warming in the last century are correct, and even those who accept the measurements don't all agree that the warming is a consequence of human-caused greenhouse gases. Some argue that it would have happened anyway, in the context of natural cycles of short-term climate change, and others point out that aerosols released by burning fossil fuels reflect sunlight and thus counter the effect of increased CO_2.

There is even more disagreement about whether global warming will continue in the future and what its consequences might be, because predictions are dependent on computer models, and not all researchers agree on how to construct these models. In the worst-case scenario, global warming will continue into the future at the present rate, so that by 2050—within the lifetime of many readers of this book—the average annual temperature will have increased in some parts of the world by 1.5° to 2.0°C (▶Fig. 23.24b). At these rates, by 2150 global temperatures may be 5° to 11° warmer than present—the warmest since the Eocene epoch, 40 million years ago. The effects of such a change are controversial, but according to some climate models, the following events might happen.

- *A shift in climate belts:* Temperate climates would move to higher latitudes, and vegetation belts would follow this trend. As a result, desert regions would expand, and the soil would dry out in the breadbasket (cereal-producing) regions (▶Fig. 23.25a, b).

FIGURE 23.25 The distribution of climate belts, indicated by vegetation type, will change if the global temperature warms. (a) Distribution of vegetation types today. (b) Distribution of vegetation types if the climate warms by just a couple of degrees.

(a)

(b)

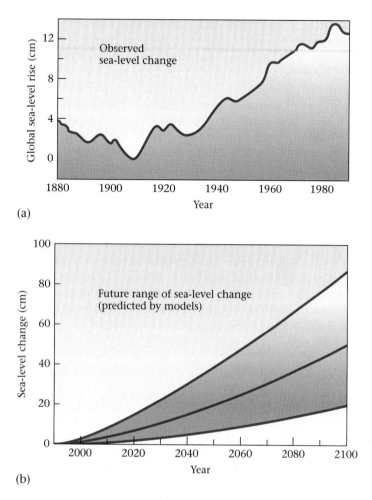

(a)

(b)

FIGURE 23.26 (a) Estimates of the average global sea level between 1880 and 1990 suggest that there has been a rise of about 12 cm. (b) All models of future sea-level change predict that there will be a rise. In the worst-case model, the sea level could rise by almost 1 m in the next century.

• *Stronger storms:* An increase in average ocean temperatures would mean that more of the ocean could evaporate when a tropical depression passed over. This evaporation would nourish stronger hurricanes. In nondesert areas, there might be more precipitation and therefore flooding.

• *A rise in sea level:* The melting of ice sheets in polar regions would make the sea level rise enough to flood coastal wetlands and communities and damage deltas. Measurements suggest that there has already been a rise of almost 12 cm in the past century (▶Fig. 23.26a). Even a change of only 1 m, which some models say may happen in little more than a century, could inundate regions of the world where 20% of the population currently lives (▶Fig. 23.26b). According to one model, the West Antarctic ice sheet could float off its substrate and break apart if the Earth warmed just a little more,

which would bring a quick change in sea level in the not-so-distant future.

• *An interruption of the oceanic heat conveyor:* Oceanic currents play a major role in transferring heat across latitudes. If global warming melts polar ice, the resulting freshwater would dilute surface ocean water at high latitudes. This water could not sink, and thus thermohaline circulation would be shut off (see Chapter 18), preventing it from conveying heat.

Models suggesting that unchecked combustion of fossil fuels will cause global warming imply that governments must work to require modifications in industrial procedures or life-styles. Because of the importance of this issue, the United Nations and the World Meteorological Association have established the Intergovernmental Panel on Climate Change (IPCC), which issues a report every five years. There have also been meetings designed to address issues of future climate change. Global warming was discussed at the 1992 Earth Summit held in Rio de Janeiro, and again at a 1997 summit in Kyoto, Japan, where 160 nations signed a treaty to slow the input of CO_2 into the atmosphere by reducing emissions to levels below those of 1990 (the treaty has not been ratified by the U.S. Senate).

23.7 THE FUTURE OF THE EARTH: A SCENARIO

Most of the discussion in this book has focused on the past, for the geology record preserved in rocks tells us of earlier times. Let's now bring this book to a close by facing in the opposite direction and speculating what the world might look like in geologic time to come.

In the geologic near term, the future of the world depends largely on human activities. Whether the Earth system undergoes a major disruption and shifts to a new equilibrium, whether a catastrophic mass-extinction event takes place, or whether society achieves **sustainable growth** (meaning an ability to prosper within the constraints of the Earth system) will depend on our own foresight and ingenuity. Projecting thousands of years into the future, we might well wonder if the Earth will return to ice-age conditions, with glaciers growing over major cities and the continental shelf becoming dry land, or if the ice age is over for good because of global warming. No one really knows for sure.

If we project millions of years into the future, it is clear the map of the planet will change significantly because of the continuing activity of plate tectonics. For example, during the next 50 million years or so, the Atlantic Ocean will probably become bigger, the Pacific Ocean will shrink, and the western part of California will migrate northward.

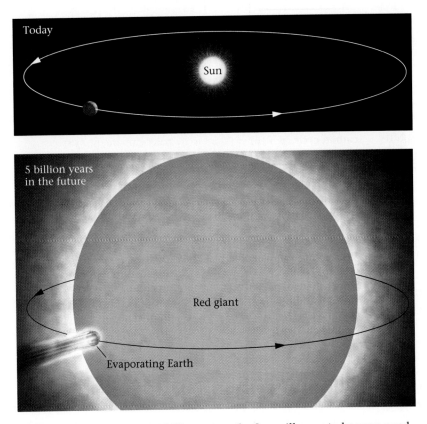

FIGURE 23.27 In about 5 billion years, the Sun will grow to become a red giant. When that happens, the Earth will first dry out, then vaporize. Initially, it may look like a giant comet. Perhaps some of its atoms will become part of a nebula that someday condenses to form a new Sun and planetary system.

future as they have in the past. We can't predict when the next strike will come, but unless the object can be diverted, Earth is in for another radical readjustment of surface conditions. In fact, one collision with a planet-sized bolide may fragment the Earth and transform it into another asteroid belt.

Astronomers predict that if such planet-destroying collisions are avoided, the end of the Earth will occur some 5 billion years from now, when the Sun begins to run out of nuclear fuel. As noted in Chapter 13, when this happens, the thermal pressure caused by fusion reactions will no longer be able to prevent the Sun from collapsing inward, because of the immense gravitational pull of its mass. Were the Sun a few times larger than it is, the collapse would trigger a supernova explosion that would blast matter of the Universe out into space to form a new nebula, perhaps surrounding a black hole. But since the Sun is not that large, the thermal energy generated when its interior collapses inward will heat the gases of its outer layers sufficiently to cause them to expand. As a result, the Sun will become a **red giant,** a huge star whose radius would grow beyond the orbit of Earth (▶Fig. 23.27). Our planet will then vaporize, and its atoms will join an expanding ring of gas—the ultimate global change. If this happens, the atoms that once made Earth and all its inhabitants through geologic time may eventually be incorporated in a future solar system, where the cycle of planetary formation and evolution will begin anew.

Eventually, Australia will crush against the southern margin of Asia, and the islands of Indonesia will be flattened in between.

Predicting the map of the Earth beyond that is hard, because we don't know for sure where subduction zones will develop. But most likely, a subduction zone will form on one side of the Atlantic Ocean, and the ocean will be consumed. As a consequence, the eastern margin of the Americas will collide with the western margin of Europe and Africa. The sites of major cities—New York, Miami, Rio de Janeiro, Buenos Aires, London—will be incorporated in a collisional mountain belt, and likely will be subjected to metamorphism and igneous intrusion before being uplifted and eroded. Shallow seas may once again cover the interiors of continents and then later retreat; it's happened in the past, so it could happen again. And if the past is the key to the future, we *Homo sapiens* might not be around to watch our cities enter the rock cycle, for biological evolution may have introduced new species to the biosphere—and there is no way to predict what these species will be like.

And what of the end of the Earth? Geologic catastrophes resulting from bolide collisions will undoubtedly occur in the

CHAPTER SUMMARY

• We refer to the global interconnecting web of physical and biological phenomena on Earth as the Earth system. Global change refers to the transformations or modifications of physical and biological components of the Earth system through time. Unidirectional change results in transformations that never repeat, while cyclic change involves repetition of the same steps over and over.

• Examples of unidirectional change include the gradual evolution of the solid Earth from a homogeneous collection of planetesimals to a layered planet on which lithosphere plates move; the formation of the oceans and the gradual change in the composition of the atmosphere; and the evolution of life.

• Examples of physical cycles that take place on Earth include the supercontinent cycle, the sea-level cycle, and the rock cycle.

• A biogeochemical cycle is the passage of a chemical among nonliving and living reservoirs. Global change occurs when something changes the relative proportions of the chemical in different reservoirs.

• During the hydrologic cycle, water cycles between the oceans, the atmosphere, glaciers, surface and groundwater, and living organisms. During the carbon cycle, carbon cycles between the oceans, the atmosphere, living organisms, fossil fuels, and limestone. CO_2 is a greenhouse gas, so changes in atmospheric concentration may affect the climate.

• Tools for documenting global climate change, the transformations or modifications of Earth's climate through time, include the stratigraphic record, paleontology, oxygen-isotope ratios, bubbles in ice, growth rings, and human history.

• Studies of long-term climate change (measured in millions of years) show that at times in the past the Earth experienced greenhouse (warmer) periods, while at other times there were ice-house (cooler) periods. Factors leading to long-term climate change include the positions of continents, volcanic activity, the uplift of land (which controls rates of chemical weathering), and the formation of materials (such as coal and limestone) that remove CO_2.

• Short-term climate change can be seen in the record of the last million years. In fact, during only the past 15,000 years, we see that the climate has warmed and cooled a few times. Causes of short-term climate change include fluctuations in solar radiation, changes in Earth's orbit and tilt, changes in reflectivity, and changes in ocean currents.

• Mass extinction, a catastrophic change in biodiversity, may be caused by a bolide impact or by intense volcanic activity associated with a superplume.

• During the last two centuries, humans have changed landscapes, destroyed ecosystems, and added pollutants to the land, air, and water at rates faster than the Earth system can process and neutralize.

• The addition of CO_2, and CH_4 to the atmosphere may be causing global warming, which could shift climate belts, cause stronger storms, lead to a rise in sea level, and shut off thermohaline oceanic currents.

• In the future, in addition to climate change, the Earth will witness a continued rearrangement of continents resulting from plate tectonics, and will likely suffer the impact of bolides. The end of the Earth may come when the Sun runs out of fuel in about 5 billion years and becomes a red giant star.

KEY TERMS

albedo (p. 724)
anthropogenic (p. 708)
biodiversity (p. 725)
biogeochemical cycle (p. 708)
biosphere (p. 710)
carbon cycle (p. 716)
catastrophic change (p. 708)
climate-change models (p. 718)
cyclic change (p. 708)
cyclothem (p. 711)
dendrochronologist (p. 719)
ecosystem (p. 727)
eustatic sea-level change (p. 712)
faint young Sun paradox (p. 722)
Gaia (p. 708)
global change (p. 708)
global warming (p. 717)
Goldilocks effect (p. 709)
gradual change (p. 708)

greenhouse period (p. 720)
hydrologic cycle (p. 713)
ice-house period (p. 720)
oxygen-isotope ratio (p. 718)
ozone hole (p. 729)
paleoclimate (p. 718)
palynologist (p. 718)
red giant (p. 733)
runaway greenhouse effect (p. 722)
slash-and-burn agriculture (p. 727)
steady-state condition (p. 713)
supercontinent cycle (p. 711)
superplume (p. 726)
sunspot cycle (p. 724)
unidirectional change (p. 708)

REVIEW QUESTIONS

1. Why do we use the term "Earth system" to describe the processes operating on this planet?

2. How have the Earth's crust and atmosphere changed since they first formed?

3. What processes control the rise and fall of sea level on Earth?

4. How does carbon cycle through the various Earth systems?

5. How do paleoclimatologists study ancient climate change?

6. Contrast ice-house and greenhouse conditions.

7. What are the possible causes of long-term climatic change?

8. What climatic changes may have been responsible for the birth and growth of the first civilizations in Europe and Asia?

9. What four factors explain short-term climatic change?

10. Give some examples of events that cause catastrophic change.

11. Give some examples of how humans have changed the Earth.

12. What is the ozone hole, and how does it affect us?

13. Describe how carbon-dioxide-induced global warming takes place, and how humans may be responsible. What effects might global warming have on the Earth's system?

14. What are some of the likely scenarios for the long-term future of the Earth?

SUGGESTED READING

Alvarez, W. 1997. *T. Rex and the Crater of Doom.* Princeton, N.J.: Princeton University Press.

Holland, H. D., and U. Petersen. 1995. *Living Dangerously: The Earth, Its Resources, and the Environment.* Princeton N.J.: Princeton University Press.

Kump, L. R., J. F. Kasting, and R. G. Crane. 1999. *The Earth System.* Upper Saddle River, N.J.: Prentice-Hall.

MacKenzie, F. T. 1998. *Our Changing Planet: An Introduction to Earth System Science and Global Environmental Change.* Upper Saddle River, N.J.: Prentice-Hall.

Officer, C., and J. Page. 1993. *Tales of the Earth: Paroxysms and Perturbations of the Blue Planet.* New York: Oxford University Press.

Silver, C. S., and R. S. DeFries. 1990. *One Earth, One Future: Our Changing Global Environment.* Washington D. C.: National Academy of Sciences Press.

Turco, R. P. 1997. *Earth Under Siege: From Air Pollution to Global Change.* Oxford, Eng.: Oxford University Press.

Van Andel, T. H. 1994. *New Views on an Old Planet: A History of Global Change.* Cambridge, Eng.: Cambridge University Press.

Wilson, E. O. 1993. *The Diversity of Life.* New York: W.W. Norton.

Metric Conversion Chart

Length

1 kilometer (km) = 0.6214 mile (mi)
1 meter (m) = 1.094 yards = 3.281 feet
1 centimeter (cm) = 0.3937 inch
1 millimeter (mm) = 0.0394 inch
1 mile (mi) = 1.609 kilometers (km)
1 yard = 0.9144 meter (m)
1 foot = 0.3048 meter (m)
1 inch = 2.54 centimeters (cm)

Area

1 square kilometer (km^2) = 0.386 square mile (mi^2)
1 square meter (m^2) = 1.196 square yards (yd^2)
= 10.764 square feet (ft^2)
1 square centimeter (cm^2) = 0.155 square inch (in^2)
1 square mile (mi^2) = 2.59 square kilometers (km^2)
1 square yard (yd^2) = 0.836 square meter (m^2)
1 square foot (ft^2) = 0.0929 square meter (m^2)
1 square inch (in^2) = 6.4516 square centimeters (cm^2)

Volume

1 cubic kilometer (km^3) = 0.24 cubic mile (mi^3)
1 cubic meter (m^3) = 264.2 gallons
= 35.314 cubic feet (ft^3)
1 liter (1) = 1.057 quarts
= 33.815 fluid ounces
1 cubic centimeter (cm^3) = 0.0610 cubic inch (in^3)
1 cubic mile (mi^3) = 4.168 cubic kilometers (km^3)
1 cubic yard (yd^3) = 0.7646 cubic meter (m^3)
1 cubic foot (ft^3) = 0.0283 cubic meter (m^3)
1 cubic inch (in^3) = 16.39 cubic centimeters (cm^3)

Mass

1 metric ton = 2,205 pounds
1 kilogram (kg) = 2.205 pounds
1 gram (g) = 0.03527 ounce
1 pound (lb) = 0.4536 kilogram (kg)
1 ounce (oz) = 28.35 grams (g)

Pressure

1 kilogram/square centimeter (kg/cm^2) = 0.96784 atmosphere (atm)
= 14.2233 pounds/square inch (lb/in^2)
= 0.098067 bar
1 bar = 0.98692 atmosphere (atm)
= 105 pascals (Pa)
= 1.02 kilograms/square centimeter (kg/cm^2)

Temperature

To change from Fahrenheit (F) to Celsius (C):
$$°C = \frac{(°F - 32°)}{1.8}$$

To change from Celsius (C) to Fahrenheit (F):
$$°F = (°C \times 1.8) + 32°$$

To change from Celsius (C) to Kelvin (K):
$$K = °C + 273.15$$

To change from Fahrenheit (F) to Kelvin (K):
$$K = \frac{(°F - 32°)}{1.8} + 273.15$$

Scientific Background: Matter and Energy

a.1 DISCOVERING THE NATURE OF MATTER

In order to understand the formation and evolution of the Universe, as well as descriptions of materials that comprise the Earth and processes that shape the Earth, we must first understand some basic facts about matter, energy, and heat (▶Fig. a.1). The following synopsis should be consulted whenever necessary.

Matter takes up space—you can feel it and see it. We use the word **matter** to refer to all the material making up the universe. The amount of matter in an object is its **mass.** An object with a greater mass than another object contains more matter—for example, a large tree contains more matter than a blade of grass. There's a subtle but important distinction between mass and weight. **Weight** depends on the amount of an object's mass but also on the strength of gravity. An astronaut has the same mass on both the Earth and the Moon, but weighs a different amount. Since the Moon has about one-sixth the mass of the Earth, an astronaut weighing 68 kilograms (150 pounds) on Earth would weigh only about 11 kilograms (25 pounds) on the Moon. Thus, lunar explorers have no trouble jumping great distances even when burdened by a space suit and oxygen tanks.

What is matter made of? Early philosophers deduced that the Earth and the plants and animals on it must form from simpler components, just as bread is made from a measured mixture of primary ingredients. They initially thought that the "primary ingredients" making up matter included only earth, air, fire, and water. Then a philosopher named Democritus (ca. 460–370 B.C.E.) argued that if you were able to keep dividing matter into progressively smaller pieces, you would eventually end up with nothing, and since it doesn't seem possible to make something out of nothing, there must be a smallest piece of matter that can't be subdivided further. He proposed the name **atom** for these smallest pieces, based on the Greek word *atomos*, which means "indivisible." Democritus also speculated, incorrectly, that different materials displayed different properties

FIGURE a.1 A lightning storm over mountains. The solid rock, the air, and the clouds consist of matter. The lightning, a rapidly moving current of electrons, is a manifestation of energy.

because of the shapes of the atoms composing them: for example, water consisted of smooth, spherical atoms, while fire was made of atoms with sharp edges.

Our modern classification of elements didn't take hold until the seventeenth century, when chemists like Robert Boyle (1627–1691) recognized that certain substances, like hydrogen, oxygen, carbon, and sulfur, cannot break down into other substances, while others, like water and salt, *can*

FIGURE a.2 A compound, salt, can be subdivided to form two elements, sodium metal and chlorine gas. Neither sodium nor chlorine can be divided further.

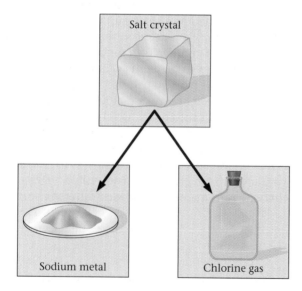

Salt crystal

Sodium metal

Chlorine gas

break down into other substances (▶Fig. a.2). For example, water breaks down into oxygen and hydrogen. Substances that can't be broken down came to be known as **elements,** while those that can be broken down came to be known as **compounds.** An English schoolteacher, John Dalton (1766–1844), adopted the word "atom" for the smallest piece of an element that maintains the property of the element, and suggested that compounds consisted of combinations of different atoms. Then in 1869, a Russian chemist named Dmitri Mendeléev (1834–1907) realized that groups of elements share similar characteristics. Mendeléev organized the elements into a chart we now call the **periodic table of the elements** (▶Fig. a.3). In the figure, elements within each column of the table behave similarly; for example, all elements in the right-hand column are **inert gases,** meaning that they can't combine with other elements to form compounds. But Mendeléev and his contemporaries didn't know what caused the similarities and differences. An understanding of the cause would have to wait until twentieth-century physicists discovered the internal structure of the atom.

a.2 A MODERN VIEW OF MATTER

Atoms

In modern terminology, an **element** is a substance composed only of atoms of the same kind. Ninety-two different elements occur naturally on Earth, but physicists have created more than a dozen new elements using "atom smashers," machines that fuse smaller atoms together to make larger ones. Each element has a name (for example, nitrogen, hydrogen, sulfur) and a **symbol,** an abbreviation of its English or Latin name (N = nitrogen, H = hydrogen, Fe = iron, Ag = silver).

As noted earlier, an atom is the smallest identifiable piece of an element. Work in the late 1800s and early 1900s demonstrated that, contrary to the view of Democritus, atoms actually can be divided. Ernest Rutherford, a British physicist, made this key discovery in 1910 when he shot a beam of atoms at a gold foil and found, to his amazement, that only a tiny fraction of the mass bounced back; most of the mass in the beam passed through the foil as if it were invisible. This result could mean only one thing. Most of the mass in an atom clusters in a dense ball at the atom's center, and this ball is surrounded by a cloud that contains very little mass. It is the surface of the whole atom that defines the top surface of this low-density cloud, so an atom as a whole consists mostly of empty space.

Physicists now refer to the dense ball at the center of the atom as the **nucleus,** and the low-density cloud sur-

FIGURE a.3 The modern periodic table of the elements. The columns group elements with related properties. For example, inert gases are listed in the column on the right. Metals are found in the central and left parts of the chart.

Legend:
- **Alkali metals**
- **Symbol** He 2 / **Name** Helium / **Atomic weight** 4.002 — with *Atomic number*
- **Inert gases** (right column)
- **Nonmetals**
- **Transition elements (metals)**

1	2	3	4	5	6	7	8	9	10	11	12	13	14	15	16	17	18
H 1 Hydrogen 1.007																	He 2 Helium 4.002
Li 3 Lithium 6.941	Be 4 Beryllium 9.0121											B 5 Boron 10.811	C 6 Carbon 12.011	N 7 Nitrogen 14.006	O 8 Oxygen 15.999	F 9 Fluorine 18.998	Ne 10 Neon 20.179
Na 11 Sodium 22.989	Mg 12 Magnesium 24.305											Al 13 Aluminum 26.981	Si 14 Silicon 28.085	P 15 Phosphorus 30.973	S 16 Sulfur 32.066	Cl 17 Chlorine 35.452	Ar 18 Argon 39.948
K 19 Potassium 39.098	Ca 20 Calcium 40.078	Sc 21 Scandium 44.955	Ti 22 Titanium 47.88	V 23 Vanadium 50.941	Cr 24 Chromium 51.996	Mn 25 Manganese 54.938	Fe 26 Iron 55.847	Co 27 Cobalt 58.933	Ni 28 Nickel 58.693	Cu 29 Copper 63.546	Zn 30 Zinc 65.39	Ga 31 Gallium 69.723	Ge 32 Germanium 72.61	As 33 Arsenic 74.921	Se 34 Selenium 78.96	Br 35 Bromine 79.904	Kr 36 Krypton 83.80
Rb 37 Rubidium 85.467	Sr 38 Strontium 87.62	Y 39 Yttrium 88.905	Zr 40 Zirconium 91.224	Nb 41 Niobium 92.906	Mo 42 Molybdenum 95.94	Tc 43 Technetium 98.907	Ru 44 Ruthenium 101.07	Rh 45 Rhodium 102.905	Pd 46 Palladium 106.12	Ag 47 Silver 107.868	Cd 48 Cadmium 112.411	In 49 Indium 114.82	Sn 50 Tin 118.710	Sb 51 Antimony 121.757	Te 52 Tellurium 127.60	I 53 Iodine 126.904	Xe 54 Xenon 131.29
Cs 55 Cesium 132.905	Ba 56 Barium 137.327	La 57 Lanthanum 138.905	Hf 72 Hafnium 178.49	Ta 73 Tantalum 180.947	W 74 Tungsten 183.85	Re 75 Rhenium 186.207	Os 76 Osmium 190.2	Ir 77 Iridium 192.22	Pt 78 Platinum 195.08	Au 79 Gold 196.966	Hg 80 Mercury 200.59	Tl 81 Thallium 204.383	Pb 82 Lead 207.2	Bi 83 Bismuth 208.980	Po 84 Polonium 208.982	At 85 Astatine 209.987	Rn 86 Radon 222.017
Fr 87 Francium 223.019	Ra 88 Radium 226.025	Ac 89 Actinium 227.027															

Lanthanides:

Ce 58 Cerium 140.115	Pr 59 Praseodymium 140.907	Nd 60 Neodymium 144.24	Pm 61 Promethium 144.912	Sm 62 Samarium 150.36	Eu 63 Europium 151.965	Gd 64 Gadolinium 157.25	Tb 65 Terbium 158.925	Dy 66 Dysprosium 162.50	Ho 67 Holmium 164.930	Er 68 Erbium 167.26	Tm 69 Thulium 168.934	Yb 70 Ytterbium 173.04	Lu 71 Lutetium 174.967

Actinides:

Th 90 Thorium 232.038	Pa 91 Protactinium 231.035	U 92 Uranium 238.028	Np 93 Neptunium 237.048	Pu 94 Plutonium 244.064	Am 95 Americium 243.061	Cm 96 Curium 247.070	Bk 97 Berkelium 247.070	Cf 98 Californium 251.079	Es 99 Einsteinium 252.083	Fm 100 Fermium 257.095	Md 101 Mendelevium 258.10	No 102 Nobelium 259.100	Lr 103 Lawrencium 262.11

rounding the nucleus as the **electron cloud.** Further study in the first half of the twentieth century led to the conclusion that the nucleus contains two types of **subatomic particles: neutrons,** which have a neutral electrical charge, and **protons,** which have a positive electrical charge (▶Fig. a.4a–d). The electron cloud consists of negatively charged particles, **electrons,** which are only about $\frac{1}{1,836}$ times as massive as protons and move around the nucleus at high speed. (For simplicity, think of a positive charge as the "+" end of a battery and a negative charge as the "−" end.) The mass of a neutron is approximately equal to the sum of the mass of a proton and the mass of an electron. It's as if a neutron consists of a proton and electron stitched together. In the past few decades, physicists have found that protons and neutrons are made up of a myriad of even smaller particles, the smallest of which is called a **quark.** Since quarks appear to be so much smaller than protons and neutrons, it seems that protons and neutrons are themselves mostly empty space.

Electron clouds have a complex internal structure. Each electron within the cloud moves within a specific range of distances from the nucleus. This range is variously called an **orbital, energy level,** or **shell.** Some shells have a spher-ical shape, whereas others resemble dumbbells, rings, or groups of balls—for simplicity, we portray shells as circles in cross section (▶Fig. a.5). Successively higher shells lie at progressively greater distances from the nucleus. Each shell can only contain a specific number of electrons at a given instant: the lowest shell can contain only two electrons, while the next several shells each can hold 8. Electrons fill the lowest shells first, so that atoms with a small number of electrons only have the innermost shells; the outer shells do not exist unless there are electrons to fill them.

The outermost shell of electrons, in effect, defines the outer edge of an atom, so an atom with many occupied electron shells is larger than one with few occupied shells (argon, for example, with 18 electrons, contains more occupied electron shells than helium, with 2 electrons, so an argon atom is bigger than a helium atom). If we picture the nucleus of a carbon atom as an orange, the electrons of the outermost shell would lie at a distance of over 1 km from the orange—again, an atom is mostly open space. But if so, you might say, then solids are also mostly empty space, so why doesn't this book sink through your desk, pass through your legs, and fall to the center of the Earth? The answer is that the electron cloud acts as a carapace, or shield, isolat-

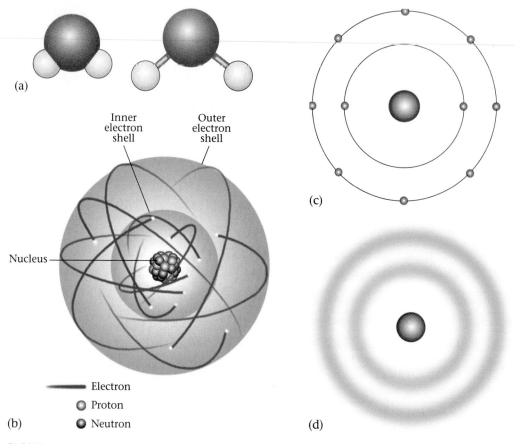

(a)

Inner electron shell

Outer electron shell

Nucleus

—— Electron

○ Proton

● Neutron

(b)

(c)

(d)

FIGURE a.4 (a) Two ways of portraying a water molecule. The large ball is oxygen, and the small ones are hydrogen. (b) An image of an atom with a nucleus surrounded by electrons. (c) This diagram shows the number of electrons in the inner shells. (d) An alternative depiction of electron shells, implying that the electrons comprise a cloud. In reality, electrons do not follow simple circular orbits.

FIGURE a.5 This schematic drawing of a neon atom shows the two complete electron shells. The inner shell contains two electrons, the outer one eight. The "shells" merely represent the most likely location for an election to be.

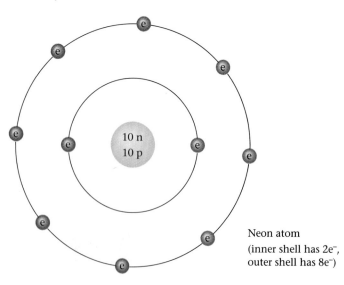

10 n
10 p

Neon atom
(inner shell has 2e⁻,
outer shell has 8e⁻)

ing the interior of an atom from its surroundings, so no two atoms can occupy the same space at the same time.

Atoms are so small that they can't be seen with even the strongest light microscopes. (However, by using some clever techniques, scientists have been able in recent years to create images of very large atoms; ▶Fig. a.6.) Atoms are so small, in fact, that 1 gram (0.036 ounces) of helium contains 6.02×10^{23} atoms; in other words, a quantity of helium weighing little more than a postage stamp contains 602,000,000,000,000,000,000,000 atoms! The number of atoms of helium in a small balloon is approximately the same as the number of balloons it would take to replace the whole of Earth's atmosphere.

Atomic Number, Atomic Mass, and Isotopes

We distinguish atoms of different elements from one another by their **atomic number,** the number of protons in their nucleus. For example, hydrogen's atomic number is 1, oxygen's is 8, lead's is 82, and uranium's is 92. We write the atomic number as a subscript to the left of the element's

FIGURE a.6 An image of atoms taken with a special microscope that uses electrons that illuminate the surface.

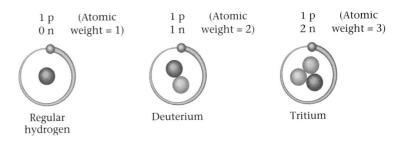

FIGURE a.7 The three isotopes of hydrogen: regular hydrogen, deuterium, and tritium.

Ions: Atoms with a Charge

If the number of electrons (negatively charged particles) exactly equals the number of protons (positively charged particles) in an atom, then the atom is **electrically neutral.** But if the numbers aren't equal, then the atom has a net electrical charge and is called an **ion.** For example, if an atom has two fewer electrons than protons, then we say it has a charge of +2, and if it has two more electrons than protons, it has a charger of –2. We write the charge as a superscript to the right of the symbol (for example, Na^+). Ions

FIGURE a.8 Common ions making up materials in rocks at the Earth's surface.

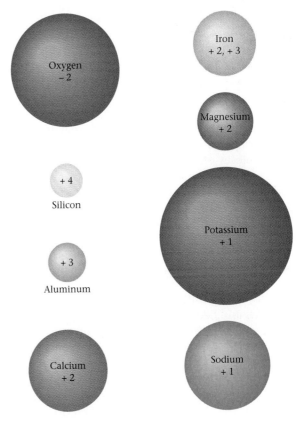

symbol (for example $_1H$, $_8O$, $_{82}Pb$, $_{92}U$). With the exception of most hydrogen nuclei, all atomic nuclei also contain neutrons. In smaller atoms, the number of neutrons equals the number of protons, but in larger atoms, the number of neutrons exceeds the number of protons. Subatomic particles are held together in a nucleus by **nuclear bonds.**

Atomic weight (or **atomic mass**) defines the amount of matter in a single atom. For a given element, the atomic weight *approximately* equals the number of protons plus the number of neutrons. (Precise atomic weights are actually slightly greater than this sum, because neutrons have slightly more mass than protons, and because the electrons have mass.) For example, helium contains 2 protons and 2 neutrons, and thus has an atomic weight of about 4, while oxygen, which contains 8 protons and 8 neutrons, has an atomic weight of about 16. We indicate the weight of an atom by a superscript to the left of the element's symbol (^4He, ^{16}O).

Some elements occur in more than one form, and these differ in atomic weight. For example, uranium 235 ($^{235}_{92}U$) contains 92 protons and 143 neutrons, while uranium 238 ($^{238}_{92}U$) contains 92 protons and 146 neutrons. Note that both forms of uranium have the *same* atomic number—they must, if they are to be considered the same element; they differ only in atomic weight. Multiple versions of the same element, which differ only in atomic weight (the number of neutrons), are called **isotopes** of the element (▶Fig. a.7).

with a negative charge are **anions** (pronounced ANN-eye-ons), and ions with a positive charge are **cations** (pronounced CAT-eye-ons). Ions form because atoms "prefer" to have a complete outer electron shell. Thus, an atom may give up or take on electrons in order for its outer shell to contain the proper number of electrons. As is the case with neutral atoms, the size of an ion depends on the number of shells containing electrons. Note that oxygen ions are larger than silicon ions, even though silicon has a larger atomic number, because oxygen that has gained electrons uses more electron shells than silicon that has lost electrons (▶Fig. a.8).

Molecules and Chemical Bonds

Most of the materials we deal with in everyday life—oxygen, water, plastic—are not composed of isolated atoms. Rather, most atoms except for the inert gases tend to stick, or **bond,** to other atoms; two or more atoms stuck together comprise a **molecule.** Hydrogen gas, for example, consists of H_2 molecules; note that a hydrogen molecule consists of two of the same kind of atom. But many materials contain different kinds of atoms bonded together. Such materials are called **compounds.** For example, common table salt is a compound with sodium and chlorine atoms bonded together. Compounds generally differ markedly from the elements that make them up—salt bears no resemblance at all to pure sodium (a shiny metal) or pure chlorine (a noxious gas) (Fig. a.2). A molecule is the smallest identifiable piece of a compound, containing the correct proportion of the compound's elements. For example, a molecule of water consists of two atoms of hydrogen and one atom of oxygen. We represent water by the **chemical formula** H_2O, a concise recipe that indicates the relative proportions of different elements in the molecule (Fig. a.4a).

Chemical bonds act as the "glue" that holds atoms together to form molecules and holds molecules together to form larger pieces of a material. Chemical bonding results from the interaction among the electrons of nearby atoms, and can take place in four different ways. (Note that chemical bonds are not the same as the nuclear bonds that hold together protons and neutrons in a nucleus.)

• *Ionic bonds:* As an inviolate rule of nature, "like" electrical charges repel (two positive charges push each other away), while "unlike" electrical charges attract (a negative charge sticks to a positive charge). Thus, if one atom gives up electrons and becomes a cation, it attracts an atom that has gained electrons to become an anion, because positive charges attract negative charges. Bonds that form in this way are called **ionic bonds** (▶Fig. a.9). For example, in a molecule of salt, positively charged sodium ions (Na⁺) attract negatively charged chloride (Cl⁻) ions ("chloride" is the name given to ions of chlorine). Note

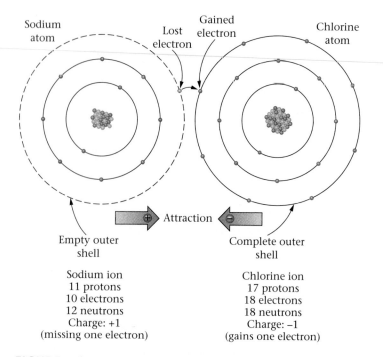

Sodium atom — Lost electron — Gained electron — Chlorine atom

Empty outer shell — Attraction — Complete outer shell

Sodium ion
11 protons
10 electrons
12 neutrons
Charge: +1
(missing one electron)

Chlorine ion
17 protons
18 electrons
18 neutrons
Charge: –1
(gains one electron)

FIGURE a.9 The sodium atom has an unfilled outer shell—the shell has room for eight electrons but only has one. The chlorine atom also has an unfilled outer shell—it's missing one electron. The sodium atom gives up its outer electron, and thus has one more proton than electron (that is, a net positive charge), while the chlorine gains an electron and thus has one more electron than proton (a net negative charge). The two ions attract each other.

that a visible grain of salt from a salt shaker is not one molecule of salt but rather consists of a large number of sodium and chloride ions bonded to each other in equal proportions. The chemical formula for salt, NaCl, reflects the proportion of sodium to chlorine.

• *Covalent bonds:* The atoms of carbon making up a diamond do not transfer electrons to one another, but rather share electrons. Bonding that involves the sharing of electrons is called **covalent bonding** (▶Fig. a.10). Because of the sharing, the electron shells of all the carbon atoms in a diamond are complete, and all the carbon atoms have a neutral charge. Water molecules also exist because of covalent bonding: in a water molecule, two hydrogen atoms are covalently bonded to one oxygen atom. By sharing electrons, the outer shell of the oxygen atom can hold the requisite 8 electrons and the outer shell of the hydrogen atoms can contain the requisite 2 electrons, to make the individual atoms neutral.

• *Metallic bonds:* Atoms in metals pack together because they share electrons in the inner shells, but electrons of the outer shells move easily from atom to atom. We call

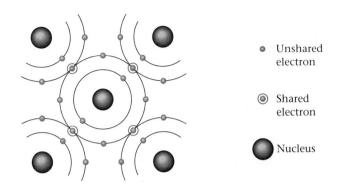

FIGURE a.10 Covalent bonding. Carbon only has four electrons in its outer shell, which has a capacity of eight. Thus, the carbon shares electrons with four other carbon atoms—it is covalently bonded to the other atoms.

• Unshared electron

◉ Shared electron

● Nucleus

Bonds resulting from the polarity of molecules are important in many geologic materials. For example, liquid water consists of molecules in which two hydrogen atoms covalently bond to one oxygen atom. The hydrogen atoms both lie on the same side of the oxygen atom, and the oxygen nucleus, which contains 8 protons, exerts more attraction to the shared electrons than do the hydrogen nuclei, each of which contains only one proton. The shared electrons concentrate closer to the oxygen side of the molecule, making the oxygen side more negative than the hydrogen side. Thus, the water molecules are polar, and they attract each other. This attraction is called a **hydrogen bond** (▶Fig. a.12). The polarity of water molecules makes water a good solvent (it can dissolve substances), because the polar water molecules attract and surround ions of soluble materials and pull them apart.

this type of bonding **metallic bonding** (▶Fig. a.11). Because outer-shell electrons move so freely, metals conduct electricity easily—when you connect a metal wire to an electrical circuit, a current of electrons flows through the metal.

• *Bonds resulting from the polarity of atoms or molecules:* Chemists long recognized that some materials break or split so easily that they must be held together by particularly weak chemical bonds. Eventually, they realized that these bonds are due to the permanent or temporary polarity of atoms or molecules. **Polarity** means that the atom or molecule has a positive charge on one side and a negative charge on the other.

FIGURE a.11 In a metallically bonded material, nuclei and their inner shells of electrons float in a "sea" of free electrons. Sometimes the electrons orbit the nuclei, but at other times they stream through the metal.

FIGURE a.12 Water is a polar molecule, because the two hydrogen atoms lie on the same side of the oxygen atom. Thus, water molecules tend to attract one another (this attraction creates surface tension, and causes water to form drops). Similarly, the polar molecules surround chlorine and sodium ions when salt dissolves in water.

A polar water molecule

Hydrogen bonding in water (→▶◀— = hydrogen bond)

Water molecules surrounding and bonding to chlorine and sodium ions in dissolved salt

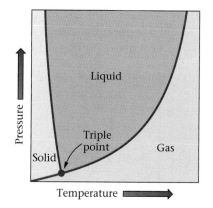

FIGURE a.13 The states of matter. (a) A solid block sitting in a bottle retains its shape regardless of the size of the container. Atoms within the block have an orderly arrangement. (b) A liquid conforms to the shape of the container. If the container changes shape, the liquid also changes shape, so that it stays constant in density (mass per unit volume). Note that it is possible for a liquid to fill only part of the container. In the liquid, there may be clumps of atoms or molecules, but no long-range order. (c) A gas fills the whole container, and the atoms or molecules comprising it have no order at all. A gas will expand to fill whatever volume it occupies, and thus can change density if the volume changes. (d) The Sun contains plasma, a gas-like material that is so hot that electrons have been stripped from the nuclei.

Johannes van der Waals (1837–1923), a Dutch physicist, discovered another type of weak chemical bonding that depends on polarity. This type, now known as **van der Waals bonding,** links one covalently bonded molecule to another. The bonds exist because electrons temporarily cluster on one side of each molecule, giving it a polarity and making it attract molecules with the opposite polarity.

The Forms of Matter

Matter exists in one of four states: **solids,** which can maintain their shape for a long time without the restraint of a container; **liquids,** which flow fairly quickly and assume the shape of the container they fill (it's possible for a liquid to fill only part of its container); **gases,** which expand to fill the entire container they have been placed in and will disperse in all directions when not restrained; and **plasma,** an unfamiliar, gas-like mixture of positive ions and free electrons that exists only at very high temperatures (►Fig. a.13a–d). Which state of matter exists at a location depends on the pressure and temperature (►Fig. a.14).

To understand the differences at an atomic scale between solids, liquids, and gases—the common forms of matter—imagine a large group of students getting ready to take a 9:00 A.M. exam in a windowless auditorium. The students represent atoms. It's 8 o'clock, and some students are in their rooms, some are in the library, some are walking to the auditorium, and some are at the cafeteria. In other words, everyone is scattered all over the place; there is no order to the distribution of people, and they are too far apart to interact and bond to one another. Such a distribution is characteristic of a gas.

Ten minutes before nine, most students have arrived at the auditorium. Some are already in their seats, but many are walking in or out of the room. Such a distribution characterizes the state of a liquid, which in most of the particles are in the container, except for a slow exchange with the atmosphere (by evaporation or condensation). There is short-range order, represented by little clumps of students in their seats or chatting in the aisles, but overall there is still a lot of motion and no overall organization, or long-range order.

Suddenly the lights go off, when someone bumps against the switch, and everyone stops in place to avoid stumbling. Even though motion has stopped, there is still no long-range order in the room. Such a situation character-

FIGURE a.14 A material such as water can "change state" (from gas to liquid, liquid to solid, or solid to liquid), depending on the pressure and temperature. Water can exist simultaneously in all three states at a unique pressure and temperature called the triple point.

izes a special kind of solid called a **glass.** Glass has a disordered arrangement of atoms, like a liquid, but cannot flow and thus behaves like solid.

The lights go on again, and the examiner finally enters. Everyone takes a seat according to a specified plan—alternate rows and alternate seats. Even though individual students are not motionless, they are no longer moving out of an orderly arrangement. The situation now corresponds to that in a solid.

a.3 THE FORCES OF NATURE

In our everyday experience, we constantly see or feel the effect of force. Forces can squash objects, speed them up and slow them down, tear, stretch, spin, and twist them, and make them float or sink. Isaac Newton, the great British scientist who effectively founded the field of physics, was the first to describe the way forces work. He defined a **force** as simply the push or pull that causes the velocity (speed) of a mass to change in magnitude and/or direction.

We can distinguish between two categories of force. The first, which includes force applied by the movement of a mass (a hand, a hammer, the wind, waves), is called **mechanical force,** or **contact force.** When you push a block across the floor, you are applying a mechanical force. The second category, which includes force resulting from the action of an invisible agent, is called a **field force,** or **noncontact force.** If you drop a book, the invisible force of gravity pulls the book toward the floor.

Physicists further recognize four types of field forces: gravity, electromagnetic, strong nuclear, and weak nuclear. **Gravity** is a force of attraction between any two masses, and can act over large distances. The magnitude of gravitational attraction depends on the size of the masses and the distance between them. We feel a much stronger gravitational pull to the planet we walk on than we do to a baseball. **Electromagnetic force,** the force associated with electricity and magnetism, is stronger than gravity, but only operates between materials that have electrical charges or are magnetic, and only operates over short distances. Like gravity, electromagnetic force depends on the distance between objects, but unlike gravity it can be either attractive or repulsive—as mentioned previously, like charges repel and unlike charges attract. A **strong nuclear force** holds together the protons and neutrons of an atomic nucleus, while a **weak nuclear force** affects the even smaller particles making up protons and neutrons. Nuclear forces act only at tiny distances within atoms.

a.4 ENERGY

We frequently use the word "energy" in a variety of contexts. For example, since the oil embargo of the early 1970s, people have frequently talked about "conserving energy," and when you finish a 10-km race, you say, "I've run out of energy!" To a physicist, **energy** has a more specific meaning: it refers to the ability to do work, or, in other words, the ability to apply a force that moves a mass by some distance (**work** = force × distance). According to this definition, gasoline serves as an energy source because we can burn gasoline to move heavy cars and trucks (a type of work), and after you run out of energy in a race, you no longer have the ability to do work.

Kinetic and Potential Energy

Formally, we can classify energy into two basic types: **kinetic energy,** the energy of motion, and **potential energy,** the energy stored in a material. A boulder sitting at the top of a hill has potential energy, because gravity could make it topple down. If the boulder does start falling down the hill, its potential energy converts to kinetic energy (▶Fig. a.15). This kinetic energy can do work if the boulder strikes another one and sets it rolling. Similarly, the gas in a car's tank has potential energy, which converts to kinetic energy when it burns and starts the cylinders in the engine moving up and down. Now we briefly look at where energy comes from.

Energy from Chemical Reactions

A **chemical reaction** refers to the process whereby one or more compounds (the **reactants**) come together or break apart to form new compounds (the **products**). Many

FIGURE a.15 A boulder sitting on a ledge in a gravitational field has potential energy. When the boulder starts to roll, this potential energy converts to kinetic energy.

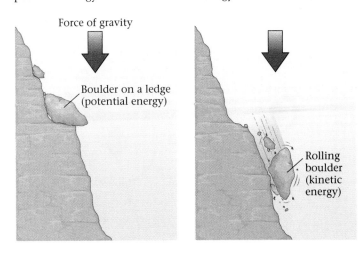

Force of gravity

Boulder on a ledge (potential energy)

Rolling boulder (kinetic energy)

chemical reactions produce energy. In the case of burning gasoline, gasoline molecules react with the oxygen molecules in the air to form carbon dioxide molecules and water molecules plus energy. We can describe this reaction using a **chemical formula,** a kind of shorthand that uses symbols for the molecules:

$$2C_8H_{18} + 25O_2 \rightarrow 16CO_2 + 18H_2O + \text{heat energy.}$$

Note that **reactants** appear on the left side of the formula (C_8H_{18} is a typical compound in gasoline), and **products** on the right side. The arrow indicates the direction in which the reaction proceeds. The energy comes from the breaking of chemical bonds—more energy is stored in the chemical bonds of gasoline and oxygen than in the chemical bonds of carbon dioxide and water.

Energy from Field Forces

When an object is placed in a field force, the force acts on the object, and the object, unless it is restrained, will move. A restrained object in a field force has potential energy, while a moving object in a field force has kinetic energy. As described earlier, a boulder resting on top a hill has potential energy (it is restrained by the ground), but when the boulder starts rolling, it has kinetic energy. Similarly, an iron bar taped to a table near a magnet has potential energy, but a bar that is moving toward the magnet has kinetic energy.

Fission and Fusion: Energy from Breaking or Making Atoms

Nuclear forces come into play in the production of nuclear energy. Two processes, which physicists first understood only in the twentieth century, produce nuclear energy.

During **nuclear fission,** a large atomic nucleus splits into two or more pieces (▶Fig. a.16). Such splitting involves one of the following processes: (1) the ejection of an electron from one of the neutrons in the nucleus, transforming the neutron into a proton; (2) the ejection of a couple of protons or neutrons from the nucleus; or (3) the separation of the nucleus into smaller atoms. Each of these processes creates new atoms with different atomic numbers from that of the original atom. In other words, one element is transformed into another. In effect, the ultimate goal of medieval alchemists, to turn lead into gold, can now be achieved in modern atom smashers, though at a significant cost.

When fission occurs, some of the matter comprising the atom transforms into a huge amount of energy (heat and electromagnetic radiation), as defined by Einstein's famous equation: $E = mc^2$, where E is energy, m is mass, and c is the speed of light. The realization that fission releases vast

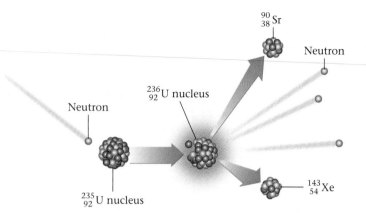

FIGURE a.16 A uranium atom splits during nuclear fission.

quantities of energy led to the rush to build atomic weapons during and after World War II. Nuclear fission provides the energy in atomic bombs, nuclear power plants, and nuclear submarines.

Nuclear fusion results when two nuclei slam together at such high velocity that they get close enough for the nuclear forces to bind them together, thereby creating a new and larger atom of a different element (▶Fig. a.17). In

FIGURE a.17 (a) In the Sun, four hydrogen atoms fuse to form a helium atom. (b) In a hydrogen bomb, a deuterium and a tritium atom fuse to form a helium atom plus a free neutron.

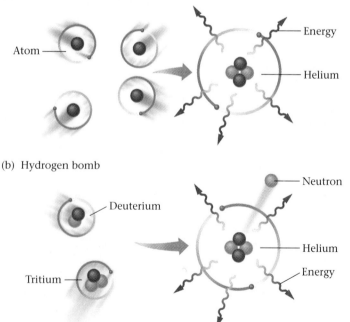

this regard, fusion is the opposite of fission. Fusion reactions also produce huge amounts of energy, because during fusion some of the matter converts into energy, according to Einstein's equation. Fusion reactions generate the heat in stars; in the Sun, for example, hydrogen atoms fuse together to form helium. Fusion reactions also generate the explosive energy of a hydrogen bomb (a "thermonuclear" device). Such reactions can only take place at extremely high temperatures (over 1,000,000°C). In fact, in order to get a hydrogen bomb to blow up, you need to use an atomic bomb as the trigger, because only an atomic explosion can generate high enough temperatures to start fusion. So far, no one has figured out how to economically sustain a controlled fusion reaction on Earth for the purpose of generating electricity.

a.5 HOT AND COLD

Thermal Energy and Temperature

The atoms and molecules that make up an object do not stay rigidly fixed in place, but rather jiggle and jostle with respect to one another. This vibration creates **thermal energy**—the faster the atoms move, the greater the thermal energy and the hotter the object. In simple terms, the thermal energy in a substance represents the sum of the kinetic energy of all the substance's atoms (this includes the back-and-forth displacements that an atom makes as it vibrates as well as the movement of an atom from one place to another).

When we say that one object is hotter or colder than another, we are describing its temperature. It represents the average kinetic energy of atoms in the material. **Temperature** is a measure of warmth relative to some standard. In everyday life, the standard we generally use is the freezing or boiling point of water at sea level. When using the **Celsius (centigrade) scale,** we arbitrarily set the freezing point of water (at sea level) as 0°C and the boiling point as 100°C; whereas in the **Fahrenheit scale,** we set the freezing point as 32°F and the boiling point as 212°F.

The coldest a substance can be is the temperature at which its atoms or molecules stand still. We call this temperature **absolute zero,** or 0K, where **K** stands for **Kelvin** (after Lord Kelvin [1824–1907], a British physicist), another unit of temperature. You simply can't get colder than absolute zero, meaning that you can't extract any thermal energy from a substance at 0K (−273.15°C). Degrees in the Kelvin scale are the same increment as degrees in the Celsius scale. On the Kelvin scale, ice melts at 273.15K and water boils at 373.15K. ▶Figure a.18 shows the range of temperatures we observe in the Universe today, as specified by the Kelvin scale.

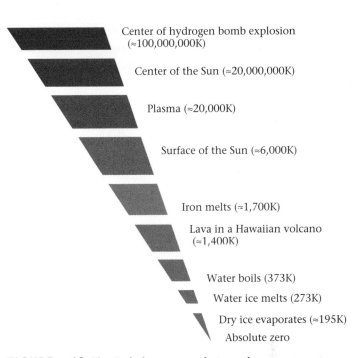

FIGURE a.18 Physical phenomena that we observe occur at a wide range of temperatures (here, as specified by the Kelvin scale).

Heat and Heat Transfer

"Heat" refers to the thermal energy transferred from one object to another. The unit of heat is the **calorie.** One thousand calories can heat 1 kilogram of water by 1°C. There are three ways in which heat transfer takes place in the Earth system: conduction, convection, and advection.

Conduction takes place when you stick the end of an iron bar in a fire (▶Fig. a.19a). The iron atoms at the fire-licked end of the bar start to vibrate more energetically; they gradually incite atoms farther from the flame to start jiggling, and these atoms in turn set atoms even farther along in motion. In this way, heat slowly flows down the bar until you feel it with your hand. Note that even though the iron atoms vibrate more as the bar heats up, the atoms remain locked in their position within the solid. Thus, **conduction** is the transfer of heat without the actual movement of atoms from one place to another. If you place two bars of different temperatures in contact with each other, heat conducts from the hot bar into the cold one until both end up at the same temperature.

Convection takes place when you set a pot of water on a stove (▶Fig. a.19b). The heat from the stove warms the water at the base of the pot (it makes the molecules of water vibrate faster and move around more). As a consequence, the density of the water at the base of the pot decreases, for as you heat a liquid, the atoms move away

FIGURE a.19 The three processes of heat transfer. (a) Conduction occurs when you heat the end of an iron bar in a flame. Heat flows from the hot region toward the cold region, as vibrating atoms cause their neighbors to vibrate. (b) Convection takes place when moving fluid carries heat with it. Hot water at the bottom of the pot rises, while cool water sinks, setting up a convective cell. (c) During advection, a hot liquid (such as molten rock) rises into cooler material. Heat conducts from the hot liquid into the cooler material.

from each other and the liquid expands. For a time, cold water remains at the top of the pot, but eventually the warm, less dense water becomes buoyant relative to the cold, dense water. In a gravitational field, a buoyant material rises (like a styrofoam ball in a pool of water) if the material above it is weak enough to flow out of the way. Since liquid water can easily flow, the hot water rises. When this happens, the cold water sinks to take its place. The new volume of cold water heats up and then rises itself. Thus, during **convection,** heat is carried by the actual flow of the material itself. The trajectory of flow defines **convective cells;** in a convection cell, water follows a loop, with warm water rising and cold water sinking. Convection occurs in the atmosphere, the ocean, and the interior of the Earth.

Advection, a less familiar process, happens when heat is carried with a moving fluid that is flowing through cracks and pores into a solid material. The heat brought by the fluid conductively heats up the material the fluid passes through. Advection takes place, for example, if you pass hot water through a sponge and the sponge itself gets hot. In the Earth, advection occurs where molten rock rises through the crust beneath a volcano and heats up the crust in the process.

Flow Charts for Identifying Minerals

Geologists use a **mineral-identification flow chart** to identify unknown mineral specimens (▶Fig. b.1a, b). A flow chart is simply a series of questions organized in such a way that by answering them, you quickly identify the mineral. The concept is simple. Start by observing the specimen's luster. Let's say the mineral has a metallic luster, so you follow the path on the chart for metallic-luster minerals (Fig. b.1a). The first thing you have to decide is whether the specimen is magnetic or nonmagnetic. If it is nonmagnetic, you follow the path for nonmagnetic minerals, where you then must identify the mineral as silvery or bronze. Again, depending on your answer, you proceed along the flow chart.

Notice that one of the flow-chart questions in Figure b.1b is whether the specimen reacts with hydrochloric acid (HCl) or not. Only calcite and dolomite do (calcite reacts with acid to form carbon dioxide gas), so the question allows a definitive identification of these minerals. Another question pertains to striations, parallel lines on cleavage planes created by the arrangement of atoms in the crystal lattice. Only plagioclase has striations.

Mineral-Identification Flow Chart (metallic or dark-colored nonmetallic)

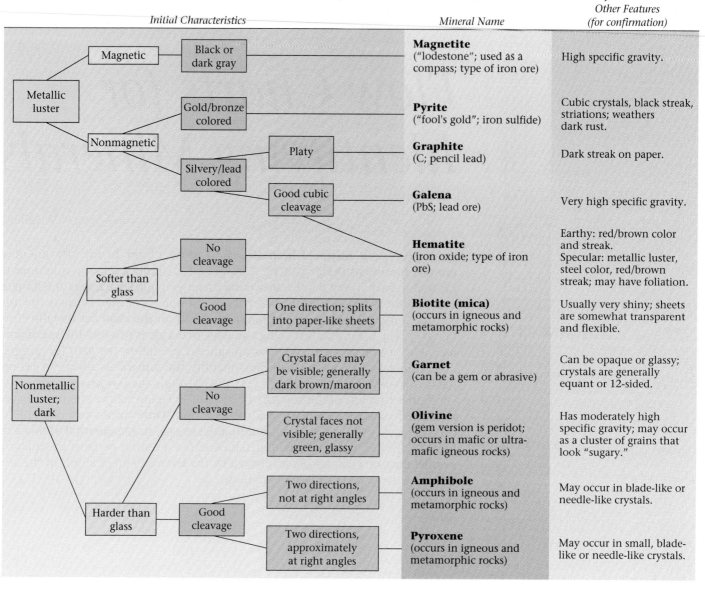

Initial Characteristics	Mineral Name	Other Features (for confirmation)
Magnetic → Black or dark gray	**Magnetite** ("lodestone"; used as a compass; type of iron ore)	High specific gravity.
Metallic luster → Nonmagnetic → Gold/bronze colored	**Pyrite** ("fool's gold"; iron sulfide)	Cubic crystals, black streak, striations; weathers dark rust.
Silvery/lead colored → Platy	**Graphite** (C; pencil lead)	Dark streak on paper.
Silvery/lead colored → Good cubic cleavage	**Galena** (PbS; lead ore)	Very high specific gravity.
Softer than glass → No cleavage	**Hematite** (iron oxide; type of iron ore)	Earthy: red/brown color and streak. Specular: metallic luster, steel color, red/brown streak; may have foliation.
Softer than glass → Good cleavage → One direction; splits into paper-like sheets	**Biotite (mica)** (occurs in igneous and metamorphic rocks)	Usually very shiny; sheets are somewhat transparent and flexible.
Nonmetallic luster; dark → Harder than glass → No cleavage → Crystal faces may be visible; generally dark brown/maroon	**Garnet** (can be a gem or abrasive)	Can be opaque or glassy; crystals are generally equant or 12-sided.
No cleavage → Crystal faces not visible; generally green, glassy	**Olivine** (gem version is peridot; occurs in mafic or ultra-mafic igneous rocks)	Has moderately high specific gravity; may occur as a cluster of grains that look "sugary."
Harder than glass → Good cleavage → Two directions, not at right angles	**Amphibole** (occurs in igneous and metamorphic rocks)	May occur in blade-like or needle-like crystals.
Good cleavage → Two directions, approximately at right angles	**Pyroxene** (occurs in igneous and metamorphic rocks)	May occur in small, blade-like or needle-like crystals.

Mineral-Identification Flow Chart (light-colored nonmetallic)

Initial Characteristics			*Mineral Name*	*Other Features (for confirmation)*
Softer than a fingernail	Poor cleavage	"Greasy" feel; makes distinct mark on cloth	**Talc** (pulverized to make talcum powder)	Usually milky white, yellowish, or greenish.
	Good cleavage	One direction; splits into thick sheets	**Gypsum** (used in drywall)	Usually clear.
		One direction; splits into thick sheets	**Muscovite (mica)** (occurs in igneous and metamorphic rocks)	Usually light tan or brown; sheets are transparent and flexible; some varieties are harder than a fingernail.
Softer than glass; harder than a fingernail	Good cubic cleavage		**Halite** (common table salt)	Tastes salty; generally clear or light gray; transparent or translucent.
	Good cleavage (noncubic)	Reacts with HCl	**Calcite** (CaCO₃; found in limestone, shells)	Breaks into white to clear blocks whose sides are ≈75° to one another; double refraction.
		Powder reacts with HCl	**Dolomite** (CaMg[CO₃]₂)	May occur as rhombus-shaped crystals.
		No reaction with HCl	**Fluorite** (state mineral of Illinois)	Generally clear or purple crystals; may be cubic or trapezoidal; cleavage in 4 directions.
Harder than glass	Good cleavage	May have striations	**Plagioclase** (variety of feldspar)	Commonly off-white, brick-shaped crystals with a milky luster.
		No striations	**Orthoclase** (also known as K-feldspar)	Commonly milky luster, pinkish to off-white, brick-shaped crystals.
	No cleavage		**Quartz** (SiO₂; common component of sandstone)	Can be clear, white, pink, smoky, or purple.

Glossary

aa A lava flow with a rubbly surface.

abandoned meander A meander that dries out after it gets cut off.

ablation The removal of ice at the toe of a glacier by melting, sublimation (the evaporation of ice into water vapor), and/or calving.

abrasion The process in which one material (such as sand-laden water) grinds away at another (such as a stream channel's floor and walls).

absolute age Numerical age (the age specified in years).

absolute plate velocity The movement of a plate relative to a fixed point in the mantle.

abyssal plain A broad, relatively flat region of the ocean that lies at least 4.5 km below sea level.

Acadian orogeny A convergent mountain-building event that occurred around 400 million years ago in which continental slivers accreted to the eastern edge of the North American continent.

accreted terrane A block of land that collided with a continent at a convergent margin and stayed attached to the continent.

accretionary coast A coastline that receives more sediment than erodes away.

accretionary orogen An orogen formed by the attachment of numerous buoyant slivers of crust to an older, larger continental block.

accretionary prism A wedge-shaped mass of sediment and rock scraped off the top of a downgoing plate and accreted onto the overriding plate at a convergent plate margin.

acid mine runoff A dilute solution of sulfuric acid, produced when sulfur-bearing minerals in mines react with rainwater, that flows out of a mine.

acid rain Precipitation in which air pollutants react with water to make a weak acid that then falls from the sky.

active continental margin A continental margin that coincides with a plate boundary.

active fault A fault that has moved recently or is likely to move in the future.

active sand The top layer of beach sand, which moves daily because of wave action.

active volcano A volcano that has erupted within the past few centuries and will likely erupt again.

adiabatic cooling The cooling of a body of air or matter without the addition or subtraction of thermal energy (heat).

adiabatic heating The warming of a body of air or matter without the addition or subtraction of heat.

aerosols Tiny solid particles or liquid droplets that remain suspended in the atmosphere for a long time.

aftershocks The series of smaller earthquakes that follow a major earthquake.

air The mixture of gases that make up the Earth's atmosphere.

air-fall tuff Tuff formed when ash settles gently from the air.

air mass A body of air, about 1,500 km across, that has recognizable physical characteristics.

air pressure The push that air exerts on its surroundings.

albedo The reflectivity of a surface.

Alleghenian orogeny The convergent orogenic event that occurred about 270 million years ago when Africa collided with North America.

alloy A metal containing more than one type of metal atom.

alluvial fan A gently sloping apron of sediment dropped by an ephemeral stream at the base of a mountain in arid or semi-arid regions.

alluvium Sorted sediment deposited by a stream.

alluvium-filled valley A valley whose floor fills with sediment.

amber Hardened (fossilized) ancient sap or resin.

amplitude The height of a wave from crest to trough.

Ancestral Rockies The late Paleozoic uplifts of the Rocky Mountain region; they eroded away long before the present Rocky Mountains formed.

angiosperm A flowering plant.

angle of repose The angle of the steepest slope that a pile of uncemented material can attain without collapsing from the pull of gravity.

angularity The degree to which grains have sharp or rounded edges or corners.

angular unconformity An unconformity in which the strata below were tilted or folded before the unconformity developed; strata below the unconformity therefore have a different tilt than strata above.

anhedral grains Crystalline mineral grains without well-formed crystal faces.

Antarctic bottom water mass The mass of cold, dense water that sinks along the coast of Antarctica.

antecedent stream A stream that cuts across an uplifted mountain range; the stream must have existed before the range uplifted and must then have been able to downcut as fast as the land was rising.

anthracite coal Shiny black coal formed at temperatures between 200° and 300°C.

anticline A fold with an arch-like shape in which the limbs dip away from the hinge.

anticyclone The clockwise flow of air around a high-pressure mass.

Antler orogeny The Late Devonian mountain-building event in which slices of deep-marine strata were pushed eastward, up and over the shallow-water strata on the western coast of North America.

anvil cloud A large cumulonimbus cloud that spreads laterally at the tropopause to form a broad, flat top.

aphanitic A textural term for fine-grained igneous rock.

apparent polar-wander path A path on the globe along which a magnetic pole appears to have wandered over time; in fact, the continents drift, while the magnetic pole stays fairly fixed.

aquiclude Sediment or rock that transmits no water.

aquifer Sediment or rock that transmits water easily.

aquitard Sediment or rock that does not transmit water easily and therefore retards the motion of the water.

Archaeobacteria A kingdom of "old bacteria," now commonly found in extreme environments like hot springs. (Also called "Archaea.")

Archean The middle Precambrian eon.

Archimedes' principle The mass of the water displaced by a block of material equals the mass of the whole block of material.

arête A residual knife-edge ridge of rock that separates two adjacent cirques.

argillaceous sedimentary rock Sedimentary rock that contains abundant clay.

arroyo The channel of an ephemeral stream; dry wash; wadi.

artesian well A well in which water rises on its own.

ash fall Ash that falls to the ground out of an ash cloud.

ash flow An avalanche of ash that tumbles down the side of an explosively erupting volcano.

assimilation The process of magma contamination in which blocks of wall rock fall into a magma chamber and dissolve.

asthenosphere The layer of the mantle that lies between 100–150 km and 350 km deep; the asthenosphere is relatively soft and can flow when acted on by force.

atm A unit of air pressure that approximates the pressure exerted by the atmosphere at sea level.

atmosphere A layer of a mixture of gases (air) that surrounds the Earth.

atoll A coral reef that develops as a circular reef surrounding a lagoon.

atomic number The number of protons in the nucleus of a given element.

atomic weight The number of protons plus the number of neutrons in the nucleus of a given element. (Also known as atomic mass.)

aurora australis The same phenomenon as the aurora borealis, but in the Southern Hemisphere.

aurora borealis A ghostly curtain of varicolored light that appears across the night sky in the Northern Hemisphere when charged particles from the Sun interact with the ions in the ionosphere.

avalanche A turbulent cloud of debris mixed with air that rushes down a steep hill slope at high velocity; the debris can be rock and/or snow.

avalanche chute A downslope hillside pathway along which avalanches repeatedly fall, consequently clearing the pathway of mature trees.

avulsion The process in which a river overflows a natural levee and begins to flow in a new direction.

axial plane The imaginary surface that encompasses the hinges of successive layers of a fold.

axial trough A narrow depression that runs along a mid-ocean ridge axis.

backscattered light Atmospheric scattered sunlight that returns back to space.

backshore zone The zone of beach that extends from a small step cut by high-tide swash to the front of the dunes or cliffs that lie farther in shore.

backswamp The low marshy region between the bluffs and the natural levees of a floodplain.

backwash The gravity-driven flow of water back down the slope of a beach.

bajada An elongate wedge of sediment formed by the overlap of several alluvial fans emerging from adjacent valleys.

Baltica A Paleozoic continent that included crust that is now part of today's Europe.

banded-iron formation (BIF) Iron-rich sedimentary layers consisting of alternating gray beds of iron oxide and red beds of iron-rich chert.

bar (1) A sheet or elongate lens or mound of alluvium; (2) a unit of air-pressure measurement approximately equal to 1 atm.

barchan dune A crescent-shaped dune whose tips point downwind.

barrier island An offshore sand bar that rises above the mean high-water level, forming an island.

barrier reef A coral reef that develops offshore, separated from the coast by a lagoon.

basal sliding The phenomenon in which meltwater accumulates at the base of a glacier, so that the mass of the glacier slides on a layer of water or on a slurry of water and sediment.

basalt A fine-grained mafic igneous rock.

base level The lowest elevation a stream channel's floor can reach at a given locality.

basement Older igneous and metamorphic rocks making up the Earth's crust beneath sedimentary cover.

basement uplift Uplift of basement rock by faults that penetrate deep into the continental crust.

base metals Metals that are mined but not considered precious, such as copper, lead, zinc, or tin.

basin A fold or depression shaped like a right-side-up bowl.

Basin and Range Province A broad, Cenozoic continental rift that has affected a portion of the western United States in Nevada, Utah, and Arizona; in this province, tilted fault blocks form ranges, and alluvium-filled valleys are basins.

batholith A vast composite, intrusive, igneous rock body up to several hundred km long and 100 km wide, formed by the intrusion of numerous plutons in the same region.

bathymetric map A map illustrating the shape of the ocean floor.

bathymetric profile A cross section showing ocean depth plotted against location.

bathymetry Variation in depth.

bauxite A residual mineral deposit rich in aluminum.

baymouth bar A sandspit that grows across the opening of a bay.

beach drift The gradual migration of sand along a beach.

beach erosion The removal of beach sand caused by wave action and longshore currents.

beach face A steep concave part of the foreshore zone formed where the swash of the waves actively scours the sand.

bedding Layering or stratification in sedimentary rocks.

bed load Large particles, such as sand, pebbles, or cobbles, that bounce or roll along a stream bed.

bedrock Rock still attached to the Earth's crust.

Bergeron process Precipitation involving the growth of ice crystals in a cloud at the expense of water droplets.

berm A horizontal or landward-sloping terrace in the backshore zone of a beach that receives sediment during a storm.

big bang A cataclysmic explosion that scientists suggest represents the formation of the Universe; before this event, all matter and all energy were packed into one volumeless point.

biochemical sedimentary rock Sedimentary rock formed from material (such as shells) produced by living organisms.

biodiversity The number of different species that exist at a given time.

biogeochemical cycle The exchange of chemicals between living and nonliving reservoirs in the Earth system.

bioremediation The injection of oxygen and nutrients into a contaminated aquifer to foster the growth of bacteria that will ingest or break down contaminants.

biosphere The region of the Earth and atmosphere inhabited by life; this region stretches from a few km below the Earth's surface to a few km above.

bioturbation The mixing of sediment by burrowing animals such as clams and worms.

bituminous coal Dull black coal formed at temperatures between 100° and 200°C.

black-lung disease Lung disease contracted by miners from the inhalation of too much coal dust.

black smoker The cloud of suspended minerals formed where hot water spews out of a vent along a mid-ocean ridge; the dissolved sulfide components of the hot water instantly precipitate when the water mixes with seawater and cools.

blind fault A fault that does not intersect the ground surface.

blocking temperature The temperature below which isotopes in a mineral are no longer free to move, so the radiometric clock starts.

blowout A deep, bowl-like depression scoured out of desert terrain by a turbulent vortex of wind.

blue shift The phenomenon in which a source of light moving toward you appears to have a higher frequency.

body waves Seismic waves that pass through the interior of the Earth.

bog A wetland dominated by moss and shrubs.

bolide A solid extraterrestrial object such as a meteorite, comet, or asteroid.

bornhardt An inselberg with a loaf geometry, like that of Uluru (Ayers Rock) in central Australia.

Bowen's reaction series The sequence in which different silicate minerals crystallize during the progressive cooling of a melt.

braided stream A sediment-choked stream consisting of entwined subchannels.

breaker A water wave in which water at the top of the wave curves over the base of the wave.

breakwater An offshore wall, built parallel or at an angle to the beach, that prevents the full force of waves from reaching a harbor.

breccia Coarse sedimentary rock consisting of angular fragments; or rock broken into angular fragments by faulting.

breeder reactor A nuclear reactor that produces its own fuel.

brine Water that is not fresh but is less salty than seawater; brine may be found in estuaries.

brittle deformation The cracking and fracturing of a material subjected to stress.

brittle-ductile transition (brittle-plastic transition) The depth above which materials behave brittlely and below which materials behave ductilely (plastically); this transition typically lies between a depth of 10 and 15 km in continental crustal rock, and 60 m deep in glacial ice.

buoyancy The upward force acting on a less dense object immersed or floating in denser material.

butte A medium-sized, flat-topped hill in an arid region.

caldera A large circular depression with steep walls and a fairly flat floor, formed after an eruption as the center of the volcano collapses into the drained magma chamber below.

caliche A solid mass created where calcite cements the soil together.

calving The breaking off of chunks of ice at the edge of a glacier.

Cambrian explosion of life The remarkable diversification of life, indicated by the fossil record, that occurred at the beginning of the Cambrian period.

Canadian Shield A broad, low-lying region of exposed Precambrian rock in the Canadian interior.

canyon A trough or valley with steeply sloping walls, cut into the land by a stream.

capillary fringe The thin subsurface layer in which water molecules seep up from the water table by capillary action to fill pores.

carbonate rocks Rocks containing calcite and/or dolomite.

carbon-14 dating A radiometric dating process that can tell us the age of organic material containing carbon originally extracted from the atmosphere.

cast Sediment that preserves the shape of a shell it once filled before the shell dissolved or mechanically weathered away.

catabatic winds Strong winds that form at the margin of a glacier where the warmer air above ice-free land rises and the cold, denser air from above the glaciers rushes in to take its place.

catastrophic change Change that takes place either instantaneously or rapidly in geologic time.

catchment *Drainage network.*

cement Mineral material that precipitates from water and fills the spaces between grains, holding the grains together.

cementation The phase of lithification in which cement, consisting of minerals that precipitate from groundwater, partially or completely fills the spaces between clasts and attaches each grain to its neighbor.

Cenozoic The most recent era of the Phanerozoic eon, lasting from 65 Ma up until the present.

chalk Very fine-grained limestone consisting of weakly cemented plankton shells.

change of state The process in which a material changes from one phase (liquid, gas, or solid) to another.

channel A trough dug into the ground surface by flowing water.

channeled scablands A barren, soil-free landscape in eastern Washington, scoured clean by a flood unleashed when a large glacial lake drained.

chatter marks Wedge-shaped indentations left on rock surfaces by glacial plucking.

chemical sedimentary rocks Sedimentary rocks made up of minerals that precipitate directly from water solution.

chemical weathering The process in which chemical reactions alter or destroy minerals when rock comes in contact with water solutions and/or air.

chert A sedimentary rock composed of very fine-grained silica (cryptocrystalline quartz).

Chicxulub crater A circular excavation buried beneath younger sediment on the Yucután peninsula; geologists suggest that a bolide landed there 65 Ma.

chimney (1) A conduit in a magma chamber in the shape of a long vertical pipe through which magma rises and erupts at the surface; (2) an isolated column of strata in an arid region.

cinder cone A subaerial volcano consisting of a cone-shaped pile of tephra whose slope approaches the angle of repose for tephra.

cinders Fragments of glassy rock ejected from a volcano.

cirque A bowl-shaped depression carved by a glacier on the side of a mountain.

cirrus cloud A wispy cloud that tapers into delicate, feather-like curls.

clastic (detrital) sedimentary rock Sedimentary rock consisting of cemented-together detritus derived from the weathering of preexisting rock.

cleavage (1) The tendency of a mineral to break along preferred planes; (2) a type of foliation in low-grade metamorphic rock.

cleavage planes A series of surfaces on a crystal that form parallel to the weakest bonds holding the atoms of the crystal together.

cliff (or scarp) retreat The movement of the position of a cliff face caused by erosion.

climate The average weather conditions, along with the range of conditions, of a region over a year.

cloud A mist of tiny water droplets.

coal rank A measurement of the carbon content of coal; higher-rank coal formed at higher temperatures.

coal reserve The quantities of discovered coal in sedimentary rock of the continents.

coal swamp A swamp whose oxygen-poor water allows thick piles of woody debris to accumulate; this debris transforms into coal upon deep burial.

coastal plain A flat stretch of coastal land that merges with the continental shelf along a passive margin.

cold front The boundary at which a cold air mass pushes underneath a warm air mass.

collision The process of two buoyant pieces of lithosphere converging and squashing together.

columnar jointing A type of fracturing that yields roughly hexagonal columns of basalt; columnar joints form when a dike, sill, or lava flow cools.

comet A ball of ice and dust, probably remaining from the formation of the solar system, that orbits the Sun.

compaction The phase of lithification in which the pressure of the overburden on the buried rock squeezes out water and air that was trapped between clasts, and the clasts press tightly together.

composite volcano *Stratovolcano.*

compositional banding A type of metamorphic foliation defined by alternating bands of light and dark minerals.

compressibility The degree to which a material's volume changes in response to squashing.

compression A push or squeezing felt by a body.

compressional waves Waves in which particles of material move back and forth parallel to the direction in which the wave itself moves.

conchoidal fractures Smoothly curving, clamshell-shaped surfaces along which materials with no cleavage planes tend to break.

condensation The process of gas molecules linking together to form a liquid.

condensation nuclei Preexisting solid or liquid particles, such as aerosols, onto which water condenses during cloud formation.

cone of depression The downward-pointing, cone-shaped surface of the water table in a location where the water table is experiencing drawdown because of pumping at a well.

confined aquifer An aquifer that is separated from the Earth's surface by an overlying aquitard.

conglomerate Very coarse-grained sedimentary rock consisting of rounded clasts.

consuming boundary *Convergent plate boundary.*

contact The boundary surface between two rock bodies (as between two stratigraphic formations, between an igneous intrusion and adjacent rock, between two igneous rock bodies, or between rocks juxtaposed by a fault).

contact metamorphism *Thermal metamorphism.*

contaminant plume A cloud of contaminated groundwater that moves away from the source of the contamination.

continental crust The crust beneath the continents.

continental divide A highland separating drainage that flows into one ocean from drainage that flows into another.

continental-drift hypothesis The hypothesis that continents have moved and are still moving slowly across the Earth's surface.

continental glacier A vast sheet of ice that spreads over thousands of square km of continental crust.

continental-interior desert An inland desert that develops because by the time air masses reach the continental interior, they have lost all of their moisture.

continental lithosphere Lithosphere topped by continental crust; this lithosphere reaches a thickness of 150 km.

continental margin A continent's coastline.

continental rift A linear belt along which continental lithosphere stretches and pulls apart.

continental rifting The process by which a continent stretches and splits along a belt; if successful, rifting separates a larger continent into two smaller continents separated by a divergent boundary.

continental rise The sloping sea floor that extends from the lower part of the continental slope to the abyssal plain.

continental shelf A broad, shallowly submerged region of a continent along a passive margin.

continental slope The slope at the edge of a continental shelf, leading down to the deep sea floor.

continental volcanic arc A long curving chain of subaerial volcanoes on the margin of a continent adjacent to a convergent plate boundary.

contour lines Lines on a map along which a parameter has a constant value; for example, all points along a contour line on a topographic map are at the same elevation.

control rod Rods that absorb neutrons in a nuclear reactor and thus decrease the number of collisions between neutrons and radioactive atoms.

convection Heat transfer that results when warmer, less dense material rises while cooler, denser material sinks.

convergence zone A place where two surface air flows meet so that air has to rise.

convergent margin *Convergent plate boundary.*

convergent plate boundary A boundary at which two plates move toward each other so that one plate sinks (subducts) beneath the other; only oceanic lithosphere can subduct.

coral reef A mound of coral and coral debris forming a region of shallow water.

core The dense, iron-rich center of the Earth.

core-mantle boundary An interface 2,900 km below the Earth's surface separating the mantle and core.

Coriolis effect The deflection of objects, winds, and currents on the surface of the Earth owing to the planet's rotation.

cornice A huge, overhanging drift of snow at the crest of a mountain ridge built up by strong winds.

correlation The process of defining the age relations between the strata at one locality and the strata at another.

cosmic rays Nuclei of hydrogen and other elements that bombard the Earth from deep space.

cosmology The study of the overall structure of the Universe.

country rock (wall rock) The preexisting rock into which magma intrudes.

crater (1) A circular depression at the top of a volcanic mound; (2) a depression formed by the impact of a bolide.

craton A long-lived block of durable continental crust commonly found in the stable interior of a continent.

cratonic platform A province in the interior of a continent in which Phanerozoic strata bury most of the underlying Precambrian rock.

creep The gradual downslope movement of regolith.

crevasse A large crack that develops by brittle deformation in the top 60 m of a glacier.

critical mass A sufficiently dense and large mass of radioactive atoms in which a chain reaction happens so quickly that the mass explodes.

cross section A diagram depicting the geometry of materials underground as they would appear on an imaginary vertical slice through the Earth.

crude oil Oil extracted directly from the ground.

crust The rock that makes up the outermost layer of the Earth.

crustal root Low-density crustal rock that protrudes downward beneath a mountain range.

crystal A single, continuous piece of a mineral bounded by flat surfaces that formed naturally as the mineral grew.

crystal form The geometric shape of a crystal, defined by the arrangement of crystal faces.

crystal habit The general shape of a crystal or cluster of crystals that grew unimpeded.

crystal lattice The orderly framework within which the atoms or ions of a mineral are fixed.

crystalline Containing a crystal lattice.

cuesta An asymmetric ridge formed by tilted layers of rock, with a steep cliff on one side cutting across the layers and a gentle slope on the other side; the gentle slope is parallel to the layering.

cumulonimbus cloud A rain-producing puffy cloud.

cumulus cloud A puffy, cotton-ball-shaped cloud.

current (1) A well-defined stream of ocean water; (2) the moving flow of water in a stream.

cut bank The outside bank of the channel wall of a meander, which is continually undergoing erosion.

cutoff A straight reach in a stream that develops when erosion eats through a meander neck.

cyanobacteria Blue-green algae; a type of Archaeobacteria.

cycle A series of interrelated events or steps that occur in succession and can be repeated, perhaps indefinitely.

cyclone (1) The counterclockwise flow of air around a low-pressure mass; (2) the equivalent of a hurricane in the Indian Ocean.

cyclothem A repeated interval within a sedimentary sequence that contains a specific succession of sedimentary beds.

Darcy's law A mathematical equation stating that a volume of water, passing through a specified area of material at a given time, depends on the material's permeability and hydraulic gradient.

daughter isotope The decay product of radioactive decay.

day The time it takes for the Earth to spin once on its axis.

debris avalanche An avalanche in which the falling debris consists of rock fragments and dust.

debris flow A downslope movement of mud mixed with larger rock fragments.

debris slide A sudden downslope movement of material consisting only of regolith.

decompression melting The kind of melting that occurs when hot mantle rock rises to shallower depths in the Earth so that pressure decreases while the temperature remains unchanged.

deep current An ocean current at a depth greater than 100 m.

deep-focus earthquake An earthquake that occurs at a depth between 300 and 670 km; below 670 km, earthquakes do not happen.

deflation The process of lowering the land surface by wind abrasion.

deformation A change in the shape, position, or orientation of a material, by bending, breaking, or flowing.

dehydration Loss of water.

delta A wedge of sediment formed at a river mouth when the running water of the stream enters standing water, the current slows, the stream loses competence, and sediment settles out.

delta plain The low, swampy land on the surface of a delta.

delta-plain flood A flood in which water submerges a delta plain.

dendritic network A drainage network whose interconnecting streams resemble the pattern of branches connecting to a deciduous tree.

dendrochronologist A scientist who uses tree rings to determine the geologic age of features.

density Mass per unit volume.

denudation The removal of rock and regolith from the Earth's surface.

deposition The process by which sediment settles out of a transporting medium.

depositional landform A landform resulting from the deposition of sediment where the medium carrying the sediment evaporates, slows down, or melts.

desert A region so arid that it contains no permanent streams except for those that bring water in from elsewhere, and has very sparse vegetation cover.

desertification The process of transforming nondesert area into desert.

desert pavement A mosaic-like stone surface over the ground in a desert.

desert varnish A dark, rusty-brown coating of iron oxide and magnesium oxide that accumulates when water seeps into a rock, dissolves iron and magnesium ions, and carries the ions back to the surface of the rock.

detachment fault A nearly horizontal fault at the base of a fault system.

detritus The chunks and smaller grains of rock broken off outcrops by physical weathering.

dewpoint temperature The temperature at which air becomes saturated so that dew can form.

differential stress The condition in which a material experiences a push or pull in one direction of a greater magnitude than the push or pull in another direction; in some cases, differential stress can result in shearing.

differential weathering The condition in which different rocks in an outcrop undergo weathering at different rates.

diffraction The splitting of light into many tiny beams that interfere with one another.

dike A tabular (wall-shaped) intrusion of rock that cuts across the layering of country rock.

dimension stone An intact block of granite or marble to be used for architectural purposes.

dipole A magnetic field with a north and south pole, like that of a bar magnet.

dipole field (for Earth) The part of the Earth's magnetic field, cause by the flow of liquid iron alloy in the outer core, that can be represented by an imaginary bar magnet with a north and south pole.

dip-slip fault A fault in which sliding occurs up or down the slope (dip) of the fault.

dip slope A hill slope underlain by bedding parallel to the slope.

disappearing stream A stream that intersects a crack or sinkhole leading to an underground cavern, so that the water disappears into the subsurface and becomes an underground stream.

discharge The volume of water in a conduit or channel passing a point in a second.

discharge area A location where groundwater flows back up to the surface, and may emerge at springs.

disconformity An unconformity parallel to the two sedimentary sequences it separates.

displacement (or offset) The amount of movement or slip across a fault plane.

disseminated deposit A hydrothermal ore deposit in which ore minerals are dispersed throughout a body of rock.

dissolution A process during which materials dissolve in water.

dissolved load Ions dissolved in a stream's water.

distillation column A vertical pipe in which crude oil is separated into several components.

distributaries The fan of small streams formed where a river spreads out over its delta.

divergence zone A place where sinking air separates into two flows that move in opposite directions.

divergent plate boundary A boundary at which two lithosphere plates move apart from each other; they are marked by mid-ocean ridges.

diversification The development of many different species.

DNA (deoxyribonucleic acid) The complex molecule, shaped like a double helix, containing the code that guides the growth and development of an organism.

doldrums Very slow winds along the equator.

dome Folded or arched layers with the shape of an overturned bowl.

Doppler effect The phenomenon in which the frequency of wave energy appears to change when a moving source of wave energy passes an observer.

dormant volcano A volcano that has not erupted for hundreds to thousands of years but does have the potential to erupt again in the future.

downcutting The process in which water flowing through a channel cuts into the substrate and deepens the channel relative to its surroundings.

downdraft Downward-moving air.

downgoing plate (or slab) A lithosphere plate that has been subducted at a convergent margin.

downslope force The component of the force of gravity acting in the downslope direction.

downslope movement The tumbling or sliding of rock and sediment from higher elevations to lower ones.

downwelling zone A place where near-surface water sinks.

drag fold A fold that develops in layers of rock adjacent to a fault during or just before slip.

drainage divide A highland or ridge that separates one watershed from another.

drainage network (or basin) An array of interconnecting streams that together drain an area.

drawdown The phenomenon in which the water table around a well drops because the users are pumping water out of the well faster than it flows in from the surrounding aquifer.

drilling mud A slurry of water mixed with clay that oil drillers use to cool a drill bit and flush rock cuttings up and out of the hole.

dripstone Limestone (travertine in a cave) formed by the precipitation of calcium carbonate out of groundwater.

drop stone A rock that drops to the sea floor once the iceberg that was carrying the rock melts.

drumlin A streamlined, elongate hill formed when a glacier overrides glacial till.

dry-bottom (polar) glacier A glacier so cold that its base remains frozen to the substrate.

dry wash The channel of an ephemeral stream when empty of water.

dry well (1) A well that does not supply water because the well has been drilled into an aquitard or into rock that lies above the water table; (2) a well that does not yield oil, even though drilled into an anticipated reservoir.

ductile (plastic) deformation The bending and flowing of a material (without cracking and breaking) subjected to stress.

dune A pile of sand generally formed by deposition from the wind.

dust storm An event in which strong winds hit unvegetated land, strip off the topsoil, and send it skyward to form rolling dark clouds that block out the Sun.

dynamic metamorphism Metamorphism that occurs as a consequence of shearing alone, with no change in temperature or pressure.

dynamo A power plant generator in which water or wind power spins an electrical conductor around a permanent magnet.

dynamothermal metamorphism Metamorphism that involves heat, pressure, and shearing.

earthquake A vibration caused by the sudden breaking or frictional sliding of rock in the Earth.

earthquake belt A relatively narrow, distinct belt of earthquakes that defines the position of a plate boundary.

earthquake engineering The design of buildings that can withstand shaking.

earthquake zoning The determination of where land is relatively stable and where it might collapse because of seismicity.

Earth system The global interconnecting web of physical and biological phenomena involving the solid Earth, the hydrosphere, and the atmosphere.

ebb tide The falling tide.

eccentricity cycle The cycle of the gradual change of the Earth's orbit from a more circular to a more elliptical shape; the cycle takes around 100,000 years.

ecliptic The plane defined by a planet's orbit.

ecosystem An environment and its inhabitants.

eddy An isolated, ring-shaped current of water.

effusive eruption An eruption that yields mostly lava, not ash.

Ekman spiral The change in flow direction of water with depth, caused by the Coriolis effect.

Ekman transport The overall movement of a mass of water, resulting from the Eckman spiral, in a direction 90° to the wind direction.

elastic strain A change in shape of a material that disappears instantly when stress is removed.

electromagnet An electrical device that produces a magnetic field.

electron microprobe A laboratory instrument that can focus a beam of electrons on a small part of a mineral grain in order to create a signal that defines its chemical composition.

El Niño The flow of warm water east from the Pacific Ocean that reverses the upwelling of cold water along the west coast of South America and causes significant global changes in weather patterns.

embayment A low area of coastal land.

emergent coast A coast where the land is rising relative to sea level or sea level is falling relative to the land.

end moraine (terminal moraine) A low, sinuous ridge of till that develops when the terminus (toe) of a glacier stalls in one position for a while.

energy The capacity to do work.

energy resource Something that can be used to produce work; in a geologic context, a material (such as oil, coal, wind, flowing water) that can be used to produce energy.

eon The largest subdivision of geologic time.

epeirogenic movement The gradual uplift or subsidence of a broad region of the Earth's surface.

epeirogeny An event of epeirogenic movement; the term is usually used in reference to the formation of broad mid-continent domes and basins.

ephemeral (intermittent) stream A stream whose bed lies above the water table, so that the stream flows only when the rate at which water enters the stream from rainfall or meltwater exceeds the rate at which water infiltrates the ground below.

epicenter The point on the surface of the Earth directly above the focus of an earthquake.

epicontinental sea A shallow sea overlying a continent.

epoch An interval of geologic time representing the largest subdivision of a period.

equant A term for a grain that has the same dimensions in all directions.

equatorial low The area of low pressure that develops over the equator because of the intertropical convergence zone.

equinox One of two days out of the year (September 22 and March 21) in which the Sun is directly overhead at noon at the equator.

era An interval of geologic time representing the largest subdivision of the Phanerozoic eon.

erg Sand seas formed by the accumulation of dunes in a desert.

erosion The grinding away and removal of Earth's surface materials by moving water, air, or ice.

erosional coast A coastline where sediment is not accumulating and wave action grinds away at the shore.

erosional landform A landform that results from the breakdown and removal of rock or sediment.

erratic A boulder that was picked up by a glacier and deposited hundreds of kilometers away from the outcrop from which it detached.

esker A ridge of sorted sand and gravel that snakes across a ground moraine; the sediment of an esker was deposited in subglacial meltwater tunnels.

estuary An inlet in which seawater and river water mix, created when a coastal valley is flooded because of either rising sea level or land subsidence.

Eubacteria The kingdom of "true bacteria."

euhedral crystal A crystal whose faces are well formed and whose shape reflects crystal form.

eukaryotic cell A cell with a complex internal structure, capable of building multicellular organisms.

eustatic sea-level change A global rising or falling of the ocean surface.

evaporate To change from liquid to vapor.

evapotranspiration The sum of evaporation from bodies of water and the ground surface and transpiration from plants and animals.

exfoliation The process by which an outcrop of rock splits apart into onion-like sheets along joints that lie parallel to the face of the outcrop.

exhumation The process (involving uplift and erosion) that returns deeply buried rocks to the surface.

exotic terrane A block of land that collided with a continent along a convergent margin and attached to the continent; the term "exotic" implies that the land was not originally part of the continent to which it is now attached.

expanding Universe theory The theory that says that the whole Universe must be expanding because galaxies in every direction seem to be moving away from us.

explosive eruptions Violent volcanic eruptions that produce clouds and avalanches of pyroclastic debris.

external process A geomorphologic process—such as downslope movement, erosion, or deposition—that is the consequence of gravity or the interaction between the solid Earth and its fluid envelope (air and water). Energy for these processes come from gravity and sunlight.

extinction The death of the last members of a species so that there are no parents to pass on their genetic traits to offspring.

extinct volcano A volcano that was active in the past but has now shut off entirely and will not erupt in the future.

extraordinary fossil A rare fossilized relict, or trace, of the soft part of an organism.

extrusive igneous rock Rock that forms by the freezing of lava above ground, after it flows or explodes out (extrudes) onto the surface and comes into contact with the atmosphere or ocean.

eye The relative calm in the center of a hurricane.

eye wall A rotating vertical cylinder of clouds surrounding the eye of a hurricane.

fault A fracture on which one body of rock slides past another.

fault-block mountains An outdated term for a narrow, elongate range of mountains that develops in a continental-rift setting as normal faulting drops down blocks of crust, or tilts blocks.

fault breccia Fragmented rock in which angular fragments were formed by brittle fault movement; fault breccia occurs along a fault.

fault creep Gradual movement along a fault that occurs in the absence of an earthquake.

fault gouge Pulverized rock consisting of fine powder that lies along fault surfaces; gouge forms by crushing and grinding.

faulting Slip events along a fault.

fault scarp A small step on the ground surface where one side of a fault has moved vertically with respect to the other.

fault system A grouping of numerous related faults.

fault trace (or **line**) The intersection between a fault and the ground surface.

Ferrel cell The name given to the middle-latitude convection cells in the atmosphere.

fetch The distance across a body of water along which a wind blows to build waves.

fine-grained A textural term for rock consisting of many fine grains or clasts.

firn Compacted granular ice (derived from snow) that forms where snow is deeply buried; if buried more deeply, firn turns into glacial ice.

fission track A line of damage formed in the crystal lattice of a mineral by the ejection of an atomic particle during the decay of a radioactive isotope.

fissure A conduit in a magma chamber in the shape of a long crack through which magma rises and erupts at the surface.

fjord A deep, glacially carved, U-shaped valley flooded by rising sea level.

flank eruption An eruption that occurs when a secondary chimney, or fissure, breaks through the flank of a volcano.

flash flood A flood that occurs during unusually intense rainfall or as the result of a dam collapse, during which the floodwaters rise very fast.

flexing The process of folding in which a succession of rock layers bends and slip occurs between the layers.

flocculation The clumping together of clay suspended in river water into bunches that are large enough to settle out.

flood An event during which the volume of water in a stream becomes so great that it covers areas outside the stream's normal channel.

flood basalt Vast sheets of basalt that spread from a volcanic vent over an extensive surface of land; they may form where a rift develops above a continental hot spot, and lava is particularly hot and has low viscosity.

floodplain The flat land on either side of a stream that becomes covered with water during a flood.

floodplain flood A flood during which a floodplain is submerged.

flood stage The stage when water reaches the top of a stream channel.

flood tide The rising tide.

floodway A mapped region likely to be flooded, in which people avoid constructing buildings.

flow fold A fold that forms when the rock is so soft that it behaves like weak plastic.

flowstone A sheet of limestone that forms along the wall of a cave when groundwater flows along the surface of the wall.

fluvial deposit Sediment deposited in a stream channel, along a stream bank, or on a floodplain.

flux Flow.

focus The location where a fault slips during an earthquake (hypocenter).

fog A cloud that forms at ground level.

fold A bend or wrinkle of rock layers or foliation; folds form as a consequence of ductile deformation.

fold axis An imaginary line that, when moved parallel to itself, can trace out the shape of a folded surface.

fold-thrust belt An assemblage of folds and related thrust faults that develop above a detachment fault.

foliation Layering formed as a consequence of the alignment of mineral grains, or of compositional banding in a metamorphic rock.

foraminifera Microscopic plankton with calcitic shells, components of some limestones.

foreland sedimentary basin A basin located under the plains adjacent to a mountain front, which develops as the weight of the mountains pushes the crust down, creating a depression that traps sediment.

foreshocks The series of smaller earthquakes that precede a major earthquake.

foreshore zone The zone of beach regularly covered and uncovered by rising and falling tides.

formation *Stratigraphic formation.*

fossil The remnant, or trace, of an ancient living organism that has been preserved in rock or sediment.

fossil assemblage A group of fossil species found in a specific sequence of sedimentary rock.

fossil correlation A determination of the stratigraphic relation between two sedimentary rock units, reached by studying fossils.

fossil fuel An energy resource such as oil or coal that comes from organisms that lived long ago, and thus stores solar energy that reached the Earth then.

fossiliferous limestone Limestone consisting of abundant fossil shells and shell fragments.

fossilization The process of forming a fossil.

fractional crystallization The process by which a magma becomes progressively more silicic as it cools, because early-formed crystals settle out.

fracture zone A narrow band of vertical fractures in the ocean floor; fracture zones lie roughly at right angles to a mid-ocean ridge, and the actively slipping part of a fracture zone is a transform fault.

fresh rock Rock whose mineral grains have their original composition and shape.

friction Resistance to sliding on a surface.

fringing reef A coral reef that forms directly along the coast.

front The boundary between two air masses.

frost wedging The process in which water trapped in a joint freezes, forces the joint open, and may cause the joint to grow.

fuel rod A metal tube that holds the nuclear fuel in a nuclear reactor.

Fujita scale A scale that distinguishes among tornadoes based on wind speed, path dimensions, and possible damage.

Ga Billion years ago (abbreviation).

gabbro A coarse-grained, intrusive mafic igneous rock.

Gaia The term used for the Earth system, with the implication that it resembles a complex living entity.

galaxy An immense system of hundreds of billions of stars.

gene An individual component of the DNA code that guides the growth and development of an organism.

genetics The study of genes and how they transmit information.

geocentric Universe concept An ancient Greek idea suggesting that the Earth sat motionless in the center of the Universe while stars and other planets and the Sun orbited around it.

geochronology The science of dating geologic events in years.

geode A cavity in which euhedral crystals precipitate out of water solutions passing through a rock.

geographical pole The locations (north and south) where the Earth's rotational axis intersects the planet's surface.

geologic column A composite stratigraphic chart that represents the entirety of the Earth's history.

geologic history The sequence of geologic events that has taken place in a region.

geologic map A map showing the distribution of rock units and structures across a region.

geologic time The span of time since the formation of the Earth.

geologic time scale A scale that describes the intervals of geologic time.

geology The study of the Earth, including our planet's composition, behavior, and history.

geotherm The change in temperature with depth in the Earth.

geothermal energy Heat and electricity produced by using the internal heat of the Earth.

geothermal gradient The rate of change in temperature with depth.

geothermal region A region of current or recent volcanism in which magma or very hot rock heats up groundwater, which may discharge at the surface in the form of hot springs and/or geysers.

geyser A fountain of steam and hot water that erupts periodically from a vent in the ground in a geothermal region.

glacial abrasion The process by which clasts embedded in the base of a glacier grind away at the substrate as the glacier flows.

glacial advance The forward movement of a glacier's toe when the supply of snow exceeds the rate of ablation.

glacial drift Sediment deposited in glacial environments.

glacial incorporation The process by which flowing ice surrounds and incorporates debris.

glacial marine Sediment consisting of ice-rafted clasts mixed with marine sediment.

glacial outwash Coarse sediment deposited on a glacial outwash plain by meltwater streams.

glacially polished surface A polished rock surface created by the glacial abrasion of the underlying substrate.

glacial plowing The process by which flowing ice bulldozes and moves loose sediment.

glacial plucking (or quarrying) The process by which a glacier breaks off and carries away fragments of bedrock.

glacial rebound The process by which the surface of a continent rises back up after an overlying continental ice sheet melts away and the weight of the ice is removed.

glacial retreat The movement of a glacier's toe back toward the glacier's origin; glacial retreat occurs if the rate of ablation exceeds the rate of supply.

glacial subsidence The sinking of the surface of a continent caused by the weight of an overlying glacial ice sheet.

glacial till Sediment transported by flowing ice and deposited beneath a glacier or at its toe.

glaciation   A period of time during which glaciers grew and covered substantial areas of the continents.

glacier   A river or sheet of ice that slowly flows across the land surface and lasts all year long.

glass   A solid in which atoms are not arranged in an orderly pattern.

glassy igneous rock   Igneous rock consisting entirely of glass, or of tiny crystals surrounded by a glass matrix.

glide horizon   The surface along which a slump slips.

global change   The transformations or modifications of both physical and biological components of the Earth system through time.

global circulation   The movement of volumes of air in the paths that ultimately take it around the planet.

global climate change   Transformations or modifications in Earth's climate over time.

global cooling   A fall in the average atmospheric temperature.

global positioning system (GPS)   A satellite system people can use to measure rates of movement of the Earth's crust relative to one another, or simply to locate their position on the Earth's surface.

global warming   A rise in the average atmospheric temperature.

gneiss   A compositionally banded metamorphic rock typically composed of alternating dark and light layers.

Gondwana   A supercontinent that consisted of today's South America, Africa, Antarctica, India, and Australia. Also called Gondwanaland.

graben   A down-dropped crustal block bounded on either side by a normal fault dipping toward the basin.

gradualism   The theory that evolution happens at a constant, slow rate.

grain   A fragment of a mineral crystal or group of crystals.

grain rotation   The process by which rigid, inequant mineral grains distributed through a soft matrix may rotate into parallelism as the rock changes shape owing to differential stress.

granite   A coarse-grained intrusive silicic igneous rock.

gravitational spreading   A process of lateral spreading that occurs in a material because of the weakness of the material; gravitational spreading causes continental glaciers to grow and mountain belts to undergo orogenic collapse.

graywacke   An informal term used for sedimentary rock consisting of sand-sized up to small-pebble-sized grains of quartz and rock fragments all mixed together in a muddy matrix; typically, graywacke occurs at the base of a graded bed.

greenhouse conditions (greenhouse period)   Relatively warm global climate leading to the rising of sea level for an interval of geologic time.

greenhouse effect   The trapping of heat in the Earth's atmosphere by carbon dioxide and other greenhouse gases, which absorb infrared radiation; somewhat analogous to glass in a greenhouse.

greenhouse gases   Atmospheric gases, such as carbon dioxide and methane, that regulate the Earth's atmospheric temperature by absorbing infrared radiation.

greenschist facies   The lowest metamorphic grade, in which chlorite has formed.

greenstone   A low-grade metamorphic rock formed from basalt; if foliated, the rock is called greenschist.

Greenwich mean time (GMT)   The time at the astronomical observatory in Greenwich, England; time in all other time zones is set in relation to GMT.

Grenville orogeny   The orogeny that occurred about 1 billion years ago and yielded the belt of deformed and metamorphosed rocks that underlie the eastern fifth of the North American continent.

groin   A concrete or stone wall built perpendicular to a shoreline in order to prevent beach drift from removing sand.

ground moraine   A thin, hummocky layer of till left behind on the land surface during a rapid glacial recession.

groundwater   Water that resides under the surface of the Earth, mostly in pores or cracks of rock or sediment.

group   A succession of stratigraphic formations that have been lumped together, making a single, thicker stratigraphic entity.

growth ring   A rhythmic layering that develops in trees, travertine deposits, and shelly organisms as a consequence of seasonal changes.

gusher   A fountain of oil formed when underground pressure causes the oil to rise on its own out of a drilled hole.

guyot   A seamount that had a coral reef growing on top of it, so that it is now flat-crested.

gymnosperm   A plant whose seeds are "naked," not surrounded by a fruit.

gyre   A large circular flow pattern of ocean surface currents.

Hadean   The oldest of the Precambrian eons; the time between Earth's origin and the formation of the first rocks that have been preserved.

Hadley cell   The name given to the low-latitude convection cells in the atmosphere.

hail   Falling ice balls from the sky, formed when ice crystallizes in turbulent storm clouds.

hail streak   An approximately 2-by-10-km stretch of ground, elongate in the direction of a storm, onto which hail falls.

half-graben   A wedge-shaped basin in cross section that develops as the hanging-wall block above a normal fault slides down and rotates; the basin develops between the fault surface and the top surface of the rotated block.

half-life   The time it takes for half of a group of a radioactive element's isotopes to decay.

halocline   The boundary in the ocean between surface-water and deep-water salinities.

hamada   Barren rocky highlands in a desert.

hanging valley   A glacially carved tributary valley whose floor lies at a higher elevation than the floor of the trunk valley.

hanging wall   The rock or sediment above an inclined fault plane.

hard water   Groundwater that contains dissolved calcium and magnesium, usually after passing through limestone or dolomite.

head   (1) The elevation of the water table above a reference horizon; (2) the edge of ice at the origin of a glacier.

headland   A place where a hill or cliff protrudes into the sea.

head scarp   The distinct step along the upslope edge of a slump where the regolith detached.

headward erosion   The process in which a stream channel lengthens up its slope as the flow of water increases.

headwaters   The beginning point of a stream.

heat   Thermal energy resulting from the movement of molecules.

heat capacity   A measure of the amount of heat that must be added to a material to change its temperature.

heat flow   The rate at which heat rises from the Earth's interior up to the surface.

heat-transfer melting   Melting that results from the transfer of heat from a hotter magma to a cooler rock.

heliocentric Universe concept   An idea proposed by Greek philosophers around 250 B.C.E. suggesting that all heavenly objects including the Earth orbited the Sun.

Hercynian orogen   The late Paleozoic orogen that affected parts of Europe; a continuation of the Alleghenian orogen.

heterosphere   A term for the upper portion of the atmosphere in which gases separate into distinct layers based on composition.

hiatus   The interval of time between deposition of the youngest rock below an unconformity and deposition of the oldest rock above the unconformity.

high-altitude westerlies   Westerly winds at the top of the troposphere.

high-grade metamorphic rocks   Rocks that metamorphose under relatively high temperatures.

high-level waste   Nuclear waste containing greater than 1 million times the safe level of radioactivity.

hinge   The portion of a fold where curvature is greatest.

hogback   *Cuesta.*

Holocene   The period of geologic time since the last glaciation.

Holocene climatic optimum   The period from 5,000 to 6,000 years ago, when Holocene temperatures reached a peak.

homosphere   The lower part of the atmosphere, in which the gases have stirred into a homogenous mixture.

hoodoo   The local name for the brightly colored shale and sandstone chimneys found in Bryce Canyon National Park in Utah.

horn   A pointed mountain peak surrounded by at least three cirques.

hornfels   Rock that undergoes metamorphism simply because of a change in temperature, without being subjected to differential stress.

horse latitudes   The region of the subtropical high in which winds are weak.

horst   The high block between two grabens.

hot spot   A location at the base of the lithosphere, at the top of a mantle plume, where temperatures can cause melting.

hot-spot track   A chain of now-dead volcanoes transported off the hot spot by the movement of a lithosphere plate.

hot-spot volcano   An isolated volcano not caused by movement at a plate boundary, but rather by the melting of a mantle plume.

hot spring   A spring that emits water ranging in temperature from about 30° to 104°C.

hummocky surface   An irregular and lumpy ground surface.

hurricane   A huge rotating storm, resembling a giant spiral in map view, in which sustained winds blow over 119 km per hour.

hurricane track   The path a hurricane follows.

hyaloclastite   A rubbly extrusive rock consisting of glassy debris formed in a submarine or sub-ice eruption.

hydration   The absorption of water into the crystal structure of minerals; a type of chemical weathering.

hydraulic conductivity   The coefficient K in Darcy's law; hydraulic conductivity takes into account the permeability of the sediment or rock as well as the fluid's viscosity.

hydraulic gradient   The slope of the water table.

hydrocarbon   A chain-like or ring-like molecule made of hydrogen and carbon atoms; petroleum and natural gas are hydrocarbons.

hydrologic cycle   The continual passage of water from reservoir to reservoir in the Earth system.

hydrolysis   The process in which water chemically reacts with minerals and breaks them down.

hydrosphere   The Earth's water, including surface water (lakes, rivers, and oceans), groundwater, and liquid water in the atmosphere.

hydrothermal deposit   An accumulation of ore minerals precipitated from hot-water solutions circulating through a magma or through the rocks surrounding an igneous intrusion.

hypsometric curve   A graph that plots surface elevation on the vertical axis and the percentage of the Earth's surface on the horizontal axis.

ice age   An interval of time in which the climate was colder than it is today, glaciers occasionally advanced to cover large areas of the continents, and mountain glaciers grew; an ice age can include many glacials and interglacials.

iceberg   A large block of ice that calved off the front of a glacier and dropped into the sea.

icehouse period   A period of time when the Earth's temperature was cooler than it is today and ice ages could occur.

ice-margin lake   A meltwater lake formed along the edge of a glacier.

ice-rafted sediment   Sediment carried out to sea by icebergs.

ice sheet   A vast glacier that covers the landscape.

ice shelf   Sea ice that surrounds and protrudes outward from land.

ice stream   A portion of a glacier that travels much more quickly than adjacent portions of the glacier.

ice tongue   The portion of a valley glacier that flowed out into the sea.

igneous rock   Rock that forms when hot molten rock (magma or lava) cools and freezes solid.

ignimbrite   Rock formed when deposits of pyroclactic flows solidify.

inactive fault   A fault that last moved in the distant past and probably won't move again in the near future, yet is still recognizable because of displacement across the fault plane.

inactive sand   The sand along a coast that is buried beneath a layer of active sand and moves only during severe storms or not at all.

incised meander   A meander that lies at the bottom of a steep-walled canyon.

index minerals   Minerals that serve as good indicators of metamorphic grade.

induced seismicity   Seismic events caused by people.

industrial minerals   Minerals that serve as the raw materials for manufacturing chemicals, concrete, and wallboard, among other products.

inequant   A term for a mineral grain whose length and width are different lengths.

inertia   The tendency of an object at rest to remain at rest.

infiltrate   Seep down into.

injection well   A well in which a liquid is pumped down into the ground under pressure so that it passes from the well back into the pore space of the rock or regolith.

inner core   The inner section of the core 5,155 km deep to the Earth's center at 6,371 km, and consisting of solid iron alloy.

inselberg   An isolated mountain or hill in a desert landscape created by progressive cliff retreat, so that the hill is surrounded by a pediment or an alluvial fan.

insolation   Exposure to the Sun's rays.

interglacial   A period of time between two glaciations.

interior basin   A basin with no outlet to the sea.

interlocking texture   The texture of crystalline rocks in which mineral grains fit together like pieces of a jigsaw puzzle.

internal process   A process in the Earth system, such as plate motion, mountain building, or volcanism, ultimately caused by Earth's internal heat.

intertidal zone   The area of coastal land across which the tide rises and falls.

intertropical convergence zone   The equatorial convergence zone in the atmosphere.

intraplate earthquakes   Earthquakes that occur away from plate boundaries.

intrusive contact   The boundary between country rock and an intrusive igneous rock.

intrusive igneous rock   Rock formed by the freezing of magma underground.

ionosphere The interval between 50 and 400 km distance from the Earth containing abundant positive ions.

iron catastrophe The proposed event very early in Earth history when the Earth partly melted and molten iron sank to the center to form the core.

isobar A line on a map along which the air has a specified pressure.

isograd A line along a pressure-temperature graph along which all points are taken to be at the same metamorphic grade.

isostasy (or isostatic equilibrium) The condition that exists when the buoyancy force pushing lithosphere up equals the gravitational force pulling lithosphere down.

isostatic compensation The process in which the surface of the crust slowly rises or falls to reestablish isostatic equilibrium after a geologic event changes the density or thickness of the lithosphere.

isotherm Lines on a map or cross section along which the temperature is constant.

isotopes Different species of a given element that have the same atomic number but different atomic weights.

jet stream A fast-moving current of air that flows at high elevations.

jetty A manmade wall that protects the entrance to a harbor.

joints Naturally formed cracks in rocks.

joint set A group of systematic joints.

Jovian A term used to describe the outer gassy, Jupiter-like planets (gas-giant planets).

kame A stratified sequence of lateral-moraine sediment that's sorted by water flowing along the edge of a glacier.

karst landscape A region underlain by caves in limestone bedrock; the collapse of the caves creates a landscape of sinkholes separated by higher topography, or of limestone spires separated by low areas.

kerogen The waxy molecules into which the organic material in shale transforms on reaching 100°C.

kettle hole A circular depression in the ground made when a block of ice calves off the toe of a glacier, becomes buried by till, and later melts.

knob-and-kettle topography A land surface with many kettle holes separated by round hills of glacial till.

K-T boundary event The mass extinction that happened at the end of the Cretaceous period, 65 million years ago, possibly because of the collision of a bolide with the Earth.

lag deposit The coarse sediment left behind in a desert after wind erosion removes the finer sediment.

lagoon A body of shallow seawater separated from the open ocean by a barrier island.

lahar A thick slurry formed when volcanic ash and debris mix with water, either in rivers or from rain or melting snow and ice on the flank of a volcano.

landslide A sudden movement of rock and debris down a nonvertical slope.

landslide-potential map A map that ranks regions according to the likelihood that a mass movement will occur.

land subsidence Sinking elevation of the ground surface; the process may occur over an aquifer that is slowly draining and decreasing in volume because of pore collapse.

La Niña Years in which the El Niño event is not strong.

lapilli Marble-to-plum-sized fragments of pyroclastic debris.

Laramide orogeny The mountain-building event that lasted from about 80 Ma to 40 Ma, in western North America; in the United States, it formed the Rocky Mountains as a result of basement uplift and the warping of the younger overlying strata into large monoclines.

latent heat of condensation The heat released during condensation, which comes only from a change in state.

lateral moraine A strip of debris along the margins of a glacier.

laterite soil Soil formed over iron-rich rock in a tropical environment, consisting primarily of a dark-red mass of insoluble iron and/or aluminum oxide.

Laurentia A continent in the early Paleozoic era composed of today's North America and Greenland.

Laurentide ice sheet An ice sheet that spread over northeastern Canada during the Pleistocene ice age.

lava Molten rock that has flowed out onto the Earth's surface.

lava dome A dome-like mass of rhyolitic lava that accumulates above the eruption vent.

lava flows Sheets or mounds of lava that flow onto the ground surface or sea floor in molten form and then solidify.

lava lake A large pool of lava produced around a vent when lava fountains spew forth large amounts of lava in a short period of time.

lava tube The empty space left when a lava tunnel drains; this happens when the surface of a lava flow solidifies while the inner part of the flow continues to stream downslope.

leach To dissolve and carry away.

leader A conductive path stretching from a cloud toward the ground, along which electrons leak from the base of the cloud, and which provides the start for a lightning flash to the ground.

lightning flash A giant spark or pulse of current that jumps across a gap of charge separation.

light year The distance that light travels in one Earth year (about 6 trillion miles or 9.5 trillion km).

lignite Low-rank coal that consists of 50% carbon.

limb The side of a fold, showing less curvature than at the hinge.

limestone Sedimentary rock composed of calcite.

liquification The process in which clay flakes in wet sediment unstick from one another in clay-rich sediments so that the sediment becomes a slurry of mud and water; liquification may be triggered by earthquake vibrations.

lithification The transformation of loose sediment into solid rock through compaction and cementation.

lithologic correlation A correlation based on similarities in rock type.

lithosphere The relatively rigid, nonflowable, outer 100–150-thick layer of the Earth; comprising the crust and the top part of the mantle.

little ice age A period of cooler temperatures between 1500 and 1800 C.E.

local base level A base level upstream from a drainage network's mouth.

lodgment till A flat layer of till smeared out over the ground when a glacier overrides an end moraine as it advances.

loess Layers of fine-grained sediments deposited from the wind; large deposits of loess formed from fine-grained glacial sediment blown off outwash plains.

longitudinal (seif) dune A dune formed when there is abundant sand and a strong, steady wind, and whose axis lies parallel to the wind direction.

longitudinal profile A cross-sectional image showing the variation in elevation along the length of a river.

longshore current A current that flows parallel to a beach.

lower mantle The deepest section of the mantle, stretching from 670 km down to the core-mantle boundary.

low-grade metamorphic rocks Rocks that metamorphose under relatively low temperatures.

low-velocity zone The asthenosphere underlying oceanic lithosphere in which seismic waves travel more slowly, probably because rock has partially melted.

luster The way a mineral surface scatters light.

L-waves Surface seismic waves that cause the ground to ripple back and forth, creating a snake-like movement.

Ma Million years ago (abbreviation).

macrofossil A fossil large enough to be seen with the naked eye.

mafic A term used in reference to magmas or igneous rocks that are relatively poor in silica and rich in iron and magnesium.

magma Molten rock beneath the Earth's surface.

magma chamber A space below ground filled with magma.

magma contamination The process in which flowing magma incorporates components of the country rock through which it passes.

magmatic deposit An ore deposit formed when sulfide ore minerals accumulate at the bottom of a magma chamber.

magnetic anomaly The difference between the expected strength of the Earth's dipole field at a certain location and the actual measured strength of the field at that location.

magnetic declination The angle between the direction a compass needle points at a given location and the direction of true north.

magnetic field The region affected by the force emanating from a magnet.

magnetic field lines The trajectories along which magnetic particles would align, or charged particles would flow, if placed in the field.

magnetic force The push or pull exerted by a magnet.

magnetic inclination The angle between a magnetic needle free to pivot on a horizontal axis and a horizontal plane parallel to the Earth's surface.

magnetic reversal The change of the Earth's magnetic polarity; when a reversal occurs, the field flips from normal to reversed polarity, or vice versa.

magnetic-reversal chronology The history of magnetic reversals through geologic time.

magnetization The degree to which a material can exert a magnetic force.

magnetometer An instrument that measures the strength of the Earth's magnetic field.

magnetosphere The region protected from the electrically charged particles of the solar winds by Earth's magnetic field.

magnetostratigraphy The comparison of the pattern of magnetic reversals in a sequence of strata, with a reference column showing the succession of reversals through time.

manganese nodules Lumpy accumulations of manganese-oxide minerals precipitated onto the sea floor.

mantle The thick layer of rock below the Earth's crust and above the core.

mantle plume A column of very hot rock rising up through the mantle.

marble A metamorphic rock composed of calcite and transformed from a protolith of limestone.

mare The broad darker areas on the Moon's surface, which consist of flood basalts that erupted over 3 billion years ago and spread out across the Moon's lowlands.

marginal sea A small ocean basin created when sea-floor spreading occurs behind an island arc.

maritime tropical air mass A mass of air that originates over tropical or subtropical oceanic regions.

marsh A wetland dominated by grasses.

mass-extinction event A time when vast numbers of species abruptly vanish.

mass movement (or mass wasting) The gravitationally caused transport of rock, regolith, snow, or ice downslope.

matrix Finer-grained material surrounding larger grains in a rock.

meander A snake-like curve along a stream's course.

meandering stream A reach of stream containing many meanders (snake-like curves).

meander neck A narrow isthmus of land separating two adjacent meanders.

mean sea level The average level between the high and low tide over a year at a given point.

mechanical weathering *Physical weathering.*

medial moraine A strip of sediment in the interior of a glacier, parallel to the flow direction of the glacier, formed by the lateral moraines of two merging glaciers.

medieval warm period A period of high temperatures in the Middle Ages.

melt Molten (liquid) rock.

meltdown The melting of the fuel rods in a nuclear reactor that occurs if the rate of fission becomes too fast and the fuel rods become too hot.

melting curve The line defining the range of temperatures and pressures at which a rock melts.

melting temperature The temperature at which the thermal vibration of the atoms or ions in the lattice of a mineral is sufficient to break the chemical bonds holding them to the lattice, so a material transforms into a liquid.

meltwater lake A lake fed by glacial meltwater.

Mercalli intensity scale An earthquake characterization scale based on the amount of damage that the earthquake causes.

mesa A large, flat-topped hill (with a surface area of several square km) in an arid region.

mesopause The boundary that marks the top of the mesosphere.

mesosphere The cooler layer of atmosphere overlying the stratosphere.

Mesozoic The middle of the three Phanerozoic eras; it lasted from 245 Ma to 65 Ma.

metal A solid composed almost entirely of atoms of metallic elements; it is generally opaque, shiny, smooth, malleable, and can conduct electricity.

metallic bond A chemical bond in which the outer atoms are attached to each other in such a way that electrons flow easily from atom to atom.

metamorphic aureole The region around a pluton, stretching tens to hundreds of meters out, in which heat transferred into the country rock and metamorphosed the country rock.

metamorphic facies A set of metamorphic mineral assemblages indicative of metamorphism under a specific range of pressures and temperatures.

metamorphic foliation A fabric defined by parallel surfaces or layers that develop in a rock as a result of metamorphism; schistocity and gneissic layering are examples.

metamorphic mineral assemblage A group of minerals that form in a rock as a result of metamorphism.

metamorphic rock Rock that forms when preexisting rock changes into new rock as a result of an increase in pressure and temperature and/or shearing under elevated temperatures; metamorphism occurs without the rock first becoming a melt or a sediment.

metamorphic zone The region between two metamorphic isograds, typically named after an index mineral found within the region.

metamorphism The process by which one kind of rock transforms into a different kind of rock.

metasomatism The process by which a rock's overall chemical composition changes during metamorphism because of reactions with hot water that bring in or remove elements.

meteoric water Water that falls to Earth from the atmosphere as either rain or snow.

meteorite A piece of rock or metal alloy that fell from space and landed on Earth.

micrite Limestone consisting of lime mud (that is, very fine-grained limestone).

microfossil A fossil that can be seen only with a microscope or an electron microscope.

mid-latitude (wave) cyclone The circulation of air around large, low-pressure masses.

mid-ocean ridge A 2-km-high submarine mountain belt that forms along a divergent oceanic plate boundary.

migmatite A rock formed when gneiss is heated high enough so that it begins to partially melt, creating layers, or lenses, of new igneous rock that mix with layers of the relict gneiss.

Milankovitch cycles Climate cycles that occur over tens to hundreds of thousands of years, because of changes in Earth's orbit and tilt.

mine A site at which ore is extracted from the ground.

mineral A homogenous, naturally occurring, solid inorganic substance with a definable chemical composition and an internal structure characterized by an orderly arrangement of atoms, ions, or molecules in a lattice.

mineral classes Groups of minerals distinguished from each other based on chemical composition.

mineral resources The minerals extracted from the Earth's upper crust for practical purposes.

Mississippi Valley–type (MVT) ore An ore deposit, typically in dolostone, containing lead- and zinc-bearing minerals that precipitated from groundwater that had moved up from several km depth in the upper crust; such deposits occur in the upper Mississippi Valley.

Moho The seismic-velocity discontinuity that defines the boundary between the Earth's crust and mantle.

Mohs hardness scale A list of ten minerals in a sequence of relative hardness, with which other minerals can be compared.

mold A cavity in sedimentary rock left behind when a shell that once filled the space weathers out.

monocline A fold whose shape resembles that of a carpet draped over a stair step.

monsoon A seasonal reversal in wind direction that causes a shift from a very dry season to a very rainy season in some regions of the world.

moraine A sediment pile composed of till deposited by a glacier.

mountain front The boundary between a mountain range and adjacent plains.

mountain (alpine) glacier A glacier that exists in or adjacent to a mountainous region.

mountain ice cap A mound of ice that submerges peaks and ridges at the crest of a mountain range.

mouth The outlet of a stream where it discharges into another stream, a lake, or a sea.

mudflow A downslope movement of mud at slow to moderate speed.

mud pot A viscous slurry that forms in a geothermal region when hot water rises into soils rich in volcanic ash and clay.

mudstone Very fine-grained sedimentary rock that will not easily split into sheets.

mylonite Rock formed during dynamic metamorphism and characterized by foliation that lies roughly parallel to the fault (shear zone) involved in the shearing process; mylonites have very fine grains formed by the nonbrittle subdivision of larger grains.

native metal A naturally occurring pure mass of a single metal in an ore deposit.

natural arch An arch that forms when erosion along joints leaves narrow walls of rock; when the lower part of the wall erodes while the upper part remains, an arch results.

natural levees A pair of low ridges that appear on either side of a stream and develop as a result of the accumulation of sediment deposited naturally during flooding.

natural selection The process by which the fittest organisms survive to pass on their characteristics to the next generation.

neap tide An especially low tide that occurs when the angle between the direction of the Moon and the direction of the Sun is 90°.

nebula A cloud of gas or dust in space.

nebula theory of planet formation The concept that planets grow out of planetary nebula.

negative anomaly An area where the magnetic field strength is less than expected.

negative feedback Feedback that slows a process down or reverses it.

Nevadan orogeny A convergent-margin mountain-building event that took place in western North America during the Late Jurassic period.

nonconformity A type of unconformity at which sedimentary rocks overlie basement (older intrusive igneous rocks and/or metamorphic rocks).

nonflowing artesian well An artesian well in which water rises on its own up to a level that lies below the ground surface.

nonfoliated metamorphic rock Rock containing minerals that recrystallized during metamorphism, but which has no foliation.

nonmetallic mineral resources Mineral resources that do not contain metals; examples include building stone, gravel, sand, gypsum, phosphate, and salt.

nonplunging fold A fold with a horizontal hinge.

nonrenewable resource A resource that nature will take a long time (hundreds to millions of years) to replenish or may never replenish.

nonsystematic joints Short cracks that come in a range of orientations and are randomly placed and oriented.

nor'easter A large, mid-latitude North American cyclone; when it reaches the East Coast, it produces strong winds that come out of the northeast.

normal fault A fault in which the hanging-wall block moves down the slope of the fault.

normal force The component of the gravitational force acting perpendicular to a slope.

normal polarity Polarity in which the paleomagnetic dipole has the same orientation as it does today.

normal stress The push or pull that is perpendicular to a surface.

North Atlantic deep-water mass The mass of cold, dense water that sinks in the north polar regions.

northeast tradewinds Surface winds that come out of the northeast and occur in the region between the equator and 30°N.

nuclear fuel Pellets of concentrated uranium oxide or a comparable radioactive material that can provide energy in a nuclear reactor.

nuclear fusion The process by which the nuclei of atoms fuse together, thereby creating new, larger atoms.

nuclear reactor The part of a nuclear power plant where the fission reactions occur.

nuée ardente *Pyroclastic flow.*

oasis A verdant region surrounded by desert, occurring at a place where natural springs provide water at the surface.

oblique-slip fault A fault in which sliding occurs diagonally along the fault plane.

obsidian An igneous rock consisting of a solid mass of volcanic glass.

occluded front A front that no longer intersects the ground surface.

oceanic crust The crust beneath the oceans; composed of gabbro and basalt, overlain by sediment.

oceanic lithosphere Lithosphere topped by oceanic crust; it reaches a thickness of 100 km.

Oil Age The period of human history, including our own, so named because the economy depends on oil.

oil field A region containing a significant amount of accessible oil underground.

oil reserve The known supply of oil held underground.

oil shale Shale containing kerogen.

oil trap A geologic configuration that keeps oil underground in the reservoir rock and prevents it from rising to the surface.

oil window The narrow range of temperatures under which oil can form in a source rock.

olistotrome A large, submarine slump block, buried and preserved.

ophiolite A slice of oceanic crust that has been thrust onto continental crust.

ordinary well A well whose base penetrates below the water table, and can thus provide water.

ore Rock containing native metals or a concentrated accumulation of ore minerals.

ore deposit An economically significant accumulation of ore.

ore minerals Minerals that have metal in high concentrations and in a form that can be easily extracted.

organic carbon Carbon that has been incorporated in an organism.

organic chemical A carbon-containing compound that occurs in living organisms, or that resembles such compounds; it consists of carbon atoms bonded to hydrogen atoms along with varying amounts of oxygen, nitrogen, and other chemicals.

organic coast A coast along which living organisms control landforms along the shore.

organic sedimentary rock Sedimentary rock (such as coal) formed from carbon-rich relicts of plants.

organic shale Lithified, muddy, organic-rich ooze that contains the raw materials from which hydrocarbons eventually form.

orogen (or orogenic belt) A linear range of mountains.

orogenic collapse The process in which mountains begin to collapse under their own weight and spread out laterally.

orogeny A mountain-building event.

orographic barrier A landform that diverts air flow upward or laterally.

outcrop An exposure of bedrock.

outer core The section of the core, between 2,900 and 5,150 km deep, that consists of liquid iron alloy.

outwash plain A broad area of gravel and sandbars deposited by a braided stream network, fed by the meltwater of a glacier.

overburden The weight of overlying rock on rock buried deeper in the Earth's crust.

overriding plate (or slab) The plate at a subduction zone that overrides the downgoing plate.

oversaturated solution A solution that contains so much solute (dissolved ions) that precipitation begins.

oversized stream valley A large valley with a small stream running through it; the valley formed earlier when the flow was greater.

oxbow lake A meander that has been cut off yet remains filled with water.

oxidation reaction A reaction in which an element loses electrons; an example is the reaction of iron with air to form rust.

ozone O_3, an atmospheric gas that absorbs harmful ultraviolet radiation from the Sun.

ozone hole An area of the atmosphere, over polar regions, from which ozone has been depleted.

pahoehoe A lava flow with a surface texture of smooth, glassy, rope-like ridges.

paleoclimate The past climate of the Earth.

paleomagnetism The record of ancient magnetism preserved in rock.

paleopole The supposed position of the Earth's magnetic pole in the past, with respect to a particular continent.

paleosol Ancient soil preserved in the stratigraphic record.

Paleozoic The oldest era of the Phanerozoic eon.

Pangaea A supercontinent that assembled at the end of the Paleozoic era.

Pannotia A supercontinent that may have existed sometime between 800 Ma and 600 Ma.

parabolic dunes Dunes formed when strong winds break through transverse dunes to make new dunes whose ends point upwind.

parallax The apparent movement of an object seen from two different points not on a straight line from the object (for example, from your two different eyes).

parallax method A trigonometric method used to determine the distance from the Earth to a nearby star.

parent isotope A radioactive isotope that undergoes decay.

partial melting The melting in a rock of the minerals with the lowest melting temperatures, while other minerals remain solid.

passive margin A continental margin that is not a plate boundary.

passive-margin basin A thick accumulation of sediment along a tectonically inactive coast, formed over crust that stretched and thinned when the margin first began.

patterned ground A polar landscape in which the ground splits into pentagon or hexagon shapes.

pause An elevation in the atmosphere where temperature stops decreasing and starts increasing, or vice versa.

peat Compacted and partially decayed vegetation accumulating beneath a swamp.

pedalfer soil A temperate-climate soil formed on granite and characterized by well-defined soil horizons and an organic A-horizon.

pediment The broad, nearly horizontal bedrock surface at the base of a retreating desert cliff.

pedocal soil Thin soil formed in desert climates and containing very little organic matter.

pegmatite A coarse-grained igneous rock containing crystals of up to tens of centimeters across and occurring in dike-shaped intrusions.

peidmont glacier A fan or lobe of ice that forms where a valley glacier emerges from a valley and spreads out into the adjacent plain.

pelagic sediment Microscopic plankton shells and fine flakes of clay that settle out and accumulate on the deep-ocean floor.

Pelé's hair Droplets of basaltic lava that mold into long glassy strands as they fall.

Pelé's tears Droplets of basaltic lava that mold into tear-shaped glassy beads as they fall.

peneplain A nearly flat surface that lies at an elevation close to sea level; thought to be the product of long-term erosion.

perched water table A quantity of groundwater that lies above the regional water table because an underlying lens of impermeable rock or sediment prevents the water from sinking down to the regional water table.

percolation The process in which groundwater meanders through tiny, crooked channels in the surrounding material.

peridotite A coarse-grained ultramafic rock.

periglacial environment A region with widespread permafrost but without a blanket of snow or ice.

period An interval of geologic time representing a subdivision of a geologic era.

permafrost Permanently frozen ground.

permanent magnet A special material that behaves magnetically for a long time all by itself.

permanent stream A stream that flows year-round because its bed lies below the water table, or because more water is supplied from upstream than can infiltrate into the ground.

permeability The degree to which a material allows fluids to pass through it via an interconnected network of pores and cracks.

permineralization The fossilization process in which plant material becomes transformed into rock by the precipitation of silica from groundwater.

petrified A term used by geologists to describe plant material that has transformed into rock by permineralization.

petroglyph Drawings formed by chipping into the desert varnish of rocks to reveal the lighter rock beneath.

phaneritic A textural term used to describe coarse-grained igneous rock.

Phanerozoic eon The most recent eon, an interval of time from 545 Ma to the present.

phenocryst A large crystal surrounded by a finer-grained matrix in an igneous rock.

photochemical smog Brown haze that blankets a city when exhaust from cars and trucks reacts in the presence of sunlight.

photosynthesis The process by which chlorophyll-containing plants remove carbon dioxide from the atmosphere, form tissues, and expel oxygen back to the atmosphere.

phreatomagmatic eruption An explosive eruption that occurs when water is introduced into the magma chamber.

phyllite A fine-grained metamorphic rock with a foliation caused by the preferred orientation of very fine-grained muscovite or chlorite.

phyllitic luster A silk-like sheen characteristic of phyllite, a result of the rock's fine-grained mica.

phylogenetic tree A chart representing the ideas of paleontologists showing which groups of organisms radiated from which ancestors.

physical weathering The process in which intact rock breaks into smaller grains or chunks.

pillow basalt Glass-encrusted basalt blobs formed when magma is extruded on the sea floor and cools very quickly.

placer deposit Concentrations of metal grains in stream sediment developed when rocks containing native metals erode and create a mixture of sand grains and metal fragments; the moving water of the stream carries away lighter mineral grains.

planetesimal Tiny, solid pieces of rock and metal that collect in a planetary nebula and eventually accumulate to form a planet.

plankton Tiny plants and animals that float in sea or lake water.

plastic deformation The deformational process in which mineral grains behave like plastic and, when compressed or sheared, become flattened or elongate without cracking or breaking.

plate One of about twenty distinct pieces of the relatively rigid lithosphere.

plate boundary The border between two adjacent lithosphere plates.

plate-boundary earthquakes The earthquakes that occur along and define plate boundaries.

plate-boundary volcano A volcanic arc or mid-ocean ridge volcano, formed as a consequence of movement along a plate boundary.

plate interior A region away from the plate boundaries that consequently experiences few earthquakes.

plate tectonics *Theory of plate tectonics.*

playa The flat, typically salty lake bed that remains when all the water evaporates in drier times; forms in desert regions.

Pleistocene ice age The period of time from about 2 Ma to 14,000 years ago, during which the Earth experienced an ice age.

plunge pool A depression at the base of a waterfall scoured by the energy of the falling water.

plunging fold A fold with a tilted hinge.

pluton An irregular or blob-shaped intrusion; can range in size from tens of m across to tens of km across.

pluvial lake A lake formed to the south of a continental glacier as a result of enhanced rainfall during an ice age.

point bar A wedge-shaped deposit of sediment on the inside bank of a meander.

polar cell A high-latitude convection cell in the atmosphere.

polar easterlies Prevailing winds that come from the east and flow from the polar high to the subpolar low.

polar front The convergence zone in the atmosphere at latitude 60°.

polar glacier *Dry-bottom glacier.*

polar high The zone of high pressure in polar regions created by the sinking of air in the polar cells.

polarity The orientation of a magnetic dipole.

polarity chron The time interval between polarity reversals of Earth's magnetic field.

polarity subchron The time interval between magnetic reversals if the interval is of short duration (less than 200,000 years long).

polarized light A beam of filtered light waves that all vibrate in the same plane.

polar wander The phenomenon of the progressive changing through time of the position of the Earth's magnetic poles relative to a location on a continent; significant polar wander probably doesn't occur—in fact, poles seem to remain fairly fixed, while continents move.

polar-wander path The curving line representing the apparent progressive change in the position of the Earth's magnetic pole, relative to a locality X, assuming that the position of X on Earth has been fixed through time (in fact, poles stay fixed while continents move).

pollen Tiny grains involved in plant reproduction.

polymorphs Two minerals that have the same chemical composition but a different crystal lattice structure.

pore A small open space within sediment or rock.

pore collapse The closer packing of grains that occurs when groundwater is extracted from pores, thus eliminating the support holding the grains apart.

porosity The total volume of empty space (pore space) in a material, usually expressed as a percentage.

porphyritic A textural term for igneous rock that has phenocrysts distributed throughout a finer matrix.

positive anomaly An area where the magnetic field strength is stronger than expected.

positive-feedback mechanism A mechanism that enhances the process that causes the mechanism in the first place.

potentiometric surface The elevation to which water in an artesian system would rise if unimpeded; where there are flowing artesian wells, the potentiometric surface lies above ground.

pothole A bowl-shaped depression carved into the floor of a stream by a long-lived whirlpool carrying sand or gravel.

Precambrian The interval of geologic time between Earth's formation about 4.6 billion years ago and the beginning of the Phanerozoic eon 545 million years ago.

precession The gradual conical path traced out by Earth's spinning axis; simply put, it is the "wobble" of the axis.

precious metals Metals (like gold, silver, and platinum) that have high value.

precipitation (1) The process by which atoms dissolved in a solution come together and form a solid; (2) rainfall or snow.

preferred mineral orientation The metamorphic texture in which platy grains lie parallel to one another and/or elongate grains align in the same direction.

pressure Force per unit area, or the "push" acting on a material in cases where the push is the same in all directions.

pressure gradient The rate of pressure change over a given horizontal distance.

pressure solution The process of dissolution at points of contact where pressure is greatest, producing ions that then precipitate elsewhere, where compression is less.

prevailing winds Belts in which surface winds generally flow in a consistent direction.

primary porosity The space that remains between solid grains or crystals immediately after sediment accumulates or rock forms.

principal aquifer The geologic unit that serves as the primary source of groundwater in a region.

principle of baked contacts When an igneous intrusion "bakes" (metamorphoses) surrounding rock, the rock that has been baked must be older than the intrusion.

principle of cross-cutting relations If one geologic feature cuts across another, the feature that has been cut is older.

principle of fossil succession In a stratigraphic sequence, different species of fossil organisms appear in a definite order; once a fossil species disappears in a sequence of strata, it never reappears higher in the sequence.

principle of inclusions If a rock contains fragments of another rock, the fragments must be older than the rock containing them.

principle of original continuity Sedimentary layers, before erosion, formed fairly continuous sheets over a region.

principle of original horizontality Layers of sediment, when originally deposited, are fairly horizontal.

principle of superposition In a sequence of sedimentary rock layers, each layer must be younger than the one below, for a layer of sediment cannot accumulate unless there is already a substrate on which it can collect.

principle of uniformitariansim The physical processes we observe today also operated in the past in the same way, and at comparable rates.

prograde metamorphism Metamorphism that occurs as temperatures and pressures are increasing.

Proterozoic The most recent of the Precambrian eons.

protocontinent A block of crust composed of volcanic arcs and hot-spot volcanoes sutured together.

protolith The original rock from which a metamorphic rock formed.

protoplanet A body that grows by the accumulation of planetesimals but has not yet become big enough to be called a planet.

protoplanetary nebula A ring of gas and dust that surrounded the newborn Sun, from which the planets were formed.

protostar A dense body of gas that is collapsing inward because of gravitational forces and may eventually become a star.

pumice A glassy igneous rock that forms from frothy lava and contains abundant (over 50%) pore space.

punctuated equilibrium The hypothesis that evolution takes place in fits and starts; evolution occurs very slowly for quite a while and then, during a relatively short period, takes place very rapidly.

P-waves Compressional seismic waves that move through the body of the Earth.

P-wave shadow zone A band between 103° and 143° from an earthquake epicenter, as measured along the circumference of the Earth, inside which P-waves do not arrive at seismograph stations.

pycnocline The boundary between layers of water of different densities.

pyroclastic debris Fragmented material that sprayed out of a volcano and landed on the ground or sea floor in solid form.

pyroclastic flow A fast-moving avalanche formed when hot volcanic ash and debris mix with air and flows down the side of a volcano.

pyroclastic rock Rock made from fragments blown out of a volcano during an explosion that were then packed or welded together.

quarry A site at which stone is extracted from the ground.

quartzite A metamorphic rock composed of quartz and transformed from a protolith of quartz sandstone.

quenching A sudden cooling of molten material to form a solid.

quick clay Clay that behaves like a solid when still (because of surface tension holding the water-coated clay flakes together), but that flows like a liquid when shaken.

radial network A drainage network in which the streams flow outward from a cone-shaped mountain, and define a pattern resembling spokes on a wheel.

radioactive decay The process by which a radioactive atom undergoes fission or releases particles.

radioactive isotope An unstable isotope of a given element.

radiometric dating The science of dating geologic events in years by measuring the ratio of parent atoms to daughter atoms in a rock's radioactive elements.

rain band A spiraling arm of a hurricane radiating outward from the eye.

rain shadow The inland side of a mountain range, arid because the mountains block rain clouds from reaching the area.

range (for fossils) The interval of a sequence of strata in which a specific fossil species appears.

rapids Particularly turbulent stream water that develops where water flows over a bed with clasts whose diameter approaches the water depth.

reach A specified segment of a stream's path.

recessional moraine The end moraine that forms when a glacier stalls for a while as it recedes.

recharge area A location where water enters the ground and infiltrates down to the water table.

recrystallization The process in which ions or atoms in minerals rearrange to form new minerals.

rectangular network A drainage network in which the streams join each other at right angles because of a rectangular grid of fractures that breaks up the ground and localizes channels.

recurrence interval The average time between successive geologic events.

red giant A huge red star that forms when Sun-sized stars start to die and expand.

red shift The phenomenon in which a source of light moving away from you very rapidly shifts to a lower frequency; that is, toward the red end of the spectrum.

reef bleaching The death and loss of color of a coral reef.

reflected ray A ray that bounces off a boundary between two different materials.

refracted ray A ray that bends as it passes through a boundary between two different materials.

refraction The bending of a ray as it passes through a boundary between two different materials.

reg A vast stony plain in a desert.

regional metamorphism *Dynamothermal metamorphism;* metamorphism of a broad region, usually the result of deep burial during an orogeny.

regolith Any kind of unconsolidated debris that covers bedrock.

regression The seaward migration of a shoreline caused by a lowering of the sea level.

relative age The age of one geologic feature with respect to another.

relative humidity The ratio between the measured water content of air and the maximum possible amount of water the air can hold at a given condition.

relative plate velocity The movement of one lithosphere plate with respect to another.

relief The difference in elevation between adjacent high and low regions on the land surface.

renewable resource A resource that can be replaced by nature within a short time span relative to a human life span.

reservoir rock Rock with high porosity and permeability, so it can contain an abundant amount of easily accessible oil.

residence time The average length of time that a substance stays in a particular reservoir.

residual mineral deposit Soils in which the residuum left behind after leaching by rainwater is so concentrated in metals that the soil itself becomes an ore deposit.

resurgent dome The new mound, or cone, of igneous rock that grows within a caldera as an eruption begins anew.

retrograde metamorphism Metamorphism that occurs as pressures and temperatures are decreasing; for retrograde metamorphism to occur, water must be added.

return stroke An upward-flowing electric current from the ground that carries positive charges up to a cloud during a lightning flash.

reversed polarity Polarity in which the paleomagnetic dipole points north.

reverse fault A steeply dipping fault on which the hanging-wall block slides up.

Richter magnitude scale A scale that defines earthquakes based on the amplitude of the largest ground motion recorded on a seismogram.

ridge axis The crest of a mid-ocean ridge; the ridge axis defines the position of a divergent plate boundary.

right-lateral strike-slip fault A strike-slip fault in which the block on the opposite fault plane from a fixed spot moves to the right of that spot.

rip current A strong, localized seaward flow of water perpendicular to a beach.

riprap Loose boulders or concrete piled together along a beach to absorb wave energy before it strikes a cliff face.

roche moutonnée A glacially eroded hill that becomes elongate in the direction of flow and asymmetric; glacial rasping smoothes the upstream part of the hill into a gentle slope, while glacial plucking erodes the downstream edge into a steep slope.

rock A coherent, naturally occurring solid, consisting of an aggregate of minerals or a mass of glass.

rock burst A sudden explosion of rock off the ceiling or wall of an underground mine.

rock cycle The succession of events that results in the transformation of Earth materials from one rock type to another, then another, and so on.

rock flour Fine-grained sediment produced by glacial abrasion of the substrate over which a glacier flows.

rock glacier A slow-moving mixture of rock fragments and ice.

rock slide A sudden downslope movement of rock.

rocky coast An area of coast where bedrock rises directly from the sea, so beaches are absent.

Rodinia A proposed Precambrian supercontinent that existed around 1 billion years ago.

rotational axis The imaginary line through the center of the Earth around which the Earth spins.

R-waves Surface seismic waves that cause the ground to ripple up and down, like water waves in a pond.

sabkah A region of once-flooded coastal desert in which stranded seawater has left a salt crust over a mire of organic-rich mud.

salinity The degree of concentration of salt in water.

saltation The movement of a sediment in which grains bounce along their substrate, knocking other grains into the water column (or air) in the process.

salt dome A rising bulbous dome of salt that bends up the adjacent layers of sedimentary rock.

salt wedging The process in arid climates by which dissolved salt in groundwater crystallizes and grows in open pore spaces in rocks and pushes apart the surrounding grains.

sand spit An area where the coastline indents landward so the beach stretches out into open water.

sandstone Coarse-grained sedimentary rock consisting almost entirely of quartz.

sand volcano (or sand blow) A small mound of sand produced when sand layers below the ground surface liquify as a result of seismic shaking, causing the sand to erupt through cracks or holes in overlying clay layers.

saprolite A layer of rotten rock created by chemical weathering in warm, wet climates.

Sargasso Sea The center of North Atlantic Gyre, named for the tropical seaweed sargassum, which accumulates in its relatively noncirculating waters.

saturated solution Water that carries as many dissolved ions as possible under given environmental conditions.

saturated zone The region below the water table, where pore space is filled with water.

scattering The dispersal of energy that occurs when light interacts with particles in the atmosphere.

schist A medium-to-coarse-grained metamorphic rock that possesses schistosity.

schistosity Foliation caused by the preferred orientation of large mica flakes.

scientific method A sequence of steps for systematically analyzing scientific problems in a way that leads to verifiable results.

scoria A glassy igneous rock containing abundant air-filled holes; scoria differs from pumice in that holes account for less than 50% of the rock volume.

scouring A process by which running water removes loose fragments of sediment from a stream bed.

sea arch An arch of land protruding out into the sea and connected to the mainland by a narrow bridge.

sea-floor spreading The gradual widening of an ocean basin as new oceanic crust forms at a mid-ocean ridge axis and then moves away from the axis.

sea ice Ice formed by the freezing of the surface of the sea.

seal A relatively impermeable rock such as shale, salt, or unfractured limestone, which lies above a reservoir rock and stops the oil from rising further.

seam A sedimentary bed of coal interlayered with other sedimentary rocks.

seamount An isolated submarine mountain.

seasonal floods Floods that appear almost every year during seasons when rainfall is heavy or when winter snows start to melt.

seasonal well A well that provides water only during the rainy season when the water table rises below the base of the well.

sea stack An isolated piece of land just offshore, disconnected from the mainland by the collapse of a sea arch.

seawall A wall of riprap built on the landward side of a backshore zone in order to protect shore cliffs from erosion.

second The basic unit of time measurement, now defined as the time it takes for the magnetic field of a cesium atom to flip polarity 9,192,631,770 times, as measured by an atomic clock.

secondary enrichment The process by which a new ore deposit forms from metals that were dissolved and carried away from preexisting ore minerals.

secondary porosity New pore space in rocks, created some time after a rock first forms.

secondary recovery technique A process used to extract the quantities of oil that will not come out of a reservoir rock with just simple pumping.

sediment An accumulation of loose mineral grains, such as boulders, pebbles, sand, silt, or mud, that are not cemented together.

sedimentary basin A depression, created as a consequence of subsidence, that fills with sediment.

sedimentary rock Rock that forms either by the cementing together of fragments broken off preexisting rock or by the precipitation of mineral crystals out of water solutions at or near the Earth's surface.

sedimentary sequence A grouping of sedimentary units bounded on top and bottom by regional unconformities.

sediment budget The proportion of sand supplied to sand removed.

sediment load The total volume of sediment carried by a stream.

sediment maturity The degree to which a sediment has evolved from a crushed-up version of the original rock into a sediment that has lost its easily weathered minerals and become well-sorted and rounded.

sediment sorting The segregation of sediment by size.

seep A place where oil-filled reservoir rock intersects the ground surface, or where fractures connect a reservoir to the ground surface, so that oil flows out onto the ground on its own.

seiche Rhythmic movement in a body of water caused by ground motion.

seismic belts (or zones) The relatively narrow strips of crust on Earth under which most earthquakes occur.

seismicity Earthquake activity.

seismic-moment magnitude scale A scale that defines earthquake size using calculations involving the amount of slip, length of rupture, depth of rupture, and rock strength.

seismic ray The changing position of an imaginary point on a wave front as the front moves through rock.

seismic-reflection profile A cross-sectional view of the crust made by measuring the reflection of artificial seismic waves off boundaries between different layers of rock in the crust.

seismic tomography Analysis by sophisticated computers of global seismic data in order to create a three-dimensional image of variations in seismic-wave velocities within the Earth.

seismic velocity The speed at which seismic waves travel.

seismic-velocity discontinuity A boundary in the Earth at which seismic velocity changes, abruptly.

seismic (earthquake) waves Waves of energy emitted at the focus of an earthquake.

seismogram The record of an earthquake produced by a seismograph.

seismograph (seismometer) An instrument that can record the ground motion from an earthquake.

semipermanent pressure cell A somewhat elliptical zone of high or low atmospheric pressure that lasts much of the year; it forms because high-pressure zones tend to be narrower over land than over sea.

Sevier orogeny A mountain-building event that affected western North America between about 150 Ma and 80 Ma, a result of convergent margin tectonism; a fold-thrust belt formed during this event.

shale Very fine-grained sedimentary rock that breaks into thin sheets.

shatter cones Small, cone-shaped fractures formed by the shock of a meteorite impact.

shear strain A change in shape of an object that involves the movement of one part of a rock body sideways past another part.

shear stress A stress that moves one part of a material sideways past another part.

shear waves Seismic waves in which particles of material move back and forth perpendicular to the direction in which the wave itself moves.

shear zone A fault in which movement has occurred ductilely.

sheetwash A film of water less than a few mm thick that covers the ground surface.

shield An older, interior region of a continent.

shield volcano A subaerial volcano with a broad, gentle dome, formed either from low-viscosity basaltic lava or from large pyroclastic sheets.

shocked quartz Grains of quartz that have been subjected to intense pressure such as that of an immense bolide impact.

shoreline The boundary between the water and land.

shortening The process during which a body of rock or a region of crust becomes shorter.

short-term climate change Climate change that takes place over hundreds to thousands of years.

Sierran arc A large continental volcanic arc along western North America that initiated at the end of the Jurassic period and lasted until about 80 million years ago.

silica SiO_2.

silicate minerals Minerals composed of silicon-oxygen tetrahedra linked in various arrangements; most contain other elements too.

silicate rock Rock composed of silicate minerals.

siliceous sedimentary rock Sedimentary rock that contains abundant quartz.

silicic Rich in silica with relatively little iron and magnesium.

sill A nearly horizontal table-top-shaped tabular intrusion which injects between the layers of country rock.

siltstone Fine-grained sedimentary rock.

sinkhole A circular depression in the land that forms when an underground cavern collapses.

slab-pull force The force that downgoing plates (or slabs) apply to oceanic lithosphere at a convergent margin.

slate Fine-grained, low-grade metamorphic rock, formed by the metamorphism of shale.

slaty cleavage The foliation typical of slate and reflective of the preferred orientation of slate's clay minerals, which allows slate to be split into thin sheets.

slickensides The polished surface of a fault caused by slip on the fault; lineated slickensides also have groves that indicate the direction of fault movement.

slip face The lee side of a dune, which sand slides down.

slip lineations Linear marks on a fault surface created during movement on the fault; some slip lineations are defined by grooves, some by aligned mineral fibers.

slope failure The downslope movement of material on an unstable slope.

slumping Downslope movement in which a mass of regolith detaches from its substrate along a spoon-shaped sliding surface and slips downward semicoherently.

smelting The heating of a metal-containing rock to high temperatures in a fire so that the rock will decompose to yield metal plus a nonmetallic residue (slag).

snotite A long gob of bacteria that slowly drips from the ceiling of a cave.

snow line The boundary above which snow remains all year.

soda straw A hollow stalactite in which calcite precipitates around the outside of a drip.

soil Sediment that has undergone changes at the surface of the Earth, including reaction with rainwater and the addition of organic material.

soil erosion The removal of soil by wind and runoff.

soil horizon Distinct zones within a soil, distinguished from each other by factors like chemical composition and organic content.

soil moisture Underground water that wets the surface of the mineral grains and organic material making up soil, but lies above the water table.

soil profile A vertical sequence of distinct zones of soil.

solar wind A stream of particles with enough energy to escape from the Sun's gravity and flow outward into space.

solid-state diffusion The slow movement of atoms or ions through a solid.

solifluction The type of creep characteristic of tundra regions; during the summer, the uppermost layer of permafrost melts, and the soggy, weak layer of ground then flows slowly downslope in overlapping sheets.

solstice A day in which the Sun is visible from one of the poles for the full twenty-four hours, while the opposite pole is in darkness for those twenty-four hours.

Sonoma orogeny A convergent-margin mountain-building event that took place on the western coast of North America in the Late Permian and Early Triassic periods.

sorting (1) The range of clast sizes in a collection of sediment; (2) the degree to which sediment has been separated by flowing currents into different-size fractions.

source rock A rock (organic-rich shale) containing the raw materials from which hydrocarbons eventually form.

southeast tradewinds Tradewinds in the Southern Hemisphere, which start flowing northward, deflect to the west, and end up flowing from southeast to northwest.

southern oscillation The oscillating of pressure cells back and forth across the Pacific Ocean, associated with El Niño.

specific gravity A number representing the density of a mineral, as specified by the ratio between the weight of a volume of the mineral and the weight of an equal volume of water.

speleothem A formation that grows in a limestone cave by the accumulation of travertine precipitated from water solutions dripping in a cave or flowing down the wall of a cave.

sphericity The measure of the degree to which a clast approaches the shape of a sphere.

spreading boundary *Divergent plate boundary.*

spreading rate The rate at which sea floor moves away from a mid-ocean ridge axis, as measured with respect to the sea floor on the opposite side of the axis.

spring A natural outlet from which groundwater flows up onto the ground surface.

spring tide An especially high tide that occurs when the Sun is on the same side of the Earth as the Moon.

stable air Air that does not have a tendency to rise rapidly.

stable slope A slope on which downward sliding is unlikely.

stalactite An icicle-like cone that grows from the ceiling of a cave as dripping water precipitates limestone.

stalagmite An upward-pointing cone of limestone that grows when drips of water hit the floor of a cave.

standing wave A wave whose crest and trough remain in place as water moves through the wave.

star dune A constantly changing dune formed by frequent shifts in wind direction; it consists of overlapping crescent dunes pointing in many different directions.

stick-slip behavior Stop-start movement along a fault plane caused by friction, which prevents movement until stress builds up sufficiently.

stone rings Ridges of cobbles between adjacent bulges of permafrost ground.

stoping A process by which magma intrudes; blocks of wall rock break off and then sink into the magma.

storm An episode of severe weather in which winds, precipitation, and in some cases lightning become strong enough to be bothersome and even dangerous.

storm-center velocity A storm's (hurricane's) velocity along its track.

storm surge Excess seawater driven landward by wind during a storm; the low atmospheric pressure beneath the storm allows the sea level to rise locally, increasing the surge.

strain The change in shape of an object in response to deformation (that is, as a result of the application of a stress).

stratified drift Glacial sediment that has been redistributed and stratified by flowing water.

stratigraphic column A cross-section diagram of a sequence of strata summarizing information about the sequence.

stratigraphic formation A recognizable layer of a specific sedimentary rock type or set of rock types, deposited during a certain time interval, that can be traced over a broad region.

stratigraphic sequence An interval of strata deposited during periods of relatively high sea level, and bounded above and below by regional unconformities.

stratopause The temperature pause that marks the top of the stratosphere.

stratosphere The stable, stratified layer of atmosphere directly above the troposphere.

stratovolcano A large, cone-shaped subaerial volcano consisting of alternating layers of lava and tephra.

stratus cloud A thin, sheet-like, stable cloud.

streak The color of the powder produced by pulverizing a mineral on an unglazed ceramic plate.

stream A ribbon of water that flows in a channel.

stream bed The floor of a stream.

stream capacity The total quantity of sediment a stream carries.

stream capture (or piracy) The situation in which headward erosion causes one stream to intersect the course of another, previously independent stream, so that the intersected stream starts to flow down the channel of the first stream.

stream competence The maximum particle size that a stream can carry.

stream gradient The slope of a stream's channel in the downstream direction.

stream rejuvination The renewed downcutting of a stream into a floodplain or peneplain, caused by a relative drop of the base level.

stress The push, pull, or shear that a material feels when subjected to a force; formally, the force applied per unit area over which the force acts.

stretching The process during which a layer of rock or a region of crust becomes longer.

striations Linear scratches in rock.

strike-slip fault A fault in which one block slides horizontally past another (and therefore parallel to the strike line), so there is no relative vertical motion.

strip mining The scraping off of all soil and sedimentary rock above a coal seam in order to gain access to the seam.

stromatolite Layered mounds of sediment formed by cyanobacteria; cyanobacteria secrete a mucuous-like substance to which sediment sticks, and as each layer of cyanobacteria gets buried by sediment, it colonizes the surface of the new sediment, building a mound upward.

structural control The condition in which geologic structures, such as faults, affect the distribution and drainage of water or the shape of the land surface.

subaerial Above sea level or the land surface.

subduction The process by which one oceanic plate bends and sinks down into the asthenosphere beneath another plate.

subduction zone The region along a convergent boundary where one plate sinks beneath another.

sublimation The evaporation of ice directly into vapor without first forming a liquid.

submarine canyon A narrow, steep canyon that dissects a continental shelf and slope.

submarine fan A wedge-shaped accumulation of sediment at the base of a submarine slope; fans usually accumulate at the mouth of a submarine canyon.

submarine slump The underwater downslope movement of a semicoherent block of sediment along a weak mud detachment.

submergent coast A coast at which the land is sinking relative to sea level.

subpolar low The rise of air where the surface flow of a polar cell converges with the surface flow of a Ferrel cell, creating a low-pressure zone in the atmosphere.

subsidence The vertical sinking of the Earth's surface in a region relative to a reference plane.

substrate A general term for material just below the ground surface.

subtropical high (subtropical divergence zone) A belt of high pressure in the atmosphere at 30° latitude formed where the Hadley cell converges with the Ferrel cell, causing cool, dense air to sink.

subtropics Desert climate regions that lie on either side of the equatorial tropics between the lines of 20° and 30° north or south of the equator.

summit eruption An eruption that occurs in the summit crater of a volcano.

sunspot cycle The cyclic appearance of large numbers of sunspots (black spots thought to be magnetic storms on the Sun's surface) every 9 to 11.5 years.

supercontinent cycle The process of change during which supercontinents develop and later break apart, forming pieces that may merge once again in geologic time to make yet another supercontinent.

supernova A short-lived, very bright object in space that results from the catacylsmic explosion marking the death of a very large star; the explosion ejects large quantities of matter into space to form new nebulae.

superplume A huge mantle plume.

superposed stream A stream whose geometry has been laid down on a rock structure and is not controlled by the structure.

surface current An ocean current in the top 100 m of water.

surface waves Seismic waves that travel along the Earth's surface.

surface westerlies The prevailing surface winds in North America and Europe, which come out of the west or southwest.

surf zone A region of the shore in which breakers crash onto the shore.

surge (glacial) A pulse of rapid flow in a glacier.

suspended load Tiny solid grains carried along by a stream without settling to the floor of the channel.

swamp A wetland dominated by trees.

swash The upward surge of water that flows up a beach slope when breakers crash onto the shore.

S-waves Seismic shear waves that pass through the body of the Earth.

S-wave shadow zone A band between 103° and 180° from the epicenter of an earthquake inside of which S-waves do not arrive at seismograph stations.

swelling clay Clay possessing a mineral structure that allows it to absorb water between its layers and thus swell to several times its original size.

symmetry The condition in which the shape of one part of an object is a mirror image of the other part.

syncline A trough-shaped fold whose limbs dip toward the hinge.

systematic joints Long planar cracks that occur fairly regularly throughout a rock body.

tabular intrusions Sheet intrusions that are planar and of roughly uniform thickness.

Taconic orogeny A convergent mountain-building event that took place around 400 million years ago, in which a volcanic island arc collided with eastern North America.

tailings pile A pile of waste rock from a mine.

talus A sloping apron of fallen rock along the base of a cliff.

tar Hydrocarbons that exist in solid form at room temperature.

tarn A lake that forms at the base of a cirque on a glacially eroded mountain.

tar sand Sandstone reservoir rock in which less viscous oil and gas molecules have either escaped or been eaten by microbes, so that only tar remains.

taxonomy The study and classification of the relationships between different forms of life.

tension A stress that pulls on a material and could lead to stretching.

tephra Unconsolidated accumulations of pyroclastic grains.

terminal moraine The end moraine at the farthest limit of glaciation.

terminator The boundary between the half of the Earth that has daylight and the half experiencing night.

terrace The elevated surface of an older floodplain into which a younger floodplain had cut down.

terrestrial A term used to describe the inner, Earth-like planets.

thalweg The deepest part of a stream's channel.

theory A scientific idea supported by an abundance of evidence that has passed many tests and failed none.

theory of plate tectonics The theory that the outer layer of the Earth (the lithosphere) consists of separate plates that move with respect to one another.

thermal metamorphism Metamorphism caused by heat conducted into country rock from an igneous intrusion.

thermocline A boundary between layers of water with differing temperatures.

thermohaline circulation The rising and sinking of water driven by contrasts in water density, which is due in turn to differences in temperature and salinity; this circulation involves both surface and deep-water currents in the ocean.

thermosphere The outermost layer of the atmosphere containing very little gas.

thin section A 3/100-mm-thick slice of rock that can be examined with a petrographic microscope.

thin-skinned deformation A distinctive style of deformation characterized by displacement on faults that terminate at depth along a subhorizontal detachment fault.

thrust fault A gently dipping reverse fault; the hanging-wall block moves up the slope of the fault.

tidal bore A visible wall of water that moves inland with the rising tide in quiet waters.

tidal flat A broad, nearly horizontal plain of mud and silt, exposed or nearly exposed at low tide but totally submerged at high tide.

tidal reach The difference in sea level between high tide and low tide at a given point.

tide The daily rising or falling of sea level at a given point on the Earth.

tide-generating force The force, caused in part by the gravitational attraction of the Sun and Moon, and in part by the centrifugal force created by the Earth's spin, that generates tides.

till A mixture of unsorted mud, sand, pebbles, and larger rocks deposited by glaciers.

tillite A rock formed from hardened ancient glacial deposits and consisting of larger clasts distributed through a matrix of sandstone and mudstone.

toe (terminus) The leading edge or margin of a glacier.

tombolo A narrow ridge of sand that links a seastack to the mainland.

topographical map A map that uses contour lines to represent variations in elevation.

topography Variations in elevation.

topsoil The top soil horizons, which are typically dark and nutrient-rich.

tornado A near-vertical, funnel-shaped cloud in which air rotates extremely rapidly around the axis of the funnel.

tornado swarm Dozens of tornadoes produced by the same storm.

tower karst A karst landscape in which steep-sided residual bedrock towers remain between sinkholes.

transform fault A fault marking a transform plate boundary; along mid-ocean ridges, transform faults are the actively slipping segment of a fracture zone between two ridge segments.

transform plate boundary A boundary at which one lithosphere plate slips laterally past another.

transgression The inland migration of shoreline resulting from a rise in sea level.

transition zone The middle portion of the mantle from 400 to 670 km deep, defined by jumps in seismic velocity.

transpiration The release of moisture as a metabolic byproduct.

transverse dune A simple, wave-like dune that appears when enough sand accumulates for the ground surface to be completely buried, but only moderate winds blow.

travel-time curve A graph that plots the time since an earthquake began on the vertical axis, and the distance to the epicenter on the horizontal axis.

trellis network A drainage system that develops across a landscape of parallel valleys and ridges so that major tributaries flow down the valleys and join a trunk stream that cuts through the ridge; the resulting map pattern resembles a garden trellis.

trench A deep elongate trough bordering a volcanic arc; a trench defines the trace of a convergent plate boundary.

triangulation The method for determining the map location of a point from knowing the distance between that point and three other points; this method is used to locate earthquake epicenters.

tributary A smaller stream that flows into a larger stream.

triple junction A point where three lithosphere plate boundaries intersect.

tropical depression A tropical storm with winds reaching up to 61 km per hour; such storms develop from tropical disturbances, and may grow to become hurricanes.

tropical disturbance Cyclonic winds that develop in the tropics.

tropopause The temperature pause marking the top of the troposphere.

troposphere The lowest layer of the atmosphere, where air undergoes convection and where most wind and clouds develop.

truncated spur A spur (elongate ridge between two valleys) whose end was eroded off by a glacier.

trunk stream The single larger stream into which an array of tributaries flow.

tsunami A large wave along the sea surface triggered by an earthquake or large submarine slump.

tuff A pyroclastic igneous rock composed of volcanic ash and fragmented pumice, formed when accumulations of the debris cement together.

tundra A cold, treeless region of land at high latitudes, supporting only species of shrubs, moss, and lichen capable of living on permafrost.

turbidite A graded bed of sediment built up at the base of a submarine slope and deposited by turbidity currents.

turbidity current A submarine avalanche of sediment and water that speeds down a submarine slope.

turbulence The chaotic twisting, swirling motion in flowing fluid.

typhoon The equivalent of a hurricane in the western Pacific Ocean.

ultimate base level Sea level; the level below which a trunk stream cannot cut at its mouth.

ultramafic A term used to describe igneous rocks or magmas that are rich in iron and magnesium and very poor in silica.

unconfined aquifer An aquifer that intersects the surface of the Earth.

unconformity A boundary between two different rock sequences representing an interval of time during which new strata were not deposited and/or were eroded.

unconsolidated Consisting of unattached grains.

undercutting Excavation at the base of a slope that results in the formation of an overhang.

undersaturated A term used to describe a solution capable of holding more dissolved ions.

unsaturated zone The region of the subsurface above the water table.

unstable air Air that is significantly warmer than air above and has a tendency to rise quickly.

unstable ground Land capable of moving downslope in a matter of seconds to weeks.

unstable slope A slope on which sliding will likely happen.

updraft Upward-moving air.

upper mantle The uppermost section of the mantle, reaching down to a depth of 400 km.

upwelling zone A place where deep water rises in the ocean, or hot magma rises in the asthenosphere.

U-shaped valley A steep-walled valley shaped by glacial erosion into the form of a U.

vacuum Space that contains very little matter in a given volume (for example, a region in which air has been removed).

valley A trough with sloping walls, cut into the land by a stream.

valley glacier A river of ice that flows down a mountain valley.

Van Allen radiation belts Belts of solar wind particles and cosmic rays that surround the Earth, trapped by Earth's magnetic field.

varve A pair of thin layers of glacial lake-bed sediment, one consisting of silt brought in during the spring floods, and the other of clay deposited during the winter when the lake's surface freezes over and the water is still.

vascular plant A plant with woody tissue and seeds and veins for transporting water and food.

vein A seam of minerals formed when dissolved ions carried by water solutions precipitate in cracks.

vein deposit A hydrothermal deposit in which the ore minerals occur in veins that fill cracks in preexisting rocks.

velocity-versus-depth curve A graph that shows the variation in the velocity of seismic waves with increasing depth in the Earth.

ventifact (faceted rock) A desert rock whose surface has been faceted by the wind.

vesicles Open holes in igneous rock formed by the preservation of bubbles in magma as the magma cools into solid rock.

viscosity The resistance of material to flow.

volatiles Elements or compounds such as H_2O and CO_2 that evaporate easily and can exist in gaseous forms at the Earth's surface.

volatility A specification of the ease with which a material evaporates.

volcanic arc A curving chain of active volcanoes formed adjacent to a convergent plate boundary.

volcanic ash Tiny glass shards formed when a fine spray of exploded lava freezes instantly upon contact with the atmosphere.

volcanic bomb A large piece of pyroclastic debris thrown into the atmosphere during a volcanic eruption.

volcanic-danger-assessment map A map delineating areas that lie in the path of potential lava flows, lahars, debris flows, or pyroclastic flows of an active volcano.

volcanic gas Elements or compounds that bubble out of magma or lava in gaseous form.

volcanic island arc The volcanic island chain that forms where one oceanic plate subducts beneath another oceanic plate.

volcano (1) A vent from which melt from inside the Earth spews out onto the planet's surface; (2) a mountain formed by the accumulation of extrusive volcanic rock.

V-shaped valley A valley whose cross-sectional shape resembles the shape of a V; the valley probably has a river running down the point of the V.

Wadati-Benioff zone A sloping band of seismicity defined by intermediate- and deep-focus earthquakes that occur in the downgoing slab of a convergent plate boundary.

wadi The name for a dry wash in the Middle East and North Africa.

warm front A front in which warm air rises slowly over cooler air in the atmosphere.

waste rock Rock dislodged by mining activity yet containing no ore minerals.

waterfall A place where water drops over an escarpment.

water gap An opening in a resistant ridge where a trunk river has cut through the ridge.

watershed The region that collects water that feeds into a given drainage network.

water table The boundary, approximately parallel to the Earth's surface, that separates substrate in which groundwater fills the pores from substrate in which air fills the pores.

wave base The depth, approximately equal in distance to half a wavelength in a body of water, beneath which there is no wave movement.

wave-cut bench A platform of rock, cut by wave erosion, at the low-tide line that was left behind a retreating cliff.

wave-cut notch A notch in a coastal cliff cut out by wave erosion.

wave erosion The combined effects of the shattering, wedging, and abrading of a cliff face by waves and the sediment they carry.

wave front The boundary between the region through which a wave has passed and the region through which it has not yet passed.

wavelength The horizontal difference between two adjacent wave troughs or two adjacent crests.

wave refraction (ocean) The bending of waves as they approach a shore so that their crests make no more than a 5° angle with the shoreline.

weather Local-scale conditions as defined by temperature, air pressure, relative humidity, and wind speed.

weathered rock Rock that has reacted with air and/or water at or near the Earth's surface.

weathering The processes that break up and corrode solid rock, eventually transforming it into sediment.

weather system A specific set of weather conditions, reflecting the configuration of air movement in the atmosphere, that affects a region for a period of time.

welded tuff Tuff formed by the welding together of hot volcanic glass shards at the base of pyroclastic flows.

well A hole in the ground dug or drilled in order to obtain water.

Western Interior Seaway A north-south-trending seaway that ran down the middle of North America during the Late Cretaceous period.

wet-bottom (temperate) glacier A glacier with a thin layer of water at its base, over which the glacier slides.

wetted perimeter The area in which water touches a stream channel's walls.

wind abrasion The grinding away at surfaces in a desert by windblown sand and dust.

wind gap An opening through a high ridge that developed earlier in geologic history by stream erosion, but is now dry.

xenolith   A relict of wall rock surrounded by intrusive rock when the intrusive rock freezes.

yardang   A mushroom-like column with a resistant rock perched on an eroding column of softer rock; created by wind abrasion in deserts where a resistant rock overlies softer layers of rock.

yazoo stream   A small tributary that runs parallel to the main river in a floodplain because the tributary is blocked from entering the main river by levees.

Younger Dryas   An interval of cooler temperatures, that took place 4,500 years ago, during a general warming/glacier-retreat period.

zone of ablation   The area of a glacier in which ablation (melting, sublimation, calving) subtracts from the glacier.

zone of accumulation   (1) The layer of regolith in which new minerals precipitate out of water passing through, thus leaving behind a load of fine clay; (2) the area of a glacier in which snowfall adds to the glacier.

zone of aeration   *Unsaturated zone.*

zone of leaching   The layer of regolith in which water dissolves ions and picks up very fine clay; these materials are then carried downward by infiltrating water.

Credits

UNNUMBERED PHOTOS AND ART

ii Galen Rowell/Corbis; **2** Stephen Marshak; **12** Courtesy Johnson Space Center/NASA; **14** R. Williams (ST Scl)-NASA; **26–27** original artwork by Gary Hincks; **34** David Turnley/Corbis; **42–43** original artwork by Gary Hincks; **50** Courtesy of Thomas Taylor, University of Kansas, Lawrence; **68–69** original artwork by Gary Hincks; **76** Johnson Space Center/NASA; **80–81** original artwork by Gary Hincks; **106** Corbis; **108** © 1998 Jeff Scovil; **129** © J. C. Leacock—Photographers/Aspen; **136** © Science VU; **140–41** original artwork by Gary Hincks; **162** © 1992 Tom Bean; **192–93** original artwork by Gary Hincks; **202** Stephen Marshak; **218–19** original artwork by Gary Hincks; **227 (left)** © Rob Simpson/Visuals Unlimited; **227 (top right)** © Martin Miller; **227 (bottom right)** © Martin Miller; **230–31** original artwork by Gary Hincks; **234** © DigitalVision/PictureQuest; **236** Bettmann/Corbis; **248–49** original artwork by Gary Hincks; **268** U.S. Geological Survey; **276–77** original artwork by Gary Hincks; **318** Stephen Marshak; **342–43** original artwork by Gary Hincks; **350** Stephen Marshak; **364** Stephen Marshak; **378–79** original artwork by Gary Hincks; **394** Stephen Marshak; **404–05** original artwork by Gary Hincks; **424** © DigitalVision/PictureQuest; **426** © Tomas Muscionico/Contact Press Images/PictureQuest; **452** © Janine Pestel/Visuals Unlimited; **470** Tashi Tenzing/ImageState; **474–75** original artwork by Gary Hincks; **478** Robert L. Schuster/U.S. Geological Survey; **488–89** original artwork by Gary Hincks; **504** © Lowell Georgia/Corbis; **528–29** original artwork by Gary Hincks; **542** Joel W. Rogers/Corbis; **578** © Yann Arthus-Bertrand/Corbis; **600–01** original artwork by Gary Hincks; **608** Johnson Space Center, NASA; **642** © Christine Osborne/Corbis; **658–59** original artwork by Gary Hincks; **664** © 2000 Loren Taft/Alaskan Images; **684–85** original artwork by Gary Hincks; **704** © DigitalVision/PictureQuest; **712–13** original artwork by Gary Hincks.

NUMBERED PHOTOS AND ART

Prelude: P.2, P.3, P.4 Stephen Marshak; **P.5** Associated Press HURRIYET; **P.9** Courtesy Peter Fiske.

Chapter 1: 1.1 Modified from Snow, 1987; **1.2A** Rare Books Division, The New York Public Library, Astor, Lenox and Tilden Foundations; **1.3** © Richard E. Hill/Visuals Unlimited; **1.7** Anglo-Australian Observatory; **1.13** J. Hester and P. Scowen/NASA; **1.14** Courtesy Palomar Observatory, California Institute of Technology; **1.15** NASA.

Chapter 2: 2.2 Modified from Laing; **2.3** Johnson Space Center/NASA; **2.9** Tate Gallery, London/Art Resource, NY.

Chapter 3: 3.1 Alfred Wegener Institute for Polar and Marine Research; **3.2** Wegener A., *The Origin of the Continents and Oceans* (New York: Dover, 1966; trans. from 1929 German ed.); **3.4** Modified from American Association of Petroleum Geologists; **3.5** Modified from Motz; **3.6A** Modified from Hurley; **3.15A** Modified from McElhinney; **3.22** Rothé, J. R. 1954. La zone séismique médian Indo-Atlantique: Proceedings of the Royal Society of London, Series A, v. 222, p. 388; **3.24B** Modified from Mason, 1955; **3.27** Modified from Cox, Dalrymple, and Doell, in Hamblin and Christiansen, 1998; **3.29C** Modified from Cox, Dalrymple, and Doell.

Chapter 4: 4.9 Photo by Dudley Foster, Woods Hole Oceanographic Institute; **4.10** Modified from Sloss, NOAA; **4.17B** R. E. Wallace (228), U.S. Geological Survey; **4.18A** Ric Ergenbright/Corbis; **4.18B** © Oddur Sigurdsson/Visuals Unlimited; **4.28** Modified from Cox and Hardt, 1986.

Chapter 5: 5.1 © Crown copyright: Historic Royal Palaces; **5.2A** © Ken Lucas/Visuals Unlimited; **5.2B** © G. R. Roberts; **5.3A** © Jay Schomer; **5.3B** © Wally McNamee/Corbis; **5.4** Modified from Wicander and Monroe; **5.9D** Courtesy of John A. Jaszczak, Michigan Technological University; **5.10A–B** Courtesy F. Hilgers; **5.12A** © 1996 Jeff Scovil; **5.14** © Richard P. Jacobs/JLM Visuals; **5.15** © Richard P. Jacobs/JLM Visuals; **5.16A** © Breck P. Kent/JLM Visuals; **5.16B** © Richard P. Jacobs/JLM Visuals; **5.17A** Copyright © 1995–1998 by Amethyst Galleries, Inc., http://mineral.galleries.com; **5.17B** © 1997 Jeff Scovil; **5.17C** © 1993 Jeff Scovil; **5.18A** © Richard P. Jacobs/JLM Visuals; **5.18B** © 1992 Jeff Scovil; **5.18C** © Martin Miller; **5.18D–F, 5.19** © Richard P. Jacobs/JLM Visuals; **5.23** © 1996 Smithsonian Institution; **5.24A** © 1985 Darrel Plowes; **5.24B** © Ken Lucas/Visuals Unlimited; **5.25** © A. J. Copley/Visuals Unlimited; **5.27** © 1994 Chip Clark.

Interlude A: A.1A © Richard P. Jacobs/JLM Visuals; **A.1B** Courtesy David W. Houseknecht, U.S. Geological Survey; **A.2A** © Mark A. Schneider/Visuals Unlimited; **A.2B** Courtesy of Dr. Kent Ratajeski, Department of Geology and Geophysics, University of Wisconsin, Madison; **A.3A** G. R. Roberts; **A.3B–D** Stephen Marshak; **A.4A** © Corbis; **A.4B** © John Elk III/Stock, Boston/PictureQuest; **A.4C** © David Muench/Corbis; **A.5C** Modified from AGI data sheet; **A.6A** © 1999 Tom Bean; **A.6B** H. Johnson; **A.8A** Stephen Marshak; **A.8B** Courtesy of Bradley Hacker, Department of Geology, University of California, Santa Barbara.

Chapter 6: 6.1A Jim Sugar Photography/Corbis; **6.1B** Photo by J. D. Griggs/U.S. Geological Survey; **6.1C** Roger Ressmeyer/Corbis; **6.8B** Stephen Marshak; **6.9C** Stephen Marshak; **6.11A** Stephen Marshak; **6.11B** © 1998 Tom Bean; **6.11C** Paul Hoffmann; **6.13A** Stephen Marshak; **6.14C** Stephen Marshak; **6.16A,B** Courtesy of Dr. Kent Ratajeski, Department of Geology and Geophysics, University of Wisconsin, Madison; **6.16C** © Richard P. Jacobs/JLM Visuals; **6.16D** © Richard P. Jacobs/JLM Visuals; **6.16E** © John S. Shelton; **6.16F** © Richard P. Jacobs/JLM Visuals; **6.16G** © Dane S. Johnson/Visuals Unlimited; **6.16H** © Doug Sokell/Visuals Unlimited; **6.17A** Mineral proportions after Hamblin and Howard; **6.19** Stephen Marshak; **6.20B** Tony Stone Images; **6.21B** Stephen Marshak; **6.21C** © Peter Kresan.

Chapter 7: 7.2A Stephen Marshak; **7.3** Stephen Marshak; **7.5A** © Martin Miller; **7.5B–C** Stephen Marshak; **7.6B** © 1998 Tom Bean; **7.6C** Stephen Marshak; **7.7C** Stephen Marshak; **7.7D** Courtesy Carlo Giovanella; **7.10C–D** Stephen Marshak; **7.12A** © John D. Cunningham/Visuals Unlimited; **7.14** © John C. Coulter/Visuals Unlimited; **7.17A (left)** Stephen Marshak; **7.17A (right)** © G. R. Roberts; **7.17B–C** Stephen Marshak; **7.17D** From R. L. Kugler and J. C. Pashin, Reservoir Heterogeneity in Carter Sandstone, North Blowhorn Creek Oil Unit and Vicinity, Black Warrior Basin, Alabama. Geological Survey of Alabama, Bulletin 159, 1994; **7.19A** U.S. Geological Survey; **7.19B** Stephen Marshak; **7.19C** Courtesy Mark A. Wilson, The College of Wooster; **7.20A** Stephen Marshak; **7.20B** © Steve McCutcheon/Visuals Unlimited; **7.21B** © Gerald and Buff Corsi/Visuals Unlimited; **7.21D** John S. Shelton, courtesy of Morton Salt Company; **7.22A** © Martin Miller; **7.22B** © 1998 M. W. Schmidt; **7.22C** Stephen Marshak; **7.24** Stephen Marshak; **7.25D** Stephen Marshak; **7.26B** © Martin G. Miller/Visuals Unlimited; **7.27A–D** Stephen Marshak; **7.28A** G. R. Roberts; **7.28B** Stephen Marshak; **7.28C** © Martin Miller; **7.29A** © John S. Shelton; **7.29C** Stephen Marshak; **7.31A** Yann Arthus-Bertrand/Corbis; **7.32A** Scripps Institution of Oceanography, University of California, San Diego; **7.32B** G. R. Roberts.

Chapter 8: 8.1A Courtesy of Yong Il Lee, School of Earth and Environmental Sciences, Seoul National University; **8.1B** Courtesy Kurt Friehauf, Kutztown University of Pennsylvania; **8.2** Kevin Schafer/Corbis; **8.3A** © L. S. Stepanowicz/Visuals Unlimited; **8.3B** Courtesy Carlo Giovanella; **8.6** © 1996 Bruce Clendenning/Visuals Unlimited; **8.12B** Ludovic Maisant/Corbis; **8.14** Stephen Marshak; **8.15A** Stephen Marshak; **8.15B** Pamela J. W. Gore, Georgia Perimeter College; **8.16** © 1999 Stephen Weaver; **8.17A** Stephen Marshak; **8.17B** Diego Lezama Orezzoli/Corbis; **8.29B** Stephen Marshak; **8.29C** David Muench/Corbis.

Chapter 9: 9.1 Stephen Marshak; **9.3** Stephen Marshak; **9.4** Stephen Marshak; **9.5B** D. W. Peterson/Corbis; **9.5C** © Ken Wagner/Visuals Unlimited; **9.5D** Stephen Weaver; **9.5E** Associated Press AP; **9.6A** Roger Ressmeyer/Corbis **9.6B** © 2001 Stephen and Donna O'Meara, Volcano Watch International; **9.6C** Associated Press *Antigua Sun*; **9.7** U.S. Geological Survey; **9.8** Stephen Marshak; **9.9C** Stephen Marshak; **9.11C** Yann Arthus-Bertrand/Corbis; **9.11D** © Will Troyer/Visuals Unlimited; **9.12B** © Martin Miller/Visuals Unlimited; **9.12C** © 1985 Tom Bean; **9.12E** Roger Ressmeyer/Corbis; **9.14A** Stephen Marshak; **9.14B** © S. Jonasson/FLPA; **9.16D–E** Gary Braasch/Corbis; **9.18B–C** Stephen Marshak; **9.19** Stephen Marshak; **9.20B** © Oddur Sigurdsson; **9.21** Stephen Marshak; **9.22A** Stephen Marshak; **9.22B** J. D. Griggs/U.S. Geological Survey; **9.22C** © Philippe Bourseiller; **9.22D** AP/Wide World Photos; **9.22E** Photo by J. Marso, U.S. Geological Survey; **9.23** Peter Turnley/Corbis; **9.25** Neil Ryder Hoos; **9.26** Fact Sheet by Scott, K. M., Wolfe, E. W., and Driedger, C. L., Hawaiian Volcano Observatory, U.S. Geological Survey; **9.27A** Bettmann/Corbis; **9.27B** photo by Sigurgeir Jonasson © Frank Lane Picture Agency/Corbis; **9.28** Joseph Mallord William Turner. *Slave Ship*, 1840. Museum of Fine Arts, Boston. Henry Lillie Pierce Fund, 99.22. Courtesy, Museum of Fine Arts, Boston. Reproduced with permission. © 2000 Museum of Fine Arts, Boston. All rights reserved.; **9.29A** Gail Mooney/Corbis; **9.29B** © 1997 Birke Schreiber; **9.30** JPL/NASA.

Chapter 10: 10.2 J. Dewey, U.S. Geological Survey; **10.5A** National Geophysical Data Center/NOAA; **10.6G** Photo courtesy of Paul "Kip" Otis-Diehl, USMC, 29 Palms, CA; **10.20** Adapted from Bolt, 1978; **10.21** Modified from Dolgoff, 1996; **10.25A** Adapted from U.S. Geological Survey map of New Madrid seismicity; **10.25B** Courtesy State Historical Society of Missouri, Columbia; **10.28A** Associated Press HURRIYET; **10.28B** M. Celebi, U.S. Geological Survey; **10.28C** J. Dewey, U.S. Geological Survey; **10.28D** © Woodrow L. Higdon/Geo-Tech Imagery International; **10.28E** Associated Press; **10.29** Associated Press AP; **10.30C** National Geophysical Data Center/NOAA; **10.32A** George Hall/Corbis; **10.32B** U.S. Geological Survey; **10.34A** Cecilio Licos, Yasuki Arakaki Collection/Pacific Tsunami Museum; **10.34B** Pacific Tsunami Museum; **10.34C** NOAA; **10.36A** Adapted from U.S. Geological Survey; **10.36B** Adapted from Wesson and Wallace, 1985; **10.36C** Adapted from Nishenko, 1989 (U.S. Geological Survey); **10.39B** Associated Press AP.

Interlude C: C.5A From "Patterns of Nature," Department of Physics and Astronomy, Arizona State University, Arizona Board of Regents; **C.11A** From Su, W. J., and Dziewonski, A. M., 1994. Degree 12 model of shear velocity heterogeneity in the mantle. *Journal of Geophysical Research,* v. 99, pp. 6945–80. This image was produced by X. Song; **C.11B** Steve Grand, University of Texas; **C.12A–B** © Jon Q. Thompson, courtesy of Dawson Geophysical Company; **C.12C** Courtesy Gregory Mountain, Lamont-Doherty Earth Observatory; **C.12D** Courtesy John Diebold, Lamont-Doherty Earth Observatory; **C.12E** Image courtesy of Magic Earth, Houston.

Chapter 11: 11.1 © 2001 Corbis; **11.2** National Geophysical Data Center/NOAA; **11.5** Stephen Marshak; **11.8C–D** Stephen Marshak; **11.12A** Galen Rowell/Corbis; **11.12B–C** Stephen Marshak; **11.13A** John S. Shelton; **11.13B** Stephen Marshak; **11.15A** Stephen Marshak; **11.15C–F** Stephen Marshak; **11.18A** Stephen Marshak; **11.20** Stephen Marshak; **11.24C** Stephen Marshak; **11.24E** Stephen Marshak; **11.26** Stephen Marshak; **11.28A** Stephen Marshak; **11.28B** © Art Wolfe; **11.31B** Adapted from Coney et al., 1980; **11.34** U.S. Geological Survey; **11.35A** Adapted from Geologic Map of the United States, U.S. Geological Survey.

Interlude D: D.1 Stephen Marshak; **D.2A** Photo by William L. Jones from the Stones & Bones Collection, http://www.stonesbones.com; **D.4A** © Saint-Petersburg Zoological Museum; **D.4B** Doug Lundberg (www.ambericawest.com); **D.4C–D** Stephen Marshak; **D.4E** Department of Earth and Space Sciences, UCLA; **D.4F** Kevin Schafer/Corbis; **D.4G–H** Stephen Marshak; **D.5** Photo by Dennis Greig and Al Melillo, Chevron, USA; **D.6A** Courtesy of Senckenberg, Messel Research Department; **D.6B** Humboldt-Universität zu Berlin, Museum für naturkunde. Photo by W. Harre; **D.6C** Courtesy of Smithsonian Institution.

Chapter 12: 12.2 © A. J. Copley/Visuals Unlimited; **12.4A–B** Stephen Marshak; **12.4D–E** Stephen Marshak; **12.6** Layne Kennedy/Corbis; **12.8A** Stephen Marshak; **12.10A** Stephen Marshak; **12.16 (top)** Adapted from Coney, Zion National History Association, 1975; **12.16 (bottom)** Adapted from U.S. Geological Survey; **12.16B–C** Stephen Marshak; **12.16E** © Charles Preitner/Visuals Unlimited; **12.16G** Stephen Marshak; **12.18** Courtesy Jim Connelly, University of Texas at Austin; **12.20A–B** © Michael Pole/Corbis; **12.20C** Lonnie G. Thompson, Byrd Polar Research Center, The Ohio State University, Columbus; **12.22A** Courtesy of Yong Il Lee, School of Earth and Environmental Sciences, Seoul National University; **12.25A** Courtesy of the Royal Ontario Museum © ROM; **12.25B** NASA.

Index

NOTE: Italicized page numbers refer to pictures, tables, and figures. Bold page numbers refer to key words.